ALLE · ZEIT · WACH · 1842

Hans Flohr (Editor)

Post-Lesion Neural Plasticity

With 276 Figures

Springer-Verlag Berlin Heidelberg New York
London Paris Tokyo

Professor Dr. Hans Flohr

Dept. of Neurobiology
University of Bremen, NW2
Postfach 330440
D-2800 Bremen 33
Federal Republic of Germany

ISBN-13:978-3-642-73851-7 e-ISBN-13:978-3-642-73849-4
DOI: 10.1007/978-3-642-73849-4

Library of Congress Cataloging in Publication Data
Post-lesion neural plasticity/edited by H. Flohr.
p. cm. "Proceedings of a symposium held at the University of Bremen in August 1987 as a satellite meeting of the Second World Congress of Neuroscience" – P. Includes index.
ISBN-13:978-3-642-73851-7 (U.S.): DM 198.00 (W. Germany).
1. Neuroplasticity – Congresses. 2. Nervous system – Regeneration – Congresses. I. Flohr, H. (Hans). 1936–.
QP363.5.P67 1988 599'.0188–dc19 88-28158

Softcover reprint of the hardcover 1st edition 1988

Typesetting: Brühlsche Universitätsdruckerei, Gießen

2131/3020-543210 – Printed on acid-free paper

Preface

Neural networks are not rigidly wired, but rather highly plastic structures, the functional architecture of which can be actively reorganized in response to external or internal events. Such networks react to lesions with plastic changes, whereby the system is "rewired" and "reprogrammed", eventually bringing about a partial or complete restoration of the initially disrupted function.

This form of neural plasticity has been the subject of increasing attention over the last decade. There would appear to be two main reasons for this growing interest. Firstly, the once dominant, rather pessimistic, attitude as to the clinical possibilities of treatment of lesions in the central nervous system has given way to the hope that a better understanding of the physiology of post-lesion plasticity might have concrete therapeutic consequences. Secondly, the idea has become attractive that the study of lesion-induced events is an ideal approach for understanding neural networks as dynamic and self-organizing structures. It is implicitly assumed that the mechanisms and principles underlying post-lesion reorganization of neural nets are the same as or similar to those responsible for the original development and subsequent maintenance of such dynamic structures.

This volume provides an overview of the latest trends and findings in this rapidly expanding field of research. It contains the proceedings of a symposium held at the University of Bremen in August 1987 as a satellite meeting of the Second World Congress of Neuroscience. The financial support of the Volkswagenstiftung and the University of Bremen is gratefully acknowledged. I am also indebted to the staff of the Department of Neurobiology at the University of Bremen for their valuable assistance. In particular, I would like to thank Ms Helga Kortmann for her tireless and painstaking efforts in organizing the conference and in preparing this volume. The cooperation of Springer-Verlag in the publication of these proceedings is much appreciated.

Bremen, July 1988 Hans Flohr

Contents

I Cellular and Molecular Mechanisms of Lesion-Induced Plastic Processes

Multiple Sites and Modes for the Regulation of Axonal Growth
S. Rotshenker 3

Models, Mechanisms and Kinetics of Neuromuscular Synapse Elimination in Reinnervated Adult Skeletal Muscle
R. R. Ribchester 11

Elimination of a Single Axon Changes Synaptic Characteristics of the Other Axons Innervating a Muscle Fibre of Lobster
J. Dudel, I. Parnas 25

Nerve Growth Factor and the Reinnervation of Skin After Peripheral Nerve Lesions
J. Diamond 35

Modulation of Glial Cell Response to Injury and CNS Regeneration
M. Schwartz, A. Cohen, A. Harel, A. Solomon, M. Belkin 49

Structural Plasticity in Lesioned Motoneurons
G.W. Kreutzberg, A.K. Engel, M.B. Graeber, W. Tetzlaff, L. Tóth 57

The Influence of Magnetic Field on the Sheath Cell-Axon Interactions
K. Shibib, M. Brock, G. Gosztonyi, S.N. Erné, H.-D. Hahlbohm, G. Schoknecht 65

Putative Role of Apolipoprotein E and Lipoproteins in Peripheral Nerve Repair
H.W. Müller, T. Rothe 85

Protein Kinase C and Protein F1: Potential Molecular Mediators of Lesion-Induced Synaptic Plasticity Recapitulate Developmental Plasticity
A. Routtenberg 93

The Kinase C Substrate Protein B-50 (GAP43) and its Function in Axonal Repair Mechanisms
J. Verhaagen, C.O.M. Van Hooff, W.H. Gispen 103

Neuronal Metabolic Basis of the Conditioning Lesion Effect
I.G. McQuarrie 113

Postsynaptic Receptor Elimination During Synaptic Competition
M.M. Rich, J.W. Lichtman 123

Changes in Dendrites of Adult Rat Ganglion Cells Regenerating Axons into Peripheral Grafts
S. Thanos, A.J. Aguayo 129

Compensatory Sprouting of Retinofugal Axon Trees After Early Tectal Lesions: New Evidence for the Principle of "Conservation of Total Axonal Arborizations"
B.A. Sabel, M. Donicht, G.E. Schneider 139

II CNS Development and Lesions

The Effect of Early and Late Hemidecortication on Vision and Locomotion in the Rabbit
M.W. Van Hof, W.H. De Vos-Korthals, J.F. Hobbelen 157

Post-Lesion Neural Plasticity in Cerebral Cortex: The Response of Rat Sensorimotor Cortex to Neonatal Ablation of the Opposite Hemisphere
P.R. Huttenlocher, R. Raichelson 165

Reduction of Plasticity in the Primary Visual Cortex of the Rat
A. Wree, K. Zilles, A. Schleicher 173

Split Brain Surgically Performed in Developing and in Adult Cats: Physiological Properties and Recovery of Visual Cortex Neurons
U. Yinon, M. Chen 187

Neuronal Remodeling After Early Cerebellar Hemispherectomy in Rats
A. Gramsbergen 203

Neonatal Versus Adult Hemicerebellectomy: A Behavioral and Anatomical Analysis
L. Petrosini, M. Molinari, T. Gremoli, A. Granato 213

III Lesions to Specific CNS Subsystems: Spinal Cord

Does Collateral Sprouting from Corticospinal Fibers Participate in Motor Recovery After Spinal Hemisection in Monkeys?
M. Aoki, Y. Fujito, I. Kosaka, H. Satomi 223

Pathfinding and Synaptic Specificity of Regenerating Spinal Axons in the Lamprey
M.E. Selzer, D. Lurie, S.A. Mackler 233

Altered Kinematic Patterns and Synapses in Recovery of Motor Behavior After Hindlimb Deafferentation in Cats
M.E. Goldberger 249

Influence of the Noradrenergic Coeruleospinal System on Recurrent Inhibition in the Spinal Cord and its Role During Postural Reflexes
O. Pompeiano 259

IV Lesions to Specific CNS Subsystems: Vestibular System

Plastic, Adaptive Changes in the Vestibuloocular Reflex and Their Role in Recovery from Labyrinthine Lesions
B.W. Peterson, J.F. Baker, V. Matsuo, S.A. Rude, X.N. Sahyouni 281

Vestibulo-Ocular Reflex Compensation in the Vertical and Horizontal Planes Following Unilateral Peripheral Vestibular Deficit in Man
J.H.J. Allum, M. Yamane, C.R. Pfaltz 293

Recovery from Unilateral Labyrinthectomy in Primates
M. Fetter, D.S. Zee 305

Vestibular Compensation: Aspects of Time Course and De-Compensation
K.-P. Schaefer, D.L. Meyer 313

Cervico-Ocular Reflex After Labyrinthine Damage
V.E. Pettorossi, P. Errico, A. Ferraresi, R. Fedeli, R.M. Santarelli 321

Compensation of Cerebellar-Vestibular Lesions and Microgravity
O.G. Gazenko, R.A. Grigorian, Yu.V. Kreidich, G.S. Aizikov, A.S. Markin 331

HRP Study on Structural Changes in the Commissural Fiber System of *Rana temporaria* Following Labyrinthectomy
U. Will, H. Kortmann, H. Flohr 345

Multimodal Sensory Substitution Process in Vestibular Compensation
C. Xerri, M. Lacour, L. Borel 357

Electrophysiological Changes in Vestibular Cortex After Labyrinthine Input Deprivation
G.B. Azzena, M.A. Caria, F. Melis 371

Optokinetic and Vestibular Oculomotor Responses and Their Interaction Following Hemilabyrinthectomy in the Cat
C. Maioli 381

"Acute" Vestibular Compensation in the Goldfish: A Visual Substitution Process?
A. Burt, H. Flohr 393

V Lesions to Specific CNS Subsystems: Cerebellum and Oculomotor System

Recovery from Smooth Pursuit Impairments After Successive Unilateral and Bilateral Chemical Lesions in the Dorsolateral Pontine Nucleus of the Monkey
J.G. May III, E.L. Keller 413

Regeneration of Cerebellofugal Projection in Kittens
S. Kawaguchi 421

Compensatory Mechanisms at the Level of the Vestibular Nuclei Following Post-Natal Degeneration of Specific Cerebellar Cell Classes and Ablation of the Cerebellum in Mutant Mice
U. Grüsser-Cornehls 431

Functional Recovery of the Primate Oculomotor Pursuit System Following Cerebellar and Cerebral Cortical Lesions
R. Eckmiller 443

VI Lesions to Specific CNS Subsystems: Visual System

The Effect of Monocular Pattern Deprivation and Open-Loop Stimulation on Optokinetic Nystagmus in Squirrel Monkeys (*Saimiri sciureus*)
F. Behrens, O.-J. Grüsser 455

Return of Function After Optic Tract Lesions in Adult Rats: Spontaneous Axonal Regeneration?
A.P. Foerster 473

Lesion-Induced Relative Supersensitivity to an Excitatory Amino Acid Agonist in the Cat Dorsal Lateral Geniculate Nucleus
U.Th. Eysel, H.-Chr. Pape, T. Salt, A. Sillito 481

Path- and Homefinding of Regenerating Retinal Axons in Goldfish
C.A.O. Stuermer 489

Regeneration of the Retinotectal Projection in Goldfish: Selective Stabilization of Retinotopic Synapses by Correlated Activity
J. Schmidt 499

VII Lesions to Specific CNS Subsystems: Somatosensory Cortex, Motor Cortex, Basal Ganglia

Plasticity in Nonprimary Somatosensory Cortex of Adult Monkeys
T.P. Pons, P.E. Garraghty, M. Mishkin 511

Plasticity and Function of Associational Input to the Motor Cortex
H. Asanuma, M. Bornschlegl 519

Anatomical and Electrophysiological Studies of Aberrant Corticorubral Fibers Induced by Lesions of the Cerebral Cortex in Kitten
F. Murakami, S. Higashi, E. Kosar, Y. Fujito 527

Parallels in Behavioral and Neural Plasticity Induced by Unilateral Vibrissae Removal and Unilateral Lesion of the Substantia Nigra
J.P. Huston, H. Steiner, R.K.W. Schwarting, S. Morgan 537

VIII Lesions to Specific CNS Subsystems: Auditory System, Olfactory System

Axonal Regeneration After Transection of the Commissure of the Inferior Colliculus
I. Taniguchi, H. Tanaka 555

Post-Lesion Plasticity in the Auditory System of the Cricket
K. Schildberger, F. Huber 565

Peripheral and Central Post-Lesion Plasticity in the Olfactory System of the Goldfish: Behavior and Morphology
H.P. Zippel, D.L. Meyer, M. Knaust 577

IX Pharmacology of CNS Injury

Melanocortins: A Potential Role in the Therapy of Neurological Disorders
P. De Koning, P.M. Edwards, W.H. Gispen 595

Sprouting Pattern and B-50 Phosphorylation in Regenerating Sciatic Nerve Respond to ACTH Peptides
F.L. Strand, L.A. Zuccarelli, B. Kirschenbaum, R.E. Frischer 605

ACTH(4–10) and ACTH(1–24) Influence Neurite Outgrowth and Neural Development of Fetal Rat Cerebral Cultures
C. Richter-Landsberg, I. Bruns, H. Flohr 615

Comparative Effect of Some Neurotropic Agents on Balance Compensation After Unilateral and Bilateral (Two-Staged) Labyrinthectomy in Squirrel Monkeys
M. Igarashi, M. Ishii, K. Ishikawa, T. Himi 627

Glucocorticoids and Lesion-Induced Plasticity
E.R. De Kloet, J.A.M. Van Eekelen, M.A. Seger 635

Ganglioside Treatment and Post-Lesion Neural Plasticity. A Unifying Theory on Ganglioside Mechanism of Action
A. Gorio, A.M. Di Giulio 645

Role of Gangliosides in Functional Recovery
of Damaged Nervous System
A. Consolazione, G. Vantini, S. Seren, G. Toffano, A. Leon 655

Central and Cerebellar Norepinephrine Depletion and Vestibulo-Ocular
Reflex (VOR) Adaptation
J.G. McElligott, W. Freedman 661

Sensorimotor Activity and Metabolic Factors in Vestibular Compensation
M. Lacour, L. Ez-Zaher, C. Xerri 675

Role of the Ca^{2+} Entry Blocker Flunarizine in Vestibular Compensation
E. Tolu, O. Mameli, F. Melis, M.A. Caria 687

Effect of Substance P on Vestibular Compensation
U. Lüneburg, H. Flohr 699

List of Contributors

Aguayo, A.J. 129
Aizikov, G.S. 331
Allum, J.H.J. 293
Aoki, M. 223
Asanuma, H. 519
Azzena, G.B. 371
Baker, J.F. 281
Behrens, F. 455
Belkin, M. 49
Borel, L. 357
Bornschlegl, M. 519
Brock, M. 65
Bruns, I. 615
Burt, A. 393
Caria, M.A. 371, 687
Chen, M. 187
Cohen, A. 49
Consolazione, A. 655
Diamond, J. 35
Donicht, M. 139
Dudel, J. 25
Eckmiller, R. 443
Edwards, P.M. 595
Van Eekelen, J.A.M. 635
Engel, A.K. 57
Erné, S.N. 65
Errico, P. 321
Eysel, U.Th. 481
Ez-Zaher, L. 675
Fedeli, R. 321
Ferraresi, A. 321
Fetter, M. 305
Flohr, H. 345, 393, 615, 699
Foerster, A.P. 473
Freedman, W. 661
Frischer, R.E. 605
Fujito, Y. 223, 527
Garraghty, P.E. 511
Gazenko, O.G. 331
Gispen, W.H. 103, 595
Di Giulio, A.M. 645
Goldberger, M.E. 249
Gorio, A. 645
Gosztonyi, G. 65
Graeber, M.B. 57
Gramsbergen, A. 203
Granato, A. 213
Gremoli, T. 213
Grigorian, R.A. 331
Grüsser, O.-J. 455
Grüsser-Cornehls, U. 431
Hahlbohm, H.-D. 65
Harel, A. 49
Higashi, S. 527
Himi, T. 627
Hobbelen, J.F. 157
Van Hof, M.W. 157
Van Hooff, C.O.M. 103
Huber, F. 565
Huston, J.P. 537
Huttenlocher, P.R. 165
Igarashi, M. 627
Ishii, M. 627
Ishikawa, K. 627
Kawaguchi, S. 421
Keller, E.L. 413
Kirschenbaum, B. 605
De Kloet, E.R. 635
Knaust, M. 577
De Koning, P. 595
Kortmann, H. 345
Kosaka, I. 223
Kosar, E. 527
Kreidich, Yu.V. 331
Kreutzberg, G.W. 57
Lacour, M. 357, 675
Leon, A. 655
Lichtman, J.W. 123
Lüneburg, U. 699
Lurie, D. 233
Mackler, S.A. 233
Maioli, C. 381
Mameli, O. 687
Markin, A.S. 331
Matsuo, V. 281
May III, J.G. 413
McElligott, J.G. 661
McQuarrie, I.G. 113
Melis, F. 371, 687
Meyer, D.L. 313, 577
Mishkin, M. 511
Molinari, M. 213
Morgan, S. 537
Müller, H.W. 85
Murakami, F. 527
Pape, H.-Chr. 481
Parnas, I. 25
Peterson, B.W. 281
Petrosini, L. 213
Pettorossi, V.E. 321
Pfaltz, C.R. 293
Pompeiano, O. 259
Pons, T.P. 511
Raichelson, R. 165
Ribchester, R.R. 11
Rich, M.M. 123
Richter-Landsberg, C. 615
Rothe, T. 85
Rotshenker, S. 3
Routtenberg, A. 93
Rude, S.A. 281
Sabel, B.A. 139
Sahyouni, X.N. 281
Salt, T. 481
Santarelli, R.M. 321
Satomi, H. 223
Seger, M.A. 635
Selzer, M.E. 233
Seren, S. 655
Shibib, K. 65
Sillito, A. 481
Solomon, A. 49
Schaefer, K.-P. 313
Schildberger, K. 565
Schleicher, A. 173

Schmidt, J. 499
Schneider, G.E. 139
Schoknecht, G. 65
Schwarting, R.K.W. 537
Schwartz, M. 49
Steiner, H. 537
Strand, F.L. 605
Stuermer, C.A.O. 489
Tanaka, H. 555
Taniguchi, I. 555
Tetzlaff, W. 57
Thanos, S. 129
Toffano, G. 655
Tolu, E. 687
Tóth, L. 57
Vantini, G. 655
Verhaagen, J. 103
De Vos-Korthals, W.H. 157
Will, U. 345
Wree, A. 173
Xerri, C. 357, 675
Yamane, M. 293
Yinon, U. 187
Zee, D.S. 305
Zilles, K. 173
Zippel, H.P. 577
Zuccarelli, L.A. 605

I Cellular and Molecular Mechanisms of Lesion-Induced Plastic Processes

Regulation of Axonal Growth

S. Rotshenker

Department of Anatomy and Embryology, Hebrew University-Hadassah Medical School, Jerusalem 91010, Israel

1 Introduction

A major manifestation of neural plasticity is the ability of adult nerve cells to regulate the growth of their axons. For example, peripheral nerve injury can be followed by the regeneration of the severed axons and the sprouting of intact axons. These two modes of growth may result in new synapse formation and therefore bear functional significance. Normally, severed axons regenerate into denervated tissues. Intact axons, on the other hand, can sprout into either deneravted or innervated target tissues. For example, after cutting some of the axons innervating a muscle, the remaining intact axons grow new processes that innervate denervated muscle fibers (Edds 1953; Brown et al. 1981). Sprouting and synapse formation in normally innervated muscles may take place after denervating muscles on the contralateral side of the body (e.g., Rotshenker 1979; Rotshenker and Tal 1985).

To account for the different modes of neural plasticity I suggest three major groups of mechanisms for the regulation of axonal growth and synapse formation: transneuronal, central, and peripheral. The transneuronal mechanism involves the induction of axonal sprouting of intact motoneurons by other nerve cells within the spinal cord (motor, sensory, or others). The central mechanism formulates that a site of regulation of axonal growth is the nerve cell body and that target muscle fibers can play a role in this regulation by means of retrograde transport of muscle-derived molecules. Peripheral mechanisms involve the direct influence of growth-promoting factors on intramuscular axons, preterminal branches and nerve endings, thus suggesting that the neurons' peripheral extensions are an additional site at which growth can be regulated.

2 Sprouting and Synapse Formation Following Contralateral Axotomy

We examined the pattern of motor innervation to individual muscle fibers of the cutaneous-pectoris and sartorius muscles of the frog, and the peroneal and extensor digitorum longus muscles of the mouse. Various degrees of sprouting and synapse formation were observed in all intact muscles of unoperated animals. Sprouting was enhanced in intact muscles after axotomizing neurons innervating contralateral, homologous muscles (contralateral axotomy).

H. Flohr (Ed.)
Post-Lesion Neural Plasticity

Frog cutaneous-pectoris muscles were stained with a combined silver cholinesterase stain that visualizes the entire neuromuscular junction: (Rotshenker 1982). The majority of muscle fibers are innervated by a single myelinated axon. In some instances, two myelinated axons innervate a single muscle fiber at a single site. Occasionally, single muscle fibers receive innervation from two different sources: a myelinated axon and a sprout that arises either from another axon or from nerve-endings of a neighboring neuromuscular junction. The incidence of muscle fibers demonstrating the latter pattern of innervation varied between different batches of unoperated frogs. Some batches exhibited an average rate of sprouting of $1.5 \pm 0.6\%$, whereas others averaged $9.7 \pm 0.7\%$. These incidences of sprouting and synapse formation increased, respectively, following contralateral axotomy to the average values of $16.9 \pm 3.0\%$ and $18.2 \pm 1.0\%$. Thus muscle fibers that were previously innervated by a single motoneuron (via one axon) became polyneuronally innervated (via a sprout).

Polyneuronally innervated muscle fibers were also identified electrophysiologically by recording multiple end-plate potentials in response to graded intensities of stimulation applied to the motor nerve (Rotshenker 1979). The incidence of polyneuronal innervation detected in intact muscles of unoperated frogs varied between the different batches of frogs. Frogs exhibiting the low rate of anatomically identified sprouting and synapse formation (1.5%) also exhibited a low incidence (16 to 20%) of electrophysiologically identified polyneuronal innervation. Frogs that showed a higher incidence of anatomical sprouting (9.7%) also had a higher percentage (28%) of muscle fibers in which multiple end-plate potentials were recorded. In all frogs, contralateral axotomy was followed by an increase in the incidence of electrophysiologically identified polyneuronal innervation (36%). The additional number of muscle fibers in which a polyneuronal pattern of innervation was detected electrophysiologically is thus similar to the additional number of muscle fibers that became supernumerary innervated by sprouts (see also Rotshenker and Reichert 1980).

All of the experiments reported below involving the cutaneous-pectoris muscle were performed in frogs exhibiting the low incidence of sprouting (anatomical) and of polyneuronal innervation (electrophysiological).

Sartorius muscles of the frog were studied with the light microscope following a combined silver cholinesterase stain (Ring et al. 1983). The incidence of sprouting and synapse formation observed in intact muscles of normal frogs was $6.0 \pm 0.8\%$. It increased more than three fold to $20.2 \pm 2.6\%$ after contralateral axotomy. In a recent study, Herrera and Scott (1985) failed to observe an increase in the incidence of sprouting in intact sartorius muscles following contralateral axotomy. This discrepancy could easily be explained by the fact that these authors used unoperated frogs exhibiting a high incidence of sprouting (20%) that we observed only after contralateral axotomy. It is possible that in these muscles sprouting had already reached its maximal capacity and therefore no additional increase was observed following contralateral axotomy.

Recent findings (Rotshenker unpublished) indicate that in some unoperated frogs a small proportion of muscle fibers are in a state of denervation. This observation could explain the high incidence of sprouting observed in some unoperated frogs.

Pyriformis muscles of the frog were examined electrophysiologically by Stefani and co-workers (Elizalde et al. 1983). This muscle is comprised of twitch and tonic muscle fibers that are innervated by fast and slow motor axons, respectively. Contralateral axotomy induced the sprouting of fast motor neurons. Consequently, there was an increase in the incidence of polyneuronal innervation of twitch muscle fibers and hyperinnervation of tonic fibers by fast motor axons.

Peroneal and extensor digitorum longus (edl) muscles of the mouse were examined with the light microscope after neuromuscular junctions were visualized by a combined silver-cholinesterase stain (Rotshenker and Tal 1985). In normal adult mice all muscle fibers are innervated by a single axon. In a small proportion of cases ($7.0 \pm 0.9\%$) axons gave rise to sprouts, and both parent axon and sprout innervated the same muscle fiber. This incidence of sprouting and synapse formation increased three fold, to an average value of $20.7 \pm 2.1\%$ after contralateral axotomy.

Peroneus longus and edl muscles of the cat were studied by Steinbach (1981) following contralateral axotomy. The mean junctional size increased in intact muscles, most probably consequent to the induction of sprouting.

3 The Transneuronal Mechanism for the Induction of Sprouting and Synapse Formation

The transneuronal mechanism was formulated to explain the induction of sprouting and synapse formation in intact muscles after contralateral axotomy (Rotshenker 1979, 1982). The transneuronal mechanism suggests that axotomy initiates a signal for growth in cell bodies of axotomized motoneurons. The signal is then transferred transneuronally across the spinal cord to intact motoneurons that respond with sprouting and synapse formation. The experimental results that led to the formulation of the transneuronal mechanism did so by ruling out alternative mechanisms, and by being in accord with predictions of the suggested mechanism.

Possible alternatives to the transneuronal mechanism are that sprouting was induced by products of degeneration and/or denervation, or by inactivity of the denervated muscle.

Products of degeneration and denervation that arise from the denervated muscle could induce neuronal sprouting after reaching the opposite muscle by local or systemic distribution. If this were the case, then the surgical removal of one cutaneous-pectoris muscle should not be followed by contralateral sprouting, since the sources for products of degeneration and denervation would be removed. Such was not the case; sprouting and synapse formation developed in intact cutaneous-pectoris muscles after removing the homologous muscle on the opposite side (Rotshenker 1979). Conversely, denervation of one hind limb, which would allow the systemic distribution of products of degeneration and denervation, produced sprouting in contralateral hind limb muscles (Elizalde et al. 1983; Ring et al. 1983; Rotshenker and Tal 1985; Rotshenker 1979; Steinbach 1981) but not in the chest cutaneous-pectoris muscles (Rotshenker 1979).

The possible role of products of degeneration and denervation was further ruled out by the observation that axotomy of the 2nd spinal nerve at its exit from the spinal column did not produce sprouting in intact opposite muscles in about one half of the animals, and that ventral rhizotomy failed to produce the contralateral effect in all animals. Sprouting was always observed after contralateral axotomies that were placed at distances of 3 mm or more from the spinal column (Ring, Sugarman and Rotshenker in preparation). We do not understand why contralateral axotomy that was placed in too close proximity to the spinal cord failed to produce sprouting. However, one conclusion is clear: products of degeneration and denervation could not produce sprouting. At the same time, this experimental result does not contradict the transneuronal mechanism. It can be argued, based on extensive data in the literature (e.g., Lieberman 1971, 1974), that the response of the motoneuron cell body to axotomy is different for injuries which are placed either very close to the cell body or at some distance from it (e.g., cell death following axotomy is more prominent after proximal axotomy). The sprouting response, which is suggested to be dependent on changes in cell bodies of contralateral axotomized motoneurons could also differ in conjunction with differences in the soma reaction to axotomy.

It is possible that inactivity of the denervated muscles could produce sprouting in opposite muscles. The experimental results that contralateral axotomy of the 2nd spinal nerve at its exit from the spinal column and ventral rhizotomy did not produce sprouting whereas more distant axotomies always did, suggest that inactivity could not be a major factor in the induction of sprouting. In other experiments, we altered the sensory input to the spinal cord by producing a tenotomy of one cutaneous-pectoris muscle (Rotshenker 1979) or by cutting the 2nd dorsal root through which sensory information from the muscle reaches the spinal cord (Ring, Sugarman and Rotshenker in preparation). Neither one of the above procedures resulted in transneuronal sprouting and synapse formation.

One prediction of the transneuronal mechanism is that axotomized motoneurons and those responding with sprouting and synapse formation should be located within the same, or at least adjacent, segments of the spinal cord. The interruption of the sciatic nerve in one hind limb induced sprouting of the intact nerve in the contralateral hind limb (Elizalde et al. 1983; Ring et al. 1983) but not of chest cutaneous-pectoris nerves that arise from cervical segments of the spinal cord (Rotshenker 1979).

A second prediction of the transneuronal mechanism is that the delay with which new synapses develop in intact muscles should decrease when the contralateral axotomy is placed closer to the spinal cord. This prediction is based on the assumption that sprouting is dependent on changes in cell bodies of contralateral axotomized motoneurons. It is known that retrograde post-axotomy changes (e.g., chromatolysis, increase in RNA production) develop more rapidly when the site of axotomy is placed closer to the soma (Lieberman 1971, 1974). The predicted results were observed both in frogs and in mice. (Rotshenker 1979, 1982; Rotshenker and Tal 1985).

A third prediction of the transneuronal mechanism relates to metabolic changes in RNA synthesis. Axotomized regenerating motoneurons exhibit an increase in RNA synthesis which is thought to be required for axonal growth. If such is the case then one may expect an increase in RNA synthesis in axotomized cutaneous-pectoris motoneurons followed by RNA synthesis increase in intact contralateral motoneurons. We examined motoneurons, by autoradiography, for the incorporation of ^{3}H-uridine into rapidly labeled RNA after unilatral axotomy (Ring, Reichert and Rotshenker in preparation). The incorporation of uridine in cell bodies of axotomized motoneurons increased 2 days after surgery, peaked at 14 days and then declined to below normal values at days 26 and 34. An increase over normal was also observed in contralateral intact motoneurons at days 14 and 21, and then a decline to below normal values at day 34.

4 The Central Mechanism for the Induction of Sprouting

The transneuronal mechanism for the induction of sprouting formulated above is composed of two steps. First, the initiation of a signal for growth by the axotomy in the injured nerve cell bodies, and second, the transneuronal transfer of the signal to intact motoneurons. We suggest that axotomy initiated a signal for sprouting in the injured nerve cell bodies by preventing the normal supply of a muscle-derived trophic substance to the motoneurons' cell body.

We tested our working hypothesis by designing two experimental conditions that are likely to inhibit the retrograde supply of a trophic substance from the periphery of the nerve cell to its soma, the inhibition of the retrograde axonal transport, on whose integrity the retrograde transport of trophic substances is dependent, and the removal of muscle fibers, which are the source of the hypothetical trophic substance, without injuring the innervating axons.

The inhibition of the retrograde axonal transport in one cutaneous-pectoris muscle was achieved by the local application of colchicine to the nerve in doses that did not produce nerve degeneration or muscle denervation (Rotshenker 1982). Anatomical and electrophysiological examination of the intact cutaneous-pectoris muscle on the opposite side revealed the development of sprouting and synapse formation 2.5 weeks after drug application to the nerve at the axilla. The development of new synapses in the intact muscle following the contralateral inhibition of axonal transport mimicked in magnitude and timing that resulting from contralateral axotomy at the same site.

The removal of muscle fibers of one cutaneous-pectoris muscle was produced by the local application of the myotoxic local anesthetic carbocaine (Tal and Rotshenker 1984). Drug application was followed by partial degeneration of the muscle. Macrophages phagocytized degenerating muscle fibers, leaving behind vacated spaces enclosed by a basal lamina sheath. Motor axons and nerve endings remained structurally intact and retained their ability to recycle synaptic vesicles, as normal (Tal and Rotshenker 1983). Thus intact motoneurons became partially deprived of their target muscle fibers and of trophic substances derived from them. Cutaneous-pectoris muscles were examined electrophysiologically. Sprouting and synapse formation developed in intact muscles about 4 weeks after the application of the drug to opposite muscles. The response was similar in magnitude and timing to that obtained after cutting the cutaneous-pectoris nerve close to the muscle. Sprouting and synapse formation developed also in the drug-treated muscles but about 2.5 weeks after drug application. These results agree well with predictions from our working hypothesis. A signal for growth is expected to develop in the cell bodies of motoneurons that have been deprived of a muscle-derived trophic substance. Therefore, these motoneurons should be the first to sprout. After the signal for growth is transferred transneuronally across the spinal cord to intact motoneurons, the recipient cells should also respond with sprouting, but with some delay. Similar observations were recently made and conclusions reached by Pecot-Dechavassine (1986).

5 Peripheral Mechanisms for the Induction of Sprouting

We define peripheral mechanisms for the induction of sprouting as those that involve the direct effect of a growth-promoting factor on the peripheral extensions of the axon. Such factors could be trophic factors (e.g., NGF), products of denervation, products of degeneration, inactivity, toxins, and others (e.g., reviews by Edds 1953; Brown et al. 1981). The evidence, in in vivo systems, that peripheral mechanisms exist is rather indirect and circumstantial. Recently, we were able to provide direct evidence for the possibility that such mechanisms exist.

In experiments previously described, colchicine was applied to one cutaneous-pectoris nerve at the axilla. Sprouting and synapse formation developed in the intact muscle on the opposite side after 2.5 weeks. Surprisingly, new synapses were also formed by the nerve that was exposed to colchicine, as little as 2 days after drug application (Rotshenker 1981, 1982, see also Riley and Fahlman 1985). The rapid response to drug application (2 days) raised the possibility that colchicine induced sprouting by affecting nerve-endings and axons directly without the immediate involvement of the cell body. To test this possibility, we examined the ability of axons that were separated from their cell bodies to sprout. In the frog, degeneration that follows axotomy is rather slow. Failure of synaptic transmission in cutaneous-pectoris muscles occurred first on the fourth day after cutting the 2nd spinal nerve close to the spine. The pattern of innervation to single muscle fibers did not change from normal during the first 3 days after surgery. Thus, it was possible to cut the 2nd spinal nerve and apply colchicine to the axillary nerve at the same time, and then test the muscle 2 and 3 days later. Anatomical and electrophysiological examination provided evidence for sprouting and synapse formation. Thus, axons that were separated from their cell bodies were still able to respond with sprouting and synapse formation.

6 Conclusions

Our studies in the amphibian and mammalian motor systems suggest that sprouting of intact motoneurons and synapse formation can be regulated by three mechanisms: peripheral, central, and transneuronal. We therefore suggest that under different physiological and pathological conditions, a different set of interactions between those mechanisms will determine whether and to what extent a motoneuron will sprout.

Peripheral mechanisms provide the means of a direct and rapid mode of interaction between the periphery of the cell and the target to determine the extent of target innervation and the size of a synapse. Such interactions could be mediated by either trophic substances, degree of activity, and others. For example, NGF is one trophic agent whose site of action as a nerve growth-promoting factor is documented to be at the periphery of nerve cell (e.g., Campenot 1977; Levi-Montalcini 1982).

The central mechanism enables target muscles to signal the cell bodies of their innervating motoneurons to regulate axonal growth. As proposed by our findings, a regulatory factor could be, for example, a muscle-derived trophic substance that under normal conditions exerts an inhibitory effect on the growth properties of motoneuron. Such a mechanism suggests that in addition to the presence of "grow" signals there are also "stop" signals. This working hypothesis could easily explain observations such as the ability of axotomized motoneurons to regenerate their axons in the absence of any target tissue, and thus most likely in the absence of any nerve growth promoting factor (Kuffler 1986, Rotshenker unpublished results), or the observation that single cultured leach neurons extend neurites in the absence of target cells while bathed in a defined medium (Chiquet and Acklin 1986; Rotshenker unpublished results). The peripheral and central mechanisms suggest, therefore, two sites at which growth can be regulated; at the periphery of the nerve cell (nerve-endings) and at its center (cell body). The target muscle could play an important role in this regulation by means of trophic substances and level of activity.

The transneuronal mechanism provides a vehicle by which the pattern of innervation of a muscle can be altered by nerve cells that themselves do not innervate the muscle. Similar transneuronal effects may also take place in other systems such as the developing oculomotor nerve of the *Xenopus* (Schonenberger et al. 1983), the red nucleus following cross-innervation of limb muscles (Tsukahara and Murakami 1986), and the visual cortex following reversed suture of an eye during the post-natal period (Levay et al. 1980).

References

Brown MC, Holland RL, Hopkins WG (1981) Motor nerve sprouting. Annu Rev Neurosci 4:17–42

Campenot RB (1977) Local control of neurite development by nerve growth factor. Proc Natl Acad Sci USA 74:4516–4518

Chiquet M, Acklin SE (1986) Attachment to con A or extracellular matrix initiates rapid sprouting by culture leech neurons. Proc Natl Acad Sci USA 83:6188–6192

Edds MV (1953) Collateral nerve regeneration. Q Rev Biol 28:260–276

Elizalde AM, Huerta M, Stefani E (1983) Selective reinnervation of twitch and tonic muscle fibers of the frog. J Physiol 340:513–524

Herrera AA, Scott DR (1985) Motor axon sprouting in frog sartorius muscles is not altered by contralateral axotomy. J Neurocytol 14:145–156

Kuffler DP (1986) Isolated satellite cells of peripheral nerve direct the growth of regenerating frog axons. J Comp Neurol 249:57–64

Levay S, Weisel TN, Hubel DH (1980) The development of ocular dominance columns in normal and visually deprived monkeys. J Comp Neurol 191:1–51

Levi-Montalcini R (1982) Developmental neurobiology and the natural history of nerve growth factor. Annu Rev Neurosci 5:341–362

Lieberman AR (1971) The axon reaction: a review of the principle features of perikaryon responses to axon injury. Int Rev Neurobiol 14:49–124

Lieberman Ar (1974) Some factors affecting retrograde neuronal responses to axonal lesions. In: Bellairs R, Gray EG (eds) Essays on the nervous system. Clarendon, Oxford, pp 71–105

Pecot-Dechavassine M (1986) Increase in polyneuronal innervation in frog muscle after muscle injury. J Physiol 371:167–177

Riley DA, Fahlman CF (1985) Colchicine-induced sprouting of endplates on fast and slow muscle fibers in rat extensor digitorum longus, soleus and tibialis anterior muscles. Brain Res 329:83–95

Ring G, Reichert F, Rotshenker S (1983) Sprouting in intact sartorius muscles of the frog following contralateral axotomy. Brain Res 60:313–316

Rotshenker S (1979) Synapse formation in intact innervated cutaneous-pectoris muscles of the frog following denervation of opposite muscle. J Physiol 292:535–547

Rotshenker S (1981) Sprouting and synapse formation by motor axons separated from their cell bodies. Brain Res 223:141–145

Rotshenker S (1982) Transneuronal and peripheral mechanisms for the induction of motor neuron sprouting. J Neurosci 2:1359–1368

Rotshenker S, Reichert F (1980) Motor axon sprouting and site of synapse of formation in intact innervated skeletal muscle of the frog. J Comp Neurol 193:413–422

Rotshenker S, Tal M (1985) The transneuronal induction of sprouting and synapse formation in intact mouse muscle. J Physiol 360:387–396

Schonenberger N, Escher G, Loos H Van Der (1983) Axon number in oculomotor nerves in *Xenopus:* removal of eye primordium affects both sides. Neurosci Lett 41:238–245

Steinbach HJ (1981) Neuromuscular junctions and α-bungarotoxin binding sites in denervated and contralateral cat skeletal muscle. J Physiol 313:513–528

Tal M, Rotshenker S (1983) Recycling of synaptic vesicles in motor nerve endings separated from their target muscle fibers. Brain Res 270:131–133

Tal M, Rotshenker S (1984) Sprouting and synapse formation produced by carbocaine. J Neurosci 4:458–463

Tsukahara N, Murakami F (1986) Correlative morphological and physiological studies on sprouting of excitatory and inhibitory synapses of red nucleus neurons. In: Gilad FM, Gorio A, Kreutzberg GW (eds) Processes of recovery from neuronal trauma. Springer, Berlin Heidelberg New York, p 109

Models, Mechanisms and Kinetics of Neuromuscular Synapse Elimination in Reinnervated Adult Skeletal Muscle

R. R. Ribchester
Department of Physiology, University Medical School, Teviot Place, Edinburgh, EH8 9AG, Great Britain

1 Introductory Overview

The elimination of polyneuronal innervation of muscle fibres in immature skeletal muscle is well documented (see review by Bennett 1983). The main purpose of this chapter is to address briefly the issue of polyneuronal innervation and its elimination in reinnervated adult muscle, especially following injury to part of its nerve supply. The repair of connections in partially denervated muscle sometimes occurs with only limited success. This topic is therefore germane to the failure of reorganization of neuronal convergence and divergence in the injured central nervous system, because it is more comparable than, say, reinnervation of completely denervated muscle, which is usually completely successful.

Anatomists working earlier this century commented on the imperfect nature of newly-formed connections in reinnervated muscle. For instance, Boeke (1921) noted "... complicated aberrant endplates with their exuberance of branches ... gradually disappear .... Superfluous expansions ... seem to be drawn in" (see Fig. 1). Similarly, Ramon y Cajal (1928) described "many duplications and errors of distribution ... incongruences [which] are progressively corrected", during reinnervation of peripheral structures. These early observations suggested that reinnervated muscle fibres receive a transient convergent innervation by motor nerve terminals arising from different motoneurons. Dual innervation of muscle fibres by sprouted and regenerated motor nerve terminals following reinnervation of partially denervated muscles was shown anatomically by Hoffmann (1951). Physiological evidence of polyneuronal innervation of muscle fibres and its subsequent elimination was obtained from recordings of tension overlap on stimulation of separate ventral roots and from intracellular recording of compound endplate potentials in reinnervated muscle (Guth 1962; McArdle 1975; Thompson 1978; Brown and Ironton 1978).

The functional significance of neuromuscular synapse elimination, in development or repair, is that it constrains movement by bringing about an adjustment in the numbers of motor nerve terminals supplied by a motoneuron; that is, motor unit size. At least three properties of regenerating axons influence the numbers of muscle fibres that they ultimately recover. These are activity, the number and size of competing motor units, and the time taken for axons to regenerate into the target muscle. Thus paralysis of reinnervated muscle promotes retention of synapses and stimulation accelerates synapse elimination (Taxt 1983; Duchen and

H. Flohr (Ed.)
Post-Lesion Neural Plasticity

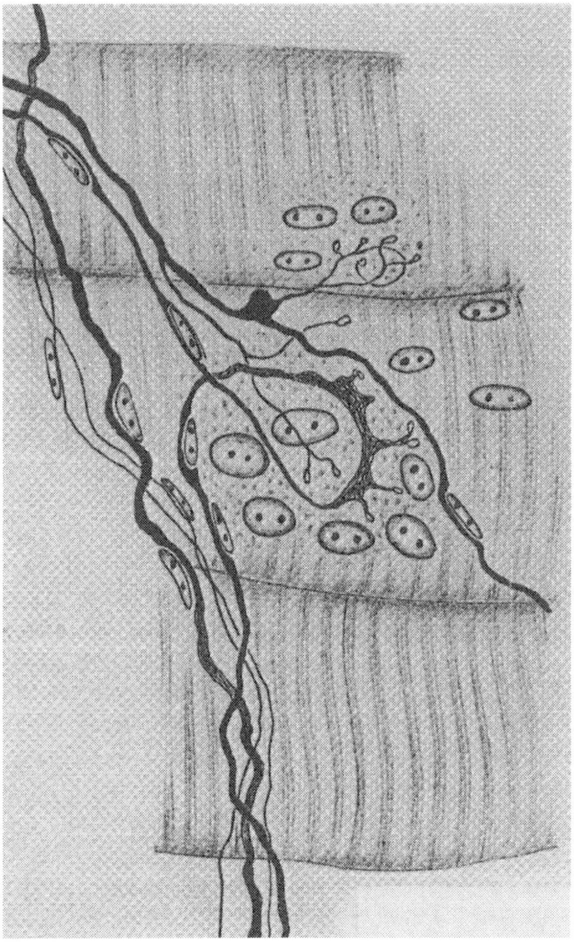

Fig. 1. Drawing made by Boeke (1916) of motor nerve terminals in a reinnervated intercostal muscle of the hedgehog. Branches of two motor axons converge on the same motor endplate. One of the axons is illustrated as forming a synapse en passant

Tonge 1977; Jansen et al. 1973; Magchielse and Meeter 1986). Selective inhibition of motor unit activity promotes stabilization of the more active nerve terminals (Ribchester 1986; but see Callaway et al. 1987). Secondly, when axons regenerate into completely innervated muscles, terminals belonging to the intact, larger motor units regress (Brown and Ironton 1978; Thompson 1978; Bixby and Van Essen 1978). Finally, the longer axons take to regrow to their targets, the smaller the size of the regenerating units (Frank et al. 1975; Thompson 1978; Brunetti et al. 1985).

A number of hypothetical mechanisms of elimination of immature synapses have been proposed (e.g. Gouzé et al. 1983; Willshaw 1981; Purves and Lichtman 1985; Nelson and Brenneman 1982; O'Brien et al. 1978, 1984). What these models have in common is the notion that synapses rely on latent external crises in order to maintain their structure and function. The idea is that the stability of the synapse in the normal state is due to a balance of forces tending to displace it and forces promoting its expansion. Nerve injury stimulates the forces leading to expansion (that is, sprouting and synapse formation), whereas ingrowth of excessive innervation favours neurite regression. Through this balance, the functional requirements – for growth at early stages of innervation, cessation of growth once connections are established, and regression of superfluous connections thereafter – is be successfully controlled.

Models of synaptic stabilization differ, however, in the supposed loci of origin and action of hypothetical stabilizing and de-stabilizing factors. For instance, Gouzé et al. (1983) postulate the existence of a muscle-derived trophic factor, μ, which is released from muscle in an activity-dependent manner: that is, denervation or paralysis enhances its release. This factor is supposed to act on convergent pre-synaptic terminals at the motor end-plate and to be preferentially taken up by the more active terminals. Inside the terminal, μ acts permissively in stimulating synthesis of a second, terminal stabilizing factor, s. Further synthesis of s depends on an autocatalytic process. Simulations using this model can predict the selective stabilization of active synapses when active and inactive terminals converge on the same motor end-plates (e.g. Ribchester and Taxt 1983). But this model does not really provide a satisfactory explanation for the formation of large numbers of stable, functioning connections in completely paralyzed muscle (e.g. Brown et al. 1982; Taxt 1983). By comparison, O'Brien et al. (1978) proposed the existence of a degrading substance, specifically a proteolytic enzyme that may be a calcium-activitated neutral peptidase (O'Brien et al. 1984), produced either by the muscle fibres or in the pre-synaptic terminals. More of this is supposed to be produced at active neuromuscular synapses than inactive ones. Formal analysis of this model (Willshaw 1981) predicts the acceleration of synapse elimination in stimulated muscle and its retardation in paralyzed muscle but the model is unsatisfactory in explaining selective stabilization during competition between active and inactive terminals converging on the same motor end-plates.

What is missing from both kinds of model is an empirical kinetic description of events during synaptic rearrangement in vivo. The potential value of kinetic descriptions is to yield insight into cellular mechanisms and suggest strategies for further investigation (see Hille 1984, for a clear illustration of this). I outline here an attempt to describe reinnervation kinetically. A full account of the experimental data on which the description is based is published elsewhere (Ribchester 1988).

2 Theoretical Considerations

Suppose muscle fibres can acquire neuronal input from two sets of motoneurons, S or L. The fibres can exist in one of three states (excluding complete denervation): innervated by S only, by L only, or receiving convergent input from S and L. Suppose there is a simple equilibrium between these states

$$\mathrm{S} \underset{k_{-1}}{\overset{k_1}{\rightleftharpoons}} \mathrm{DUAL} \underset{k_{-2}}{\overset{k_2}{\rightleftharpoons}} \mathrm{L}\,,$$

and the rate constants k_i are first order, their absolute magnitude determined by the number of motoneurons in either set. If we consider the system with all the muscle fibres initially in state S, the achievement of equilibrium should occur in a predictable way – analogous to that in sequential, reversible chemical reactions. The appropriate equations have been derived (Dawes 1972). Specifically,

$$\frac{d}{dt}[S]=k_{-1}[DUAL]-k_1[S]$$

$$\frac{d}{dt}[DUAL]=k_1[S]+k_{-2}[L]-(k_{-1}+k_2)[DUAL]$$

$$\frac{d}{dt}[L]=k_2[DUAL]-k_{-2}[L].$$

The total numbers of muscle fibres, M, is a constant so:

$$M=S+L+DUAL.$$

The explicit solutions of the differential equations are:

$$S(t) \quad =c1\cdot\exp(b1\cdot t)+c2\cdot\exp(b2\cdot t)+M(k_{-1}\cdot k_{-2})/D \tag{1}$$

$$DUAL(t)=c3\cdot\exp(b1\cdot t)+c4\cdot\exp(b2\cdot t)+M(k_1\cdot k_{-2})/D \tag{2}$$

$$L(t) \quad =c5\cdot\exp(b1\cdot t)+c6\cdot\exp(b2\cdot t)+M(k_1\cdot k_2)/D, \tag{3}$$

where $D=k_1\cdot k_2+k_{-1}\cdot k_{-2}+k_1\cdot k_{-2}$.

The constants ci, b1 and b2 are obtained from the solution of a set of simultaneous equations, viz.

$$b1+b2=-(k_1+k_{-1}+k_2+k_{-2})$$

$$b1\cdot b2=k_1\cdot k_2+k_{-1}\cdot k_{-2}+k_1\cdot k_{-2}$$

$$c1\cdot b1+c2\cdot b2=-k_1\cdot[M]$$

$$c1+c2=[M]\cdot(k_1\cdot k_2+k_{-2}\cdot k_1)/D$$

$$c3\cdot b1+c4\cdot b1=k_1\cdot[M]$$

$$c3+c4=-[M]\cdot(k_1\cdot k_{-2})/D$$

$$c5\cdot b1+c6\cdot b1=0$$

$$c5+c6=-[M]\cdot(k_1\cdot k_2)/D.$$

Example solutions are shown in Fig. 2. If the forward and backward rate constants were equal, then the system would converge toward equilibrium with equal amounts of muscle innervated by S, L or receiving DUAL innervation (Fig. 2A). If the backward rate constants were zero, then there would be a transient rise in dual innervation, but ultimately all the muscle fibres would be innervated by neurones in set L (Fig. 2B). Unequal, non-zero rate constants would predict an intermediate outcome.

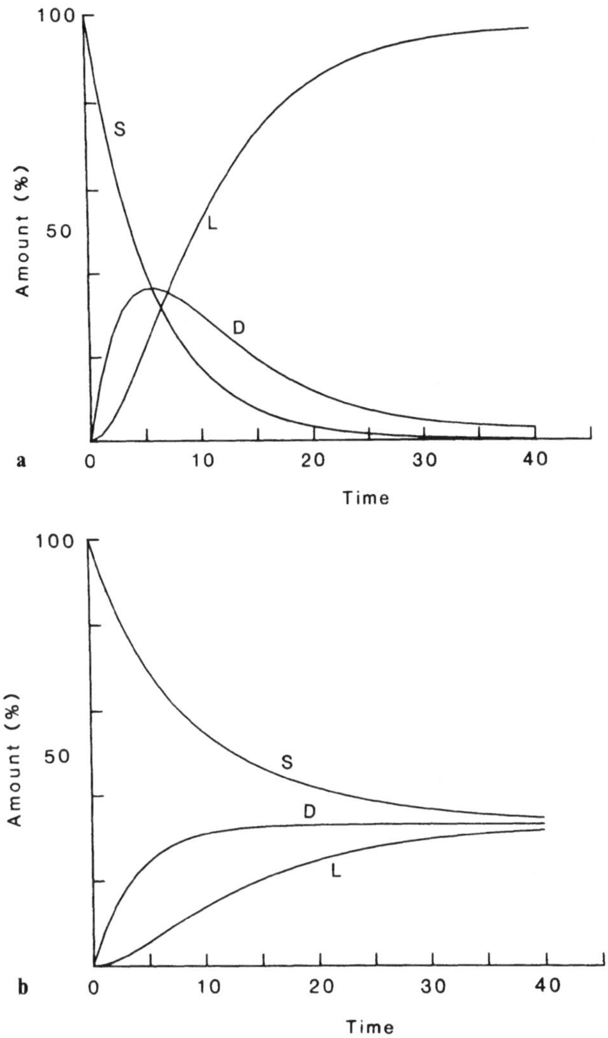

Fig. 2. Predicted regression of mononeuronal innervation of muscle fibres by a sprouted population of motoneurons, *S* in partially denervated muscle when axons belonging to a second population; *L* return. The changes in the amounts of mononeuronal innervation and the amount of convergent (*D*) innervation are calculated by Eqs. (1)–(3). In **a** the rate constants are all equal. In **b** the backward rate constants are zero

3 Experimental Design and Results

A state of complete initial innervation by one set of neurons, followed by competition with a regenerating set, was engineered using the distributed motor nerve supply to the rat hind foot. In rodents, the motoneurons innervating plantar flexor muscles, lumbricals and interossei are supplied by the medial plantar nerve, the lateral plantar nerve (LPN) and a branch of the sural nerve (SN) which forms a natural anastomosis with the LPN near the ankle (Nakanishi and Norris 1970;

Peyronnard and Charron 1982). The SN supplies between one and five of the motor motor axons innervating the fourth deep lumbrical muscle (Betz et al. 1979, 1980a). This muscle normally contains a total of about ten motor units. In normal, unoperated lumbrical muscles the tension evoked by SN or LPN stimulation depends on the number of motor units and the numbers of muscle fibres supplied by the two nerves. In muscles receiving three SN motor units, for instance, the tension produced by SN stimulation is about 30% of the total tension, while stimulation of the LPN generates about 70% of the total tension (Betz et al. 1979).

Adult female rats were anaesthetized with sodium pentobarbitone (Sagatal, 60 mg/kg; i.p.) and the hind foot was partially denervated by crushing the LPN and the medial plantar nerve with fine forceps, proximal to the SN anastomosis (see for example, Betz et al. 1980b). Isometric twitch and tetanic tension measurements were made from isolated nerve-muscle preparations 15–50 days later, according to methods described previously (Ribchester and Taxt 1983). Comparison of the total muscle tension produced by combined nerve stimulation (Fig. 3), with the arithmetic sum of the tension produced by stimulating the LPN or SN separately (the tension overlap) gives an estimate of the proportion of DUAL innervation and the proportion of the muscle innervated exclusively by LPN or SN motor axons:

$$\%\text{DUAL} = (\text{SN} + \text{LPN} - \text{COMBINED}) \cdot 100/\text{COMBINED}$$

$$\%\text{LPN only} = (1 - \text{SN}/\text{COMBINED}) \cdot 100$$

$$\%\text{SN only} = 100(\%\text{DUAL} + \%\text{LPN only}) .$$

Partial denervation of the lumbrical muscles by LPN crush leaves most muscles innervated by the few motor axons supplied by the sural nerve. Expansion of the SN motor units then occurs by motor nerve sprouting (Betz et al. 1980b). By 16 days, the LPN motor axons return to the lumbrical muscle (Ribchester 1984). Provided three or more SN motor units remain, the tension they produce is not significantly different from the total muscle tension evoked by direct muscle stimulation. Thus SN motor units are capable of sprouting to at least three times their normal size following lateral plantar nerve crush. This is a similar order of magnitude to that seen previously in partially denervated rat soleus muscle, for instance (Thompson 1978). Recovery of LPN tension is accompanied by regression of the sprouted SN motor units (Fig. 3).

Figure 4 shows the change in exclusive SN innervation, exclusive LPN innervation and DUAL innervation with time, calculated from the tension measurements in muscles containing three SN motor units. Day zero is taken as 17 days after LPN crush, the mean time found experimentally for the damaged axons to produce any tension in the muscles. These data show that the proportion of the muscle innervated by the three SN motor axons fell from 100% to about 50% over the 40 days following return of the LPN motor axons, whereas the percentage of the muscle receiving exclusive innervation by the LPN axons (mean, seven motor units) reached only about 40%, over the same period. Dual innervation reached a peak of only about 20% of the muscle fibres by 10 days after the LPN axons returned and remained at about 10% throughout the remaining period.

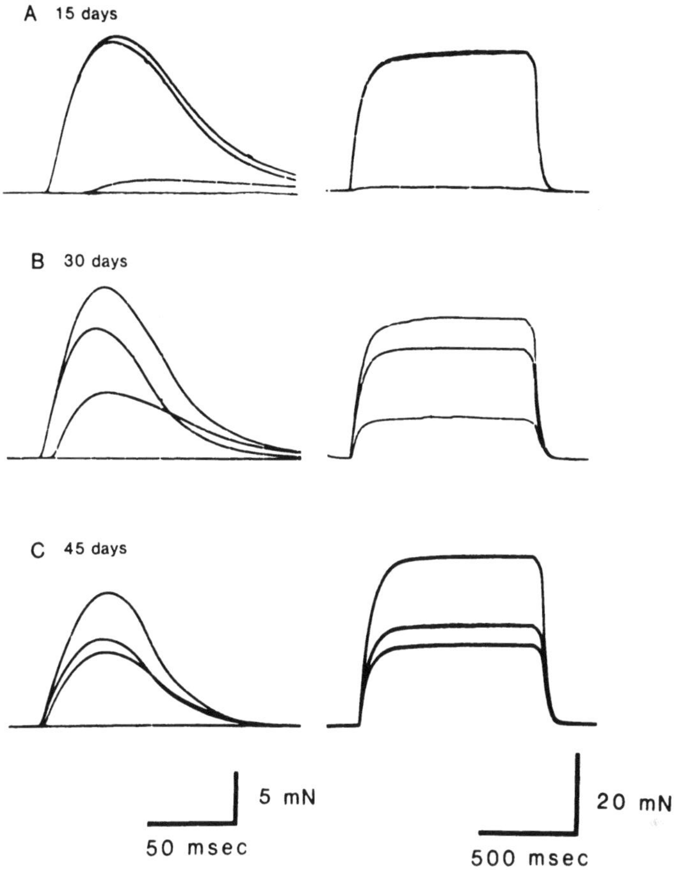

Fig. 3. Examples of isometric twitch *(left)* and tetanic *(right)* tension measurements in partially denervated lumbrical muscles after crushing the lateral plantar nerve, showing regression of sprouted motor units and recovery of regenerated motor units with time. In each case, the *lowest trace* is the response to stimulating the regenerating, LPN axons; the *middle trace* is the response to stimulating intact, SN motor axons; and the *top trace* is due to combined nerve stimulation. The recovery of the LPN responses was accompanied by a decrease in the latency of the isometric twitch, as conduction velocity in the regenerating axons also recovered. A, 15 days after LPN crush; muscle containing four SN motor units. B. 30 days post crush; three SN motor units; C. 45 days post crush; three SN motor units

This confirms that regenerating motor axons have an impaired capacity to effect functional recovery following reinnervation of a partially denervated muscle.

Thus there are three features of the reinnervated muscles that seem problematical. First, levels of dual innervation never become particularly high despite significant reinnervation by the regenerating nerve. Second, some dual innervation of muscle fibres persists until late stages of reinnervation. Third, the regression of the sprouted motor axons is incomplete and the recovery of LPN motor units is smaller than expected based on the numbers of intact and regenerating axons.

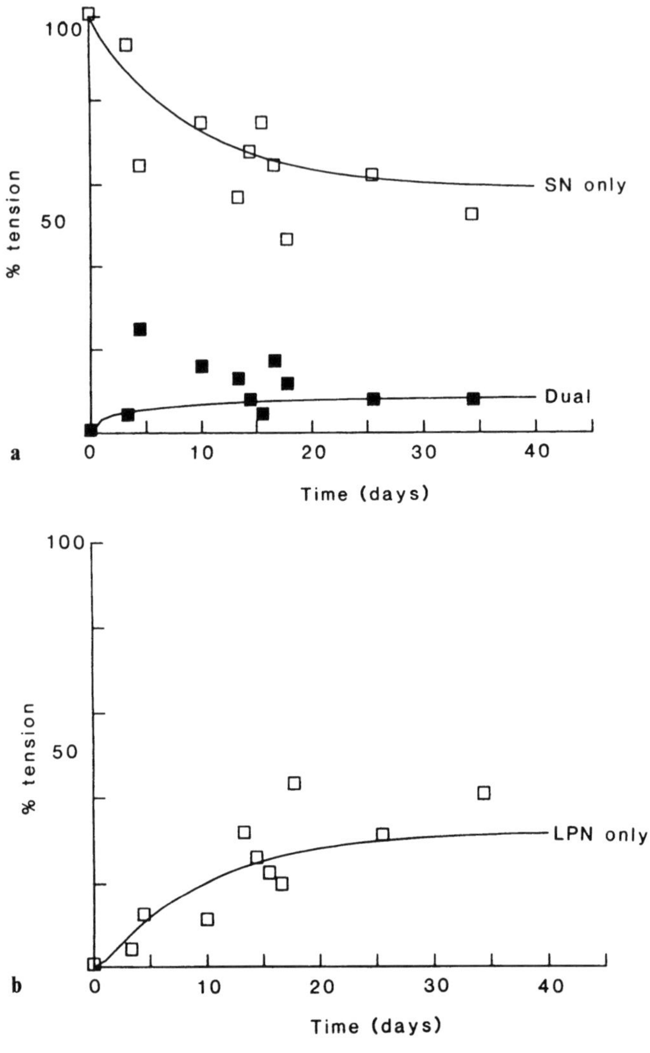

Fig. 4. Data on proportion of reinnervated lumbrical muscle fibres innervated exclusively by SN motor nerve terminals, exclusively by LPN motor nerve terminals, and receiving convergent (DUAL) innervation by SN and LPN motor axons, 17–25 days crushing the LPN. Day zero is taken as 17 days after nerve crush, the mean time found experimentally for the damaged LPN motor axons to return to the muscle. All the data are from muscles containing precisely three SN motor units. The *solid curves* were computed from Eqs. (1)–(3) with values for the rate constants $k_2 = 1.0$ days^{-1} and $k_{-2} = 0.1$ days^{-1}. The ratios k_2:k_{-1} and k_{-2}:k_1 were 7:3 – that is, similar to the ratio of the numbers of LPN to SN motor units. In **a** *open squares* are calculated data on exclusive SN innervation, *filled squares* the convergent innervation calculated from the tension overlap. In **b** *open squares* are data on exclusive LPN innervation calculated from the tension overlap

An attempt to reconcile these observations using the kinetic model is also shown in Fig. 4. The predicted variation in amounts of SN only, DUAL innervation and LPN only were obtained empirically using Eqs. (1)–(3). The sizes of rate constants that determine the transition between SN, LPN and DUAL innervated states were made proportional to the numbers of motor axons supplied by the two nerves. Thus the ratio of the rate constants k_1:k_{-2}=k_{-1}:k_2 was 3 (SN) to 7 (LPN). Despite the noise in the data (inevitable given the relatively small number of elements in the system) the model appears to provide a quantitatively-acceptable description of the main experimental findings.

4 Conclusions and Discussion

The kinetic analysis provides a quantitative description of the dynamic interactions that occur between motor nerve terminals when they converge on muscle fibres. An important implication is that although the overall distribution of motor unit tensions in the reinnervated muscles can be predicted on the basis of the numbers of motor units supplied by regenerating and intact motor nerves, at the level of single muscle fibres it may be impossible to predict which terminals are stabilized and which are subseqeuntly eliminated, or when during their interaction this might occur. The description further predicts that when regenerating and intact nerve terminals compete for exclusive innervation of a muscle fibre, there is a finite probability that the *regenerating* terminal may be eliminated. This is unexpected in light of contemporary ideas about the role of disparities in motor unit size in determining the outcome of neuromuscular synapse elimination (Brown et al. 1976; Smallheiser and Crain 1984).

It is not clear what determines the size of the forward and backward rate constants in the kinetic description. It is perhaps significant that an acceptable fit was obtained by assuming that the rate constants were proportional to the numbers of competing motor axons. This suggests that these rate constants might be determined by the levels of factors intrinsic to the motoneurons, and further that the amounts of these factors might be constant between motoneurons. The effects of either selective or non-selective changes in activity could perhaps be predicted if activity altered one or more of the rate constants in the kinetic equations.

A possible mechanism controlling synaptic competition, consistent with the kinetic description, might result from interactions involving two substances whose functions are antagonistic: a promoting substance arising from skeletal muscle which stimulates neurite growth, and a growth retarding substance which activates neurite withdrawal (Fig. 5). The latter could either be expressed on the surface, or be secreted by immature or growing motor nerve terminals. Together, these two substances could operate a "push-pull" control system analogous to that regulating other kinds of internal physiological variables (e.g. blood glucose). This kind of control would predict that in denervated, paralyzed, or newly-reinnervated muscle there would be excesses of both growth-promoting and repressing factors. As suggested by others (e.g. Henderson et al. 1984; Dohrman et al. 1986; Gurney et al. 1986), the levels of growth promoter would decline as muscle activity was restored. The balance would then tip in favour of neurite re-

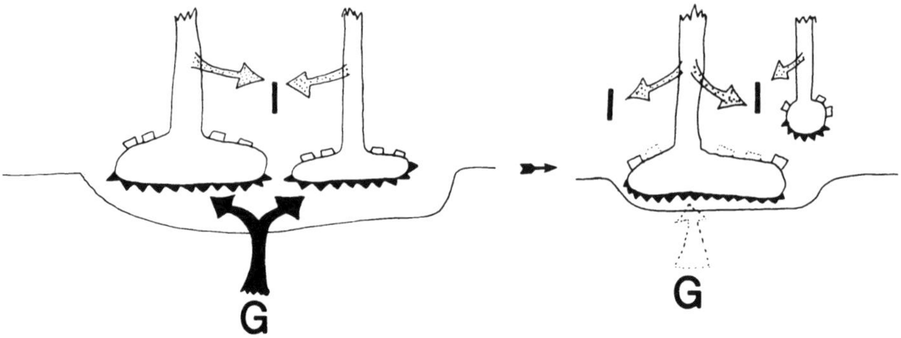

Fig. 5. Hypothetical model of neuromuscular synapse elimination consistent with the kinetics and experimental observations on the reorganization of motor units following normal reinnervation, or reinnervation combined with partial or complete paralysis. Muscle-derived growth-promoting factor, *G*, acts on receptors *(solid triangles)* on the motor nerve terminals to stimulate sprouting. Release of G is enhanced by denervation or paralysis and suppressed by normal muscle activity. Nerve-derived synaptic repressor, *I*, acts on receptors *(open rectangles)* to stimulate regression. Expression of I (which could be diffusible or a cell surface factor) may be partly activity dependent, and a mechanism should exist for inactivation of its expression and/or down regulation of receptors for I as the motor nerve terminals mature. Immature, ingrowing terminals might be potent sources of I, and their membranes loaded with receptors for G and I. The rates of regression of synapses would depend on binding kinetics of I and G for their receptors. Evidence exists for molecules with properties predicted for G, but the existence of I is speculative at present

pressor released from the growing nerve terminals. Paralyzed muscles would continue to produce the growth promoter (Henderson et al. 1986), and in stimulated muscles its production might be turned off more rapidly. The balance of the promoter and repressor substances would also be altered at motor end-plates receiving convergent input from terminals which differed in their activity.

If the growth repressor were a surface factor it might participate in some form of contact inhibition of growth, like that described for growth cones by Kapfhammer and Raper (1987). It is perhaps difficult to see how this could be activity-dependent. If the factor were diffusible, however, mechanisms would be required to insure against autolytic effects. For instance, as surviving terminals matured the production of neurite growth repressor would have to decline, or receptors for it on nerve terminals would have to be down-regulated or disappear. Alternatively, the regression factor could act in a voltage-dependent manner, such that active nerve terminals (i.e. the ones releasing regression factor) would be immune to its effects. There are a number of receptors showing voltage-dependent gating behaviour in the central nervous system: for instance, the receptor for N-methyl-D-aspartate (NMDA) is associated with a calcium channel which is subject to a voltage-dependent block by magnesium ions under normal physiological conditions (Nowak et al. 1984). In passing, it is perhaps worthy of note that the NMDA receptor has been implicated in a number of forms of synaptic plasticity (Bear et al. 1987), and NMDA receptors are expressed in motoneuron membranes (O'Brien and Fischbach 1986).

Another possible mechanism of regression is suggested by experiments in which depolarization of motor nerve terminals, produced pharmacologically or by raised extracellular potassium ion concentration, inhibits neurite growth and stimulates synapse elimination (Fishman and Nelson 1981; Campenot 1986). Vrbova et al. (1988) suggest that elevation of extracellular potassium, caused by a potassium efflux during muscle activity, could depolarize inactive nerve terminals, promote a calcium influx, activate a peptidase and precipitate synapse elimination. One problem with this idea is that estimates of potassium efflux from axons during activity indicate that the concentrations of K^+ which have an effect on neurite growth in vitro are unlikely to occur in the synaptic cleft (Attwell and Iles 1979).

It may be that we are only likely to learn more about the mechanisms of synaptic competition by studying cellular interactions continously: observing the properties of specific identified neurons and their targets over long periods of time. Studies on nerve terminal form in vivo (Lichtman et al. 1987) and recordings of synaptic potentials in the same cells on different days in vitro (e.g. Fuchs et al. 1981; Magchielse and Meeter 1986) suggest that repeated observations of identified synapses made over long periods are feasible in principle.

Acknowledgement. This work was supported by grants from the MRC and Action Research for the Crippled Child.

References

Attwell D, Iles J (1979) Synaptic transmission: ion concentrations in the synaptic cleft. Proc R Soc Lond B Biol Sci 206:115–131

Bear M, Cooper L, Ebner F (1987) A physiological theory for synapse modification. Science 237:42–48

Bennett M (1983) Development of neuromuscular synapses. Physiol Rev 63:915–1048

Betz W, Caldwell J, Ribchester R (1979) The size of motor units during postnatal development of rat lumbrical muscle. J Physiol 297:463–478

Betz W, Caldwell J, Ribchester R (1980a) The effects of partial denervation at birth on the development of muscle fibres and motor units in rat lumbrical muscle. J Physiol 303:265–279

Betz W, Caldwell J, Ribchester R (1980b) Sprouting of active nerve terminals in partially inactive muscles of the rat. J Physiol 303:281–297

Bixby J, Essen D Van (1978) Competition between foreign and original nerves in adult mammalian skeletal muscle. Nature 282:726–728

Boeke J (1916) Studien zur Nervenregeneration. I. Die Regeneration der motorischen Nervenelemente und die Regeneration der Nerven der Muskelspindeln. Verh K Acad Wet Amst 18(6):1–120

Boeke J (1921) The innervation of striped muscle fibres and Langley's receptive substance. Brain 44:1–22

Brown M, Ironton R (1978) Sprouting and regression of neuromuscular synapses in partially denervated mammalian muscles. J Physiol 278:325–348

Brown M, Jansen J, Van Essen D (1976) Polyneuronal innervation of skeletal muscle in newborn rats and its elimination during maturation. J Physiol 261:387–422

Brown M, Hopkins W, Keynes R (1982) Short and long term effects of paralysis on the motor innervation of two different neonatal mouse muscles. J Physiol 329:439–450

Brunetti O, Magni F, Pazzaglia U (1985) Neural nature of the time dependent factors involved in motor reinnervation. Exp Neurol 90:677–683

Callaway E, Soha J, Van Essen D (1987) Competiton favouring inactive over active motor neurones during synapse elimination. Nature 328:422–426

Campenot R (1986) Retraction and degeneration of sympathetic neurites in response to locally elevated potassium. Brain Res 399:357–363

Dawes E (1972) Quantitative problems in biochemistry, 5th edn. Livingstone, London, pp 163–165
Dohrmann U, Edgar D, Sendtner M, Thoenen H (1986) Muscle-derived factors that support survival and promote fiber outgrowth from embryonic chick spinal motor neurons in culture. Dev Biol 118:209–221
Duchen L, Tonge D (1977) The effects of implantation of an extra nerve on axonal sprouting usually induced by botulinum toxin in skeletal muscle of the mouse. J Anat 124:205–216
Fishman M, Nelson P (1981) Depolarization-induced synaptic plasticity at cholinergic synapses in tissue culture. J Neurosci 1:1043–1051
Frank E, Jansen J, Lømo T, Westgaard R (1975) The interaction between foreign and original motor nerves innervating the soleus muscle of rats. J Physiol 247:725–743
Fuchs P, Nicholls J, Ready D (1981) Membrane properties and selective connexions of identified leech neurones in culture. J Physiol 316:203–223
Gouzé JL, Lasry JM, Changeux JP (1983) Selective stabilization of muscle innervation during development: a mathematical model. Biol Cybern 46:207–215
Gurney M, Appattoff B, Heinrich S (1986) Suppression of terminal axonal sprouting at the neuromuscular junction by monoclonal antibodies against a muscle-derived antigen of 56000 daltons. J Cell Biol 102:2264–2272
Guth L (1962) Neuromuscular function after regeneration of interrupted nerve fibres into partially denervated muscle. Exp Neurol 6:129–141
Henderson C, Huchet M, Changeux JP (1984) Neurite promoting activities for embryonic spinal neurons and their developmental changes in the chick. Dev Biol 104:336–347
Henderson C, Benoit P, Huchet M, Guenet JG, Changeux JP (1986) Increase of neurite promoting activity for spinals neurons in muscles of paralysé mice and tenotomized rats. Dev Brain Res 25:65–70
Hille B (1984) Ionic channels in excitable membranes. Sinauer, Sunderland, pp 45–53
Hoffmann H (1951) Fate of interrupted nerve fibres regenerating into partially denervated muscle. Aust J Expl Med Sci 29:211–219
Jansen J, Lømo T, Nicolaysen K, Westgaard R (1973) Hyperinnervation of skeletal muscle fibres: dependence on muscle activity. Science 181:559–561
Kapfhammer J, Raper J (1987) Collapse of growth cone structure on contact with specific neurites in culture. J Neurosci 7:201–212
Lichtman J, Magrani L, Purves D (1987) Visualization of neuromuscular junctions over periods of several months in living animals. J Neurosci 7:1215–1222
Magchielse T, Meeter E (1986) The effect of neuronal activity on the competitive elimination of neuromuscular junctions in tissue culture. Dev Brain Res 25:211–220
McArdle J (1975) Complex end-plate potentials at the regenerating neuromuscular junction of the rat. Exp Neurol 49:629–638
Nakanishi T, Norris F (1970) Motor fibers in the rat sural nerve. Exp Neurol 26:433–435
Nelson P, Brenneman D (1982) Electrical activity of neurons and development of the brain. Trends Neurosci 5:229–232
Nowak L, Bregeskovski P, Ascher P, Herbet A, Prochiantz A (1984) Magnesium gating of glutamate-activated channels in mouse central neurons. Nature 307:462–465
O'Brien R, Fischbach G (1986) Characterization of excitatory amino acid receptors expressed by embryonic chick motoneurons in vitro. J Neurosci 6:3284–3289
O'Brien R, Ostberg A, Vrbovà G (1978) Observations on the elimination of polyneuronal innervation in developing mammalian skeletal muscle. J Physiol 282:571–582
O'Brien R, Ostberg A, Vrbovà G (1984) Protease inhibitors reduce the loss of nerve terminals induced by activity and calcium in developing rat soleus muscle. Neuroscience 12:637–646
Peyronnard JM, Charron L (1982) Motor and sensory neurones of the rat sural nerve: a horseradish peroxidase study. Muscle Nerve 5:654–660
Purves D, Lichtman J (1985) Principles of neural development. Sinauer, Sunderland, pp 295, 361
Ramon y Cajal S (1928) Degeneration and regeneration of the nervous system. Oxford University Prss, Oxford, p 279
Ribchester R (1984) Non-selective elimination of motor nerve terminals after reinnervation of partially denervated rat lumbrical muscles. J Physiol 346:25 P

Ribchester R (1986) Neural activity and the reorganization of motor units in reinnervated skeletal muscle. In: Nix W, Vrbovà G (eds) Electrical stimulation and neuromuscular disorders. Springer, Berlin Heidelberg New York, pp 53–63

Ribchester RR (1988) Activity dependent and independent synaptic interactions during reinnervation of partially denervated rat muscle. J Physiol 401:53–75

Ribchester R, Taxt T (1983) Motor unit size and synaptic competition in rat lumbrical muscles reinnervated by active and inactive motor axons. J Physiol 344:89–111

Smallheiser N, Crain S (1984) The possible role of "sibling neurite bias" in the co-ordination of neurite extension, branching and survival. J Neurobiol 15:517–529

Taxt T (1983) Local and systemic effects of tetrodotoxin on the formation and elimination of synapses in reinnervated adult rat muscle. J Physiol 340:175–194

Thompson W (1978) Reinnervation of partially denervated rat soleus muscle. Acta Physiol Scand 103:81–91

Vrbovà G Lowrie MB, Evers J (1988) Reorganization of synaptic inputs to developing skeletal muscle fibres. In: Plasticity of the neuromuscular system. Ciba Foundation Symposium No. 138 (in press)

Willshaw D (1981) The establishment and subsequent elimination of polyneuronal innervation of developing muscle: theoretical considerations. Proc R Soc Lond Biol Sci 212:233–252

Elimination of a Single Axon Changes Synaptic Characteristics of the Other Axons Innervating a Muscle Fibre of Lobster

J. Dudel[1] and I. Parnas[2]

[1] Physiologisches Institut der Technischen Universität München, Biedersteiner Straße 29, D-8000 München 40, FR Germany

[2] The Otto Loewi Center for Cellular and Molecular Neurobiology, The Hebrew University, Jerusalem, Israel

1 Introduction

The neuromuscular system of crustaceans is simple enough to allow the study of plastic changes and interactions between neurons innervating the same postsynaptic cell. Each muscle fibre may be innervated by two or more excitatory axons and one or two inhibitory axons (Atwood 1977). The excitatory transmitter is glutamate (Kawagoe et al. 1982), and the inhibitory one γ-amino butyric acid (GABA) (Otsuka et al. 1966). The axons innervate each muscle fibre at many points (multiterminal innervation) and the presynaptic as well as postsynaptic properties of the synapses were studied in great detail (Dudel and Kuffler 1961 a, b; Atwood 1977; Dudel 1981; Parnas H et al. 1982, 1986; Parnas I et al. 1982 b, c, 1986; Dudel et al. 1983; Franke et al. 1986). Modulation of release can be produced by a variety of treatments such as long-term stimulation (Atwood and Wojtowicz 1986), applications of modulators (Kravitz et al. 1985), changes occurring during development (Atwood and Kwan 1976) or after microlesions (Parnas I et al. 1982 a). Thus, the neuromuscular system of crustaceans, although being more complicated than the frog neuromuscular junction (see Pecot-Dechavassine 1986), is still simple enough to enable the study of changes that occur after partial denervation of excitatory or inhibitory inputs. For example, if the muscle is denervated by the removal of one excitatory axon, will the fibre show supersensitivity as found in the frog after denervation? If the inhibitory axon is removed, will changes occur in the remaining excitatory axons innervating the same cell? If an axon with a "strong" synaptic input is eliminated, will the weaker axon become stronger? Such questions certainly cannot be answered in the vertebrate neuromuscular system, and are very difficult to analyze in the vertebrate CNS.

While in vertebrates cutting a nerve trunk is a useful technique to produce degeneration of the peripheral stump, in crustaceans and several other invertebrates severing a nerve does not produce degeneration of the peripheral stump (Hoy et al. 1967; Hoy 1969). In some cases the central and peripheral axon sections even fuse to form the correct functional connectivity. In order to be able to study changes after partial denervation, a technique was required to eliminate single neurons with their arborizations without damage to neighbouring neuronal branches. This was achieved by intracellular injection of the proteolytic enzyme pronase (Parnas and Bowling 1977; Bowling et al. 1978), which diffuses

H. Flohr (Ed.)
Post-Lesion Neural Plasticity

throughout the cell arborizations and dissolves the injected neuron. Such an injection does not produce any detectable damage even to adjacent neuronal arborizations. Using this technique it was shown, in the leech, than when one motoneuron (the annulus erector) was removed, only homologous neurons of the adjacent segments expanded their field of innervation to restore the functional activity of the vacant territory. In a similar way, when sensory neurons were removed, only the remaining sensory neurons of the same modality expanded their fields of innervation into the vacant skin regions (Blackshaw et al. 1982). Thus, using the technique of intracellular injection of pronase, we could produce specific denervations, excitatory or inhibitory in the lobster, and check for changes in the remaining neurons innervating the same muscle fibres.

The Deep Abdominal Extensor Muscles (DEAM) of the Lobster

As our experiments were performed on the DEAM of the lobster *Panulirus penicillatus,* a brief description of the anatomical and "electrophysiological" organization is given. The DEAM are used by the lobster to extend its tail, they are found on the dorsal side above the deep flexor muscles and below the superficial extensor muscles. On each side (Fig. 1 a) the muscles are composed of the medial, M, and the two lateral muscles L_1 and L_2. In each segment five axons innervate the three muscles. The ExM, the ExL_1 and ExL_2 are specific excitatory axons innervating each muscle. In addition, the M and L_1 muscles share a common excitatory axon, ComEx, and the three muscles are also innervated by a common inhibitor, the ComIn (Parnas and Atwood 1966). Our experiments were done mainly on the L_1. The innervation pattern of this muscle is given in greater detail in Fig. 1 b. The anatomical organization of the fibres in this muscle is such that the muscle is divided by a clear line into an anterior and a posterior bundle. While the ComEx in each segment innervates both bundles, the ExL_1 innervates only the posterior bundle. However, the ExL_1 from one segment emits a posterior branch to innervate also the anterior bundle of the adjacent posterior segment. Thus, each muscle fibre in a given segment is innervated by two excitatory axons. All muscle fibres of the two bundles are innervated by the ComIn.

The development of the macropatch recording technique (Dudel 1981) enabled the quantitative characterization of the synaptic inputs to this muscle. The macropatch electrode (about 20 μm inner diameter and 40 μm outer diameter) could glide along the muscle fibre until it reached a release site. Only when the electrode was over a release site, was the synaptic current seen as an inward current with short rise time. At low extracellular Ca^{2+} concentration (6.5 mM Ca^{2+}), most of the synaptic responses obtained are composed of a single quantum and failures in release are seen. Samples of such responses are given in Fig. 2. With this technique, the number of quanta released over many trials (128–256 stimuli) can be directly counted and the average quantum content determined. It should be remembered that the quantum content is controlled by presynaptic mechanisms, while the amplitude and shape of the quantum response (rise time, decay time) reflect postsynaptic mechanisms. As mentioned, the muscles receive multiterminal innervation and the distance between release sites was found to be about 100 μm.

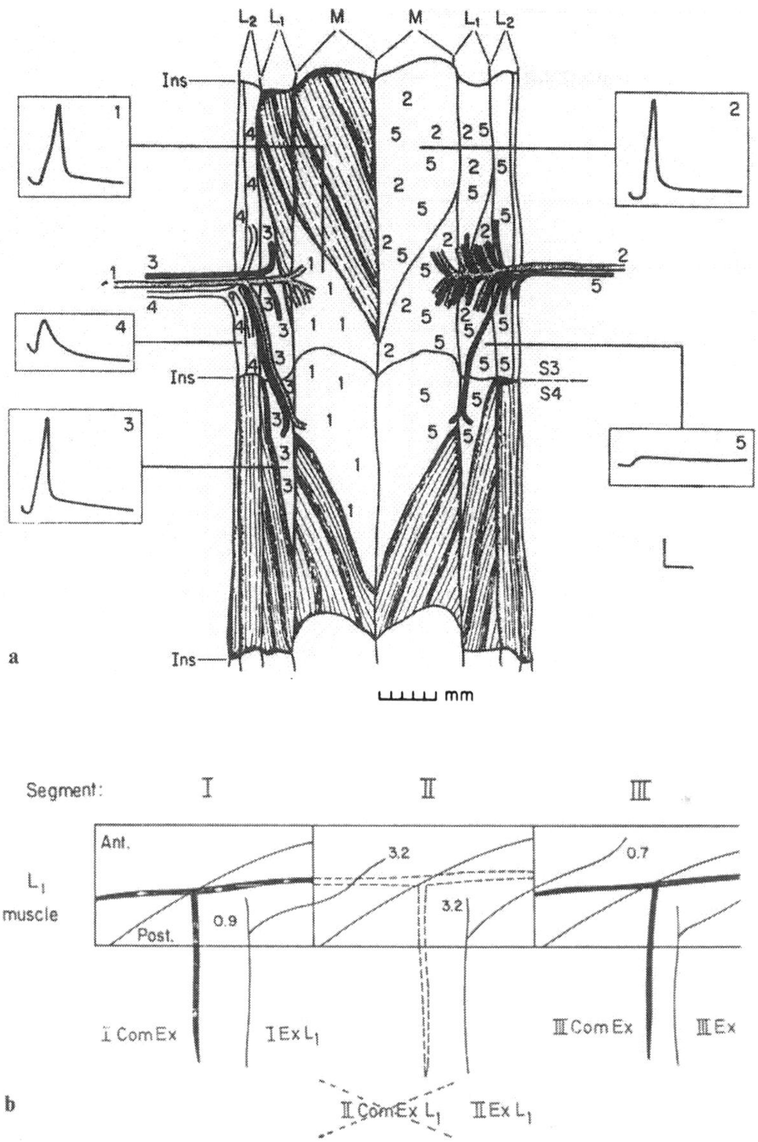

Fig. 1 a. Diagram of the innervation of the deep extensor muscles in segments 3 and 4 of the rock lobster. The subdivisions of the muscles (ventral aspect) and the main branches of the axons are indicated. Each axon is numbered, together with the parts of the musculature responding electrically to the axon. Tracings of representative electrical responses are shown for each axon. *M* medial deep extensor; L_1, L_2 sections of the lateral deep extensor; *Ins* segmental ligamentous inscriptions. Calibration for electrical recordings: Time, 20 ms; voltage, *1–4* 20 mV; *5* 4 mV. (Parnas and Atwood 1966). **b** Schematic diagram of the L_1 muscles of the three first segments; right-hand side. Each muscle consists of an anterior and a posterior bundle. The distributions of synapses of the common excitor axon (*thick lines*) and of the specific L_1 excitor axon (*thin lines*), are shown. The common excitor of segment II has been eliminated (*dashes*). The *numbers* inserted at the different muscle bundles give the average quantum content of synaptic release at single synapses. (Dudel and Parnas 1987)

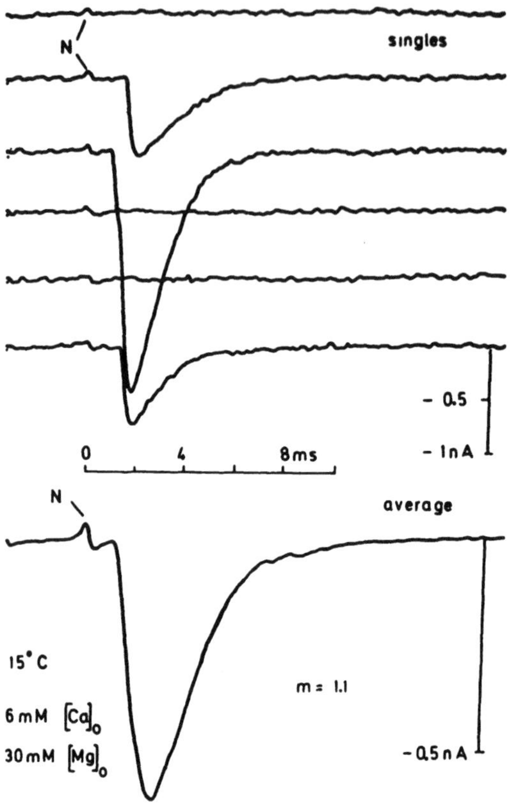

Fig. 2. Measurement of EPSCs. The *upper part* shows six single traces. In each, a record from the nerve terminal (*N*) is seen, but in three sweeps no release occurs. The second and the sixth traces show a unitary response. The lower part shows the average of 128 such traces at the same synaptic site. The quantal content (m) of this average EPSC was 1.1, as calculated from the proportion of the observed releases and failures. Amplitude scales are different in singles and in the average; time bases are the same. The EPSCs were recorded at 15° C and in low $[Ca]_0 = 6$ mM and high $[Mg]_0 = 30$ mM. (Parnas et al. 1984)

3 Effects of Removal of the Inhibitory Axon on the Remaining Excitatory Inputs

When the inhibitory axon was removed by intracellular injection of pronase, the excitatory synaptic potential (EPSP), both of ComEx and ExL_1, became longer, the decay time of the EPSP increasing by a factor of 2–4. It took 10 to 16 days after removal of the inhibitor for the prolongation of the EPSP to be expressed. This shows that the prolonged decay did not arise from a simple increase in postsynaptic membrane resistance due to the removal of postsynaptic inhibition. Further, synaptic currents (EPSCs) which do not depend on the muscle input resistance, were also prolonged by a factor of 2–4. The slowed decay phase of the synaptic current could still be fitted by a single exponent. A detailed description of these results is given in Dudel et al. (1981), and Parnas I et al. (1982a).

A longer synaptic current can be obtained by several mechanisms. For example, the synaptic current may be composed of several quanta released asynchronously. Since single quanta were prolonged like the EPSCs, this possibility is ruled out. Alternatively, the removal of glutamate may be slowed, and a glutamate molecule may have a chance to combine more than once with the postsynaptic receptor. In this case, however, the decay phase of the synaptic current would not be exponential. As mentioned above, the decay phase of the longer synaptic currents could be fitted by a single exponential. Recently Franke et al. (1986) recorded single-channel activity in excised patches from synaptic regions of the crayfish DEAM muscles. Preliminary experiments in the lobster show (mild treatment with collagenase, only 100–200 MΩ seals) that after the removal of the inhibitory axon the average open time of the excitatory channels was almost doubled. This line of experiments should be further pursued.

To our knowledge, this is the first time that "cross talk" between two axons releasing different transmitter substances is demonstrated. The mechanism of prolongation of the channel open time is not clear. It is possible that existing channels changed their properties, but it is also possible that removal of the inhibitory axon produced a signal by which new channels, with a longer open time, are inserted into the membrane. Another possibility is a higher effective glutamate concentration at excitatory terminals which would increase the average channel open time (Dudel and Franke 1987).

4 Removal of an Excitatory Axon

The L_1 muscle fibres are innervated by two excitatory axons, the ComEx and the ExL_1. The ComEx axon produces a larger synaptic potential. Thus, the synapses of the ComEx can be looked upon as "strong" while those of the ExL_1 as "weaker". The issue of "strong" and "weak" synapses located on the same neuron is complicated. The amplitude of an EPSP as recorded in the soma does not necessarily reflect the "strength" of a synapse in producing its postsynaptic effect. Distance from the soma, the magnitude of the conductance change, the driving force of the synaptic battery, the location of the synapse at a strategic point (e.g. a branch point) on the dentritic tree, activation of second messenger mechanisms, the duration of the synaptic potentials, all these and others are important in defining "synaptic strength" (see Burke 1987). In the case of the neuromuscular junction the issue is simpler. Here, the recording of the synaptic response is from a restricted region and the number of quanta is directly determined. Postsynaptic factors such as space constant or amplitude of a single quantum event do not affect the number of quanta. For our purpose we defined "synaptic strength" by he number of quanta released on average, per impulse. Other mechanisms that may affect "synaptic strength" by postsynaptic mechanisms were not studied at this stage.

At the first stage of our study we established the release levels at ExL_1 synapses, in controls, for the L_1 muscles on both sides. In controls, the quantal content varied between 0.2 and 1 (6.5 mM Ca^{2+}, 30 mM Mg^{2+}). The average was found to be 0.5 quanta/action potential (Fig. 3). After elimination of the ComEx

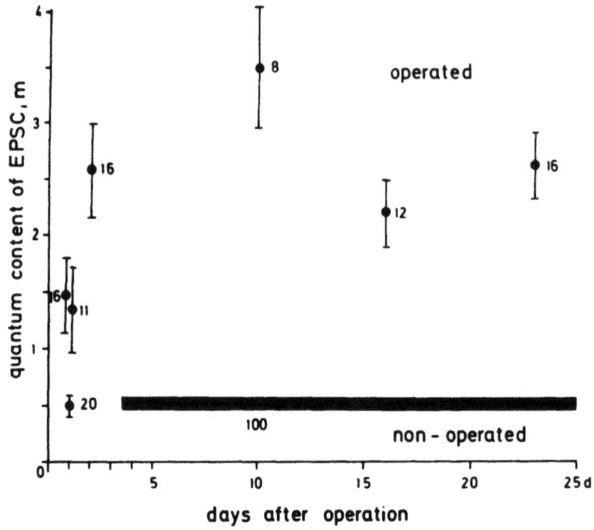

Fig. 3. Time course of increase in quantal content of EPSCs; average quantal content of EPSCs (*m* ordinate) as a function of time after operation (*abscissa*), i.e. after pronase injection into the common excitor axon. The *black bar* indicates the average m of the 100 control recordings on the nonoperated side of the L_1 muscle. The *width* of this bar represents the standard error of the mean. The *points* represent average m values with their standard errors of the mean measured at different times after the elimination of the common excitor. The *numbers at the points* indicate the number of sites at which m was measured in each preparation. Each determination of m is based on at least 128 single recordings. $[Ca]_0 = 6$ mM; $[Mg]_0 = 30$ mM. (Parnas et al. 1984)

on one side, release was doubled within 28–48 h, while the control side remained normal. Release increased by a factor of 5–10 over a period of 10–15 days (Fig. 3).

In controls the distance between release sites of ExL_1 was about 100 μm. In the first 2–5 days after elimination of ComEx the distance between release sites was still about 100 μm. Later, however, we found that the distance between release sites was smaller (less than 50 μm) and in fact it became very easy to find a release site. We interpret this finding to indicate that new release sites were formed in the existing terminals. It is also possible that these "new" release sites were present before but had a very low probability for release, becoming stronger after the elimination of the ComEx. A third effect was sprouting. The network of L_1 axon branches and terminals became much more extensive. Despite the pronounced sprouting, the newly grown branches and sprouts were limited to the original posterior bundle of the segment. ExL_1 terminals did not cross to the anterior bundle of the same segment. Similarly, sprouting was not seen in the anterior bundle of the following segment, even though it is innervated by the same ExL_1 axon (see Fig. 1).

What can be the signal or signals for these changes? In the frog, cutting of the cutaneous-pectoris nerve on one side produced sprouting and an increase in the percentage of muscle fibres with double innervation in the intact contralateral muscle (Rotshenker 1979, 1982). In this case, the signal for sprouting seems to have been mediated centrally at the level of the spinal cord, the damaged motoneurons producing a transneuronal effect. Rotshenker and his colleagues (Rotshenker and Tal 1985) demonstrated also the existence of a peripheral signal which evokes sprouting. "Synaptic strength" was not studied in great detail in the above-mentioned papers.

In the case of the removal of the ComEx, the signal for strengthening of the ExL_1 terminals could also be central. Alternatively, the mechanism could be peripheral. For example, removal of the ComEx could trigger a signal from the partially denervated muscle fibres. It is also possible that the two axons were competing for a "trophic" factor produced in a limited amount by the muscle fibres. One may also assume that the two axons release a factor or factors (modulators) which reduce release of each other (and of themselves). The removal of one axon

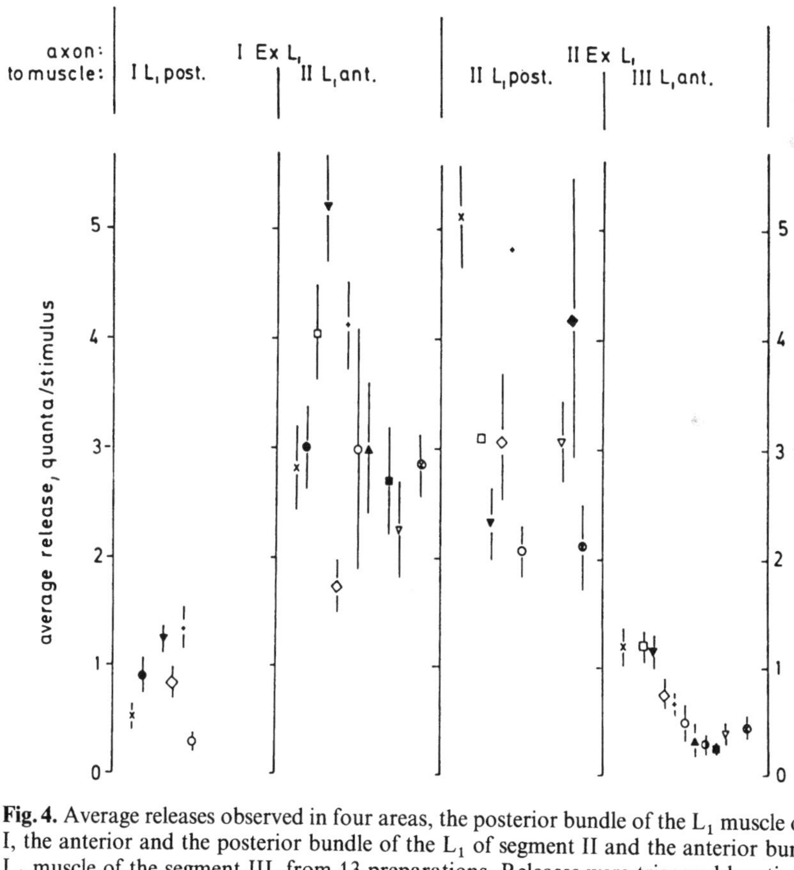

Fig. 4. Average releases observed in four areas, the posterior bundle of the L_1 muscle of segment I, the anterior and the posterior bundle of the L_1 of segment II and the anterior bundle of the L_1 muscle of the segment III, from 13 preparations. Releases were triggered by stimulation of the respective L_1 excitor axon. Average release rates from one preparation are presented by a common symbol. The *vertical lines* next to the respective symbols indicate the statistical error of the average value. The averages contain 2 to 22 determinations of release rates at different sites on the respective bundle of L_1, and for each such release rate the results of at least 128 stimulations were evaluated. No statistical errors are given if only two sites are contained in an average. In the anterior and posterior bundle of the L_1 of segment II the common excitor axon had been eliminated, and in these areas release is much higher than in the controls (the posterior bundle of the L_1 muscle of segment I and the anterior bundle of the L_1 of segment III). Segments I, II and III represented by *I*, *II* and *III* on the diagram. (Dudel and Parnas 1987)

reduces the amount of such a tonic signal, enabling the remaining axons to release more transmitter. The anatomical organization of the L_1 muscle and its innervation enables at least to distinguish whether the signal for the strengthening of release is central or peripheral.

As shown in Fig. 1b, removal of the ComEx produces partial denervation in both the anterior and posterior bundles of segment II. On the other hand, these two bundles are innervated by ExL_1 axons originating in two segments. The anterior bundle of segment II is innervated by axon terminals from segment I, and the posterior bundle is innervated by terminals originating in segment II. Thus, if the signal for synaptic strengthening were central, limited to the ganglion of segment II, it is expected that only terminals originating in this ganglion and innervating the posterior bundle of segment II and the anterior bundle of segment III (Fig. 1 b) become stronger. Terminals originating in segment I and innervating the posterior bundle of segment I and the anterior bundle of segment II should remain normal. If, however, the signal is peripheral, then, all terminals "sensing" a vacant territory in segment II should become stronger: The branches of ExL_1 from segment I and the branches of ExL_1 from segment II should become stronger in both bundles of segment II, while terminals of these two axons on the posterior bundle of segment I and on the anterior bundle of segment II should remain normal. The result was clear. Only the terminals innervating segment II became stronger (Figs. 1 b and 4).

The signal for synaptic strengthening is not only peripheral, it must be restricted to the "vacant territory" of the ComEx. The posterior branch of ExL_1, from segment I, on its way to segment II innervates fibres in segment I. It was thus possible to measure release properties of this branch at different regions in segments I and II. It was found that the fibres in segment I received normal innervation, but as soon as the electrode crossed the border to segment II (about 100 μm distance) release increased sharply (Fig. 4). Similarly, the terminals of ExL_1 from segment II became stronger in segment II, but the branches of this same axon were found to be normal, when the electrode was moved to segment III (Fig. 4) (for more details see Dudel and Parnas 1987).

5 Is Electrical Activity the Signal for the Synaptic Strengthening?

Electrical activity of the postsynaptic cell was found to be an important signal for receptor organization at the frog neuromuscular junction (Lømo and Slater 1978). The removal of the ComEx by intracellular injection of pronase, although not producing complete block of electrical activity in the muscle, certainly reduces it. The question was whether the reduced electrical activity plays a role as a signal for synaptic strengthening of the remaining ExL_1 axon on the DEAM system.

As mentioned above, cutting a nerve in crustaceans does not produce degeneration of the peripheral stump (Hoy 1969). This was found to be true for the DEAM system as well. Thus, by cutting the nerve, the peripheral stump survives but it is electrically silent (unless electrically stimulated). Cutting of the nerve in segment II (all five axons) produced "weakening" of the severed ComEx and ExL_1 terminals in segment II. Although the peripheral stump did not degenerate, the quantal content of EPSCs was drastically reduced, and release sites were found only at the region where the nerve enters and crosses the L_1 muscle.

The ExL_1 axon from segment I was not damaged by such a treatment. It was interesting to find that under these conditions the terminals of ExL_1 of segment I which innervate the anterior bundle of segment II, remained normal and did not become stronger. Evidently, the presence of the weaker and electrically silent ComEx terminal in segment II was sufficient to prevent the signal for the strengthening of the ExL_1 terminals. The results of this experiment indicate two conclusions. Electrical or lack of electrical activity is not the signal for synaptic strengthening. The weak but viable ComEx terminals are sufficient to prevent the signal from being expressed.

The molecular events and signals involved in the strengthening of synaptic release are not clear. The changes described are in accord with the concept of competition between neurons innervating the same target for some trophic factor or factors released by the muscle fibres (Njå and Purves 1978; Csillik et al. 1985; Purves 1986; Thoenen and Barde 1980).

References

Atwood HL (1977) Crustacean neuromuscular systems: past, present and future. In: Hoyle G (ed) Identified neurons and behaviour of arthropods, vol 6. Plenum, New York, pp 9–29

Atwood HL, Kwan I (1976) Synaptic development in the crayfish opener muscle. J Neurobiol 7:289–312

Atwood HL, Wojtowicz JM (1986) Short-term and long-term plasticity and physiological differentiation of crustacean motor synapses. Int Rev Neurobiol 28:275–362

Blackshaw SE, Nicholls JG, Parnas I (1982) Expanded receptive fields of cutaneous mechanoreceptor cells after single neuron deletion in leech central nervous system. J Physiol (Lond) 326:261–268

Bowling D, Nicholls JG, Parnas I (1978) Destruction of a single cell in the central nervous system of the leech as a means of analysing its connexions and functional role. J Physiol (Lond) 282:169–180

Burke RE (1987) Synaptic efficacy and the control of neuronal input-output relations. TINS 10:42–45

Csillik B, Schwab ME, Thoenen ME (1985) Transganglionic regulation of central terminals of dorsal root ganglion cells by nerve growth factor (NGF). Brain Res 331:11–15

Dudel J (1981) The effect of reduced calcium on quantal unit current and release at the crayfish neuromuscular junction. Pflügers Arch 391:35–40

Dudel J, Franke C (1987) Single glutamate-gated synaptic channels at the crayfish neuromuscular junction. II. Dependence of channel open time on glutamate concentration. Pflügers Arch 408:307–314

Dudel J, Kuffler SW (1961 a) The quantal nature of transmission and spontaneous miniature potentials at the crayfish neuromuscular junction. J Physiol (Lond) 155:514–529

Dudel J, Kuffler SW (1961 b) Presynaptic inhibition at the crayfish neuromuscular junction. J Physiol (Lond) 155:543–562

Dudel J, Parnas I (1987) Augmented synaptic release by one excitatory axon in regions in which a synergistic axon was removed, in lobster muscle. J Physiol (Lond) 390:189–199

Dudel J, Grossman Y, Parnas I (1981) Synaptic transmission in crustacean muscle: effects of elimination of the inhibitor fibre on excitatory transmission. In: Feudis FV de, Mandel P (eds) Amino acid neurotransmitters. Raven, New York, pp 359–364

Dudel J, Parnas I, Parnas H (1983) Neurotransmitter release and its facilitation in crayfish muscle. VI. Release determined by both, intracellular calcium concentration and depolarization of the nerve terminal. Pflügers Arch 399:1–10

Franke C, Hatt H, Dudel J (1986) The excitatory glutamate-activated channel recorded in cell attached and excised patches from membranes of tail, leg and stomach muscles of crayfish. J Comp Physiol A Sens Neural Behav Physiol 159:579–589
Hoy RR (1969) Degeneration and regeneration in abdominal flexor motor neurons in the crayfish. J Exp Zool 172:219–232
Hoy RR, Bittner GO, Kennedy D (1967) Regeneration in crustacean motor neurons: evidence for axonal fusion. Science 156:251–252
Kawagoe R, Onodera K, Takeuchi A (1982) On the quantal release of endogenous glutamate from the crayfish neuromuscular junction. J Physiol (Lond) 322:529–539
Kravitz EA, Beltz B, Glusman S et al. (1985) The well-modulated lobster: the roles of serotonin, octopamine, and proctolin in the lobster nervous system. In: Selverston a (ed) Model neural networks and behavior. Plenum, New York, pp 339–360
Lømo T, Slater CR (1978) Control of acetylcholine sensitivity and synapse formation by muscle activity. J Physiol (Lond) 275:391–402
Njå A, Purves D (1978) The effects of nerve growth factor and its antiserum on synapses in the superior cervical ganglion of the guinea pig. J Physiol (Lond) 252:429–463
Otsuka M, Iversen LL, Hall ZW, Kravitz EA (1966) Release of gamma-aminobutyric acid from inhibitory nerves of lobster. Proc Natl Acad Sci USA 56:1110–1115
Parnas H, Dudel J, Parnas I (1982) Neurotransmitter release and its facilitation in crayfish. I. Saturation kinetics of release, and of entry and removal of calcium. Pflügers Arch 393:1–14
Parnas H, Dudel J, Parnas I (1986) Neurotransmitter release and its facilitation in crayfish. VII. Another voltage dependent process besides Ca entry controls the time course of phasic release. Pflügers Arch 406:121–130
Parnas I, Atwood HL (1966) Phasic and tonic neuromuscular systems in the abdominal extensor muscles of the crayfish and rock lobster. Comp Biochem Physiol 18:701–723
Parnas I, Bowling D (1977) Killing of single neurons by intracellular injection of proteolytic enzymes. Nature 270:626–628
Parnas I, Dudel J, Grossman Y (1982a) Chronic removal of inhibitory axon alters excitatory transmission in a crustacean muscle fibre. J Neurophysiol 47:1–10
Parnas I, Parnas H, Dudel J (1982b) Neuotransmitter release and its facilitation in crayfish. II. Duration of facilitation and removal processes of calcium from the terminal. Pflügers Arch 393:232–236
Parnas I, Parnas H, Dudel J (1982c) Neurotransmitter release and its facilitation in crayfish muscle. V. Basis for synapse differentiation of the fast and slow type in one axon. Pflügers Arch 395:261–270
Parnas I, Dudel J, Cohen I, Franke C (1984) Strengthening of synaptic contacts of an excitatory axon on elimination of a second excitatory axon innervating the same target. J Neurosci 4:1912–1923
Parnas I, Parnas H, Dudel J (1986) Neurotransmitter release and its facilitation in crayfish. VIII. Modulation of release by hyperpolarizing pulses. Pflügers Arch 406:131–137
Pécot-Dechavassine M (1986) Increase in polyneuronal innervation in frog muscle after muscle injury. J Physiol (Lond) 371:167–177
Purves D (1986) The trophic theory of neural connections. TINS 9:486–489
Rotshenker S (1979) Synapse formation in intact innervated cutaneous-pectoris muscles of the frog following denervation of the opposite muscle. J Physiol (Lond) 292:535–547
Rotshenker S (1982) Transneuronal and peripheral mechanism for the induction of motor neuron sprouting. J Neurosci 2:1359–1368
Rotshenker S, Tal M (1985) The transneuronal induction of sprouting and synapses formation in intact mouse muscle. J Physiol (Lond) 360:387–396
Thoenen H, Barde YA (1980) Physiology of nerve growth factor. Physiol Rev 60:1284–1335

Nerve Growth Factor and the Reinnervation of Skin After Peripheral Nerve Lesions

J. Diamond

Department of Neurosciences, Mcmaster University, 1200 Main Street West
Hamilton, Ontario, Canada L8N 3Z5

1 Reinnervation of Deprived Skin: Regeneration and Collateral Sprouting Contrasted

The reinnervation of any tissue following an injury of its nerves can be achieved only by (1) regenerating fibres originating from the damaged nerves (or other cut nerves appropriately redirected to the deprived region), and/or (2) the collateral sprouting of *intact* axons that were spared when the original lesion occurred. The two growth behaviours have a lot in common; they are both initiated as a consequence of axonal damage (though of different sets of nerves), they are both able to restore lost physiological functions, and both will readily occur along degenerating pathways (see below). It seems unlikely that they would utilize different cellular mechanisms of axonal extension. What then can explain the obvious differences that often are observed between regeneration and collateral sprouting? For example, the large diameter, low threshold, mechanosensory nerves supplying the skin of adult mammals will readily regenerate after they are crushed, but when intact they do not sprout into adjacent, denervated, territory (reviewed in Diamond 1987). In the adult rat this applies even within the confines of a single, but dually innervated, touch dome: after elimination of one of the axons supplying such a dome, the other fails to sprout functionally into the deprived region, even as little as 50–100 μm; nevertheless if this remaining axon is crushed, it regenerates to occupy much or all of the dome (Yasargil et al. 1988). In very young animals, however, these same low threshold axons show a variety of sprouting behaviours. For example, axons of the dorsal cutaneous nerves of rats *can* sprout collaterally during an early "window" of time which ends at about 20 days of age (Jackson and Diamond 1984); with a different rat population we now find that in some animals of a litter *no sprouting occurs at all* (Mills, Macintyre, Holmes and Diamond in preparation). Moreover in none of these very young (or older) rats have we ever obtained sprouting of the equivalent nerves supplying the low threshold mechanoreceptors to the paws (Mills 1987), although they too regenerate without difficulty. We have found no explanation of these variations. Interestingly, collateral sprouting is a constant feature of the low threshold nerves supplying the skin of adult salamanders (Aguilar et al. 1973). Another intriguing finding comes from young rats as they continue their normal growth; the low threshold cutaneous axons clearly not only elongate, but exhibit a proportionate increase in the size of their terminal arbors within the expanding skin area, even

H. Flohr (Ed.)
Post-Lesion Neural Plasticity

though they are incapable of collateral sprouting into adjacent skin if this is denervated.

In contrast to low threshold nerves, the *high* threshold, nociceptive, nerves of mammals are able both to regenerate and, at all ages, and in all the regions examined, to sprout vigorously into adjacent denervated skin (Devor et al. 1979, Nixon et al. 1984; Doucette and Diamond 1987; Mills 1987). Similarly, in the adult mammalian *central* nervous system (CNS), collateral sprouting of spared axons is a readily evoked growth behaviour (Tsukahara 1981); however, in the CNS it is the regeneration of cut axons that is problematical, perhaps occurring only under special conditions (see Foerster this Vol., So and Aguayo 1985). There are two further interesting contrasts between sprouting and regenerating axons. Firstly, while collateral sprouting seems usually to be expressed at and within target tissues, nerves will regenerate along peripheral pathways independently of the presence of their target tissues. Secondly, when both are in competition for occupancy of the same target tissue, regenerating nerves will almost always "displace" (at least functionally) already resident ones that had sprouted earlier from neighbouring intact axons (Diamond 1982; Diamond et al. 1987).

2 Extrinsic Influences: the Role of Growth Factors

Much of this almost bewildering variability in the expression of collateral sprouting and axonal regeneration could reflect a variability in extrinsic influences that initiate and affect the two growth behaviours. Could some of their contrasting character derive from differences in the growth factors that evoke them? The evidence from developmental neurobiology suggests not. During the development of the sympathetic and at least one population of peripheral sensory systems a single growth factor (Nerve Growth Factor, or NGF) seems to be responsible for all of the observed neuronal growth (Levi-Montalcini and Angeletti 1968). It is worth noting, however, that these developmental studies almost never allowed a distinction to be made between axonal elongation and axonal arborization; if elongation failed to occur, for example, the possibility that axonal arborization (i.e., collateral sprouting) within the target area might be a differently regulated phenomenon could not be examined. In adult animals the situation has seemed somewhat different: NGF is still required for the long-term viability of sympathetic neurons (Angeletti et al. 1971), and has been implicated in the reconstitution of their terminal arborizations after these were destroyed by chemical means (Bjerre et al. 1974; but see Diamond et al. 1987); however, sensory neurons will survive and grow independently of NGF soon after birth (Levi-Montalcini and Angeletti 1968; Schwartz et al. 1982). This fact, and the result of a study that failed to implicate NGF in the regeneration of nociceptive nerves in the adult (Rich et al. 1984), together with the indications that prolonged NGF deprivation leads to a reduction of substance P levels in sensory neurons (Schwartz et al. 1982), led to the view that in the adult sensory system the continued role of NGF must presumably be one of general metabolic support, perhaps specifically directed to maintenance of transmitter levels.

That NGF might play a role in the growth behaviours of adult sensory nerves was not excluded, however, by the findings described above, and we therefore examined this possibility. We had already studied how after peripheral nerve lesions the denervated back skin in the adult rat can become functionally reinnervated, either by regenerating axons, or by collateral sprouts originating in axons that were spared at the initial denervation procedure (Jackson and Diamond 1984; Nixon et al. 1984; Doucette and Diamond 1987), and had demonstrated that regenerating low threshold axons, after reinnervating their former territory, can *continue to extend* somewhat beyond their former boundaries (Jackson and Diamond 1984; Yasargil et al. 1988). We since showed that regenerating *nociceptive* axons can also do this (see below), and they then appear to be behaving exactly as if they had not been damaged in the first instance, but are simply innervating the surrounding skin by collateral sprouting. Thus we looked for an involvement of NGF in axonal regeneration along the peripheral neural pathway, in the reestablishment of the original functional terminal arbor in the skin, and also in the extension of this arbor *beyond* the boundaries of the original field. The results were somewhat unexpected.

3 The Axonal Regeneration and Collateral Sprouting Paradigm

We isolate the field of a selected cutaneous nerve in the back skin of adult rats by eliminating all the nerves supplying the surrounding skin (Nixon et al. 1984). There remains then an "island" of innervation, effectively surrounded by a "sea" of totally denervated skin. Any collateral sprouting that occurs from the intact axons supplying the isolated field is easily recognized functionally, and when required can be demonstrated morphologically, by the progressive expansion of the denervated field into the surrounding skin. When *regeneration* is the growth behaviour selected for study, the intact nerve supplying the island is crushed immediately following after the isolation of the field; its axons then regenerate along the degenerating nerve distal to the region of crush. The principal difference between the reinnervation achieved by regeneration and that by collateral sprouting is that the latter is initiated from undamaged axons supplying an adjacent cutaneous nerve field, while in the case of regeneration the development of a cutaneous arborization has been preceded by regrowth of initially damaged fibres along the (degenerating) peripheral nerve pathway.

The techniques for measuring regeneration and collateral sprouting are essentially the same. To follow the reinnervation achieved by nociceptive nerves, we utilize their ability to evoke a reflex contraction of a cutaneous skeletal muscle (the *cutaneus trunci* muscle or CTM) which in innervated back skin of the rat produces a brisk bilateral wrinkling when the skin is locally pinched or heated, that is sustained during the period of stimulation (Nixon et al. 1984; Doucette and Diamond 1987). The reappearance of this reflex response in initially denervated skin was confirmed as being based upon new neuritic growth by correlative morphological studies, involving both light and electron microscopy (see Diamond et al. 1987): these studies were needed to rule out an alternative explanation of the progressive expansion of an isolated nociceptive field *as measured behav-*

iourally; this is the development of post-lesion changes in central excitability, which cause previously ineffective inputs from "outlying" nociceptive endings (which are not detected by the behavioural testing) now to become effective in evoking the reflex response (see Doucette and Diamond 1987). The presence of *low*-threshold mechanosensory nerve endings is revealed by the electrophysiological recording of the impulses evoked by light stroking of the skin (Jackson and Diamond 1984).

An expansion of the isolated mechano-nociceptive area by collateral sprouting of the Aδ fibres (Nixon et al. 1984), and of the heat-nociceptive area by sprouting of the C fibres (Doucette and Diamond 1987) (Fig. 1), becomes first detectable at about 10–14 days after isolation of the fields, and reaches a maximum at 24–25 days; at no time is there any change in the size of the light touch field. After nerve crush, a recovery of sensitivity by *regenerating* fibres is first detected in the skin some 8 to 25 days later, depending on the location of the crush (Diamond et al. 1987). Often the first responses to light touch can be recorded somewhat earlier (by a few days) than the first reflex responses to pinch, and the latter always precede those to heat; interestingly, the mechano-nociceptive field almost always enlarges more rapidly than the others, overtaking the light touch field in both rate and extent, and always outstripping the heat-nociceptive field. In older animals especially, the latter often does not exceed its original area, but the mechano-nociceptive fields eventually always extend beyond their former boundaries (see below).

The characteristics of the two growth behaviours we selected for a test of the influence of NGF were thus well established for three distinct sensory modalities.

4 Anti-NGF Treatment Distinguishes Between the Two Growth Responses of Sensory Nerves

Daily subcutaneous injection into operated adult rats of anti-serum to NGF was begun immediately after the selected nerve field was isolated; when regeneration was to be studied the nerve supplying this field was crushed some 1–2 cm from the skin (Diamond et al. 1987).

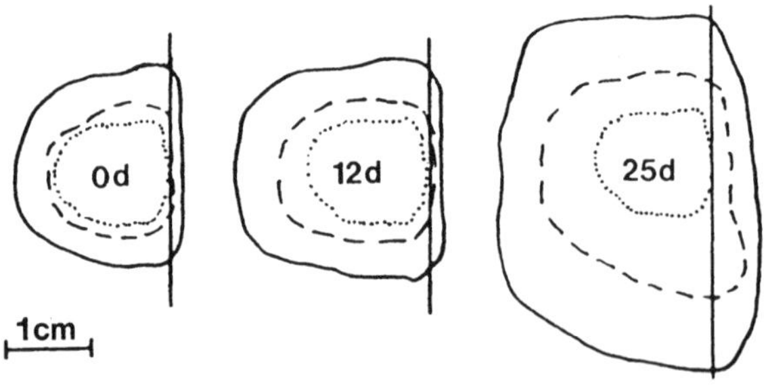

Fig. 1. Collateral sprouting of cutaneous nociceptive fields. The midline is indicated by the *vertical straight line*, the light touch field (the smallest) by the *dotted line*, the heat-nociceptive field by the *dashed line*, and the mechano-nociceptive field (the largest) by the *continuous line*. The fields were measured at the indicated times (*d* days) after their isolation. Note that the touch field did not change, while the two nociceptive ones expanded. The incision line for the operation (not shown) was to the right of the midline. (Diamond et al. 1987)

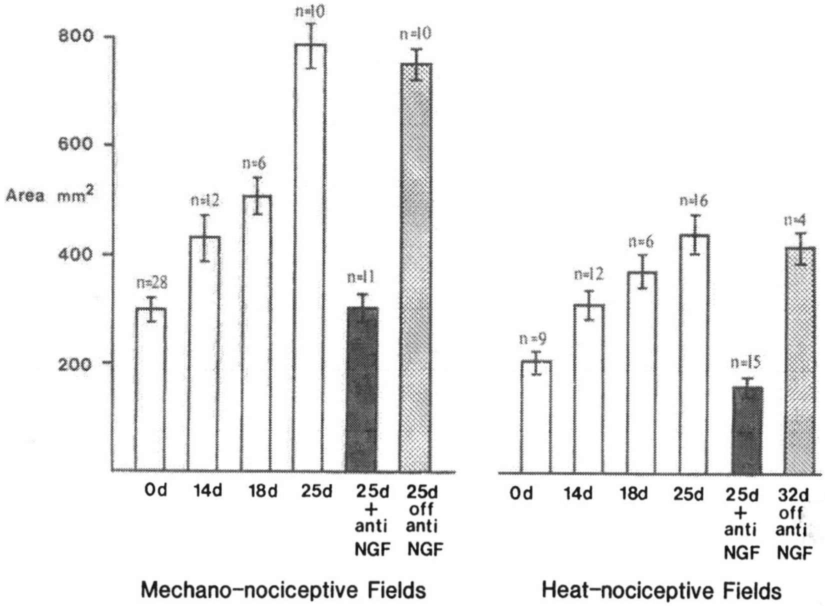

Fig. 2. Histograms of isolated mechano-nociceptive and heat-nociceptive fields in various groups of rats. Abscissae are not scaled, but shown below are the various times (in days) when the measurements were done after the initial field isolation. The number of animals in each group is indicated by the n values. Error bars indicates ± SEM. The normal rate of field expansion is seen from the first four histograms (*open*) to the left for both series. Expansion was maximum at about 24–25 days. The *filled* histograms (25 days + anti-NGF) came from animals that were treated daily for the entire period; clearly there had been no field expansions. The treatment was then discontinued and 25 days later for the mechano-nociceptive fields (the 25 days off anti-NGF histogram, *stippled*) and somewhat longer, 32 days later, for the heat-nociceptive fields (32 days off anti-NGF) the normal expansion had occurred. (Diamond et al. 1987)

4.1 Collateral Sprouting is Prevented

Figure 2 shows that the collateral sprouting of nociceptive nerves was totally prevented by the anti-NGF treatment, in comparison with groups of control operated animals that either received no injections, or (not illustrated) that received non-immune serum, and that the effect was fully reversible. We also found that if the anti-NGF injections were begun *after* the collateral sprouting was already in progress, the expansions of the nociceptive fields were brought rapidly to a halt. Thus endogenous NGF seems to be continuously required, and is continuously available, during such sensory field expansions; the growth factor does not act simply to trigger the phenomenon.

We next examined the possibility that the anti-NGF treatment did not actually prevent collateral sprouting as such, but instead prevented the functional maturation of the newly-grown sprouts. Figure 3 shows the morphological findings in skin taken from an animal that had received 34 d of anti-NGF treatment; the silver staining used to reveal axons within the dermal perineurial tubes (Nixon et al. 1984), and the electron microscopy (EM) used here specifically to reveal the

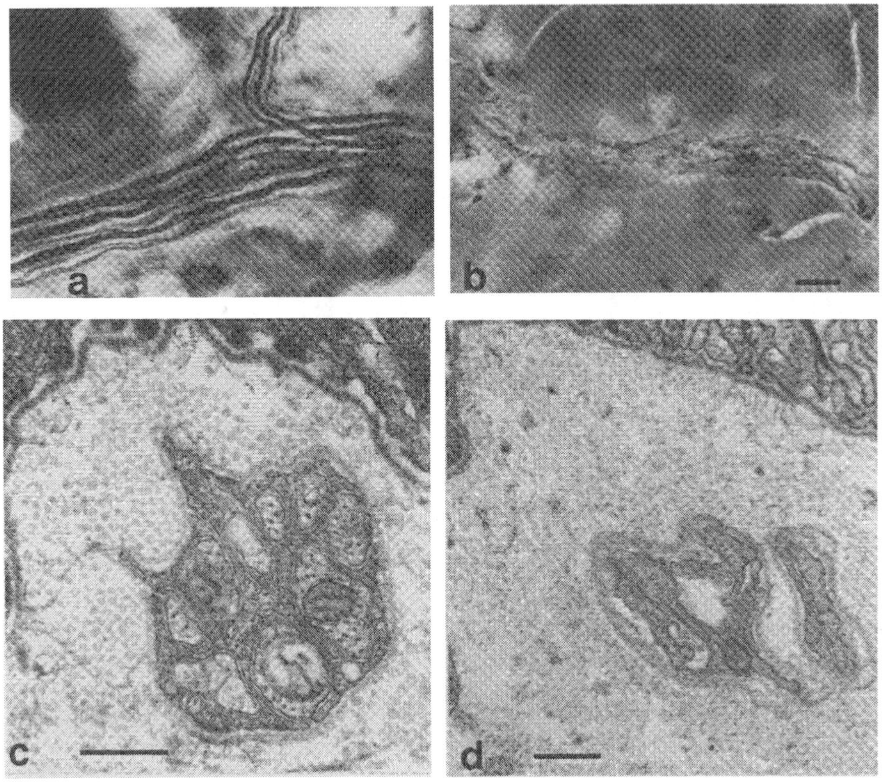

Fig. 3 a, b. Silver-stained sections of rat skin showing perineurial tubes within the dermis. **a** Skin from centre of a behaviourally-mapped nociceptive field, showing the typical parallel bundles of axons. The nerve field had been isolated 34 days earlier but prevented from expanding by daily anti-NGF treatment. **b** Insensitive skin immediately surrounding this field. In the absence of the anti-NGF treatment such "empty" perineurial tubes (which were typical of denervated, insensitive skin), would have contained sprouted fibers. The calibration is 20 μm.
Fig. 3 c, d. Electron micrographs of portions of the subepidermal horizontal fiber network in rat skin, each showing a transverse section across an epidermal "gutter" containing a Schwann tube surrounded by a basal lamina. In **c** (the same skin region that provided **a**) the tube contains a number of unmyelinated axon profiles, while in **d** (the same skin region that provided **b**) only an "empty" (axon-free) Schwann tube is present. Had expansion been allowed, C fibers would have been revealed (by EM) both in the dermal perineurial tubes and in some of these subepidermal Schwann tubes. Calibrations = 0.5 μm. (Diamond et al. 1987)

C fibres within the subepidermal horizontal network (Cauna 1969; Doucette and Diamond 1987) showed that nerve fibres were present in the centre of the isolated nerve field, but absent from the insensitive region immediately outside it. Thus the failure of the nociceptive field to expand was correlated with an absence of collateral sprouting into the surrounding skin; when expansions of the mechano- and the heat-nociceptive fields occur (i.e. in the absence of anti-NGF treatment) axons reappear in these two cutaneous networks (Nixon et al. 1984; Doucette and Diamond 1987).

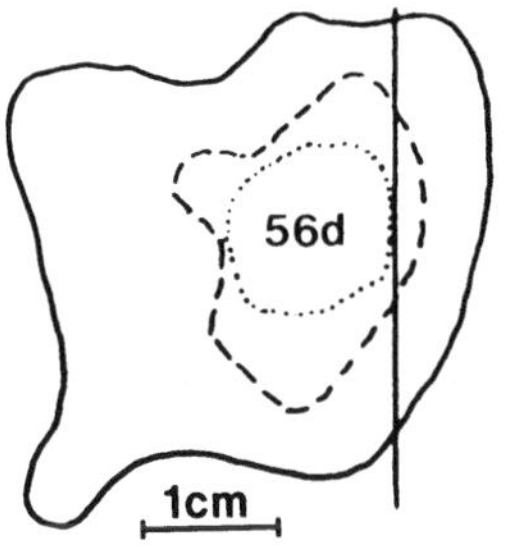

Fig. 4. Sensory recovery achieved by regeneration, 56 days after the nerve supplying the isolated field was crushed, and during daily anti-NGF administration; the initial fields prior to crushing resembled those shown at day 0 in Fig. 1. The light touch field (*dotted line*) did not exceed its original size but both the mechano-nociceptive (*continuous line*) and heat-nociceptive (*dashed line*) fields had encroached into "foreign" territory. (Diamond et al. 1987)

4.2. Regeneration is Unaffected

Unexpected results came when *regeneration* was similarly studied. Our experimental protocol allowed us not only to re-examine the reported failure of anti-NGF treatment to prevent the regrowth of nociceptive axons along a degenerating distal nerve trunk (Rich et al. 1984), but also to study the ability of regenerating axons to reconstruct a functional terminal arbor in the skin, and to extend into the surrounding "foreign" territory in the manner achieved by collaterally sprouting fibres. Since anti-NGF treatment prevented the latter arborization, we anticipated that it would similarly disallow the former, *even though axonal elongation along the peripheral pathway might still occur*. In the event, we were wrong. The regenerating axons *did* succed in producing functional arbors in the skin during continuous anti-NGF treatment, and these arbors *did* sometimes exceed their original extent (Fig. 4).

4.3 Further Examination of the NGF-Independence of Regeneration (Diamond et al. 1987)

Because the anti-NGF treatment we used routinely may have been simply not enough to block regeneration, we increased the antiserum injections to 3 × the amount that always *totally* prevented collateral sprouting; we observed no obvious differences between the ability of regenerating axons to establish functional nociceptive (and incidentally, light touch) fields in control animals, and in animals receiving the augmented anti-NGF treatment. Secondly, because unobserved differences may have developed between the animals that were used to produce collateral sprouting and those used in the regeneration paradigm, we did an experiment in which the two growth behaviours were tested in *the same* animal, utilizing a midline incision to create field isolations on both the right and left sides (necessarily sacrificing the small medial portions of the fields which extended across the midline). The nerve on the right side was crushed to evoke regeneration, while that on the left was untouched so that collateral sprouting would be possible; the standard anti-NGF treatment was then begun. By the time the experiment was terminated 35 d later, the nerves on the right had regenerated to reestablish essentially normal sensory fields, but the left side showed no evidence of collateral sprouting from the isolated field, even though this would normally be maximum by this time. Thus the differential effect of the treatment on sprouting and regeneration was expressed in one and the same animal.

Could there be sources of endogenous NGF available to regenerating nerves, but not (adequately) to collaterally sprouting ones, in a region which was relatively inaccessible to the injected anti-NGF antibodies? It turns out that both at a crush site *and along the distal pathway of the regenerating axons* there is a virtual disappearance of the normal hindrance to circulating macromolecules that defines the "blood nerve barrier" (see Sparrow and Kiernan 1981), a disruption that persists for periods much longer than those needed for functional regeneration in the present experiments. These observations effectively discount explanations requiring that regenerating axons are inaccessible to the administered antibodies, at least within the distal nerve trunk. What of the regenerating nerve pathways *in the skin itself*? Could the growing axons be following unusual pathways within the dermis that are less accessible than the pathways used by collaterally sprouting ones? Our morphological studies indicated not; regenerating axons used the same perineurial and horizontal Schwann cell systems that were followed by collaterally sprouting fibres, and we rarely observe axons outside these two pathways in either of the two reinnervation situations.

These findings make it highly unlikely that there could be an inaccessible *peripheral* source of NGF that was available to regenerating nerves but not to collaterally sprouting ones. One source of NGF could well be the Schwann cells of the degenerating nerve pathways (e.g. Finn et al. 1986; Assouline et al. 1987); Taniuchi et al. (1986) found that these cells rapidly begin to express low-affinity NGF receptors, and suggested that they might bring about a direct transfer of NGF to the high-affinity receptors of the axonal membranes (Greene and Shooter 1980). As we now see, this mechanism could not explain the *differential* effect of our anti-NGF treatment (though it could well be involved in collateral sprouting).

Another possible way that NGF could be hypothesized as being especially mobilized to support regeneration would be its uptake and transport by the *central* projections of dorsal root ganglion (DRG) neurons (Johnson and Yip 1985); this potential supply of NGF would be inacessible to our anti-NGF treatment because of the blood brain barrier. We disallowed this possibility by cutting the dorsal roots of the same DRG's whose peripheral axons were crushed to evoke regeneration. Electrophysiological studies revealed that the large myelinated mechanosensory axons of the decentralized DRG neurons continued to regenerate into denervated skin (Diamond et al. 1987), confirming an earlier report (Horch 1979); the problem, however, was to examine the regeneration of *nociceptive* axons after such dorsal rhizotomy, since this procedure precluded the behavioural testing of nociceptive function. However, the presence of *unmyelinated* sensory endings (those which subserve heat-nociception [Doucette et al. 1987]) can be detected by the exquisitely sensitive Evans blue extravasation test (Brenan 1982), in which antidromic excitation of sensory C fibres leads to the visible extravasation of the dye in the skin (Jancso and Jancso-Gabor 1980). By this technique we showed that dorsal rhizotomy did not prevent crushed C fibres from regenerating to the skin, and that this regeneration (and also that of the large light touch fibres) was unaffected by concomitant anti-NGF treatment. Although the mechano-nociceptive Aδ fibres were not separately studied, it seems unlikely that they would have failed to regenerate similarly in these experiments. Thus central sources of NGF are clearly not responsible for the peripheral nerve regeneration that is unaffected by anti-NGF treatment.

5 Significance of the Findings

5.1 Two Growth Behaviours: at Least Two Growth Factors

The results described here indicate that the two axonal growth behaviours which are capable of restoring an innervation to deprived skin occur in response to different endogenous growth factors. One of these, NGF, evokes the *collateral sprouting* of undamaged nociceptive fibres into adjacent skin regions deprived of their innervation. Presumably the denervation of these areas leads to an increase in their available NGF levels, either because a proportion of the axons competing for a constant supply of it has been eliminated, as was suggested to explain the (indirect) indications of nerve sprouting in the partially denervated iris (e.g. Ebendal et al. 1983; Kessler 1985), or because of changes in other mechanisms that could affect NGF levels, as discussed by Shelton and Reichardt (1986), who also showed that the levels of mRNA for NGF in the iris in vivo do not increase after its denervation. *Regenerating* nociceptive axons, however, seem not to require NGF for their continuing growth, and we have ruled out the possibility that this finding depended upon their having access to sources of NGF which were not available to collaterally sprouting nerves, and were inaccessible to the anti-NGF treatment. Especially compelling, if an ongoing anti-NGF treatment that had successfully prevented intact axons from collateral sprouting for 3–4 weeks was stopped, those axons would quickly begin to sprout; 3–4 weeks is about the same time after a nerve lesion that *regenerating* axons would be entering the skin, using *the same intradermal nerve pathways* as those followed by sprouting nerves. Since sprouting within these pathways is also rapidly brought to a halt by anti-NGF treatment, while regeneration is unaffected, then the latter is clearly independent of the NGF manufactured by cells associated with these pathways (Richardson and Ebendal 1982). One or more of the various *non*-NGF growth promoting influences that have been demonstrated in degenerating pathways, and/or in Schwann or other cells normally associated with these pathways (Longo et al. 1984; Windebank and Poduslo 1986), could be responsible for maintaining axonal regeneration. Conversely, these influences were clearly inadequate to evoke collateral sprouting of undamaged axons during anti-NGF treatment.

5.2 Implications of Different Growth Factors Being Responsible for Collateral Sprouting and for Regeneration

Although other reasons can be proposed to explain why, for example, low threshold sensory nerves regenerate but do not sprout, or why central axons sprout but do not (usually) regenerate, one possibility that must now be considered is that one or other of the growth factors responsible for the respective growth behaviours is absent or markedly reduced, or alternatively that the appropriate axonal growth factor receptors may be missing. Considerations such as these serve to emphasize the point that the ability of regeneration and collateral sprouting to provide an apparently identical functional reinnervation of deprived tissues should *not* be taken to imply that these two growth behaviours *have the same functional significance*. Collateral sprouting, particularly as evoked and recognized experimentally, could well be a dramatic expression of a normally more subtly expressed, ongoing, growth behaviour in the nervous system, namely the continuous sprouting of new nerve endings, probably associated with a continual regression of endings (Diamond 1982). According to this viewpoint, in the adult organism nerve growth factors would play an important role in the *plasticity* of the nervous system, specifically in an ongoing regulation of the turnover of nerve endings, and thus of the terminal fields of axonal projections (Diamond et al. 1987). NGF, for example, continues to be synthesized in adult animals (Shelton and Reichardt 1984; Heumann et al. 1984), and is retrogradely transported in axons after its uptake by specific receptors, both in the peripheral and central nervous systems (e.g. Richardson and Riopelle 1984; Seiler and Schwab 1984). The parent neurons of these same axons are those known or believed to require NGF for their normal development (Levi-Montalcini and Angeletti 1968; Honegger and Lenoir 1982; Hefti et al. 1985). Presumably the numerous neuronal systems that are not NGF-dependent in very early life utilize their own original growth factors in the adult nervous system, in an analogous manner. A continuous turnover of axonal endings within target tissues has previously been suggested on the basis of morphological indications of apparently spontaneously degenerating or "regressing" terminals in both the peripheral and the central nervous system (referenced in Diamond et al. 1987). In the skin, too, we see not infrequent profiles in electron micrographs of the subepidermal horizontal network suggestive of regressing C fibre terminals, always in the vicinity of perfectly normal endings (Diamond and Visheau unpublished observations). The proposals (1) that endogenous NGF regulates a continual turnover of nerve endings, and (2) that this could provide one basis for plasticity in the sensory nervous system, could apply to nerve growth factors generally in the adult, although this does not disallow roles for these same growth factors in other general metabolic functions of neurons.

Why would regeneration of an axon *not* be dependent on the same growth factor that is responsible for its collateral sprouting? We propose that the biological role postulated above for nerve growth factors in the adult nervous system implies a *selectivity* of action; only the subsets of neurons having a particular sensitivity to a given growth factor would be affected by its levels within target tissues. As was pointed out earlier, regeneration can occur independently of the presence of target tissues, and moreover virtually *any* nerves, peripheral (Simpson and Young 1945) or central (So and Aguayo 1985), seem able to regenerate along available degenerating peripheral nerve pathways. Regeneration, viewed in this way, is a growth behaviour that contrasts with collateral sprouting in being able to be evoked and directed in a relatively non-selective manner. Such a mechanism could have evolved as an effective global means of optimizing the regrowth of axons towards appropriate target regions. Once there, however, one would predict that *selective* mechanisms would be involved in the final targeting of the regenerating terminals. Paradoxically, we found this not to be so in the skin, i.e. the achievement of a functional arbor by regenerating axons continued to be NGF-independent. It would be interesting to know if this result applies only to a target tissue such as skin, which lacks specific target cells for the nociceptive axons that are able to undergo collateral sprouting. Obviously the ability of growing axons to recognize appropriate appropriate target cells or sites will also be an important factor in the re-establishment of a functional reinnervation in a deprived target tissue. In this regard one wonders to what extent the regrowth into and along correct *terminal* pathways is critical to the achieving of an appropriate functional end-result. Such guidance might be relatively easy to achieve in the skin, but difficult perhaps in the CNS, where the functionally successful penetration of target tissue by nerves that regenerated there by way of a peripheral nerve "bridge" is proving difficult to attain (Vidal-Sanz et al. 1987).

5.3 Clinical Considerations

A concern that arises from the present studies relates to the therapeutic use of growth factors in promoting the reinnervation of deprived tissues. As is implied above, *in the adult organism* a true sprouting agent, defined by its action either in vivo or in vitro, will probably not be the agent of choice for promoting regeneration along long pathways, including artificially constructed ones. On the other hand, there may be traumatic or disease conditions when the collateral sprouting of spared axons (or axons that have regenerated to appropriate target regions via, e.g. implants of peripheral nerve) would be the desired growth of choice, and then identified sprouting agents ought to be administered. Although NGF, studied in vitro, can cause both elongation and sprouting of axons (Levi-Montalcini and Angeletti 1968; Campenot 1982), the experiments have generally utilized immature nerve cells, and thus may not indicate what would happen with adult neurons in vivo. The distinction between agents that promote regeneration (or axonal elongation), and those characterized purely as sprouting factors, should be an important consideration whenever identified growth factors, or even tissues capable of generating these, are being proposed for therapeutic use in conditions when one or other of the two growth behaviours is to be encouraged.

References

Aguilar CE, Bisby MA, Cooper E, Diamond J (1973) Evidence that axoplasmic transport of trophic factors is involved in the regulation of peripheral nerve fields in salamanders. J Physiol (Lond) 234:449–464

Angeletti PU, Levi-Montalcini R, Caramia F (1971) Analysis of the effects of the antiserum to the nerve growth factor in adult mice. Brain Res 27:343–355

Assouline JG, Bosch P, Lim R, Kim IS, Jensen R, Pantazis NJ (1987) Rat astrocytes and Schwann cells in culture synthesize nerve growth factor-like neurite-promoting factors. Dev Brain Res 31:103–118

Bjerre B, Bjorklund A, Edwards DC (1974) Axonal regeneration of peripheral adrenergic neurons: effects of antiserum of antiserum to nerve growth factor in mouse. Cell Tissue Res 148–441–476

Brenan A (1982) Plasma extravasation following antidromic stimulation of cutaneous nerves in the rat. J Physiol (Lond) 327:7

Campenot RB (1982) Development of sympathetic neurons in compartmentalized culture. I. Local control of neurite growth by nerve growth factor. Dev Biol 93:1–12

Cauna N (1969) The fine morphology of the sensory receptor organs in the auricle of the rat. J Comp Neurol 136:81–98
Devor M, Schonfeld D, Seltzer Z, Wall PD (1979) Two modes of cutaneous reinnervation following peripheral nerve injury. J Comp Neurol 185:211–220
Diamond J (1982) Modelling and competition in the nervous system: clues from the sensory innervation of skin. In: Moscona AA, Monroy A (eds) Current topics in developmental biology, vol 17. Academic Press, London, pp 147–205
Diamond J (1987) Plasticity and stability in the sensory innervation of skin. In: Pubols LM, Sessle BJ (eds) Neurology and neurobiology, vol 30. Alan R Liss, New York, pp 43–48
Diamond J, Coughlin M, Macintyre L, Holmes M, Visheau B (1987) Evidence that endogenous β-nerve growth factor is responsible for the collateral sprouting but not the regeneration of nociceptive axons in adult rats. Proc Natl Acad Sci USA 84:6956–6600
Doucette R, Diamond J (1987) Normal and precocious sprouting of heat nociceptors in the skin of adult rats. J Comp Neurol 261:592–603
Doucette R, Theriault E, Diamond J (1987) Regionally selective elimination of cutaneous thermal nociception in rats by neonatal capsaicin. J Comp Neurol 261:583–591
Ebendal T, Olson L, Seiger A (1983) The level of nerve growth factor (NGF) as a function of innervation: a correlative radio-immunoassay and bioassay study of the rat iris. Exp Cell Res 148:311–317
Finn PJ, Ferguson IA, Renton FJ, Rush RA (1986) Nerve growth factor immunohistochemistry and biological activity in the rat iris. J Neurocytol 15:169–176
Greene LA, Shooter EM (1980) The nerve growth factor: biochemistry, synthesis, and mechanism of action. Annu Rev Neurosci 3:353–402
Hefti F, Hortikka J, Eckenstein F, Gnahn H, Heumann R, Schwab M (1985) Nerve growth factor increases choline acetyltransferase but not survival or fiber outgrowth of cultured fetal septal cholinergic neurons. Neuroscience 14:55–68
Heumann R, Korsching S, Scott J, Thoenen J (1984) Relationship between levels of nerve growth factor (NGF) and its messenger RNA in sympathetic ganglia and peripheral targt tissues. EMBO J 3:3183–3189
Honegger P, Lenoir D (1982) Nerve growth factor (NGF) stimulation of cholinergic telencephalic neurons in aggregating cell cultures. Dev Brain Res 3:229–238
Horch KW (1979) Guidance of regrowing sensory axons after cutaneous lesions in the cat. J Neurophysiol 42:1437–1449
Jackson P, Diamond J (1984) Temporal and spatial constraints on the collateral sprouting of low-threshold mechanosensory nerves in the skin of rats. J Comp Neurol 226:336–345
Jancso G, Jancso-Gabor A (1980) Effects of capsaicin on morphine analgesia: possible involvement of hypothalamic structures. Naunyn Schmiedebergs Arch Pharmacol 211:285–288
Johnson EM, Yip HK (1985) Central nervous system and peripheral nerve growth factor provide trophic support critical to mature sensory neuronal survival. Nature 314:751–752
Kessler JA (1985) Parasympathetic, sympathetic, and sensory interactions in the iris: nerve growth factor regulates cholinergic ciliary ganglion innervation in vivo. J Neurosci 5:2719–2725
Levi-Montalcini R, Angeletti PU (1968) Nerve growth factor. Physiol Rev 48:534–569
Longo RM, Hayman EG, Davis GE, Ruoslahti E, Engvall E, Manthorpe M, Varon S (1984) Neurite-promoting factors and extracellular matrix components accumulating in vivo within nerve regeneration chambers. Brain Res 309:105–117
Mills L (1987) Interactions between Merkel cells and nerve endings in the rat. Ph D Thesis, McMaster University, Hamilton
Nixon BJ, Doucette R, Jackson PC, Diamond J (1984) Impulse activity evokes precocious sprouting of nociceptive nerves into denervated skin. Somatosens Res 2:97–126
Rich KM, Yip HK, Osborne PA, Schmidt RE, Johnson EM (1984) Role of nerve growth factor in the adult dorsal root ganglia neuron and its response to injury. J Comp Neurol 230:110–118
Richardson PM, Ebendal T (1982) Nerve growth activities in the rat peripheral nerve. Brain Res 246:57–64
Richardson PM, Riopelle RJ (1984) Uptake of nerve growth factor along peripheral and spinal axons of primary sensory neurons. J Neurosci 4:1683–1689

Schwartz JP, Pearson J, Johnson EM (1982) Effect of exposure to anti-NGF on sensory neurons of adult rats and guinea pigs. Brain Res 244:378–381

Seiler M, Schwab ME (1984) Specific retrograde transport of nerve growth factor (NGF) from neocortex to nucleus basalis in the rat. Brain Res 300:33–39

Shelton D, Reichardt L (1984) Expression of the β-nerve growth factor gene correlates with the density of sympathetic innervation in effector organs. Proc Natl Acad Sci USA 81:7951–7955

Shelton DL, Reichardt LF (1986) Studies on the regulation of beta-nerve growth factor gene expression in the rat iris: the level of mRNA-encoding nerve growth factor is increased in irises placed in explant cultures in vitro, but not in irises deprived of sensory or sympathetic innervation in vivo. J Cell Biol 102:1940–1948

Simpson SA, Young JZ (1945) Regeneration of fibre diameter after cross-unions of visceral and somatic nerves. J Anat 79:48–65

So KF, Aguayo AJ (1985) Lengthy regrowth of cut axons from ganglion cells after peripheral nerve transplantation into the retina of adult rats. Brain Res 328:349–354

Sparrow JR, Kiernan JA (1981) Endoneurial vascular permeability in degenerating and regenerating peripheral nerves. Acta Neuropathol 53:181–188

Taniuchi M, Clark HB, Johnson EM (1986) Induction of nerve growth factor receptor in Schwann cells after axotomy. Proc Natl Acad Sci USA 83:4094–4098

Tsukahara N (1981) Synaptic plasticity in the mammalian central nervous system. Annu Rev Neurosci 4:351–379

Vidal-Sanz M, Bray GM, Villegas-Perez MP, Thanos S, Aguayo AJ (1987) Axonal regeneration and synapse formation in the superior colliculus by retinal ganglion cells in the adult rat. J Neurosci 7:2894–2909

Windebank AJ, Poduslo JF (1986) Neuronal growth factors produced by adult peripheral nerve after injury. Brain Res 385:197–200

Yasargil GM, Macintyre L, Doucette R, Visheau B, Holmes M, Diamond J (1988) Axonal domains within shared touch domes in the rat: a comparison of their fate during conditions favoring collateral sprouting and following axonal regeneration. J Comp Neurol 270:301–312

Modulation of Glial Cell Response to Injury and CNS Regeneration

M. Schwartz, A. Cohen, A. Harel, A. Solomon, and M. Belkin
Department of Neurobiology, The Weizmann Institute of Science, Rehovot, Israel

1 Introduction

Axonal injury initiates a cellular response in the non-neuronal cells surrounding the axons. This response includes proliferation of certain non-neuronal cells and infiltration of scavenger cells such as macrophages. For years it was accepted that this cellular response impedes regeneration due to the mechanical barrier created by the proliferating cells which make up scar tissue. The experimental evidence in this direction is not conclusive and it is not clear whether the lack of growth evolves from a barrier or from a deficiency in supportive elements.

Recent observations from our laboratory and others have offered new clues as to why the adult CNS fails to regenerate (Aguayo et al. 1978a; Hadani et al. 1984; Kao et al. 1977; Neuman et al. 1983; Nieto-Sampedro et al. 1983; Schwartz et al. 1985; Williams and Varon 1985). These studies have demonstrated that the environment of injured neurons determines the ability of the axon to regenerate following injury.

The first observation that the environment can either facilitate or inhibit axonal regeneration was made by Ramon Y Cajal (1928). More recently nerve transplantation experiments demonstrated that the neuronal environment should be conducive for regeneration to occur. For example, spinal and brain axons regenerate readily through columns of Schwann cells in peripheral nerves but usually fail to enter grafted CNS nerve segments (Aguayo et al. 1978a, b; David and Aguayo 1981; Richardson et al. 1982; Richardson and Issa 1984). These in vivo studies were confirmed lately by in vitro experiments in which the growth of neurites showed preference towards co-cultured sciatic nerves (Schwab and Thoenen 1985) and regenerating fish optic-nerves (Carbonetto et al. 1987), rather than optic nerves of adult mammals (Schwab and Thoenen 1985). Among the elements that the environment may provide are diffusible substances, cell surface molecules, and extracellular matrix components (David and Aguayo 1981; Daniloff et al. 1986; Longo et al. 1982, 1983; Lundborg et al. 1982; Manthorpe et al. 1982, 1983; Martini and Schachner 1986; Nieto-Sampedro et al. 1982; Richardson et al. 1982; Richardson and Issa 1984; Williams and Varon 1985).

Our studies were initiated by suggestions drawn from the available data that environmental elements, both soluble and nonsoluble, determine the fate of the neuron after injury. The visual systems of lower and higher vertebrates served us as models for CNS of regenerating and nonregenerating systems, respectively. It

H. Flohr (Ed.)
Post-Lesion Neural Plasticity

is suggested from our studies that the proliferating glial cells at the site of injury in nonregenerating systems do not necessarily impede growth. Moreover, if modulated properly at the appropriate time they can even support growth.

To obtain an insight into the nature of these environmental elements, we began by studying soluble substances originating from the regenerating fish visual system in the form of conditioned media (CM) derived from either injured or noninjured fish optic nerves (Rachailovich and Schwartz 1984). We reasoned that these soluble substances in the CM would be largely derived from the surrounding non-neuronal cells with the contribution of axoplasmic proteins leaking from the axon being, at most, minute. Such CM are likely to be comprised of multiple constituents, some of which may participate in the regenerative response to injury, by acting in concert or in a temporally synchronized fashion, as occurs in other growth processes. Such components may include neuronal factors (Mizrachi et al. 1985), glial factors (Giulian et al. 1986), proteases and their inhibitors (Monard et al. 1983), and other substances which may neutralize inhibitors. Our results provide evidence that injury does indeed cause changes both in the type and amount of substances originating from the non-neuronal cells (Harel et al. 1987 b; Rachailovich and Schwartz 1984; Schwartz et al. 1985).

In this chapter we show the changes in soluble substances occurring after injury in a regenerating system and their possible physiological relevance. We wish to show the results which led us to the suggestion that substances originating from growing systems (developing or regenerating) can modulate the glial and the neuronal response to injury in a nonregenerative system.

2 Apolipoprotein in Response to Injury – Putative Scavenger of Degradation Products

Among the prominent changes in soluble substances (i.e., conditioned media, CM) derived from non-neuronal cells, which occur following optic nerve injury in the fish, is the increased labeling of a 28 kDa polypeptide in CM of regenerating nerves (Rachailovich and Schwartz 1984). In the rat sciatic nerves, a 37 kDa polypeptide, showing increased labeling under similar conditions (Muller et al. 1985; Skene and Shooter 1983) has been identified as apolipoprotein-E (apo-E) (Ignatius et al. 1986; Snipes et al. 1986). This polypeptide was found to be accumulated in the regenerating PNS in contrast to its increased synthesis without accumulation in the nonregenerating CNS (Muller et al. 1985).

We hypothesized that the fish 28 kDa polypeptide may also be an apolipoprotein. We found that fish plasma contains a 28 kDa polypeptide which we identified as apo-A-I based on its molecular weight, amino acid composition and microheterogeneity (apo-E is absent from fish plasma, Harel et al. 1987 a). Antibodies were raised against the isolated and purified fish plasma apolipoprotein, as no immunological cross-reactivity could be detected between fish and rodent apolipoprotein. Using these antibodies we succeeded in identifying 28 kDa polypeptides in media conditioned by injured and noninjured fish optic nerves as apo-A-I isoforms. Moreover, injury seemed to be accompanied by an increased amount of these polypeptides manifested more in the least acidic of the three isoforms.

3 Glial Modulating Factors Counteracting Regeneration Deficits

In the absence of an appropriate control mechanism, the non-neuronal cells in mammalian CNS will form a glial scar which may impede or at least will not support axonal regrowth. An appropriate glial response should involve a well-controlled process to activate the glial cells to acquire the regeneration-supportive properties at the proper time. This response may include production of soluble factors, expression of an appropriate cell adhesion molecule and formation of an appropriate extracellular matrix needed for neuronal survival, neurite growth and elongation.

Both the available data and our own results indicated a correlation between the ability of a nerve to grow and the composition of its environment. Therefore, we reasoned that both regenerating and growing nerves possess the machinery to stimulate in the neuron, as well as in the environment, the appropriate response for regeneration. Accordingly, we developed an experimental model using the normally nonregenerating adult rabbit visual system, which allowed application of exogenous soluble substances to crush-injured rabbit optic nerves (Solomon et al. 1985). As an exogenous source for soluble substances we used CM of regenerating fish optic nerves, assuming that the molecules involved in growth stimulation may not be species-specific.

Our previous studies have shown that the application of these active substances to injured optic nerve of adult rabbit results in biochemical changes, reminiscent of regeneration, in the corresponding rabbit retinas (Hadani et al. 1984; Schwartz et al. 1985; Solomon et al. 1985; Stein-Izsak et al. 1985). These include a general increase in protein synthesis and elevation in labeling of a few polypeptides. Similar activity was observed when media conditioned by developing mammalian optic nerves (i.e., newborn rabbit optic nerves) were examined.

These biochemical changes in the retinas were accompanied also by morphological changes in the nerves, as manifested by the survival of injured fibers and the appearance of growth cones in the treated injured nerves. The growth cones were observed adjacent to glial surfaces in regions in which large extracellular spaces were seen (Lavie et al. 1987). Similar morphological relationship between glial organization, growth cones and extracellular spaces was seen in developing nerves (Schwartz et al. 1987). The apparent changes in the glial cell surface caused by the active CM stimulated us to look for the presence of supportive elements on the surface of the glial cells.

Cellular elements supportive of growth may include cell surface molecules involved in cell-cell recognition (Edelman 1984; Nieke and Schachner 1985; Kruse et al. 1985; Martini and Schachner 1986; Sanes et al. 1986; Schachner 1982); extracellular matrix components like collagen (Bunge et al. 1980); laminin (Rogers et al. 1983; Sanes 1983); fibronectin (Schachner et al. 1978); and glycosaminoglycans (Margolis and Margolis 1979), which are pivotal for growth.

In vivo and in vitro studies carried out in both the CNS and the PNS have provided evidence that laminin is one of the extracellular matrix components which has a key role in growth and regeneration (Carbonetto et al. 1983; Carbonetto et al. 1987; Evercooren et al. 1982; Schwab and Thoenen 1985). Moreover, a correlation was found between the ability of nerves to grow and the presence of an organized extracellular matrix which includes laminin. For example, in the fish visual system, similarly to peripheral nervous system of mammals, laminin is continuously expressed and injury causes elevation in its level (Hopkins et al. 1985; Liesi et al. 1984a; Liesi 1985). The absence of laminin in the CNS of adult mammals was proposed as a reason for its poor ability to regenerate (Liesi et al. 1983, 1984b). We therefore examined whether the morphological alterations which we observed in the injured implanted rabbit optic nerves, expressed by apparent changes in glial cell surface, are associated with laminin.

Our results revealed that in intact adult rabbit optic nerves, laminin immunoreactive sites are present in the epineurium and around the sparsely distributed blood vessels. Following injury, laminin immunoreactivity can be detected in the epineurium, in connective tissue surrounding blood vessels and in the scar tissue. However, after injury and implantation of media conditioned by regenerating fish optic nerves, the pattern of laminin immunoreactivity was changed and additional laminin-immunoreactive sites were detected. These appeared as a fine network coursing throughout the matrix of the nerves (Zak et al. 1987). These results agree with the assumption that the absence of laminin may be a cause for the poor inherent capacity of the mammalian CNS to regenerate. These results also suggest that factor(s) in the CM may, at least partially, counteract this deficit.

The aforementioned studies suggest that the glial cells of mammals may be amenable to activation and that the non-neuronal cells of a regenerative system, like the fish optic nerve, may contain the appropriate activators. In contrast, the injured mammalian CNS may not contain these activators or contain them only in an inaccessible form or at an inappropriate time. To test this hypothesis we have established an in vitro assay system where the ability to activate glial cells directly by active substances from fish can be tested and the isolation of the active components should thus be feasible. Our results suggest that components within the media conditioned by regenerating fish optic nerves can directly modulate the level of laminin synthesis and/or accumulation (A. Cohen and M. Schwartz in preparation).

It is therefore possible that several factors within the conditioned media are responsible for bringing the adult mammalian nonregenerative CNS after injury into a growth state. These factors should be identified and used.

4 Other Potential Growth Modulators

We have shown that low energy laser irradiation can prevent the degeneration of crushed sciatic nerves (Rochkind et al. 1986, 1987). Similar irradiation prevented, at least temporarily, the morpholigical (Schwartz et al. 1987) and electrophysiological (Rosner et al. 1987) manifestations of degeneration after optic nerve injury. The possibility exists that combining the application of the activators mentioned above with low energy laser irradiation which prolonged survival might result in complimentary or synergistic action which will induce CNS regeneration to a higher extent than achieved by either modality separately (Schwartz Solomon, Benbarsat, Harel and Belkin in preparation).

Another factor with potential use for the induction of CNS regeneration is Fibroblast Growth Factor (FGF), an angiogenic agent which stimulates mesoderm and neuroectoderm derived cells. FGF was found to cause exuberant vascular proliferation, as well as some degree of neuronal growth in the injured rabbit optic nerve (unpublished data). It is not clear whether FGF affects axons or glia directly or whether its neuronal growth-promoting effects are achieved via the provision of better blood supply to the injury site.

We believe that using the approach taken in our studies of modulating the axon and its own environment should ensure growth of fibers with CNS properties and therefore with the potention of eventually making proper reconnection.

Summary

Axonal injury stimulates two complimentary systems, the axon and its surrounding non-neuronal cells, which operate in synchrony. Our results suggest that the non-neuronal cell response to injury has an impact on the subsequent ability of the axon to grow and elongate. We found that injury to a regenerating CNS (i.e., fish visual system) causes changes in the amount and the nature of substances originating from the axon and its surrounding non-neuronal cells. One example of these changes is the increased synthesis and accumulation of a 28 kDa polypeptide in media conditioned by these injured nerves. The 28 kDa polypeptide was identified by us as apolipoprotein-A-I (apo-A-I) in analogy to a 37 kDa polypeptide in mammalian regenerating peripheral nerves which was identified as apolipoprotein E (apo-E). Both apo-A-I and apo-E may be involved in elimination of myelin degradation products and thereby may contribute to pave the way for regeneration.

In addition to the 28 kDa peptide, we found that substances within media conditioned by regenerating nerves, similarly to substances derived from developing nerves of mammalian CNS, activate neurons and may modulate the glial cell response to injury. These substances, when applied to injured adult rabbit optic nerves, cause biochemical alterations in the rabbit retinas and acquisition of a growth-supportive surface, including accumulation of laminin, by the surrounding non-neuronal cells. Our results suggest that proper modulation of the glial cells at the site of injury may be essential to the regenerative process. This process might be deficient in nonregenerating systems and may be circumvented by exogenous factors.

Identification of the various factors which can modulate the glial and neuronal response to injury and evaluation of the activities of known factors will allow us to develop therapeutic means for CNS injury. This may be achieved by supplying the injured nerves with factors which may save neurons from dying or which activate neurons to grow; or by providing the neural glial environment with factors which may modulate their response to injury, thereby bringing these cells to a stage where they support regeneration rather than impede it.

Acknowledgement. This work was supported by grants given to M.S. from the Israel Academy of Science and the US Army Medical Research and Developmental Command.

References

Aguayo A, Samuel D, Dickson R, Trecarten Y (1978 a) Axonal elongation in peripheral and central nervous system transplantations. Neurosci Lett 9:97

Aguayo A, Samuel D, Richardson P, Bray G (1978 b) Axonal elongation in peripheral and central nervous system transplantations. Adv Cell Neurobiol 3:215–221

Bunge MB, Williams AK, Wood PM, Uitto J, Jeffrey JJ (1980) Comparison of nerve cell cultures with particular emphasis on basal lamina and collagen formation. J Cell Biol 84:184–202

Carbonetto S, Gruver MM, Turner DC (1983) Nerve fiber growth in culture on fibronectin, collagen, glycosaminoglycan substrates. J Neurosci 3:2324–2335

Carbonetto S, Evans D, Cochard P (1987) New fiber growth in culture on tissue substrata from central and peripheral nervous system. J Neurosci 7:610–620

Daniloff JK, Levi G, Grumet M, Rieger F, Edelman GM (1986) Altered expression of neuronal cell adhesion molecules induced by nerve injury and repair. J Cell Biol 102:929–942

David S, Aguayo AJ (1981) Axonal elongation into peripheral nervous system "bridges" after central nervous injury in adult rats. Science 214:931–933

Edelman GM (1984) Cell adhesion molecule. Science 219:450–457

Evercooren A Baron-van, Klienman HK, Ohno S, Marangos D, Schwatz JP, Dubois-Dalcq ME (1982) Nerve growth factor, laminin, and fibronectin promote neurite growth in human fetal sensory ganglia cultures. J Neurosci Res 8:179–193

Giulian D, Allen RL, Baker TJ, Tomozawa Y (1986) Brain peptides and glial growth. I. Glia-promoting factors as regulator of gliogenesis in the developing and injured central nervous system. J Cell Biol 102:803

Hadani M, Harel A, Solomon A, Belkin M, Lavie V, Schwartz M (1984) Substances originating from the optic nerve of neonatal rabbit induce regeneration-associated response in the injured optic nerve of adult rabbit. Proc Natl Acad Sci USA 81:7965–7969

Harel A, Fainaru M, Shafer Z, Schwartz M (1987 a) Optic nerve regeneration in adult fish is associated with accumulation of apolipoprotein-A-I manifested preferentially in one isoform. (submitted)

Harel A, Rubinstein M, Solomon A, Belkin M, Schwartz M (1987 b) Isolation of a growth associated triggering factor from regenerating optic nerve. J Neurochem (Abst) (in press)

Hopkins JM, Ford-Holevnski TS, McCoy JP, Agranoff BW (1985) Laminin and optic nerve regeneration in the goldfish. J Neurosci 5:3030–3038

Ignatius MJ, Gebicke-Härter PJ, Skene JHP, Schilling JW, Weisgraber KH, Mahley RW, Shooter EM (1986) Expression of apolipoprotein E during nerve degeneration and regeneration. Proc Natl Acad Sci USA 83:1125–1129

Kao CC, Chang LW, Bloodworth JMB Jr (1977) Axonal regeneration across transected mammalian spinal cords: an electron microscopic study of delayed microsurgical grafting. Exp Neurol 54:591–615

Kruse J, Keilhauer G, Faissner A, Timpl R, Schachner M (1985) The J. glycoprotein – a novel nervous system cell adhesion molecule of the L2/HNK-1 family. Nature 316:146–148

Lavie V, Harel A, Doron A et al. (1987) Morphological response of injured adult rabbit optic nerve to implant containing neonatal derived substances. Brain Res 419:106–172)

Liesi P (1985) Laminin-immunoreactive glia distinguish regenerative adult central nervous systems form non-regenerative ones. EMBO J 4:2505–2511

Liesi P, Dahl D, Vaheri A (1983) Laminin is produced by early rat astrocytes in primary culture. J Cell Biol 96:920–924

Liesi P, Dahl D, Vaheri A (1984 a) Neurons cultured from developing rat brain attach and spread preferentially to laminin. J Neurosci Res 11:241–251

Liesi P, Kaakola S, Dahl D, Vaheri A (1984 b) Laminin is induced in astrocytes of adult brain by injury. EMBO J 3:683–686

Longo FM, Manthorpe M, Varon S (1982) Spinal cord neurotrophic factors (SCNTFs), 1. Bioassay of schwannoma and other conditioned media. Dev Brain Res 3:277

Longo FM, Skaper SD, Manthorpe M, Lundborg G, Varon S (1983) Temporal changes in neuronotrophic activities accumulate in vivo within nerve regeneration chambers. Brain Res 261:109

Lundborg G, Longo FM, Varon S (1982) Nerve regeneration model and trophic factors in vivo. Brain Res 232:157–161

Manthorpe M, Longo FM, Varon S (1982) Comparative features of spinal neurotrophic factors in fluids collected in vitro and in vivo. J Neurosci Res 8:241–250

Manthorpe M, Engrall E, Ruoslahti E, Longo FM, Davis GE, Varon S (1983) Laminin promotes neuritic regeneration from cultured peripheral and central neurons. J Cell Biol 97:1882–1890

Margolis RV, Margolis RK (1979) Complex carbohydrates of nervous tissue. Plenum, New York

Martini R, Schachner M (1986) Immunoelectron microscopic localization of neural cell adhesion molecules (L1, N-CAM, and MAG) and their shared carbohydrate epitope and myelin basic protein in developing sciatic nerve. J Cell Biol 103:2439–2448

Mizrachi Y, Rubinstein M, Kimhi Y, Schwartz M (1985) Neurotrophic factor derived from goldfish brain: characterization and purification. J Neurochem 46:1675–1682

Monard D, Niday E, Limat A, Solomon F (1983) Inhibition of protease activity can lead to neurite extension in neuroblastoma cells. Prog Brain Res 58:359–365

Muller HW, Gebicke-Harter PJ, Hangen DH, Shooter EM (1985) A specific 37,000-dalton protein that accumulates in regenerating but not in non-regenerating mammalian nerves. Science 228:499–501

Neumann D, Yerushalmi A, Schwartz M (1983) Inhibition of non-neuronal cell proliferation in the goldfish visual pathway affects the regenerative capacity of the retinas. Brain Res 272:237–245

Nieke J, Schachner M (1985) Expression of the neural cell adhesion molecule L1 and N-CAM and their common carbohydrate epitope L2/HNK-1 during development and after transection of the mouse sciatic nerve. Differentiation 30:141–151

Nieto-Sampedro M, Lewis ER, Cotman CW et al. (1982) Brain injury causes a time-dependent increase in neurotrophic activity at the lesion site. Science 217:860–861

Nieto-Sampedro M, Manthorpe M, Barbin G, Varon S, Cotman CW (1983) Injury-induced neuronotrophic activity in adult rat brain: correlation with survival of delayed implants in the wound cavity. J Neurosci 3:2219–2289

Rachailovich I, Schwartz M (1984) Molecular events associated with increased regenerative capacity of the goldfish retinal ganglion cells following X-irradiation: decreased level of axonal growth inhibitors. Brain Res 306:149–155

Ramon y Cajal S (1928) Degeneration and regeneration of the nervous system. English translation and reprint (1959) Hafner, New York

Richardson PM, Issa VMK (1984) Peripheral injury enhances central regeneration of primary sensory neurons. Nature 309:791–793

Richardson PM, McGuinness UM, Aguayo AJ (1982) Peripheral nerve autografts to the rat spinal cord: studies with axonal tracing methods. Brain Res 237:147–162

Rochkind S, Nissan M, Razon N, Schwartz M, Bartal A (1986) Electrophysiological effect of He-Ne laser on normal and injured sciatic nerve in the rat. Acta Neurochir 83:125–130

Rochkind S, Barrnea L, Razon N, Bartal A, Schwartz M (1987) Stimulatory effect of He-Ne low dose laser on injured sciatic nerves of rats. J Neurosurg 20:843–847

Rogers SL, Letourneau PC, Palm SL, McCarthy J, Furcht LT (1983) Neurite extension by peripheral and central nervous system neurons in response to substratum-bound fibronectin and laminin. Dev Biol 98:212–220

Sanes JR (1983) Roles of extracellular matrix in neural development. Annu Rev Physiol 45:581–600

Sanes JR, Schachner M, Covault J (1986) Expression of several adhesive macromolecules (N-CAM, L1, J1, NILE, uvomorulin, laminin fibronectin and heparan sulfate proteoglycan) in embryonic, adult and denervated adult skeletal muscle. J Cell Biol 102:430–431

Schachner M (1982) Glial antigens and the expression of neuroglial phenotypes. TINS, July 1982, pp 225–228

Schachner M, Schoonemaker G, Hynes RO (1978) Cellular and subcellular localization of LETS protein in the nervous system. Brain Res 158:149–158

Schwab ME, Thoenen H (1985) Dissociated neurons regenerate into sciatic but not optic nerve explants in culture irrespective of neurotrophic factors. J Neurosci 5:2415–2423

Schwartz M, Belkin M, Harel A et al. (1985) Regenerating fish optic nerves and regeneration-like response in injured optic nerve of adult rabbits. Science 228:600–603

Schwartz M, Harel A, Stein-Izsak C, Solomon A, Bawnik Y, Belkin M (1987a) Growth associated triggering factors and central nervous system response to injury. In: Stein DG, Sabel B (eds) Pharmacological Approaches to the Treatment of Brain and Spinal Cord Injuries. Plenum, New York

Schwartz M, Erlich M, Doron A, Benbasat S, Lavie V, Belkin M (1987b) The effects of low energy laser irradiation of posttraumatic degeneration of adult rabbit optic nerve. Lasers Surg Med 7:51

Skene JHP, Shooter EM (1983) Denervated sheath cells secrete a new protein after nerve injury. Proc Natl Acad Sci USA 80:4169–4173

Snipes GJ, McGuire CB, Norden JJ, Freeman JA (1986) Nerve injury stimulates the secretion of apolipoprotein E by nonneuronal cells. Proc Natl Acad Sci USA 83:1130–1134

Solomon A, Belkin M, Hadani M, Harel A, Rachailovich I, Lavie V, Schwartz M (1985) A new transorbital approach to the rabbits optic nerve. J Neurosci Methods 12:259–262

Stein-Izsak C, Harel A, Solomon A, Belkin M, Schwartz M (1985) Alterations in mRNA translation products are associated with regenerative response in the retina. J Neurochem 45:1754–1760

Williams LR, Varon S (1985) Modification of fibrin matrix formation in situ enhances nerve regeneration in silicone chambers. J Comp Neurol 231:209–220

Zak NB, Harel A, Bawnik Y, Benbasat S, Vogel Z, Schwartz M (1987) Laminin immunoreactive sites are induced by growth-associated triggering factors in injured rabbit optic nerve. Brain Res 408:263–266

Structural Plasticity in Lesioned Motoneurons

G. W. Kreutzberg, A. K. Engel, M. B. Graeber, W. Tetzlaff, and L. Tóth
Department of Neuromorphology, Max Planck Institute for Psychiatry,
Am Klopferspitz 18 A, D-8033 Planegg-Martinsried, FR Germany

1 Introduction

Neurons react to axon lesioning with a number of changes which have been known for almost a century. They are referred to as "chromatolysis", "retrograde change" or "the axon reaction". In nerve cells of the peripheral nervous system such as sensory and autonomic ganglia or motoneurons, the retrograde changes are seen as part of the regeneration program that can be triggered in these nerve cells. Under favorable conditions this leads to a regrowth of the lost axon and reinnervation of the target tissue and thus to a restitution of function. In the regenerating neuron many metabolic changes occur concerning RNA and protein synthesis, enzymatic activities, glucose consumption, neurotransmitters and receptors (for review see Lieberman 1971; Grafstein and McQuarrie 1978; Bisby 1980; Kreutzberg 1986; Tetzlaff et al. 1986). There are also changes in the microenvironment of the neuron afflicting local glial cells and the vasculature (Kreutzberg 1982). As a result of the neuronal and glial cell activation, a profound change of the synaptic arrangement on the neurons has been revealed. It shows that the axon lesion can provoke a surprising structural plasticity in the synapse-bearing elements. Several of these alterations will be described in this chapter as they have been studied in motoneurons.

2 Early Astroglial Reaction

The earliest retrograde changes in motoneurons occur within hours and lead to a significant increase in the activity of ornithine decarboxylase (300%) and glucose uptake within the first 24 h (Tetzlaff and Kreutzberg 1985; Kreutzberg and Emmert 1980). Activation of local astrocytes occurs at the same time. Taken the amount of glial fibrillary acidic protein (GFAP) as a marker for the metabolic and structural activation we have measured a significant increase within the first 24 h (Tetzlaff et al. 1988) By immunocytochemistry we have studied the occurrence and distribution of GFAP at the light and electron microscope levels (Graeber and Kreutzberg 1986). In the normal facial nucleus of a young adult rat, GFAP-immunoreactivity is very weak and almost restricted to a few perivascular structures corresponding to astrocyte foot-plates on the local capillaries. Following facial nerve transection GFAP-immunoreactivity changes dramatically within 2–3

H. Flohr (Ed.)
Post-Lesion Neural Plasticity

days. All local astrocytes in the facial nucleus become positively stained with GFAP-antibodies. At the level of electron microscopy, astrocytes show a hypertrophy and change from the protoplasmic into the fibrillary type. The cell bodies become more prominent and the processes increase in both diameter and number. They are filled with bundles of glial filaments and run in a cable-like fashion through the neuropil of the nucleus. We have not observed [^{3}H]-thymidine uptake, mitotic divisions or a resulting increase in the number of astrocytes (Graeber et al. 1988). Thus we conclude that there is only hypertrophy but not hyperplasia in astrocytes of the facial nucleus during retrograde reaction.

3 Microglia and Synaptic Stripping

The above conclusion is important since another species of glial cells does in fact react with mitotic activity and vigorous proliferation. Several (3–4) days after nerve transection local resting microglial cells must have received a signal for entering the S-phase of the mitotic cycle, as can easily be recognized by the uptake of [^{3}H]-thymidine (Kreutzberg 1966, 1968). The activated microglial cells are characterized by their morphology, by their exceptionally strong activity in the ecto-enzyme 5′-nucleotidase (Kreutzberg and Barron 1978) and by the presence of a α-D-galactose glycoconjugate on their plasmalemma which can be stained with lectin GSA-I-B_4 from *Griffonia simplicifolia* (Streit and Kreutzberg 1987). Most of the microglial cells are in a perineuronal position covering almost completely the surface of the soma and the stem dendrites of the motoneurons (Fig. 1). Thus, the synaptic terminals which are normally very dense at these sites are displaced. As a result, the motoneurons become centrally deafferented (Blinzinger and Kreutzberg 1968) a phenomenon first described in our laboratory and which is now generally accepted as "synaptic stripping" occurring in a number of different systems as a consequence of nerve lesions (for review see Cotman et al. 1981). Activated microglial cells can develop in macrophages provided the right signal is present. This seems to be the case if neuronal death occurs in situ (Streit and Kreutzberg 1988) or in vitro (Rieske et al. 1987). The option of activated microglial cells to become macrophages with potential cytotoxic properties may play a role in triggering a second or delayed reaction of the astrocytes.

4 The Delayed Astrocytic Reaction

The early hypertrophy of local astrocytes characterized by a strong increase in thick processes containing glial filaments with GFAP-immunoreactivity is fol-

Fig. 1. Regenerating rat facial nucleus 10 days after operation. A chromatolytic motoneuron (*N*) is surrounded by microglial cells (*M*). They are in close contact to the neuronal surface (*small arrows*). Only three synaptic contacts are preserved (*arrowheads*). Bar = 1.5 μm

Fig. 2. Ninety days following axotomy astrocytic lamellar processes (*arrows*) are still present. They cover the surface of the regenerating motoneurons (*N*); *A* astrocytic processes containing glial filaments (*arrowheads*); *D* dendrite. Bar = 0.2 μm

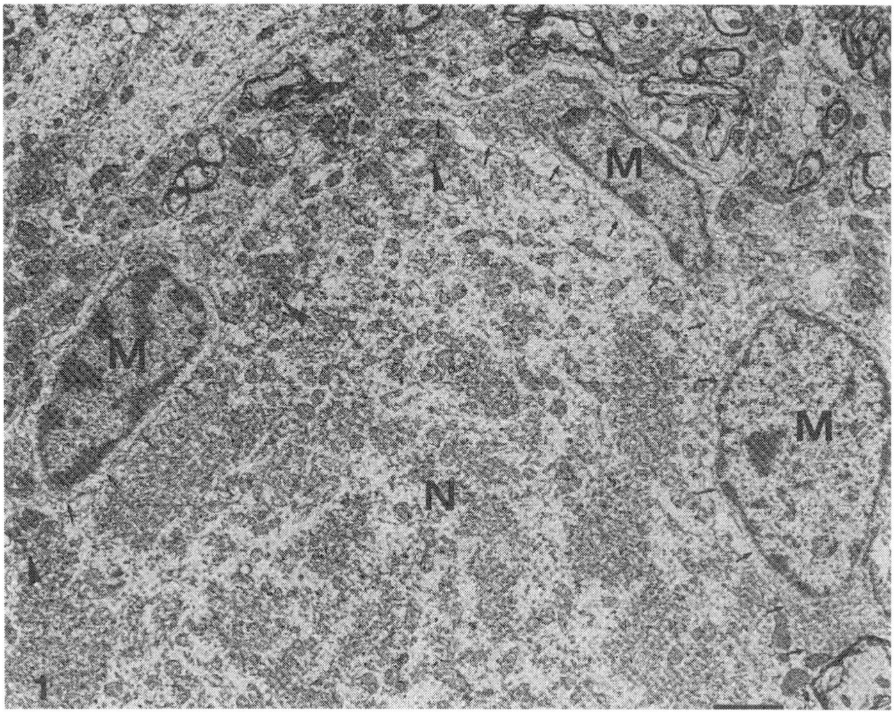
M
M
M
N
1

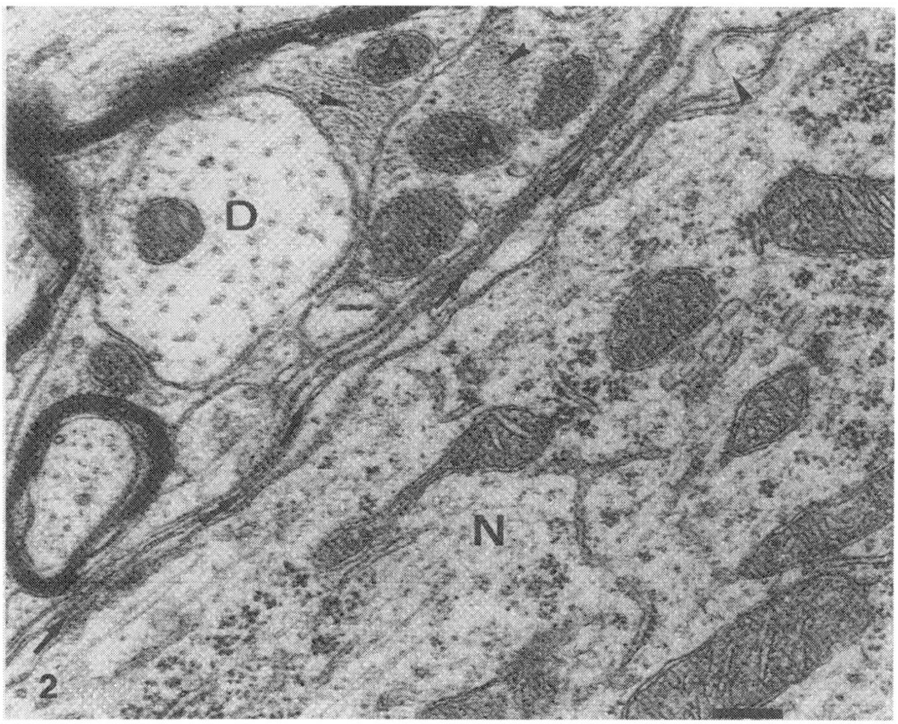
D
N
2

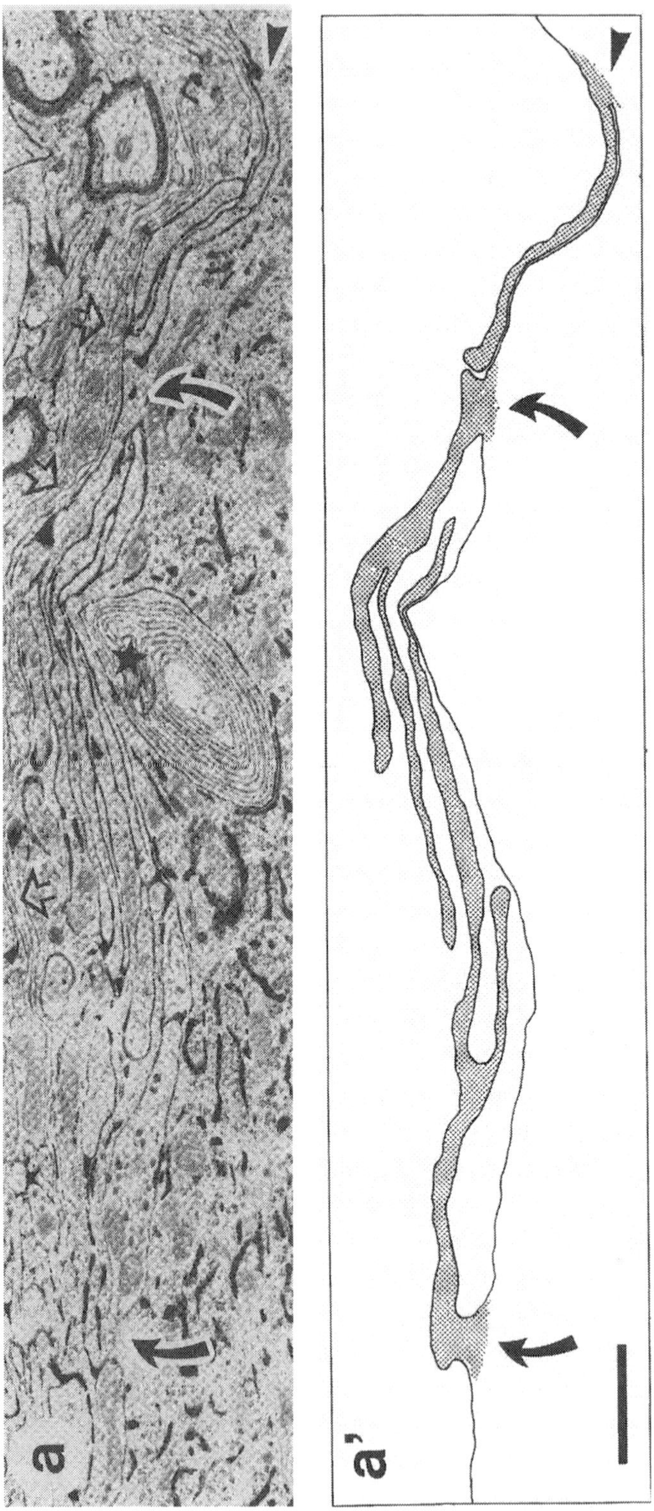

Fig. 3a

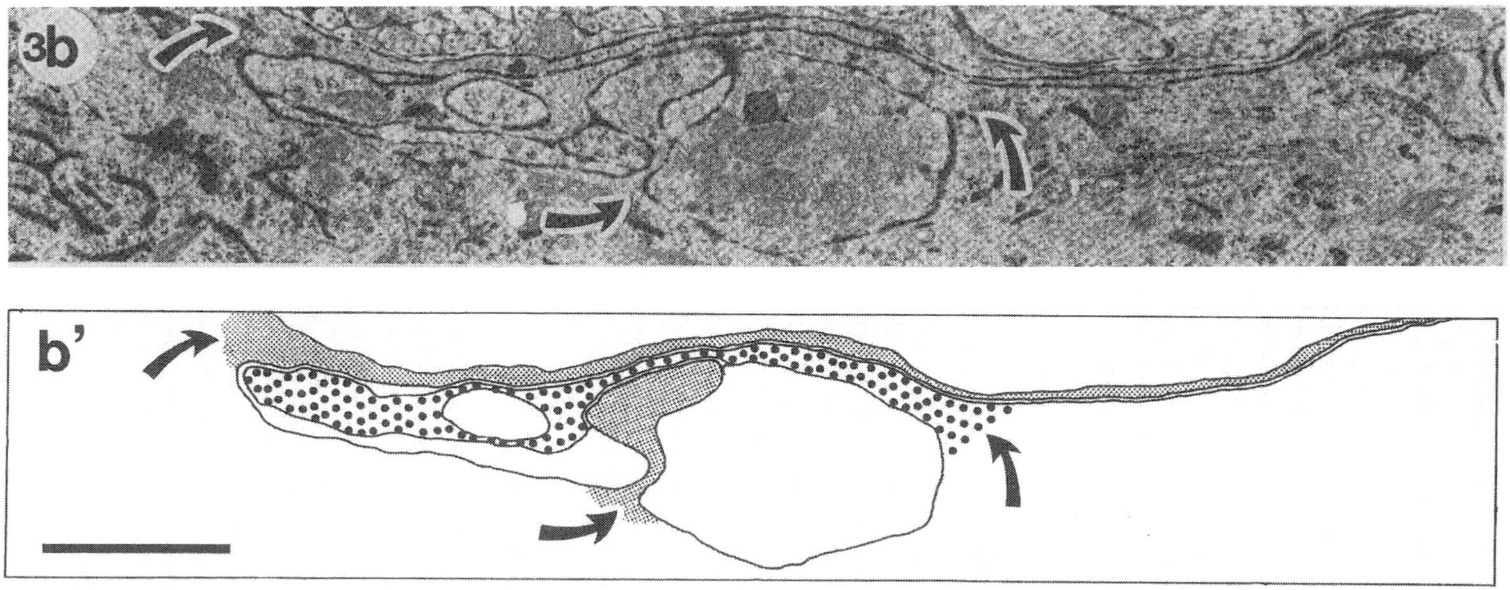

Fig. 3. Sprouting of flap-like processes from regenerating vagal motoneurons in the guinea pig. **a** Electron micrograph from the dorsal motor nucleus of vagus, 6 days after axotomy. Acetylcholinesterase (AChE) cytochemistry was used to distinguish neuronal from glial lamellae. Note that multiple neuronal processes (*solid arrows, arrowhead*) originate from the perikaryon, interdigitate and pile up to a lamellae stack that covers part of the surface. They are schematically redrawn in **a**′. Intermediate profiles which also stain positive for AChE activity can be assumed to stem from the same cell. In addition, AChE-negative glial processes wrap large parts of the neuron (*open arrows*). An astrocytic lamellae whorl (*asterisk*) has been invaginated into the neuron, indicating the extreme malleability and deformation of the neuronal surface. Bar = 1 μm. **b** Vagal motoneuron, 6 days after axotomy. AChE cytochemistry. On many occasions, the neuronal sprouts were seen to engulf presynaptic boutons. This might suggest that in particular the synaptic sites of the vagal motoneurons are subjected to plastic modifications. The neuronal sprouts are usually rich in microfilaments, pleomorphic vesicular elements and polyribosomes, thus yielding a growth cone-like appearance. Note the strong continuous staining of the plasmalemma for AChE activity. **b**′ schematic drawing of neuronal processes. Bar = 1 μm

lowed by a second phase (Graeber and Kreutzberg 1988). Between 3 weeks and 3 months following nerve lesion astrocytes form increasingly lamellar processes which can be seen in single sheets or in multiple stacks in the facial nucleus. During this period it is rather typical to observe lamellar stacks especially covering the motoneurons. Thus it seems that after the initial wrapping with microglia, astrocytes have taken over and now shield or insulate the neuronal cell bodies for a long time (Fig. 2). The astrocytic nature of the lamellae can be clearly demonstrated by following the continuity of the sheets to somewhat larger processes with glial filaments. It is also of interest that the lamellae-forming astrocytes are slowly losing GFAP-immunoreactivity. After 90 days they are weakly stained but still GFAP-positive, unlike the resting protoplasmic astrocytes of the unoperated control side. We have to realize that astrocytes are occupying synaptic sites to a great extent by their wrapping processes. Thus the motoneurons are deprived of their central input for a long period.

5 Sprouting of Pseudo-Dendritic Processes

We have observed that chromatolytic motoneurons exhibit a wavy, occasionally spiny or scalloped surface. In the dorsal motor nucleus of the vagus an extreme form of sprouting and budding is demonstrated after transection of the vagal nerve (Engel and Kreutzberg 1988) (Fig. 3 a, b). These processes arise in a flap-like form from the cell bodies of the motoneurons. They are often multiple and curl around the neuronal surface, form budlike structures, and may extend into the neuropil where they may become ensheathed and wrapped by astrocytic lamellae. The whole aspect has a highly pathological appearance as, for instance, in the Purkinje cells of Menkes disease (Purpura et al. 1976). Again, the formation of the flap-like pseudo-dendrites takes the sites of axo-somatic synaptic terminals and cannot be without consequence for the regular function of the neurons.

6 Changes in Transmitter-Related Proteins

With all these severe changes in the synapse-bearing structures of the motoneurons in mind, it is tempting to speculate that the preservation of electrical activity and synaptic function has low priority during the regeneration process. Very much in favor of such a view are also the down regulation of neurotransmitter-related proteins such as dopamine decarboxylase, tyrosine hydroxylase, choline acetyltransferase, acetylcholinesterase, and the muscarinic acetylcholine receptor as shown in Fig. 4 (for references see Grafstein and McQuarrie 1978; Kreutzberg et al. 1984). Nevertheless, the situation is not as simple as this list might suggest. The most prominent exception is seen in the guinea pig facial motoneurons, where a special molecular form of acetylcholinesterase, the 16S heavy form, is increased by 300% and is secreted from dendrites as well as from the growing axonal sprouts. The functional significance of this is unknown (Engel and Kreutzberg 1986; Engel et al. 1988).

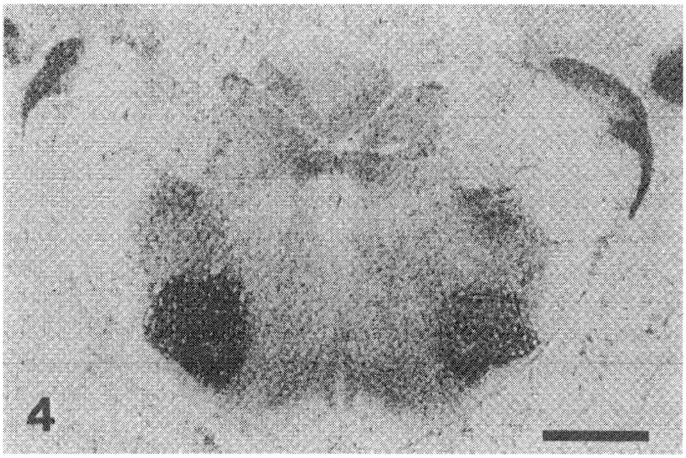

Fig. 4. Guinea pig facial nucleus. Autoradiograph of M_2 muscarinic cholinergic receptors as demonstrated by incubation with 1 nM [^{3}H]-oxotremorine. One week after axotomy, ligand binding is markedly reduced in both perikarya and neuropil of the operated side (*right*). Bar = 1 mm

7 Summary

The axotomized motoneuron represents an experimental model to study how neurons and their microenvironment can react when challenged by a lesion. As a response to axonal injury, motoneurons show several retrograde changes accompanied by glial reactions. (1) Proliferation by mitosis can be seen in local microglia cells. Within a few days the neuronal cell bodies are covered by microglial cells and the presynaptic terminals are displaced from the soma and the stem dendrites (synaptic stripping). (2) Local astrocytes become reactive and increasingly display GFAP-immunoreactivity, glial filaments, and processes. They form lamellar stacks around the regenerating neurons, thus insulating them from the covering microglial cells. With time the latter show a tendency to migrate from perineuronal to neuropil positions. The wrapping of neurons by astroglial lamellar stacks seems to be a relatively long-lasting phenomenon, since it has been observed even after gross recovery of motor function following reinnervation, e.g., after 90 days. (3) The nerve cells seem to supplement the glial wrapping by contributing their own processes. Supposedly they must originate as flap-like protrusions from the soma surfaces or from stem dendrites. They can also form lamellar stacks covering parts of the surface of the motoneurons. As a cytochemical marker, AChE activity is used. Thus the neuronal lamellar stacks can be clearly distinguished from the glial analogs. These novel findings can be discussed in terms of a protective mechanism, as a deafferentation mechanism, or as a truncated growth process.

References

Bisby MA (1980) Changes in the composition of labeled protein transported in motor axons during their regeneration. J Neurobiol 11:435–445

Blinzinger K, Kreutzberg GW (1968) Displacement of synaptic terminals from regenerating motoneurons by microglial cells. Z Zellforsch 85:145–157

Cotman CW, Nieto-Sampedro M, Harris EW (1981) Synapse replacement in the nervous system of adult vertebrates. Physiological Reviews 61:684–784

Engel AK, Kreutzberg GW (1986) Changes of acetylcholinesterase molecular forms in regenerating motor neurons. Neuroscience 18:467–473

Engel AK, Kreutzberg GW (1988) Neuronal surface changes in the dorsal vagol motor nucleus of the guinea pig in the response to axotomy. J Comp Neurol 274:4

Engel AK, Tetzlaff W, Kreutzberg GW (1988) Axonal transport of 16S acetylcholinesterase is increased in regenerating peripheral nerve in guinea pig, but not in rat. Neuroscience 24:729–738

Graeber MB, Kreutzberg GW (1986) Astrocytes increase in GFAP during retrograde changes of facial motoneurons. J Neurocytol 15:363–373

Graeber MB, Kreutzberg GW (1988) Delayed astrocyte reaction following facial nerve axotomy. J Neurocytol 17:209–220

Graeber MB, Tetzlaff W, Streit WJ, Kreutzberg GW (1988) Microglial but not astrocytes undergo mitosis following facial nerve axotomy. Neurosci Lett 85:317–321

Grafstein B, McQuarrie IG (1978) Role of the nerve cell body in axonal regeneration. In: Cotman CW (ed) Neuronal plasticity. Raven Press, New York, pp 155–195

Kreutzberg GW (1966) Autoradiographische Untersuchung über die Beteiligung von Gliazellen an der axonalen Reaktion im Facialiskern der Ratte. Acta Neuropathol 7:149–161

Kreutzberg GW (1968) Über perineuronale Mikrogliazellen (Autoradiographische Untersuchungen). Acta Neuropathol Suppl 4:141–145

Kreutzberg GW (1982) Acute neural reaction to injury. In: Nicholls JG (ed) Repair and regeneration of the nervous system. Life Sciences Research Report 24, Dahlem Konferenzen 1982. Springer, Berlin Heidelberg New York, pp 57–69

Kreutzberg GW (1986) Neurobiology of regeneration and degeneration. In: May M (ed) The facial nerve. Thieme, New York, pp 75–83

Kreutzberg GW, Barron KD (1978) 5′-Nucleotidase of microglial cells in the facial nucleus during axonal reaction. J Neurocytol 7:601–610

Kreutzberg GW, Emmert H (1980) Glucose utilization of motor nuclei during regeneration: a 14C deoxyglucose study. Exp Neurol 70:712–716

Kreutzberg GW, Tetzlaff W, Tóth L (1984) Cytochemical changes of cholinesterases in motor neurons during regeneration. In: Brzin M, Barnard EA, Sket D (eds) Cholinesterases – fundamental and applied aspects. Walter de Gruyter, Berlin New York, pp 273–288

Lieberman AR (1971) The axon reaction: a review of the principal features of perikaryal responses to axon injury. In: Pfeiffer CC, Smythies JR (eds) International review of neurobiology, vol 14. Academic Press, New York London, pp 49–124

Purpura DP, Hirano A, French JH (1976) Polydendritic Purkinje cells in X-chromosome linked copper malabsorption: a Golgi study. Brain Res. 117:125–129

Rieske-Shows E, Tetzlaff W, Czlonkowska A, Graeber M, Kreutzberg GW (1987) Microglia in culture. In: Althaus HH, Seifert W (eds) Glial-neuronal communication in development and regeneration. NATO ASI Series H, vol 2. Springer, Berlin Heidelberg New York London Paris Tokyo, pp 43–51

Streit WJ, Kreutzberg GW (1987) Lectin binding by resting and reactive microglia. J Neurocytol 16:249–260

Streit WJ, Kreutzberg GW (1988) The response of endogenous glial cells to motor neuron degeneration induced by toxic ricin. J Comp Neurol268:248–263

Tetzlaff W, Kreutzberg GW (1985) Ornithine decarboxylase in motoneurons during regeneration. Exp Neurol 89:679–688

Tetzlaff W, Graeber MB, Kreutzberg GW (1986) Reaction of motoneurons and their microenvironment to axotomy. In: Gilad G, Gorio A, Kreutzberg GW (eds) Processes of recovery from neural trauma. Exp Brain Res, Suppl 13:3–8

Tetzlaff W, Graeber MB, Bisby MA, Kreutzberg GW (1988) Increased glial fibrillary acidic protein synthesis in astrocytes during retrograde reaction of the rat facial nucleus. Glia 1:90–95

The Influence of Magnetic Field on the Sheath Cell-Axon Interactions

K. Shibib[1], M. Brock[1], G. Gosztonyi[2], S.N. Erne[3], H.-D. Hahlbohm[3], and G. Schoknecht[4]

[1] Neurosurgical Department, University Medical Center Steglitz, Free University of Berlin, Hindenburgdamm 30, D-1000 Berlin (West) 45
[2] Institute of Neuropathology, Hindenburgdamm 30, D-1000 Berlin (West) 45
[3] Physikalisch-Technische Bundesanstalt, Institut Berlin, Abbestr. 2, D-1000 Berlin (West) 21
[4] Bundesgesundheitsamt für Sozialmedizin und Epidemiologie, General-Pape-Str. 62, D-1000 Berlin (West) 42

1 Introduction

The geomagnetic field which is a part of the meteorological variables, is related to biological activities, especially to the orientation of microorganisms, insects, birds and fishes (Bamberger et al. 1978; Blackmore 1975; Brown et al. 1979; Chernyshev et al. 1978; Chervinets et al. 1979; Kirschvink 1981; Vilches-Troya et al. 1984; Wehner 1979). Variations in geomagnetic field have been correlated to congenital anomalies (Archer 1978; Archer 1979), psychological and convulsive disorders (Pokorny et al. 1966; Rajam et al. 1981), to the incidence of myocardial infarction (Bellossi et al. 1985; Knox et al. 1979), and to death rate variations (Feinleib et al. 1975). These variation can alter the growth and metabolism of cells (Kholodov 1964; Kopanev et al. 1979; Liboff et al. 1984; Shibib et al.; Sutherland et al. 1978). Because of the possible importance of biomagnetic phenomena in the nervous system, we developed experimental models to examine the effect of magnetic field variations on the cellular interactions of the developing nervous system of newborn rats.

Normally the degree of maturation of peripheral nerves in the newborn rat can be judged from the amount of myelinated fibers, and the amount of premature fibers (fetal and promyelin fibers) still present among the myelin-forming fibers (Asbury 1976; Friede et al. 1967; Robertson 1955) at a certain point in time. According to the authors mentioned above, signs of prematurity disappear to a great extent 21 days following birth and are completely absent 1 week later. At this point in time the peripheral nerves have attained adult features except for fiber diameter. Almost the same applies to maturation steps of optic nerve fibers which take place in even more regular sequence (Peters 1960; Skoff et al. 1980; Skoff 1978; Tennekoon et al. 1977).

In the immediate post-natal period, extensive migration and development of nerve cells in the CNS which might have some physical (mechanical) aspects occurs. Because of its regularity (Altman 1972, 1979; Corliss 1976; Herndon 1964; Yu et al. 1980) the cerebellar Cortex of the rat is examined.

2 Material and Method

Thirty newborn Wistar rats were used for this experiment. Three types of magnetic environment were tested.

H. Flohr (Ed.)
Post-Lesion Neural Plasticity

1. Control group: ten experimental animals were kept under normal conditions and received standard diet and care.
2. BMSR group: ten animals were kept in a cage with their mother in the Berlin Magnetically Shielded Room (BMSR), and received similar diet and care as the controls. The AC- and DC-induced magnetic fields in this room are reduced to practically zero. This group is referred to as hypomagnetic or „amagnetic".
3. Five Gauss (0.0005 T) group: Ten animals were kept under a DC-induced magnetic field of 5 Gauss, i.e., (0.0005 T) parallel to the floor (horizontal plane).

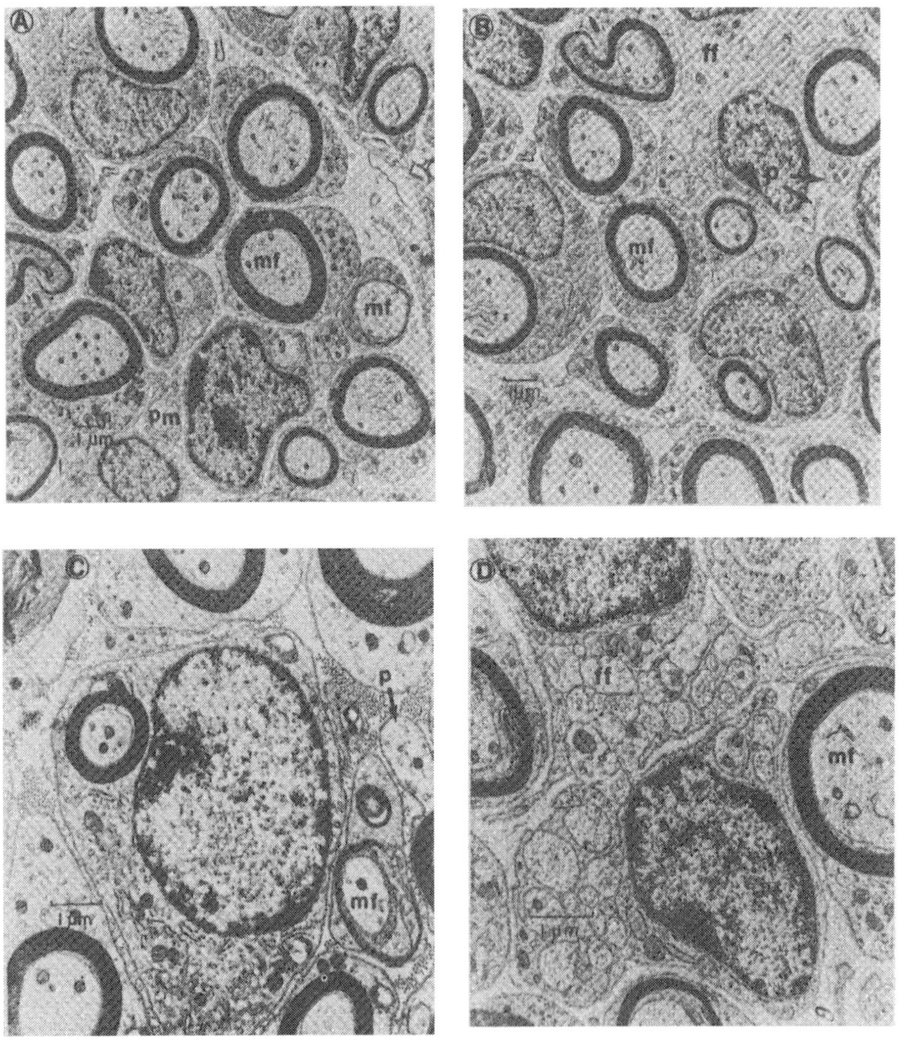

Fig. 1. Trigeminal nerve of a 10-day-old rat from the hypermagnetic group. **A**, **B** and **C** Myelin forming fiber (*mf*), promyelin fiber (*pm*), and Schwann cell processes (*p*). **D** a cluster of fetal fibers (*ff*). Nerves of the hypermagnetic and control groups revealed similar findings

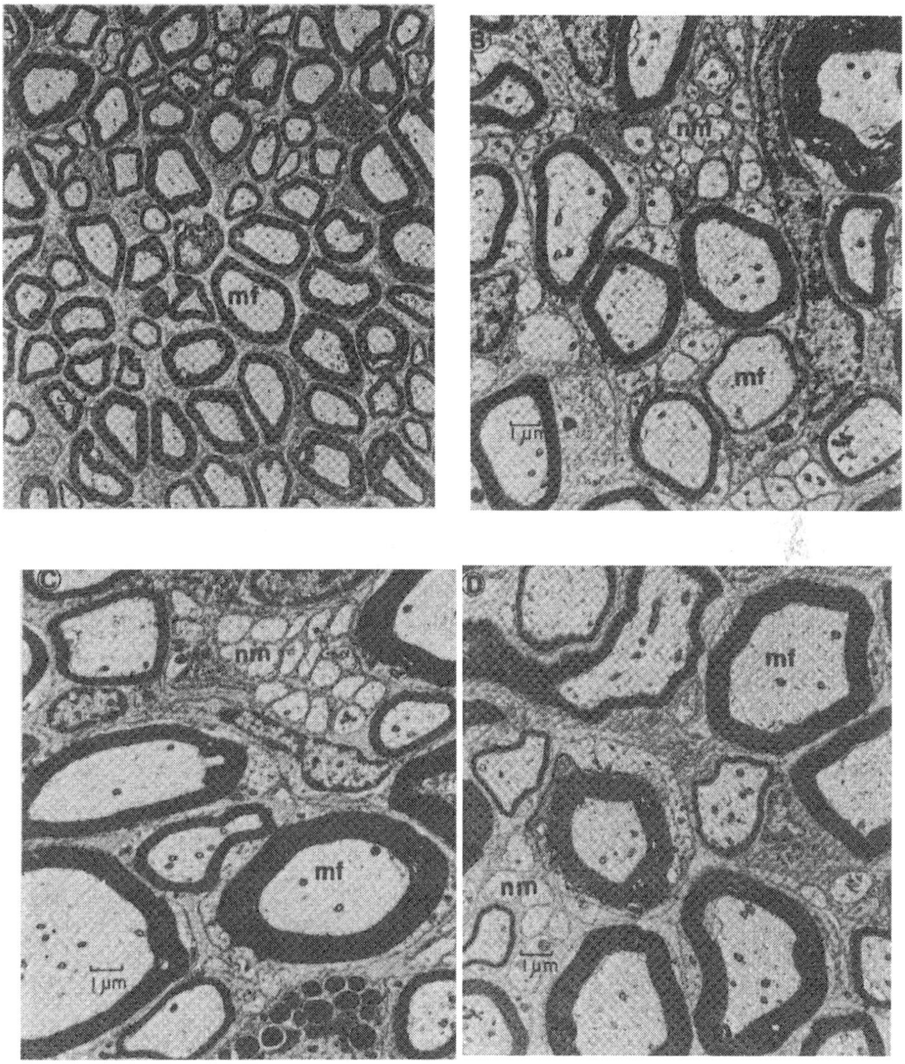

Fig. 2. Trigeminal nerve of 20-day-old rats from the control group. Myelinated (*mf*) and unmyelinated fibers (*nm*). No fetal or promyelin fibers are seen

The animals were looked after as previously mentioned. This group will be referred to as hypermagnetic.

At the end of the experiment, the remaining newborn rats were killed by intracardiac perfusion with Karnovsky solution. The sciatic and the trigeminal nerves on both sides were removed and kept in the same solution overnight. Optic nerves and chiasm, cerebellum, and specimens from other parts of the CNS were also removed and treated similarly. Parts of these specimens were processed for electron microscopy and part for paraffin.

3 **Results:** Clinical Conditon

The newborn rats of the hypomagnetic group showed sluggishness and apathy as compared to those belonging to the hypermagnetic and control groups. The sluggishness of the rats belonging to the hypomagnetic group decreased during the time elapsed between their removal from the magnetic chamber and sacrifice (2–6 h).

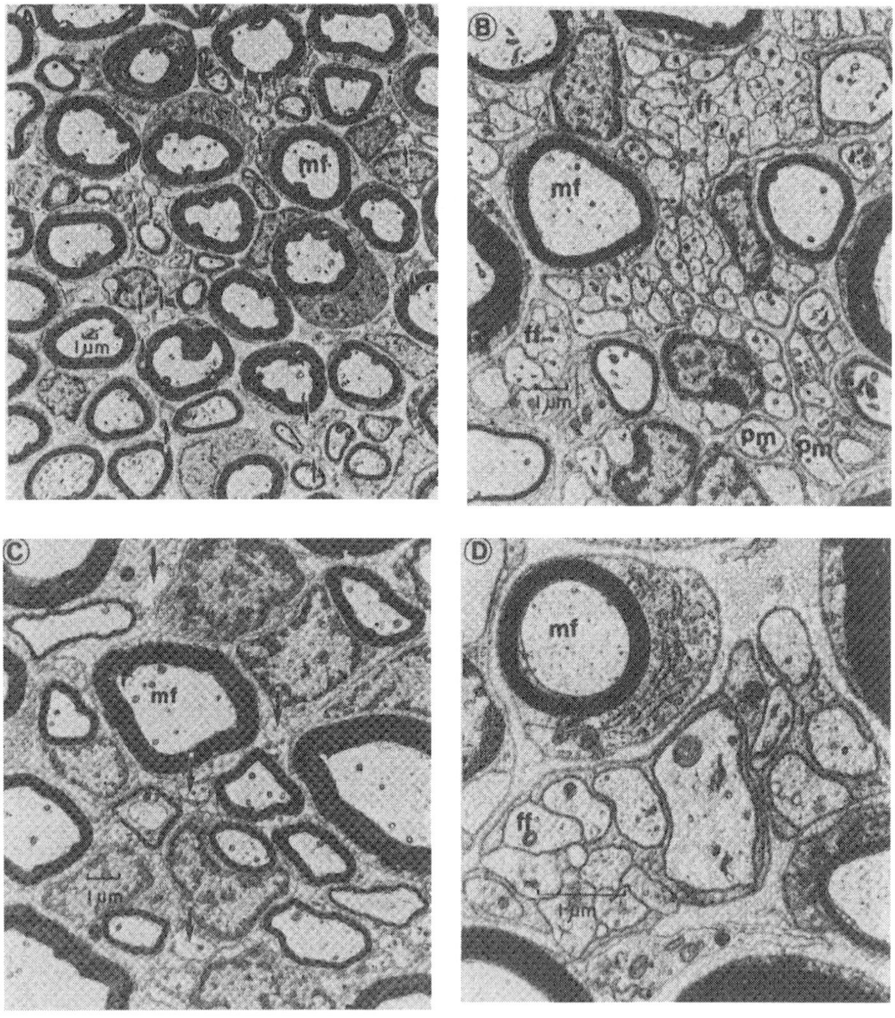

Fig. 3. Trigeminal nerve of 20-day-old rats from the hypomagnetic group. **A** and **C** In addition to the myelin forming fibers (*mf*), there is widespread distribution of promyelin (*pm*) fibers in all sections (*arrows*). **B** and **D** Some clusters of fetal fibers (*ff*) are also seen

3.1 Examination of Peripheral Nerves

At the 10th postnatal day, electron microscopy of both the sciatic and trigeminal nerves of 9 newborn rats (3 from each group) showed no noticeable differences between the two nerves as well as between the three groups (Fig. 1) Many fibers were myelinated. Myelin forming fibers (mf) and embryonal fibers, i.e., promyelin fibers (pm) and clusters of fetal fibers exist beside each other. Schwann cell processes persist at this stage in all groups.

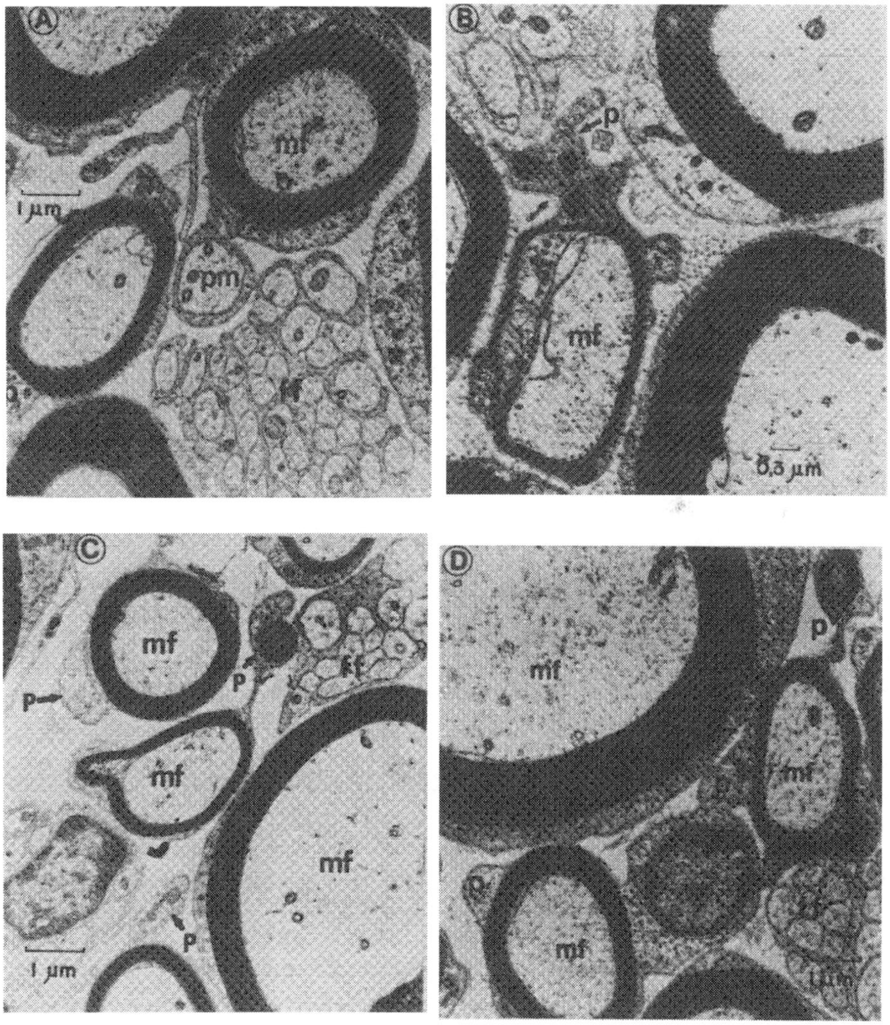

Fig. 4: Trigeminal nerve of 20-day-old rats from the hypermagnetic group. In addition to the myelin forming fibers (*mf*), there is widespread distribution of fetal fibers (*ff*) (**A**, **C** and **D**). **D** Persistence of Schwann cell processes (*p*). **C** and **D** Schwann cell processes (*p*) extending from one myelinated fiber (*mf*) to ensheath and myelinate another myelin forming fiber. In addition, other processes of the same cell still protrude into the extracellular space

At the 20th day, promyelin fibers (pm) and fetal fibers are seen in both hypo- and in the hypermagnetic groups, while in the control group fetal fibers are extremely rare (Fig. 2). The incidence of promyelin fibers (pm) is higher in the hypomagnetic (Fig. 3 A, C) than in the hypermagnetic (Fig. 4 A) group. On the other hand, the fetal fibers (ff) are more common in the hypermagnetic (Fig. 4 A, C, D, and Fig. 5 B) than in the hypomagnetic (Fig. 3) group.

The free Schwann cells and those in early stages of development show a tendency to sprout in the environment. Following ensheathment and myelination of the axon, these processes or sprouts tend to regress, and the Schwann cell attains its oval shape with regular surface (Fig. 6). The Schwann cells show a higher ten-

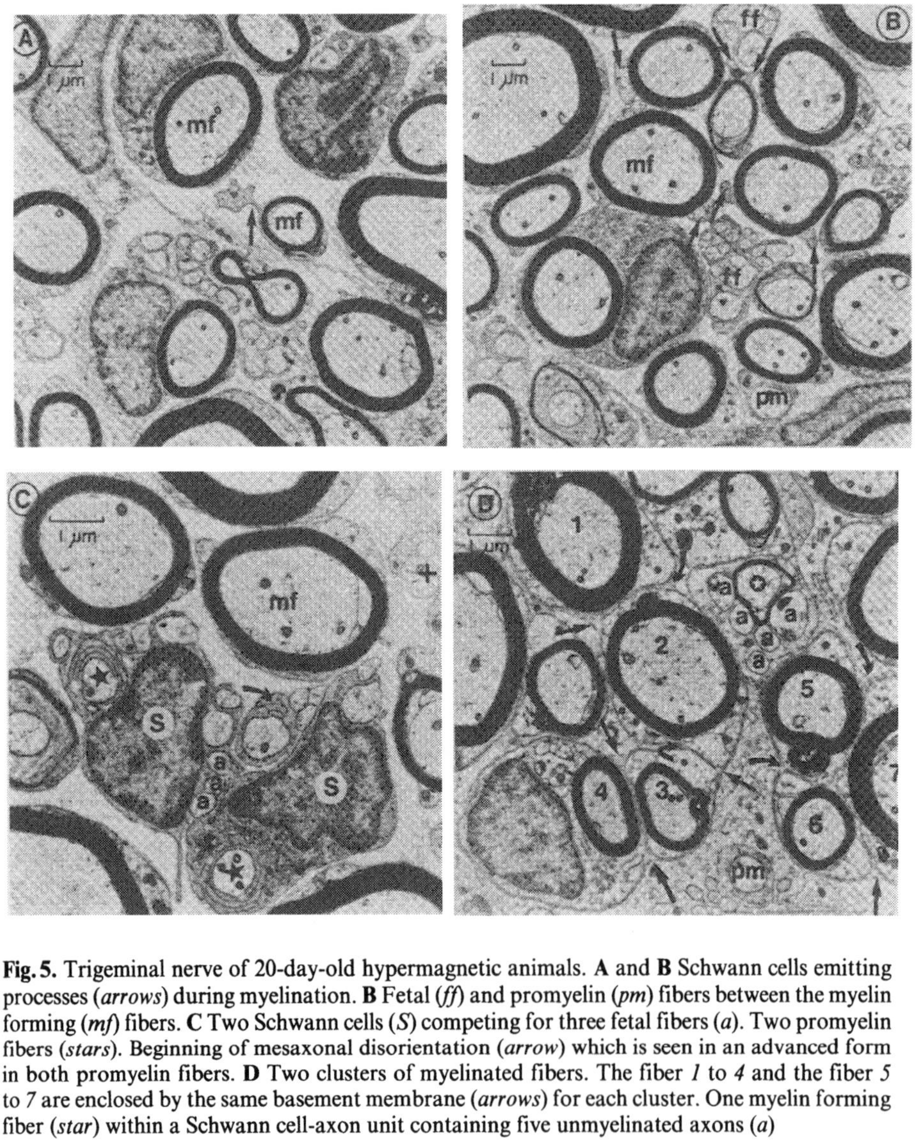

Fig. 5. Trigeminal nerve of 20-day-old hypermagnetic animals. **A** and **B** Schwann cells emitting processes (*arrows*) during myelination. **B** Fetal (*ff*) and promyelin (*pm*) fibers between the myelin forming (*mf*) fibers. **C** Two Schwann cells (*S*) competing for three fetal fibers (*a*). Two promyelin fibers (*stars*). Beginning of mesaxonal disorientation (*arrow*) which is seen in an advanced form in both promyelin fibers. **D** Two clusters of myelinated fibers. The fiber *1* to *4* and the fiber *5* to *7* are enclosed by the same basement membrane (*arrows*) for each cluster. One myelin forming fiber (*star*) within a Schwann cell-axon unit containing five unmyelinated axons (*a*)

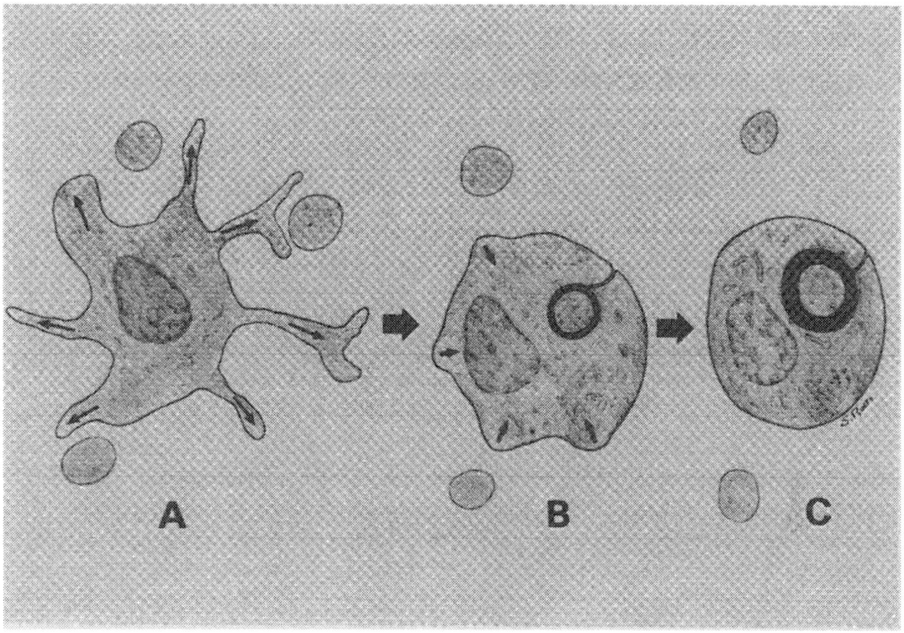

Fig. 6. Schematic illustration of the "normal" Schwann cell-axon interaction. **A** Initial stage of migration and sprouting before establishing contact with an axon. **B** The Schwann cell is myelinating the axon, migration stops and other processes retract towards the cell body. **C** No further contact is established with other axons. The Schwann cell has become "indifferent" towards its environment

dency to emit processes in the hypermagnetic group as compared to the other two groups (Fig. 4). Consequently, the existence of more than one myelinated axon, or of myelinated and unmyelinated axons in one Schwann cell was common in this group and absent in both other groups (Fig. 5 B, C, D and Fig. 7 D).

Computer-assisted morphometry of the myelinated nerve fibers in the trigeminal nerve showed that while the relation between myelin thickness and fiber diameter (fiber index) decreases during development in both control and in the hypomagnetic group. In the hypermagnetic group – on the other hand – this index increases (Table 1).

3.2 Examination of the Optic Chiasm

The examination of optic chiasms of nine animals at the 10th postnatal day revealed significant differences between the groups. In all sections of animals belonging to the control „geomagnetic" group, many nerve fibers were myelinated. Myelin had a normal appearance (compact, uniform, and lammelated as usual) (Fig. 8 C, D). Most of the fibers in this group were, however, still from the fetal type.

Few myelinated fibers were found altogether throughout the entire cross section of the optic chiasm of one animal belonging to the hypermagnetic group

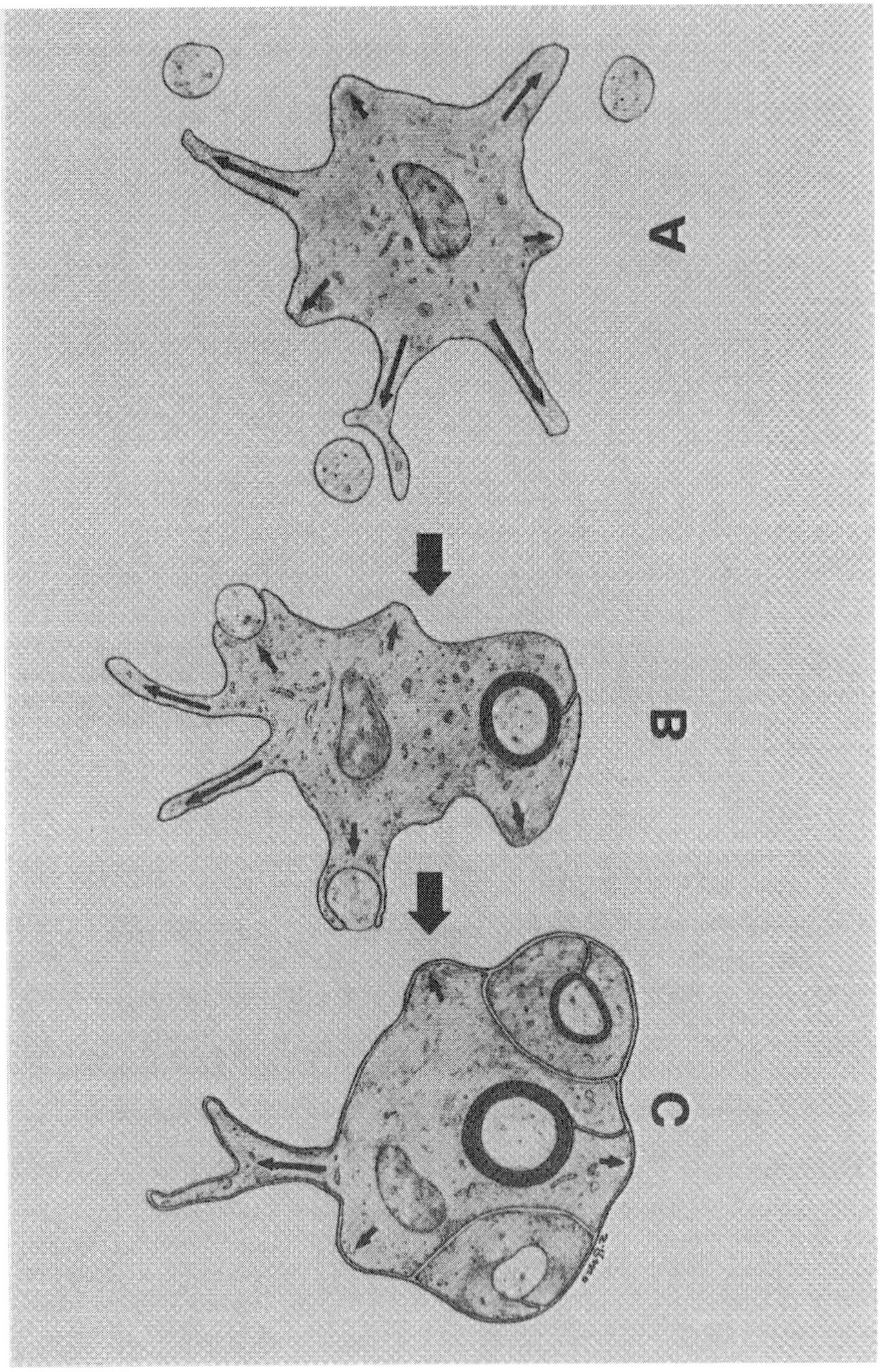
A
B
C

Table 1. Morphometry of the myelinatec trigeminal nerve fibers. While the fiber index (relation between myelin thickness to total fiber diameter) decreases between the 10th and the 20th day in the controls (and to a lesser extent in the hypomagnetic group), it remains almost constant or rather increases in the hypermagnetic group. The larger standard deviation in both hypomagnetic and hypermagnetic may reflect individual variations. The T-test demonstrates a decrease of the fiber index between the 10th and the 20th day in the control ($P=0.0327$) and hypomagnetic ($P=0.05829$) groups and an increase in the fiber index in the hypermagnetic group ($P=0.4762$)

		Control group		Hypomagnetic group (BMSR)		Hypermagnetic group (5 Gauss)	
		10 days	20 days	10 days	20 days	10 days	20 days
Fiber maximal diameter	Mean	59.74	80.40	52.59	73.18	59.82	65.54
	SD	3.5938	3.9388	4.1224	22.0400	4.1214	7.9464
Fiber minimal diameter	Mean	45.83	57.82	39.67	55.69	44.93	50.04
	SD	4.9427	1.7632	5.4942	17.7868	3.4603	6.4806
Myelin diameter	Mean	4.13	4.41	3.53	3.85	3.22	4.05
	SD	0.6408	0.2836	0.9383	0.6665	1.0174	0.2768
Fiber index	Mean	0.07801	0.06392	0.07585	0.05953	0.06960	0.0704
	SD	0.0075	0.0061	0.0124	0.0069	0.0166	0.0337

(Fig. 10D). In all other animals of both the hypomagnetic and hypermagnetic groups, no signs of regular myelination could be detected. The electron microscopic picture of these sections is dominated by fibers of the fetal type (Fig. 9 and Fig. 10B, C). Folding of membranes of the glial processes around clusters of fetal fibers was seen (Fig. 9 and Fig. 10A) but no compact myelin sheaths.

At the 20th day time point most fibers were myelinated. No significant differences could be detected between the control group the hypomagnetic group, and the hypermagnetic group (Fig. 11 and Table 2).

3.3 Cerebellar Cell Migration

At the 10th postnatal day light microscopy of paraffin sections of the cerebellum of animals belonging to all three groups showed persistence of nerve cells in the outer layer of the cerebellar cortex (Fig. 12). The majority of the Purkinje cells had already attained the internal granular layer. On the other hand, the overwhelming part of the granular nerve cells was still located in the external granular layers. Very few nerve cells were seen in the molecular layer.

At the 20th postnatal day, in all three groups complete migration of the nerve cells from the outer layer to the internal granular layers had taken place. Since

Fig. 7. Schematic illustration of the Schwann cell-axon interaction in a disturbed magnetic field. **A** Migration and sprouting of the Schwann cell before establishing contact with an axon. **B** Contact can be initiated with one or more axons. Myelination of one axon and contact with another. **C** Myelination of more than one axon, contact with other and further sprouting into the environment

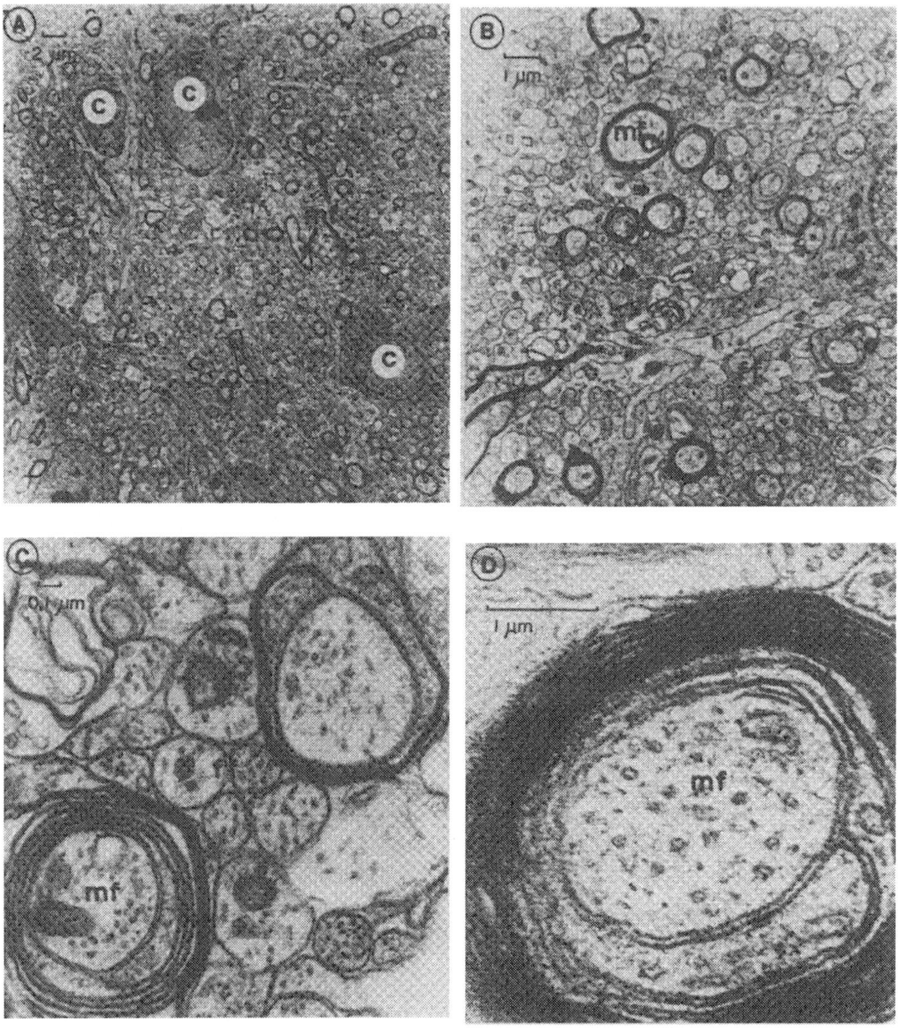

Fig. 8. Optic chiasm of 10-day-old rats of the central group. **A** and **B** Scattered myelin forming (*mf*) fibers in all sectors. **A** Oligodendrocytes (*c*) also scattered in the field. **B** Myelin forming (*mf*) and promyelin fibers (*arrows*). **C** Myelin forming fiber with still uncompact myelin. **D** Myelin forming fiber (*mf*) showing 11 myelin lammelae. The axon is about 2 µm in diameter

the majority of the granular cells were still in the external granular layer at the 10th post-natal day, the migration of granular cells was striking. The migration of Purkinje cell was completed at this time. The external granular layer was totally devoid of nerve cells (Fig. 13). The cerebellar cells showed no significant structural differences between the three groups. At 10 days the Purkinje cells in the external granular layer were smaller and less differentiated than those in the internal layer. The differentiation of the Purkinje cells at the 20th post-natal day was expressed by the increase in cytoplasmic volumen development of a distinct dendritic tree and axonal extension (Fig. 14).

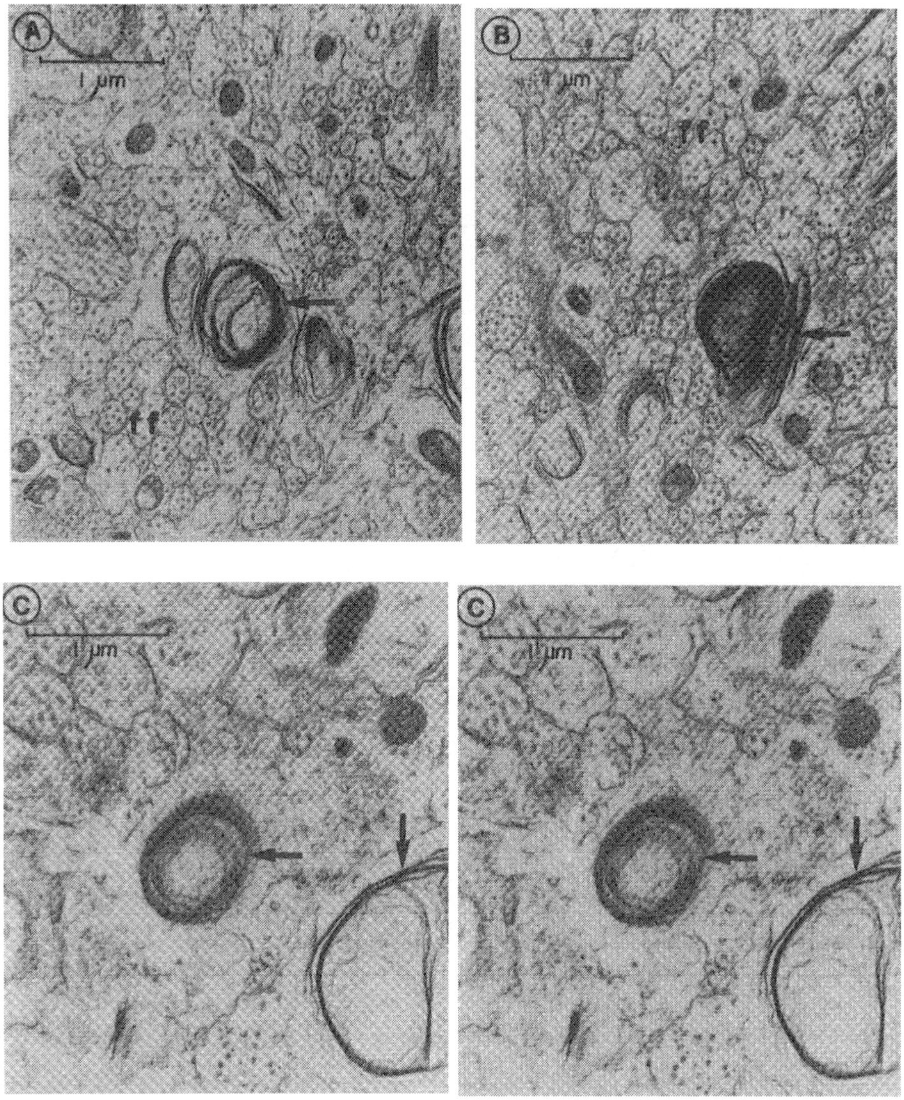

Fig. 9. Optic chiasm of 10-day-old rats o the hypomagnetic group. **A**, **B**, **C** and **D** In all nerve sections no evidence of "normal" myelin. Fetal fibers (*ff*) dominate the picture. Opaque structures probably resulting from abnormal membrane folding (*arrows*) are scattered throughout all sections

Electron microscopy of the cerebellar folia of animals belonging to all groups in the time from 10th to 20th postnatal days showed no significant differences between the cerebellar neurons of the three groups of animals (Fig. 15).

Because animals belonging to the control group are subjected to magnetic fields of unknown nature and intensity in addition to the geomagnetic field, this group is reexamined. The "new" control group is subjected to a DC-induced magnetic field generated inside the Berlin Magnetically Shielded Room of similar in-

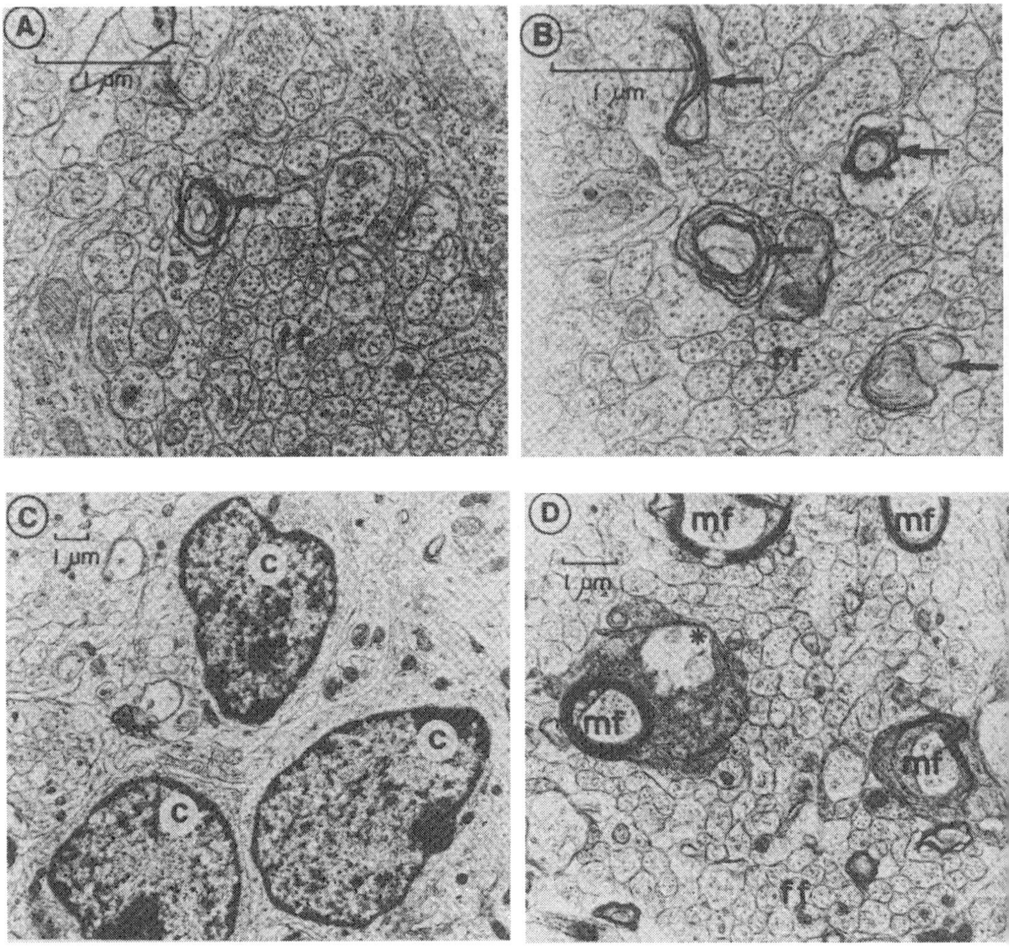

Fig. 10. Optic chiasm of 10-day-old rats of the hypermagnetic group. **A** and **B** Membranous folding scattered throughout all sections (*arrows*), which contain numerous fetal fibers (*ff*). **C** Cluster of oligodendrocytes (*c*). **D** In only one animal a few myelin forming fibers (*mf*) are seen in the center of the section which is also dominated by fetal fibers (*ff*). In the vicinity of the myelin forming fibers, and inside the same glia processes that enclose them, other, probably degenerating unmyelinated axons, are seen (*asterisks*)

tensity as the geomagnetic field. The animals developed normally and the electron microscope examination was similar to that of the control group of the first experiments.

4 Discussion

The present study was motivated by the following considerations:

1. There is a transport of electrical charges – flowing current – along the fibers of the Central and Peripheral Nervous System. From the physical point of

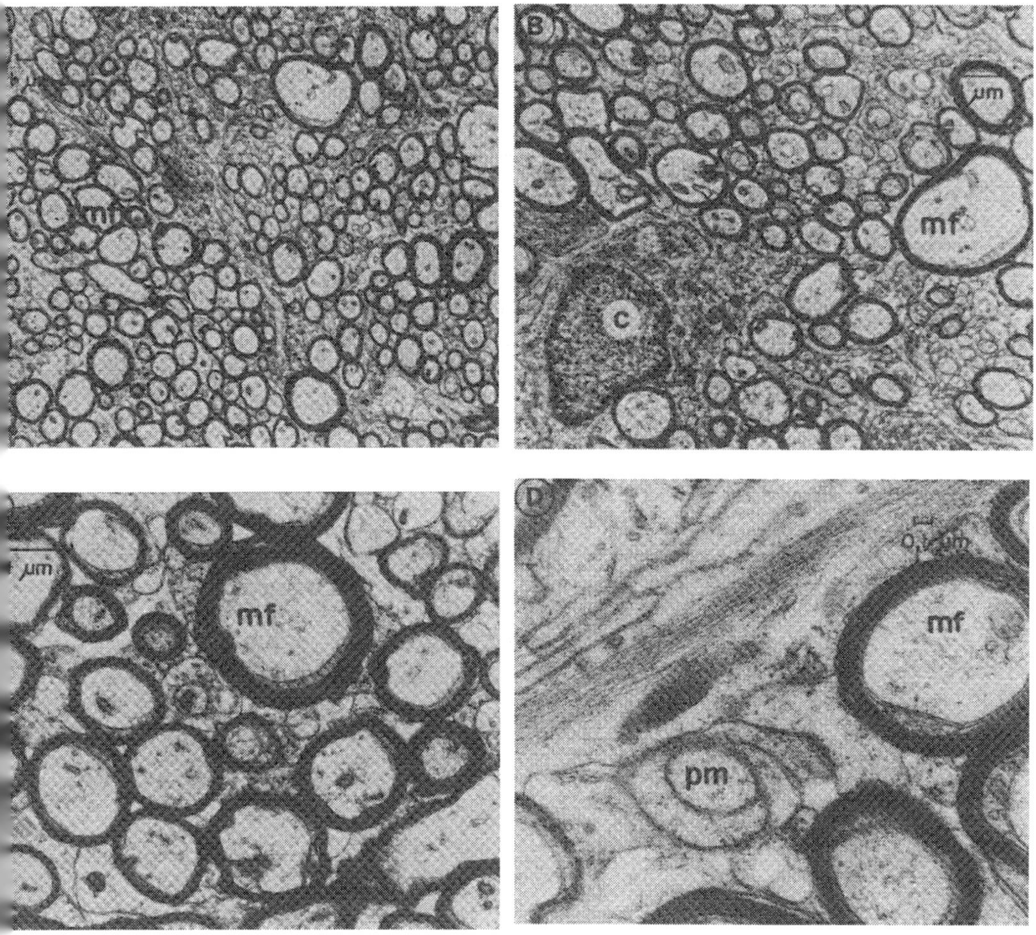

Fig. 11. Optic chiasm in 20-day-old rats of the control group. **A**, **B**, and **C** Intensive myelination of the majority of fibers (*mf*). **B** Astrocyte (*C*). **D** Promyelin fiber (*pm*) seen between the myelinated nerve fibers

view, this flow of current may be considered a dipole giving origin to a magnetic field around these fibers (Wikswo et al. 1980). As is known, the external magnetic field acts on objects with current dipole behavior and may cause their displacement for short distances.

2. Phenomena such as myelination, remyelination, demyelination, and Wallerian degeneration result from the axon-sheath cell interaction. This takes place in the immediate vicinity of the axon, and within the small magnetic field generated by the nerve impulse (Skoff et al. 1980), or by other forms of electricity which might pass along the axons.
3. It can be speculated that the mechanical element (movement) of the abovementioned interactions is influenced either by the interaction of the external magnetic field with the cells (axons, and probably, sheath cells) (Robertson

Table 2. Morphometry of the myelinated fibers of the optic chiasm. In 10-day-old animals, the axonal count and morphometric characters of myelinated fibers in animals belonging only to the control group is performed. In the other two groups, no evident axonal myelination can be detected under HPF of 1000. In the 20-day group, no significant differences are detected between the three different groups. The number of premature fibers cannot be examined under light microscopy

		Control group		Hypomagnetic group (BMSR)		Hypermagnetic group (5 Gauss)	
		10 days	20 days	10 days	20 days	10 days	20 days
Maximal fiber diameter	Mean	37.30	58.72	–	51.15	–	44.85
	SD	2.9459	12.2612	–	6.8956	–	2.1101
Minimal fiber diameter	Mean	19.66	18.27	–	37.34	–	34.28
	SD	5.4960	7.8186	–	4.7335	–	1.1717
Myelin thickness	Mean	2.28	3.21	–	3.24	–	2.75
	SD	0.1662	1.1088	–	0.6046		0.2374
Fiber index	Mean	0.06786	0.06651	–	0.07308	–	0.06944
	SD	0.0157	0.0060	–	0.0088	–	0.0039

1955) having a current dipole behavior, or through magnetization of these cells by ponderomotoric field-induced forces (Rajam et al. 1981).

In both the absence and presence of the additional magnetic field, the sheath cell-axon interaction appears to be disturbed. This disturbance consists not only in a retardation of development but also in changes in the character of the Schwann cell, especially in a hypermagnetic environment.

In a disturbed magnetic field, especially under hypermagnetic conditions, this tendency of the Schwann cell to emit processes into its environment persists despite establishment of contact with one axon.

The transient disturbance of development of the optic nerve fibers and the undisturbed migration and differentiation of the nerve cells in the cerebellar cortex and in other parts of the CNS might indicate that the disturbance of myelination noticed in the central and peripheral nerve fiber is caused mainly by the effect of magnetic field variation on the sheath cells. It appears permissable to assume that magnetic fields act mainly on the sheath cells. The sensitivity of the sheath cells to external magnetic field might indicate that the sheath cell-axon interaction is partly electromagnetic since the source of magnetism nearest to a sheath cell is that generated by the flow of current within the axon. The nerve cell which conducts electrical current(s) and is sensitive to electrical changes in its environment (Hinkle et al. 1981; Jaffe et al. 1979; Shibib) seems to be insensitive to magnetic field changes in the ranges examined.

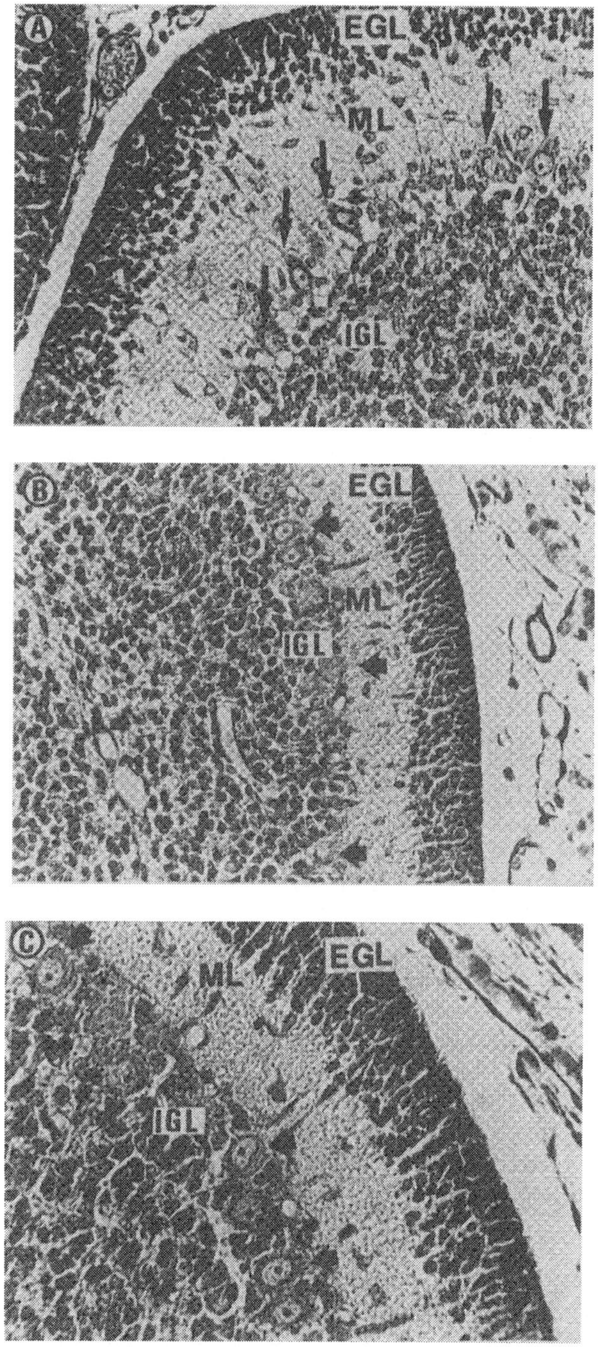

Fig. 12. Transverse paraffin section (7 μm) of a cerebellar folium of 10-day-old rats. The external granular layer (*EGL*) is crowded with cells of the granular type. In the internal granular layer (*IGL*) well-differentiated Purkinje cells are seen among granular cells in **A**, **B**, and **C**. In comparison, the molecular layer (*ML*) appears almost featureless on light microscopy. (Nissl-fast blue stain). **A** Animal of the hypomagnetic group. (Magnification 12.5 × 16). **B** Animal of the hypermagnetic group. (Magnification 12.5 × 16). **C** Animal of the control group. (Magnification 12.5 × 24)

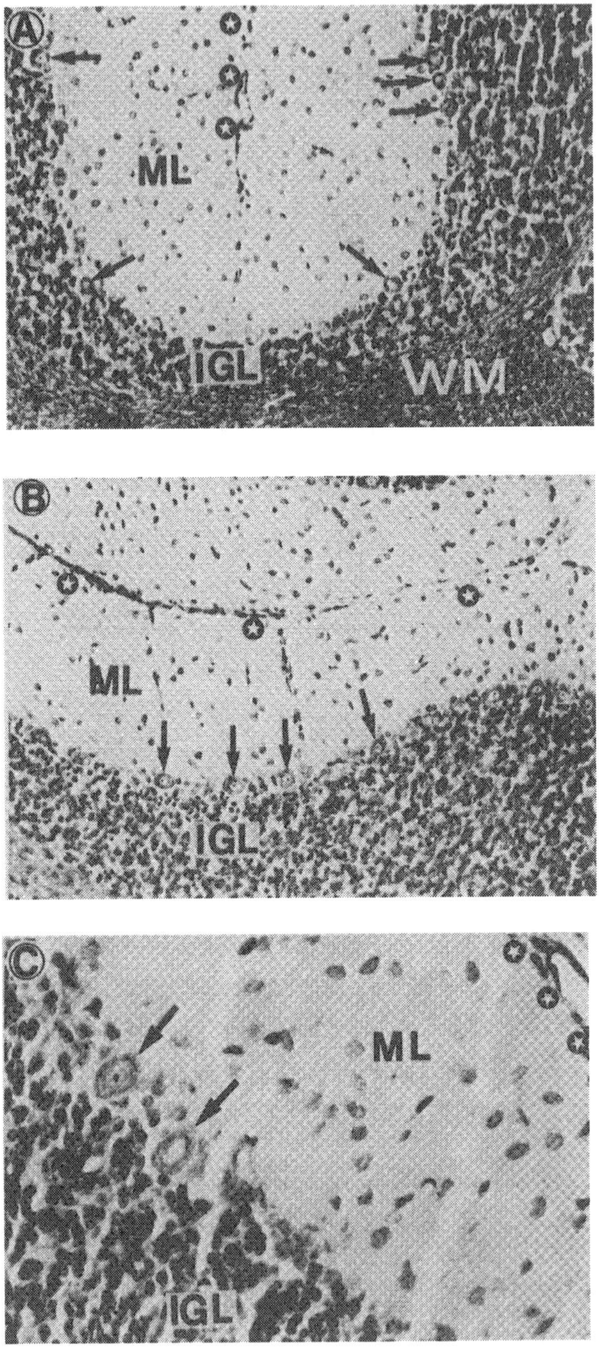

Fig. 13. Transverse paraffin section of cerebellar folium of 20-day-old rats. The external granular layer disappeared almost completely (*asterisks*). The molecular layer (*ML*) now extends to the outermost part of the cortex. The Purkinje cells occupy exclusively the deepest part of the molecular layer and form a monocellular layer (*arrows*). In the subcortical white matter (*WM*), myelination is seen (Nissl-fast blue stain). **A** Animal of the hypomagnetic group. (Magnification 12.5 × 6.3). **B** Animal of the hypermagnetic group. (Magnification 12.5 × 6.3). **C** Animal of the control group. (Magnification 12.5 × 16)

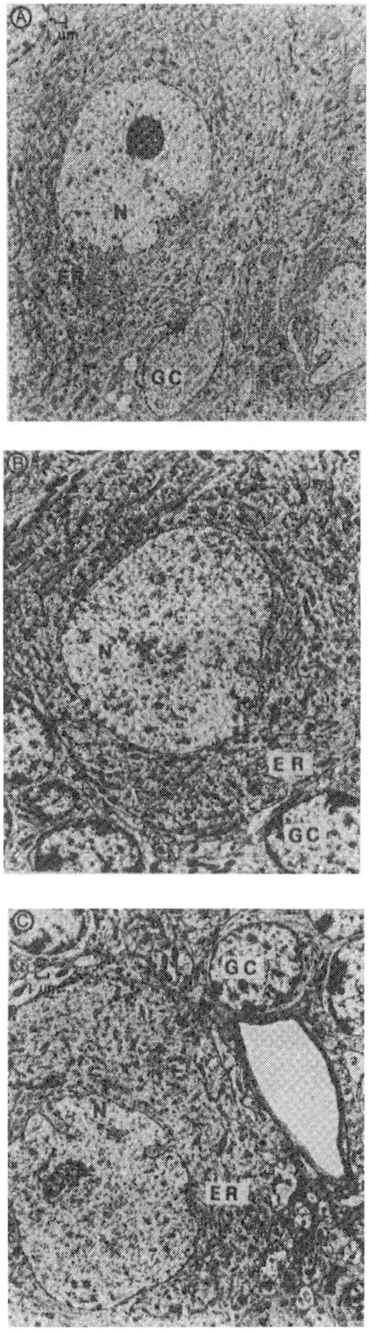

Fig. 14. Transverse paraffin section of the deeper part of the molecular layer (*ML*) of a cerebellar folium of 20-day-old rats. The Purkinje cells have attained good differentiation and have large cell bodies and dendritic trees with clear polarity (*white arrows*) Nissl-fast blue stain. (Magnification 12.5 × 60). **A** Animal of the hypomagnetic group. **B** Animal of the hypermagnetic group. **C** Animal of the control group

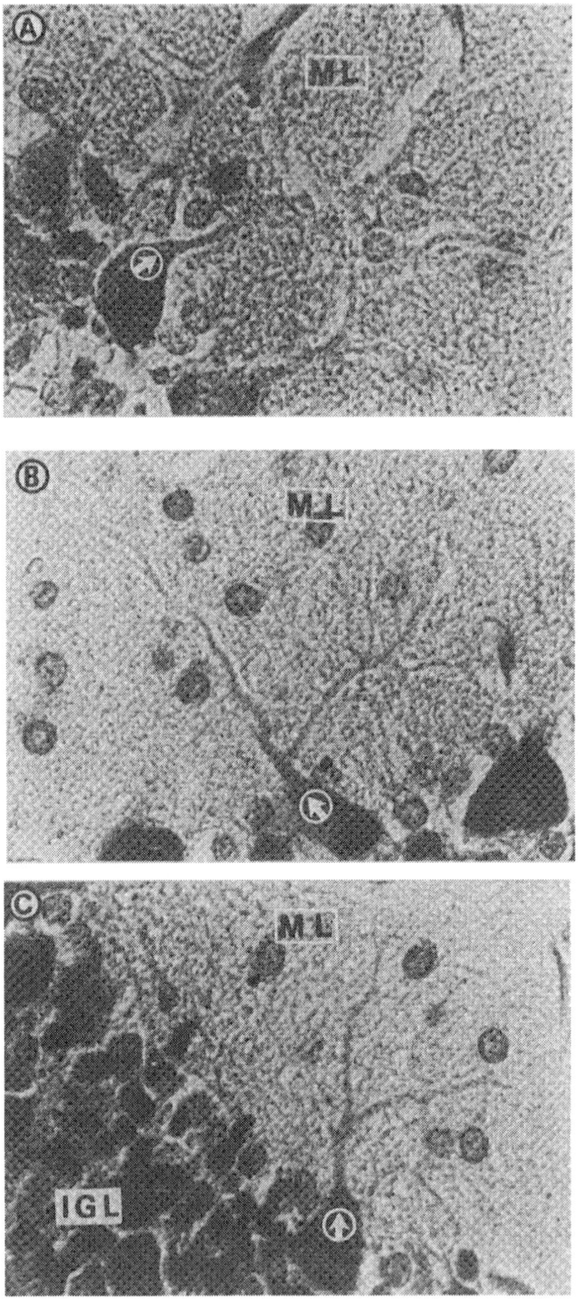

Fig. 15. The deeper part of the molecular layer in cerebellar folium of a 20-day-old rat. Mature nucleus (*N*), bulky cytoplasm and well-differentiated endoplasmic reticulum (*ER*) of the Purkinje cells. **A** Animal of the hypomagnetic group. **B** Animal of the hypermagnetic group. **C** Animal of the control group

References

Altman J (1972) Postnatal development of cerebellar cortex in the rat. I. The external germinal layer in the transitional molecular layer. J Comp Neurol 145:353–394

Altman J (1979) Autoradiographic and histological studies of postnatal neurogenesis. III. Dating the time of production and onset of differentiation of the cerebellar microneurons. J Comp Neurol 136:269–294

Archer VE (1978) Geomagnetism, cancer, weather and cosmic radiation. Health-Phys 34(3):237–247

Archer VE (1979) Anencephalus, drinking water, geomagnetism and cosmic radiation. Am J Epidemiol 109(1):88–97

Asbury AK (1976) Schwann cell proliferation in developing mouse sciatic nerve. J Cell Biol 34:735–743

Bamberger S, Valet G, Stroch F, Ruhenstroth-Bauer G (1978) Electromagnetically induced fluid streaming as a possible mechanism of the biomagnetic orientation of organisms. Z Naturforsch (C) 33(1–2):159–160

Bellossi A, de-Certaines J, Bernard AM (1985) Is there an association between myocardial infarction and geomagnetic activity? Int J Biometeorol 29(1):1–6

Blackmore R (1975) Magnetotactic bacteria. Science 190(4212):377–379

Brown HR, Ilyinsky OB, Muravejko VM, Corshkov ES (1979) Evidence that geomagnetic variations can be detected by lorenzinian ampullae. Nature 277(5698):648–649

Chernyshev VB, Afonina VM (1978) Effect of electric field on the behaviour of *Drosophila melangaster*. Biol Bull Acad Sci USSR 5(5):577–584

Chervinets VM, Novitskii-Yui, Pavloich SA (1979) Magnetic susceptibility of microorganisms. J Hyg Epidemiol Microbiol Immunol 23(3):256–260

Corliss CE (1976) CNS elements of clinical development. In: Patten's human embryology. McGraw-Hill Inc, pp 195–226

Feinleib M, Rogot E, Sturrock PA (1975) Solar activity and mortality in the United States. Int J Epidemiol 4(3):227–229

Friede RL, Samorajski T (1968) Myelin formation in the sciatic nerve of the rat. A quantitative electron microscopic, histochemical and radioautographic study. J Neuropathol Exp Neurol 27:546–570

Friede RL, Samorajski T (1967) Relation between the number of myelin lammelae and axon circumference in fiber of the vagus and sciatic nerve of mice. J Comp Neurol 130:223–232

Herndon RM (1964) The fine structures of the rat cerebellum. II. The stellate neurons, granule cells and glia. J Cell Biol 23:277–293

Hinkle I, McCaig CD, Robinson KR (1981) The direction of growth of differentiating neurons and myoblasts from frog embryos in an electric field. J Physiol 314:121–135

Jaffe LF, Poo Mu-ming (1979) Neurites grow faster towards the cathode than the anode in a steady field. J Exp Zool 209:115–128

Kirschvink JL (1981) The horisontal magnetic dance of the honeybee is comatable with a single-domain ferromagnetic magnetoreceptor. Biosystems 14(2):193–203

Kholodov YuA (1964) Effects on the central nervous system. In: Barnothy MF (ed) Biological effects of magnetic fields. Plenum Press, New York, pp 196–208

Knox EG, Armstrong E, Lancashire R, Wall M, Haynes R (1979) Heart attacks and geomagnetic activity. Nature 281(5732):564–565

Kopanev VI, Efimenko GD, Shakula AV (1979) Biological effect of hypomagnetic environment on the organism. Biol Bull Acad Sci USSR 6:289–298

Liboff AR, Williams T Jr, Strong DM, Wistar R Jr (1984) Time-varying magnetic field: effect on DNA synthesis. Science 223(4638):818–820

Peters A (1960) The formation and structure of myelin sheaths in the Central Nervous System. J Biophys Biochem Cytol 8:431–451

Pokorny AD, Mefferd RB Jr (1966) Geomagnetic fluctuations and disturbed behaviour. J Nerv Ment Dis 143(2):140–151

Rajam M, Mitra S (1981) Correlation between convulsive seizure and geomagnetic activity. Neurosci Lett 24(2):187–191

Robertson D (1955) The ultrastructure of adult vertebrate myelinated peripheral nerves in relation to myelogenesis. J Biophysic and Biochem Cytol 1(4):271–286

Schwan HP (1982) Nonthermal cellular effects of electromagnetic fields: AC-field induced ponderomotoric forces. Br J Cancer 45:220–224

Shibib K, Brock M, Buljat G, Gosztonyi G, Schoknecht G. Polarization of nerve regeneration. Submitted for publication

Skoff RP, Toland D, Nast E (1980) Pattern of myelination and distribution of neuroglial cells along the developing optic system of the rat and rabbit. J Comp Neurol 191:237–253

Skoff RP (1978) The pattern of myelination along the developing rat optic nerve. Neurosci Lett 7:191–196

Sutherland RM, Marton JP, McDonald JC, Howell RL (1978) Effect of weak magnetic field on growth of cells in tissue culture. Physiol Chem Phys 10(2):125–131

Tennekoon GI, Cohen SR, Price DL, McKhann G (1977) Myelogenesis in optic nerve. J Cell Biol 72:604–616

Vilches-Troya J, Dunn RF, Leary DP (1984) Relationship of the vestibular hair cells to magnetic particles in the otolith of guitarfish sacculus. J Comp Neurol 226(4):489–494

Wehner R, Labhart T (1979) Perception of geomagnetic field in the fly *Drosophila melanogaster*. Experientia 26(9):967–968

Wikswo JP, Barach JP, Freeman JA (1980) Magnetic field of a nerve impulse: a first measurement. Science 208:53–55

Yu MC, Yu WH (1980) Effect of hypoxia on cerebellar development. Morphologic and radioautographic study. Exp Neurol 70:652–664

Putative Role of Apolipoprotein E and Lipoproteins in Peripheral Nerve Repair

H. W. Müller and T. Rothe

Molecular Neurobiology Laboratory, Department of Neurology, University of Düsseldorf, Moorenstr. 5, D-4000 Düsseldorf, FR Germany

1 Introduction

On the molecular level one of the most prominent effects following crush injury of a mature rat peripheral nerve is the stimulation of synthesis and the accumulation of endoneurial apolipoprotein E (apo E, M_r 37kDa) in the degenerating distal nerve stump (Skene and Shooter 1983; Müller et al. 1985; Müller et al. 1986; Ignatius et al. 1986; Snipes et al. 1986). The rate of apo E synthesis, which normally declines to low control levels by 8 weeks following sciatic nerve crush as the nerve regenerates, remains elevated if the sciatic nerve is prevented from regenerating (Müller et al. 1985). Interestingly, during the first 3 weeks of postnatal development of rat sciatic nerve, when young axons are known to be myelinated by Schwann cells, apo E synthesis is significantly enhanced compared to the mature nerve (Müller et al. 1986; Müller and Shooter 1987). It has been shown that apo E is expressed by (nonresident) macrophages in peripheral nerve (Stoll and Müller 1986; Müller and Minwegen 1987).

Apo E in serum is known to be associated with various classes of lipoproteins and serves as a ligand for the binding of these lipoproteins to the LDL (low density lipoprotein) receptor thus mediating lipid uptake into a variety of cells (Mahley and Innerarity 1983; Brown and Goldstein 1983). It has been suspected previously that apo E in the nervous system could be involved in mobilization of myelin and axonal membranes during Wallerian degeneration and in the redistribution of lipid precursors to Schwann cells for myelination of regenerating axons (Stoll and Müller 1986; Müller and Shooter 1987; Ignatius et al. 1987 a).

Support for this hypothesis has come from biochemical studies on cholesterol metabolism in normal and injured nerve. During the degradation of myelin and axonal membranes in degenerating sciatic nerve substantial amounts of esterified cholesterol are retained within the nerve and incorporated into regenerating myelin sheaths (Rawlins et al. 1970, 1972; Yao et al. 1980).

Cholesterol esters and apo E are complexed in plasma lipoproteins, including high density lipoprotein (HDL). Similar apo E-containing lipoproteins are formed in degenerating rat sciatic nerve (Ignatius et al. 1987 b). The same authors further reported that lipoprotein complexes are incorporated into neuronal growth cones and cell bodies in vitro (Ignatius et al. 1987 b). However, uptake of lipoproteins into Schwann cells, a prerequisite for the reutilization of lipids retained in degenerating peripheral nerve for myelination of regenerating axons, has thus far not been demonstrated.

H. Flohr (Ed.)
Post-Lesion Neural Plasticity

In the present communication, we investigated lipoprotein uptake into Schwann cells, using a well-characterized cell culture model of peripheral nerve regeneration. We provide evidence that Schwann cells incorporate various lipoprotein complexes but, in contrast to macrophages, do not express detectable amounts of apo E immunoreactivity.

2 Cell Culture Model of Peripheral Nerve Regeneration

An in vitro model system to study cellular and molecular aspects of peripheral nerve repair has to fulfill a variety of structural and functional requirements characterizing critical steps in Wallerian degeneration and nerve regeneration. Such a model system ideally consists of a culture containing primary peripheral neurons, Schwann cells, and macrophages as well as myelinated and nonmyelinated axons and depositions of extracellular matrix and basal lamina components. Injury of nerve fibers should further initiate degeneration of axons, Schwann cell proliferation, phagocytosis of myelin and axonal debris and, subsequently, regeneration and remyelination of neurites. At present, explant cultures of dorsal root ganglia (DRG) are one of the best in vitro model systems of peripheral nerve to investigate the expression and function of defined molecules as well as cell interactions in nerve regeneration.

Cultures of embryonic (day E 16–18) rat DRG were prepared essentially as described by Wood (1976). Briefly, the DRG were initially grown for 2 weeks in anti-mitotic medium (to suppress fibroblast growth) containing human placental serum and then switched to serum-free N2 medium (Bottenstein and Sato 1979) supplemented with fetuin (1 mg/ml) and ascorbic acid (50 μg/ml) according to Carey and Todd (1987). All media contained nerve growth factor (7S NGF, 75 ng/ml).

Figure 1 a represents a section of a DRG culture in serum-free medium showing regenerating neurites identified by neurofilament immunoreactivity (Fig. 1 b) and laminin expressing Schwann cells located along axons (Fig. 1 c). In the presence of ascorbic acid and fetuin Schwann cells provided myelination of small and large nerve fibers (Carey and Todd 1987), as indicated by the expression of myelin basic protein immunoreactivity (Fig. 1 e, f) and the intense "railroad-track"-like Sudan black staining pattern (Fig. 2 d).

3 Expression of Apolipoprotein E in Cultures of Dorsal Root Ganglia

A small number of cells expressing apo E immunoreactivity could be identified in DRG cultures as visualized by immunofluorescence microscopy using anti-apo E antibodies (Fig. 2). The apo E-positive cells coexpressed the macrophage-specific ED1 antigen (Dijkstra et al. 1985) (data not shown). This result indicates that apo E is expressed by the same cell type (macrophage) in DRG cultures as it is in peripheral nerve (Stoll and Müller 1986; Müller and Minwegen 1987). In contrast to degenerating nerve in vivo, however, the number of intrinsic apo E-immunoreactive cells in DRG cultures did not increase for at least 4 weeks follow-

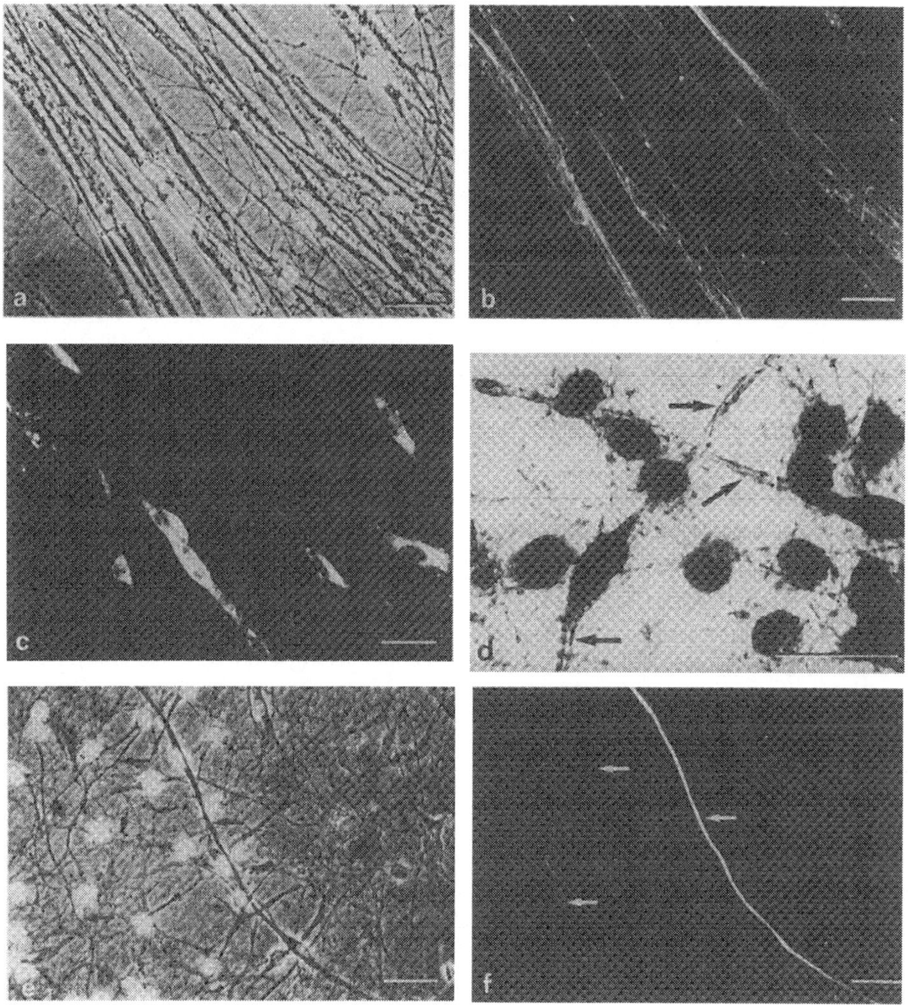

Fig. 1. Embryonic rat DRG in culture: **a–c** Culture maintained for 1 week in anti-mitotic medium containing NGF and human placental serum (Wood 1976). **a** Phase-contrast image; **b, c** indirect immunofluorescence double labeling of the same culture section after fixation in ethanol/acetic acid (95/5% v/v, 10 min, −20° C), **b** with monoclonal anti-neurofilament antibody (RT97; 1:10, generous gift of Dr. B. Anderton, London) as first antibody and fluorescein-conjugated goat anti-mouse IgG (GAM-FITC, 1:25) as second antibody, **c** with rabbit anti-laminin antiserum (1:25, Bethesda Research Laboratories) and rhodamine-conjugated goat anti-rabbit IgG (GAR-TRITC, 1:25), respectively. **d–f** Myelin formation in DRG cultures maintained in serum-free medium for 3 weeks. **d** Sudan black staining. Note the "railroad-track"-like staining pattern of myelinated axons (*arrows*); **e** Phase-contrast image with DAPI nuclear staining; **f** indirect immunofluorescence labeling of the same culture section as in **e** using a mouse monoclonal antibody directed to myelin basic protein (1:10, Hybritech) and GAM-FITC (1:25). *Arrows* mark myelinated axons. Bars = 10 μm

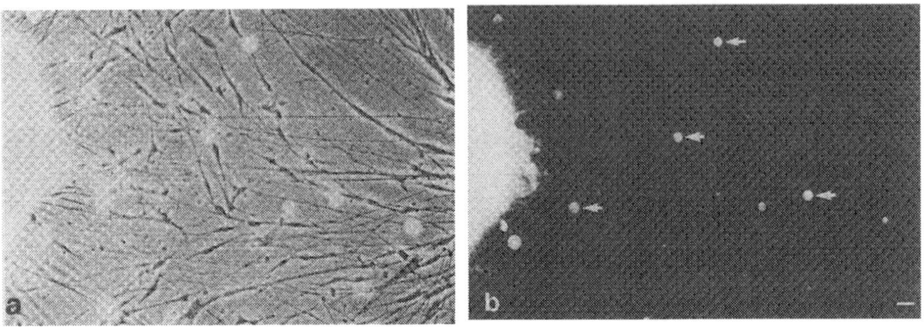

Fig. 2. DRG culture maintained for 1 week in vitro. **a** Phase contrast image; **b** Indirect immunofluorescence microscopy using rabbit anti-apo E antiserum (1 : 25) and GAR-FITC (1 : 25). Apo E positive cells (*arrows*) in these cultures further express the macrophage specific ED1 antigen (not shown). Staining of the ganglion is nonspecific. Note that both Schwann cells and neurites did not stain with anti-apo E antibodies. Bar = 10 μm

ing axonal damage. This finding is in agreement with a similar observation recently reported for explant cultures of rat sciatic nerve (Crang and Blakemore 1986), and further suggests that most of the apo E-expressing macrophages which could be identified in a peripheral nerve during Wallerian degeneration (Stoll and Müller 1986; Müller and Minwegen 1987) are nonresident cells, presumably of hematogenous origin. Metabolically ^{35}S-methionine labeled apo E secreted into the medium could not be detected by gel electrophoresis and fluorography due to the small number of macrophages in our DRG cultures.

4 Lipoprotein Uptake and Phagocytosis by Schwann Cells

Lipoprotein uptake by Schwann cells is one of the necessary, yet unproven, prerequisites of the hypothesis that lipids derived from myelin debris are redistributed in peripheral nerve for reutilization by Schwann cells to myelinate regenerating axons (Rawlins et al. 1970, 1972). To test the ability of the Schwann cell to incorporate lipoproteins we added various lipoprotein complexes to a monolayer of primary Schwann cells which was prepared from a DRG culture by excision of the ganglion as described (Wood 1976). Removal of the ganglion initiated the degeneration of myelinated and nonmyelinated neurites, whereas most of the typically spindle-shaped Schwann cells still remained aligned as they were prior to removal of the ganglion (Fig. 3 a). Virtually all Schwann cells incorporated significant amounts of rhodamine-labeled cholesterol containing lipoproteins such as low density lipoprotein (LDL), acetylated LDL (ac-LDL), and HDL of human origin (generous gift from Dr. G. Schmitz, Münster) within 6 or more hours of incubation at 37° C, as shown for the LDL complex in Fig. 3 b. Uptake of lipoproteins was inhibited when the incubation temperature was decreased to 0–4° C. These results suggest that mechanisms are provided by the Schwann cell to take up different lipoprotein complexes from their extracellular environment. On the other hand, macrophages accumulated fluorescent ac-LDL more effectively than

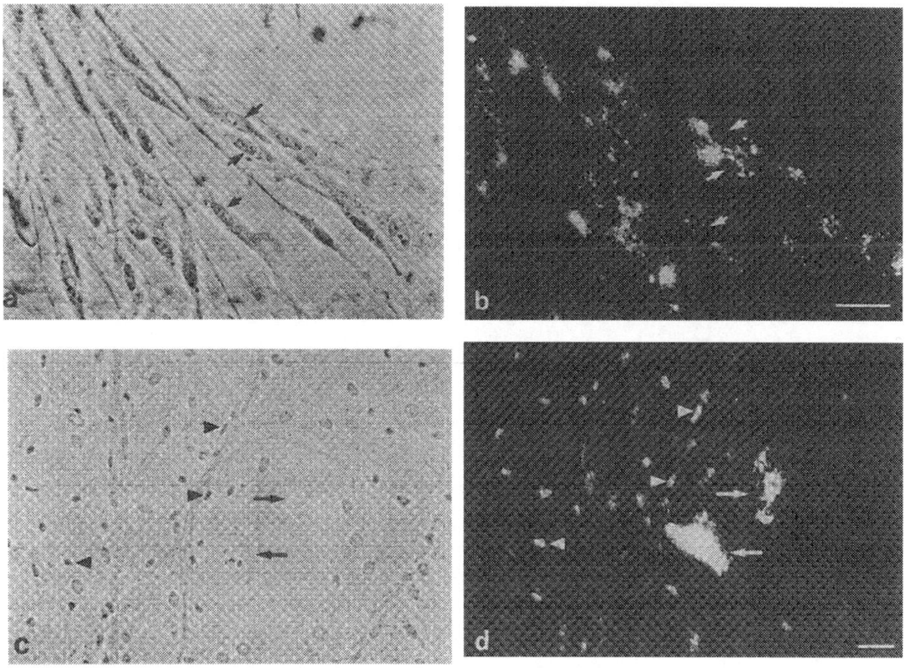

Fig. 3. a, b Monolayer of Schwann cells derived from a serum-free DRG culture containing myelinated axons by excision of the ganglion 1 week prior to the addition of rhodamine-labeled human low density lipoprotein (LDL, 50 µg protein/ml culture medium, 18 h incubation at 37° C). **a** Phase contrast image of the typically spindle shaped Schwann cells. **b** Rhodamine fluorescence of internalized LDL. *Arrows* indicate nuclei spared by fluorescence. **c, d** Uptake of rhodamine-conjugated acetylated LDL (50 µg protein/ml of medium, 18 h incubation) by Schwann cells and macrophages. **c** Phase contrast image of a DRG culture; **d** Rhodamine fluorescence associated with Schwann cells (*arrowheads*) and macrophages (*arrows*). Bars = 10 µm

Schwann cells in a DRG culture (Fig. 3c, d) probably via endocytosis mediated by specific high-affinity ac-LDL receptors (Brown and Goldstein 1983; Mahley and Innerarity 1983). The mechanism underlying the internalization of lipoproteins into Schwann cells, however, remains unknown. It should be noted that primary Schwann cells in DRG cultures express phagocytic activity as demonstrated by the incorporation of rhodamine-labeled polystyrene latex beads into their cytoplasm (Fig. 4). Thus it is possible that phagocytosis represents one mechanism to internalize endoneurial lipoproteins into Schwann cells in vivo.

5 Discussion: Putative Role of Macrophages and Apolipoprotein E in Lipid Redistribution to Myelinating Schwann Cells

We have shown that primary Schwann cells in a peripheral nerve regeneration model take up various cholesterol containing lipoprotein complexes (LDL, ac-LDL, HDL). The mechanisms by which Schwann cells internalize these lipoproteins are temperature-sensitive but nonselective for the distinct complexes. Recep-

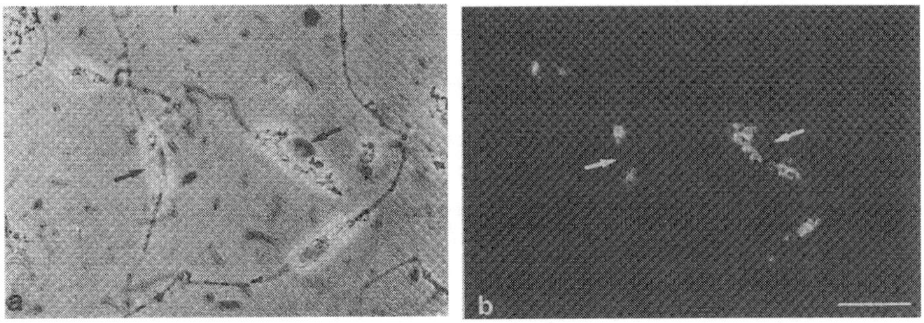

Fig. 4. Phagocytosis of rhodamine-labeled polystyrene latex beads (0.4 μm diameter, Serva) by Schwann cells within 12 h of incubation. Noninternalized adherent beads had been removed from the cell surface by repetitive rinses with phosphate-buffered saline. **a** Phase contrast image of primary Schwann cells. Aggregates of latex beads are associated with the cells; **b** Rhodamine fluorescence of the same culture section. Note that the Schwann cell nuclei (*arrows*) are spared by fluorescent beads which are located within the cytoplasm. Bar = 10 μm

tor-mediated endocytosis of LDL via the apo (B,E)-LDL receptor is possible but unlikely, since the internalization of rhodamine-conjugated LDL into Schwann cells is rather slow. Comparable amounts of fluorescent ac-LDL, a lipoprotein which does not bind to the apo (B,E)-LDL receptor (Mahley and Innerarity 1983; Brown and Goldstein 1983), seem to accumulate in Schwann cells at a similar slow rate as LDL and HDL. Therefore, we suggest that cholesterol-containing lipoproteins accumulating in the environment of Schwann cells are internalized by these cells mainly via phagocytosis rather than a regulated receptor-mediated pathway. Phagocytic activity of Schwann cells could be demonstrated (see Fig. 4). Furthermore, Orekhov et al. (1987) have shown that LDL is phagocytized by subendothelial intimal cells from arterial walls when complexed with extracellular matrix components such as fibronectin or heparin. In fact, immunocytochemical

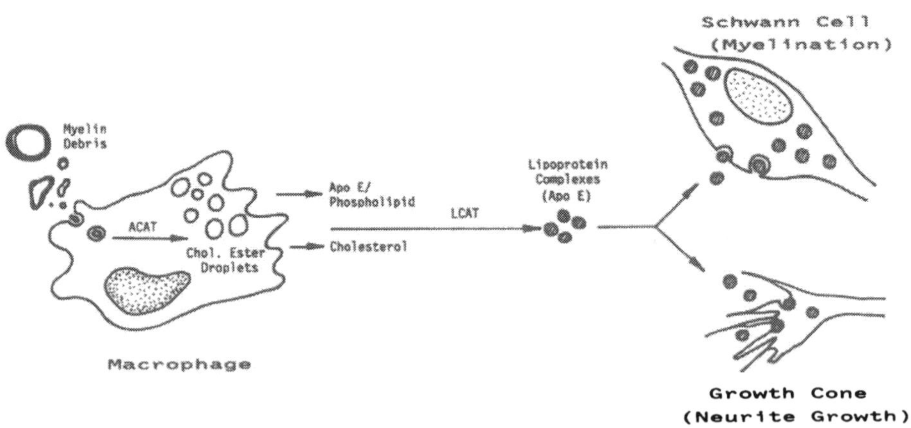

Fig. 5. Proposed function(s) of apo E and lipoproteins in lipid storage and redistribution in nerve regeneration. *ACAT* Acyl-CoA: cholesterol-acyltransferase; *LCAT* Lecithin: cholesterol-acyltransferase

studies demonstrated that apo E (or apo E-containing lipoproteins) is located at or near the basal lamina in rat sciatic nerve and in other tissues (Stoll and Müller 1986) probably linked via its specific heparin binding sites (Cardin et al. 1986; Weisgraber et al. 1986). These findings strengthen the possibility that apo E-containing lipoproteins may be complexed with basal lamina components in injured nerve prior to phagocytosis by the Schwann cell or other cells.

Our present data, together with previous observations from other laboratories, support the hypothesis that lipoproteins, which accumulate in injured peripheral nerve, may participate in the redistribution of lipid to Schwann cells for myelination of regenerating axons. The diagram in Fig. 5, which is an extension of a previously proposed model (Ignatius et al. 1987a), describes the putative role of apo E in mediating cholesterol transfer from macrophages ("foam cells") to Schwann cells and neuronal growth cones via lipoprotein complexes in peripheral nerve regeneration.

Acknowledgments. The authors are grateful to PD Dr. G. Schmitz and T. Brüning (Münster) for rhodamine labeled human lipoproteins. This work was supported in part by the Deutsche Forschungsgemeinschaft SFB 200 (C6).

References

Bottenstein JE, Sato GH (1979) Growth of a rat neuroblastoma cell line in serum-free supplemented medium. Proc Natl Acad Sci USA 76:514–517

Brown MS, Goldstein JL (1983) Lipoprotein metabolism in the macrophage: implications for cholesterol deposition in artherosclerosis. Ann Rev Biochem 52:223–261

Cardin AD, Hirose N, Blankenship DT, Jackson RL, Harmony AK (1986) Binding of a high reactive heparin to human apolipoprotein E: identification of two heparin-binding domains. Biochem Biophys Res Commun 134:783–789

Carey DJ, Todd MS (1987) Schwann cell myelination in a chemically defined medium: demonstration of a requirement for additives that promote Schwann cell extracellular matrix formation. Dev Brain Res 32:95–102

Crang AJ, Blakemore WF (1986) Observations on Wallerian degeneration in explant cultures of cat sciatic nerve. J Neurocytol 15:471–482

Dijkstra CD, Döpp EA, Joling P, Kraal G (1985) The heterogeneity of mononuclear phagocytes in lymphoid organs: distinct macrophage subpopulations in the rat recognized by monoclonal antibodies ED1, ED2 and ED3. Immunology 54:589–599

Ignatius MJ, Gebicke-Härter PJ, Skene JHP, Schilling JW, Weisgraber KH, Mahley RW, Shooter EM (1986) Expression of apolipoprotein E during nerve degeneration and regeneration. Proc Natl Acad Sci USA 83:1125–1129

Ignatius MJ, Gebicke-Härter PJ, Pitas RE, Shooter EM (1987a) Apolipoprotein E in nerve injury and repair. Prog Brain Res 71:177–184

Ignatius MJ, Shooter EM, Pitas RE, Mahley RW (1987b) Lipoprotein uptake by neuronal growth cones in vitro. Science 236:959–962

Mahley RW, Innerarity TL (1983) Lipoprotein receptors and cholesterol homeostasis. Biochim Biophys Acta 737:197–222

Müller HW, Minwegen P (1987) Nonresident macrophages in peripheral nerve of rat: effect of silica on migration, myelin phagocytosis, and apolipoprotein E expression during Wallerian degeneration. J Neurosci Res 18:222–229

Müller HW, Shooter EM (1987) Molecular approach to the role of macrophages and astrocytes in nerve growth and regeneration. In: Althaus HH, Seifert W (eds) Glial-neuronal communication in development and regeneration, NATO ASI Series H (Cell Biology), vol 2. Springer, Berlin Heidelberg New York, pp 630–640

Müller HW, Gebicke-Härter PJ, Hangen OH, Shooter EM (1985) A specific 37000-Dalton protein that accumulates in regenerating but not in nonregenerating mammalian nerves. Science 228:499–501

Müller HW, Ignatius MJ, Hangen OH, Shooter EM (1986) Expression of specific sheath cell proteins during peripheral nerve growth and regeneration in mammals. J Cell Biol 102:293–402

Orekhov AN, Tertov VV, Mukhin DN, Koteliansky VE, Glukhova MA, Khashimov KA, Smirnov VN (1987) Association of low-density lipoprotein with particulate connective tissue matrix components enhances cholesterol accumulation in cultured subendothelial cells of human aorta. Biochim Biophys Acta 928:251–258

Rawlins FA, Hedley-Whyte ET, Villegas GM, Uzman BG (1970) Reutilization of cholesterol-1,2-H^3 in the regeneration of peripheral nerve: an autoradiographic study. Lab Invest 22:237–240

Rawlins FA, Villegas GM, Hedley-Whyte ET, Uzman BG (1972) Fine structural localization of cholesterol-1,2-^3H in degenerating and regenerating mouse sciatic nerve. J Cell Biol 52:615–625

Skene JHP, Shooter EM (1983) Denervated sheath cells secrete a new protein after nerve injury. Proc Natl Acad Sci USA 80:4169–4173

Snipes GJ, McGuire CB, Norden JJ, Freeman JA (1986) Nerve injury stimulates the secretion of apolipoprotein E by nonneuronal cells. Proc Natl Acad Sci USA 83:1130–1134

Stoll G, Müller HW (1986) Macrophages in the peripheral nervous system and astroglia in the central nervous system of rat commonly express apolipoprotein E during development but differ in their response to injury. Neurosci Lett 72:233–238

Weisgraber KH, Rall SC, Mahley RW, Milne RW, Marcel YL, Sparrow JT (1986) Human apolipoprotein E: determination of the heparin binding sites of apolipoprotein E3. J Biol Chem 261:2068–2076

Wood PM (1976) Separation of functional Schwann cells and neurons from normal peripheral nerve tissue. Brain Res 115:361–375

Yao JK, Natarajan V, Dyck PJ (1980) The sequential alterations of endoneurial cholesterol and fatty acid in Wallerian degeneration and regeneration. J Neurochem 35:933–940

Protein Kinase C and Protein F1: Potential Molecular Mediators of Lesion-Induced Synaptic Plasticity Recapitulate Developmental Plasticity

A. Routtenberg
Cresap Neuroscience Laboratory, Northwestern University, Evanston, IL 60201, USA

Introduction

This review is focused on the suggestion that a potential molecular mechanism involving protein kinase C and its substrate protein F1 may be directly involved in the process of mediating lesion-induced neuronal plasticity. As several chapters in this Volume illustrate, after damage to the nervous system, intact neurons show growth activities such as axonal sprouting. Under certain conditions this can lead to recovery of function, in whole or in part. The mechanisms for this process are not known, though some recent clues from biochemistry suggest particular proteins that may be essential for the process. In this review I wish to emphasize one essential point: that the input-dependent molecular mechanism related to synaptic growth of intact synapses is also recruited by injury-produced growth factors that stimulate the response to nerve injury.

This same specific molecular machinery related to mechanisms of plasticity of synapses in the adult brain has recently been co-identified with neurite outgrowth in developing brain. Since both embryonic and adult brain exhibit plasticity in respose to injury, it is instructive to review the possibility that this machinery at both stages may regulate the success or failure of the lesion-induced neuronal plasticity process. Understanding this molecular machinery in the context of lesion-induced neuronal plasticity may provide quite specific agents to facilitate, or retard if necessary, the outgrowth and integration of the sprouting axons.

Our interest in this area has grown out of studies on synaptic plasticity in adult brain beginning in 1974. These studies led to the identification of a protein termed protein F1, that was phosphorylated in relation to learning and long-term synaptic potentiation (Routtenberg et al. 1975, 1983). Protein F1 is phosphorylated by protein kinase C (PKC); the activation of the kinase and the phosphorylation of F1 are closely linked now to models of information storage in the nervous system (for reviews see Routtenberg 1984, 1985, 1986). In a recent series of collaborative studies we have demonstrated that protein F1 is identical to a protein associated with growth of axons in development (pp46) and during lesion-induced axonal regeneration (GAP43). This had led to the view that information storage in the adult may be linked to presynaptic terminal growth. It is intriguing to think that both developmental growth and adult plasticity regulated by protein kinase C/protein F1 (PKC/F1) are playing an important role in the lesion-induced neuronal plasticity phenomenon. This would predict that it may be possi-

H. Flohr (Ed.)
Post-Lesion Neural Plasticity

ble to regulate neurite growth by activation and/or inhibition of the phosphotransferase enzyme, protein kinase C.

Additionally, regulation of PKC substrates such as protein F1, may also prove to be effective in regulating neurite growth. The present chapter charts the lines of evidence that have led to this new proposal.

The link between growth of axons and the formation and plasticity of synapses emerged as a result of independent lines of study by four different laboratories that, as we now believe, were studying the identical brain protein. We have reported that protein F1 discovered in our laboratory now over 10 years ago (Ehrlich and Routtenberg 1974; Routtenberg and Ehrlich 1975) appears to be identical to pp46, GAP-43 and B-50, as originally proposed by Nelson and Routtenberg (1985). In what follows we review some of the evidence for these identities and then consider the implications of these new findings for lesion-induced neuronal plasticity.

2 Input-Dependent Regulation of Protein Kinase C/Protein F1: Molecular Mechanism for Synaptic Growth in Adult Rat Brain?

Increased phosphorylation of the neuronal membranebound protein F1 and translocation of its kinase, the Ca^{2+} and phospholipid stimulated protein kinase C (PKC), have been related to long-term increases in adult synaptic efficacy in a number of reports from our laboratory (Routtenberg et al. 1985; Lovinger et al. 1985; Akers et al. 1986). Converging lines of evidence from a number of laboratories indicate that activation of protein kinase C can decrease potassium conductance, reveal cryptic calcium channels and increase the duration of and enhance EPSP. A high-frequency train of stimulation applied to any of several fiber pathways in the hippocampus or other specific fiber pathways in brain results in long-term potentiation (LTP), a prolonged change in trans-synaptic communication (measured physiologically as an enhanced population spike or population EPSP in the post-synaptic cells; Bliss and Lomo 1973). Several properties of LTP – including the briefness of stimulation necessary to induce it, its persistence (measured for months in chronically recorded animals; Douglas and Goddard 1975), its similarities to neuronal events measured during learning (Berger 1984), and its property of associativity (i.e., in some instances LTP can only be induced if concomitant activation of nearby synapses occurs; Levy and Steward 1983; Larson and Lynch 1986) – have made LTP an attractive model for use in studying how information storage might occur. Thus environmental events involved in learning generate an input-dependent activation of hippocampus that may parallel the events generated by LTP.

For theoretical reasons reviewed elsewhere (Routtenberg 1982), we have chosen to study changes in protein phosphorylation following induction of LTP in an effort to uncover biochemical events which may accompany information storage. Following high-frequency stimulation of the perforant path (an afferent fiber system to the hippocampal formation), we assayed different rat brain regions and found a selective increase in the in vitro phosphorylation of a single substrate termed protein F1 (M_r 47 kDa, pI 4.5) in animals killed 5 min but not

1 min after the LTP-inducing stimulation (Routtenberg et al. 1985; Nelson and Routtenberg 1985). This increase in phosphorylation at 5 min was found only in dorsal hippocampal formation, the region containing the potentiated synapses, and did not appear after low-frequency control stimulation (which does not produce LTP). The degree of LTP achieved (measured as the increase in population spike amplitude) was directly correlated with the extent of increase in protein F1 phosphorylation across animals, suggesting a direct relationship between protein F1 phosphorylation and LTP.

Related studies exploring longer durations (hours, days) of LTP suggest that protein F1 phosphorylation is associated with maintenance of the potentiated response rather than with initial enhancement of the response (Lovinger et al. 1985). This hypothesis is further supported by the observation that protein F1 phosphorylation has no direct relationship with the magnitude of potentiation seen immediately after high-frequency stimulation.

Since our evidence suggested that protein F1 is phosphorylated by the Ca^{2+} and phospholipid stimulated protein kinase C (PKC; Kikkawa et al. 1982; Akers and Routtenberg 1985), we chose to examine whether LTP was accompanied by an increase in membrane PKC activity. PKC is normally distributed in both cytosol and membrane fractions. Recent evidence has shown that the PKC-activation, tumor-promoting phorbol esters (Castagna et al. 1982), as well as elevated Ca^{2+} levels, can activate PKC by translocating it from the cytosolic to the membrane-associated state where co-factors for PKC activity reside (Kraft and Andersen 1983; Wolf et al. 1985). Because protein F1 is a membrane-bound protein (Nelson and Routtenberg 1985), we were intrigued by the idea that translocation of PKC to the membrane might be responsible for the increase in protein F1 phosphorylation we observed following LTP. In order to test this hypothesis, we induced LTP in the dentate gyrus of anesthetized animals, and assayed for PKC activity in both cytosolic and membrane fractions from dorsal hippocampus after LTP.

In agreement with the changes we found in protein F1 phosphorylation following LTP, we detected an increase in membrane-associated PKC activity at 1 h but not 1 min following the high-frequency stimulation. We also discovered a corresponding decrease in cytosolic PKC activity (Akers et al. 1986). The sum of PKC activity in the two fractions did not change significantly after induction of LTP, suggesting that PKC was physically transferred from the cytosol to the membrane. Thus, increased phosphorylation of membrane-bound protein F1 following induction of LTP may be due to activation of PKC through its movement to the membrane.

As discussed in the next sections, increased synthesis of protein F1, as well as its post-translational modification by phosphorylation, may be important for neurite outgrowth in developing brain. Since our LTP data show that protein kinase C and protein F1 also appear to be important for modifying synaptic communication between adult neurons, it may be suggested that certain components of the developmental programs of growth and differentiation are preserved for the purpose of regulating synaptic plasticity in the adult nervous system. Under physiological conditions such plasticity occurs as a result of neuronal activity generated by environmental stimulation, growth factors, and so forth. This raises the possibility that the environmental context of the animal, activating the nervous system in a particular manner, might regulate recovery of function after a lesion. Such evidence is already available (Will and Kelche 1979) and may involve a convergence of environmental input-dependent processes on a molecular mechanism involved with axonal growth and recovery from injury. Such regulation would most likely involve both translational and post-translational events expressing both PKC and F1.

3 Developmental Regulation of Protein Kinase C/Protein F1: Role in Neurite Outgrowth

In collaboration with K. Pfenninger and co-workers, we found that protein F1 appears to be the same as a major growth cone phosphoprotein termed pp46 (Katz et al. 1985), on the basis of identical molecular weight, isoelectric point, microheterogeneity on two-dimensional gels, phosphorylation by PKC, cAMP-independent endogenous phosphorylation, membrane enrichment, and two-dimensional phosphopeptide maps following limited proteolysis with *S. aureus* V8 protease (Nelson et al. 1985). Potentially more interesting than this co-identification are the quantitative differences in protein F1 between growth cone and synaptosomal preparations.

The endogenous phosphorylation of protein F1 was 20-fold higher in growth cones than in a crude adult synaptosome preparation. In fact, protein F1 was the most highly labeled phosphoprotein detected in growth cones in the range of phosphoproteins assayed (30–100 kDa), in contrast to adult brain, where there are many phosphoproteins more prominent than protein F1. This higher phosphorylation in growth cones is due in part to a higher ratio of protein F1 to total protein in the growth cones vs. crude adult synaptosomes as measured qualitatively by protein staining of two-dimensional gels. It is interesting and relevant that the lesion-induced plasticity occurs with near virulence in the developing hippocampus.

Of four major growth cone phosphoproteins detectable in the 30–100 kDa range, three, including protein F1, could be phosphorylated by exogenously added purified protein kinase C, suggesting that PKC activity might be relatively more important in growth cone function than activity of other protein kinases. Burgess et al. (1986) have recently arrived at a similar conclusion. Evidence from other laboratories has also suggested that PKC might play an important role in promoting neurite outgrowth and synaptogenesis. PKC can be bound with high affinity by 12-O-tetradecanoylphorbol-13-acetate (TPA), a potent tumor promoter (Niedel et al. 1983). When labeled TPA has been used to determine PKC levels in particular tissue types, the highest levels of TPA binding have been found in growth cone-rich regions of fetal rat brain (Murphy et al. 1983). The role that PKC might occupy in neuronal function has also been explored using TPA since this substance, besides directly binding to PKC, is known to stimulate PKC activity strongly (Castagna et al. 1982). In cultured dorsal root ganglia cells or neuroblastoma, TPA induces neurite outgrowth (Hsu et al. 1984), suggesting that PKC and its substrates in growth cones play a crucial role in neurite extension.

Because TPA also strengthens the synaptic response of LTP in adults (Routtenberg et al. 1986) it was suggested that it may do so by promoting growth of axonal arbors in the synaptic region. It may be of interest to study the effects of phorbol ester on facilitating neurite outgrowth after lesions of the central nervous system.

Two recent reports indicate an association between neurite outgrowth of rat neurons in culture and synthesis of the protein F1/protein kinase C module. Burgess et al. (1986) observed that there was an increase both in protein kinase C (as indexed by phorbol ester binding) and in PKC-substrate phosphorylation correlated with the age of cultured neurons. One acidic PKC substrate was identified as protein F1 by Burgess et al. (1986). Thus, the elaboration of processes of cells in culture correlated with increase in both PKC/F1 substrate amount and phosphorylation. A similar conclusion was drawn by Perrone-Bizzozero et al. (1986) studying cortical neurons in culture and demonstrating an increase in 35S-methionine incorporation into protein F1 in direct correlation to neurite outgrowth.

It is suggested from the foregoing that intact nerve cells, following response to injury to other nerve cells, extend their axons and axon terminals in relation to the increased presence of certain identifiable proteins. In this chapter we have focused on the PKC/F1 module as a central mechanism for this growth. It is of direct relevance to the link between the PKC/F1 module and the lesion-induced neuronal plasticity phenomenon that only early neonatal neurons are capable of more rapid axonal regrowth. The decrease in protein F1 amount and F1 phosphorylation in the adult suggests two potential mechanisms for this reduced capacity: decrease in the synthesis of protein F1 and decreased phosphorylation by PKC. To begin addressing the former possibility we have initiated studies on gene regulation and mRNA expression of protein F1 during development.

In collaboration with Rosenthal, Ullrich, Menzel and others at Genentech, we have recently cloned the cDNA for protein F1 (Rosenthal et al. 1987) and described the nucleotide and amino acid sequence for protein F1. Northern blots for protein F1 mRNA indicate that transcription of protein F1 DNA in rat brain is below detection levels at embryonic day 15, but can be observed at E17, precisely the day used for our growth cone study. Interestingly, the levels of F1 mRNA decrease significantly from week 1 to week 5 postnatally. This corresponds pari passu with decreased requirement for extensive neuritic growth. At present the linking between expression of protein F1 and axonal growth is indirect, although primary cell culture studies correlating neurite outgrowth with increased protein F1 are suggestive. By defining the complete structure of the F1 gene, it is hoped that the regulatory events in the nucleus and ribosome leading to protein F1 expression will be understood. The regulation mechanisms might then be directly manipulated to test their relationship to aspects of neurite growth.

4 Selective Down-Regulation of Protein F1 and its Relation to Capacity for Axonal Regeneration

In lesion-induced neuronal plasticity experiments the brain is damaged and uninjured axons and intact synapses adjacent to degenerating ones tend to grow into the region vacated by the phagocytosed tissue. Why the damaged tissue cannot regenerate is a difficult question to answer, though it might well be related to the decreased synthesis of protein F1. This conclusion is strengthened by recent evidence which indicates that in those situations where regeneration can occur protein F1 synthesis is increased severalfold. We have co-identified protein F1 with the growth-associated protein (GAP43; Skene and Willard 1981 a, b, c; Benowitz and Lewis 1983), using identical molecular weight, isoelectric point, microheterogeneity on two-dimensional gels, phosphorylation by PKC, membrane-enrichment, and immunological cross-reactivity of protein F1 with an antibody raised against GAP43 as criteria (Snipes et al. 1987). Protein F1 synthesis and the amount of material transported by subsequent fast-axonal transport is greatly increased during the axonal sprouting which follows nerve crush or axotomy. This increase in protein F1 synthesis typically persists until axons have reached their targets and only occurs in nervous systems where successful nerve regeneration occurs, i.e., the CNS and PNS of anamniotic creatures such as fishes and amphibians, but only the PNS of amniotes such as reptiles, birds, and mammals (Skene 1984).

Protein F1 synthesis, as pp46 and GAP43, increases in vertebrates during axonal development prior to synaptogenesis. One therefore might predict that CNS of higher vertebrates would be capable of axonal regeneration during nervous system development to the extent that expression of the gene for this protein is important in the capacity for axonal regeneration. This prediction is confirmed by studies demonstrating regrowth of CNS fiber pathways in neonatal rats for limited period following birth, after which the same pathways will fail to regenerate (Kalil and Reh 1979; Bernstein and Stelzner 1983). Thus, with regard to induction of protein F1 synthesis, the regenerative state of the neuron recapitulates initial axonal development.

5 Protein F1: Cellular and Subcellular Localization

PKC is a multifunctional enzyme with a long list of putative physiological substrates (Nishizuka 1986). This raises the issue of how PKC might play a specific role in axonal and synaptic plasticity without simultaneously altering a host of other cellular events. One means of achieving this specificity of action would be to isolate changes in PKC activity to a defined cellular compartment, and to limit the distribution of PKC substrates important for synaptic plasticity to this same compartment. Collaborative studies on protein F1 in our laboratory and in the laboratory of Gispen and co-workers on B-50 protein indicate that they are most likely identical (Gispen et al. 1986; Nelson and Routtenberg 1985) and that phosphorylation and immunoreactivity of the protein are restricted to brain (Kristjansson et al. 1982; Oestreicher et al. 1986). Within the nervous system, the protein has a heterogeneous distribution, being highest in such areas as septum, hippocampus, and cerebral cortex. Within the neuron itself, the evidence to date, both from ultrastructural immunocytochemical studies, and from our studies of the growth cone, suggest that protein F1 concentration is highest in the presynaptic terminal (Gispen et al. 1985a; Nelson et al. 1985). Finally, within the terminal, protein F1 is tightly associated with the membrane (Skene and Willard 1981b; Nelson and Routtenberg 1985). It is of interest in this regard that PKC activity, despite its wide distribution among tissues and across species, has its highest enrichment in neural plasma membranes (Kikkawa et al. 1982), strongly suggesting that it has an important role in regulating protein F1 function, and, having F1 as a neuron-specific kinase C substrate, suggests more generally that it has an important role in functions specific to nervous system.

We have recently obtained the first information on the expression of protein F1 in cell types within particular structures using in situ hybridization of protein F1 mRNA (Rosenthal et al. 1987). It is striking that a good deal of selectivity is apparent within a given structure. For example, protein F1 is expressed in the pyramidal cells of the hippocampus, but granule cells appear to be devoid of F1 transcripts. The significance of this differential expression of protein F1 may lie in the capacity of the terminals of these cells to demonstrate plasticity. It will be of interest to determine whether those cell types that demonstrate plasticity in the adult are also the ones which demonstrate the most clearcut plasticity following nerve damage.

6 Conclusions and Prospect

A recent search of the biomedical data base indicates that the present proposal to investigate the linkage between lesion-induced plasticity on the one hand and protein kinase C and its substrates on the other has not been explored. The indirect evidence suggesting such a link and the potential that this mechanism might facilitate recovery of function after brain damage is potentially promising, and therefore worthwhile to determine in what manner these molecular processes contribute to the success of lesion-induced neuronal plasticity.

The present review suggests potential interventions based on the molecular mechanisms underlying lesion-induced neuronal plasticity. Empirical analysis will first necessarily involve the characterization of protein kinase C and protein F1 (e.g., mRNA expression, F1 phosphorylation, PKC subcellular distribution) in both the damaged and the intact, re-growing neurons. It should then be possible to manipulate the activity of protein kinase C with activators and inhibitors and determine whether the facilitated neuritic outgrowth observed in culture by protein kinase C activators can be mimicked in the case of lesion-induced neuronal plasticity.

If protein kinase C and protein F1 in the intact cells play a critical role in regulating the extent of the plastic response, then manipulation of gene expression of cells prior to lesion-induced neuronal plasticity, using the transgenic mouse preparation, for example, could provide direct evidence on the role of these proteins in lesion-induced plasticity. Amplification of expression of those proteins related to axonal growth and regeneration could significantly influence the manner in which the nervous system responds to injury.

References

Akers RF, Routtenberg A (1985) Kinase C phosphorylates a protein involved in synaptic plasticity. Brain Res 334:147–151

Akers, RF, Lovinger D, Colley P, Linden D, Routtenberg A (1986) Translocation of protein kinase C activity may mediate hippocampal long-term potentiation. Science 231:587–589

Benowitz LI, Lewis ER (1983) Increased transport of 44000- to 49000-dalton acidic proteins during regeneration of the goldfish optic nerve: a two-dimensional gel analysis. J Neurosci 3:2153–2163

Berger TW (1984) Long-term potentiation of hippocampal synaptic transmission affects rate of behavioral learning. Science 224:627–630

Bernstein E, Stelzner D (1983) Plasticity of the corticospinal tract following mid-thoracic spinal injury in post-natal rat. J Comp Neurol 221:382–400

Bliss TVP, Lomo T (1973) Long lasting potentiation of synaptic transmission in the dentate area of the anesthetized rabbit following stimulation of the perforant path. J Physiol 232:357–374

Burgess SK, Sahyoun N, Blanchard SG, LeVine H III, Chang KJ, Cuatrecasas P (1986) Phorbol ester receptors and protein kinase C in primary neuronal cultures: development and stimulation of endogenous phosphorylation. J Cell Biol 102:312–319

Castagna M, Takai Y, Kaibuchi K, Sano K, Kikkawa U, Nishizuka Y (1982) Direct activation of calcium-activated, phospholipid-dependent protein kinase by tumor-promoting phorbol esters. J Biol Chem 257:7847–7851

Douglas RM, Goddard GV (1975) Long-term potentiation of the perforant path-granule cell synapse in the rat hippocampus. Brain Res 86:205–215

Gispen WH, Leunissen JLM, Oestreicher AB, Verkleij AJ, Zwiers H (1985a) Presynaptic localization of B-50 phosphoprotein: the (ACTH)-sensitive protein kinase substrate involved in rat brain polyphosphoinositide metabolism. Brain Res 328:381–385

Gispen WH, Dongen CJ van, Graan PNE de, Oestreicher AB, Zwiers H (1985b) The role of phosphoprotein B50 in phosphoinositide metabolism in brain synaptic plasma membranes. In: Bleasdale JE, Hauser G, Eichberg J (eds) Inositol and phosphoinositides. Humana, Dallas, pp 399–414

Gispen WH, Graan PNE de, Chan SY, Routtenberg A (1986) Comparison between the neural acidic proteins B50 and F1. In: Gispen WH, Routtenberg A (eds) Phosphoproteins in neuronal function, Prog Brain Res, vol 69. Elsevier, Amsterdam, pp 383–386

Harreveld A van, Fifkova E (1975) Swelling of dendritic spines in the fascia dentata after stimulation of the perforant fibers as a mechanism of post-tetanic stimulation. Exp Neurol 49:736–739

Hsu L, Natyzak D, Laskin JD (1984) Effects of the tumor promoter 12-O-tetradecanoylphorbol-13-acetate on neurite outgrowth from chick embryonic sensory ganglia. Cancer Res 44:4607–4614

Kalil K, Reh T (1979) Regrowth of severed axons in the neonatal CNS; establishment of normal connections. Science 205:1158–1161

Katz F, Ellis L, Pfenninger KH (1985) Nerve growth cones isolated from fetal rat brain: calcium-dependent protein phosphorylation. J Neurosci 5:1402–1411

Kikkawa U, Takai Y, Minakuchi R, Inohara S, Nishizuka Y (1982) Calcium-activated, phospholipid-dependent protein kinase from rat brain. J Biol Chem 257:13341–13348

Kraft AS, Andersen WB (1983) Phorbol esters increase the amount of calcium, phospholipid-dependent protein kinase associated with the plasma membrane. Nature 301:621–623

Kristjansson GI, Zwiers H, Oestreicher AB, Gispen WH (1982) Evidence that the synaptic phosphoprotein B50 is localized exclusively in nerve tissue. J Neurochem 39:371–378

Larson J, Lynch G (1986) Induction of synaptic potentiation in hippocampus by patterned stimulation involves two events. Science 232:985–988

Levy WB, Steward O (1983) Temporal contiguity requirements for long-term associative potentiation/depression in the hippocampus. Neuroscience 8:791–797

Lovinger DM, Akers RF, Nelson RB, Barnes CA, McNaughton BL, Routtenberg A (1985a) A selective increase in the phosphorylation of protein F1, a protein kinase C substrate, directly related to three day growth of long term synaptic enhancement. Brain Res 343:137–143

Macara IG (1985) Oncogenes, ions, and phospholipids. Am J Physiol 248:C3–C11
Murphy KMM, Gould RJ, Oster-Granite ML, Gearheart JD, Snyder SH (1983) Phorbol esters receptors: autoradiographic identification in the developing rat. Science 222:1036–1038
Nelson RB, Routtenberg A (1985) Characterization of the 47kD protein F1 (pI 4.5), a kinase C substrate directly related to neural plasticity. Exp Neurol 89:213–224
Nelson RB, Routtenberg A, Hyman C, Pfenninger KH (1985) A phosphoprotein, F1, directly related to neuronal plasticity in adult rat brain may be identical to a major growth cone membrane protein. Soc Neurosci Abstr 11:927
Niedel JE, Kuhn LJ, Vandenbark GR (1983) Phorbol diester receptor copurifies with protein kinase C. Proc Natl Acad Sci USA 80:36–40
Nishizuka Y (1986) Studies and perspectives of protein kinase C. Science 233:305–312
Oestreicher AB, Dekker LV, Gispen WH (1986) A radioimmunoassay for the phosphoprotein B50: distribution in rat brain. J Neurochem 46:1366–1369
Perrone-Bizzozero NI, Finklestein SP, Benowitz LI (1986) Synthesis of a growth-associated protein by embryonic rat cerebrocortical neurons in vitro. J Neurosci 6:3721–3730
Rosenthal A, Chan SY, Henzel W et al. (1987) Primary structure and mRNA localization of protein F1, a growth-related protein kinase C substrate, associated with synaptic plasticity. EMBO J 6:3641–3646
Routtenberg A (1982) Memory formation as a posttranslational modification of brain proteins. In: Marsden CA, Matthies H (eds) Mechanisms and models of neural plasticity. Proc VIth Intl Neurobiol IBRO Symposium on Learning and Memory. Raven, New York, pp 17–24
Routtenberg A (1984) Brain phosphoproteins kinase C and protein F1: Protagonists of plasticity in particular pathways. In: Lynch G, McGaugh J, Weinberger N (eds) Neurobiology of learning and memory. The Guilford Press, New York, pp 479–490
Routtenberg A (1985) Protein kinase C activation leading to protein F1 phosphorylation may regulate synaptic plasticity by presynaptic terminal growth. Behav Neural Biol 44:186–200
Routtenberg A (1986) Synaptic plasticity and protein kinase C. In: Gispen WH, Routtenberg A (eds) Phosphoproteins in the nervous system. Elsevier, Amsterdam, pp 211–234
Routtenberg A, Ehrlich YH (1975) Endogenous phosphorylation of four cerebral cortical membrane proteins: role of cyclic nucleotides, ATP and divalent cations. Brain Res 92:415–430
Routtenberg A, Ehrlich YH, Rabjohns R (1975) Effect of a training experience on phosphorylation of a specific protein in neocortical and subcortical membrane preparations. Fed Proc 34:293
Routtenberg A, Lovinger D, Cain S, Akers R, Steward O (1983) Effects of long-term potentiation of perforant path synapses in the intact hippocampus on in vitro phosphorylation of a 47 KD protein (F-1). Fed Proc 42:755
Routtenberg A, Lovinger D, Steward O (1985) Selective increase in the phosphorylation of a 47 kD protein (F1) directly related to long-term potentiation. Behav Neural Biol 43:3–11
Skene JHP (1984) Growth-associated proteins and the curious dichotomies of nerve regeneration. Cell 37:697–700
Skene JHP, Willard M (1981 a) Changes in axonally transported proteins during axon regeneration in toad retinal ganglion cells. J Cell Biol 89:86–95
Skene JHP, Willard M (1981 b) Axonally transported proteins associated with axon growth in rabbit central and peripheral nervous system. J Cell Biol 89:96–103
Skene JHP, Willard M (1981 c) Characteristics of growth-associated polypeptides in regenerating toad retinal ganglion cell axons. J Neurosci 1:419–426
Snipes GJ, Chan S, McGuire CB, Costello BR, Routtenberg A, Norden JJ, Freeman JA (1987) Evidence that GAP-43, a growth-related protein, and F1, a synaptic plasticity-associated protein, are identical. J Neurosci 7:4066–4075
Will BE, Kelche CR (1979) Effects of different postoperative environments on the avoidance behavior of rats with hippocampal lesions: recovery or improvement of function? Behav Neural Biol 27:96
Wolf MP, Cuatrecasas P, Sahyoun N (1985) Interaction of protein kinase C with membranes is regulated by Ca^{++}, phorbol esters, and ATP. J Biol Chem 260:15718–15722

The Kinase C Substrate Protein B-50 (GAP43) and its Function in Axonal Repair Mechanisms

J. Verhaagen [1], C. O. M. Van Hooff [2], and W. H. Gispen [2]

[1] Laboratory for Neuromuscular Diseases, Academic Hospital and
[2] Division of Molecular Neurobiology, Rudolf Magnus Institute for Pharmacology and Institute of Molecular Biology and Medical Biotechnology, University of Utrecht, Padualaan 8, NL-3584 CH Utrecht, The Netherlands

1 Introduction

Severing of a nerve is accompanied by an assortment of morphological and biochemical changes in the neuronal cell bodies, the axonal processes and in the neuronal interstitium. In the peripheral nervous system (PNS) axons normally respond to damage by axonal regeneration, while in the central nervous system (CNS) axons generally regenerate poorly or not at all. The successful regeneration of axons appears to be dependent on the establishment of a neuronal micro-environment which is favorable for axonal regrowth. The importance of trophic factors and extracellular surface components in sustained axonal regrowth has been demonstrated widely (Berg 1984). The difference in the regenerative capacities of CNS and PNS neurons seems largely based on the failure of CNS tissue to provide the damaged axon with a milieu appropriate for axonal regeneration. This has, for instance, been shown in experiments in which damaged CNS axons were exposed to grafts of sciatic nerve tissue. One month after implantation of these PNS grafts, a large number of CNS axons had grown into the graft for distances of 3 cm or more. In the absence of the PNS graft the severed CNS axons would only have exhibited some abortive sprouting (Aguayo et al. 1982). Thus, CNS axons are not necessarily unable to regenerate, but it seems that the CNS environment is not permissive for axonal sprout formation.

The study of the various interactions among the components of the nervous system, active in the process of successful axonal regeneration, is in the center of interest. Unraveling of the complex molecular events underlying these interactions will be useful in the development of new therapies improving the recovery process following nerve injury. In this chapter we will summarize current research concerning the role of a small subset of fast transported proteins in the process of axonal regeneration. These proteins have been designated "growth-associated proteins" (GAP's) since they were found to be present in growing (developing/regenerating) axons while in mature non-growing neurons they are present at much lower levels (Benowitz et al. 1981; Skene and Willard 1981 a). On basis of this behavior of the GAP's, Skene and Willard (1981 a–c; Skene 1984) formulated the "GAP hypothesis". This hypothesis implicates that the expression of GAP's is a crucial step in the functional switch of a neuron from a non-growing state to one of growth and active sprout formation.

H. Flohr (Ed.)
Post-Lesion Neural Plasticity

Discovery of the GAP's revealed that the synthesis of a relatively small number of neuronal gene products is altered during axonal regeneration. Recently it has become apparent that the characteristics of a prominent GAP, GAP43, correspond closely with the properties of the neuron-specific phosphoprotein B-50 (Zwiers et al. 1980). The notion that GAP43 and B-50 are identical proteins may provide a clue to the function of GAP43 in axonal growth. The neuron-specific B-50 protein has been studied extensively in our laboratory for more than 10 years. Neurochemical evidence suggests that B-50 may be involved in a feedback control on receptor-mediated breakdown of polyphosphoinositides by phospholipase C (PL-C) (Gispen 1986). The growth-associated expression of B-50/GAP43 and the involvement of B-50 in polyphosphoinositide metabolism invites the speculation that B-50/GAP43 might play a role in the regulation of the sensitivity of regenerating axons to environmental signals (e.g., growth factors) which interact with PL-C linked receptor systems.

2 Changes in Synthesis and Transport of Proteins Associated with Regeneration

In mature intact neurons there is a continuous flow of proteins synthesized in the cell bodies and transported to axonal processes. Virtually no axonal proteins are derived from local protein synthesis. Since the source of axonal proteins is the cell body, axonal proteins can be radioactivity labeled by injecting a radioactive amino acid near the cell bodies of a nerve. The injected amino acid is incorporated into newly synthesized proteins and subsequently these labeled proteins are conveyed to the axons by axoplasmic transport. Electrophoretic separation of the labeled proteins has revealed the existence of hundreds of axonally transported proteins. These proteins can be divided in five distinct groups on basis of their difference in transport velocity (Lasek 1981; Wilson 1984).

Analysis of pulse labeled proteins by gel electrophoresis and autoradiography has been employed to compare synthesis and transport of proteins in intact and damaged nerves. The rationale behind this approach is that if a protein serves a specific function in the reconstruction of axons it would be labeled and transported at substantially elevated levels in damaged axons as compared to intact axons. The following conclusions can be drawn from these studies: (1) in mammalian species no overall increase in protein synthesis has become apparent. In some cold-blooded animals (goldfish and toad) a three-to-five fold increase in total protein synthesis was found; (2) no new polypeptides were detected; (3) decreases in some enzymes involved in neurotransmitter metabolism, e.g., acetylcholine esterase and choline acetyl transferase (Frizell and Sjöstrand 1974) and a few unidentified proteins (MW 28 kDa and 13 kDA) have been observed (Bisby 1980; Redshaw and Bisby 1985); and (4) in goldfish and toad enhanced production of the structural proteins, tubulin and actin, have been reported (Heacock and Agranoff 1976; Giulian et al. 1980). In mammals the picture that emerges with regard to these constituents of the cytoskeleton is less clear: both increased and normal unchanged synthesis have been reported for actin and tubulin during regeneration. A small number of fast transported proteins showed significant increases in their metabolic labeling. The most reproducible changes have been detected in protein spots with relative molecular weights of 50, 43–49 and 22–24 kDa (Theiler and McClure 1978; Bisby 1980; Benowitz et al. 1981; Skene and Willard 1981 a–c). These proteins were first referred to as GAP50, GAP43 and GAP24 by Skene and Willard (1981 a).

GAP50 is a basic glycoprotein (pI 8.9) induced during regenerative growth in the toad optic nerve. No counterpart of GAP50 has been identified in mammals and its induction in response to nerve damage has not been corroborated by observations in other cold-blooded species. GAP43 and GAP24 have been shown to be present in the nervous systems of both cold-blooded and mammalian species. In the developing and regenerating rabbit hypoglossal nerve and in the developing rabbit optic nerve the metabolic radiolabeling of GAP43 increases 10–100 fold. The increased GAP43 labeling remains high during the period of axon elongation but decreases readily to basal levels when the growing axons reach their target cells. In neurons exhibiting no regenerative sprouting (rabbit optic neurons and rat and guinea pig corticospinal neurons) also no increased labeling of GAP43 was seen following nerve crush (Skene and Willard 1981 b; Kalil and Skene 1986; Reh et al. 1987). Nearly identical patterns of labeling have been demonstrated for GAP24. One major exception is that GAP24 labeling remains high in adult rabbits hypoglossal nerve cells and no additional labeling of this protein is observed following lesioning of this nerve (Skene and Willard 1981 b).

The relatively minor changes in the expression of neuronal gene products in response to nerve damage, as observed in these pulse-labeling experiments, have lead to the notion that successful axonal regeneration is not dependent on the expression of a large number of proteins that are unique to the process of regeneration. However, the continued supply of the proteins synthesized both in intact and damaged nerves to the site of injury of the damaged axon is not in itself sufficient to permit a severed neuron to repair its axon. Successful regeneration seems to require major alterations in the synthesis of a small subset of neuronal gene products, the above mentioned GAP's (Willard et al. 1985).

The function of GAP's in the process of axonal repair is not known. Recent cross-laboratory studies have provided evidence that GAP43 is identical to growth cone phosphoprotein pp46, protein F_1 and protein B-50. With respect to the function of GAP43 the latter correspondence is of particular interest, since B-50 is a biochemically well-characterized polypeptide with a putative role in membrane-associated signal transduction processes.

3 The Neuron-Specific Phosphoprotein B-50 and Axonal Regeneration

B-50 is an acidic phosphoprotein, exclusively present in neurons (Zwiers et al. 1980; Kristjansson et al. 1982). B-50 immunoreactivity is widely distributed in rat brain, where it is concentrated in synaptic dense areas (Oestreicher and Gispen 1986). Electron microscopic analysis indicates that B-50 is associated with presynaptic membranes (Gispen et al. 1985). When synaptosomal plasma membranes (SPM), isolated from adult rat brain, are subjected to endogenous phosphorylation with [γ-^{32}P]-ATP, B-50 is seen as a prominent phosphoprotein (Fig. 1). B-50 phosphorylation in SPM is mediated by the Ca^{2+}/phospholipid-dependent pro-

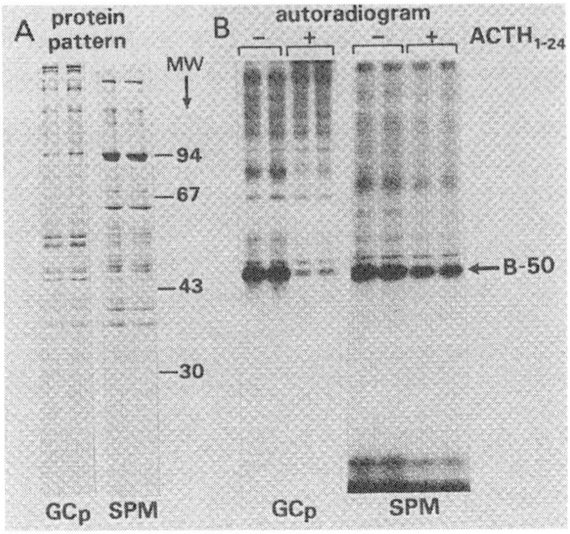

Fig. 1. Protein-staining patterns (**A**) and autoradiograms (**B**) of nerve growth cone particulate material (GCp) and SPM subjected to SDS-PAGE separation, after endogenous phosphorylation with [γ-^{32}P]-ATP in the absence or presence of ACTH(1–24) (3×10^{-6} M). Molecular weight markers (MW) in kilodaltons

tein kinase C (Aloyo et al. 1982, 1983) and can be inhibited by addition of ACTH(1–24) (Fig. 1; Zwiers et al. 1978). On two-dimensional gel electrophoresis B-50 could be resolved in four distinct spots in the isoelectric focusing dimension of the gel. These spots are in part interconventable by exhaustive phosphorylation and dephosphorylation. These results indicate that B-50 has at least two phosphorylation sites (Zwiers et al. 1985). B-50 has been purified to complete homogeneity from rat and calf brain employing subcellular fractionation, DEAE-cellulose chromatography, ammonium sulphate precipitation and Sephadex flat bed isoelectric focusing (Zwiers et al. 1980; Oestreicher et al. 1983). In addition to this isolation procedure, an alternative protocol has been described (Oestreicher et al. 1983). In this procedure the subcellular fractionation and DEAE-chromatography steps have been replaced by alkaline extraction of B-50 present in a crude brain homogenate. Subsequently, B-50 is subjected to ammonium sulfate precipitation and flat bed isoelectric focusing as described in the original purification procedure. Using the simplified isolation procedure, mg quantities of B-50 can be purified within relatively short time.

In 1985 it became apparent that B-50 shares several characteristics with a prominent phosphoprotein in growth cones, designated pp46 (De Graan et al. 1985; Katz et al. 1985). This has been achieved by showing that: (1) purified B-50 co-migrates with pp46 in growth cone particulate material on two-dimensional gels; (2) phosphorylation of the growth cone protein is inhibited by ACTH(1–24), a typical property of B-50 in SPM (Fig. 1), and (3) affinity-purified anti-B-50 antibodies cross-react with the growth cone phosphoprotein on immunoblots and immunoprecipitate this phosphoprotein from growth cone particulate material (De

Graan et al. 1985; Van Hooff et al. 1988). GAP43, as identified on two-dimensional gel prepared from regenerating toad optic nerve and from developing rat brain, also comigrates with both B-50 (Jacobson et al. 1986) and pp46 (Meiri et al. 1986). Comparison of GAP43 and B-50 has revealed the following additional similarities: GAP43 is phosphorylated upon exposure to [γ-^{32}P]-ATP (Jacobson et al. 1986), both proteins exhibit a peculiar acrylamide concentration-dependent shift in relative molecular weight on SDS gels (Gower and Rodnight 1982; Oestreicher et al. 1982; Jacobson et al. 1986; Benowitz et al. 1987) and antibodies raised against GAP43, isolated from neonatal rat brain, cross-react with B-50 partially purified from adult rat brain (Jacobson et al. 1986).

In order to reinforce the growth-associated behavior of B-50/GAP43, actual B-50 levels have been compared in crushed and intact rat sciatic nerves, using a sensitive B-50 radioimmunoassay and immunohistochemistry (Verhaagen et al. 1986). In these studies virtually no B-50 immunoreactivity was seen in control nerves but bright immunofluorescence appeared in regenerating sprouts that grow into the distal nerve portion following nerve crush. The amount of B-50 levels per mg nerve protein in the crushed nerves 6 days following the crush, increase 5.3-fold as compared to non-crushed controls. Normal intact neuromuscular junctions of the soleus muscle, one of the target muscles of sciatic nerve axons, do not contain detectable amounts of B-50, but during the period of reinnervation, neuromuscular junctions contain abundant amounts of the B-50 protein (Fig. 2A–D). In the newly formed terminals, B-50 is associated with presynaptic vesicular structures and with the presynaptic plasma membrane (Fig. 2E). After completion of reinnervation, B-50 levels in the nerve decrease to control values at the crush site as well as at the neuromuscular junction (Verhaagen et al. 1987). The latter observation may indicate that B-50 is subject to regulation at two different levels of organization: (1) the expression of the B-50 gene is restored to control values upon interaction of the growing axons with their target cells. This could for instance be achieved by uptake and retrograde transport of a signal molecule from the target cells to the neuronal cell bodies. Evidence for target-dependent synthesis of GAP43 and GAP24 has been presented previously (Redshaw and Bisby 1984, 1985). On the other hand, it has been shown that in regenerating goldfish optic axons GAP43 levels decreased after a fixed time period independent of the reconnection of the growing axons with their target tissue (Yoon et al. 1986). (2) The decrease of the high amounts of B-50 observed in nerve terminals following reinnervation may be due to the activation of a (B-50 specific?) protease in the newly formed synaptic contacts. Zwiers et al. (1982) described a protease that cleaves B-50 in B-60 (MW 46 kDa) and a small molecular weight peptide (MW 2 kDa). Since B-60 is also recognized by anti-B-50 antibodies the conversion of B-50 into B-60 will not result in the observed decrease in B-50 immunoreactivity. Thus it may be expected that other proteases exist in nerve terminals which are active in degrading B-50 to non-immunoreactive material.

B-50 was also shown to be present in an established model system to study the mechanism of neuronal differentiation, the rat pheochromocytoma cell line PC12. Upon exposure to nerve growth factor (NGF) PC12 cells acquire a phenotype resembling sympathetic neurons, including neurite outgrowth and cessation of proliferation. Culturing PC12 cells in the presence of 100 ng/ml NGF for two

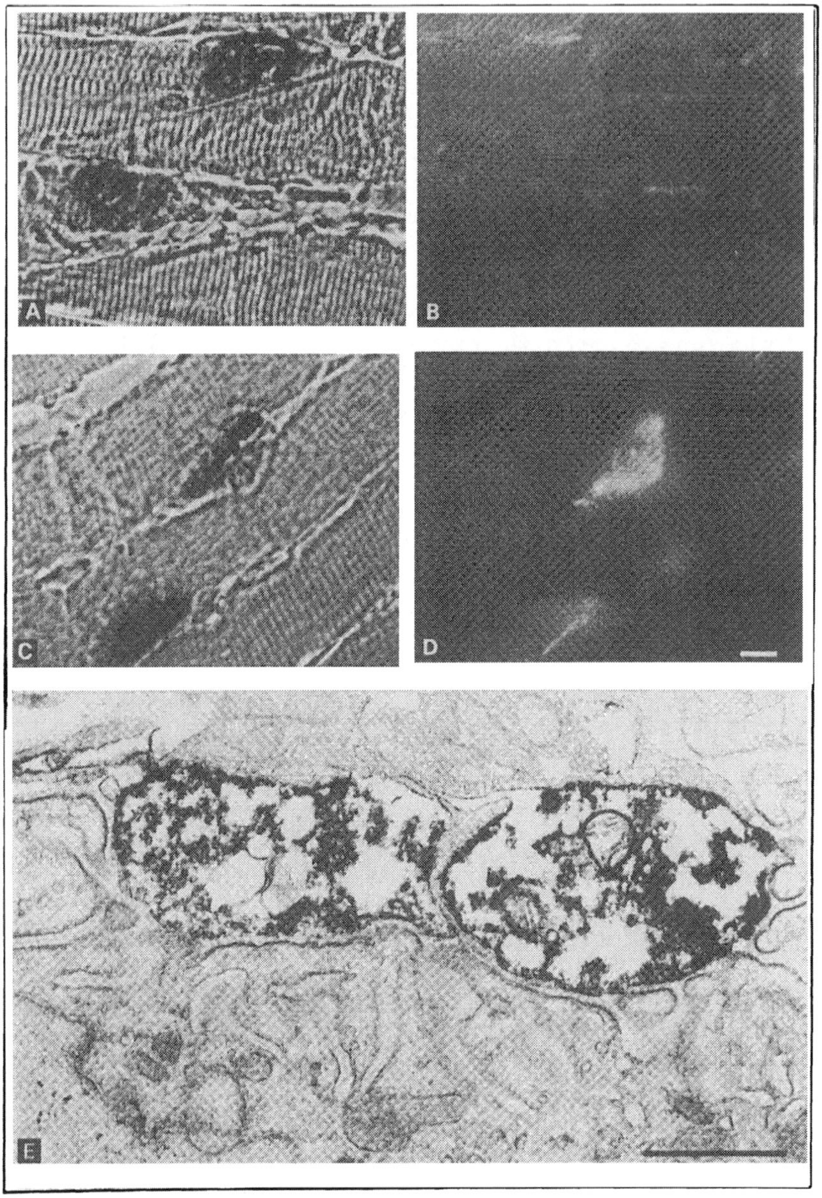

Fig. 2. Cryostat sections of a control non-denervated muscle (**A**, **B**) and a muscle 18 days following nerve crush lesion (**C**, **D**). Phase contrast micrograph (**A**, **C**) of the AchE-stained end plates and corresponding immunostaining (**B**, **D**) of the sections with anti-B-50 antibodies are shown. Note that the newly innervated neuromuscular junctions strongly react with the anti-B-50 antibodies while no B-50-immunoreactive material is present in intact neuromuscular junctions. **E** Immuno-electronmicrograph of a reinnervated neuromuscular junction 18 days after nerve crush. The B-50-immunoreactivity is associated with presynaptic vesicles and with the presynaptic plasma membrane

Table 1. Effect of NGF exposure on B-50 levels in PC12 cells

Treatment	ng B-50/mg protein
NGF−	35.8 ± 2.0 (n = 4)
NGF+	89.6 ± 7.4* (n = 4)

P12 cells were grown for two days in the presence (+) or absence (−) of 100 ng/ml NGF. B-50 levels of total cell homogenates were assayed by B-50 RIA. Data are expressed as mean ± SEM.
* Significantly different ($p < 0.001$).

days increased the B-50 levels of the cell homogenates 2,5-fold (Table 1). In these differentiated cells, B-50 immunoreactivity was most pronounced at the neurite-like processes and their growth cones (Van Hooff et al. 1986).

4 The Function of B-50

In the rat CNS B-50 may be involved in the regulation of the activity of the phosphatidyl 4-phosphate (PIP) kinase (Jolles et al. 1980, 1981; Oestreicher et al. 1983; Van Dongen et al. 1985). The phosphorylation state of B-50 has been connected to the modulation of the synthesis of phosphatidylinositol 4,5-bisphosphate (PIP_2) from PIP: phosphorylation of B-50 in SPM results in concomitant diminished formation of PIP_2 while in the presence of dephospho-B-50 more PIP_2 is formed (Van Dongen et al. 1985; Gispen 1986). This inverse relationship has recently also been demonstrated in growth cone membranes (Van Hooff et al. 1988). The hydrolysis of PIP_2, mediated by receptor activation, results in the formation of the second messengers DG and inositol trisphosphate (IP_3). The transient rise in the concentrations of these second messengers results in the simultaneous activation of protein kinase C and the release of calcium from endoplasmic reticulum stores. The synergistic activation of protein-kinase C and the release of Ca^{2+} facilitates the fusion of vesicles with the plasma membrane. If a predominant number of B-50 molecules in growth cones is in the dephospho-state one might anticipate that this could result in high cytoplasmic IP_3 and Ca^{2+} concentrations. This in turn may cause sustained fusion of vesicles with the plasma membrane resulting in the forward movement of the growth cone. The phospho state of B-50 in growing axons is not yet known, although one report appeared suggesting that a substantial amount of the B-50 molecules in cultured embryonic cortex neurons is indeed in the dephosphorylated state (Perrone-Bizzozero et al. 1986). Phosphorylation of B-50 is shown to be inhibited by ACTH(1–24), a neuropeptide-stimulating peripheral nerve regenerating (Bijlsma et al. 1981). Although it is tempting to speculate that this neurotrophic property of ACTH(1–24) is based on its ability to inhibit the phosphorylation of B-50, the link between the neurotrophic action of ACTH and the degree of phosphorylation of B-50 is problematic. Peptides with profound stimulating effects on axonal regeneration, like αMSH and ORG.2766 [a metabolically stable ACTH(4–9) analog] do not inhibit the B-50 phosphorylation in growth cone membranes (Van Hooff et al. 1988).

Thus, although we presume that the phosphorylation state of B-50 may be an important variable in the process of axonal growth, more work is necessary to elucidate the precise role of this protein in axonal growth.

5 Perspectives

Currently the field of growth-associated proteins is in a rapid flux. Several cDNA clones encoding for B-50/GAP43 have now been isolated (Karns et al. 1987; Basi et al. 1987; Neve et al. 1987; Nielander et al. 1987). Studies using these clones revealed that the expression of GAP43/B-50 mRNA is higher during periods of neurite growth, thus corroborating the observations at the protein level reviewed in this paper. The available cDNA clones will be employed to study the regulation of the GAP43/B-50 gene in detail. Important issues with regard to the regulation of the GAP43/B-50 gene expression are (1) do known growth factors (e.g., NGF, CNTF, melanocortin neuropeptides) directly induce the expression of the GAP43/B-50 gene? (2) what is the relationship between the growth related appearance of B-50 and B-50 found in adult mature synapses?

The revelation of the primary structure of B-50 is essential to further investigate the function of this protein. The amino acid sequence of B-50/GAP43 is abundant in negatively charged amino acids (matching with the acid nature of the protein) and contain no potential hydrophobic membrane spanning regions. A short N-terminal hydrophobic tail could probably serve as a membrane "anchor" for B-50/GAP43 (Basi et al. 1987). This would be in line with the observation that B-50 is a membrane protein localized at the inner surface of neuronal membranes (Gispen et al. 1985). In order to study the phosphorylation state of B-50 in vivo it will be necessary to generate antibodies which can distinguish between phosphorylated and dephosphorylated B-50. With the help of the currently available sequence data the production of these antibodies is now feasible. Such region specific antibodies may provide important tools to collect more information concerning the function of the B-50/GAP43 protein and the role of its phosphorylation in neuronal growth.

References

Aguayo A, David S, Richardson P, Bray G (1982) Axonal elongation in peripheral and central nervous system transplants. Adv Cell Neurobiol 3:215–233

Aloyo VJ, Zwiers H, Gispen WH (1982) B-50 protein kinase and kinase C in rat brain. Prog Brain Res 56:303–315

Aloyo VJ, Zwiers H, Gispen WH (1983) Phosphorylation of B-50 protein by calcium-activated, phospholipid-dependent protein kinase and B-50 protein kinase. J Neurochem 41:649–643

Basi GS, Jacobson RD, Virag I, Schilling J, Skene P (1987) Primary structure and transcriptional regulation of GAP43, a protein associated with nerve growth. Cell 49:785–791

Benowitz LJ, Yoon MG, Lewis ER (1981) Specific changes in rapidly transported proteins during regeneration of the goldfish optic nerve. J Neurosci 1:300–307

Benowitz LI, Perrone-Bizzozero NI, Finklestein SP (1987) Molecular properties of the growth-associated protein GAP43 (B-50). J Neurochem 48:1640–1647

Berg DK (1984) New neuronal growth factors. Annu Rev Neurosci 7:149–170

Bijlsma WA, Jennekens FGI, Schotman P, Gispen WH (1981) Effects of corticotropin (ACTH) on recovery of sensorimotor function in the rat: structure-activity study. Eur J Pharmacol 76:73–79
Bisby MA (1980) Changes in the composition of labelled protein transported in motor axons during their regeneration. J Neurobiol 11:435–445
De Graan PNE, Hooff COM Van, Tilly BC, Oestreicher AB, Schotman P, Gispen WH (1985) Phosphoprotein B-50 in nerve growth cones from fetal rat brain. Neurosci Lett 61:235–241
Frizell M, Sjöstrand J (1974) Transport of protein, glycoproteins and cholinergic enzymes in regenerating hypoglossal nerve. J Neurochem 22:845–850
Gispen (1986) Phosphoprotein B-50 and phosphoinositides in brain synaptic plasma membranes: a possible feedback relationship. Trans Biochem Soc UK 14:163–165
Gispen WH, Leunissen JLM, Oestreicher AB, Verkleij AJ, Zwiers H (1985) Presynaptic localization of B-50 phosphoprotein: the ACTH-sensitive protein kinase substrate involved in rat brain phosphoinositide metabolism. Brain Res 328:381–385
Giulian D, Ruisseaux H Des, Cowburn D (1980) Biosynthesis and intra-axonal transport of proteins during neuronal regeneration. J Biol Chem 255:6494–6501
Gower H, Rodnight R (1982) Intrinsic protein phosphorylation in synaptic plasma membrane fragments from the rat. Biochim Biophys Acta 716:45–52
Heacock AM, Agranoff BW (1976) Enhanced labelling of a retinal protein during regeneration of optic nerve in goldfish. Proc Natl Acad Sci USA 73:828–832
Jacobson RD, Virag I, Skene JHP (1986) A protein associated with axon growth, GAP43, is widely distributed and develop mentally regulated in rat CNS. J Neurosci 6:1843–1855
Jolles J, Zwiers H, Dongen CJ Van, Schotman P, Wirtz KWA, Gispen WH (1980) Modulation of brain polyphosphoinositide metabolism by ACTH-sensitive protein phosphorylation. Nature 286:623–625
Jolles J, Zwiers H, Dekker A, Wirtz KWA, Gispen WH (1981) Corticotropin-(1-24)-tetracosapeptide affects protein phosphorylation and polyphosphoinositide metabolism in rat brain. Biochem J 194:283–291
Kalil K, Skene JHP (1986) Elevated synthesis of an axonally transported protein correlates with axon outgrowth in normal and injured pyramidal tracts. J Neurosci 6:2563–2570
Karns LR, Ng SC, Freeman JA, Fishman MC (1987) Cloning of complementary DNA for GAP43, a neuronal growth-related protein. Science 236:597–600
Katz F, Ellis L, Pfenninger RH (1985) Nerve growth cones isolated from fetal rat brain III. Calcium-dependent phosphorylation. J Neurosci 5:1402–1411
Kristjansson GI, Zwiers H, Oestreicher AB, Gispen WH (1982) Evidence that the synaptic phosphoprotein B-50 is localized exclusively in nerve tissue. J Neurochem 39:371–378
Lasek RJ (1981) Axonal transport: a dynamic view of neuronal structures. TINS 1:87–91
Levine J, Skene THP, Willard H (1981) GAP's and fodrin, novel axonally transported proteins. TINS 2:273–277
Meiri K, Pfenninger K, Willard M (1986) Growth-associated protein, GAP43, a polypeptide that is induced when neurons extend axons, is a component of growth cones and corresponds to a major polypeptide enriched in growth cones. Proc Natl Acad Sci USA 83:3537–3541
Neve RL, Perrone-Bizzozero NI, Finklestein S, Zwiers H, Bird E, Kurnit DM, Benowitz LI (1987) The neuronal growth associated protein GAP43 (B-50, F1): neuronal specificity, developmental regulation and regional distribution of the human and rat mRNAs. Mol Brain Res 2:177–183
Nielander HB, Schrama LH, Van Rozen AJ, Kasperaitis M, Oestreicher AB, De Graan PNE, Gispen WH, Schotman P (1987) Primary structure of the neuron-specific phosphoprotein B-50 is identical to growth-associated protein GAP43. Neurosci Res Commun 1:163–172
Oestreicher AB, Gispen WH (1986) Comparison of the immunocytochemical distribution of the phosphoprotein B-50 in the cerebellum and hippocampus of immature and adult rat brain. Brain Res 375:267–279
Oestreicher AB, Zwiers H, Gispen WH, Roberts S (1982) Characterization of infant rat cerebral cortical membrane proteins phosphorylated in vivo: identification of the ACTH-sensitive phosphoprotein B-50. J Neurochem 39:683–692
Oestreicher AB, Dongen CJ Van, Zwiers H, Gispen WH (1983) Affinity-purified anti-B-50 protein antibody: interference with the function of the phosphoprotein B-50 in synaptic plasma membranes. J Neurochem 41:331–340

Perrone-Bizzozero NI, Finklestein S, Benowitz L (1986) Synthesis of a growth-associated protein by embryonic rat cerebrocortical neurons in vitro. J Neurosci 6:3721–3730
Reh IA, Redshaw JD, Bisby MA (1987) Axons of the pyramidal tract do not increase their transport of growth-associated proteins after axotomy. Mol Brain Res 2:1–6
Redshaw JD, Bisby MA (1984) Proteins of fast axonal transport in the regenerating hypoglossal nerve of the rat. Can J Physiol Pharmacol 62:1387–1393
Redshaw JD, Bisby MA (1985) Comparison of the effects of sciatic nerve crush or resection on the proteins of fast axonal transport in rat dorsal root ganglion cell axons. Exp Neurol 88:437–446
Skene JHP (1984) Growth-associated proteins and the curious dichotomies in nerve regeneration. Cell 37:697–700
Skene JHP, Willard M (1981 a) Changes in axonally transported proteins during regeneration in toad retinal ganglion cells. J Cell Biol 89:86–95
Skene JHP, Willard M (1981 b) Axonally transported proteins associated with axon growth in rabbit central and peripheral nervous systems. J Cell Biol 89:96–103
Skene JHP, Willard M (1981 c) Characteristics of growth-associated polypeptides in regenerating toad retinal ganglion cell axons. J Neurosci 1:419–426
Theiler RF, McClure WO (1978) Rapid axoplasmic transport of proteins in regenerating sensory nerve fibers. J Neurochem 31:433–447
Van Dongen CJ, Zwiers H, Graan PNE De, Gispen WH (1985) Modulation of the activity of purified phosphatidylinositol 4-phosphate kinase by phosphorylated and dephosphorylated B-50 protein. Biochem Biophys Res Commun 128:1219–1227
Van Hooff COM, Graan PNE De, Boonstra J, Oestreicher AB, Schmidt-Michels MH, Gispen WH (1986) Nerve growth factor enhances the level of the protein kinase C substrate B-50 in pheochromocytoma PC12 cells. Biochem Biophys Res Commun 2:644–651
Van Hooff COM, Graan PNE De, Oestreicher AB, Gispen WH (1988) B-50 phosphorylation and polyphosphoinositide metabolism in nerve growth cone membranes. J Neurosci 8:1789–1795
Verhaagen J, Hooff COM Van, Edwards PM (1986) The kinase C substrate protein B-50 and axonal regeneration. Brain Res Bull 17:737–741
Verhaagen J, Edwards PM, Gispen WH (1987) Damaged rat peripheral nerves do not contain detectable amounts of α-MSH. J Neurosci Res 19:14–18
Willard MB, Meiri K, Glicksman M (1985) Changes of state during neuronal development: regulation of axon elongation. In: Edelman GM, Gall WE, Cowan MW (eds) Molecular bases of neural development. Wiley, New York, pp 341–361
Wilson DL (1984) Molecular events associated with peripheral nerve regeneration. In: Elam JS, Cancelon P 8eds) Axonal transport in neuronal growth and regeneration. Plenum, New York, pp 137–145
Yoon MG, Benowitz LI, Baker FA (1986) The optic tectum regulates the transport of specific proteins in regenerating optic fibres of goldfish. Brain Res 382:339–351
Zwiers H, Wiegant VM, Schotman P, Gispen WH (1978) ACTH-induced inhibition of endogenous rat brain protein phosphorylation in vitro: structure activity. Neurochem Res 3:455–463
Zwiers H, Schotman P, Gispen WH (1980) Purification and some characteristics of an ACTH-sensitive protein kinase and its substrate protein in rat brain membranes. J Neurochem 34:1689–1699
Zwiers H, Gispen WH, Kleine L, Mahler HR (1982) Specific proteolyis of a brain membrane phosphoprotein (B-50): effects of calcium and calmodulin. Neurochem Res 7:127–137
Zwiers H, Verhaagen J, Dongen CJ Van, Graan PNE De, Gispen WH (1985) Resolution of rat brain synaptic phosphoprotein B-50 into multiple forms by two-dimensional electrophoresis: evidence for multi-site phosphorylation. J Neurochem 44:1083–1090

Neuronal Metabolic Basis of the Conditioning Lesion Effect

I. G. McQuarrie

Division of Neurosurgery and Department of Developmental Genetics and Anatomy, School of Medicine, Case Western Reserve University Cleveland, Ohio 44106, USA
Veterans Administration Medical Center, Cleveland, Ohio, 44106, USA

1 The Cell Body Response to Axonal Injury

The cell body response to axonal interruption is a fundamental reaction of the neuron. This activates the "growth mode" of neuronal metabolism, a genomic induction that leads to the accelerated turnover of messenger ribonucleic acid (mRNA) in association with the production of new messenger RNA and additional nucleolar DNA (Torvik and Skjorten 1974; Watson 1974; Grafstein and McQuarrie 1978; Hoffman et al. 1987). The "growth mode" gives first priority to the synthesis and axonal transport of structural and regulatory proteins that are used to assemble microtubules and microfilaments: tubulin, actin, microtubule associated proteins (MAPs), and calmodulin (Grafstein and McQuarrie 1978; McQuarrie 1983, 1988; Tetzlaff et al. 1986; Hoffman et al. 1987). Increases in transport are also seen for a trace protein, GAP-43, that appears to catalyze membrane assembly (Skene and Willard 1981 a; Jacobson et al. 1986; Perry et al. 1987). There is a concurrent reduction in metabolic activities that support the production of neurotransmitter molecules and neurofilaments (Grafstein and McQuarrie 1978; Hoffman and Lasek 1980; Tetzlaff et al. 1986; Hoffman et al. 1987; Wong and Oblinger 1987). The reduction in neurofilament transport causes a thinning of parent axons; radial regrowth does not occur until reconnection of the daughter axon with a functionally matched target organ (Cragg and Thomas 1961; Zalewski 1970; Hoffman et al. 1985, 1987; Wong and Oblinger 1987).

1.1 Supply of Proteins to the Growing Axon

Since almost all proteins needed for axonal outgrowth are synthesized in the nerve cell body, and outgrowth typically involves a tenfold increase in neuronal volume, axonal transport can be expected to influence regeneration (McQuarrie 1983, 1986). Membranous organelles and membrane proteins are carried by the fast component (FC) of axonal transport, whereas fibrillar macromolecules (neurofilaments, microtubules, microfilaments) are conveyed through the axon with related structural and regulatory proteins by the slow component (SC) of axonal transport (Grafstein and Forman 1980; Lasek et al. 1984; McQuarrie et al. 1986).

The integrity of FC is necessary to sustain outgrowth (McQuarrie and Grafstein 1983; Grafstein et al. 1987), possibly because FC carries GAP-43, a protein

H. Flohr (Ed.)
Post-Lesion Neural Plasticity

that exhibits a marked increase in transport during outgrowth (Skene and Willard 1981a; Perry et al. 1987; Redshaw and Bisby 1987; Reh et al. 1987). This membrane protein is widely distributed in axonal terminals and growth cones, and exhibits increased phosphorylation when the axonal arbor expands during either development or post-tetanic potentiation (Katz et al. 1985; Routtenberg et al. 1985; Jacobson et al. 1986).

While fast axonal transport is needed to sustain outgrowth, it is doubtful that it plays a rate-limiting role. The outgrowth rate is two orders of magnitude slower than FC, and there is a growing body of work suggesting that it is a function of SC (Grafstein and McQuarrie 1978; Lasek et al. 1981; McQuarrie and Grafstein 1982b; McQuarrie 1983, 1984, 1985a, 1986; McQuarrie and Lasek 1989). A rate-limiting factor can be identified by using methods to retard or accelerate the rate under investigation. A particular cell function can then be shown to influence the rate by its behavior under controlled conditions: the function undergoes a major change in the appropriate direction *before* the rate changes.

A study of axonally transported proteins that show increased transport may help to identify a rate-limiting factor. The best-documented candidates are GAP-43 in FC (Skene and Willard 1981a, b; Perry et al. 1987; Reh et al. 1987) and tubulin in SC (Giulian et al. 1980; Heacock and Agranoff 1982; Hoffman et al. 1985, 1987). Axonal outgrowth can be retarded by using cycloheximide to reduce protein synthesis during the cell body response (McQuarrie and Grafstein 1983; McQuarrie 1985b), and accelerated by using a conditioning axotomy to increase protein synthesis during the cell body response (McQuarrie and Grafstein 1982b). It has been shown that a conditioning axotomy does not produce a sustained increase in transport (after a subsequent testing axotomy) for either FC or GAP-43 (McQuarrie and Grafstein 1982b; Redshaw and Bisby 1987). Therefore, it is unlikely that FC plays a role in governing the rate of axonal outgrowth. The role of SC will be examined in detail after further consideration of the conditioning lesion method.

1.2 Suboptimal Supply of Structural Proteins to Distally Injured Axons

The "growth mode" does not initially provide increased support for the construction of daughter axons from the proximal (parent) axon stump. This is because key structural elements (tubulin, actin, MAP's, calmodulin) travel with SCb – the faster subcomponent of SC – at only 2–6 mm/day in both daughter and parent axons of rodent motor neurons (McQuarrie 1983; Lasek et al. 1984; McQuarrie et al. 1986; McQuarrie and Lasek 1989). In these neurons, axonal outgrowth begins after a 1–2 day "initial delay" and proceeds at the same rate as SCb (McQuarrie 1985a; McQuarrie 1988; McQuarrie and Lasek 1989). Thus, if the lesion is not made very near the nerve cell body, any additional structural elements carried with SCb (in response to axotomy) will not reach the growing axon tip until after it has reached the target organ (Hoffman and Lasek 1980). For lesions at greater distances from the cell body, structural elements can only come from the parent axon stump – through ongoing axonal transport and the depolymerization of microtubules and microfilaments that accumulate there (McQuarrie 1988; McQuarrie and Lasek 1989).

2 Method of Analyzing Cell Body Support for the Regenerating Axon

The magnitude of the cell body response and its importance for axonal regrowth can be analyzed by making a distal conditioning lesion to initiate the cell body response, followed by a testing lesion that is timed and located to allow the augmented supply of structural elements (carried by SCb) to reach the daughter axon sprouts. The effect of a conditioning lesion on the initial delay (latent period during which axons bridge the lesion), outgrowth rate, and SC (rate, amount and composition) is measured. The conditioning lesion causes earlier and/or faster outgrowth after the testing lesion (McQuarrie 1984). The hypothesis that this is due to an augmented supply of structural elements carried by SCb has been examined in three types of neurons: goldfish retinal ganglion cells (McQuarrie 1985b; McQuarrie and Grafstein 1981, 1982a, b), newt motor neurons (Maier and McQuarrie 1988), and rat motor neurons (Hoffman and Lasek 1980; Hoffman et al. 1985; McQuarrie 1986, 1988; McQuarrie and Lasek 1989).

3 Characteristic Features of the Conditioning Lesion Effect

When the outgrowth of daughter axons at a testing lesion begins earlier or proceeds more rapidly because of a conditioning lesion, the phenomenon is termed the conditioning lesion effect (McQuarrie et al. 1977). The characteristic features of this effect vary with respect to the type of neuron that is conditioned, the extrinsic factors that regulate the effect, the conditioning event, and the conditioning interval (McQuarrie 1984).

3.1 Neurons That Can Be Conditioned

There now appear to be no restrictions as to the type of neuron that can be conditioned: all that have been tested to date exhibit a shorter initial delay, faster outgrowth, or both. Some neurons, e.g., retinal ganglion cells of goldfish, exhibit a marked conditioning lesion effect (McQuarrie 1985b; Grafstein et al. 1987), while others, e.g., dorsal root ganglion cells of rodents, exhibit a modest effect (McQuarrie 1984). The latter have two axonal branches, central and peripheral, that have distinctly different responses to a conditioning lesion. Examination of these can be used to analyze the mechanism by which the cell body response is regulated. One branch can be used for the conditioning lesion and the other for the testing lesion. Axonal outgrowth from a testing lesion of the peripheral branch is accelerated by a peripheral conditioning lesion, whereas, neither a peripheral nor central branch conditioning lesion will accelerate outgrowth from a dorsal root testing lesion (Oblinger and Lasek 1984). However, a peripheral conditioning lesion improves the regrowth of central axons by ultimately producing a 100-fold increase in the number of dorsal column axons that will enter a nerve autograft of the cervical spinal cord, placement of which represents the testing lesion (Richardson and Issa 1984).

3.2 Extrinsic Factors that Regulate the Conditioning Lesion Effect

Factors extrinsic to the neuron influence both the cell body reaction and the conditioning lesion effect (McQuarrie 1984). For example, the acceleration of outgrowth is greater if the testing lesion is made at the same distal site as the conditioning lesion, rather than more proximally (Bisby and Pollock 1983; Bisby 1985). This could be because axons entering a pre-degenerated nerve encounter greater amounts of a local growth factor than those entering a distal nerve that had been normal until the testing lesion (Lubinska 1952, 1959; Brown and Hopkins 1981). Although this putative factor appears to act peripherally (on growth cone function), a central site of action is equally plausible. The cell body reaction may therefore be controlled, rather than merely influenced, by a change in the supply of a factor arriving from the periphery either via retrograde axonal transport or the blood stream.

Cragg's (1970) hypothesis that the cell body reaction to axotomy is triggered by the absence of an extrinsic factor has been neither proven nor disproven. Recently, however, Richardson and Verge (1986) have developed a novel method for testing it that takes advantage of the separate axonal branches of dorsal root ganglion neurons. They find that the hypothesis holds for these neurons, and that the peripheral factor reaches the cell body via retrograde transport. Any reduction in delivery of this factor stimulates regeneration. Their experimental design eliminates the possibility that the cell body reaction is triggered by a "positive" signal, either a new factor (produced by the injury) or a decreased turnaround time of fast-transported materials (due to axons being shortened by the lesion).

The "negative" signal results from a loss of axonal contact with Schwann cells: it is not related to atrophy of the target organ or loss of the distal axon. When loss of access to the Schwann cell factor is prolonged, by using nerve transection or ligation instead of using a crush lesion or colchicine (which permit rapid regeneration), the cell body response is more marked. (This may explain the faster growth of axons in pre-degenerated nerve). Richardson and Verge (1986) have not characterized the negative factor, aside from establishing that it is not NGF (nerve growth factor).

One feature of the cell body response that cannot be explained by loss of access to a Schwann cell factor is the earlier onset and greater magnitude of the cell body response that occur when lesions are made close to the cell body (Watson 1968; Richardson and Verge 1986). For every 4–5 mm more distally that a lesion of the hypoglossal nerve is placed, the onset of the increase in nucleolar (ribosomal) RNA is delayed by 1 day (Cragg 1970; Watson 1968). Since the retrograde transport rate of 200–400 mm/day is the rate that corresponds to the delay for signalling the onset of the cell body reaction (Kristensson and Olsson 1975; Singer et al. 1982), two different factors are responsible for a complete cell body response to axotomy. One is extrinsic to the neuron and signals a loss of connection with Schwann cells in the peripheral nerve, while the other is intrinsic to the neuron and signals the extent of axonal loss.

The extrinsic stimulus for the cell body reaction in somatic and autonomic motor neurons may differ from that in sensory neurons. For axotomized somatic motor neurons, artificial continuation of electrical activity in the denervated muscle may increase the available amount of neuronotrophic factor (Sebille and Bondoux-Jahan 1980). Whether denervated muscle or nerve is the source of neuronotrophic factor remains to be determined. For aminergic autonomic neurons, the neuronotrophic factor is NGF produced by Schwann cells (Stoeckel and Thoenen 1975; Richardson and Riopelle 1984). As in sensory neurons, the cell body response is due to lack of access to the peripheral factor. Thus, anti-NGF induces sprouting in neonatal animals (Stoeckel and Thoenen 1975; Hulsebosch et al. 1984; Richardson and Riopelle 1984).

3.3 Conditioning Event and Conditioning Interval

A conditioning event is any perturbation of the nervous system that produces a growth mode of neuronal metabolism (McQuarrie 1984). By this definition, a neuron that is developing its axonal arborization for the first time (either in the embryo or in vitro) has been conditioned (Collins and Lee 1982), as is a neuron that has expanded its axonal arbor to innervate target organs that have been vacated by axons from other neurons (Sparrow and Grafstein 1983; McQuarrie 1984). However, for most experimental applications the conditioning event is a distally placed axotomy.

Effective conditioning intervals range from 2–28 days, depending on the neuron being studied, the proximity of the conditioning lesion to the soma, and the extent to which the conditioning lesion prevents daughter axons from promptly reentering the distal nerve stump (Forman et al. 1980, 1981; Edwards et al. 1981; McQuarrie 1984; Bisby 1985; Bondoux-Jahan and Sebille 1986; Redshaw and Bisby 1987).

4 Experimental Methods

A convenient approach for evaluating the effect of metabolic changes on axonal outgrowth is to directly examine the influence these have on the daughter axon. That is, what are the changes in flux of identified proteins and macromolecules reaching the daughter axon via SC, and how is axonal outgrowth affected by these changes? With respect to structural elements that are neither degraded nor augmented during transport, flux (F) is the mass of an element traversing a cross-section of the axon in a unit of time (Lasek et al. 1983): $F = A(C \times R)$, where A is the cross-sectional area of the axon, C is the concentration of the element, and R is the rate of transport. Therefore, two basic questions need to be answered when reviewing the results of experiments examining SC in relation to axonal outgrowth: (1) is the rate, instantaneous amount, or composition of SC changed by axotomy, and (2) do these changes affect axonal outgrowth?

4.1 Methods for Measuring Slow Transport During Axonal Regeneration

Axonal transport is most effectively studied by labeling newly synthesized proteins at the nerve cell body. This is done by microinjecting radioactive amino acids near a population of neurons. Five distinct waves of radiolabeled proteins can than be examined while advancing distally in the axon. Each of these rate-components moves at a different rate from the others and has a nearly unique protein composition. Thus, a rate-component can be identified by removing nerve segments for gel electrophoresis and fluorography, and analyzing the labeled proteins that have traveled a certain distance after a certain labeling interval (e.g., Skene and Willard 1981 a; McQuarrie et al. 1986).

In mammalian neurons, FC is the fastest rate-component (250–400 mm/day) and SCa (0.2–2.0 mm/day) is the slowest (Grafstein and Forman 1980; McQuarrie et al. 1986). Axons must be long enough to separate the peaks of labeling for each rate-component, and it is necessary to label many neurons if individual proteins are to have measureable amounts of radioactivity (McQuarrie et al. 1986). These constraints effectively eliminate the possibility of studying axonal transport in cultured neurons, thus necessitating chronic animal studies (Forman 1983).

4.2 Methods for Measuring Axonal Outgrowth Distances

A variety of morphological, electrophysiological, behavioral, and biochemical methods are available for measuring the length of daughter axons (McQuarrie 1984). Currently, the most widely used method is to label FC with tritiated amino acids shortly before killing the animal. The nerve is then serially cross sectioned into 2–3-mm pieces; these are dissolved for liquid scintillation spectrometry to detect the total radioactivity in each piece (Forman and Berenberg 1978; McQuarrie 1978). The distribution of labeling shows two peaks, which correlate with the presence of growth cones. One peak is where sprouts are forming immediately proximal to the lesion; the other is distal to the lesion and locates the main population of regenerating axon tips (Griffin et al. 1976). Beyond the second peak, radioactivity falls off in a log-linear fashion to intersect with the tissue background level of radioactivity. This intersection marks the maximum outgrowth distance, while the steepest point on the log-linear descent has been used as an approximation of mean outgrowth distance (McQuarrie 1978).

The radioactivity levels in each nerve from a group of similarly treated animals can be normalized by assigning the total amount of axonally transported radioactivity distal to the lesion site a value of 100%. Then the mean fraction of that amount located in each nerve segment distal to the lesion is plotted (Forman et al. 1981; Redshaw and Bisby 1987). This method is suitable for statistical analysis of differences between two different groups of animals having the same injection-kill and lesion-kill intervals (Forman et al. 1981). Finally, a variant of the FC method has been developed to produce morphological data: a whole-mount of the nerve is prepared with a fluorographic reagent and exposed to X-ray film to obtain a fluorogram (Edstrom et al. 1986).

While the FC method is the most reliable, it is often helpful to have a method for estimating the maximum outgrowth distance when SC has been labeled. This is done by plotting the distribution of total radioactivity, or radioactivity in a heavily labeled protein such as tubulin, and denoting the most distal value that is clearly greater than tissue background as the maximum outgrowth distance (Black and Lasek 1979; McQuarrie 1986). Since the radioactivity declines gradually instead of steeply (as in the FC method), the SC method underestimates the maximal outgrowth distances by 10% (McQuarrie 1986). Recently, however, it has been possible to modify the SC method to closely approximate the distance obtained by the FC method. In heavily labeled nerves with low tissue background, there is a small distal peak of radioactivity contained in MAP's and actin; the position of this peak correlates closely with the maximum outgrowth distance obtained by the FC method (McQuarrie and Lasek 1989).

5 The Flux of Cytoskeletal Elements Governs Outgrowth Rate

As mentioned above, axonal transport studies of the conditioning lesion effect have been carried out in three types of neuron: retinal ganglion cells of goldfish, ventral motor neurons that serve the forelimb musculature of newts, and ventral motor neurons that serve the hindlimb musculature of rats (McQuarrie and Grafstein 1982b; McQuarrie 1986, 1988; Maier and McQuarrie 1987). The evidence that the conditioning lesion acts to shorten the initial delay and/or accelerate the outgrowth rate in goldfish retinal ganglion cells has previously been reviewed (McQuarrie 1984). In studies on the newt, limb amputation serves as the testing lesion. Since limb regeneration is a neurotrophic phenomenon, the end point for evaluating axonal outgrowth is limb regrowth. This begins 3–4 days earlier as a result of a nerve conditioning lesion, then proceeds at the rate expected following amputation alone (Maier et al. 1984). For motor axon regeneration in the rat sciatic nerve, a conditioning lesion accelerates the outgrowth of most axons after a testing lesion but the leading axons appear to show only a 10% increase in outgrowth rate (McQuarrie 1978, 1986).

The flux (overall transport) of SC structural elements is considerably increased in all three types of neuron following axotomy (McQuarrie and Grafstein 1982a, b; Hoffman et al. 1985; Maier and McQuarrie 1987; McQuarrie and Lasek 1989). There is a decrease in the transport of neurofilaments (carried by SCa) in rat motor neurons, which causes a reduction in axonal diameter that begins at the nerve cell body and extends distally with time (Hoffman et al. 1985). This appears to be a general response of neurons to axotomy (Hoffman et al. 1987; Wong and Oblinger 1987). There also appears to be an increase in the synthesis (and transport via SCb) of other structural elements: tubulin/microtubules, actin/microfilaments, MAP's, and calmodulin (Hoffman and Lasek 1980; Heacock and Agranoff 1982; Hoffman et al. 1985, 1987; McQuarrie 1983, 1988; Maier and McQuarrie 1987; Wong and Oblinger 1987; McQuarrie and Lasek 1989).

The current working hypothesis is that the SCb transport of structural elements is increased because of the conditioning lesion. An increased supply of these acts to accelerate the formation of daughter axons that emerge following the testing lesion (McQuarrie 1985a). To the extent that this hypothesis has been tested, it has been substantiated (Maier and McQuarrie 1987; McQuarrie 1986, 1988). However, it has not yet been demonstrated that a testing lesion made ahead of an enhanced wave of SCb structural elements will produce daughter axons which fail to exhibit accelerated outgrowth. The alternate hypothesis, that accelerated outgrowth is due to a reduction in neurofilament transport, also has not been tested. However, this is a less likely explanation since (1) short conditioning intervals are fully effective, which means outgrowth is accelerated at testing lesion sites located ahead of the wave of axonal thinning caused by decreased neurofilament transport, and (2) there is little reason to think that the minimal packing density of neurofilaments is reduced by axotomy-induced thinning, given the information that it is not reduced by drug-induced thinning (Monaco et al. 1988). Finally, the hypothesis that axonal acceleration is caused by a reduction in neurofilament density does not provide a reason for the dramatic increases in the SCb transport of MAP's and calmodulin that are seen following axotomy (McQuarrie 1988). We can tentatively conclude that the arrival of unusually large amounts of tubulin, actin, MAP's, and calmodulin at the testing lesion site is the apparent basis of the conditioning lesion effect. The corollary is that the rate of axonal outgrowth is limited by the availability of these structural elements.

References

Black MM, Lasek RJ (1979) Slowing of the rate of axonal regeneration during growth and maturation. Exp Neurol 63:616–623
Bisby MA (1985) Enhancement of the conditioning lesion effect in rat sciatic motor axons after superimposition of conditioning and test lesions. Exp Neurol 90:385–394
Bisby MA, Pollock B (1983) Increased regeneration rate in peripheral nerve axons following double lesions: enhancement of the conditioning lesion phenomenon. J Neurobiol 14:467–472
Bondoux-Jahan M, Sebille A (1986) Electrophysiological study of conditioning lesion effect on rat sciatic nerve regeneration following either prior section or freeze. I. Intensity and time course. Brain Res 382:39–45
Brown MC, Hopkins WG (1981) Role of degenerating axon pathways in regeneration of mouse soleus motor axons. J Physiol 318:365–373
Collins F, Lee MR (1982) A reversible developmental change in the ability of ciliary ganglion neurons to extend neurites in culture. J Neurosci 2:424–430
Cragg BG (1970) What is the signal for chromatolysis? Brain Res 23:1–21
Cragg BG, Thomas PK (1961) Changes in conduction velocity and fibre size proximal to peripheral nerve lesions. J Physiol (Lond) 157:315–327
Edstrom A, Sjoberg J, Kanje M (1986) The use of whole-mount preparations of nerves labelled with axonally transported radioactive proteins to study regeneration. J Neurosci Meth 16:19–27
Edwards DL, Alpert RM, Grafstein B (1981) Recovery of vision in regeneration of goldfish optic axons: enhancement of axonal outgrowth by a conditioning lesion. Exp Neurol 72:672–686
Forman DS (1983) Axonal transport and nerve regeneration: A review. In: Kao CC, Bunge RP, Reier PJ (eds) Spinal cord reconstruction. Raven, New York, pp 75–86
Forman DS, Berenberg RA (1978) Regeneration of motor axons in the rat sciatic nerve studied by labeling with axonally transported radioactive proteins. Brain Res 156:213–225
Forman DS, McQuarrie IG, Labore FW, Wood DK, Stone LS, Braddock CH, Fuchs DA (1980) Time course of the conditioning lesion effect on axonal regeneration. Brain Res 182:180–185
Forman DS, McQuarrie IG, Grafstein B, Edwards DL (1981) Effect of a conditioning lesion on axonal regeneration and recovery of function. In: Flohr H, Precht W (eds) Lesion-induced neuronal plasticity in sensorimotor systems. Springer, Berlin Heidelberg New York, pp 103–113
Giulian D, Ruisseaux H des, Cowburn D (1980) Biosynthesis and intra-axonal transport of proteins during neuronal regeneration. J Biol Chem 255:6494–6501
Grafstein B, Forman DS (1980) Intracellular transport in neurons. Physiol Rev 60:1167–1283
Grafstein B, McQuarrie IG (1978) Role of the nerve cell body in axonal regeneration. In: Cotman CW (ed) Neuronal plasticity. Raven, New York, pp 155–195
Grafstein B, Burmeister DW, McGuinness CM, Perry GW, Sparrow JR (1987) Role of fast axonal transport in regeneration of goldfish optic axons. In: Seil FJ, Herbert E, Carlson BM (eds) Neural regeneration. Elsevier, Amsterdam, pp 113–120 (Progress in Brain Research, vol 71)

Griffin JW, Drachman DB, Price DL (1976) Fast axonal transport in motor nerve regeneration. J Neurobiol 7:355–370

Heacock AM, Agranoff BW (1982) Protein synthesis and transport in the regenerating goldfish visual system. Neurochem Res 7:771–788

Hoffman PN, Lasek RJ (1980) Axonal transport of the cytoskeleton in regenerating motor neurons: constancy and change. Brain Res 202:317–333

Hoffman PN, Thompson GW, Griffin JW, Price DL (1985) Changes in neurofilament transport coincide temporally with alterations in the caliber of axons in regenerating motor fibers. J Cell Biol 101:1332–1340

Hoffman PN, Cleveland DW, Griffin JW, Landes PW, Cowan NJ, Price DL (1987) Neurofilament gene expression: a major determinant of axonal caliber. Proc Natl Acad Sci USA 84:3472–3476

Hulsebosch CE, Coggeshall RE, Perez-Polo JR (1984) Increased numbers of thoracic dorsal root axons in rats given antibodies to nerve growth factor. Science 225:525–526

Jacobson RD, Virag I, Skene JHP (1986) A protein associated with axon growth, GAP-43, is widely distributed and developmentally regulated in rat CNS. J Neurosci 6:1843–1855

Katz F, Ellis L, Pfenninger KH (1985) Nerve growth cones isolated from fetal rat brain. III. Calcium dependent protein phosphorylation. J Neurosci 5:1402–1411

Kristensson K, Olsson Y (1975) Retrograde transport of horseradish peroxidase in transected axons. II. Relations between rate of transfer from the site of injury to the perikaryon and the onset of chromatolysis. J Neurocytol 4:653–661

Lasek RJ, McQuarrie IG, Wujek JR (1981) The central nervous system regeneration problem: neuron and environment. In: Gorio A, Millesi H, Mingrino S (eds) Posttraumatic peripheral nerve regeneration. Raven, New York, pp 59–70

Lasek RJ, Oblinger MM, Drake PF (1983) Molecular biology of neuronal geometry: expression of neurofilament genes influences axonal diameter. Cold Spring Harbor Symp Quant Biol 48:731–744

Lasek RJ, Garner JA, Brady ST (1984) Axonal transport of the cytoplasmic matrix. J Cell Biol 99:212 s–221 s

Lubinska L (1952) The influence of the state of the peripheral stump on the early stages of nerve regeneration. Acta Biol Exp (Warsaw) 16:55–63

Lubinska L (1959) Region of transition between preserved and regenerating parts of myelinated nerve fibers. J Comp Neurol 113:315–335

Maier CE, McQuarrie IG (1987) Acceleration of slow transport in axons of regrowing newt limbs by a nerve conditioning lesion made prior to amputation. Soc Neurosci Abstr 13:973

Maier CE, McQuarrie IG, Singer M (1984) A nerve conditioning lesion accelerates limb regeneration. J Exp Zool 232:181–186

McQuarrie IG (1978) The effect of a conditioning lesion on the regeneration of motor axons. Brain Res 152:597–602

McQuarrie IG (1983) Role of the axonal cytoskeleton in the regenerating nervous system. In: Seil FJ (ed) Nerve, organ and tissue regeneration: research perspectives. Academic Press, London, pp 51–88

McQuarrie IG (1984) Effect of a conditioning lesion on axonal transport during regeneration: The role of slow transport. In: Elam J, Cancalon P (eds) Axonal transport in neuronal growth and regeneration. Plenum, New York, pp 185–209 (Advances in neurochemistry, vol 6)

McQuarrie IG (1985 a) Effect of a conditioning lesion on axonal sprout formation at nodes of Ranvier. J Comp Neurol 231:239–249

McQuarrie IG (1985 b) Stages of axonal regeneration following optic nerve crush in goldfish: contrasting effects of conditioning nerve lesions and intraocular acetoxycycloheximide injections. Brain Res 333:247–253

McQuarrie IG (1986) Structural protein transport in elongating motor axons after sciatic nerve crush: effect of a conditioning lesion. Neurochem Pathol 5:153–164

McQuarrie IG (1988) Transport of cytoskeletal proteins into axonal sprouts during nerve regeneration. In: Gordon T, Stein RB (eds) The current status of peripheral nerve regeneration. Alan R Liss, New York, pp 25–34

McQuarrie IG, Grafstein B (1981) Effect of a conditioning lesion on optic nerve regeneration in goldfish. Brain Res 216:253–264

McQuarrie IG, Grafstein B (1982 a) Protein synthesis and fast axonal transport in regenerating goldfish retinal ganglion cells. Brain Res 235:213–223
McQuarrie IG, Grafstein B (1982 b) Protein synthesis and axonal transport in goldfish retinal gangliom cells during regeneration accelerated by a conditioning lesion. Brain Res 251:25–31
McQuarrie IG, Grafstein B (1983) Effect of acetoxycycloheximide and dibutyryladenosine cyclic 3′,5′-monophosphate on axonal regeneration in the goldfish optic nerve. Brain Res 279:377–381
McQuarrie IG, Lasek RJ (1989) Slow transport of cytoskeletal elements from parent axons into regenerating daughter axons. J Neurosci (in press)
McQuarrie IG, Grafstein B, Gershon MD (1977) Axonal regeneration in the rat sciatic nerve: effect of a conditioning lesion and of dbcAMP. Brain Res 132:443–453
McQuarrie IG, Brady ST, Lasek RJ (1986) Diversity in the axonal transport of structural proteins: major differences between optic and spinal axons in the rat. J Neurosci 6:1593–1605
Monaco S, Autilio-Gambetti L, Lasek RJ, Katz MJ, Gambetti P (1988) Drug-increased neurofilament transport rate: decreases in neurofilament number and axon diameter. J Neuropath Exp Neurol (in press)
Oblinger MM, Lasek RJ (1984) A conditioning lesion of the peripheral axons of dorsal root ganglion cells accelerates regeneration of only their peripheral axons. J Neurosci 4:1736–1744
Perry GW, Burmeister DW, Grafstein B (1987) Fast axonally transported proteins in regenerating goldfish optic axons. J Neurosci 7:792–806
Redshaw JD, Bisby MA (1987) Proteins of fast axonal transport in regenerating rat sciatic sensory axons: a conditioning lesion does not amplify the characteristic response to axotomy. Exp Neurol 98:212–221
Reh TA, Redshaw JD, Bisby MA (1987) Axons of the pyramidal tract do not increase their transport of growth-associated proteins after axotomy. Mol Brain Res 2:1–6
Richardson PM, Issa VM (1984) Peripheral injury enhances central regeneration of primary sensory neurones. Nature 309:791–793
Richardson PM, Riopelle RJ (1984) Uptake of nerve growth factor along peripheral and spinal axons of primary sensory neurons. J Neurosci 4:1683–1689
Richardson PM, Verge VMK (1986) The induction of a regenerative propensity in sensory neurons following peripheral axonal injury. J Neurocytol 15:585–594
Routtenberg A, Lovinger D, Steward O (1985) Selective increase in phosphorylation of a 47 kDa protein (F1) directly related to long-term potentiation. Behav Neural Biol 43:3–11
Sebille A, Bondoux-Jahan M (1980) Effects of electric stimulation and previous nerve injury on motor function recovery in rats. Brain Res 193:562–565
Singer PA, Mehler S, Fernandez HL (1982) Blockade of retrograde axonal transport delays the onset of metabolic and morphologic changes induced by axotomy. J Neurosci 2:1299–1306
Skene JHP, Willard M (1981 a) Changes in axonally transported proteins during regeneration in toad retinal ganglion cells. J Cell Biol 89:86–95
Skene JHP, Willard M (1981 b) Axonally transported proteins associated with axon growth in rabbit central and peripheral nervous systems. J Cell Biol 89:96–103
Sparrow JR, Grafstein B (1983) Prior collateral sprouting enhances axonal regeneration. Brain Res 269:133–136
Stoeckel K, Thoenen H (1975) Retrograde axonal transport of nerve growth factor: specificity and biological importance. Brain Res 85:337–341
Tetzlaff W, Bisby MA, Kreutzberg GW (1986) Protein changes in the axotomized facial and hypoglossal nucleus of the rat. Soc Neurosci Abstr 12:277
Torvik A, Skjorten F (1974) The effect of actinomycin D upon normal neurons and retrograde nerve cell reaction. J Neurocytol 3:87–97
Watson WE (1968) Observations on the nucleolar and total cell body nucleic acid of injured nerve cells. J Physiol (Lond) 196:655–676
Watson WE (1974) The binding of actinomycin D to the nuclei of axotomized neurones. Brain Res 65:317–322
Wong J, Oblinger MM (1987) Changes in neurofilament gene expression occur after axotomy of dorsal root ganglion neurons: an *in situ* hybridization study. Metabol Brain Dis 2:291–303
Zalewski AA (1970) Effects of reinnervation on denervated skeletal muscle by axons of motor, sensory, and sympathetic neurons. Am J Physiol 219:1675–1679

Postsynaptic Receptor Elimination During Synaptic Competition

M. M. Rich and J. W. Lichtman

Department of Anatomy and Neurobiology, Washington University School of Medicine, 660 South Euclid Avenue, St. Louis, MO 63110, USA

1 Introduction

The way in which synaptic connections are maintained between axon terminals and the target cells they innervate is a central yet poorly understood aspect of the behavior of the nervous system. Understanding synaptic maintenance is especially important for a full understanding of how the nervous system compensates following injury. At present, however, even basic questions about whether synaptic connections are stable or constantly turning over have just begun to be approached directly.

An interesting aspect of this problem is how a target cell can selectively maintain one axon's terminals while at the same time losing connections with other axons. The competitive mechanism underlying selective synaptic maintenance occurs in both development and during reinnervation governing the elimination of multiple innervation at the neuromuscular junction. In other parts of the nervous system this competitive mechanism regulates the number of axons maintaining synapses on individual neurons (for review, see Purves and Lichtman 1985, Chap. 12). Work from a number of systems indicates that during development more axons converge on individual target cells than in adults. This extra innervation is eliminated, leaving target cells with fewer innervating axons. It is important to realize that this process is not entirely regressive: the axon or axons that remain in contact with a target elaborate further synaptic endings at the very same time the other synapses (from other axons) are being eliminated (see, for example, Lichtman 1977). Thus synapse formation and elimination are going on concurrently.

Our lack of understanding about selective synaptic maintenance stems from two general problems. The first problem is that it is unclear what are the relative roles of the pre- and postsynaptic cell in the selective maintenance of some terminals and elimination of others. Thus do axons compete for a passive target cell; or alternatively is the postsynaptic cell actively involved dictating which axonal terminals will be maintained and which will be eliminated? There are several lines of evidence arguing that the postsynaptic cell does play a role in determining the outcome of synaptic interactions. At the neuromuscular junction of developing rat and chick, blockade of postsynaptic activity slows synapse elimination, while increasing postsynaptic activity increases the rate of synapse elimination, indicating that the target regulates the pace of competition (Duxson 1982; Srihari and

H. Flohr (Ed.)
Post-Lesion Neural Plasticity

Vrobova 1978; Thompson 1983). Further evidence for the importance of postsynaptic activity comes from the fact that neuromuscular junctions on muscle fibers which fire action potentials undergo synapse elimination, whereas junctions on nonfiring tonic muscle fibers maintain multiple innervation (Lichtman et al. 1985). Postsynaptic activity also appears to be important in regulating synaptic competition in the central nervous system. In the developing visual cortex, blocking postsynaptic activity drastically affects the segregation of occular dominance columns (Reiter and Stryker personal communication). This evidence argues that in order to understand the interactions which lead to selective synaptic maintenance and elimination, both the presynaptic and postsynaptic cells may have to be studied.

A second difficulty in understanding the rules that govern synaptic maintenance and elimination is that such dynamic events are not easily studied by techniques which give static information. Thus most evidence for synaptic change is indirect. Images of different synaptic elements, each obtained at a different time, are used to argue what the progression of change would look like if an individual element was followed over the time course. The conclusions drawn are wrought with problems of interpretation. For example, the individual static images themselves give no information as to whether a synapse is being formed or eliminated. Thus a thin branch emanating from an established synapse could be interpreted as a newly formed sprout or as an old process in the midst of being resorbed.

To get around these problems we have focused on changes which occur at individual neuromuscular junctions in living mice in which we can view the same junction at multiple times. The neuromuscular junction is less difficult to visualize than other synapses and therefore an ideal preparation for study. Through the use of new flourescent probes to stain nerve terminals (Magrassi et al. 1987; Lichtman et al. 1987) and alpha bungarotoxin to stain receptors we can follow individual neuromuscular junctions over time in vivo. In the studies we report here, we have followed endplates during the synaptic competition which occurs between axons following reinnervation of the neuromuscular junction. Our studies have focused on the relation between the selective maintenance and elimination of presynaptic nerve terminals and the distribution of receptors in the postsynaptic membrane (Rich and Lichtman 1988a).

2 Multiple Innervation and its Elimination Following Reinnervation

Following denervation in adult muscle, the damaged axons regenerate back to the muscle to reinnervate muscle fibers. By viewing individual endplate sites before and after reinnervation by nerve crush we have found that the site at which a muscle fiber is reinnervated is always the original endplate site present before denervation. Moreover, we find that the reinnervation of the old endplate sites is itself quite accurate: by reoccupying the old Schwann cell tubes and the previous synaptic gutters the nerve terminal obtains a branching pattern very similar to that seen prior to denervation (see also Letinsky et al. 1976). However, the axons not only reestablish neuromuscular junctions, but also sprout over the surface of the muscle fibers. Many of these sprouts travel along the length of a single muscle fiber ending blindly, but some of the sprouts grow from one endplate site across the muscle fiber into nearby endplates. These endplates are thus multiply innervated. Other multiply innervated endplates are innervated by more than one axon following the original pathway up the old Schwann cell tube and others are multiply innervated by one or more sprouts which probably are new branches from the nerve. Thus all the multiple innervation is confined to the original endplate site. The location of multiple inputs within the same endplate means that the competition between axons during synapse elimination occurs at close range. By intracellular recording we have found muscle fibers to be physiologically multiply innervated at the time sprouts can be seen innervating endplates. Because both anatomical and physiological signs of multiple innervation disappear with a similar time course we believe that these sprouts are the source of the temporary multiple innervation.

An obvious question is, for what do axons compete at the old synaptic site that enables one axon to be maintained while another is eliminated? Because it is possible that the postsynaptic cell plays a role in synaptic competition it is necessary to understand what is happening to postsynaptic sites during the process. For example, do axons compete for the same postsynaptic sites during synapse elimination? In order to study postsynaptic sites we followed the acetylcholine receptor distribution of individual endplates in vivo during reinnervation (Rich and Lichtman 1988a). We were suprised to find that following reinnervation of the muscle a fraction of the endplates lost regions of high acetylcholine receptor density that were present prior to denervation. We only rarely saw addition of new postsynaptic sites to endplates. When the nerve terminals of endplates which had lost receptors from postsynaptic sites were visualized, we found that nerve terminal had also been eliminated from precisely the same areas. Thus areas of the endplate which lose acetylcholine receptors also lose overlying nerve terminal.

Although synapse elimination could be responsible for loss of endplate regions, we have examined other possibilities as well. We have found that the loss of postsynaptic sites seen following denervation is not due to damage to the postsynaptic sites caused by the binding or visualization of the receptor distribution with rhodaminated bungarotoxin. In endplates followed with alpha bungarotoxin in normally innervated muscle no postsynaptic sites were lost.

Another possibility is that the loss of postsynaptic sites is simply the muscle fiber's reaction to being denervated. To test this idea we followed the acetylcholine receptor distribution of endplates in permanently denervated muscle. Despite the severe atrophy of permanently denervated muscle, no loss of acetylcholine receptor sites in the endplates was observed. Thus loss of postsynaptic sites does not occur in the absence of nerve (and in fact only commences once the nerve has returned – see below).

What is it that reinnervating nerve terminals do that causes elimination of synaptic sites? One possibility is that upon return to the muscle, axons do not completely reoccupy the vacated postsynaptic sites; unoccupied postsynaptic sites might then be eliminated. To test this idea we came back to the muscle at early times following reinnervation before postsynaptic sites are eliminated. We viewed endplates at this time to see whether all the postsynaptic sites are occupied by nerve terminal. We found that at most endplates the nerve terminal had completely reoccupied postsynaptic sites and many of these sites were subsequently eliminated. In the few cases where unoccupied postsynaptic sites were present, over the subsequent days the nerve terminal did come to occupy those sites as well. Thus incomplete reoccupation of postsynaptic sites does not appear to cause loss of synaptic sites.

Perhaps the most interesting possibility is that loss of synaptic sites is related to loss of nerve terminals during synapse elimination. Consistent with this idea is the finding that the time at which postsynaptic sites are lost is coincident with the time during which synapse elimination occurs. Thus 3 to 4 days after the return of the nerve to the muscle, endplates begin to lose postsynaptic sites and synapse elimination begins. After 15 days, when synapse elimination is almost over, very few endplates lose postsynaptic sites. Thus the two processes start and stop at the same time.

To directly determine whether synapse elimination is the driving force for the loss of synaptic sites, we prospectively followed endplates which appeared multiply and singly innervated at early times during reinnervation before any sites had been lost. From our work we knew that the multiply innervated endplates would become singly innervated, whereas the singly innervated endplates would remain singly innervated. We found that a signficantly higher percentage of the multiply innervated endplates lost postsynaptic sites rich in acetylcholine receptors. Thus it appears that multiple innervation and its subsequent elimination leads to loss of synaptic sites both presynaptically and postsynaptically. This clearly means that during synaptic competition, axons are not competing for fixed postsynaptic sites because these are apparently eliminated along with the overlying nerve terminal.

Because the loss of synaptic sites involves the elimination of both nerve terminal and underlying receptor, the question arises as to which element is eliminated first. Is it that the eliminated nerve terminal leaves postsynaptic sites vacated, and those sites are subsequently lost? Or is it that the postsynaptic site is first eliminated and this is then followed by (and perhaps causes) the elimination of the overlying nerve terminal? A third possibility is that the synaptic basal lamina which lies between nerve terminal and receptor is actually the element which changes first. To see which element is eliminated first we followed nerve terminal, acetylcholine receptors and synaptic basal lamina (marked by cholinesterase) during reinnervation at times when endplates first begin to change (Rich and Lichtman 1988 a). We found that 3 to 4 days after the nerve returns to the muscle the first changes in receptor staining are seen. These changes consist of a decrease in the staining intensity of the receptor areas to be entirely eliminated over the next 11 days. Interestingly, at early times we found that the lightly staining receptor areas were usually still occupied by a nerve terminal. Views of the same endplates at later times showed that both nerve terminal and receptors were totally missing from these sites. The temporal order of changes suggests that the loss of receptor precedes the loss of nerve terminal. When we followed receptors and cholinesterase staining (a synaptic basal lamina marker) following reinnervation, we found that acetylcholine receptors were eliminated in sites before any change was detectable in cholinesterase staining. Thus the loss of acetylcholine receptors appears to be the first morphological change seen in a synaptic site to be eliminated, preceding nerve terminal loss.

This data suggests the interesting possibility that loss of the postsynaptic receptors, or some closely associated molecule, is the cause of loss of presynaptic nerve terminal. Consistent with this idea is an incidental finding we have made in which we followed sprouts growing away from endplates during reinnervation. The only sprouts not resorbed are those rare sprouts which cause new synaptic sites (acetylcholine receptor clusters) to be formed under them.

The loss of postsynaptic sites and subsequent loss of overlying nerve terminal may mean that one nerve terminal (the winner) may dislodge postsynaptic molecules from beneath another axon (the loser) followed by its elimination. It is well known from work in tissue culture that neurites are capable of causing receptor clusters nearby them to dissagregate (Anderson and Cohen 1977; Frank and Fischbach 1979). Thus one axon may eliminate another axon innervating the same endplate by causing the elimination of postsynaptic sites underlying the losing axon. Once the losing axon has been eliminated, the postsynaptic region which was eliminated is no longer attractive to the winning axon and is never reoccupied.

To test this hypothesis and directly see whether postsynaptic sites are in fact necessary to maintain nerve terminals, we have followed individual nerve terminals in the absence of the muscle fibers they synapse upon (Rich and Lichtman 1988b). The muscle fibers were killed by cutting them on either side of the endplate band. Care was taken not to damage the nerve in any way. Once the muscle fibers have degenerated and been cleared, the nerve terminals are left sitting on basal lamina ghosts in which most postsynaptic molecules including the acetylcholine receptor are much diminished. We viewed nerve terminals before and several days after muscle fiber death to see whether the loss of postsynaptic sites had any effect on nerve terminal morphology. To our suprise, we found that within 3 days of muscle fiber damage, much of the innervation at every endplate was no longer stainable. Regions of nerve terminal which had previously been long continuous arms were now broken up into a number of small, poorly staining spots. Apparently in the absence of postsynaptic sites, the nerve terminal rapidly begins to retract. We believe that this finding is consistent with the idea that losing axons are eliminated during competition due to the loss of postsynaptic sites underlying them.

Another interesting finding relating to the loss of postsynaptic sites is that once synapse elimination is over, the endplates which have lost postsynaptic sites become stable in their new configuration. This means that the changes in endplate structure caused by a transient event such as synapse elimination, are subsequently maintained in a stable way for long periods of time. If such a mechanism in which a rapid change in synaptic connectivity is followed by a long period of stability also occurs at synapses elsewhere in the nervous system, it might be a potential means of information storage.

Acknowledgement. This work was supported by grants from the NIH, Muscular Dystrophy Association, and a Research Career Development Award (J. W. L.) from the McKnight Foundation

References

Anderson MJ, Cohen MW (1977) Nerve-induced and spontaneous redistribution of acetylcholine receptors on cultured muscle cells. J Physiol (Lond) 268:757–773

Duxson MJ (1982) The effect of postsynaptic block on development of the neuromuscular junction in postnatal rats. J Neurocytol 11:395–408

Frank E, Fischbach GD (1979) Early events in neuromuscular junction formation in vitro: induction of acetylcholine receptor clusters in the postsynaptic membrane and morphology of newly formed synapses. J Cell Biol 83:143–158

Letinsky MS, Fischbeck KH, McMahan UJ (1976) Precision of reinnervation of original postsynaptic sites in frog muscle after a nerve crush. J Neurocytol 5:671–718

Lichtman JW (1977) The reorganization of synaptic connexions in the submandibular ganglion of the rat. J Physiol 273:155–177

Lichtman JW, Wilkinson RS, Rich MM (1985) Multiple innervation of tonic endplates revealed by activity-dependent uptake of fluorescent probes. Nature 314:357–359

Lichtman JW, Magrassi L, Purves D (1987) Visualization of neuromuscular junctions over periods of several months in living mice. J Neurosci 7:1215–1222

Magrassi L, Purves D, Lichtman JW (1987) Fluorescent probes that stain living nerve terminals. J Neurosci 7:1207–1214

Purves D, Lichtman JW (1985) Principles of neural development. Sinauer, Sunderland, MA

Rich MM, Lichtman JW (1988 a) Transient multiple innervation of original endplate sites during reinnervation of adult mouse muscle (in preparation)

Rich MM, Lichtman JW (1988 a) In vivo visualization of pre- and postsynaptic changes during synapse elimination in reinnervated mouse muscle. J Neurosci (submitted)

Rich MM, Lichtman JW (1988 b) Elimination of postsynaptic receptordense regions during synaptic competition (in preparation)

Rich MM, Lichtman JW (1988 c) Postsynaptic receptor elimination precedes nerve terminal loss during synaptic competition (in preparation)

Rich MM, Lichtman JW (1988 b) Motor nerve terminal retraction following removal of target myofibers (in preparation)

Srihari T, Vrbova G (1978) The role of muscle activity in the differentiation of neuromuscular junctions in slow and fast chick muscles. J Neurocytol 7:529–540

Thompson W (1983) Synapse elimination in neonatal rat muscle is sensitive to pattern of muscle use. Nature 302:614–615

Changes in Dendrites of Adult Rat Ganglion Cells Regenerating Axons into Peripheral Grafts

S. Thanos [1] and A. J. Aguayo [2]

[1] Max-Planck Institut für Entwicklungsbiologie, D-7400 Tübingen, FR Germany
[2] Montreal General Hospital and McGill University, Montreal, Quebec, Canada

1 Introduction

Axotomized neurons in the retina and other regions of the central nervous system (CNS) of adult rodents can regenerate long axons into segments of peripheral nerve (PN) grafted near their cell somata (Leoz and Arcante 1914; So and Aguayo 1985; Ramon y Cajal 1928; Aguayo 1985; Politis and Spencer 1986; Vidal-Sanz et al. 1987). When such PN grafts are used as "bridges" to replace the optic nerve (ON) and join again the eye and the superior colliculus (SC), some of the retinofugal fibers regrow along the graft and re-enter the SC where they can form synaptic contacts (Vidal-Sanz et al. 1987). However, since axonal regeneration into PN grafts requires that both the axotomy and the graft be near the somata of the retinal ganglion cells (RGC's) (Aguayo et al. 1983; Aguayo 1985), the grafting procedure also results in the rapid or protracted death of a large number of RGC's (Richardson et al. 1982; Villegas-Perez et al. 1988). Because axotomy to neurons other than RGC's is known to also cause a spectrum of functional and anatomical alterations that affect the cell body (Misantone et al. 1984) and dendrites of surviving cells (James 1933; Watson 1974; Mendell 1984; Nja and Purves 1978; Villegas-Perez et al. 1988), we have studied anatomic changes in somata and dendritic arbors of RGC's from 1 to 4 months after axotomy and grafting of a segment of PN whose most distal end was left blind-ended under the scalp, unconnected to the SC. The studies reported here are an early attempt to investigate the severity and duration of retrograde responses of injured RGC's during periods of rapid regrowth – 1 and 2 months after PN grafting – and at 4 months after the insertion of the graft, when most axons can be assumed to have stopped extending actively along these blind-ended grafts (Aguayo et al. 1986; Trecarten et al. 1986). It is anticipated that improved grafting techniques and the use of agents that can enhance RGC survival (Sievers et al. 1987) and axonal penetration into the SC (Vidal-Sanz et al. 1987) may eventually permit additional studies of the effects of synaptic terminal connectivity on neuronal survival and dendritic morphology, circumstances which for cells such as spinal motor neurons and sympathetic neurons, among others, can lead to the persistence of cells and the reorganization of altered dendritic trees (Summer and Watson 1971; Standler and Bernstein 1982; Jones and Thomas 1962).

The present studies were aided by the use of sensitive retrogradely transported neuronal marker DiI (1,1-dioctadecyl-3,3,3′,3′-tetramethylindocarbocyanine

H. Flohr (Ed.)
Post-Lesion Neural Plasticity

perchlorate; Molecular Probes, Junction City, Oregon, USA) (Honig and Hume 1986; Thanos and Bonhoeffer 1987). When applied to the PN graft, this fluorescent dye labels intensely the soma and dendrites of RGC's, allowing the assessment and measurement of their changes.

2 Materials and Methods

2.1 Transplantation Techniques

For these experiments, 31 adult female Sprague-Dawley rats weighing 220–225 g were anesthetized by injecting chloral hydrate intraperitoneally (0.42 mg/g body weight). The left optic nerve was transected completely within the orbit, approximately 1 mm behind the optic disc, and a 2- to 3-cm segment of autologous sciatic nerve was removed from the leg and sutured, end to end, to the ocular stump of the ON. The remaining portion of the PN graft was extended posteriorly along the cranium and its free, distal end sutured between the scalp and the occipital bone (Vidal-Sanz et al. 1987).

2.2 Labeling of Ganglion Cells

Rats with PN grafts were examined 1, 2, and 4 months after grafting. Two days prior to killing, these animals were anesthetized as above and the middle portion of the graft was surgically exposed from under the scalp. A small longitudinal slit was made through the connective tissue and perineurium surrounding the graft to insert locally small crystals of DiI. The precise site of DiI application was 5 to 6 mm away from the ON-PN graft junction, a distance that was considered optimal to obtain a clear visualization of the dendrites while avoiding the diffusion of the dye to the optic nerve stump. DiI was then diluted by carefully pipeting into the slit approximately 0.5 µl of pure ethanol. Rats were perfused with a saline solution followed by 4% paraformaldehyde in 0.1 M phosphate buffer. The eyes and their corresponding grafts were dissected and separated from each other at the ON-PN graft junction. Retinas were prepared as flattened whole mounts, immersed for 30 min in fixative solution, and mounted in glycerol for fluorescence microscopy. Three of the grafts were frozen and examined in serial transverse sections to investigate dye diffusion along the grafted PN segment.

The retinas of 23 normal rats were used as controls after a similar application of DiI to the optic nerve 2 mm from the posterior pole of the eye, 2 days later the animals were perfused and the retinas processed as described.

2.3 Morphometry

Flat-mounted whole retinas of control and experimental animals were examined under a Leitz Orthoplan fluorescent microscope using a rhodamine filter for the optimal visualization of DiI-labeled cells (excitation: 550–560 nm; suppression: 580 nm). Labeled RGC's were photographed at least twice under a ×25 objective on a Tri-X-Panchromatic film after focusing firstly the perikarya and then their dendrites in the inner plexiform layer. Photographic negatives were magnified in a Leitz light box and projected onto a screen. The density and size of RGC somata was determined by counting the cells within selected retinal squares at distances of about 1 mm from the optic disc.

Single dendritic arbors were drawn on a superimposed transparent paper and the perimeter of each dendritic field was measured by connecting the most peripheral tips of its dendrites. The measurements were digitized with the help of an IBAS-1 computer and their diameters calculated and analyzed statistically by means of a t-test.

The branching of RGC dendrites was examined by counting the number of bifurcations within the dendritic field of each cell. The number of dendritic branches of RGC's in control and experimental animals was related to: (1) the size of their somata (large, medium, and small) and (2) the size and shape of the dendritic tree (Fukuda 1977; Perry 1979).

3 Results

3.1. Labeling with DiI

After fixation of the graft, the site of DiI application was easily recognized as a small red discoloration. In the three grafts examined in serial cross-sections, the dye was found to have diffused locally into glia and connective tissues only within a maximal radius of 3 mm from the site of DiI application. Because DiI was always applied 5–6 mm away from the ON-PN graft junction, we have assumed that: (1) the dye had not diffused to the ON and labeled RGC's with nonregenerating cut axons confined to the remaining ON stump; and (2) DiI was incorporated only into the RGC regenerated axons that extended into the PN grafts.

In all 31 PN grafted rats examined, labeled neurons were observed throughout the retina, their density being greatest near the optic disc, most likely as a reflection of the normal distribution of RGC's (Perry 1979). The number of labeled RGC's located within 1 mm of the optic disc varied among animals and postsurgical periods. The average densities were 86 ± 22 RGC/mm^2, 1 months after axotomy and grafting ($n = 8$ retinas); 83 ± 25 RGC/mm^2 at 2 months ($n = 8$) and by 4 months labeled cells averaged 64 ± 21 RGC/mm^2 (five rats). The densities of RGC's giving rise to regenerated axons entering the graft calculated in these experiments after DiI application 5–6 mm from the ON-PN graft junction resembles closely those obtained previously by Villegas-Perez et al. (1988) using other retrograde labels applied to the distal end of PN grafts that were approximately 3 cm long.

3.2 Perikaryal and Dendritic Morphology and Size

In whole mounts of the 23 control retinas examined under the fluorescent microscope after the application of DiI to the optic nerve, it was possible to visualize clearly the brightly fluorescing soma and many fine details of the dendritic arborization of labeled RGC's (Fig. 1 A). By analyzing the drawn silhouettes (Fig. 2) of these RGC's, the labeled neurons were grouped within the three cell types described in normal rodents by Fukuda (1977) and Perry (1979). Ganglion cells of

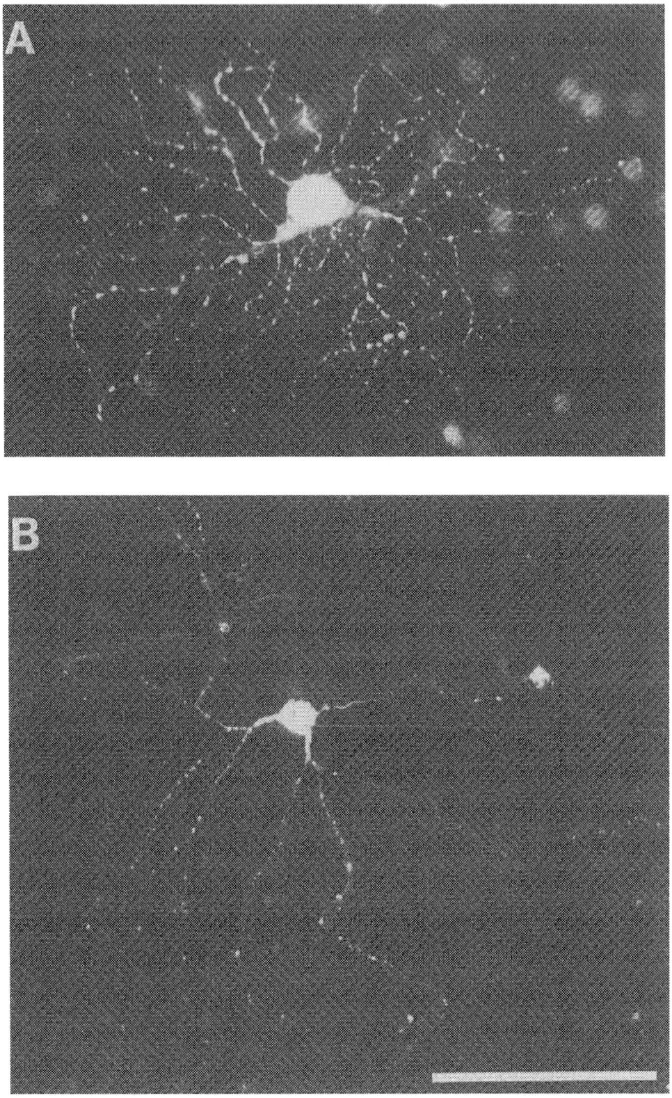

Fig. 1. Fluorescence photomicrographs showing two retrogradely labeled RGC's in the retina of a control normal rat (**A**) and in the regenerating retina of a rat 4 months post axotomy and PN grafting (**B**). The cell in **A** was identified as a type II cell. The identification of the cell in **B** has proved to be impossible. The photographs demonstrate the ability of the fluorescent dye to outline the dendritic trees of RGC's and illustrate the patterns of arborization observed in the normal and regenerated retina. Bar = 100 μm

these three types differ in soma size, number of primary dendrites, and dendritic branching and lenght. Type I RGC's are large cells with extensive dendritic trees (Fig. 2A), type II RGC's have a small to medium-sized perikarya and small, highly branched, almost bushy (Figs. 1A and 2B) dendrites, while those in type III have small to medium-sized somata with large dendrites that branch less frequently than those of types I and II (Fig. 2C).

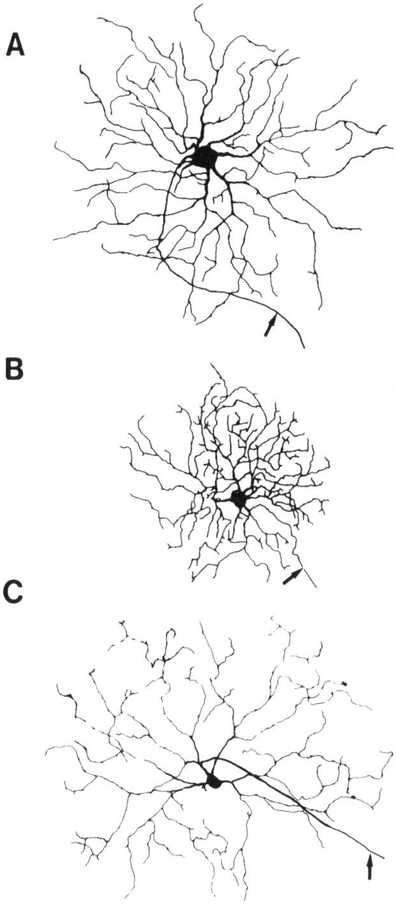

Fig. 2. Ink drawings of representative retrogradely filled type I (**A**), type II (**B**) and type III (**C**) cells in the normal rat retina. Note the differences in the dendritic organization between the three types of cells. Bar = 100 μm

Changes in the dendritic arborization of the axotomized regenerating RGC's made it impossible to discriminate reliably the incidence of axonal regeneration among the three types of RGC's in the experimental animals. Although these three cell types could not be discerned, the mean diameter of RGC somata in the population of DiI-labeled neurons regenerating axons into the grafts tended to be greater than that of the average population of controls (types I+II+III pooled together, $p < 0.01$ in the t-test) or the mean diameter of RGC's belonging to each of the three normal types of control retinal neurons (Table 1). This finding could be interpreted as an indication that either the larger RGC's (type I) were the main source of regenerated axons or that under the present experimental conditions DiI labels the larger RGC's predominantly. Although neither of these possibilities can be fully excluded, it is of interest that previous studies using HRP or fluorescent dyes other than DiI (So and Aguayo 1985; Vidal-Sanz et al. 1987; Villegas-Perez et al. 1988) have shown similar increases in mean soma size in larger populations of regenerating RGC's which included cells of all sizes and locations within the retina. It is conceivable that, as in the retinas of fish and am-

Table 1. Perikaryal and dendritic diameters and branching frequencies of normal and regenerated retinal ganglion cells (mean ± SD)

RGC type	Perikarya [μm]	Dendritic trees [μm]	Branching frequency
Normal retina			
I (n=38) range:	21.9 ± 3.4 14.6 − 32.1	318 ± 55	60.8 ± 19.6
II (n=162) range:	16.7 ± 2.5 11.2 − 24.7	187 ± 40	88.1 ± 19.7
III (n=35) range:	15.9 ± 2.5 10.0 − 22.0	299 ± 63	40.9 ± 10.6
I + II + III (n=235) range:	17.3 ± 3.3 10.0 − 32.1	193 ± 38	69.9 ± 26.6
Grafted retina			
1 month (n=41) [all types] range:	25.3 ± 5.0 17.0 − 41.0	252 ± 42	14.9 ± 5.8
2 months (n=63) [all types] range:	26.6 ± 5.1 15.6 − 36.8	266 ± 52	14.2 ± 5.2
4 months (n=45) [all types] range:	25.1 ± 4.8 15.3 − 34.0	262 ± 56	14.9 ± 6.1

phibia where regeneration is the rule, the enlargement of RGC somata in the rat is part of the regenerative response to injury (Aguayo 1985).

The most striking and novel observation made in these studies was the presence and persistence of marked alterations in the branching patterns of the RGC dendrites after axotomy, during axonal regeneration, and, finally, at a time when most axons can be assumed to have reached the end of the PN grafts. In comparison to the dendrites of the control RGC's of type I and II, the axotomized regenerating RGC's never showed the complex arbors that normally characterize these cells (Figs. 2 and 3). The bifurcations of third, fourth, and fifth order, easily recognized in normal RGC's (Fig. 2), were either absent or very few in the experimental animals (Figs. 1 B and 3). Measurements of the branching frequency of dendrites documented a nearly fivefold reduction in the overall incidence of branches along indvidual dendrites arising from RGC's with axons regenerating into the grafts. Assuming that some of these cells are type I RGC's, the loss of branches would approximate fourfold, while for types II and III these figures would near six- and threefold, respectively (Table 1).

In spite of the marked reduction in dendritic branches observed in these studies, the diameter of all dendritic trees from our experimental animals was greater than the median of all control retinas grouped together (Table 1). However, because the marked changes in dendrites seen in the PN grafted retinas made it impossible to relate dendritic field sizes to specific RGC types, the significance of this finding remains unclear. One of many possible explanations is that the poorly branched dendrites retain and perhaps extend as part of a compensatory response to the loss of many of their surrounding RGC neighbors (Perry and Linden 1982), a loss always observed after ON transection near the eye (Berry et al. 1986; Aguayo et al. 1986; Villegas-Perez et al. 1988).

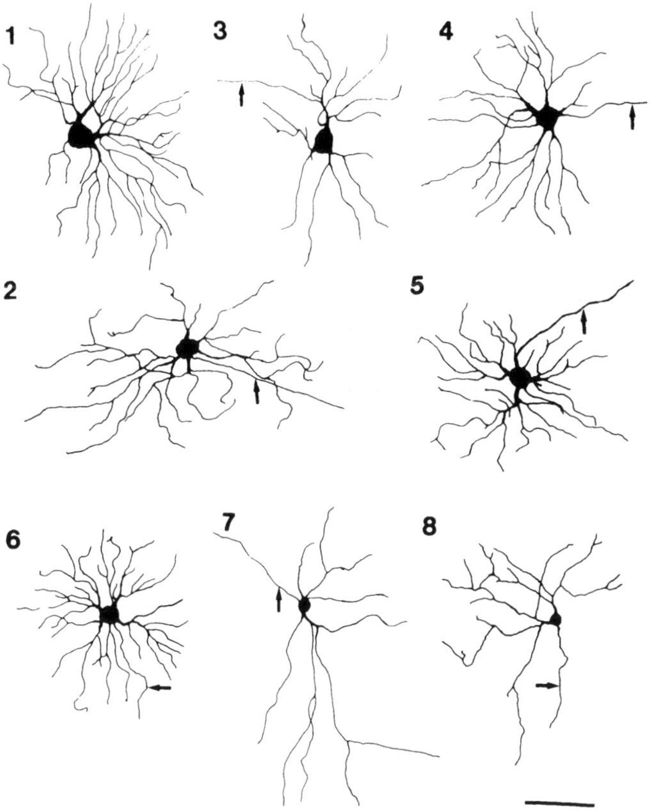

Fig. 3. Ink drawings of fluorescent images demonstrating the morphology of regenerated RGC's 1 month (**1, 2**), 2 months (**3, 4**) and 4 months (**5, 6, 7, 8** cells) after ON transection and grafting of a PN segment. Note the simplified morphology of the dendritic trees in comparison with the normal RGC's shown in Fig. 2. *Arrows* indicate the axons. Bar = 100 μm

4 Discussion

The interruption of axons in the peripheral and central nervous system of mammals initiates a series of events that affect neurons and their non-neuronal environment. In addition to the well-known anterograde changes associated with Wallerian degeneration, axotomy can lead to alterations in the proximal stump and perikaryon that often culminate in neuronal death. Neurons surviving axotomy may shed their synaptic inputs and undergo marked changes in their dendritic arbors (Mendell 1984). While it has been determined in certain axotomized peripheral neurons that some of the retrograde alterations of function and structure are only transient and reversed with axonal regrowth and the re-establishment of terminal contacts with target tissues or the administration of specific trophic molecules (Nja and Purves 1978) little is known of the extent and evolution of retrograde changes undergone by axotomized neurons in the adult mammalian CNS whose axons are induced or permitted to regenerate because, characteristically, axons do not regrow after injury in the CNS. However, recent experiments in which PN grafts were inserted into the retina or optic nerve of adult rodents have demonstrated that axotomized RGC's can successfully regenerate long axons when their central environment is substituted by that of the PNS (Aguayo 1985). It has also become apparent that, in spite of the extensive loss of RGC's that follows the transection of the optic nerve within the orbit of these rats, there is no major associated loss of cells among the various classes of neurons that constitute the intrinsic circuity of the retina. Moreover, when RGC responses to light sought in axons teased from grafts implanted in the retina (Keirstead et al. 1985), some of the RGC's showed apparently normal responses. These findings suggest that some of the RGC's regenerating axons into the PN grafts may have retained or regained a complement of synaptic inputs from cells within the retina to make possible these responses to standard forms of light stimuli. However, it was also found in these animals that only few of many viable axons capable of responding to electrical stimulation in the graft responded normally to light. The present demonstration of marked and persistent changes in the RGC dendrites following axotomy and during axonal regeneration into the PN grafts suggests a possible explanation for these functional alterations. Although PN grafts are also capable of enhancing the survival of axotomized RGC's in PN-grafted rat retinas, it wold appear from these studies that interactions with components of these grafts do not result in a full correction of the dendritic abnormalities. Intraretinal diffusion of neuroanatomic tracers and other technical difficulties determined by the shortness of the optic nerve stump of axotomized but non-PN-grafted retinas have limited additional studies of dendritic changes exclusively associated with injury. We do not know, therefore, if the PN grafts have had any effect on the dendritic alterations generated by axotomy. Although the reductions in dendritic branching remained unchanged during the time period studied, it also remains to be determined if the reestablishment of synaptic connections with tectal neurons will lead to the restoration of normal dendritic arbors. Future studies of these effects may become possible by the recent finding that synapses can be made by regenerating RGC axons that are guided back to the tectum across a PN "bridge" linking the eye and the superior colliculus (Vidal-Sanz et al. 1987).

Summary

A proportion of retinal ganglion cells (RGC's) in the adult rat regrow paradoxically lengthy axons into autologous transplanted peripheral nerve grafts. The aim of the present work was to investigate the organization of the dendritic trees among the population of cells which survived the axotomy and regrew their axons. This objective was tackled by means of the strongly fluorescent carbocyanine dye DiI.

Application of DiI into the optic nerve (ON) of normal control rats or into the PN segments in experimental animals resulted in an uptake of the dye from the RGC axons, in its retrograde transport into the perikarya of RGC's and from there to its orthograde transport into the dendritic arbors. Morphologically, the branching pattern of the regenerating RGC dendritic trees became uniformly simplified during the first month and kept this morphology until 4 months post surgery. The identification and classification of RGC types in accordance with morphological criteria derived from the normal rat retina, proved difficult due to the uniform appearance of the dendrites. The dendritic diameters of regenerated RGC's were 252 ± 42 μm, 266 ± 52 μm and 262 ± 56 μm at the post surgery periods of 1, 2 and 4 months and appeared larger than that of the small type II RGC's (187 ± 70 μm) and smaller than that of type I (318 ± 55 μm) and type III (299 ± 63 μm) cells. The branching frequencies of dendrites whose cell bodies were axotomized appeared to decrease dramatically (by about factor of 4.7) during the first weeks after axotomy. The attenuation of the dendritic branches remained stable over the period of axonal growth within the PN segments. The data suggest that axotomy of RGC's mediates regressive alterations in the organization of their dendritic trees. These changes are not restored during the period of axonal regrowth.

Acknowledgements. The authors thank Dr. Vidal-Sanz for introduction into his experimental model and Dr. S. Chang for critical reading of the manuscript. The Medical Research Council, the Multiple Sclerosis Society, and the Muscular Dystrophy Association of Canada provided research grants to A. J. A., S. T. was supported by the Max-Planck Gesellschaft.

References

Aguayo AJ (1985) Axonal regeneration from injured neurons in the adult mammalian nervous system. In: Cotman CW (ed) Synaptic Plasticity. Guilford, New York, pp 457–484

Aguayo AJ, Benfey M, David S (1983) A potential for axonal regeneration in neurons of the adult mammalian nervous system. In: Haber B, Perez-Polo JR, Hashim GA, Stella AMG (eds) Nervous system regeneration. Alan R Liss, New York, pp 327–340

Aguayo AJ, Vidal-Sanz M, Villegas-Perez MP, Keirstead SA, Raminsky M, Bray GM (1986) Axonal regrowth and connectivity from neurons in the adult rat retina. In: Agardh E, Ehinger B (eds) Retinal signal systems, degenerations and transplants. Elsevier, Amsterdam, pp 257–270

Berry M, Rees L, Sievers J (1986) Unequivocal regeneration of rat optic nerve axons into sciatic nerve isografts. In: Das GD, Wallace RB (eds) Neural transplantation and regeneration. Springer, Berlin Heidelberg New York

Fukuda Y (1977) A three-group classification of rat retinal ganglion cells: histological and physiological studies. Brain Res 119:327–344

Honig M, Hume RI (1986) Fluorescent carbocyanine dyes allow living neurons of identified origin to be studied in long-term cultures. J Cell Biol 103:171–187

James GR (1933) Degeneration of ganglion cell following axonal injury. Arch Ophthal 9:338–343

Jones WH, Thomas DB (1962) Changes in the dendritic organization of neurons in the cerebral cortex following deafferentation. J Anat 96:375–381

Keirstead SA, Vidal-Sanz M, Rasminsky M, Aguayo AJ, Levesque M, So K-F (1985) Responses to light of retinal neurons regenerating axons into peripheral nerve grafts in the rat. Brain Res 359:402–406

Leoz Ortín G, Arcaute LR (1914) Procesos regenerativos del nervo óptico y retina con ocasión de ijertos nerviosos. Trab del Lab Invest Biol 11:239–254

Mendell LM (1984) Modifiability of spinal synapses. Physiol Rev 64:260–323

Misantone LJ, Gershenbaum M, Murray M (1984) Viability of retinal ganglion cells after optic nerve crush in adult rats. J Neurocytol 13:449–465

Nja A, Purves D (1978) The effects of nerve growth factor and its antiserum on synapses in the superior cervical ganglion of the guinea pig. J Physiol 277:53–75

Perry VH (1979) The ganglion cell layer of the retina of the rat: a Golgi study. Proc R Soc Lond Biol Sci 204:363–375
Perry VH, Linden R (1982) Evidence for dendritic competition in the developing retina. Nature 297:683–685
Politis MJ, Spencer PS (1986) Regeneration of rat optic axons into peripheral nerve grafts. Exp Neurol 91:52–59
Ramón y Cajal S (1928) Degeneration and regeneration of the nervous system (May RM, Trans.) Oxford University Press, London New York
Richardson PM, Issa VMK, Shemie S (1982) Regeneration and retrograde degeneration of axons in the rat optic nerve. J Neurocytol 11:949–966
Sievers J, Hausmann B, Unsicker K, Berry M (1987) Fibroblast growth factors promote the survival of adult rat retinal ganglion cells after transection of the optic nerve. Neurosci Lett 76:157–162
So KF, Aguayo AJ (1985) Lengthy regrowth of cut axons from ganglion cells after peripheral nerve transplantation into the retina of adult rats. Brain Res 328:439–454
Standler NA, Bernstein JJ (1982) Degeneration and regeneration of motoneuron dendrites after ventral root crush: computer reconstruction of dendritic fields. Exp Neurol 75:600–615
Summer BEH, Watson WE (1971) Retraction and expansion of the dendritic tree of motor neurons of adult rats induced in vivo. Nature 233:273–275
Thanos S, Bonhoeffer F (1987) Axonal arborization in the developing chick retinotectal system. J Comp Neurol 261:155–164
Trecarten MJ, Villegas-Perez MP, Vidal-Sanz M, Thanos S, Aguayo AJ (1986) Growth of retinal axons along peripheral nerve system grafts inserted into the retina of adult rats. Neurosci Abstr 12:701
Vidal-Sanz M, Bray GM, Villegas-Perez M, Thanos S, Aguayo AJ (1987) Axonal regeneration and synapse formation in the superior colliculus by retinal ganglion cells in the adult rat. J Neurosci 7(9):2894–2909
Villegas-Perez M, Vidal-Sanz M, Bray GM, Aguayo AJ (1988) Influences of peripheral nerve grafts on the survival and regrowth of axotomized retinal ganglion cells in adult rats. J Neurosci 8(2):265–280
Watson WE (1974) Cellular responses to axotomy and to related procedures. Br Med Bull 30(2):112–115

Compensatory Sprouting of Retinofugal Axon Trees After Early Tectal Lesions: New Evidence for the Principle of "Conservation of Total Axonal Arborizations"

B. A. Sabel [1], M. Donicht [1], and G. E. Schneider [2]

[1] Institute of Medical Psychology, University of Munich School of Medicine, Goethestr. 31, D-8000 Munich 2, FR Germany
[2] Department of Brain and Cognitive Sciences, E25-634, Massachusetts Institute of Technology, Cambridge, MA 02139, USA

1 Introduction

The formation of axonal connections in the developing mammalian brain is determined by many factors, such as those affecting velocity and direction of growth, timing, interactions among growing axons, etc. Knowledge of these factors is of great importance not only for our comprehension of normal developmental processes but also for our understanding of the mechanisms which govern postlesion neuroplasticity.

The fact that damage occuring in the early stages of brain development can lead to abnormal neuronal projections provides a valuable approach to uncovering some of the factors underlying axonal growth. This aberrant axonal growth may lead to anomalies in the location or density of terminals. When drastic redirection of axon trajectories are not involved, the anomalous growth has been referred to as "compensatory sprouting", analogous to the collateral axon sprouting long known to occur in the peripheral nervous system (Edds 1953; Wedell et al. 1941). Although collateral sprouting has often been described in adult mammalian brain after injury (e.g., Cotman et al. 1973), it appears to be more vigerous when lesions are made early in life (Devor and Schneider 1975; Perry und Cowey 1979, 1982; So and Schneider 1978 a).

For several important reasons, including easy accessibility and well-studied neuronal connectivity, the visual system of murid and cricetid rodents provides a convenient model for study of abnormal axonal growth. In rats and hamsters, for example, aberrant retinothalamic and retinotectal projections occur after unilateral tectal or neocortical lesions (Baisinger et al. 1977; Schneider 1973, 1979; Schneider and Nauta 1969). In fact, under the appropriate conditions, optic tract axons invade also non-visual areas such as somatosensory or auditory thalamic nuclei (Frost 1982, 1984; Kalil and Schneider 1975; Schneider 1973).

The visual system of the Syrien hamster is particularly well suited for developmental studies. The hamster brain is very immature at birth (Shimada and Langmann 1970; Frost and Schneider 1976) and considerable neuroanatomical plasticity has been observed following neonatal surgical manipulations (Frost and Schneider 1979, Schneider 1973, 1979). Unilateral eye-removal, for example, leads to an abnormally dense ipsilateral retinal projection in most parts of the superior colliculus (Woo et al. 1985) and the dorsal nucleus of the lateral geniculate body (So et al. 1984). An abnormally recrossing projection has also been seen

H. Flohr (Ed.)
Post-Lesion Neural Plasticity

after unilateral superior colliculus ablations (Schneider 1973, So and Schneider 1978 a) which is even more prominent after removal of the eye ipsilateral to the tectal lesion (Schneider 1973; Finlay et al. 1979; So and Schneider 1978 a).

Based on the results of these and similar experiments, Schneider (1973; also Devor and Schneider 1975) presented a model that describes the behavior of axons after brain-injury in terms of two major factors, one extrinsic and one intrinsic to the cells whose axons show anomalous growth. The extrinsic factor was described in terms of competition for terminal space. In arguing for the intrinsic factor, Schneider observed that, similar to plants and trees, "axon trees" will respond to the removal of upper (distal) branches with compensatory growth of lower (proximal) branches. Schneider (1973) called this the "pruning effect". In one experiment, he pruned retinotectal fibers by a unilateral lesion of the superficial layers of the superior colliculus (SC) in newborn hamsters. After killing the animals at the age of 3 months, fields of abnormally dense termination were found in the ventral nucleus of the lateral geniculate body (LGv) and in the lateral posterior nucleus of the thalamus (LP). Based on this observation that the projection to LP was inversely related to the quantity of midbrain termination "the principle of conservation of total axonal arborization" was postulated (Schneider 1973; Devor and Schneider 1975). This concept implies that axons or axon

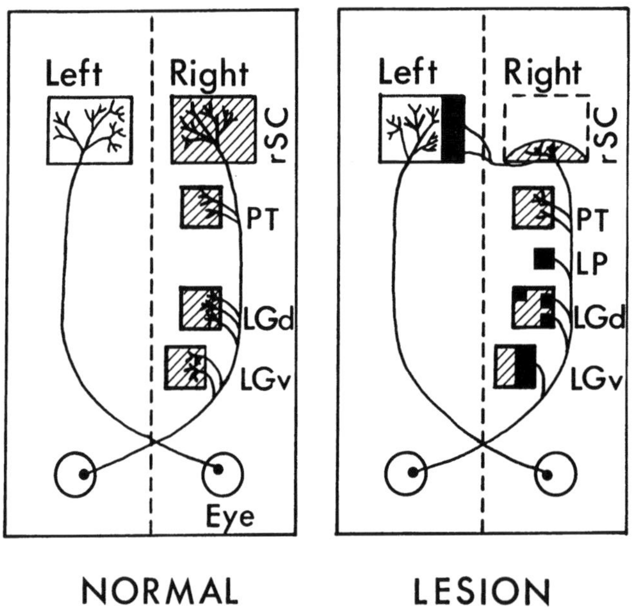

Fig. 1. Schematic representation of the hamster visual system with and without a unilateral tectal lesion. *Left panel:* in the normal hamster, retinofugal axons of the left eye cross the midline and innervate various structures in the right diencephalon and midbrain. They are: the superficial layers of the left superior colliculus (*SC*), pretectal nuclei (*PT*) as well as the ventral and dorsal nuclei of the lateral geniculate body of the thalamus (*LGd and LGv*). *Right panel:* when the right tectum is damaged on the day after birth, retinofugal axons grow into the wrong, left tectum (the re-crossing pathway), and, in addition, abnormally dense terminal fields can be found in LGv and LGd as well in the previously noninnervated lateral posterior nucleus of the thalamus (*LP*)

bundles have the tendency, presumably genetically determined, to conserve a fixed number of terminals, i.e., the loss of distal branches is compensated for by increased proximal growth (Schneider 1973; Devor and Schneider 1975).

Observations similar to those of Schneider's for the visual system and Devor's for the olfactory system were made in other brain systems as well, such as the serotoninergic and noradrenergic systems. Jonsson et al. (1978) described that application of a neurotoxin at birth produces marked and permanent 5-hydroxytryptamine (5-HT) denervation in the cerebral cortex and spinal cord which was followed by a hyperinnervation by 5-HT neurons in proximal regions. They interpreted these results as compatible with the view of "conservation of arborization" (see also Sachs et al. 1974). Unfortunately, such studies made use of neurochemical assays of local transmitter content and the possibility can not be ruled out that the increased transmitter levels may have been due to pile-up of the transmitter in nerve terminals following the lesion (Ungerstedt 1974).

Although these data are consistent with the "principle of conservation of arborization", an unequivocal demonstration of "true compensation" by the optic tract axons is still outstanding, because in all studies the increasing proximal projection was investigated only for a few selected brain areas. Since it cannot be ruled out that the increase will lead to a reduction of projections in other areas, a quantitative analysis of *all* principle terminal areas of the pruned pathway is required in order to reach a firm conclusion. Therefore, we have now studied hamsters with lesions similar to those used by Schneider (1973) and have determined the extent of compensatory sprouting in all major structures innervated by the optic tract (see Fig. 1). A more detailed report of the current findings has been presented elsewhere (Sabel and Schneider 1988 b).

2 The Quantification of Arborization

On the day after birth (postnatal day 1, P1), Syrian hamster pups received a right superior colliculus lesion with the head of a heated steel pin applied to the cartilagenous skull. By varying the heat and duration of contact, the size of the lesion was varied to obtain groups of different lesion size (small and large). In order to trace the central projections of the left retina to the diencephalon and the tectum, the left eye was removed on P36, and the animals were perfused 3 days later.

We estimated the extent of axonal arborization in the different projection areas of the optic tract with the use of a tract-tracing method (Fink and Heimer 1967) based on a technique originally developed by Nauta (1950). This method allows a semi-quantitative analysis of target structures by estimating areas of varying silver grain densities. The estimations were based on the assumption that areas with higher grain density would represent the presence of more axonal terminals than areas with lower densities. The size of these areas was measured and the volumes of the abnormal terminal fields were subsequently calculated according to the formula: $V = Am/m \times S \times S_{th}$, where V is volume, A is the area of the magnified region summed for all sections, m is the magnification factor, S is the section spacing (i.e. 5), and S_{th} is the section thickness (30 or 40 μm). For control purposes we also analyzed the terminal fields in seven otherwise intact animals (normal cases) of the same age.

3 Compensatory Sprouting in the Diencephalon and Midbrain

Figure 2 shows coronal sections through the midbrain of two representative cases, one with a small and one with a large superior-colliculus lesion. Generally, the lesion extended into the deep layers of the right superior colliculus with some, although very little, sparing of lateral portions of superficial SC in cases of smaller lesion. Unintentional damage to the left medial SC can also be seen in most cases. As for the basic anatomy of the retinofugal projection our findings are in agreement with previous observations. Retinofugal axons were found in the medial aspect of the left tectum, often with clear axon bundles crossing the midline and traveling near the surface of the midbrain (the "recrossing bundle", see Fig. 1). In the lateral posterior nucleus of the thalamus (LP) the previously known abnormal projection was observed as well as the abnormally dense innervation of the ventral nucleus of the lateral geniculate body (LGv). In addition, areas of dense innervation were seen in the dorsal nucleus of the lateral geniculate body (LGd), forming areas of dense innervation which assume, in coronal sections, the shape of "islands" or "patches". Figure 3 shows an example of these patches in one representative case and a 3-D reconstruction of the patches. Our microscopic analysis of normal hamsters (no lesions) generally did not reveal any such patterns, and terminal distribution was found to be fairly uniform, with a ventral to dorsal density gradient. It is interesting to note, however, that in two out of seven normal cases, a patch-like structure was seen in two brain sections.

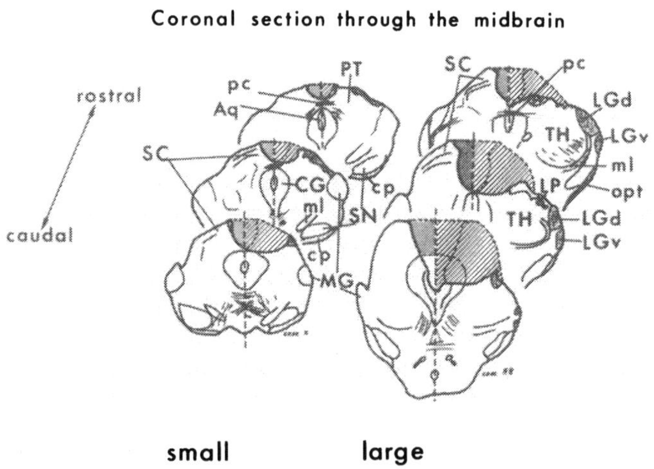

Fig. 2. Coronal midbrain sections from two cases which received unilateral tectal lesions on postnatal day 1 (P1) and which survived until P39. Each of these cases is representative for their respective group (i.e., cases closest to the group mean), receiving either a small or a large early tectal lesion. Left tectal damage is indicated by the *gray area* and right tectal damage by the *striped pattern*. The *interrupted line* shows where the right tectum would have been without any lesion, estimated from a mirror-symmetrical superimposition of the left side onto the right. Major landmarks, including some of the terminal areas are: Pretectum (*PT*), superior colliculus (*SC*), central gray (*CG*), substantia nigra (*SN*), cerebral peduncle (*CP*), medial geniculate nucleus of the thalamus (*MG*), aqueduct (*Aq*), medial lemniscus (*ml*), optic tract (*opt*), posterior commissure (*pc*), and thalamus (*TH*). (From Sabel and Schneider 1988 b)

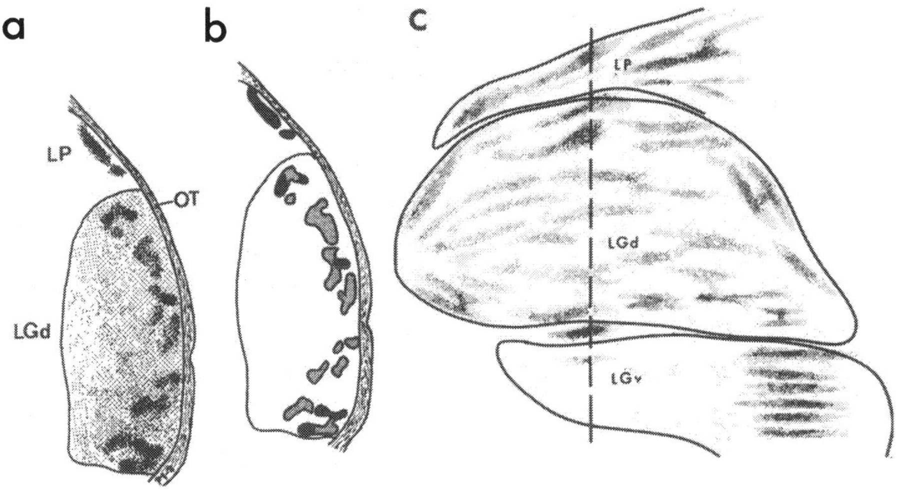

Fig. 3. When large lesions of the tectum are made early in life, areas of dense innervation (patches) are found in the diencephalon. **a** is an artist's impression of compensatory sprouting as seen through the light microscope. Note that the individual dots do not represent individual silver grains. There is a relative absence of (degenerating) terminals in the ipsilateral projection field located more medially in which no patches are found. **b** In order to quantify these areas of abnormal growth, the patches were traced on paper and their grain density was rated to be moderate (*gray area*) or heavy (*black area*). **c** A 3-D reconstruction reveals a sagittal view onto the diencephalon. The *line* shows the position of the coronal section seen in **a**. Note that the areas of dense innervation seen as patches in coronal sections derive, in reality, from columns with a rostral to caudal orientation

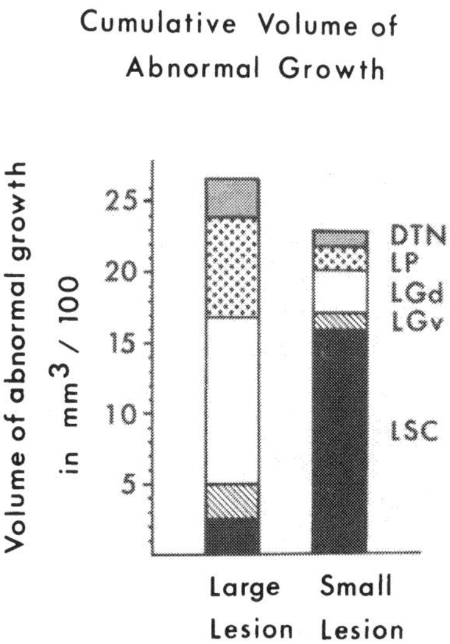

Fig. 4. The size of the abnormal terminal fields (expressed as absolute volume) indicates the extent of innervation of optic tract target structures. When the cumulative growth is compared between both groups, the total size of abnormal terminal fields is about the same for the small or large lesions group. However, most compensatory sprouting in the large lesion group occured in LGd, while in the small lesion group most sprouting was found in the left tectum. (From Sabel and Schneider 1988 b)

In hamsters with large tectal lesions the terminal field of the left SC was only 20% that of animals with small lesions. This *reduction* in the size of the abnormal left tectum innervation in hamsters with large tectal lesions coincided with a corresponding *increase* in the abnormal growth patterns in the right diencephalic and midbrain structures.

When the sum of all dense areas of abnormal growth is compared between the large and small lesion group, a surprising similarity with only 10% difference is found (Fig. 4). The evidence indicates that both groups had a comparable number of optic tract terminals; after small lesions the majority of terminals are located in the left SC, after large lesions the most extensive areas of abnormal growth are found in LGd.

4 The Degeneration Stain as an Indicator of Axons and Axon Terminals

Before discussing the findings in detail, we shall first consider the principle methodological issues and interpretative problems related to the modified Nauta silver impregnation technique. This technique allows the visualization of degenerating axons and axon terminals by silver grains after removal of the eye (Fink and Heimer 1967). These silver grains can easily be visualized by light microscopy, but we do not know the precise relationship between the number of silver grains and the number of real terminal boutons. However, the existing evidence from electron microscopic studies supports the assumption that the number of silver grains is at least proportional to the real number of axon branches and their terminals (Heimer and Peters 1968). Heimer and Peters concluded "that the Fink-Heimer silver impregnation technique is preferential in its impregnation of degenerating axons and axon terminals" (p. 346).

Furthermore, in our previous study (Sabel and Schneider 1988a) the estimated number of silver grains in the optic tectum roughly corresponded to the estimates of terminal number derived from single axon reconstructions (Sachs and Schneider 1984). Therefore, the left eye was removed and the extent of connectivity was evaluated by estimating areas of varying densities of degenerating terminals, assuming that areas with higher grain density would represent the presence of more axonal terminals than areas with lower density. One of the advantages of the modified Nauta stain is that a large number of cases and many structures can be analyzed in a reasonable period of time. This cannot be achieved as easily with electronmicroscopic methods. Furthermore, identical labelling (or nearly so) can be obtained across all the animals, a uniformity which cannot so easily be obtained with other tract-labelling techniques.

5 The "Principle of Conservation of Total Axonal Arborization"

After creating early, unilateral lesions of the superior colliculus, we have analyzed the extent of innervation in all major areas of the optic tract. We have found that the sum of anomalous growth truly compensates for the distal loss and that the sum of arborizations is roughly the same, independent of the lesion size. We also found a significant negative correlation between the size of the recrossing pathway and the sum of abnormal growth in all major proximal structures (LP, LGd, LGv, PT, $r = -0.77$, $p < 0.005$).

Based on these findings we conclude that axons have a tendency to support a fixed number of terminals. This is in agreement with earlier interpretations by Schneider (1973) who was the first to propose the "principle of conservation of total axonal arborization" after having observed abnormal growth in a few selected structures (LP and LGv). However, such an incomplete analysis was subject to the criticism that if innervation is lost in other proximal areas (such as LGd), the sprouting would not be compensatory in the true meaning of the word. Since a complete analysis of all major optic tract target structures is a necessary requirement to demonstrate clearly the "conservation principle", our study provides the first unequivocal evidence in favor of it. Comparably strong evidence for the same principle was obtained by Devor (1976) for the lateral olfactory tract axons; Devor's conclusion was based on a "compensatory stunting" phenomenon.

6 The Source of Abnormal Innervation

Given the fact that true compensation occurs, the question arises as to the source(s) of the abnormal sprouts. The following possibilities are conceivable: (a) the patches of hyperinnervation result from the maintainance of transient, exuberant projections, (b) areas of dense innervation are the consequence of rerouting, i.e., a re-direction of growing axons, or (c) a true "pruning effect", that is a sprouting of collaterals from axons which innervate simultaneously other (especially distal) structures as well.

Transient Projections. Exuberant, transient projections are a rather common observation in early development (Innocenti 1981, Schneider et al. 1987). Could our patches be due to the maintainance of such transient projections rather than the consequence of abnormal growth or compensatory sprouting? Although very small-sized patches of similar shape and location are seen in normal development (Sabel and Schneider 1988 b), the normal, transient projections are not nearly as large as the patches after early tectal lesions. This observation leads us to suspect that the areas of abnormal innervation arise from the locations of such transient projections.

Re-Routing. The alternative possibility is that axons which are deprived of their target simply grow into other areas. Thus, on the level of the single axon no compensatory, proximal sprouting of collateral branches would take place, but rather a re-direction of single, growing axons without collateral branches. We think that this possibility is also rather unlikely. First of all, most axons of the hamster optic tract have already elongated at the time of the early tectal lesion (So and Schneider 1978 a, b; Frost et al. 1979) and are therefore, in fact, cut distally. Secondly, most retinofugal axons are known to send collateral branches to the various diencephalic areas, in addition to the main branch continuing into the midbrain.

Compensatory Collateral Sprouting. Although it would certainly be desirable to investigate the pruning effect on the single-axon level (for example with single-axon tracing techniques), the arguments raised above leave us with the conclusion that the patches are formed by sprouting branches of retinofugal axon trees which send collaterals simultaneously to a number of structures (LP, LGv, LGd, PT, rSC).

7 Is the Pruning Effect Due to an Expression of Extrinsic or Intrinsic Factors?

Since the superior colliculus sends reciprocal fibers to the same areas which are innervated by the optic tract (LP, LGd, LGv), the early tectal lesion results in destruction of major inputs and thus produces vacant synaptic space in the structures where we have seen the patches. It is conceivable that the reduced competition might favor the survival of the normally transient retinal projections or their additional outgrowth (Perry und Cowey 1982; Baisinger et al. 1977; Woo et al. 1985; Frost 1984; Crain and Hall 1980). Thus, "conservation of arborizations" might not be due to pruning (an intrinsic factor) but due to additional deafferentation in hamsters with large lesions (an extrinsic factor). However, the critical projections to LP and LGv arise from the superficial gray layer and the optic fiber layer which were destroyed both in the small and the large lesion group, with only minor sparing. Also, the incidental damage to the left tectum was, in terms of absolute injury, relatively comparable (although significantly different) for both groups (see Fig. 2). It therefore seems unlikely that large differences exist between the two groups in the amount of diencephalic deafferentation and that the quantity of abnormally dense innervation is controlled by an extrinsic factor. Rather, the pruning effect appears to be due to an intrinsic program (the "growth vigor", Schneider 1981) which ensures the conservation of the number of terminal arborizations of developing axons after injury.

8 The Principle of Conservation in Other Systems

Aberrant growth has been demonstrated in many parts of the brain (visual system, olfactory system, somatosensory system, motor system, etc.). It is easily accomplished if lesions are made during development (Guillery 1972; Kalil 1972, 1973; Hickey 1975; Robson 1981; Chow et al. 1973; Jen et al. 1984; Jen and Lund 1981; Manford et al. 1984; So et al. 1984; Woo et al. 1985), but axonal sprouting in adult animals is generally less vigorous (Sakaguchi et al. 1984; Stenevi et al. 1972). In the hippocampus, where collateral sprouting has been clearly demonstrated in adult animals, the sprouting appears to take place without branch retraction elsewhere (Lund and Lund 1971; Lynch et al. 1973). This implies that the "conservation principle" is not applicable in identical fashion to different systems. The reason for these differences are not known at this point, but their elucidation may help us to better understand the conservation principle.

The conservation principle may also be applicable to plasticity events in dendritic trees. In aged human brains (nondemented cases at the average age of 79.6

years), for example, Buell and Coleman (1979) found dendritic trees, randomly chosen from layer-II pyramidal cells of the parahippocampal gyrus, which were larger and more branched than in younger cases (average age 51.2). In senile dementia (average age of 76.0 years), however, dendritic trees were less extensive than those of adult brains, largely because their terminal segments were fewer and shorter. Although Buell and Coleman postulated that during aging certain neurons die and regress and others survive and grow, it is also conceivable that neurons which show dendritic compensation might be those which have lost their axon tree.

Also Eysel and Böhm (1987) described dendritic tree plasticity in the case of retinal ganglion cells after lesion in the adult cat retina: the dendrites pointing towards the lesion were significantly shorter and smaller in number and diameter compared to normal ganglion cells, whereas the dendrites oriented away from the lesion were significantly thicker and longer (see also Eysel this Vol.). Thus, loss of membrane in dendrites adjacent to the lesion would lead to a compensatory increase of membrane in areas oriented away from the lesion.

9 The Total Number of Arborizations Can Be Altered

The principle of conservation implies not only that a neuron has a tendency to maintain a fixed number of arbors (which may shift progressively during development, Schneider 1981), but that there is also a maximum number of arbors that can be formed by an individual axon. In recent years, evidence has been presented in favor of the possibility that axonal sprouting after brain injury in adult animals can be stimulated by pharmacological intervention. In one line of experiments, animals with various types of brain lesions (such as transection of the nigro-striatal pathway) were treated with exogenous GM1 gangliosides (Toffano et al. 1984; Sabel et al. 1984; Wojcik et al. 1982) and it was found that GM1-treatment resulted in elevated levels of transmitter and a larger number of retrogradely labeled neurons in the partially damaged nigro-striatal system. The results were interpreted in terms of GM1 enhanced sprouting.

We have now also studied the effects of exogenous ganglioside treatment on sprouting in the hamster retinotectal pathway (Sabel and Schneider 1988 a). After creating an early tectal lesion, we have removed the right eye on P9, in effect deafferenting the left tectum by eliminating its normal input. In this way, the recrossing fibers were stimulated to sprout into more lateral parts of the left superior colliculus, and animals received simultaneous daily treatment with GM1 gangliosides. As Fig. 5 shows, the extent of normal and abnormal innervation of the right diencephalic structures was comparable for the control and the treated group. In contrast, the terminal field in the left tectum (as visualized by the Fink-Heimer technique) was significantly larger in the GM1 treated group. The total size of the terminal field in the GM1 cases was, in average, twice as large as that of controls, and the estimated number of silver grains in GM1-treated hamsters (more than 7×10^6 per tectum) was more than double than that of controls (about 3×10^6). This finding suggests the following: since there was no retraction of terminals from the diencephalic structures in the GM1 group, the enhanced sprouting may

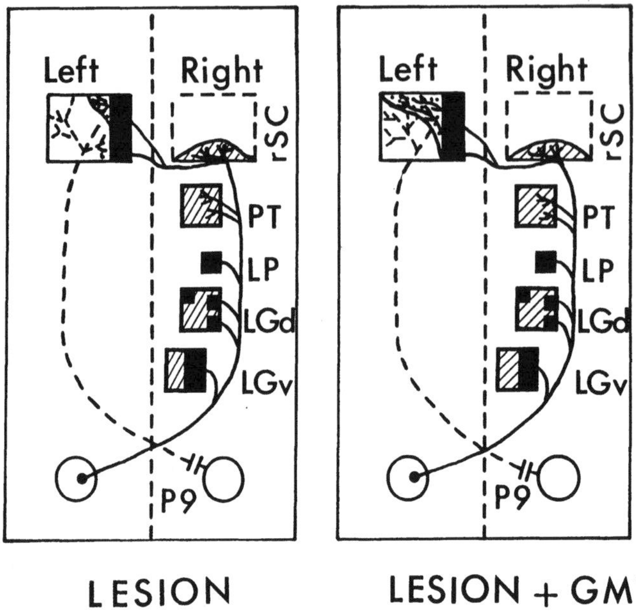

Fig. 5. *Left panel.* The axons which normally innervate the left tectum, i.e., those from the right eye, compete for terminal space with the recrossing fibers located in the medial part of the tectum (*black area*). However, if the right eye is removed (as on P9 in our study, *interrupted line*), the recrossing pathway sprouts laterally (*stippled area*), and terminals cover a larger territory. *Right panel:* if GM1-gangliosides (30 mg/kg) are administered daily for 3 weeks, commencing at the time of right eye removal, retinotectal sprouting is significantly stimulated; the total terminal field is at least twice as large as that of controls (see Sabel and Schneider 1988 a). It is interesting to note that the fields of abnormal innervation in the diencephalon are not at all affected by the treatment. They lose neither any terminals, nor do they undergo additional sprouting

be taken to indicate that the treatment increased the total number of terminals a single axon can form. It is interesting to note that if midbrain sprouting is stimulated only by neonatal right eye removal without simultaneous GM1 treatment, the size of the abnormal terminal field in the diencephalon (LP) is less (Schneider 1979).

Perhaps some products found in degenerating tissue (perhaps neurotrophic factors) are a necessary prerequisite for gangliosides to exert an effect. If so, gangliosides do not initiate a sprouting response, but rather potentiate an ongoing sprouting process. In that sense, gangliosides cannot be regarded as a neuronotrophic factor themselves, but rather as a kind of "fertilizer" of ongoing growth (see also experiments by Ferrari et al. 1983).

10 Advantages of the "Principle of Conservation of Arborization" for the Neuron and the Brain: Some Speculations

The data presented in this paper provide the best evidence to date for the validity of the "principle of conservation of total axon arborization" in the developing visual system. What are the advantage(s) of such a – presumably genetically determined – principle? How does the individual neuron or the brain profit from it?

It has long been known that neurons which fail to make synaptic contacts tend to die. One possible reason for this is that developing neurons are more dependent on target-derived neurotrophic factor for survical than are adult neurons (Hamburger and Oppenheim 1982; LaVelle and LaVelle 1984). Damage to a target would deprive neurons of trophic support, making it necessary for them to find another source of trophic factor if they are to survive. In this manner, a genetically determined intrinsic tendency to conserve the number of arborizations by compensatory sprouting would increase the probability of survival. Adult neurons survive axotomy better than developing neurons (Bleier 1969; Jonsson et al. 1984; LaVelle and LaVelle 1984), perhaps due to their greater independence from target derived trophic factor. Thus, there would be less need for adult neurons to sprout proximally. To the best of our knowledge, a pruning effect has not yet been described in adult rodents.

There is yet another possibility why the "conservation principle" may favor survival of the cell. It seems reasonable to postulate that neurons can only survive if the size of the membrane surface of perikaryon, axon tree, and dendrites is maintained within certain limits, and if the size of the membrane falls below these limits, the lack of sufficient supply of cellular nutrients may lead to cell death. We speculate that sprouting could compensate for the loss of membrane and thus maintain survival.

Assuming that survival of the axotomized, developing neuron has been accomplished, would it be reasonable to also predict that the system as a whole (the brain, the organism) would profit from the compensatory sprouting? We think that the mere survival of individual neurons or entire neuronal populations does not necessarily imply a functional benefit for the animal. The compensatory sprouts which have formed in alternative areas are perhaps good candidates to mediate behavioral recovery. But are they necessarily adaptive? We think not. If we assume, for example, that the retinofugal axons are excitatory, it is, in fact, quite likely that the abnormally dense projection fields formed after early tectal lesions in LGd might introduce noise into the spared part of the visual system, upsetting electrophysiological balance. At this point, we do not have a behavioral correlate of the abnormal innervation in LGd, but evidence deriving from other studies may provide some clues as to the behavioral impact of early lesions. (In adult brains collateral sprouting may have a more beneficial function for the organism, but again, the issue is not clear as yet, see review by Steward 1982, 1984; Ramirez and Stein 1984).

11 "Is it Better to Have Your Brain Lesion Early?"

When Hans-Lukas Teuber discussed experiments on early brain damage in baby monkeys by Margret Kennard he often introduced the issue by asking, "Is it really better to have your lesion early … that is, if you can arrange it!" (cited by Schneider 1979). Although it is quite clear that early brain lesions have generally less detrimental consequences than similar lesions in adulthood (the "Kennard principle"), it would be a deceptive oversimplification to assume that the improved behavioral recovery might be due to the enhanced compensatory sprouting or neuronal survival. Schneider (1979) has discussed several cases in the animal literature as well as human case reports that also suggest the possibility that early lesions may be responsible, at least in some cases, for behavioral abnormalities, even psychiatric disorders. It would go beyond the scope of this paper to discuss these cases in detail, and only one example of anomalous behavior will be described.

When hamsters receive early lesions of the right superior colliculus similar to the ones described in the present paper, the retinofugal pathway will form an abnormal decussation and innervate the "wrong" (left) tectum (Schneider 1973). This abnormally reorganized optic tract has been shown to mediate abnormal visual orienting responses: when a visual stimulus (such as a sunflower seed) is presented in the left visual field, the hamsters sometimes orient towards the opposite side, as if the seed was presented on the right (Schneider 1979). Whether such wrong-way orienting responses can generally be considered adaptive or malapdaptive is difficult to decide, but it probably depends on the task (Loesche and Steward 1977; Steward 1982).

Considering the converging evidence derived from neuroanatomical and neurochemical experiments with early brain lesions, it can be concluded that one of the factors which governs the growth of axon branches depends on the "principle of conservation of total axon arborization" (see also Schneider 1973; Devor and Schneider 1975). Analogous to real tree pruning, we have now shown that axon trees exhibit true compensatory sprouting of proximal branches in response to damage to the distal branches. It also appears that the maximum number of terminals an axon tree can form may be increased by providing the appropriate fertilizer (gangliosides) to the axon tree while it grows its collateral sprouts.

Acknowledgements. We thank Christina Strache for excellent histology and graphics work. Supported by NIH grants EY 00126 and EY 02621 and by the Deutsche Forschungsgemeinschaft, SFB 220/D10.

References

Baisinger J, Lund RD, Miller B (1977) Aberrant retinothalamic projections resulting from unilateral tectal lesions made in fetal and neonatal rats. Exp Neurol 54:369–382

Bleier R (1969) Retrograde transsynaptic cellular degeneration in mammilary and ventral tegmental nuclei following limbic decortication in rabbits of various ages. Brain Res 15:363–393

Buell SJ, Coleman PD (1979) Dendritic growth in the aged human brain and failure of growth in senile dementia. Science 206:854–856

Chow KL, Mathers LH, Spear PD (1973) Spreading of uncrossed retinal projection in superior colliculus of neonatally enucleated rabbits. J Comp Neurol 151:307–322

Cotman CW, Matthews DA, Taylor DA, Lynch G (1973) Synaptic rearrangement in the dentate gyrus: histochemical evidence of adjustments after lesions in immature and adult rats. Proc Natl Acad Sci (Wash) 70:3473–3477

Crain BJ, Hall WC (1980) The organization of afferents to the lateral posterior nucleus in the golden hamster after different combinations of neonatal lesions. J Comp Neurol 193:403–412

Devor M (1976) Neuroplasticity in the rearrangement of olfactory tract fibers after neonatal transections in hamsters. J Comp Neurol 166:31–48
Devor M, Schneider GE (1975) Neuroanatomical plasticity: the principle of conservation of total axonal arborization. INSERM 43:191–200
Edds MV Jr (1953) Collateral nerve regeneration. Q Rev Biol 28:260–276
Eysel UT, Böhm A (1987) Plasticity of ganglion cell dendrites at the border of photocoagulator lesions in the adult cat retina. Neuroscience, Suppl 22:5
Ferrari G, Fabris M, Gorio A (1983) Gangliosides enhance neurite outgrowth in PC 12 cells. Dev Brain Res 8:215–221
Fink RP, Heimer L (1967) Two methods for selective silver impregnation of degenerating axons and their synaptic endings in the central nervous system. Brain Res 4:369–374
Finlay BL, Wilson KG, Schneider GE (1979) Anomalous ipsilateral retinotectal projections in Syrian hamsters with early lesions: topography and functional capacity. J Comp Neurol 183:721–740
Frost DO (1982) Anomalous visual connections to somatosensory and auditory systems following brain lesions in early life. Dev Brain Res 3:627–635
Frost DO (1984) Axonal growth and target selection during development: Retinal projections to the ventrobasal complex and other "nonvisual" structures in neonatal Syrian hamsters. J Comp Neurol 230:576–592
Frost DO, Schneider GE (1976) Normal and abnormal uncrossed retinal projections in Syrian hamsters as demonstrated by Fink-Heimer and autoradiographic techniques. Neuroscience Abstr 2:812
Frost DO, So KF, Schneider GE (1979) Postnatal development of retinal projections in Syrian hamsters: a study using autoradiographic and anterograde degeneration techniques. Neuroscience 4:1649–1677
Guillery RW (1972) Experiments to determine whether retinogeniculate axons can form translaminar collateral sprouts in the dorsal lateral genicuate nucleus of the cat. J Comp Neurol 146:407–420
Hamburger V, Oppenheim R (1982) Naturally occurring neuronal death in vertebrates. Neurosci Comment 1:39–55
Heimer L, Peters A (1968) An electron microscope study of a silver stain for degeneration boutons. Brain Res 8:337–346
Hickey TL (1975) Translaminar growth of axons in the kitten dorsal lateral geniculate nucleus following removal of one eye. J Comp Neurol 161:359–382
Innocenti GM (1981) Transitory structures as substrates for developmental plasticity of the brain. In: Hof MW van, Mohn G (eds) Functional recovery from brain damage. Dev Neurosci 13:305–333
Jen LS, Lund RD (1981) Experimentally induced enlargement of the uncrossed retinotectal pathway in rats. Brain Res 211:37–57
Jen LS, So KF, Woo HH (1984) An anterograde HRP study of the retinocollicular pathways in normal hamsters and hamsters with one eye enucleated at birth. Brain Res 294:169–173
Jonsson G, Pollare T, Hallmann H, Sachs C (1978) Developmental plasticity of central serotonin neurons after 5,7-dihydroxytryptamine treatment. Ann NY Acad Sci 305:328–345
Jonsson G, Gorio A, Hallman H, Janigro D, Kojiama H, Zanoni P (1984) Effects of GM1 ganglioside on neonatally neurotoxin induced degeneration of serotonin neurons in the rat brain. Dev Brain Res 16:171–180
Kalil RE (1972) Formation of new retino-geniculate connections in kittens after removal of one eye. Anat Rec 172:339–340
Kalil RE (1973) Formation of new retino-geniculate connections in kittens: effects of age and visual experience. Anat Rec 175:353
Kalil RE, Schneider GE (1975) Abnormal synaptic connections of the optic tract in the thalamus after midbrain lesions in newborn hamsters. Brain Res 100:690–698
Kennard MA (1936) Age and other factors in motor recovery from precentral lesions in monkey. Am J Physiol 115:138–146
LaVelle A, LaVelle F (1984) Neuronal reaction to injury during development. In: Finger S, Almli R (eds) Early brain damage. Florida Academic Press, pp 3–16

Loesche J, Steward O (1977) Behavioral correlates of denervation and reinnervation of the hippocampal formation of the rat: recovery of alternation performance following unilateral entorhinal cortex lesions. Brain Res Bull 2:31–39
Lund RD, Lund JS (1971) Synaptic adjustment after deafferentation of the superior colliculus of the rat. Science 171:804–807
Lynch G, Stanfield B, Cotman CW (1973) Development differences in post-lesion axonal growth in the hippocampus. Brain Res 59:155–168
Manford M, Campbell G, Liebermann AR (1984) Postnatal development of ipsilateral retinogeniculate projections in normal albino rats and the effects of removal of one eye at birth. Anat Embryol 170:71–78
Nauta WJH (1950) Über die sogenannte Terminaldegeneration im Zentralnervensystem und ihre Darstellung durch Silberimpregnatation. Arch Neurol Psychiatr 66:353–376
Perry VH, Cowey A (1979) Changes in the retino-fugal pathways following cortical and tectal lesions in neonatal and adult rats. Exp Brain Res 35:97–108
Perry VH, Cowey A (1982) A sensitive period for ganglion cell degeneration and the formation of aberrant retino-fugal connections following tectal lesions in rats. Neuroscience 7:583–594
Ramirez JJ, Stein DG (1984) Sparing and recovery of spatial alternation performance after entorhinal cortex lesions in rats. Behav Brain Res 13:53–61
Robson JA (1981) Abnormal axonal growth in the dorsal lateral geniculate nucleus of the cat. J Comp Neurol 195:453–476
Sabel BA, Dunbar GL, Stein DG (1984) Gangliosides minimize behavioral deficits and enhance structural repair after brain injury. J Neurosci Res 12:429–443
Sabel BA, Schneider GE (1988 a) Enhanced sprouting of retinotectal fibers after early superior colliculus lesions in hamsters treated with gangliosides. Exp Brain Res (in press)
Sabel BA, Schneider GE (1988 b) The principle of "conservation of total axonal arborization": massive compensatory sprouting in the hamster subcortical visual system after early tectal lesions. Exp Brain Res (in press)
Sachs C, Pycock C, Jonsson G (1974) Altered development of central noradrenaline neurons during ontogeny by 6-hydroxydopamine. Med Biol 52:55–65
Sachs GM, Schneider GE (1984) The morphology of optic tract axons arborizing in the superior colliculus of the hamster. J Comp Neurol 230:155–167
Sakaguchi T, Shirokawa T, Nakamura S (1984) Changes in projection from locus coeruleus to lateral geniculate nucleus following ablation of visual cortex in adult rats. Brain Res 321:319–322
Schneider GE (1973) Early lesions of the superior colliculus: factors affecting the formation of abnormal retinal projections. Brain Behav Evol 8:73–109
Schneider GE (1979) Is it really better to have your brain lesion early? A revision of the "Kennard Principle". Neuropsychologia 17:557–583
Schneider GE (1981) Early lesions and abnormal neuronal connections: developmental rules can lead axons astray, with functional consequences. Trends in Neuroscience 4:187–192
Schneider GE, Jhaveri S, Davis W (1987) On the development of neuronal arbors. In: Chagas C Linden R (eds) Developmental neurobiology of mammals. Pontificiae Academiae Scientiarvm Scripta Varia 59, Vatican City, pp 31–64
Schneider GE, Nauta WJH (1969) Formation of anomalous retinal projections after removal of the optic tectum in the neonate hamster. Anat Rec 163:258
Shimada M, Langman J (1970) Cell proliferation migration and differentiation in the cerebral cortex of the golden hamster. J Comp Neurol 139:227–244
So KF, Schneider GE (1978 a) Abnormal recrossing retinotectal projections after early lesions in Syrian hamsters: age-related effects. Brain Res 147:277–295
So KF, Schneider GE (1978 b) Postnatal development of retinogeniculate projections in Syrian hamsters: an anterograde HRP study. Neuroscience Abstr 4:127
So KF, Woo HH, Jen LS (1984) The normal and abnormal postnatal development of retinogeniculate projections in golden hamsters: an anterograde horseradish peroxidase tracing study. Dev Brain Res 12:191–205
Stenevi U, Björklund A, Moore RY (1972) Growth of intact adrenergic axons in the denervated lateral geniculate body. Exp Neurol 35:290–299

Steward O (1982) Assessing the functional significance of lesion-induced neuronal plasticity. Neurobiology 23:197–254
Steward O (1984) Lesion-induced neuroplasticity and the sparing or recovery of function following early brain damage. In: Almli CR, Finger S (eds) Early brain damage, vol 1. Academic Press, London, pp 59–77
Toffano G, Agnati LF, Fuxe K, Aldinio C, Consolazione A, Valenti G, Savoini G (1984) Effect of GM1 ganglioside treatment on the recovery of dopaminergic nigro-striatal neurons after different types of lesion. Acta Physiol Scand 122:313–321
Ungerstedt U (1974) Functional dynamics of central monoamine pathways. In: Schmitt FO, Worden FG (eds) The Neurosciences. Third study program. MIT Press, Cambridge, pp 979–988
Weddell G, Gutmann L, Gutmann E (1941) The local extension of nerve fibers into denervated areas of skin. J Neurol Psychiatry 4:206–225
Wojcik M, Ulas J, Oderfeld-Nowak B (1982) The stimulating effect of ganglioside injections on the recovery of choline acetyltransferase and acetylcholinesterase activities in the hippocampus of the rat after septal lesions. Neuroscience 7:495–499
Woo HH, Jen LS, So KF (1985) The postnatal development of retinocollicular projections in normal hamsters and in hamsters following neonatal monocular enucleation: A horseradish peroxidase tracing study. Dev Brain Res 20:1–13

II CNS Development and Lesions

The Effects of Early and Late Hemidecortication on Vision and Locomotion in the Rabbit

M. W. Van Hof, W. H. De Vos-Korthals, and J. F. Hobbelen
Department of Physiology I, Erasmus Universiteit Rotterdam,
Rotterdam, The Netherlands

1 Introduction

The hemispherectomized rat and cat have been used by various investigators as a model for studying the processes underlying age-dependent recovery from brain damage.

Already in 1970, Hick and D'Amato studied motorsensory and visual behavior after hemispherectomy in newborn and mature rats and more recently Villablanca and co-workers published a series of papers on early and late hemispherectomy in cats (Villablanca et al. 1986; Burgess and Villablanca 1986; Burgess et al. 1986). This chapter is on the effects of hemispherectomy on vision and motor behavior in developing rabbits.

The main reason for studying the rabbit with respect to age-dependent recovery from hemispherectomy is that the way in which the rabbit pups are handled postnatally is quite different from that in cats and rats.

Rabbits are born after a gestation period of 31 days in a rather immature state. The eyelids are closed and motor coordination is very poor (Stuurman and van Hof 1979). Domestic rabbits make a fur-lined nest for their young. Immediately after giving birth, the doe leaves the nest and returns only once per 24 h for a nursing period of 2–4 min (Lincoln 1974).

This is in marked contrast with cats (Schneirla 1963) and rats (Grota and Adler 1969), who spend up to 80% of the first days nursing and grooming.

Newborn rabbits greatly depend on their own resources. In a series of interesting papers by Hudson and Distel (Hudson and Distel 1982, 1983, 1984; Distel and Hudson 1984, 1985) it was described that the doe behaves very passively during nursing. On the other hand, the pups are very active and find the nipples quickly by means of efficient searching, using olfactory stimuli, most likely a pheromone secreted by the skin on and around the nipples.

In other words, newborn rabbits, contrary to the cat and the rat, are relatively independent of maternal care. This makes the newborn rabbit an ideal animal for postnatal surgery and postoperative care, as was discovered already by von Gudden in the 19th century (von Gudden 1886).

H. Flohr (Ed.)
Post-Lesion Neural Plasticity

2 Unilateral Ablation of the Occipital Lobe

After a unilateral ablation of the occipital lobe, rabbits are unable to discriminate between striations of different orientation with the eye contralateral to the lesion (Van Hof and Lagers-van Haselen 1975; Stuurman and van Hof 1980; De Vos-Korthals and Van Hof 1981; Van Hof and Stuurman 1981). This phenomenon was also found in rabbits operated shortly after birth and tested at adult age (Stuurman et al. 1981). In Fig. 1 the original discrimination apparatus is shown. The presently used apparatus is essentially the same, but all manipulations are carried out by computer-controlled electropneumatic plungers.

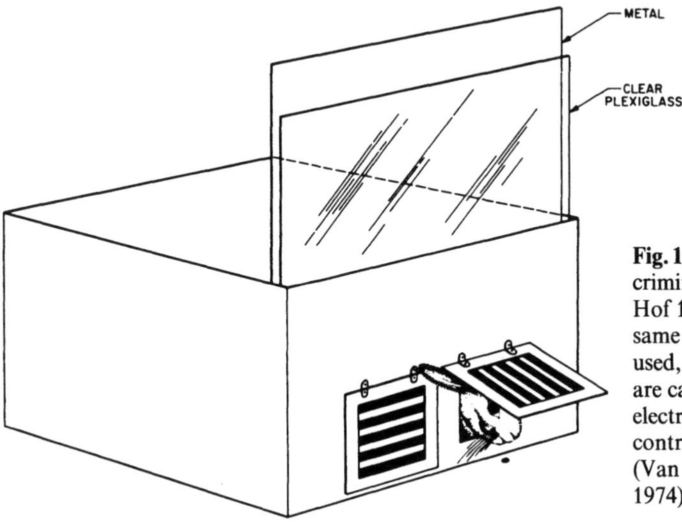

Fig. 1. Rabbit visual discrimination apparatus (Van Hof 1966). Basically the same apparatus is still being used, but all manipulations are carried out by means of electropneumatic plungers controlled by a computer. (Van der Mark and Meijer 1974)

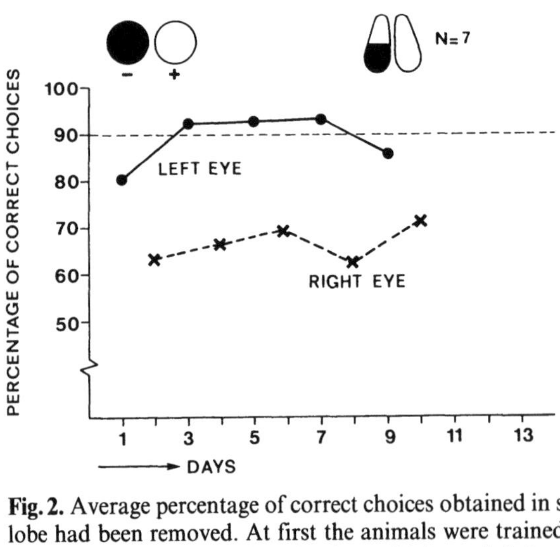

Fig. 2. Average percentage of correct choices obtained in seven rabbits in which the left occipital lobe had been removed. At first the animals were trained binocularly on a brightness discrimination task. Pushing against the door which carried the white target was food-rewarded. After the 90% correct level had been reached, training was continued monocularly. On alternate days either the right or the left eye was open. Fifty trials were given on each day. In the figure only the results obtained with monocular training are given

Recently we have found that brightness discrimination with the contralateral eye is also severely impaired. In Fig. 2 the average results obtained in seven rabbits, in which the left occipital lobe was removed, are given. At first the rabbits were trained binocularly to discriminate between two targets of different brightness. The details of the visual discrimination procedure were described earlier (Van Hof 1966, 1967; Van der Mark and Meyer 1974). After the 90% correct level had been reached, training was continued monocularly. On alternate days the animals were trained with either the left or the right eye open. On each day 50 trials were given to all animals tested. In all animals the number of correct choices was higher with the left eye than with the right eye.

3 Unilateral Ablation of the Sensorimotor Cortex

In the rabbit a unilateral ablation of the sensorimotor cortex does not affect normal walking, but the hopping reaction is impaired in the contralateral foreleg (Hobbelen and Van Hof 1986). Hopping of the forelegs was tested by means of a conveyerbelt moving at a speed of 2.5 cm/s, either to the left or to the right. The experimenter supported the animal's body and one of its legs. The other forepaw touched the surface of the conveyer belt. The animal was held at right angles to the direction of the movement of the belt. The hopping reaction has two components: a passive one in which the leg follows the conveyer belt and an active one in which the leg is lifted and returns to its starting point. The hopping reaction in the rabbit is best elicited when the conveyer belt adducts the foreleg. After a lesion of the contralateral sensorimotor cortex, the leg follows the conveyer belt passively, without active resetting. This behavioral defect was found to be age-dependent. In animals operated at adult age, the hopping reaction was still negative 4 months after the operation. A normal hopping reaction was found in rabbits operated 1 day after birth and tested 4 months later.

4 Hemidecortication

In the present investigation the behavioral effects of hemidecortication on vision and locomotion were studied in five groups of rabbits operated at different ages (0–1, 2–3, 4–5, 6–7, and 12–13 weeks old). The frontal, parietal-temporal and occipital cortex was removed on one side. The purpose of this study was twofold. First of all we wanted to re-examine the lack of age-dependent recovery in the visual system of the rabbit with a visual discrimination task which does not necessarily require an intact visual cortex. Secondly, we wanted to know whether there is a well-defined demarcation point in motor recovery after a cortical lesion.

Brightness discrimination with the left and right eye was tested, beginning 12 weeks after the operation. The hopping reaction was tested during 15 weeks, beginning 2 weeks after the operation.

The results obtained with brightness discrimination did not show any indication of age-dependent recovery. In all age groups a result was found similar to that shown in Fig. 2. The hopping reaction was found to be positive in the animals operated 0–1, 2–3, and 4–5 weeks after birth (Fig. 3).

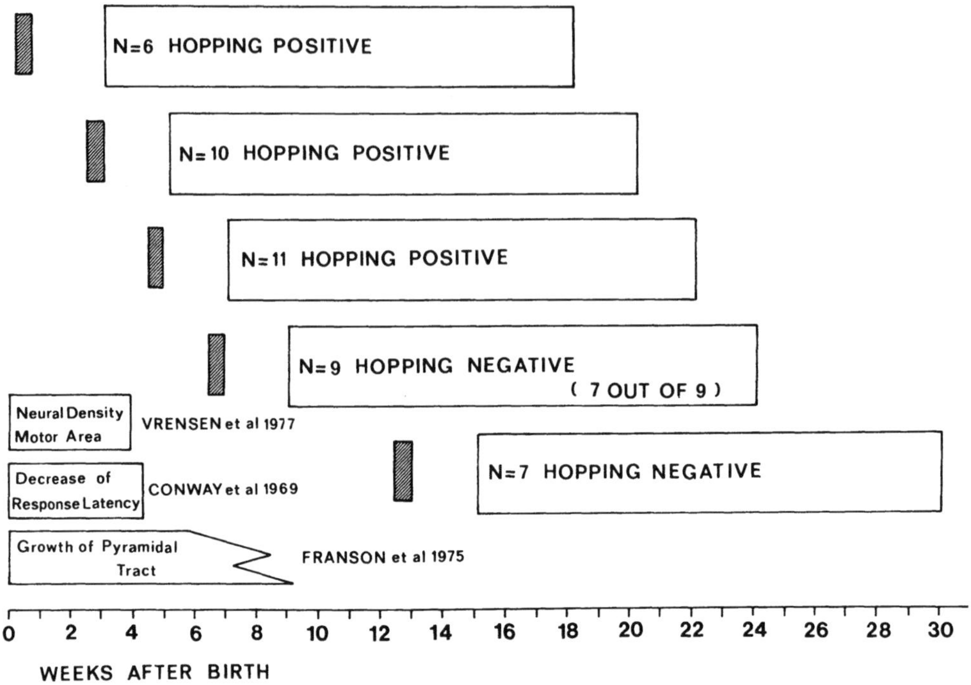

Fig. 3. The hopping reaction was tested during 15 weeks, beginning 2 weeks after the operation. The *shaded bars* indicate the postnatal weeks in which the various groups of rabbits were operated. The hopping reaction was found to be present in all animals operated before the sixth postnatal week. No hopping reaction could be elicited in seven out of nine rabbits operated in the 7th week after birth and in seven rabbits operated in the 13th week

No hopping reaction could be elicited in the animals operated 6–7 and 12–13 weeks after birth. This suggests that the demarcation point with respect to motor recovery lies in the sixth postnatal week.

The postnatal development of the motor cortex and the corticospinal connections have been studied by several authors. Vrensen et al. (1977) found that neuron density in the motor cortex increases till 27 days after birth. In a study on the maturation of the cortical surface potential evoked by stimulation of the pyramidal tract, Conway et al. (1969) found that the response latency decreased up to 30 days after birth. However, Franson and Hildebrand (1975) showed that growth of the pyramidal tract is not fully completed at the age of 30 days. Therefore the behavioral data suggest that plasticity in the motor system remains until maturation is complete.

With respect to the nature of the visual deficit, it should be remembered that there is evidence which suggests that in the rabbit, as in other animals, a brightness discrimination task is normally acquired on the basis of pattern vision mediated by the visual cortex (Le Vere and Morlock 1973; Van Hof et al. 1983). Relearning after a cortical lesion would be a less efficient process based on the discrimination of flux cues. Assuming that the superior colliculus is involved in this relearning process, the following factors will have to be taken into account in future experiments on the nature of the visual deficit after hemidecortication.

a) Contrary to the lateral geniculate nucleus (Rapisardi et al. 1975) and the visual cortex (Mathers et al. 1974), the superior colliculus of the rabbit matures relatively early (Spear et al. 1972) and could therefore show little plasticity.
b) In a study by Graham et al. (1982) on the effects of visual cortical lesions on the receptive-field properties of single units in the superior colliculus of adult rabbits, the corticotectal projection was found to be of importance for the receptive-field properties of neurons in the colliculus. In this way a cortical lesion could reduce the accuracy of the brightness discrimination via the colliculus at the side of the lesion.
c) Finally, the so-called Sprague effect has to be considered as the cause of the brightness discrimination defect. Sprague (1966) described that hemianopia after unilateral removal of the occipito-temporal cortex of the cat disappeared after ablation of the colliculus contralateral to the cortical lesion or splitting of the collicular commissure. This finding was interpreted as an abolishment of the influx of inhibition from the crossed colliculus.

Two other factors, not related to the colliculus per se, deserve attention.

1. Atrophy of retinal cells after removal of striate cortex was described in the rhesus monkey (Cowey 1974) and in the cat (Pearson et al. 1981). This was not found in the adult rat (Perry and Cowey 1979a, b). However, aberrant projections to the lateral posterior nucleus were found after removal of the striate cortex in infant rats. Similar phenomena are not likely to contribute to the above-described deficit in the rabbit. Wilkes et al. (1985) found that after unilateral ablation of the visual cortex in rabbits within 24 h after birth, no abnormality of neuron size distribution and neuron densities in whole-mounted retinas was demonstrated. In addition, studies on the effects of early enucleation in the rabbit (Chow 1981; Chow et al. 1973, 1981) indicate early plasticity to be less marked in the rabbit than in the rat (Baisden and Shen 1978; Ikeda et al. 1983).
2. In the present experiments only brightness discrimination was studied. De Vos-Korthals and van Hof (1985) found that in rabbits in which the visual cortex had been removed on both sides, in spite of a deficiency of brightness discrimination, complicated visuomotor skills like jumping from a table or jumping over a gap between obstacles returned under the influence of visual experience. This kind of behavior will have to be studied in future experiments on hemidecortication.

References

Baisden RH, Shen CL (1978) Sprouting of ipsilateral retinal projection in the optic system of the albino rat. Exp Neurol 61:549–560

Burgess JW, Villablanca JR (1986) Recovery of function after neonatal or adult hemispherectomy in cats. II. Limb bias and development, paw usage, locomotion and rehabilitative effects of exercise. Behav Brain Res 20:1–18

Burgess JW, Villablanca JR, Levine MS (1986) Recovery of functions after neonatal and adult hemispherectomy in cats. III. Complex functions: open field exploration, social interactions, maze and holeboard performances. Behav Brain Res 20:217–230

Chow KL (1981) Development and reorganization of the uncrossed retinotectal and retinogeniculate fibers in the rabbit. In: Hof MW van, Mohn G (eds) Functional recovery from brain damage. Elsevier, Amsterdam, pp 361–368

Chow KL, Mathers LH, Spear PD (1973) Spreading of uncrossed retinal projection in superior colliculus of neonatally enucleated rabbits. J Comp Neurol 151:307–322
Chow KL et al. (1981) Anomalous uncrossed retinal projections fail to activate superior colliculus neurons in rabbits unilaterally enucleated by fetal surgery. J Comp Neurol 196:189–204
Conway CJ, Wright FS, Bradley WE (1969) Electrophysiological maturation of the pyramidal tract in the post-natal rabbit. EEG and Clin Neurophys 26:565–577
Cowey A (1974) Atrophy of retinal ganglion cells after removal of striate cortex in a rhesus monkey. Perception 3:257–260
Distel H, Hudson R (1984) Nipple-search performance by rabbit pups: changes with age and time of day. Anim Behav 32:501–507
Distel H, Hudson R (1985) The contribution of the olfactory and tactile modalities to the nipple-search behaviour in newborn rabbits J Comp Physiol A 157:599–605
Franson P, Hildebrand C (1975) Postnatal growth of nerve fibres in the pyramidal tract of the rabbit. Neurobiology 5:8–22
Graham J, Berman N, Murphy EH (1982) Effects of visual cortical lesions on receptive-field properties of single units in superior colliculus of the rabbit. J Neurophysiol 47:272–286
Grota LJ, Adler R (1969) Continuous recording of maternal behaviour in Rattus Norvegiens. Anim Behav 17:722–729
Gudden B von (1886) Über die Frage der Lokalisation der Funktionen der Großgehirnrinde. Allg Z Psychiatrie Grenzgebiete 42:478–499
Hick SP, D'Amato CJ (1970) Motor-sensory and visual behavior after hemispherectomy in newborn and mature rats. Exp Neurol 29:416–438
Hobbelen JF, Hof MW van (1986) The hopping reaction in the rabbit after early and late removal of the motor cortex. Behav Brain Res 21:73–77
Hof MW van (1966) Discrimination between striated patterns of different orientation in the rabbit. Vision Res 6:89–94
Hof MW van (1967) Visual acuity in the rabbit. Vision Res 7:749–751
Hof MW van, Lagers-van Haselen GC (1975) Monocular pattern discrimination in rabbits after unilateral ablation of the visual cortex. Exp Neurol 46:257–259
Hof MW van, Stuurman PM (1981) The effect of a unilateral ablation of the occipital lobe on pattern discrimination in the rabbit. Behav Brain Res 3:401–404
Hof MW van, Stuurman PM (1984) Visual discrimination after early and late unilateral enucleation of the rabbit. Behav Brain Res 12:87–89
Hof MW van, Hof-van Duin J van, Hobbelen JF (1983) Visual discrimination after bilateral removal of the visual cortex in the rabbit. Behav Brain Res 9:257–262
Hudson R, Distel H (1982) The pattern of behaviour of rabbit pups in the nest. Behaviour 79:2–4, 255–271
Hudson R, Distel H (1983) Nipple location by newborn rabbits: behavioural evidence for pheromonal guidance. Behaviour 85:3–4, 260–275
Hudson R, Distel H (1984) Nipple-search pheromone in rabbits: dependence on season and reproductive state. J Comp Physiol A 155:13–17
Ikeda Y, Yagi F, Sakai M (1983) The role of crossed and uncrossed optic pathways mediating black-white discrimination in rats with one eye enucleated at birth. Physiol Behav 30:885–890
Lincoln DW (1974) Suckling: a time constant in the nursing behaviour of the rabbit. Physiol Behav 13:711–714
Mark F van der, Meijer JHC (1974) Automatic control of installations for experiments relating to physiological research of the visual system. Comput Programs Biomed 4:35–41
Mathers LH et al. (1974) Ontogenesis of receptive fields in the rabbit striate cortex. Exp Brain Res 19:20–35
Pearson HE et al. (1981) Transneuronal retrograde degeneration in the cat retina following neonatal ablation of visual cortex. Brain Res 212:470–475
Perry VH, Cowey A (1979a) The effects of unilateral cortical and tectal lesions on retinal ganglion cells in rats. Exp Brain Res 35:85–95
Perry VH, Cowey A (1979b) Changes in the retino-fugal pathways following cortical and tectal lesions in neonatal and adult rats. Exp Brain Res 35:97–108

Rapisardi SC, Chow KL, Mathers LH (1975) Ontogenesis of receptive field characteristics in the dorsal lateral geniculate nucleus of the rabbit. Exp Brain Res 22:295–305
Schneirla TC, Rosenblatt JS, Tobach E (1963) Maternal behavior in the cat. In: Rheingold HL (ed) Maternal behavior in mammals. Wiley, New York, pp 122–168
Spear PD et al. (1972) Ontogenesis of receptive field characteristics of superior colliculus neurons in the rabbit. Brain Res 45:67–86
Sprague JM (1966) Interaction of cortex and superior colliculus in mediation of visually guided behavior in the cat. Science 153:1544–1547
Stuurman PM, Hof MW van (1979) The postnatal development of swimming behaviour in the rabbit. Physiol Behav 23:185–186
Stuurman PM, Hof MW van (1980) Pattern discrimination in rabbits kept in environments of different complexities after unilateral removal of the occipital cortex. Behav Brain Res 1:211–226
Stuurman PM, Hof MW van, Hobbelen JF (1981) Behavioural effects of early and late unilateral ablation of the occipital lobe in the rabbit. In: Hof MW van, Mohn G (eds) Functional recovery from brain damage. Elsevier Biomedical, Amsterdam, pp 121–129
Vere TE Le, Morlock GW (1973) Nature of visual recovery following posterior neodecortication in the hooded rat. J Comp Physiol Psych 83:62–67
Villablanca JR, Burgess JW, Olmstead CE (1986) Recovery of function after neonatal or adult hemispherectomy in cats: I. Time course, movement, posture and sensorimotor tests. Behav Brain Res 19:205–226
Vos-Korthals WH de, Hof MW van (1981) Pattern discrimination after unilateral and bilateral ablation of the occipital lobe in the rabbit. Behav Brain Res 2:219–222
Vos-Korthals WH de, Hof MW van (1985) Residual visuomotor behaviour after bilateral removal of the occipital lobe in the rabbit. Behav Brain Res 15:205–209
Vrensen G, Groot D de, Nunes-Cardozo J (1977) Postnatal development of neurons and synapses in the visual and motor cortex of rabbits: a quantitative light and electron microscopic study. Brain Res Bull 2:405–416
Wilkes M, Zingaro G, Murphy EH (1985) Survival of the ganglion cell population in the rabbit retina following neonatal visual cortex ablation. Dev Brain Res 21:293–297

Post-Lesion Neural Plasticity in Cerebral Cortex: the Response of Rat Sensorimotor Cortex to Neonatal Ablation of the Opposite Hemisphere

P. R. Huttenlocher and R. Raichelson

Departments of Pediatrics and Neurology, University of Chicago, Chicago, IL, USA

Our experimental work in the rodent is concerned with the plasticity of immature cerebral cortex, and more specifically with the effects of focal cerebral lesions in infancy on the organization of the corticospinal motor system. Our interest in this problem derives from certain observations related to focal brain damage and to recovery from it in the human infant. At the outset, we would therefore like to briefly summarize relevant observations in man and point out some of the questions they raise that are not readily answerable by clinical studies, but are amenable to study in the experimental animal.

It has long been known that recovery or preservation of certain specific functions occurs after perinatal focal brain injury which is not seen after comparable injury in the adult. In man, the effect is seen most strikingly in the area of language (Basser 1962). Large lesions in the dominant hemisphere or even complete surgical removals of this hemisphere are followed by remarkable – if not entirely normal (Kiessling et al. 1983; Vargha-Khadem et al. 1985) – emergence of speech functions. Recovery occurs, provided that the destruction of the speech areas on the dominant side took place prior to a critical age of about 8 years (Woods and Teuber 1978 a).

Less complete, but nevertheless quite impressive compensations occur in other functions as well. In our present discussion we are especially interested in motor functions. Voluntary motor activities on the side opposite large hemispheral lesions of early onset, including hemispherectomy, are impaired, but not totally lost. Children who have had hemispherectomies have been observed to walk well, with only mild hemiparetic limp (Wilson 1970). While upper extremity functions are more severely impaired, some children have been reported to have voluntary motor control in the paretic hand following hemispherectomy (Ueki 1966). In a recent study of 39 children who had congenital or early acquired hemiparesis, we found that the ability to use the hand correlated highly with lesion size. However, even in the group with the largest lesions, approaching destruction of the entire hemisphere, about half of the children could reach out and grasp objects with the paretic hand (Levine et al. 1987). It has been known since the early observations of Westphal (1874) that movements of the paretic hand are frequently associated with synkinetic or mirror movements on the opposite side. Such mirror movements are much more prominent when hemiparesis is of early onset, prior to about age 1 year (Woods and Teuber 1978 b). Woods and Teuber have suggested that the presence of mirror movements implies a shift in motor control to the ip-

H. Flohr (Ed.)
Post-Lesion Neural Plasticity

silateral hemisphere, and that such a shift is more likely to occur in early lesions when the plasticity of the brain may allow for major reorganization of the motor system. This interesting hypothesis leaves, however, many unanswered questions. Is ipsilateral motor control after hemispherectomy mediated by the same cortical area as contralateral control? If so, are some pyramidal tract cells that normally are specified for control of contralateral limbs taken over for the ipsilateral connections? Or is the ipsilateral control mediated by cortical regions outside normal motor cortex? If so, what areas are involved and what other functions are displaced?

Techniques available for the study of human pathologic material do not provide answers to these questions. We therefore have utilized an animal model in which it is possible to explore the cellular basis of cortical plasticity. While extrapolations from the animal model to man are always risky, we nevertheless find it likely that general principles derived from work in lower species have some applicability to the situation in man. The model chosen for these studies is the reaction of the pyramidal motor system of the rodent to postnatal hemispherectomy. Normally, this system is an entirely crossed one in the rat. Hicks and D'Amato (1970) first reported a remarkable reaction of the corticospinal system of the rat to postnatal unilateral ablation of the motor cortex: the remaining undamaged cerebral hemisphere grows an uncrossed tract to the ipsilateral spinal cord in addition to the normal crossed one. Numerous subsequent studies have confirmed this finding (Leong and Lund 1973; Castro 1975; Leong and Lund 1976; Sharp and Evans 1983). The compensations that occur in cerebral cortex to accommodate both systems have not been studied well in the past. Retrograde labeling techniques including horseradish peroxidase and fluorescent dyes have been used widely to study the normal distribution of corticospinal neurons in the rat (Wise et al. 1979; Bates and Killackey 1984), and to look at the development of this system in infancy (Adams et al. 1983; Stanfield and O'Leary 1985). Lesion-induced changes in cerebral cortex, however, have had only limited investigation (D'Amato and Hicks 1978). One reason appears to have been the perception that unilateral labeling of axon terminals in spinal cord would be very difficult if not impossible to achieve reliably. For the studies to be presented here we have developed a method for unilateral labeling of corticospinal neurons. This has made it possible to study the distribution of neurons in cerebral cortex that make up the aberrant uncrossed corticospinal pathway as well as that of neurons forming the crossed system in hemispherectomized animals.

Unilateral retrograde labeling of corticospinal neurons has been accomplished through multiple small injections of horseradish peroxidase conjugated to wheat germ agglutinin (WGA-HRP) along the rostrocaudal extent of the cervical enlargement of spinal cord, which contains the terminals for the majority of corticospinal neurons. About 40 nl of 5% solution of WGA-HRP has been injected through a glass micropipette at each of six or seven unilateral spinal cord sites. This results in diffusion of HRP through one side of spinal cord only, with insufficient spread to the other side for labeling of cortical neurons. In addition to standard HRP histochemistry (DeOlmos et al. 1978) we have employed the method for stabilization of the HRP reaction product described by Rye et al. (1984). This has made it possible to study serial sections at leisure, and to obtain quantitative data of cell numbers and of the areas of distribution of corticospinal (CS) neurons.

Success of unilateral spinal cord injections has been confirmed by study of the brain stem nuclei, especially of the red nucleus which has a purely crossed projection to spinal cord. In control animals, the cerebral cortex ipsilateral to the spinal cord injection contains very few labeled neurons, never exceeding 2% of the number of contralateral labeled cells. Hemispherectomies were carried out between postnatal ages 5–9 days. The removals were accomplished under hypothermic anesthesia, by gentle suction of one cerebral hemisphere, including hippocampus and most of the basal ganglia. The hemispherectomized animals were allowed to live to maturity. Animals with spinal cord injections both ipsilateral and contralateral to the side of hemispherectomy were studied.

The distribution of corticospinal neurons forming the normal crossed corticospinal system was little affected by removal of the opposite cerebral hemisphere. As in control animals, labeled neurons were concentrated in three cortical regions: (1) in classic sensorimotor cortex (SM 1), (2) near the frontal pole, in a group of neurons that are separated from the more posterior SM 1 by a gap in which few corticospinal neurons are found, and (3) in a small group of cells lateral and posterolateral to SM 1, superior to the rhinal fissure. This distribution is as described by Bates and Killackey (1984) after large bilateral spinal cord injections of HRP, except for a small region near the posteromedial border of SM 1. This area contains CS neurons projecting to the lumbar cord which are not demonstrated when WGA-HRP is injected at the cervical level. In neonatally hemispherectomized rats, few neurons making up the crossed tract were ectopic to these regions, and these tended to be located in the gap that normally separates the frontal motor area from SM 1.

The findings were very different when spinal cord injections were made ipsilateral to the side of the remaining cerebral hemisphere, thus labeling the corticospinal neurons that make up the aberrant uncrossed tract. The density of these cells in normal motor areas was always quite low. However, their area of distribution was greatly enlarged. This was especially so in the region near the frontal pole, but also in the gap region and both medial and lateral to SM 1 (Fig. 1). Posteriorly, the visual areas appeared to form a barrier to expansion. No corticospinal neurons were found in visual cortex.

Quantitation of number and medial-lateral distribution of corticospinal neurons was carried out, using a computer-assisted method. Every eighth coronal section through cerebral cortex, 50 μ in thickness, was analyzed. Figure 2 provides the number of labeled neurons that were counted in control animals and in hemispherectomized animals with spinal cord WGA-HRP injections ipsilateral and contralateral to the remaining cerebral hemisphere. The number of labeled neurons in the analyzed sections is given seperately for each of the three cortical areas that contain corticospinal neurons, i.e., the region near the frontal pole, SM 1, and the small lateral cell group. In the hemispherectomized animals, each of these regions contains about as many cells that form the crossed corticospinal tract as does the cerebral cortex of the controls. The cells that make up the aberrant uncrossed tract have a rather different distribution: in sensorimotor cortex proper there are only about one third of the number of the neurons that make

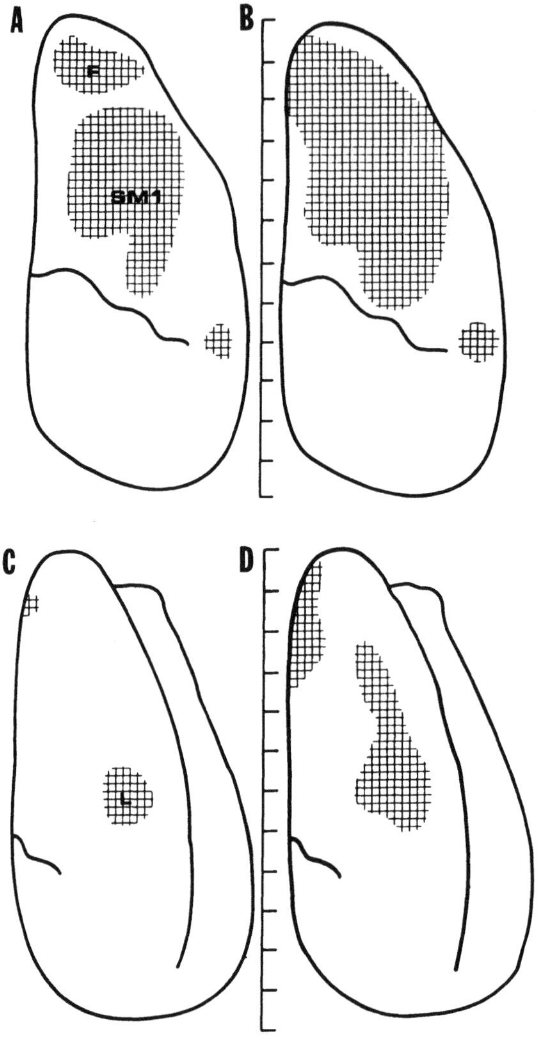

Fig. 1. Dorsal (**A** and **B**) and lateral (**C** and **D**) views of the rat cerebral hemisphere. **A** and **C** *Cross hatched* areas represent the mean distribution of CS neurons in four control animals. **B** and **D** Cross hatched areas represent the mean distribution of CS neurons in four animals that had neonatal hemispherectomies, followed by WGA-HRP injections into the cervical spinal cord on the side of the remaining cerebral hemisphere. The frontal pole is at the top of each drawing. The approximate anterior border of visual cortex is indicated by the wavy line. Scale = 1 mm. *F* Frontal motor region; *L* distribution of lateral CS neurons

up the crossed tract. Even of those, a significant number are in fringe areas lateral and medial to normal SM cortex. In contrast, compensation appears to be complete in the area of the frontal pole. Here the single hemisphere is able to accommodate about twice the normal number of CS neurons. Similarly, the lateral cell area contains about as many neurons with uncrossed connections as with contralateral ones. Normally, only about 20% of cervical CS neurons are located outside of the boundaries of SM 1. For neurons of the uncrossed tract in hemi-

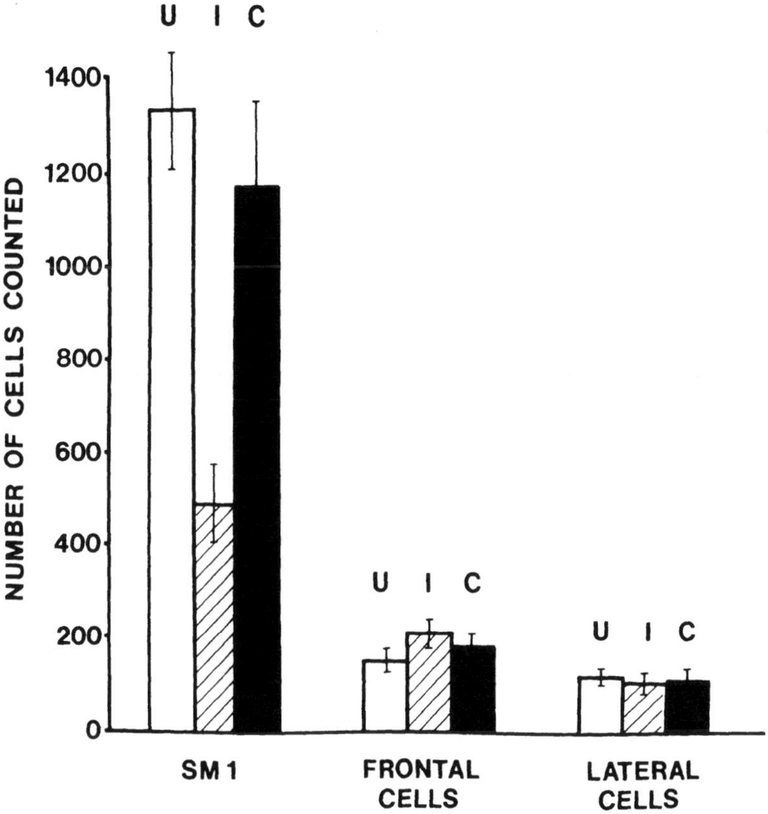

Fig. 2. Number of HRP-positive neurons counted in each of three cortical regions in unoperated control animals (*U*), hemispherectomized animals with ipsilateral spinal cord injections of WGA-HRP (*I*) and hemispherectomized animals with contralateral spinal cord injection of WGA-HRP (*C*). Cells in every eighth serial coronal section, 50 micra in thickness, were counted. Bars = 1 SEM. N = 4

spherectomized animals this number increases to almost 50%. This finding puts into doubt results of studies on corticospinal plasticity that are based on the demonstration of degenerating fibers in spinal cord after acute focal lesions in motor cortex. Such studies would fail to demonstrate the large number of connections that arise in regions outside motor cortex. An example is the study by Sloper et al. (1983) on the distribution of corticospinal fibers after unilateral motor cortex lesions in infant monkeys. No evidence for formation of an uncrossed corticospinal tract was found in these animals, but the experimental method would have missed connections from secondary motor areas or from aberrant sites.

As indicated above, the single hemisphere accommodates the extra contingent of CS neurons through expansion of the cortical region in which cells specified for this function are located. The question arises whether systems other than the corticospinal, and specifically other than the corticocervical, are displaced, or whether compensation is achieved through overlap of multiple functions in a single cortical area. We have a few preliminary observations relevant to this question. The distributions of CS neurons projecting to cervical and lumbar spinal cord in normal versus hemispherectomized animals have been compared. The cortical representation of the lumbar spinal cord is normally in a region comprising the posteromedial portion of SM cortex. After hemispherectomy, the distribution of these cells is in the same general region and the area of distribution is of the same approximate size. Superimposition of the maps for cervical and lumbar CS neurons shows some expansion of cervical corticospinal neurons into the lumbar representation, with increase in the areas of overlap.

We will now return to the questions that we posed at the outset, and will summarize how these are answered in the case of corticospinal plasticity in the rat. First, the question whether ipsilateral control of motor functions is mediated by the same cortical areas that contain the cells of the contralateral system cannot be given a simple yes or no answer. Ipsilateral corticospinal neurons occur both in normal motor areas and in certain other regions that are not normally specified for motor functions. As to the second question, whether some CS neurons normally specified for crossed connections instead send axons to ipsilateral spinal cord, we find no evidence that this occurs. Motor cortex appears to form added spinal connections from cells which do not normally send axons to cervical cord, at least in the mature state. It is not known whether these extra neurons come from a pool of cells that normally make transient connections which are later withdrawn. We are presently investigating this question in a series of infant rats who have had unilateral WGA-HRP injections at various postnatal ages. Up to now we have found no evidence for the existence of transient uncrossed corticospinal axons. The question also arises whether axonal sprouting is a factor, and whether single neurons send axons both to the ipsilateral and the contralateral side. A recent study by Reinoso and Castro (1985) failed to show evidence of double labeling of corticospinal neurons after right and left spinal cord injection of two different retrograde fluorescent tracers in hemispherectomized rats, suggesting that growth of axon collaterals is not a significant factor in the formation of the uncrossed tract.

The question as to what areas outside motor cortex are taken over for CS connections in hemispherectomized animals can now be answered in the rodent brain. The areas into which CS neurons expand are at the fringes of normal motor cortex. The largest region taken over is near the frontal pole, where the cortical area that contains CS neurons expands to approximately double the normal size. It appears that plasticity is greater in this frontal region than more posteriorly in SM 1. It is tempting to speculate that the greater plasticity in frontal cortex may be due to later development of the anterior frontal region than of SM 1. Greater functional recovery related to later development has been reported by Goldman (1972) after focal lesions in prefrontal cortex of infant monkeys. However, as far as the development of corticospinal connections in the rat is concerned, we have not been able to show that this occurs later in the frontal motor region that in SM 1. In both areas, we have found the first retrogradely labeled neurons when cervical spinal cord injections of WGA-HRP were made on postnatal day 7.

The question whether other systems are displaced by expansion of motor cortex is not answered by our studies. The frontal cortex regions into which the corticospinal system expands appear to have other motor functions normally, including eye, vibrissae and jaw movements (Hall and Lindholm 1974). It is not known whether these functions are displaced or whether the same cortical region also is able to accommodate its normal connections. The density of corticospinal neurons outside normal motor areas is always fairly low, and it is therefore quite likely that overlap of functions rather than displacement occurs. This appears to be the case at the border of corticocervical and corticolumbar regions. A possible outcome of crowding of added systems into a single hemisphere therefore is a blurring of cortical localization, with mediation of several functions by the same cortical region, rather than complete displacement. Such an adjustment would be consistent with the observation that all major functions can still be carried out by a single hemisphere. However, a disturbance in functional localization could well affect the overall accuracy and efficiency of cortical processing. It may represent the anatomical basis for observed negative effects of cortical plasticity. For example, in humans with early hemispherectomy or more restricted hemispheral lesions it has been found that there is decreased proficiency in more complex cortical functions, including reasoning and problem solving, reflected in an overall decrease in IQ scores (St. James-Roberts 1981; Levine et al. 1987), which is not seen after comparable lesions in the adult.

Our studies also indicate certain limits as to the extent to which cortical localization can be shifted. The expansion of corticospinal neurons after hemispherectomy occurs into cortical regions which normally mediate similar functions, for example jaw and vibrissae movements in the frontal areas. We have not observed corticospinal neurons in specialized sensory regions such as the barrel fields or the visual cortex. While transient axons from visual cortex to spinal cord have been described in infant rats (Adams et al. 1983; Stanfield and O'Leary 1985), these do not appear to make permanent spinal connections in hemispherectomized animals. These primary sensory areas may be genetically specified for types of connections that preclude the establishment of a spinal efferent system.

In summary, the single hemisphere that remains after neonatal hemidecortication undergoes remarkable changes in functional localization in its attempt to provide innervation to both sides of spinal cord. The areas of cortex in which corticospinal neurons occur expands, especially near the frontal pole, but also at all other margins except posteriorly, where visual cortex appears to form a barrier. The uncrossed tract appears to arise de novo, rather than from a normal transient system. Additional neurons that do not normally make CS connections are recruited. The stimuli which lead to the specification of additional neurons for corticospinal connections are presently unknown. Investigation of this problem in the pyramidal motor system is likely to lead to further insights into the question of regeneration in the central nervous system in response to injury.

Acknowledgement. Supported by grants NS 23960 and HD 04583, USPHS.

References

Adams CE, Mihailoff GA, Woodward DJ (1983) A transient component of the developing corticospinal tract arises in visual cortex. Neurosci Lett 36:243–248

Basser LS (1962) Hemiplegia of early onset and the faculty of speech with special reference to the effects of hemispherectomy. Brain 85:427–460

Bates CA, Killackey KP (1984) The emergence of a discretely distributed pattern of corticospinal projection neurons. Dev Brain Res 13:265–273

Castro AJ (1975) Ipsilateral corticospinal projections after large lesions of the cerebral hemisphere in neonatal rats. Exp Neurol 46:1–8

D'Amato CJ, Hicks SP (1978) Normal development and posttraumatic plasticity of corticospinal neurons in rats. Exp Neurol 60:557–569

Goldman PS (1972) Developmental determinants of cortical plasticity. Acta Neurobiol Exp 32:495–511

Hall RD, Lindholm EP (1974) Organization of motor somatosensory neocortex in the albino rat. Brain Res 66:23–38

Hicks SP, D'Amato JC (1970) Motor-sensory and visual behavior after hemispherectomy in newborn and mature rats. Exp Neurol 29:416–438

Kiessling LS, Denckla MB, Carlton M (1983) Evidence for differential hemispheric function in children with hemiplegic cerebral palsy. Dev Med Child Neurol 25:727–734

Kennard MA (1940) Relation of age to motor impairment in man and in subhuman primates. AMA Arch Neurol Psychiatry 48:227–240

Leong SK, Lund R (1973) Anomalous bilateral corticofugal pathways in albino rats after neonatal lesions. Brain Res 62:218–221

Leong SK, Lund R (1976) An experimental study of the corticofugal system following cerebral lesions in the albino rat. Exp Brain Res 26:235–247

Levine SC, Huttenlocher PR, Banich MT, Duda E (1987) Factors affecting cognitive functioning in hemiplegic children. Dev Med Child Neurol 29:27–35

Olmos J de, Hardy H, Heimer L (1978) The afferent connections of the main and accessory olfactory bulb formation in the rat: an experimental HRP study. J Comp Neurol 181:213–244

Reinoso BS, Castro AJ (1985) Analysis of corticospinal plasticity using retrograde fluorescent tracers in rats. Soc Neurosci Abstr 11:1286

Rye DB, Saper CB, Wainer BH (1984) Stabilization of the tetramethylbenzidine (TMB) reaction product. J Histochem Cytochem 32:1145–1153

Sharp FR, Evans KL (1983) Bilateral C14 2-deoxyglucose uptake by motor pathways after unilateral neonatal cortex lesions in the rat. Dev Brain Res 6:1–11

Sloper J, Brodal P, Powell TPS (1983) An anatomical study of the effects of unilateral removal of sensorimotor cortex in infant monkeys on the subcortical projections of the contralateral sensorimotor cortex. Brain 106:707–716

St James-Roberts I (1981) A reinterpretation of hemispherectomy data without functional plasticity of the brain. Brain Language 13:31–53

Stanfield BB, O'Leary DDM (1985) The transient corticospinal projection from the occipital cortex during the postnatal development of the rat. J Comp Neurol 238:236–248

Ueki K (1966) Hemispherectomy in the human with special reference to the preservation of function. Prog Brain Res 21:285–338

Vargha-Khadem F, O'Gorman AM, Watters GV (1985) Aphasia and handedness in relation to hemispheric side, age at injury and severity of cerebral lesion during childhood. Brain 108:677-696

Westphal C (1874) Über einige Bewegungs-Erscheinungen an gelähmten Gliedern. Arch Psychiatr 4:747–759

Wilson PJE (1970) Cerebral hemispherectomy for infantile hemiplegia: a report of 50 cases. Brain 93:147–180

Wise SP, Murray EA, Coulter JD (1979) Somatotopic organization of corticospinal and corticotrigeminal neurons in the rat. Neuroscience 4:65–78

Woods BT, Teuber H-L (1979 a) Changing patterns of childhood aphasia. Ann Neurol 3:273–280

Woods BT, Teuber H-L (1978 b) Mirror movements after childhood hemiparesis. Neurology 28:1152:1158

Reduction of Plasticity in the Primary Visual Cortex of the Rat

A. Wree, K. Zilles, and A. Schleicher
Anatomisches Institut der Universität Würzburg,
Koellikerstraße 6, D-8700 Würzburg, FR Germany

1 Introduction

The influence of the periphery on the structure and function of the representation field in the cerebral cortex has gained attention during the past century. Gudden (1870) investigated the visual cortex of rabbits which had been enucleated as neonatal animals, and found no changes in the cytoarchitectonic structure of the areas of the occipital lobes.

Since Gudden's experiments, the visual system of rodents and lagomorphs has often been used for morphological studies, because over 95% of the retinal efferents cross in the optic chiasm (Dräger and Olsen 1980; Dreher et al. 1985; Jeffery 1984; Ward and Tremblay 1982). Consequently, the effect of unilateral enucleation was assumed to be found in the hemisphere contralateral to the enucleated eye. Although much work has been done on this topic, a systematic investigation of the influence of eye removal on the structure of the primary visual cortices and the projection of the remaining eye to these areas of both hemispheres seemed necessary. This is because many parameters have not been systematically studied or quantified to date. It is particularly important, as will be shown, to quantify the results in both hemispheres after one-eye removal, and not to use the hemisphere contralateral to the remaining eye [the so called unaffected hemisphere (Fifkova 1970, Fifkova and Hassler 1969)] as intra-individual control side.

The aims of the present study are: first, to describe changes in the structure of the monocular subfield of the primary visual cortex, which were induced by neonatal enucleation of one eye, and, second, to delimit a critical period in the postnatal life of the rat. The critical period marks the age of the animal up to which enucleation still induces structural changes in the visual cortices, similar to those found in neonatally enucleated rats. Enucleations performed in older animals do not influence or only slightly influence the investigated structural parameters of the cortex. Third, the areal representation and laminar distribution of the visual afferents derived from the remaining eye were studied in layer IV of the subfields of the primary visual cortices, using quantitative autoradiography after [^{3}H]-proline injection into the remaining eye.

The primary visual cortex (occipital area 1 or Oc 1, Zilles et al. 1980) is situated in the caudal hemisphere of the rat (Adams and Forrester 1968; Montero 1973, 1981) and can be subdivided into a lateral part (Oc 1 B) and a medial part (Oc 1 M) (Wree et al. 1983; Zilles 1985; Zilles and Wree 1985; Wree et al. in

H. Flohr (Ed.)
Post-Lesion Neural Plasticity

press a). Of the primary visual areas, Oc 1 B gets a prominent input from the contralateral eye and, additionally, a small input from the ipsilateral eye (Zilles and Wree 1985; Zilles et al. 1984). Oc 1 B is binocularly driven. Oc 1 M is monocularly driven and, in the normal rat, it receives its afferents from the contralateral eye only (Zilles et al. 1984). If one eye is removed, then the Oc 1 M area ipsilateral to the remaining eye is visually deafferented and, therefore, supposedly the most affected area in the monocularly enucleated animals.

In this paper, Oc 1 M ipsilateral to the remaining eye will be the one area most thoroughly discussed. The results obtained in Oc 1 M contralateral to the remaining eye will be added to show that it is also influenced by enucleation and should not be used as an intra-individual control side. In the Oc 1 M areas of both hemispheres, we will focus on the results obtained in layer IV, since the majority of geniculo-cortical projections terminate in this layer (Dräger 1974, 1978; Ribak and Peters 1975; Zilles et al. 1984).

2 Material and Methods

Monocular enucleations (right eye) were performed in male albino rats (strain: HanWistar) ranging in age from the 23rd ontogenetic day (i.e., newborn) to the 120th ontogenetic day (i.e., young adult age). The rats were killed by perfusion fixation using Bodian's fluid (Romeis 1968) after a definite post-enucleation survival time of 120 days. The brains were embedded in paraffin and serially cut in the transverse plane at 20 μm. The parameters quantified in layers IV of the OC 1 M areas ipsilateral and contralateral to the remaining eye were: (a) the packing density of perikarya, measured in Nissl-stained sections as gray level indices according to the procedure already published (Schleicher et al. 1978, 1986; Wree et al. 1982; Zilles et al. 1980, 1984); (b) the fresh volumes (Kretschmann and Wingert 1971); (c) the optical densities of the fibers stained for myelin, using the staining protocol after Zilles (Zilles 1985; Zilles et al. 1983); and (d) the areal fraction of silver grains overlying layers IV in the autoradiographs following [^{3}H]-proline injection into the vitreous body of the remaining eye. Additionally, the same parameters, which were studied in the enucleated animals, were quantified in adult untreated control rats, in order to work out basic data, not existing in the literature, for the statistical comparison with the experimental data. Statistical differences in the values of the enucleated and control animals were tested by the Mann-Witney U-Test (Sachs 1978), $\alpha = 0.05$. The critical periods were calculated by nonlinear regression analysis.

3 Results

3.1 Packing Density of Perikarya

In the neonatally enucleated rats, the packing density of perikarya of layer IV of Oc 1 M ipsilateral to the remaining eye was increased. Furthermore, enucleations performed up to the 41st ontogenetic day (i.e., about 3 weeks after birth) also led to a significant increase in the packing density of perikarya, when compared with adult untreated control animals. In the rats enucleated from the 44th ontogenetic day onwards, the packing densities of perikarya did not differ significantly from the control values (Wree et al., in press b). In layer IV of Oc 1 M ipsilateral to the remaining eye, the critical period was found to lie between the 41st and the 44th ontogenetic day.

In layer IV of Oc 1 M contralateral to the remaining eye, a similar dependency of the packing density of perikarya on the ontogenetic age, at which the enucleation had been performed, was found. Enucleations in the early postnatal period up to the 41st ontogenetic day led to significantly increased packing densities of perikarya. Enucleations performed on the 44th ontogenetic day or later did not affect the packing density of perikarya significantly as compared with the control values. In layer IV of OC 1 M contralateral to the remaining eye, the critical period was calculated between the 41st and the 44th ontogenetic day.

Although layers IV of Oc 1 M in both hemispheres were affected with respect to the packing density of perikarya, the comparison of the data reveals that higher values were measured in the hemisphere ipsilateral to the remaining eye as compared with the contralateral one.

3.2 Fresh Volumes

Enucleations performed in neonatal rats and in early postnatal rats resulted in a reduction of the fresh volumes of layer IV of Oc 1 M ipsilateral to the remaining eye compared to the adult controls. Enucleations performed in rats older than 44 ontogenetic days did not alter the fresh volume of this layer significantly, as compared to the control values. The critical period was calculated between the 44th and 50th ontogenetic day. The fresh volumes of layer IV in OC 1 M contralateral to the remaining did not show a significant change, depending on the ontogenetic age of the animal at which the enucleation had been performed. Therefore, a critical period could not be determined in this hemisphere.

3.3 Myelin Content

Quantification of the myelin content was done by measuring the optical densities of layer IV in the myelin stain sections with a TV-image analyzer (IBAS 1+2, Kontron, Munich, FRG).

In layer IV of Oc 1 M ipsilateral to the remaining eye, the myelin content was reduced in the rats that had been enucleated in their early postnatal life. Myelin content was not changed in rats that had been enucleated as young adults. The critical period was calculated between the 60th and 70th ontogenetic day.

The myelin content of layer IV in Oc 1 M contralateral to the remaining eye did not change significantly depending on the animal's age of enucleation. The values of the enucleated animals were not significantly different from normal controls.

3.4. Geniculo-Cortical Projection

The quantification of the packing density of perikarya, the fresh volumes and the myelin content in the monocular subfields of the primary visual cortices revealed that, in rats enucleated in the first 3 weeks after birth, layer IV of OC 1 M ipsilateral to the remaining eye was more seriously affected than contralaterally. It is important, however, that the Oc 1 M areas contralaterally to the remaining eye also showed alterations in the investigated structural parameters (Fig. 4) and the interhemispheric connectivity (Wree et al. 1985, 1986) as compared to normal adult controls. Monocular enucleation led to a transneuronal reaction in both hemispheres.

In the normal rat, about 95–97% of the axons from eye cross to the contralateral hemisphere (Dräger and Olsen 1980; Dreher et al. 1985; Jeffery 1984). Shirokawa et al. (1983) could show that the normal pattern of crossing was changed by enucleation. After monocular enucleation of neonatal rats, more axons from the remaining eye than normally do not cross in the optic chiasm. In those animals, an enlarged ipsilateral projection combined with a reduced contralateral projection was shown by axon counting in the optic tracts (Shirokawa et al. 1983).

We tested this result by injecting [^{3}H]-proline into the remaining eye of enucleated rats 4 months after enucleation, using standard autoradiographic techniques (Rogers 1973). Adult control rats received similar injections into one eye. Subsequent to the transneuronal transport of the tracer (Dräger 1974, 1978), the grain densities over layer IV of the primary visual areas were measured with an image analyzer by TV-scan. These results should give a measure of the cortical areal representation and distribution of the efferents derived from the injected eye.

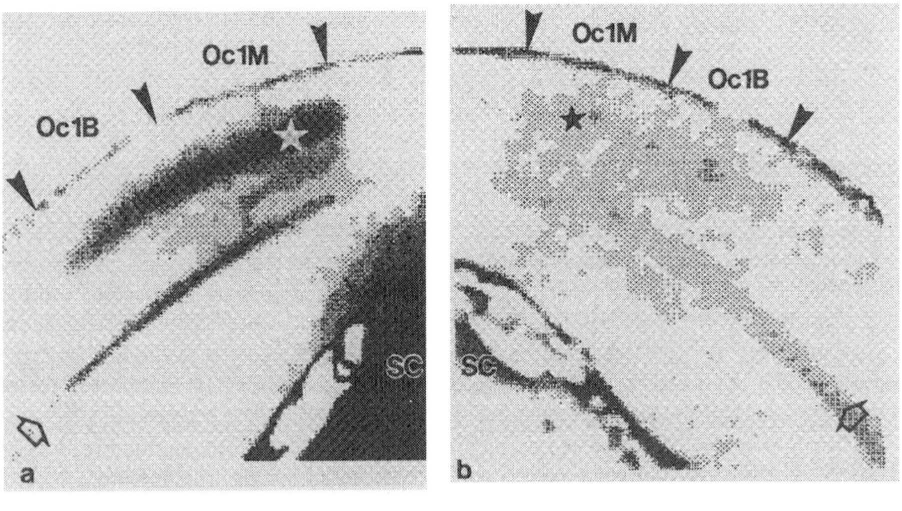

Fig. 1. Autoradiogram from a rat enucleated at the 35th ontogenetic day and injected with [^{3}H]-proline into the remaining eye 4 months post-enucleation. The grain density over the dorsal parts of the occipital hemispheres is shown after measurement by TV-scan and subsequent contrast enhancement with an image analyser (IBAS, Kontron, Munich, FRG). In the plots (**a, b**), *light areas* represent sparse label, *dark areas* dense label. **a** Hemisphere contralateral to the remaining eye; **b** hemisphere ipsilateral to the remaining eye. The extent of the primary visual cortex Oc 1 and the subfields Oc 1 M and Oc 1 B are indicated by *arrowheads*. Very dense label is visible in the superior colliculus (*SC*); in the cortex, the strongest labeling is found in layers IV (*stars*) and the optic radiations (*open arrows*)

Figure 1 shows the appearance of the primary visual cortices after [^{3}H]-proline autoradiography and after contrast enhancement with the image analyzer. Differences in label intensity over the occipital hemispheres of a rat, enucleated early postnatally, are graphically expressed using different print modes. The primary visual cortex Oc 1 is delineable due to the label of layer IV (Fig. 1 a; Zilles et al. 1984). The label intensity of the Oc 1 areas contralateral to the injected eye (Fig. 1 a) is clearly stronger than ipsilaterally (Fig. 1 b). In the hemisphere contralateral to the remaining eye (Fig. 1 a), the label of Oc 1 M is denser than in the binocular subfield (Oc 1 B). In the hemisphere ipsilateral to the injected eye (Fig. 1 b), it is important to note that a weak, but reproducibly detectable, label is seen in Oc 1 M. This label of Oc 1 M ipsilateral to the injected eye (Fig. 1 b) was only seen in enucleated animals. Oc 1 M ipsilateral to the injected eye was not labeled in normal adult control rats (Fig. 2 a; Zilles et al. 1984).

In order to obtain further insights into the areal distribution of the visual afferents, shown in Fig. 1 at low magnification, measurements of grain densities at high magnification ($\times$ 1685) were performed in layer IV of the Oc 1 M and Oc 1 B areas of both hemispheres. Using a TV-image analyzer (IBAS 1+2, Kontron, Munich FRG; for details, see Zilles et al. 1986), the areal fractions of silver grains on the quadratic measuring fields (each measuring 108 μm in size) were determined. The results are graphically represented in Fig. 2 a–d.

In Oc 1 M contralateral to the remaining eye (Fig. 2 b), the grain densities of layer IV significantly increased with the age of the animal at which the enucleation had been performed (Spearman rank correlation coefficient). The values measured in the enucleated rats were lower compared to the controls. In Oc 1 B of this hemisphere (Fig. 2 d), the grain densities over layer IV also increased with the age at which the rats' eyes had been enucleated. The values of the enucleated rats, however, did not differ significantly from the control data.

In the hemisphere ipsilateral to the remaining eye, layer IV of Oc 1 M of the enucleated animals showed a weak, but clearly reproducible label (Figs. 1 b, 2 a). In contrast, the respective area ipsilateral to the injected eye of normal adult controls was unlabeled (Zilles et al. 1984) (Fig. 2 a). Layer IV of Oc 1 B of the hemisphere ipsilateral to the injected eye was labelled in the controls, as had already been shown by Zilles et al. (1984) (Fig. 2 c), and also in the enucleated rats (Figs. 1 b, 2 c). In this layer, the grain densities measured in the enucleated rats did not differ significantly from the control data.

The calculations of the relative differences of grain densities give further arguments for the enucleation-induced changes of the retino-geniculo-cortical projections (Fig. 3). In the control animals (Fig. 3 a), the relative difference between layers IV of Oc 1 M ipsilateral to the injected eye and Oc 1 M contralaterally (Oc 1 M contralateral to the injected eye = 100%) was 99.3% ($\pm$0.2 SEM); i.e., Oc 1 M contralateral to the injected eye was heavily labeled, while Oc 1 M ipsilaterally was nearly unlabeled. In the enucleated animals, however, the relative differences were significantly smaller, compared with the controls, and increased with the age of the animals at which the enucleation had been performed (Fig. 3 a). The data indicate that, compared to normal, more retinal axons project to the ipsilateral Oc 1 M and fewer to Oc 1 M contralaterally via the dorsal lateral geniculate nuclei.

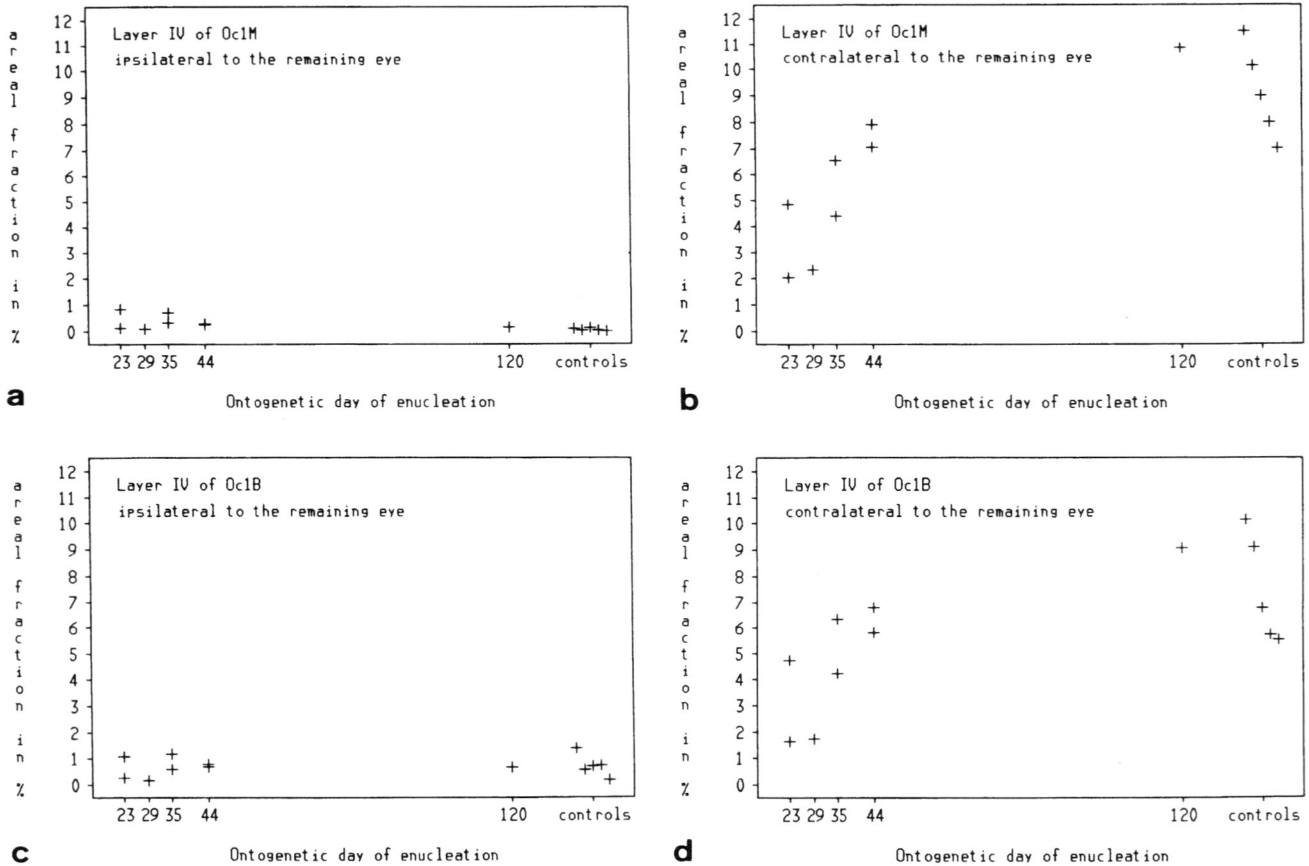

Fig. 2. Densities of silver grains overlying layers IV of the Oc 1 M (**a, b**) and Oc 1 B (**c, d**) subfields of the primary visual cortices ipsilateral (**a, c**) and contralateral (**b, d**) to the remaining (injected) eye, measured as mean areal fractions in % of the whole measuring fields with an image analyzer by TV-scan. Labeling in the autoradiograms of rats, which had been enucleated at different ontogenetic

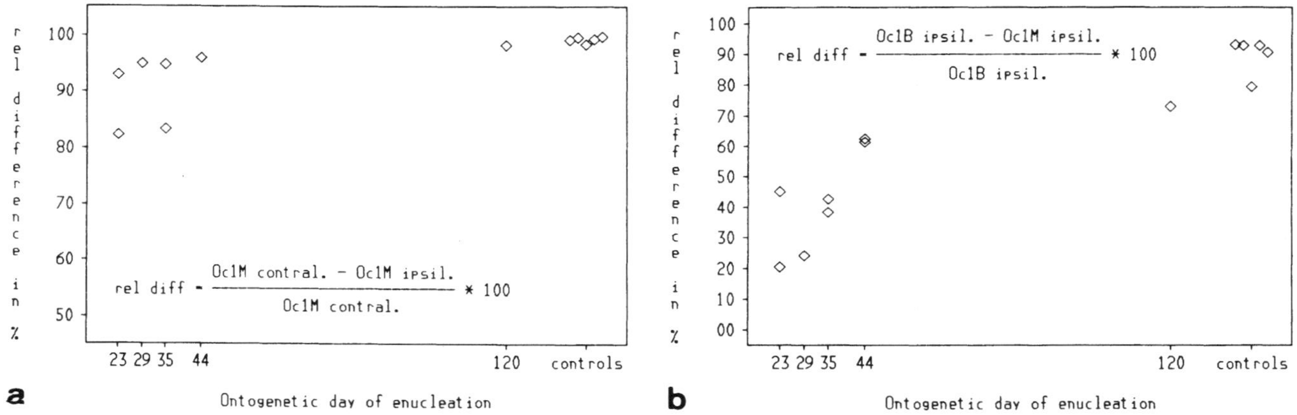

Fig. 3. Changes of the autoradiographic labeling over layers IV of the primary visual cortices subfields Oc 1 M and Oc 1 B of rats, which had been enucleated at different ontogenetic ages, calculated as relative differences (rel diff) according to the given formula and using the data given in Fig. 2. **a** In the enucleated rats, the relative differences between the Oc 1 M areas contralateral and ipsilateral to the injected eye were smaller compared with the controls; **b** in the hemispheres ipsilateral to the injected eye, the relative differences of the grain densities between the Oc 1 B and Oc 1 M subfields were large in the controls and were found to be smaller in the enucleated rats. The data indicate an enucleation-induced projection from the remaining eye to the ipsilateral Oc 1 M area

Figure 3b shows the relative differences of the grain densities between the Oc 1 B and Oc 1 M areas ipsilateral to the injected eye. In the control animals, the grain density of Oc 1 M was about 10% ot that measured in Oc 1 B (Fig. 3b). In the enucleated animals, the relative differences between the Oc 1 B and Oc 1 M areas ipsilateral to the remaining eye (grain densities of Oc 1 B = 100%) were significantly smaller, compared with the controls (Fig. 3b), and increased with the age at which the rats had been enucleated. These data are a further indication for the result that monocular enucleation in the early postnatal life of the rat led to a substantial projection to Oc 1 M ipsilateral to the remaining eye (Fig. 1b), a projection which was not seen in normal control rats (Zilles et al. 1984), leaving the projection to the directly adjoining ipsilateral Oc 1 B area unchanged.

4 Discussion

The present results demonstrate that monocular enucleation performed in neonatal rats caused changes in the structure of layers IV of the primary visual cortices' monocular subfields (Oc 1 M) of both hemispheres. The Oc 1 M area of the hemisphere ipsilateral to the remaining eye was more seriously affected than the contralateral area.

In Oc 1 M ipsilateral to the remaining eye, layers IV showed an increased packing density of perikarya, reduced fresh volumes and a reduced myelin content (Fig 4). Increased packing densities of perikarya in the primary visual cortex of monocularly enucleated mice have been reported (Heumann and Rabinowicz 1982). Enucleation-induced changes in fresh volumes and myelin content had not yet been quantified in the literature. Combining our data concerning the fresh volumes and packing density of perikarya in layers IV, it can be calculated that about 95% of the reduced cortical volume is caused by a reduction of the neuropil volume. This is in agreement with the results of Heumann and Rabinowicz (1982), who found the number of neurons in visual cortex of enucleated mice not significantly changed compared with normal controls.

Reduction of the neuropil volume could partly be due to a decreased myelin content, which could be shown in the present study. The reduced myelin content can be interpreted by a reduction of the amount of myelinated fibers entering the cortex from the severly affected dorsal lateral geniculate nucleus (DLG; Fig. 4). In neonatally enucleated rodents, the DLG was shown to be reduced in volume (Heumann and Rabinowicz 1980; Schober et al. 1977; Tsang 1937), perikarya number (Heumann and Rabinowicz 1980), perikaryal size (Schober et al. 1977; Werner et al. 1977), and synaptic (Campbell et al. 1985) and functional organization (Fukuda et al. 1983; Shirokawa et al. 1983). Due to the altered and possibly reduced geniculo-cortical projection (Fig. 4), eye removal caused a transneuronal reaction in the developing cortex, thus leading to an alteration of cortical maturation and, especially, to a deficit mainly in the neuropil development. It could be shown that in the cortex the synaptic density was reduced (Fifkova 1970), the spine densities of the pyramidal cells' apical dendrites were decreased (Globus and Scheibel 1967; Ruiz-Marcos and Valverde 1970, Valverde 1968) and dendritic trees of neurons of layers III-V were changed in shape and shortened (Valverde 1968).

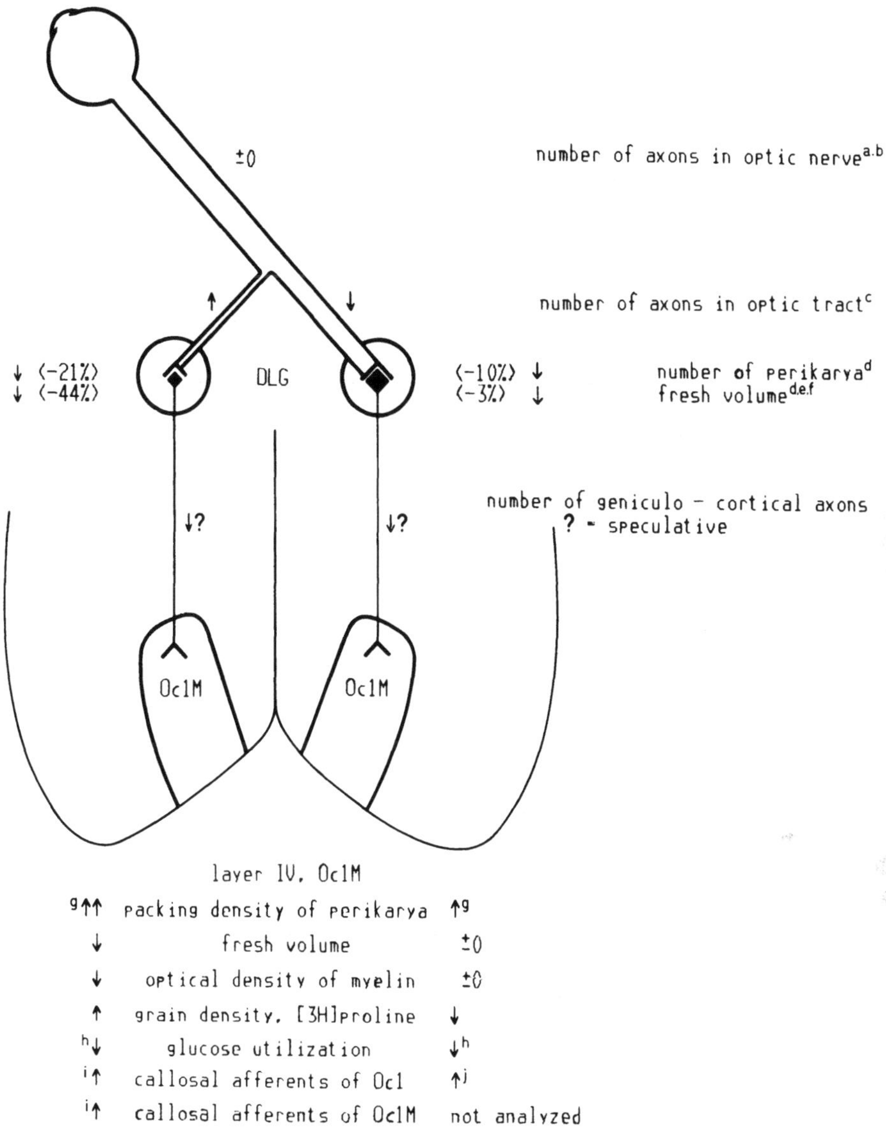

Fig. 4. Schematic diagram summarizing the changes in the retino-geniculo-cortical projections and in structural and functional parameters of the Oc 1 M areas ipsilateral and contralateral of the remaining eye in rats, which had been monocularly enucleated as newborns and investigated as adults; in the enucleated rats, the various parameters showed an increase (↑), decrease (↓), or no significant change (± 0) as compared with normal adult control animals. Data from: [a] Lam et al. 1982; – [b] Sefton and Lam 1984; – [c] Shirokawa et al. 1983; – [d] Heumann and Rabinowicz 1980; – [e] Tsang 1937; – [f] Schober et al. 1977; – [g] Heumann and Rabinowicz 1982; – [h] Wree et al. 1987; – [i] Wree et al. 1985; – [j] Wree et al. 1986. DLG – dorsal lateral geniculate nucleus; Oc 1 M – monocular subfield of the primary visual cortex Oc 1

Additionally, the results of the present study demonstrate that changes in the structural parameters of the primary visual cortex could also be induced by enucleations which had been performed up to the end of the third postnatal week. The cortical changes were similar to those seen in neonatally enucleated rats. Enucleations performed in older animals did not alter the investigated parameters significantly as compared with normal adult control rats.

As shown above, the transneuronal reaction of the cortex following neonatal enucleation mainly results in an alteration or deficit of neuropil development. If this is correct, and structurally and functionally intact afferents are essential for the normal cortical development, it can be assumed that enucleations performed throughout the time of cortical development could also interfere with the neuropil maturation. In the literature, the postnatal development of the visual cortex in normal rats and mice has been extensively described (Blue and Parnavelas 1983; Heumann et al. 1977; Leuba et al. 1977; Miller 1981; Miller and Peters 1981; Parnavelas and Uylings 1980; Parnavelas et al. 1978; Riccio and Matthews 1985; Uylings et al. 1980; Warren and Bedi 1984; Werner et al. 1981). Summarizing these reports, the cortex reached its adult-like appearance not prior to postnatal days 20 to 28, depending on the parameter studied. It can be stated that structural changes in the visual cortex could be induced in rats which had been enucleated before the cortical maturation (i.e., mainly the neuropil development) was finished. The critical periods, worked out in the present study, correspond well with the published results on the termination of the normal maturation of the rat's cortex in the postnatal ontogenesis. After that stage of development when the maturation of the visual cortex is roughly finished, enucleation no longer altered the structural parameters studied.

In Oc 1 M contralateral to the remaining eye, neonatally enucleated rats also showed a significant increase of the packing density of perikarya of layer IV as compared to normal. The packing density of perikarya in this layer, however, was significantly lower compared to Oc 1 M ipsilateral to the remaining eye. Our data on the packing density of perikarya corroborate the results of Heumann and Rabinowicz (1982), who were the first to emphazise that the Oc 1 area contralateral to the remaining eye was affected with respect to this parameter by monocular enucleation of neonatal mice. The fresh volumes and myelin content of layer IV of Oc 1 M contralateral to the remaining eye were not significantly changed compared with control animals. The fresh volumes of the whole Oc 1 area in this hemisphere, however, was reduced compared with normal controls (own unpublished observations), indicating a reduction of neuropil volume in this hemisphere also.

These results could be interpreted in the manner described above for the ipsilateral hemisphere. Neonatal monocular enucleation led to a reduction of the perikarya number and volume of the DLG contralateral to the remaining eye (Heumann and Rabinowicz 1980; Fig. 4), the latter being possibly caused by the reduced number of crossing retinal fibers reaching this DLG (Shirokawa et al. 1983; Fig. 4). The altered geniculo-cortical projection (see [^{3}H]-proline experiments) from the respective DLG might induce the transneuronal reaction seen in the Oc 1 M area contralateral to the remaining eye. In this way, one can also interpret the results of Valverde (1968), who reported a reduction of the spine density in layer IV of this area in similar experiments. Additionally, the functional activity (Wree et al. 1987) and the callosal projection pattern (Wree et al. 1986) of this Oc 1 area were found to be abnormal (Fig. 4).

Enucleations performed up to ontogenetic day 41 also caused an increase in the packing density of perikarya of layer IV of Oc 1 M contralateral to the remaining eye. Again, the interpretation is comparable to that given for the ipsilateral hemisphere. Enucleations performed before the cortical maturation had finished could lead to an alteration of cortical neuropil development, the latter possibly being caused by enucleation-induced changes of the DLG neurons and the geniculo-cortical projection. The results of the transneuronal transport experiments (Figs. 2, 3), showing a reduced transport of tracer to the cortex, corroborate this assumption.

The major results of the grain density measurements in the visual cortices following intravitreous injection of [^{3}H]-proline were that (1) Oc 1 M ipsilateral to the injected eye was labeled in the enucleated rats (Fig. 1) in contrast to normal, and (2) the labeling of Oc 1 M contralaterally was reduced compared with the respective control area (Fig. 2b).

In rodents enucleated at birth, the number of ipsilaterally projecting retinal ganglion cell axons is increased (Shirokawa et al. 1983; Fig. 4). In the respective DLG, the terminal field reached by these axons is enlarged compared to normal (Campbell et al. 1985; Lund et al. 1973; Manford et al. 1983, 1984; Shirokawa et al. 1983; So et al. 1984). Additionally, electrophysiological experiments revealed (Fukuda et al. 1983; Shirokawa et al. 1983) that the DLG relay neurons responding to stimulation of the uncrossed optic fibers were much more frequently encountered than in normal rats. It can be assumed that these relay neurons, normally receiving their afferents from the contralateral eye and projecting to the Oc 1 M of the same hemisphere, have been taken over in the enucleated rats by the enlarged number of afferents from the ipsilateral eye. Labeling of Oc 1 M ipsilateral to the remaining eye can thus be interpreted as morphological evidence for an enucleation-induced ipsilateral retino-geniculo-cortical (Oc 1 M) projection, not existing in normal rats.

Because, following monocular enucleation, the optic nerve of the remaining eye showed the same number of axons as in normal rats (Lam et al. 1982; Sefton and Lam 1984), and branching of optic nerves in the chiasm practically did not exist under these experimental conditions (Hsiao 1984), the number of retinal ganglion cell axons projecting to the contralateral DLG was reduced (Shirokawa et al. 1983; Fig. 4) compared to normal. The tracer experiments, reported in the present study, expanded the abovementioned reports onto the cortical level. The reduced crossed retino-geniculate pathway gives rise to a reduced geniculo-cortical (Oc 1 M) projection.

Acknowledgements. This work was supported by the Deutsche Forschungsgemeinschaft. We are grateful to U. Blohm for excellent technical assistance and Dr. John Sarvey for improving the English.

References

Adams AD; Forrester JM (1968) The projection of the rat's visual field on the cerebral cortex. Q J Exp Physiol 53:327–336

Blue ME, Parnavelas JG (1983) The formation and maturation of synapses in the visual cortex of the rat. II. Quantitative analysis. J Neurocytol 12:697–712

Burck HC (1973) Histologische Technik. Thieme, Stuttgart

Campbell G, So KF, Lieberman AR (1985) Identification of synapses formed by the aberrant, uncrossed retinogeniculate projection in the hamster after neonatal enucleation. Dev Brain Res 21:137–140

Dräger UC (1974) Autoradiography of tritiated proline and fucose transported transneuronally from the eye to the visual cortex in pigmented and albino mice. Brain Res 82:284–292

Dräger UC (1978) Observations on monocular deprivation in mice. J Neurophysiol 41:28–42

Dräger UC, Olsen JF (1980) Origins of crossed and uncrossed retinal projections in pigmented and albino mice. J Comp Neurol 191:383–412

Dreher B, Sefton AJ, Ni SYK, Nisbett G (1985) The morphology, number and distribution of class I cells in the retina of albino and hooded rats. Brain Behav Evol 26:10–48

Fifkova E (1970) The effects of monocular deprivation on visual centers in rats. J Comp Neurol 140:431–438

Fifkova E, Hassler R (1969) Quantitative morphological changes in visual centers in rats after monocular deprivation. J Comp Neurol 135:167–178

Fukuda Y, Sumitomo I, Hsaio CF (1983) Effects of neonatal enucleation on excitatory and inhibitory organizations of the albino rat lateral geniculate nucleus. J Neurophysiol 5:46–60

Globus A, Scheibel AB (1967) The effect of visual deprivation on cortical neurons: a Golgi study. Exp Neurol 19:331–345

Gudden B (1870) Experimentaluntersuchungen über das periphere und centrale Nervensystem. Arch Psychiatr Nervenkr 2:693–723

Heumann D, Rabinowicz T (1980) Postnatal development of the dorsal lateral geniculate nucleus in the normal and enucleated mouse. Exp Brain Res 38:75–85

Heumann D, Rabinowicz T (1982) Postnatal development of the visual cortex of the mouse after enucleation at birth. Exp Brain Res 46:99–109

Heumann D, Leuba G, Rabinowicz T (1977) Postnatal development of the mouse cerebral cortex. II. Quantitative cytoarchitectonics of visual and auditory areas. J Hirnforsch 18:483–500

Hsiao K (1984) Bilateral branching contributes minimally to the enhanced ipsilateral projection in monocular Syrian golden hamster. J Neurosci 4:368–373

Jeffery G (1984) Retinal ganglion cell death and terminal filed retraction in the developing rodent visual system. Dev Brain Res 13:81–86

Kretschmann HJ, Wingert F (1971) Computeranwendungen bei Wachstumsproblemen in Biologie und Medizin. Springer, Berlin Heidelberg New York

Lam K, Sefton A, Bennett MR (1982) Loss of axons from the optic nerve of the rat during early postnatal development. Dev Brain Res 3:487–491

Leuba G, Heumann D, Rabinowicz T (1977) Postnatal development of the mouse cerebral neocortex. I. Quantitative cytoarchitectonics of some motor and sensory areas. J Hirnforsch 18:461–481

Lund RD, Cunningham TJ, Lund JS (1973) Modified optic projections after unilateral eye removal in young rats. Brain Behav Evol 8:51–72

Manford M, Campbell G, Lieberman AR (1983) Postnatal development of ipsilateral projections from the eye to the lateral geniculate body in normal albino rats and in rats unilaterally enucleated at birth. J Anat 136:607–608

Manford M, Campbell G, Lieberman AR (1984) Postnatal development of ipsilateral retino-geniculate projections in normal albino rats and the effect of removal of one eye at birth. Anat Embryol 170:71–78

Miller M (1981) Maturation of rat visual cortex. I. A quantitative study of Golgi-impregnated pyramidal neurons. J Neurocytol 10:859–878

Miller M, Peters A (1981) Maturation of rat visual cortex. II. A combined Golgi-electron microscope study of pyramidal neurons. J Comp Neurol 203:555–573

Montero VM (1973) Evoked responses in the rat's visual cortex to contralateral, ipsilateral and restricted photic stimulation. Brain Res 53:192–196

Montero VM (1981) Comparative studies on the visual cortex. In: Woolsey CN (ed) Cortical sensory organization, vol 2. Humana, Clifton, pp 33–81

Parnavelas JG, Uylings HBM (1980) The growth of non-pyramidal neurons in the visual cortex of the rat: a morphometric study. Brain Res 193:373–382

Parnavelas JG, Bradford R, Mounty EJ, Lieberman AR (1978) The development of non-pyramidal neurons in the visual cortex of the rat. Anat Embryol 155:1–14

Ribak CE, Peters A (1975) An autoradiographic study on the projections from the lateral geniculate body of the rat. Brain Res 92:341–368

Riccio RV, Matthews MA (1985) The postnatal development of the rat primary visual cortex during optic nerve impulse blockade by intraocular tetrodotoxin: a quantitative electron microscopic analysis. Dev Brain Res 20:55–68

Rogers AW (1973) Techniques in autoradiography. 2nd edn. Elsevier, Amsterdam

Romeis B (1968) Histologische Technik. Oldenburg, München

Ruiz-Marcos A, Valverde F (1970) Dynamic architecture of the visual cortex. Brain Res 19:25–39

Sachs L (1978) Angewandte Statistik. Springer, Berlin Heidelberg New York

Schleicher A, Zilles K, Kretschmann HJ (1978) Automatische Registrierung und Auswertung eines Grauwertindex in histologischen Schnitten. Anat Anz 144:413–415

Schleicher A, Zilles K, Wree A (1986) A quantitative approach to cytoarchitectonics: software and hardware aspects of a system for the evaluation and analysis of structural inhomogeneities in nervous tissue. J Neurosci Meth 18:221–235

Schober W, Werner L, Brauer K (1977) Quantitative morphologische Veränderungen im Corpus geniculatum laterale pars dorsalis (Cgld) der Albinoratte nach unilateraler Enukleation. Anat Anz 142:385–402

Sefton AJ, Lam K (1984) Quantitative and morphological studies on developing optic axons in normal and enucleated albino rats. Exp Brain Res 57:107–117

Shirokawa T, Fukuda Y, Sugimoto T (1983) Bilateral reorganization of rat optic tract following enucleation of one eye at birth. Exp Brain Res 51:172–178

So KF, Woo HH, Jen LS (1984) The normal and abnormal postnatal development of retinogeniculate projections in golden hamsters: an anterograde horseradish peroxidase tracing study. Dev Brain Res 12:191–205

Tsang VC (1937) Visual centers in blinded rats. J Comp Neurol 66:211–261

Uylings HBM, Parnavelas JG, Walg H, Veltman WAM (1980) The morphometry of branching pattern of developing non-pyramidal neurons in the visual cortex of rats. Mikroskopie (Suppl)37:220–224

Valverde F (1968) Structural changes in the area striata of the mouse after enucleation. Exp Brain Res 5:274–292

Ward R, Tremblay L (1982) Postnatal maturation of the ipsilateral optic tract in pigmented and albino mice. J Hirnforsch 23:311–313

Warren MA, Bedi KS (1984) A quantitative assessment of the development of synapses and neurons in the visual cortex of control and undernourished rats. J Comp Neurol 227:101–108

Werner L, Schober W, Winkelmann E, Mühlig P, Kühne C (1977) Konventionelle und automatische quantitative Untersuchung histologischer Präparate nach experimenteller Beeinflussung des visuellen Systems der Ratte. Verh Anat Ges 71:127–132

Werner L, Voss K, Seifert I, Neumann E (1981) Age-related classification of pyramidal and stellate cells in the rat visual cortex: a Nissl study with the "Morphoquant". J Hirnforsch 22:397–403

Wree A, Schleicher A, Zilles K (1982) Estimation of volume fractions in nervous tissue with an image analyzer. J Neurosci Meth 6:29–43

Wree A, Zilles K, Schleicher A, Schwientek P (1983) Quantitative Cytoarchitektonik des primären visuellen Cortex der Ratte. Verh Anat Ges 77:249–250

Wree A, Kulig G, Gutmann P, Zilles K (1985) Modification of callosal afferents of the primary visual cortex ipsilateral to the remaining eye in rats monocularly enucleated at different stages of ontogeny. Cell Tissue Res 242:433–436

Wree A, Angenendt HW, Zilles K (1986) The size of the zone of origin of callosal afferents projecting to the primary visual cortex contralateral to the remaining eye in rats monocularly enucleated at different postnatal ages. Anat Embryol 174:91–96

Wree A, Schleicher A, Beck T, Zilles K (in press a) The cortical areas of the posterior cortex of the rat. Fortschr Zool 35

Wree A, Schleicher A, Beck T, Zilles K (1987) Quantitative Veränderungen im visuellen Cortex der Ratte nach Enukleation dargestellt mit der (14C)-2-Desoxyglucosetechnik. Verh Anat Ges 81:871–872

Wree A, Schleicher A, Zilles K, Beck T (in press b) Changes in the primary visual cortex caused by enucleation. Fortschr Zool 35

Zilles K (1985) The cortex of the rat. A stereotaxic atlas. Springer, Berlin Heidelberg New York

Zilles K, Wree A (1985) Cortex: Areal and laminar structure. In: Paxinos G (ed) The rat nervous system, vol 1. Academic Press, Sydney New York London, pp 375–416

Zilles K, Zilles B, Schleicher A (1980) A quantitative approach to cytoarchitectonics. VI. The areal pattern of the cortex of the albino rat. Anat Embryol 159:335–360

Zilles K, Blohm U, Koch I (1983) Neurohistologische Methoden. I. Markscheidenfärbung. MTA Praxis 29:295–297

Zilles K, Wree A, Schleicher A, Divac I (1984) The monocular and binocular subfields of the rat's primary visual cortex: a quantitative approach. J Comp Neurol 226:391–402

Zilles K, Schleicher A, Rath M, Glaser T, Traber J (1986) Quantitative autoradiography of transmitter binding sites with an image analyzer. J Neurosci Meth 18:207–220

Split Brain Surgically Performed in Developing and in Adult Cats: Physiological Properties and Recovery of Visual Cortex Neurons

U. Yinon and M.Chen

Physiological Laboratory, Maurice and Gabriela Goldschleger Eye Research Institute, Tel-Aviv University Sackler Faculty of Medicine, Chaim Sheba Medical Center, Tel-Hashomer, 52621, Israel

1 Introduction

Physiologically, several different experimental approaches have been so far adopted in order to demonstrate the involvement of visual cortex cells in interhemispheric relationships. Of these approaches the main ones will be briefly described here.

Primarily, the physiological properties of nerve fibers in the posterior corpus callosum itself, connecting interhemispherically the two visual cortical areae, were studied in cats following direct unit recording (Berlucchi et al. 1967; Hubel and Wiesel 1967; Shatz 1977). In general, the receptive field organization and other spatial visual properties of the callosal fibers in the above-mentioned studies were similar to those of the cells in the visual cortex to which they were connected in both hemispheres.

Other investigators have studied the physiological properties of callosally driven cortical cells in cats by measuring the amount of interhemispheric transfer following electrical stimulation of the contralateral hemisphere (Harvey 1980; Blakemore et al. 1983). It has been mainly proved in these studies that the corpus callosum interhemispherically connects areas 17 and 18 of the visual cortex, and especially the boundary region between them.

The effect of interhemispheric relationships on cortical cells have been also investigated following transection of the corpus callosum. When at first studied by Payne et al. (1980), it was found that binocularity in the visual cortex has been reduced following transection of the corpus callosum in adult cats. This result has been supported by the findings of Dreher and Cotte (1975); Blakemore et al. (1983), and Cynader et al. (1986), showing a considerable reduction in the binocularity level of visual cortex cells in one hemisphere following extended lesions made in the contralateral side. In contrast, other research groups (Zeki and Fries 1980; Minciacchi and Antonini 1984; Elberger and Smith 1985) claimed no changes in binocularity of cortical cells following corpus callosum transection in cats.

The involvement of visual cortex cells in interhemispheric relationships has been also physiologically studied in cats following surgical interruption of the contralateral thalamocortical pathways by midsagittal transection of the optic chiasm (Berlucchi and Rizzolatti 1968; Yinon and Hammer 1981, 1985, Cynader et al. 1981, 1986; Lepore and Guillemot 1982; Yinon et al. 1982, 1986c; Milleret

H. Flohr (Ed.)
Post-Lesion Neural Plasticity

and Buser 1984; Antonini et al. 1985). Under this condition only cortical cells with transcallosal contralateral visual input could be binocularly driven. Thus, it was possible in this way to study the properties of cortical cells interhemispherically connected with the fellow visual cortex. This has been reflected in the typical ocular dominance distribution of cells studied in the callosal projection zone of areae 17 and 18 of the visual cortex. In the above-mentioned studies either all or the majority of the cortical cells were limited in their response to stimulation of the ipsilateral eye. Only a minor proportion of cells were found to possess a contralateral visual input, indicating the presence of an interhemispheric transfer.

Another experimental approach in which the interhemispheric transfer of visual input to cortical cells has been studied, had consisted of unilateral transection of the optic tract in cats (Yinon et al. 1982, 1984, 1986c, 1987; Lepore et al. 1983; Podell et al. 1984; Yinon and Podell 1987). In these studies a certain proportion of cortical cells in the deafferented hemisphere, albeit small, had a visual input. These cells were therefore interhemispherically driven from the contralateral hemisphere since there was no other source of visual input available to them. However, the majority of the cells in the deafferented hemisphere remained visually unresponsive.

In view of this involvement of visual cortex cells in the interhemispheric relationships, in the present study their effectivity was studied in the absence of these relationships. For this purpose visual split brain has been induced in cats by interrupting both the callosal and the contralateral thalamocortical pathways to the visual cortex. Furthermore, the prevention of interhemispheric transfer to visual cortex cells in this way is assumed to induce temporary effects if produced during the developmental period. A plasticity-dependent mechanism might be involved in the recovery of cortical cells despite the complete absence of visual input from the fellow hemisphere. This hypothesis was studied in the present experimental model. Preliminary results on the subject have already been previously reported by us (Yinon et al. 1986a; Yinon and Chen 1987).

2 Methods

2.1 Animal Groups

The following groups of animals were studied:

1. The main experimental group, constituting five kittens (556 cortical cells were characterized) were operated both for callosal and chiasm transection (visual split brain, OCCCK).
2. Cats similarly operated for split brain during adulthood (OCCCA; four animals; 201 cells).
3. Kittens in which the posterior corpus callosum was sectioned (CCK; five animals; 159 cells).
4. Cats operated for posterior callosal transection during adulthood (CCA1; seven animals; 331 cells).
5. Cats operated for complete transection of the corpus callosum (CCA2; five animals; 435 cells).

6. Kittens operated for transection of the optic chiasm (OCK; five animals; 406 cells).
7. Cats operated for the transection of the optic chiasm during adulthood (OCA; six animals; 338 cells).
8. Normal intact adult cats (NORMAL; 15 animals; 523 cells).

2.2 *Surgical Procedures*

In the present study cats and kittens in which visual split brain was induced were primarily studied. This condition was achieved by separation of the cortical hemispheres, following transection of the corpus callosum. Furthermore, the binocular retinal projection to each cortical hemisphere was dissociated by interrupting the contralateral pathways in the optic chiam; this was done in order to be able to limit each cortical hemisphere to its own eye (although impractical, this could be also achieved by sectioning the ipsilateral pathways in the chiasm). Furthermore, splitting the chiasm in this way prevents the possible masking of callosal input originated from the contralateral hemisphere by a direct thalamic input from the contralateral eye. For comparison, animals which were subjected independently to the effects of corpus callosum or optic chiasm transections were examined.

The callosal operation on adult cats (CCA1, CCA2, OCCCA) was made at an unknown age and on kittens (CCK, OCCCK) when they were 6–9 weeks old. It was carried out under sterile conditions, following general anesthesia with a combination of Ketalar (15 mg/kg, i.m.) and Rompun (3 mg/kg, i.m.). Craniotomy was then made unilaterally to the sagittal suture and anteroposteriorly, with the zero (interaural) line in its center, using a David Kopf stereotaxic apparatus. The dura mater was removed and the falx cerebri and the corpus callosum exposed under Zeiss operating microscope. The entire depth of the corpus callosum was midsagittally sectioned from its posterior edge, using a microsurgical blade. In the CCA1 and OCCCA adult operated cats the transections had included approximately one third to one half of the callosal length (4–8 mm) and in the CCA2 adult operated cats the corpus callosum was completely (15–17 mm) sectioned. The callosal transections in the CCK and OCCCK kittens had ranged between 3 to 7 mm at the age of the surgery; this had consisted of approximately 50% of the total callosal length at that time. In the operated kittens and adult cats the callosal transections had thus included all the fibers connecting the visual areae in the two hemispheres (Luttenberg and Marsala 1963; Innocenti 1980). Following the transection, the cranial opening was covered with Surgicel (sterilized oxidized cellulose) serving as substitution for the removed dura mater and with the original bone flap. Antibiotic treatment was daily given for a week. Postoperative time for all the above described callosally operated cats and kittens was 3–39 months.

The chiasm operation on the adult cats (OCA, OCCCA) was made at an unkown age and on kittens (OCK, OCCCK) when they were 6–9 weeks old. The transection was midsagittally performed in the transbuccal approach (Myers 1955). Other details regarding the anesthesia, surgical procedures, and postoperative maintenance were in general as above mentioned and as previously described (Yinon and Hammer 1981, 1985). The completeness of the chiasm transection was examined during the operation itself, in the preserved brain after the electrophysiological session and in the histological preparations (see Sect. 2.3). Postoperative time for all the above-described chiasm operated cats and kittens was 1–22.5 months.

Visual split brain was induced in cats and in kittens by midsagittal transection of both the corpus callosum and the optic chiasm, as above described. The chiasm in the OCCCK kittens was sectioned during the 7–8th postnatal week and the callosum nearly one week later. In the adult cats (OCCCA) the operation was made at an unknown age with an interval of 1–6 weeks between the chiasm and the callosal transections.

2.3 Histology

The length of the callosal and the completeness of the chiasm transections, the location of the electrode penetrations within the cortical layers and the condition of the retinae (in the chiasm operated animals) were histologically examined in all groups of cats after the experiments. Blocks containing separately the callosum, the chiasm, and the visual cortex were prepared from the preserved brains following fixation in 10% buffered Formalin. Embedding was made in paraffin and staining of the sections (10 μ) in Cresyl violet (Kluver and Barrera 1953). The eyes were similarly prepared and their sections (6 μ) were stained with Hematoxylin Eosin.

The histological studies revealed the exact amount of callosal transections, as above described (see Sect. 2.2). Furthermore, it is important to note here that no difference was histologically found between the side of the cortex exposed at the time of craniotomy (during the callosotomy) and the intact fellow hemisphere (this has been also physiologically confirmed, as subsequently reported in Sect. 3).

As to the chiasm transections, they were complete in all the operated animals of the present study, as inferred from the histological cross-sections. This has been further confirmed by the absence of ganglion cells noticed in the nasal retinae in both eyes of the split chiasm cats.

The position of the electrode penetrations in the visual cortex of the various groups of cats were identified following electrical lesions which were made along the penetrations during the recording sessions. It was found that they were mostly located in the appropriate sites with regard to both cortical topography (areae 17, 18 and their boundary region) and layering.

2.4 The Electrophysiological Session

Prior to the electrophysiological sessions the cats were initially anesthetized with Ketalar and Rompun as above described, and fixed on a stereotaxic apparatus. They were then paralyzed for fixation of the extraocular eye muscles (Flaxedil, 7.5 mg/kg/h, i.v.) and artificially respired with air. Light anesthesia was continually maintained during the experiments with Thiopental Sodium (1 mg/kg/h, i.v.). Throughout the experiments the rectal body temperature (38–39 °), expiratory CO_2 level (4–5%), ECG rate and EEG (recorded with surface electrode on the visual cortex) were constantly monitored and controlled.

Craniotomy was carried out above the visual cortex and unit recording was made in both cortical hemispheres. Cells were studied mainly in the boundary region between areae 17 and 18, the callosal projection zone and where the vertical meridian is represented (Hubel and Wiesel 1967; Harvey 1980; Innocenti 1980). Our recording sites in this region were morphologically selected by using cortical landmarks (Otsuka and Hassler 1962; Tusa et al. 1979) and histologically validated in the cortical sections which had contained the penetration tracks (see Sect. 2.3). The recordings were extracellularly made using tungsten microelectrodes (2–4 Mohm). Sampling of the cells was randomly carried out from all cortical layers along each penetration (usually 4–6 mm long), with a minimum interval of 200 μ between recording sites.

2.5 Optics and Stimulation

The eyes of the cats were refracted and optically corrected; artificial pupils were then applied (the cat's pupils were maximally dilated). Following this, the spatial position of the vertical meridian was determined according to the location of the area centralae and optic disks projected for each animal on a tangent screen, 2 m in front of the cat's eyes. In the majority of the cats the actual position of the vertical meridian was in correspondence with the theoretical value given by Vakkur et al. (1963) for cats. It has also been confirmed by the positions of the receptive fields, the majority of which in all our operated and normal cats were located 4–5 ° within each side of the actual vertical meridian, as previously found for cortical cells with callosal input (Harvey 1980).

Stimulation and receptive field mapping were carried out mainly with moving light slits of various sizes. Classification of cortical cells according to their receptive field properties was made as previously described (Hubel and Wiesel 1962; Blakemore and Van Sluyters 1975; Fregnac and Imbert 1978).

Other details regarding optical considerations, various experimental procedures, as well as analysis of the data were previously described (Yinon and Hammer 1981, 1985).

3 Results

The split brain condition comprises two components, the first is the separation of the two hemispheres by corpus callosum transection and the second is the dissociation of the binocular retinal inputs to the cortex by splitting the optic chiasm. In this section we will describe the physiological consequences resulting from each of these surgical manipulations and their effects on cortical cells. Finally, we will describe the combined effect of these operations, i.e., the effect of visual split brain.

In the callosally sectioned cats, the results show a very different distribution of cortical cells according to their ocular dominance in comparison to that of the normal control cats. This difference was mainly expressed in the proportion of

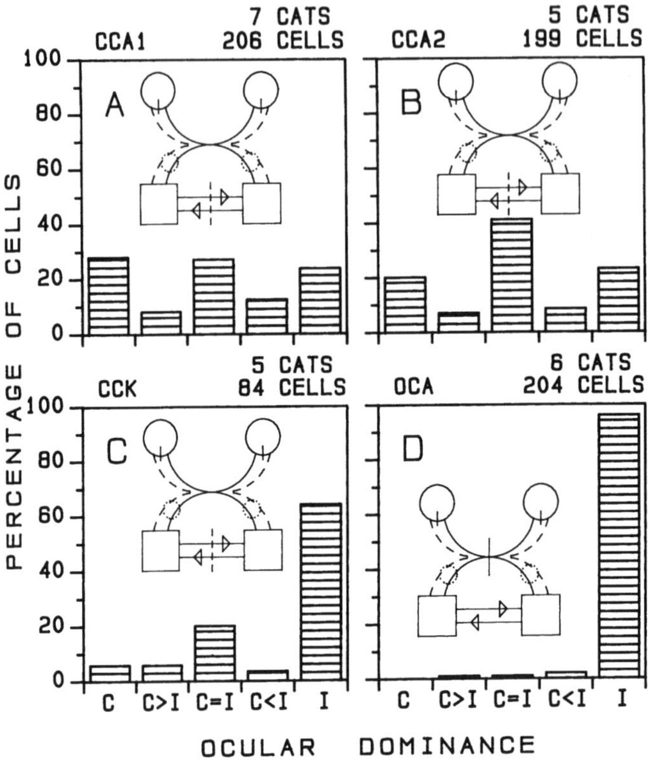

CCA1
7 CATS
206 CELLS
CCA2
5 CATS
199 CELLS
CCK
5 CATS
84 CELLS
OCA
6 CATS
204 CELLS
A
B
C
D
PERCENTAGE OF CELLS
C
C>I
C=I
C<I
I
OCULAR DOMINANCE

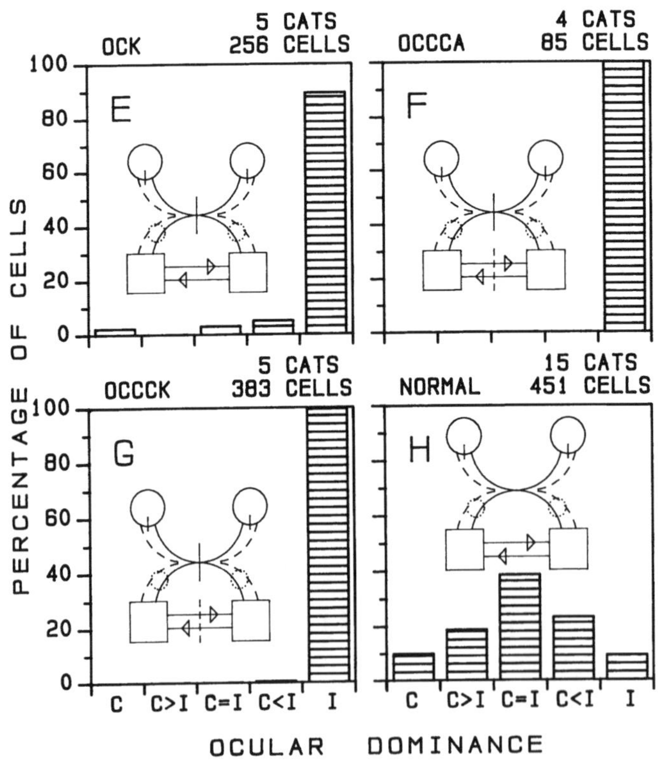

OCK
5 CATS
256 CELLS
OCCCA
4 CATS
85 CELLS
OCCCK
5 CATS
383 CELLS
NORMAL
15 CATS
451 CELLS
E
F
G
H
PERCENTAGE OF CELLS
C
C>I
C=I
C<I
I
OCULAR DOMINANCE

binocularly driven cells. In the adult cats with posterior callosal transection (CCA1) a remarkable reduction in binocularity was found (48.0% of the cells were binocularly driven; Fig. 1 A). The proportion of these cells in the adult cats which have undergone complete transection of the corpus callosum (CCA2) was 56.7% (Fig. 1 B). Furthermore, in the later group of cats (CCA2) the proportion of binocularly driven cells was almost identical in the right (57.9%) and in the left (55.0%) hemisphere. In comparison, in the cats similarly operated during development (CCK) the smallest percentage (29.7%) of binocular cells was found (Fig. 1 C). The proportion of binocularly driven cells in our normal control cats was 80.7% (Fig. 1 H), similarly to that which has been previously obtained by other investigators (Hubel and Wiesel 1962; Blakemore and Van Sluyters 1975; Fregnac and Imbert 1978) and by us (Podell et al. 1984; Yinon and Hammer 1985) for normal cats.

While in the CCA1 cats the proportion of cells with contralateral eye input was 76.2% and the ipsilateral ones 71.9%, their proportion was 90.7% for the contralateral and 90.0% for the ipsilateral eye in the normal control cats. The results were similar for the CCA1 and CCA2 cats with regard to this component of the ocular dominance distribution. However, a different result was obtained for animals callosally sectioned as kittens (CCK); in this group only 35.7% of the cells had a contralateral eye input, while 94.1% had an ipsilateral input.

The above described differences in the ocular dominance distributions of cells among the various groups of callosotomized cats are also reflected in the statistical analysis. A significant difference was obtained between the CCA1 ($Chi^2 = 74.72$; d.f. = 4; $p < 0.001$), the CCA2 ($Chi^2 = 59.12$; d.f. = 4; $p < 0.001$) and the normal control cats. It is also important to mention here that the observed difference between the two groups of callosally sectioned adult cats ($Chi^2 = 10.35$; d.f. = 4; $0.02 < p < 0.05$) is of marginal significance.

In addition to these changes in the ocular dominance distribution of cells in the callosotomized cats, changes had also taken place in the responsiveness level

Fig. 1. A–H Ocular dominance distributions of cells in the visual cortex (areae 17–18 and their boundary) of cats that underwent midsagittal transection of the posterior (**A** *CCA1*) and the entire (**B** *CCA2*) corpus callosum during adulthood. The results are also presented for cats with posterior callosal transection during the developmental period (**C** *CCK*). The distributions of cells are also given for cats following transection of the optic chiasm during adulthood (**D** *OCA*) and during development (**E** *OCK*). The data for visually split brain cats (in which transections of both optic chiasm and corpus callosum were concurrently performed) operated during adulthood (**F** *OCCCA*) and during development (**G** *OCCCK*) are also shown. Comparison is made with the results of normal control adult cats (**H** *NORMAL*). The distributions of cortical cells are calculated as percentage of the total number of responsive cells in both hemispheres. Ocular dominance groups C and C>I represent cells either monocularly or predominantly driven through the contralateral eye, respectively; groups I and C<I represent cells either monocularly or predominantly driven through the ipsilateral eye, respectively; group C=I represents cells equally binocularly driven through the contralateral and the ipsilateral eye. The numbers of cats and responsive cells studied are demonstrated in the *upper right corner* of each histogram. A general scheme of the various operations and of the visual system is presented in each histogram

(visually unresponsive cells were recognized according to the presence of action potentials and a typical spontaneous discharge). While in our normal cats 86.2% of the cells were visually responsive, their proportions were remarkably smaller in the CCA1 (62.2%) and in the CCA2 (45.7%) cats. Similarly, a considerable diminution in the responsiveness level was found in the developmentally operated (CCK) cats (52.8%).

For the split chiasm cats, the main difference in comparison to the normal control cats is indicated by the proportions of ipsilaterally driven cells. The proportion of cells driven monocularly through the ipsilateral eye in the OCA cats was 98.5% in the right hemisphere and 94.9% in the left hemisphere (Fig. 1 D). The ipsilateral eye dominance had occurred even after long periods of postoperative recovery (>6 months) given to our chronically operated adult cats, as previously reported (Yinon and Hammer 1985). In cats operated during development (OCK) the proportion of the ipsilaterally driven cells was 87.8% in the right hemisphere and 90.6% in the left hemisphere (Fig. 1 E). Thus, the general pattern of responsiveness following chiasm split indicates a symmetry between the two hemispheres regarding the amount of ipsilateral activation of the cells.

Although the ipsilateral eye remarkably dominates the cells in the visual cortex of all split chiasm cats, there is a difference between the OCA and the OCK cats as already indicated by the results above. Some reduction is seen in the proportion of cells with ipsilateral eye input (in both hemispheres) in the OCK cats (97.7%) in comparison to the OCA cats (100.0%). The difference was, however, greater when the ipsilaterally monocularly driven cells were compared; they constituted 96.1% of the cells in the OCA cats and 89.5% in the OCK cats. In comparison, while in the normal control cats the proportion of all cells with ipsilateral eye input was 90.0%, the monocularly driven ones had constituted 9.3% of the cells.

The results of the cats operated for split chiasm during adulthood (OCA) are clearly distinguished from those of cats similarly operated during development (OCK). The difference is mainly reflected in the proportion of monocularly and binocularly driven cells with input from the contralateral eye (in both hemispheres). The proportion of these cells was 4.0% in the OCA cats while in the OCK cats they constituted 10.5% of the cells. Considering the individual cats in which these contralaterally driven cells were present, the appropriate proportions ranged between 1.7 and 15.6% for the OCA cats and 9.7 to 26.5% for the OCK cats. This tendency for an increased contralateral eye input in the OCK cats is in keeping with the finding that 2.3% of their cortical cells were monocularly driven through the contralateral eye, while in the OCA cats not one cell was found in this category. In comparison, in the normal control cats the proportion of cells with contralateral eye input was 90.7% while 10.0% of the cells were monocularly driven through the contralateral eye.

The differences between the split chiasm and the control cats were also reflected in the proportions of binocularly driven cells. While a small proportion of these cells (4.0%) was found in the OCA cats a higher proportion (8.2%) was found in the OCK cats. In comparison, in the normal controls the binocularly driven cells constituted the majority of the responsive cells (80.7%).

These differences between the operated and the normal control cats were also reflected in the statistical analysis. A significant difference, albeit small, was found between the ocular dominance distributions of cortical cells in the OCK and the OCA cats ($Chi^2 = 13.2$; D.F. = 4; $0.01 < P < 0.02$). As expected, the difference between the OCA ($Chi^2 = 457.4$; D.F. = 4; $P < 0.001$), the OCK ($Chi^2 = 447.2$; D.F. = 4; $P < 0.001$) and the normal control cats is highly significant. Furthermore, there was no significant difference between OCA cats studied shortly (1–1.5 months) and long time (6.5–7 months) after the operation ($Chi^2 = 7.4$; D.F. = 3; $0.05 < P < 0.1$).

We also found considerable changes in the excitability level of the cortex in the split chiasm cats, as reflected in the proportion of visually responsive cells. The proportion of responsive cells was remarkably smaller in the OCA (60.3%) and in the OCK (63.1%) cats in comparison to the normal control cats (86.2%).

In the visually split brain cats a full separation of the two hemispheres was found, as reflected by the presence of only ipsilaterally driven cells. This was the result (100% ipsilateral cells) in both hemispheres of the split brain adult cats (OCCCA; Fig. 1 F). This result is practically the same as that (99.5% cells) found for the split brain OCCCK cats (Fig. 1 G). Thus, each hemisphere has been found to be limited to its own eye, independently of the other hemisphere.

The general level of cortical excitability has been also found to be affected in the split brain cats, as indicated by the proportion of visually responsive cells found. While the proportion of responsive cells was 86.2% in the normal control cats, it constituted 42.3% of the cells in the OCCCA cats. In comparison, the proportion of responsive cells in the OCCCK cats was 68.9%, indicating a remarkable amount of recovery. However, it must be noted here that the proportion of responsive cells in our split brain cats, no matter whether operated as kittens or during adulthood, is much smaller than the expected one, considering the absence of only one source of visual input (only the contralateral input is absent in these cats; the remaining ipsilateral pathway was not affected by the surgical procedure). Therefore, the amount of reduction in responsiveness as presently found in the split brain cats, reflects additional cortical inactivation induced by the elimination of the callosal pathway.

4 Discussion

It has been concluded from the findings of the present study that the binocularity level of cortical cells has been considerably reduced following callosotomy, whether it was performed in kittens or in adult cats. Our findings in adult operated cats, although obtained from areae 17 and 18, are in keeping with previous ones from area 17 of similarly operated adult cats (Payne et al. 1980, 1984 a, b). More specifically, the binocularity level found by Payne et al. (1984 a) in their chronic cats (58.0% binocular cells; survival time: 8–10 months) is almost equal to our finding in chronic cats (CCA2) with full callosal transection (56.7% cells; survival time: 3–12 months). It is important to mention here that even the general profile of the ocular dominance distribution of cells in the studies of Payne et al.

(1980, 1984a) is similar to that one presently obtained by us. This similarity is expressed by the characteristic appearance of a depression in the ocular dominance distribution representing cells unequally binocularly driven by the two eyes. It is conceivable that the input from the nondominant eye to these cells had disappeared under this condition, probably due to the induction of a diminished cortical excitability. That such a diminution takes place has been already presently shown by the massive existence of nonresponsive cells following callosotomy.

It is worth mentioning here that in view of the anatomical organization of the visual pathways, the condition induced by corpus callosum transection is not expected to affect cortical binocularity. However, a remarkable effect, as indicated by the above reduction in cortical binocularity, was found in cats operated for callosotomy either when young or during adulthood. Whatever the extent of the effect of callosotomy on binocularity, it is not yet explainable, although it is conceivable in view of the many binocular functions attributed to the corpus callosum.

In contrast to the finding of binocularity diminution following callosotomy (Payne et al. 1980, 1984a, b) as well as those presently shown, different results have been obtained by other investigators (Zeki and Fries 1980; Minciacchi and Antonini 1984; Elberger and Smith 1985). They showed that chronically produced callosal transection in adult cats does not disrupt binocularity of cells in cortical areae 17 and 18. Furthermore, Elberger and Smith (1985) found no change in binocularity in cats operated after the first postnatal month, in contrast to our present results in cats operated during development. We do not know yet how to explain the contradiction between the above described results with regard to the binocularity level in callosotomized kittens and adult cats.

As to the condition of optic chiasm transection, it has been presently found to drastically affect the ocular dominance of cortical cells in areae 17–18 of cats operated both during development and in adulthood. An almost complete disappearance of binocularly driven cortical cells was found in these animals, resulting from the dissociation of the two eyes under this condition. Thus, the visual cortex in each hemisphere was found to be practically monocular in its visual responsiveness. However, this monocular dominance is attributed not only to the interocular dissociation but also to the disappearance of cells with contralateral eye input, i.e., transcallosally driven cells. This had finally resulted in a great amount of hemispheric separation.

It was expected to find out that in the absence of the direct bilateral contralateral thalamic pathways following chiasm split, the corpus callosum would serve as an indirect alternative pathway for the ipsilateral thalamic inputs. An input to cortical cells from such an alternative pathway would restore binocularity, especially in areae 17, 18 boundary, which represents the vertical meridian of the visual field. On the contrary, it was found in the present study that cortical cells in cats operated for chiasm transection during adulthood were not compensated for, by and large, by an input via the corpus callosum. Furthermore, even the regular interhemispheric callosal activity had almost vanished in these animals. The amount of interhemispheric separation thus induced in these cats is similar to the separation found in our cats which have undergone callosal transection.

The minor callosal transfer found in our cats operated for chiasm split during adulthood (OCA), is in contrast to the findings of other investigators who showed a considerable transfer in similarly operated cats (Berlucchi and Rizzolatti 1968; Cynader et al. 1981; Lepore and Guillemot 1982; Antonini et al. 1985). In contrast, no such transfer has been obtained by Milleret and Buser (1984) and a small degree of transfer was found by us (Yinon and Hammer 1981, 1985; Yinon et al. 1982, 1986c) in areae 17–18 of similarly operated cats. The later results are also supported by recent behavioral findings of Lepore et al. (1986), showing no transfer in cats under these conditions, as evidenced by the absence of stereoperception. The differences among these investigators with regard to the interhemispheric transfer in cats chiasm sectioned at adulthood are not yet explainable.

Unlike in our adult cats operated for split chiasm (OCA), we have new evidence that recovery, although partial, takes place in the interhemispheric transfer following split chiasm operation in developing kittens (OCK). This has been preliminarily reported previously by us (Yinon et al. 1986b). The recovery is expressed by the considerably greater proportion of cortical cells driven via the corpus callosum in the OCK in comparison to the OCA cats. Furthermore, in contrast to our results and those of the previously mentioned investigators in cats operated during adulthood, a "pure" interhemispheric transfer was found in the OCK cats. This was revealed by the presence of cortical cells, albeit a small proportion, which had monocular input from the contralateral eye. The findings in animals operated as kittens thus indicate the involvement of a plasticity-induced mechanism during the developmental period which enhances the interhemispheric transfer of information between homolateral visual cortical areae. The anatomical substrate for this enhancement has been already given by Innocenti et al. (1977), who found a considerably exuberant projection into the corpus callosum of newborn kittens.

The present results in the OCK cats are not only evidence for plasticity-induced processes accountable for the enhancement of an interhemispheric transfer; these results serve also as an indication for less competition between callosal and thalamic fiber terminals. The thalamic fibers conventionally driven by the contralateral eye are visually inactive in our kittens due to the chiasm split. They are therefore ineffective in competing for synaptic space in comparison with the intact callosal fibers that are driven by visual input. Despite this disadvantage of the thalamic fibers, only a minor increase in the callosal input was found in our cats chiasm sectioned as kittens, indicating a very small competitive interaction, if at all, between these pathways.

Only a few studies have been so far performed in which cancelation of the direct contralateral visual inputs, by transection of the optic chiasm, was induced during the developmental period. However, in these studies the optic chiasm was sectioned in kittens concurrently with the induction of monocular deprivation. In one of these studies, based on histological evidence, only an ipsilateral effect was found as expressed by a shrinkage of cells in the deprived A1 laminae of the LGN (lateral geniculate nucleus) (Garraghty et al. 1984). The nondeprived A1 laminae had the same cell size as that of the A1 laminae of the normal control cats. No retrograde cortical effect on LGN cells was found, as would be expected in the case of extensive involvement of the corpus callosum in corticothalamic relationships. This study therefore shows the absence of a callosal effect and thus a complete independency of the two hemispheres in these kittens. In comparison, in other studies on split chiasm and monocularly deprived kittens, the properties of cortical cells were physiologically studied (Cynader et al. 1981; Yinon et al. 1985). In these studies each hemisphere was found to be mostly dominated by its own, ipsilateral eye. However, some interhemispheric transfer has been found by both investigators, as reflected by the invasion of visual input from the "experienced" hemisphere (ipsilaterally to the nondeprived eye) to the "inexperienced" hemisphere (ipsilaterally to the deprived eye) via the corpus callosum.

Although it is difficult to explain it on the basis of the anatomical organization of the visual system, the interocular dissociation caused by chiasm transection in both kittens and adult cats had induced a great amount of interhemispheric separation. This separation is a consequence of inactivation of the corpus callosum, probably related to the reduction in the excitation level of cortical cells in the callosal cell zone or an increase in their sensitivity threshold. This is a very plausible explanation in view of the remarkable amount of visually unresponsive cortical cells presently found in animals chiasm sectioned as kittens and as adults.

Using a different experimental approach, it has already been proved by us that the indirect callosal input is unable to substitute for the absent direct visual input of the thalamocortical pathway in adult operated cats. This was proved by the induction of deafferentation of the visual cortex following unilateral optic tract transection. Only a few callosally driven cells were found in areae 17–18 and their boundary in the deafferented hemisphere (Yinon et al. 1982, 1984, 1986c, 1987; Podell et al. 1984; Yinon and Podell 1987). These results are therefore analogous to our present findings in chiasm sectioned cats; under both conditions the corpus callosum, although being intact, was found to be visually inactive. A further analogy was found between the findings of the present study in animals chiasm sectioned as kittens (OCK) and previous results in cats in which cortical deafferentation has been developmentally produced (Yinon and Weiser 1985). This is reflected in the enhancement of callosal transfer found in both experimental paradigms. Thus, in both paradigms the same mechanisms are involved, either in the inactivation of the callosal transfer during adulthood or in the partial recovery of the callosal transfer during the developmental period.

The experimental group of animals presently studied are those in which visual split brain has been surgically performed by sectioning both the corpus callosum and the optic chiasm. So far no study was carried out in which the effect of visual split brain on the physiological properties of cortical cells was examined, except for our own preliminary reports (Yinon et al. 1986a; Yinon and Chen 1987). The present findings show a complete interocular dissociation in cats in which split brain was induced either developmentally or during adulthood. This dissociation was expressed in absolute ipsilateral eye dominance and has been mainly attributed to the chiasm transection in the split brain cats. The ipsilateral eye dominance has been further enhanced by the reduction of binocularity following callosotomy itself, as presently found. The complete hemispheric separation in cats operated for split brain as kittens and as adults, has been caused both by the callosal and chiasm transections (the separation was complete, despite the fact that in these animals the posterior callosal region was sectioned, enabling transfer via its more anterior part). It is further important to note here that in both kittens and adult cats subjected to split brain, the visual cortex has not been compensated for, by and large, by a supplementary transfer via an alternative commissural pathway and thus a consistent hemispheric division has been produced. It is thus conceivable that following this operation, the brain of these animals would serve as an appropriate substrate for the differential processing of visual information in each hemisphere. That this is the case has already been proven previously in behavioral studies (Myers 1955; Sperry et al. 1956).

The finding of complete interocular dissociation in our split brain cats is in keeping with a recent behavioral study showing that split brain similarly induced in cats caused the entire loss of the ability to use disparity information (Lepore et al. 1986). The later investigators have found little behavioral effect if callosotomy was induced in their cats after optic chiasm transection. This is again in agreement with our present physiological findings showing a similar pattern of cortical activity after split brain and following chiasm transection alone, indicating a relatively small effect of callosotomy under these conditions.

The diminished excitability presently found in the visual cortex of split brain cats is in keeping with the loss of visual responsiveness found in our cats following either corpus callosum or optic chiasm transection. This has been the case, although from connectivity point of view cancellation of callosal input in the split brain cats does not seem to necessarily affect the responsiveness of cortical cells. On the other hand, interruption of the thalamic pathway is of crucial importance since it blocks nearly two thirds of the input from the nasal retinae to the higher visual centers (Polyak 1957). Despite this, the diminution in excitability as presently found seems to be equally vigorous whether callosal or chiasm transection were produced.

Considering the level of neuronal excitability, a recovery was found in the split brain cats in which the operation was developmentally induced. This was expressed in the remarkable increase in the proportion of visually responsive cells found in the visual cortex of these cats in comparison to the split brain adult cats. The presently obtained difference between cats operated during development or as adults is attributed to the involvement of a plasticity mechanism during growth, compensating for the diminution in general responsiveness.

It is important to mention here that our present findings in split brain animals (as well as in the callosum and chiasm sectioned cats) are related to cortical areae 17 and 18 and the boundary region between them. Therefore, these findings do not necessarily reflect the situation in other visual cortical areae. In addition, the condition of split brain as presently established is obviously limited to the visual system and it is assumed that as such has no any implication to other sensory systems. However, it has been proved by us that it can serve as a suitable experimental model for studying various visual cortical functions independently of interhemispheric interactions.

5 Summary

Visual split brain has been experimentally performed during development in order to study the responsiveness pattern of cortical cells in the absence of the normal interhemispheric relationships throughout growth. For this purpose the interhemispheric transfer to the visual cortex has been surgically prevented by midsagittal transection of the corpus callosum in 2-month-old kittens. In order to limit each hemisphere to its own visual input, we have completely dissociated the retinal projections from the right and the left eyes to the cortex by surgically splitting the optic chiasm along the midsagittal plane, concurrently with the callosal transection. For comparison, the corpus callosum and the optic chiasm have been independently sectioned in two other groups of kittens, serving as controls. All the above-described surgical manipulations have been also performed on adult cats in order to study the effect of split brain on cortical cells in the absence of the developmental process. The results were compared to those obtained from the visual cortex of intact normal adult cats. Unit recording was made in visual areae 17 and 18, mainly from the callosal projection zone and where the vertical meridian is represented. Almost all cells in the cats operated for split brain as kittens and as adults had reacted ipsilaterally, indicating the absence of interhemispheric interaction. This has been also the result in the split chiasm operated adult cats; however, some recovery was found in cats chiasm sectioned during development, as indicated by a partial interhemispheric transfer. As to the callosal operation, a general loss in binocularity was found. However, a different profile of ocular dominance distribution of cells, as indicated by an ipsilateral eye dominance was obtained in animals operated as kittens in comparison to the callosally operated adult cats. A remarkable recovery in the level of cortical excitability was found in cats developmentally operated for split brain in comparison to cats similarly operated during adulthood, as expressed by the increased proportion of visually responsive cells. Similarly, there has been some improvement in responsiveness in cats chiasm sectioned as kittens in comparison to cats similarly operated during adulthood, but not in cats callosotomized as kittens when compared to the analogous group of adult cats. The split brain condition is thus a summation of these individual conditions. It has been further concluded that in the split brain animal some intracortical reorganization takes place during development, despite the absence of the bilateral hemispheric interactions normally available for cortical cells.

Acknowledgements. Research supported by the Israel Institute for Psychobiology, Charles E. Smith Family Foundation (Jerusalem) grant no. 1/84 to U. Yinon. We are indebted to the Alexander von Humboldt-Stiftung (Bonn, West Germany) for the support given to U. Yinon. Many thanks to R. Feldman for his excellent assistance in the various histological procedures and to E. Zimmerman for the maintenance of the electronic and optical equipment.

References

Antonini A, Stefano M Di, Minciacchi D, Tassinari G (1985) Interhemispheric influences on area 19 of the cat. Exp Brain Res 59:171–184

Berlucchi G, Rizzolatti G (1968) Binocularly driven neurons in visual cortex of split-chiasm cats. Science 159:308–310

Berlucchi G, Gazzaniga MS, Rizzolatti G (1967) Microelectrode analysis of transfer of visual information by the corpus callosum. Arch Ital Biol 105:583–596

Blakemore C, Sluyters RC Van (1975) Innate and environmental factors in the development of the kitten's visual cortex. J Physiol (Lond) 248:663–716

Blakemore C, Diao Y, Pu M, Wang Y, Xiao Y (1983) Possible functions of the interhemispheric connexions between visual cortical areas in the cat. J Physiol (Lond) 337:331–349

Cynader M, Lepore F, Guillemot J-P (1981) Inter-hemispheric competition during postnatal development. Nature 290:139–140

Cynader M, Gardner J, Dobbins A, Lepore F, Guillemot JP (1986) Interhemispheric communication and binocular vision: Functional and developmental aspects. In: Lepore F, Ptito M, Jasper HH (eds) Two hemispheres – one brain: functions of the corpus callosum. Alan R Liss, New York, pp 189–209

Dreher B, Cotte LJ (1975) Visual receptive-field properties of cells in area 18 of cat's cerebral cortex before and after acute lesions in area 17. J Neurophysiol 38:735–750

Elberger AJ, Smith III EL (1985) The critical period for corpus callosum section to affect cortical binocularity. Exp Brain Res 57:213–223

Fregnac Y, Imbert M (1978) Early development of visual cortical cells in normal and dark-reared kittens: relationship between orientation selectivity and ocular dominance. J Physiol (Lond) 278:27–44

Garraghty PE, Salinger WL, Hickey TL (1984) Monocular deprivation with concurrent sagittal transection of the optic chiasm. Dev Brain Res 14:292–294

Harvey AR (1980) A physiological analysis of subcortical and commissural projections of area 17 and 18 of the cat. J Physiol (Lond) 302:507–534

Hubel DH, Wiesel TN (1962) Receptive fields, binocular interaction and functional architecture in the cat's visual cortex. J Physiol (Lond) 160:106–154

Hubel DH, Wiesel TN (1967) Cortical and callosal connections concerned with the vertical meridian of visual fields in the cat. J Neurophysiol 30:1561–1573

Innocenti GM (1980) The primary visual pathway through the corpus callosum: morphological and functional aspects in the cat. Arch Ital Biol 118:124–188

Innocenti GM, Fiore L, Caminiti R (1977) Exuberant projection into the corpus callosum from the visual cortex of newborn cats. Neurosci Lett 4:237–242

Kluver H, Barrera E (1953) A method for the combined staining of cells and fibers in the nervous system. J Neuropathol Exp Neurol 12:400–403

Lepore F, Guillemot JP (1982) Visual receptive field properties of cells innervated through the corpus callosum in the cat. Exp Brain Res 46:413–424

Lepore F, Samson A, Molotchnikoff S (1983) Effects on binocular activation of cells in visual cortex of the cat following the transection of the optic tract. Exp Brain Res 50:392–396

Lepore F, Ptito M, Lassonde M (1986) Stereoperception in cats following section of the corpus callosum and/or the optic chiasma. Exp Brain Res 61:258–264

Luttenberg J, Marsala J (1963) The topography of the commissural fibers in the corpus callosum of the cat's brain. Cslka Morf 11:156–176

Milleret C, Buser P (1984) Receptive field sizes and responsiveness to light in area 18 of the adult cat after chiasmotomy. Postoperative evolution; role of visual experience. Exp Brain Res 57:73–81

Minciacchi D, Antonini A (1984) Binocularity in the visual cortex of the adult cat does not depend on the integrity of the corpus callosum. Behav Brain Res 13:183–192

Myers RE (1955) Interocular transfer of pattern discrimination in cats following section of crossed optic fibers. J Comp Physiol Psychol 48:470–473

Otsuka R, Hassler R (1962) Über Aufbau und Gliederung der corticalen Sehsphäre bei der Katze. Arch Psych Zeit Neurol 203:212–234

Payne BR, Elberger AJ, Berman N, Murphy EH (1980) Binocularity in the cat visual cortex is reduced by sectioning the corpus callosum. Science 207:1097–1099

Payne BR, Pearson HE, Berman N (1984a) Role of corpus callosum in functional organization of the striate cortex. J Neurophysiol 52:570–594

Payne BR, Pearson HE, Berman N (1984b) Deafferentation and axotomy of neurons in cat striate cortex: time course of changes in binocularity following corpus callosum transection. Brain Res 307:201–215

Podell M, Yinon U, Hammer A (1984) Properties of visual cortical cells of the intact and the deafferented hemisphere of unilateral optic tract sectioned acute and chronic adult cats. Exp Brain Res 55:91–96
Polyak S (1957) The vertebrate visual system. University of Chicago Press, Chicago, p 788
Shatz C (1977) Abnormal interhemispheric connections in the visual system of Boston Siamese cats: a physiological study. J Comp Neurol 171:229–246
Sperry RW, Stamm JS, Miner N (1956) Relearning tests for interocular transfer following division of the optic chiasma and corpus callosum in cats. J Comp Physiol Psychol 49:529–533
Tusa RJ, Rosenquist AC, Palmer LA (1979) Retinotopic organization of areas 18 and 19 in the cat. J Comp Neurol 185:657–678
Vakkur GJ, Bishop PO, Kozak W (1963) Visual optics in the cat, including posterior nodal distance and retinal landmarks. Vision Res 3:289–314
Yinon U, Chen M (1987) Differentiation between disuse and binocular competition effects in visual cortex cells of monocularly deprived split brain kittens. Soc Neurosci Abstr 13:592
Yinon U, Hammer A (1981) Physiological mechanisms underlying responsiveness of visual cortex neurons following optic chiasm split in cats. In: Flohr H, Precht W (eds) Lesion-induced neuronal plasticity in sensorimotor systems. Springer, Berlin Heidelberg New York, pp 360–368
Yinon U, Hammer A (1985) Optic chiasm split and binocularity diminution in cortical cells of acute and of chronic operated adult cats. Exp Brain Res 58:552–558
Yinon U, Podell M (1987) Unilateral visual cortex deafferentation induces changes in receptive field properties of cortical cells in the intact hemisphere of normal and of monocularly deprived cats. Dev Brain Res 33:205–213
Yinon U, Weiser ZD (1985) Deafferentation of the visual cortex and monocular deprivation effects in kittens. Neurosci Lett Suppl 22:S63
Yinon U, Hammer A, Podell M (1982) The hemispheric dominance of cortical cells in the absence of direct visual pathways Brain Res 232:187–190
Yinon U, Podell M, Goshen S (1984) Deafferentation of the visual cortex: the effect on cortical cells in normal and in early monocularly deprived cats. Exp Neurol 83:486–494
Yinon U, Hammer A, Weiser ZD (1985) Split chiasm and monocular deprivation during development: a model of noncompetitive mechanism in visual cortex neurons. Soc Neurosci Abstr 11:462
Yinon U, Chen M, Hammer A (1986a) Split brain induces monocular dominance of cells in the visual cortex of adult cats. Soc Neurosci Abstr 12:785
Yinon U, Chen M, Hammer A (1986b) Plasticity induced transfer in the corpus callosum of developing kittens following optic chiasm split and its effect on visual cortex cells. In: Bullier J, Kennedy H (eds) Symposium on Structure and Function in the Visual System, 10–12 Sept 1986. Lyon, France, pp 44
Yinon U, Hammer A, Podell M (1986c) Physiological changes in cortical cells following partial and complete visual cortex deafferentation in cats. In: Gilad G, Kreutzberg GW, Gorio A (eds) Processes of recovery from neural trauma. Springer, Berlin Heidelberg New York, pp 150–161
Yinon U, Achiron A, Podell M, Weiser Z (1987) The deafferented visual cortex: Neuronal activity and visual evoked potentials. Int J Neurosci 33:85–91
Zeki S, Fries W (1980) A function of the corpus callosum in the Siamese cat. Proc R Soc Lond Biol Sci 207:249–258

Neuronal Remodeling After Early Cerebellar Hemispherectomy in Rats

A. Gramsbergen
Department of Developmental Neurology,
Oostersingel 59, NL-9713 EZ Groningen
The Netherlands

1 Introduction

Numerous experimental studies have demonstrated that brain lesions at early stages of maturation lead to a greater degree of neuronal rearrangement than such lesions at adulthood. These morphological changes have been regarded as the substrate for a greater compensational capacity of the immature brain. Margaret Kennard has shown that after unilateral removal of the motor and premotor cortex in newly born rhesus monkeys no motor disturbances developed, while after such lesions at later ages monkeys remained impaired (Kennard 1942). However, more recent experiments with lesions of other brain areas raised doubts that compensation is always complete. On the contrary, even bizarre behavioral anomalies may occur, particularly after lesions at young ages. Schneider and Jhaveri (1974), for instance, made unilateral lesions in the optic tectum of Syrian hamsters at birth. On testing, these animals as adults carried out abnormal head movements away from visual stimuli. In addition, the authors demonstrated retinotectal fibers recrossing to the remaining colliculus which might explain these abnormal movements. These two examples may illustrate the complicated relationship between neuronal remodeling after early lesions and the capacity of functional compensation.

In recent years newly developed neuroanatomical and neurophysiological techniques have enabled considerable expansion of our knowledge on the extent and nature of neuronal remodeling. The new results, together with data from meticulous investigations into the behavioral effects after early lesions, are a firm basis to reconsider the compensational mechanisms of the CNS. In particular, recent studies into the effects of cerebellar hemispherectomy have attributed to new insights into the nature of post-lesion neural plasticity. In this chapter I will briefly review our own data and those of others on the morphological effects of cerebellar hemispherectomy in young rats. In the perspective of these results I will then discuss the development of locomotor behavior in these early lesioned rats.

Rats are born at an early stage of brain maturation. Therefore, it is the experimental animal of choice to perform cerebellar lesions during the early postnatal period. In addition, afferent and efferent cerebellar connections are relatively well documented in this animal.

H. Flohr (Ed.)
Post-Lesion Neural Plasticity

2 Efferent Cerebellar Projections: Rerouting of Cerebellorubral and Cerebellothalamic Connections

Normally, the cortex of the cerebellar hemisphere is connected with the lateral (dentate) and the anterior and posterior interpositus nuclei. These nuclei project onto the red nucleus and several thalamic nuclei at the contralateral side (for details see Voogd et al. 1985; Angaut et al. 1986).

Lim and Leong (1975) removed in rats at their first day of life one cerebellar hemisphere including its central nuclei. They demonstrated with the degeneration technique at adult age a bilateral projection from the nuclei in the remaining cerebellar hemisphere onto both red nuclei and the ventro-lateral thalamic nuclei at both sides (for a schematic drawing of these and other aberrant projections see Fig. 1). Leong (1978 a) expanded this study and found additionally aberrant fiber projections onto the ipsilateral zona incerta, the ventral part of the lateral geniculate nucleus, and the following thalamic nuclei: ventralis medialis, ventralis anterior, ventralis lateralis, medialis ventralis, and centralis lateralis. Another finding was that these aberrant projections occurred only when the rats were lesioned until the 10th day; lesioning at the 15th day did not lead to such projections. Essentially similar resuls were obtained independently by Castro (1978). Haroian and Campellone (1986) recently reinvestigated the cerebellothalamic projections in normal and lesioned rats with the sensitive HRP technique. They found also in normal rats cerebellar fibers which project onto the ipsilateral thalamus. These fibers arose mainly from the medially located fastigial nucleus but to a minor extent from the interpositus nuclei as well (see also Bentivoglio and Molinari 1981; Naus et al. 1984). They demonstrated that after hemicerebellectomy on the first day of life this ipsilateral projection had increased considerably (see also Castro

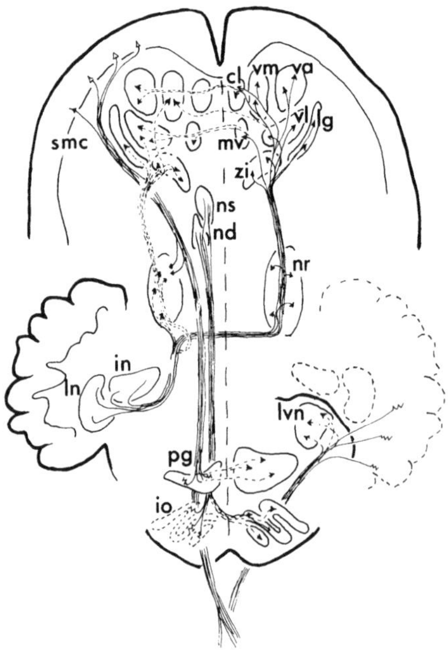

Fig. 1. Schematic drawing of aberrant fiber connections after cerebellar hemispherectomy until the 10th day in rats. *Continuous lines* normally occurring projections; *discontinuous lines* aberrant projections. *cl* n. centralis lateralis thalami; *in* n. interpositus cerebelli; *io* n. olivaris inferior; *l* n. geniculate lateralis; *ln* n. lateralis cerebelli; *lvn* n. vestibularis lateralis; *mv* n. medialis ventralis thalami; *nd* n. Darkschewitsch; *nr* n. ruber; *ns* n. subparafascicularis; *pg* pontine nuclei; *smc* sensori motor cortex; *va* n. ventralis anterior thalami; *vl* n. ventralis lateralis thalami; *vm* n. ventralis medialis thalami

1978). Summarizing the evidence from different investigations on this point, the aberrant cerebellothalamic fibers seem fo follow partly an ipsilateral course (via the red nucleus) and partly the normal contralateral route, after which they recross in the thalamus.

Aberrant cerebellothalamic fibers have also been demonstrated by electrophysiological techniques. Yamamoto and coworkers (1981) studied one group of rats hemicerebellectomized until their 4th day of life, a second group lesioned between the 8th and 13th day, as well as a control group. At adult age electrodes were inserted into the ventro-lateral thalamic nuclei and neurons in the cerebellar nuclei were stimulated antidromically. In the group of rats hemicerebellectomized until the 4th day they found that 23.3% of the neurons could be activated from the ipsilateral side. In the second group of four lesioned rats, this value decreased to 11.2%, whereas in four control rats this value was less than 1%. Yamamoto's results are important in that they suggest that these aberrant fibers are functional.

An intriguing question is, whether the aberrant fibers project at random or in orderly fashion onto their aberrant target area. Naus et al. (1984) have studied this problem in rats lesioned at the first 2 days of life. They injected anterogradely transported HRP-WGA either in the lateral or in the interposed nuclei at the age of 2 months and studied the red nucleus and thalamus. They demonstrated that the projection patterns of the aberrant fibers mirror those of normal contralateral fibers (see also Haroian and Campellone 1986). These exciting results suggest that similar ordering principles still govern the ipsilaterally outgrowing axons in their aberrant target area.

3 Synaptic Rearrangement in the Red Nucleus

In mammals the red nuclei play a key role in matching cerebellar and cortical inputs (e.g., Massion 1967). The red nucleus in rats receives afferents from the contralateral cerebellar lateral and interposed nuclei; afferents from the sensorimotor cortex stem from the ipsilateral side and project only onto the rostrally located parvocellular part of the red nucleus (Flumerfelt and Hrycyshyn 1985).

After early cerebellar hemispherectomy, the red nucleus contralateral to the lesion is obviously devoid of its normal input from the contralateral cerebellum (it does receive afferents, however, from the ipsilateral side; Sect. 2). As a consequence of the early deafferentation the denervated red nucleus is smaller and contains fewer large cells (Naus et al. 1984). Leong (1978 b) has studied the projections from the sensorimotor cortex onto this red nucleus in rats which were lesioned within the first 2 days of life. He did not find any abnormal projections, either from the ipsi- or from the contralateral sensorimotor cortex. Recently, Shieh and coworkers (1986) investigated cortical and cerebellar afferents on red nucleus neurons at an ultrastructural level. Cortical afferents of normal rats appeared to terminate upon spines and shafts of distal dendrites, whereas cerebellar afferents contacted proximal dendrites. After cerebellar hemispherectomy at the 3rd or 4th day, they found an increased number of afferents from the cerebral cortex onto proximal dendrites and a decreased number of afferents on distal dendrites. This result indicates that lesioning of one afferent system leads to synaptic rearrangement in another afferent system. Results from neurophysiological experiments after pedunculotomy in kittens (Tsukahara et al. 1983) indicated that such remodeling in the red nucleus may have important functional consequences.

4 Neuronal Rearrangement in the Pontine Nuclei

In normal rats, afferents from the lateral and interposed nuclei terminate in the contralateral pontine nuclei, upon medium-sized proximal dendrites (Ho and Leong 1977; Leong 1980). Afferents from the sensorimotor cortex terminate mainly in the ipsilateral pontine nuclei and only a sparse projection crosses to the contralateral paramedian pontine nucleus (Gwyn and Flumerfelt 1974; Leong 1978 b). Cortical afferents synapse upon distal dendrites (Leong 1980).

After cerebellar hemispherectomy in the first 2 or 3 days of life, a substantial loss of neurons in the pontine gray contralateral to the lesion occurs (Leong 1978 b; Castro and Mihailoff 1983). Fibers from the sensorimotor cortex ipsilateral to the hypotrophied pontine nuclei were in part rerouted to the contralateral side. Leong (1980) studied the cortical afferents in the pontine nuclei at an ultrastructural level. He found that terminals from cortical axons in lesioned rats contacted larger dendrites (0.2–2.4 μm instead of 0.2–1.2 μm in normal rats). Provided that dendritic diameters are normal (which has not been proven) these data would indicate that cortical afferents occupy the synaptic sites vacated by cerebellar afferents.

5 Aberrant Projections onto the Inferior Olive

In normal rats an important source of afferents to the inferior olivary nuclei stems from the subparafascicular nucleus and the nucleus of Darkschewitsch in the caudal diencephalon and mesencephalon (Flumerfelt and Hrycyshyn 1985). This projection is predominantly ipsilateral. After neonatal cerbellar hemispherectomy the inferior olivary complex, contralateral to the lesion degenerates (e.g., Castro 1978).

Swenson and Castro (1982) studied the mesodiencephalic projection onto the degenerated inferior olivary nuclei with [^{3}H]-leucine autoradiography. They reported a pronounced increase in the number of labeled terminals in the contralateral inferior olive (which was still present). These data might reflect increased axonal sprouting of the sparse contralateral projection, normally present. However, they might also indicate a rerouting of fibers which would normally terminate in the degenerated inferior olivary nuclei.

6 Rerouted Afferents to the Ablated Cerebellar Hemisphere

Even a brief outline of the normal afferents to the cortex of the cerebellar hemisphere and its lateral and interposed nuclei would go beyond the scope of this paper. Here, only a few of the most important sources of afferent projections are mentioned. *Mossy fibers* stem from pontine nuclei, the lateral reticular nucleus and spinal sources; all mossy fiber systems project bilaterally. *Climbing fibers* stem from the inferior olive. The olivocerebellar fibers project predominantly onto the contralateral hemisphere with only a small ipsilateral component. During their course to the cerebellar cortex, collaterals are given off to the deep cerebellar nuclei (Voogd et al. 1985).

After neonatal hemicerebellectomy, several of these sources of afferent projections degenerate: the contralateral inferior olivary complex, the ipsilateral lateral reticular nucleus, the ipsilateral lateral cuneate nucleus, and a considerable part of the contralateral pontine nuclei (Castro 1978). It seems possible that fibers emanating from the still present precerebellar nuclei are rerouted onto parts of the cerebellar cortex on which they normally do not project. However, virtually nothing is known yet concerning such remodeling. Data are available on spinocerebellar fibers only. Castro and Smith (1979) ablated one cerebellar hemisphere in neonatal rats and studied at adult age the terminal patterns of these spinal fibers. They found a remarkable increase in projections onto the lateral vestibular nucleus, ipsilateral to the cerebellar lesion. These results might be interpreted as a rerouting of fibers from their normal trajectory onto the removed cerebellar hemisphere.

7 Age at Lesioning in Relation to the Extent of Neuronal Remodeling

Many of the investigators mentioned so far aimed primarily at studying the main effects of cerebellar hemispherectomy. Therefore, they chose an early age for lesioning in order to maximize the effects. Data on the relation between extent of remodeling and age at lesioning are available only on cerebellorubral and cerebellothalamic projections. Leong (1978a) studied the effects of hemispherectomy performed at the 2nd, 3rd, 5th, 10th, and 15th day after birth. He reported that the density of the anomalous projections was greatest in animals operated until the 5th day, much less in those operated at the 10th day and absent in rats lesioned at the 15th day.

In order to obtain more detailed insight into this quantitative relationship, we studied the cerebellorubral projection with the double labeling technique. In this technique two retrogradely transported fluorescent tracers are injected into two projection areas which label different features of the parent cells: True Blue labeling the soma in blue and Nuclear Yellow labeling the nucleus in bright yellow. We studied six groups of four rats each: one control group and groups hemicerebellectomized on the 2nd, 5th, 10th, 20th, and 30th day respectively. At adult age the fluorescent tracers were injected stereotaxically in the left and right red nucleus. After appropriate survival times we found that the presence of aberrantly projecting fibers was clearly age-dependent, their relative numbers in both lateral and interposed nuclei decreasing from 16–33% after lesioning at the 2nd day, to 5–15% after lesioning at the 5th or 10th day and to less than 2% in rats operated at the 20th or 30th day (Fig. 2). This latter value corresponds to what we have found in normal adults (Gramsbergen and IJkema-Paassen 1982). These results

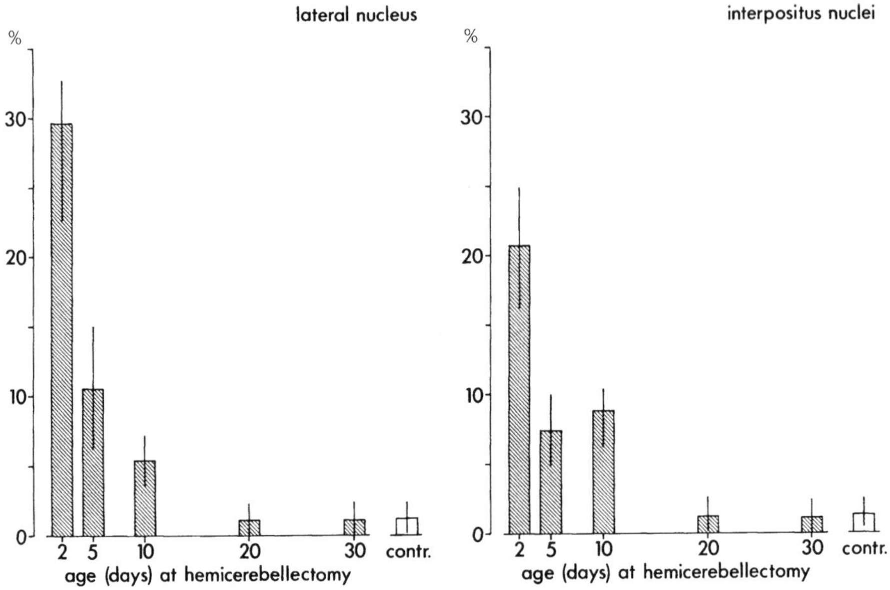

Fig. 2. Percentages of cells in lateral and interpositus nuclei labeled retrogradely by axons directed to the ipsilateral red nucleus. *Vertical scale* mean values (columns) and ranges in percentages

demonstrate a clearcut relationship between age at lesioning and the relative numbers of aberrant fibers.

8 Nature and Origin of Aberrant Cerebellorubral Fibers

An important problem concerns the origin of the aberrant cerebellorubral fibers. Theoretically, three possibilities seem to exist:

1. Ipsilaterally projecting cerebellorubral fibers are remnants of a bilateral projection which is normally present only at very early ages.
2. A sparse ipsilateral projection occurs normally which proliferates after early hemicerebellectomy.
3. The aberrant projection develops de novo.

We studied the first possibility by injecting the anterogradely transported tracer [^{3}H]-leucine into one lateral cerebellar nucleus of intact rats on their first day of life. An early bilateral cerebellorubral projection would lead to labeling in both red nuclei. However, after a survival period of 7 days we only found contralateral labeling in all six rats investigated (Gramsbergen and IJkema-Paassen 1984). Further evidence making this hypothesis unlikely is that efferent fibers from the cerebellar nuclei in rats reach the red nuclei only on the 3rd–5th day after birth (Povlishock et al. 1978; Haroian and Campellone 1986). Having excluded this possibility, the most plausible explanation is that the aberrant projection is the result of a directional change of outgrowing fibers (for a general treatise on the etiology of aberrant fiber projections see Bentivoglio et al. 1986).

Another problem is whether the aberrant fibers are collaterals from normally projecting axons. We studied this problem with the double labeling technique (Gramsbergen and IJkema-Paassen 1982; see also Sect. 7). Recall that two retrogradely transported fluorescent tracers were injected in the left and right red nucleus. In case the aberrant fibers are collaterals we should have detected double-labeled neurons in the cerebellar nuclei. However, we found double labeling only in less than 1% of the neurons. This implies that the aberrant cerebellorubral fibers stem from separate parent cells.

This result raised the question whether the number of the contralaterally projecting neurons remains normal after early lesions. In that case the total number of (contra- and ipsilaterally projecting) neurons would be increased. Such an increase in cell number might be caused by a disturbed numerical matching process in the early development of the cerebellorubral connections. Neuronal death is the major factor contributing to numerical matching of neuronal systems at early stages of development (Rager and Rager 1978; Oppenheim 1981). For instance, reducing the target area of motoneurons in the chick spinal cord by amputating a limb bud at day 3 of incubation leads to an increase in neuronal death and consequently to a decrease in cell number. On the other hand, enlarging the target area by implanting an additional limb reduces neuronal death. Theoretically, one might consider the red nucleus contralateral to the cerebellar lesion and devoid of its own contralateral input as an additional target area. With this in mind, we investigated cell death patterns in the deep cerebellar nuclei.

In one group of rats (N = 21) one cerebellar hemisphere was removed at the 2nd day of life and data from this group were compared to those of a group of unoperated controls (N = 21). Histological material was collected at different ages

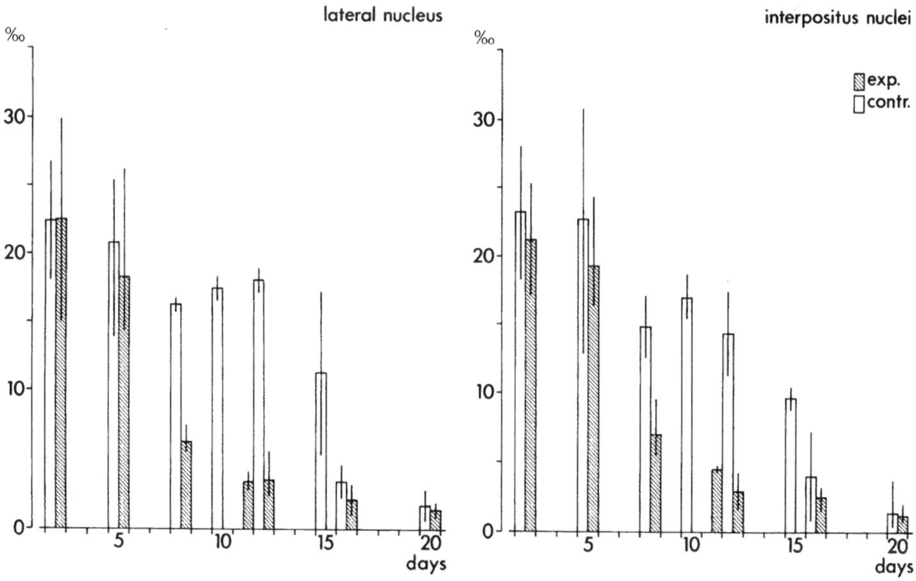

Fig. 3. Live to pyknotic cell ratios in the lateral and interpositus nuclei of control rats and rats lesioned at the 2nd day. *Vertical axis:* mean values (columns) and ranges per 1000 live cells; *horizontal axis:* ages in days

between the 2nd and 20th day and live and pyknotic cells were counted (for details on histological methods, counting criteria and the correction procedures see Gramsbergen and IJkema-Paassen 1987 and Gramsbergen et al. 1987). Pyknotic cells were relatively common in the cerebellar nuclei of intact rats until around the 15th day. (Values ranged between 14 and 28 pyknotic cells/1000 live cells; Fig. 3.) Around this age a change occurred and from the 16th day onwards all rats had less than three pyknotic cells/1000 live cells.

In lesioned rats, cell death ratios were identical to those of control rats until the 5th day. However, already at the 8th day the cell death ratio decreased abruptly in these animals. Results from this study strongly suggest that a considerable number of cells in the cerebellar nuclei are prevented from dying because they find a convenient synaptic target in the ipsilateral red nucleus. An investigation in which we study the consequences of this marked decrease in cell death for the ultimate number of neurons is in progress. However, preliminar findings indicate that the number of neurons in hemicerebellectomized rats at the 20th day is increased indeed. It seems possible that disturbed numerical matching processes in other efferent and also in afferent fiber systems have similar consequences for cell number (e.g., cell number in thalamic nuclei, precerebellar nuclei, the remaining cerebellar cortex).

9 Behavioral Sequelae

Considering the results on the extensive neuronal remodeling after early cerebellar hemispherectomy, the obvious question emerges whether the morphological changes subserve behavioral compensation. We studied the effects of such lesions on locomotor behavior (Gramsbergen 1981, 1982; Gramsbergen and IJkema-Paassen 1984). Rats were lesioned at the 5th, 10th, 20th, or 30th day and we followed their reflex development, locomotor development and specific abnormalities such as ataxia, postural tremor, hypermetria, gang width, until they reached the age of 360 days. The main conclusions from these studies were that rats lesioned at the 5th or 10th day have distinctly more disturbed locomotor patterns and an increased incidence of locomotor handicaps compared to animals operated at the 20th or 30th day. Smith et al. (1974) studied open field behavior and performance in balancing and climbing tests of rats which were hemicerebellectomized at the 10th, 15th, and 21st day. They also came to the conclusion that the rats operated at the 15th and 21st day recovered more completely and more rapidly than those lesioned at the 10th day.

Bizarre abnormalities may occur in rats lesioned at early ages. We observed in some of the rats, operated at the 5th or 10th day limb abduction at the *contralateral* side and in some other rats the abnormality shifted between the ipsilateral and contralateral side to and from. It is well known that hemicerebellectomy at adult age leads to limb abduction at the *ipsilateral* side and in our rats lesioned at the 20th and 30th day we consistently ascertained this. O'Donoghue et al. (1986) demonstrated in early hemicerebellectomized rats that microstimulation in the sensorimotor cortex at both sides leads to *ipsilateral* forelimb movements even at low current intensities. Such lesions do not lead to anomalously projecting corticospinal or corticobulbar projections (O'Donoghue et al. 1987) and therefore, to the authors, these abnormal reactions should be attributed to alterations in local spinal circuits. Our findings on limb abduction at the "wrong" side in early lesioned rats may similarly be explained by such alterations.

To summarize, hemicerebellectomy at early ages leads to extensive neuronal remodeling in efferent and afferent cerebellar connections and in the precerebellar relay nuclei. Behavioural investigations have revealed that lesioning until the 10th day leads to more impairment in locomotor behavior than similar lesions at a later age. We conclude that neuronal remodeling is a causal factor in locomotor impairment.

References

Angaut P, Batini C, Billard JM, Daniel H (1986) The cerebellorubral projection in the rat: retrograde anatomical study. Neurosci Lett 68:63–68

Bentivoglio M, Molinari M (1981) Axonal branches of the same cerebellar neurons terminate bilaterally in the thalamus. Neurosci Lett 23:291–296

Bentivoglio M, Minciacchi D, Molinari M (1986) Structural mechanisms of postlesional remodelling in the central nervous system. Ital J Neurol Sci, Suppl 5:29–36

Castro AJ (1978) Projections of the superior cerebellar peduncle in rats and the development of new connections in response to neonatal hemicerebellectomy. J Comp Neurol 178:611–628

Castro AJ, Mihailoff GA (1983) Corticopontine remodelling after cortical and/or cerebellar lesions in newborn rats. J Comp Neurol 219:112–123

Castro AJ, Smith DE (1979) Plasticity of spinovestibular projections in response to hemicerebellectomy in newborn rats. Neurosci Lett 12:69–74

Flumerfelt BA, Hrycyshyn AW (1985) Precerebellar nuclei and red nucleus. In: Paxinos G (ed) The rat nervous system, vol 2. Academic Press, London, pp 221–250

Gramsbergen A (1981) Locomotor behaviour after cerebellar lesions in the young rat. In: Flohr H, Precht W (eds) Lesion-induced neuronal plasticity in sensorimotor systems. Springer, Berlin Heidelberg New York, pp 324–336

Gramsbergen A (1982) The effects of cerebellar hemispherectomy in the young rat. I. Behavioural sequelae. Behav Brain Res 6:85–92

Gramsbergen A, IJkema-Paassen J (1982) CNS plasticity after hemicerebellectomy in the young rat. Quantitative relations between aberrant and normal cerebello-rubral projections. Neurosci Lett 33:129–134

Gramsbergen A, IJkema-Paassen J (1984) The effects of early cerebellar hemispherectomy in the rat: behavioural, neuroanatomical and electrophysiological sequelae. In: Finger S, Almli CR (eds) Early brain damage, vol 2. Aademic Press, London, pp 155–177

Gramsbergen A, IJkema-Paassen J (1987) Do early lesions affect cell death in the central nervous system? A study on the effects of early cerebellar hemispherectomy in rats. J Comp Neurol 255:617–624

Gramsbergen A, Kok LP, Poortema K, Schaafsma W (1987) Determining nucleolar multiplicity and cell number from sectional data. J Microsc 145:69–87

Gwyn DG, Flumerfelt BA (1974) A comparison of the distribution of cortical and cerebellar afferents in the red nucleus of the rat. Brain Res 69:130–134

Haroian AJ, Campellone AD (1986) A quantitative analysis of the ipsilateral cerebellothalamic projection following hemicerebellectomy in neonatal rats. A retrograde HRP study. Dev Brain Res 26:69–78

Ho CP, Leong SK (1977) A cerebellar projection onto the pontine nuclei in the albino rat. Exp Brain Res 30:149–154

Kennard MA (1942) Cortical reorganization of motor function: studies on series of monkeys of various ages from infancy to maturity. Arch Neurol Psychiat 48:227–240

Leong SK (1978 a) Plasticity of cerebellar efferents after neonatal lesions in albino rats. Neurosci Lett 7:281–289

Leong SK (1978 b) Effects of deafferenting cerebellar or cerebral inputs to the pontine and red nuclei in the albino rat. Brain Res 155:357–361
Leong SK (1980) A qualitative electron microscopic study of the corticopontine projections after neonatal cerebellar hemispherectomy. Brain Res 96:133–154
Lim KH, Leong SK (1975) Aberrant bilateral projections from the dentate and interposed nucleus in albino rats after neonatal lesions. Brain Res 96:306–309
Massion J (1967) The mammalian red nucleus. Physiol Rev 47:383–436
Naus CG, Flumerfelt BA, Hrycyshyn AW (1984) Topographic specificity of aberrant cerebello rubral projections following neonatal hemicerebellectomy in the rat. Brain Res 309:1–15
O'Donoghue DL, Kartje-Tillotson G, Neafsey EJ, Castro AJ (1986) A study of forelimb movements evoked by intracortical microstimulation after hemicerebellectomy in newborn, young and adult rats. Brain Res 385:311–320
O'Donoghue DL, Kartje-Tillotson G, Castro AJ (1987) Forelimb motor cortical projections in normal rats and after neonatal hemicerebellectomy: an anatomical study based upon the axonal transport of WGA/HRP. J Comp Neurol 256:274–284
Oppenheim RW (1981) Neuronal cell death and some related regressive phenomena during neurogenesis: a selective historical review and progress report. In: Cowan WM (ed) Studies in developmental neurobiology: essays in honor of Viktor Hamburger. Oxford UP, Oxford, pp 74–133
Povlishock JT, Taylor JT, Seiber HR (1978) Electron-microscopic study of the maturing rat red nucleus. I. The large-neuron polulation, Acta Anat 102:129–146
Rager G, Rager U (1978) Systems matching by degeneration. I. A quantitative electron microscopic study of the generation and degeneration of retinal ganglion cells in the chicken. Exp Brain Res 33:65–78
Schneider GE, Jhaveri SR (1974) Neuroanatomical correlates of spared or altered function after brain lesions in the newborn hamster. In: Stein DG, Rosen JJ, Butters N (eds) Plasticity and recovery of function in the central nervous system. Academic Press, London, pp 65–109
Shieh JY, Leong SK, Wong WC (1985) An electron microscopic study of the corticorubral fibres after neonatal deep cerebellar nuclear lesions in albino rats. Brain Res 335:201–206
Smith RL, Parks T, Lynch G (1974) A comparison of the role of the motor cortex in recovery from cerebellar damage in young and adult rats. Behav Biol 12:177–198
Swenson RS, Castro AJ (1982) Plasticity of meso-diencephalic projections to the inferior olive following neonatal hemicerebellectomy in rats. Brain Res 244:169–172
Tsukahara N, Fujito Y, Kubota M (1983) Specificity of the newly-formed corticorubral synapses in the kitten red nucleus. Exp Brain Res 51:45–56
Voogd J, Gerrits NM, Marani E (1985) Cerebellum. In: Paxinos G (ed) The rat nervous system, vol 2. Academic Press, London, pp 251–291
Yamamoto T, Kawaguchi S, Samejima A (1981) Electrophysiological studies on plasticity of cerebello thalamic neurons in rats following neonatal hemicerebellectomy. Jpn J Physiol 31:217–224

Neonatal Versus Adult Hemicerebellectomy: A Behavioral and Anatomical Analysis

L. Petrosini [1], M. Molinari [2], T. Gremoli [1], and A. Granato [2]
Institutes of Human Physiology [1] and of Neurology [2], Catholic University
Largo F. Vito, 1 I-00168 Rome, Italy

In spite of several reports on infant lesion effects, the degree to which brain damage occurring early in life results in sparing or greater recovery of motor function than that seen after comparable damage in adults is still uncertain (Schneider 1979). On one hand, following neonatal spinal transection (Bregman and Goldberger 1983) or cortical ablations (D'Amato and Hicks 1980), there is greater recovery of some postural reflexes when compared with the same lesions sustained in adulthood. On the other hand, coexistent with these examples of greater recovery, are examples of equivalent or even greater deficits or poorer motor performances (Finger et al. 1978). Namely, it has been reported that neonatally hemicerebellectomized rats exhibit a more impaired locomotion in comparison with rats hemicerebellectomized in adulthood (Gramsbergen and IJkema-Paassen 1984). These conflicting findings are complicated by the influence of normal development on motor behavior, by the stage of maturity of the pathways at the time of the lesion, and finally by the nature of the motor behaviors, since different aspects of the same behavior may respond differently to the same lesion. Considering these numerous variables, this study is focused on comparing various motor behaviors of rats hemicerebellectomized neonatally with those of rats hemicerebellectomized at weaning or in adulthood, to analyze the problem of whether cerebellar lesions made during different maturation stages provoke different effects. Particular care was taken to obtain adequate experimental conditions for comparisons of the three age-at-lesion groups; all cerebellar lesions were very similar, and repeated and extremely fine assessments of motor function using similar tests were performed at comparable long periods after lesion.

To this aim, a right hemicerebellectomy (HCb) was performed under hypothermic or Nembutal (40 mg/kg, i.p.) anesthesia on the 1st, 21st, and 90th postnatal day. All animals were regularly tested for their postural and locomotor behavior and for triggered movements. In all groups, motor performances were evaluated by means of qualitative (behavioral testing) and by kinematic quantitative assessment for 4 months. The present report deals with the behavior of the lesioned animals only after the complete stabilization of the symptomatology. After completion of behavioral testing, the animals underwent an anatomical tract tracing study of the spinal afferents, using a multiple retrograde fluorescent tracer technique. Bilateral spinal injections of fast blue and diamidino yellow were performed under barbiturate anesthesia. After the appropriate survival time, the animals were killed, brains were cut coronally and the sections were studied under

H. Flohr (Ed.)
Post-Lesion Neural Plasticity

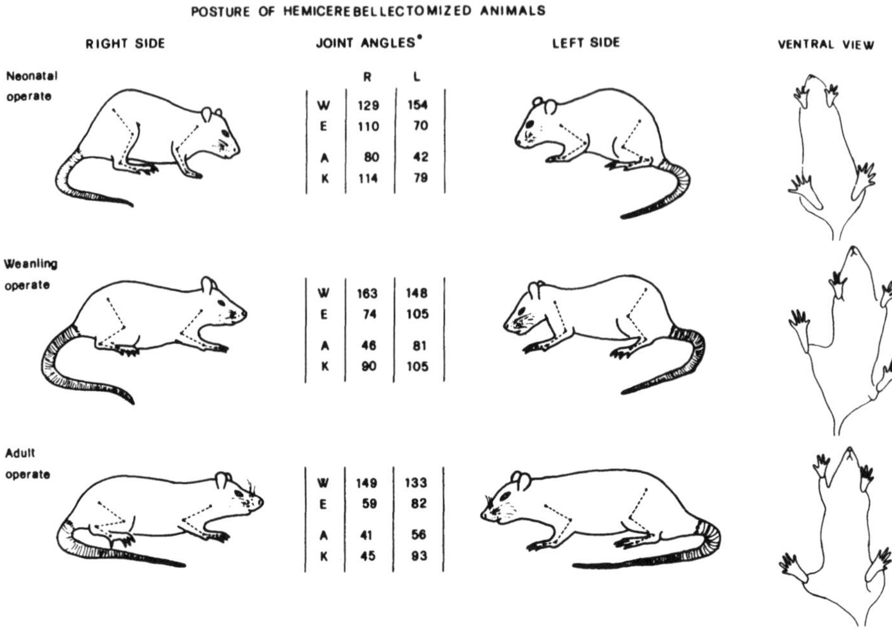

Fig. 1. Effects of right hemicerebellectomy on posture in the three experimental groups. Joint angles of wrist (*W*), elbow (*E*), ankle (*A*) and knee (*K*) of right (*R*) and left (*L*) limbs are indicated

fluorescent light. The retrograde labeling in the vestibular nuclei, the red nucleus, and the reticular formation was studied bilaterally, and chosen sections were charted with the aid of an X–Y computerized plotter. Cases were included only if (by histological examination of counterstained sections) the right cerebellar hemisphere appeared to be totally removed with a complete sparing of the left cerebellar nuclei and of extracerebellar surrounding structures. These conditions were met in seven neonatal, five weanling and six adult operated animals.

In the majority of neonatal operated animals unilateral cerebellar ablation induced head and body tilt to the left, that is contralateral to the lesioned side (Fig. 1) with a slight extension of the right hindlimb. Vice versa, weanling and adult operated animals, as classically described (Manni and Dow 1963), exhibited the extensor hypotonia to the right, then ipsilateral to the lesioned side. The body tilt to the left exhibited by the neonatal operated animals was accomplished by the reduction of elbow, knee, and ankle angles of the left limbs; the closure of these joint angles was necessarily accompanied by an extension of wrist angle. A mirror pattern of these joint angles during standing was displayed by the remaining two groups (Fig. 1).

Locomotion recovered after adult HCb and developed after neonatal lesion. During stepping, all lesioned rats demonstrated many abnormal or exaggerated movements, such as hindlimb hyperflexion and waddling gait; yet the three experimental groups exhibited striking differences among them. In comparison to neonatal operated animals, the weanling and adult operated animals appeared to be more severely impaired in their locomotor behavior. At 4 months after the lesion, some of them still tended to fall to the lesioned side, most of them had a lowered center of gravity which induced a collapsing onto their belly. To change direction, the right hindlimb was used as a pivot, remaining flexed under body weight. Wide hindlimb stance was a rather typical pattern, which stabilized their balance during locomotion and tended to prevent fore-hindpaw collisions. Both weanling and adult operated animals displayed typical cerebellar ataxia, that is a marked swaying with large amplitude and low frequency. On the other hand, in the neonatal operates the most common symptom was an unduly hard placement of the feet, resulting in a sort of slapping, which produced a clearly audible locomotion, this characteristic gait was rarely observed in the other two groups (Figs. 2 and 3). Kinematic analysis of step cycles revealed in all experimental groups during swing phase a marked hyperflexion in both hindlimbs. Regarding the stance phase, in neonatal operated animals, both hindlimbs exhibited an increase of joint angles, which resulted in a hyperextension of limbs and then in the "tiptoe" gait above described. In weanling operated animals at the end of stance a clear extension of the ankle and the knee of the right hindlimb, related to a hypermetric excursion of the limb was observed. Regarding the adult group, the very impressive finding was the exaggerated closure of knee angles of the right hindlimb accompanied by an excessive opening of ankle angles. This joint pattern caused the body to be lower than normal and also caused the outward rotation of the right hindpaw. Also in this lesioned group there was a hypermetric response in the right hindlimb. In neonatal operated rats both threshold to forward stepping and total excursion of the step decreased, causing a hypometric response in both hindlimbs. In the weanling and adult operated animals, an asymmetrical pattern was present: while in the left hindlimb a hypometric excursion was displayed, in the right hindlimb both β and ϕ angles increased, resulting in a hypermetric displacement (Fig. 4).

The motility of the forelimbs was examined by suspending the rat from a wire (Fig. 5). All lesioned rats exhibited gross deficits in dynamic postural adjustments which the test demands. All of them showed a loss of muscular force which provoked falling from the wire before a cut-off time of 120 sec. Suspended rats attempted to pull up and to provide support with one or both hindlimbs. The percentage of neonatal operated animals successful in hindlimb support was slightly higher in comparison with the other two groups, but the difference was not statistically significant. However, all lesioned animals were impaired in comparison to the controls. Furthermore, this test evidenced in all experimental groups dysmetric movements of the forelimbs and the presence of tremor always associated with muscular exertion that is superimposed on voluntary or reflex motor actions but never present at rest.

In agreement with our previous findings (Petrosini 1984) when the animals were put in water, they swam without directionality, with good coordination, maintaining their noses above the water level. The only abnormal sign was the body tilt toward the lesioned side displayed by weanling and adult operated animals.

All lesioned animals displayed deficiencies such as hyperflexion of the joints during the swing phase, increased fluctuations of amplitude of joint angles, dys-

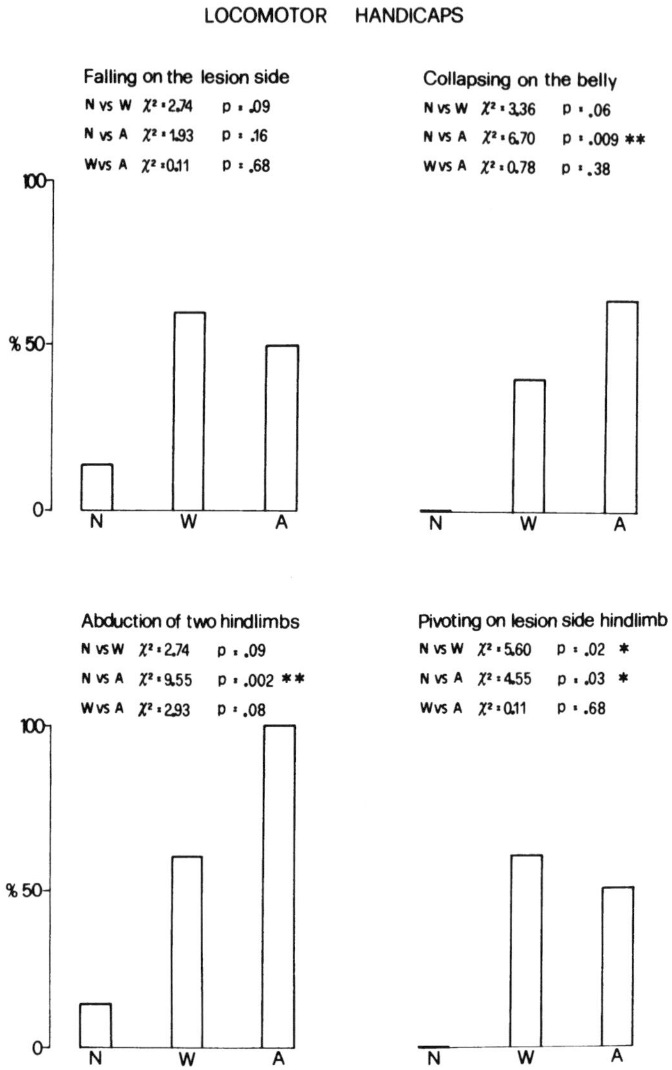

Fig. 2. Locomotion in hemicerebellectomized rats: neonatal (*N*), weanling (*W*) and adult (*A*) operated animals exhibiting the various locomotor handicaps. Comparisons between groups were made with the nonparametric χ^2-test. (*$p<0.05$; **$p<0.01$)

metria, and tremor. But by comparing the effects in the three experimental groups, a clear side difference of postural tone was observed. Ipsilaterally to the lesioned side, the slight extension bias following neonatal lesion is in contrast with the marked flexor bias after adult lesion. However, the comparison of locomotor patterns between neonatal and adult operated groups shows the existence of different motor strategies in the two groups with better performances in neonatal hemicerebellectomized animals. These data indicate that the two syndromes are not mirror-like and that the neonatal one is less impairing. Differences in time of lesioning and in animal strains employed might explain conflicting results re-

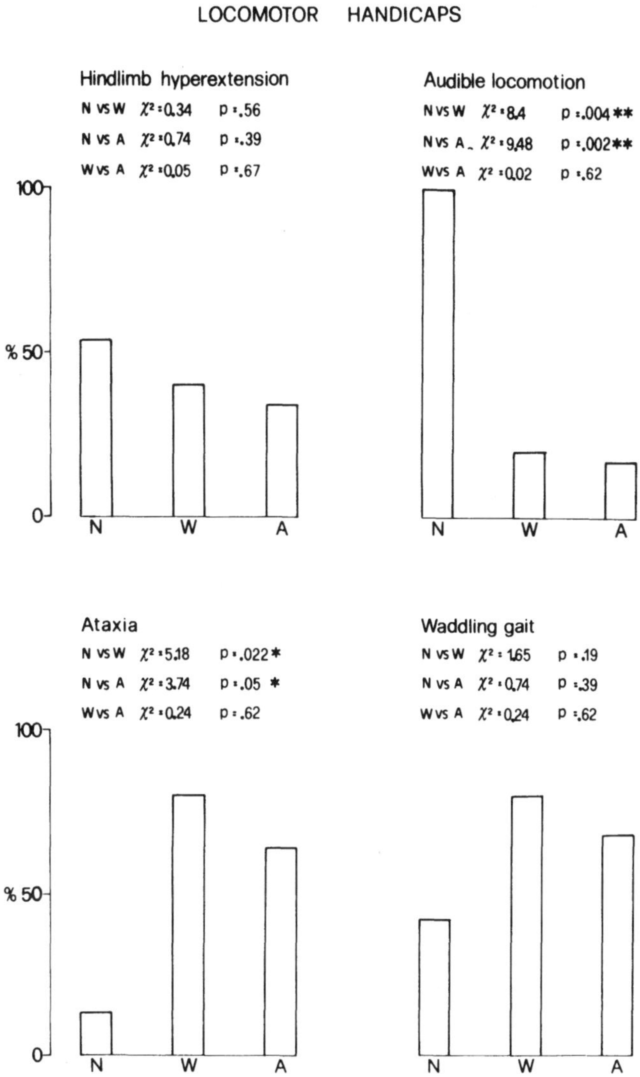

Fig. 3. Locomotion in hemicerebellectomized rats. Abbreviations and statistical comparisons as in Fig. 2

ported in the literature (Gramsbergen and IJkema-Paassen 1984). In fact the time of surgery is critical as it has been shown that the CNS undergoes dramatic anatomical and physiological changes during its early postnatal period (Altman 1982; Bovolenta et al. 1984). Because of these rapid postnatal changes even similar lesions, performed on days slightly apart, may provoke different results (Leonard and Goldberger 1987).

Why in the neonatal and adult operated animals motor performances are different? It is generally accepted that an anatomical reorganization follows neonatal lesions. Regarding cerebellar circuitries, several reports have demonstrated the

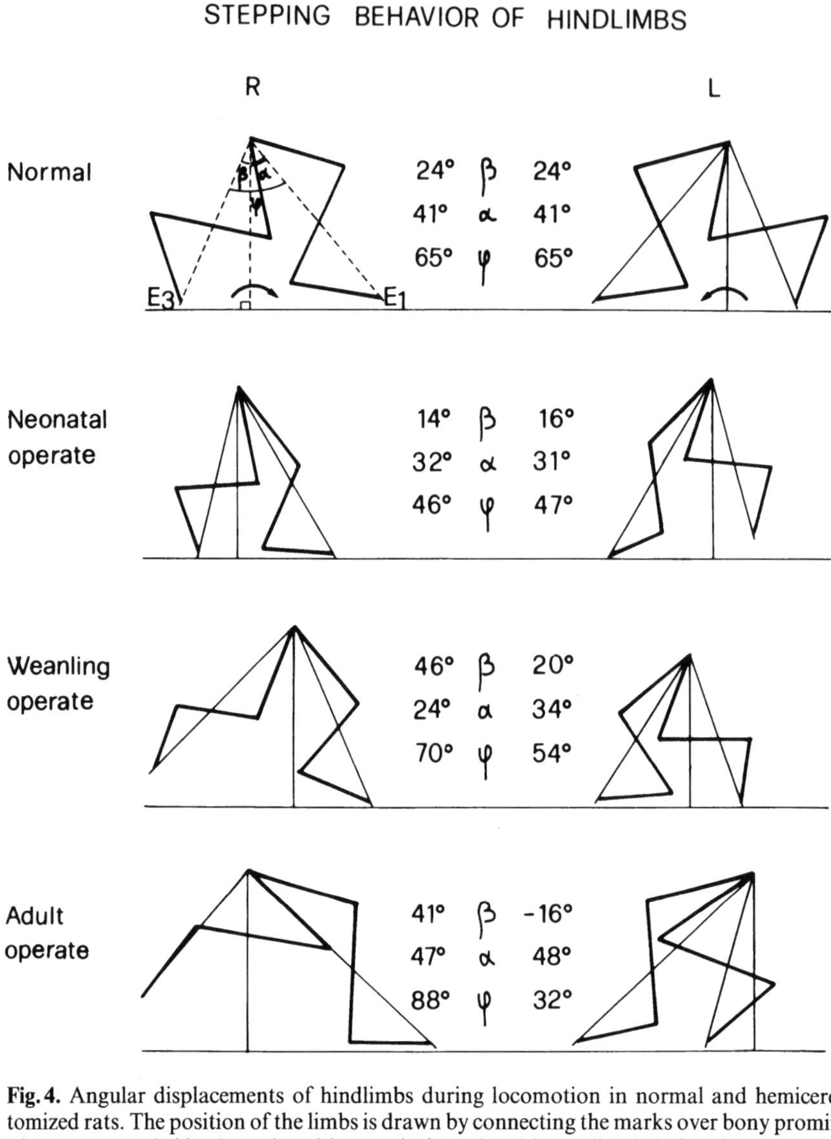

Fig. 4. Angular displacements of hindlimbs during locomotion in normal and hemicerebellectomized rats. The position of the limbs is drawn by connecting the marks over bony prominences. The most extended backward position (end of E3 phase) immediately before foot lift-off and the most extended forward position (end of E1 phase) immediately after foot placement are indicated. β posterior angular displacement; α anterior angular displacement; ϕ total angular displacement

development of aberrant neural projections after cerebellar lesions in newborn rats (Castro 1978; Leong 1977). Following neonatal HCb, aberrant ipsilateral cerebello-rubral (Gramsbergen and IJkema-Paassen 1982; Naus et al. 1984) and cerebello-thalamic (Molinari et al. 1986) projections develop that maintain the topographic specificity of the normal contralateral projection (Naus et al. 1984). Furthermore, an anomalous increase in crossed sensorimotor cortico-pontine (O'Donoghue et al. 1987) and diencephalo-olivary projections after neonatal

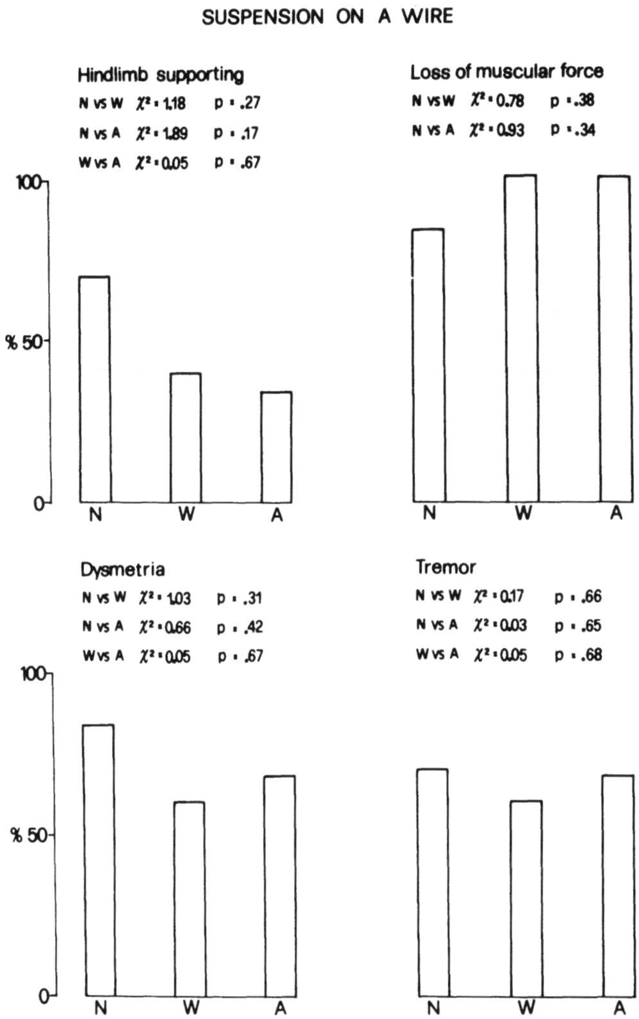

Fig. 5. Motor handicaps after suspension on a wire in hemicerebellectomized rats. Abbreviations and statistical comparisons as in Fig. 2

HCb is reported. An increased spinal projection to the ipsilateral Deiters' nucleus is also described (Castro and Smith 1979). Taking into account both these anatomical reports and our behavioral findings the study of descending spinal tracts seemed worthwhile. Regarding the rubro-spinal and vestibulo-spinal pathways, no difference was found in the ratio between ipsilateral and contralateral components of these descending projections in all hemicerebellectomized animals, independently from the age of lesioning. Also regarding the reticulo-spinal system, a gross rearrangement was not evident. Since the most important descending tracts, such as the rubro-spinal, the vestibulo-spinal and possibly the reticulo-spinal pathways, appear to maintain their normal projections, without displaying remodelling, the anatomical substrate for the better performances of neonatal operated animals, the so-called infant lesion effect (Bregman and Goldberger 1982) must be sought elsewhere. Although a synaptic reorganization at the level of spinal microcircuitry cannot be dismissed, being beyond the resolution of our neuroanatomical technique, the structural basis of these behavioral findings seems to lie in the above described anatomical reorganization at the level of the red nuclei, the thalamic nuclei, and the corticofugal projections. However, further studies are needed to elucidate the relationships between anatomical and behavioral post-lesion effects.

Acknowledgements. LP and MM want to thank Doctor Mario Galli who allowed them to discover common roots. This work was supported by grants of CNR and MPI.

References

Altman J (1982) Morphological development of the rat cerebellum and some of its mechanisms. In: Palay SL, Chan-Palay V (eds) The cerebellum. New vistas. Springer, Berlin Heidelberg New York, pp 8–49 (Experimental Brain Research, suppl 6)

Bovolenta P, Liem RKH, Mason CA (1984) Development of cerebellar astroglia: transitions in form and cytoskeletal content. Dev Biol 102:248–259

Bregman BS, Goldberger ME (1982) Anatomical plasticity and sparing of function after spinal cord damage in neonatal cats. Science 217:553–555

Bregman BS, Goldberger ME (1983) Infant lesion effect. II. Sparing and recovery of function after spinal cord damage in newborn and adult cats. Dev Brain Res 9:119–135

Castro AJ (1978) Projections of the superior cerebellar peduncle in rats and the development of new connections in response to neonatal hemicerebellectomy. J Comp Neurol 178:611–628

Castro AJ, Smith DE (1979) Plasticity of spinovestibular projections in response to hemicerebellectomy in newborn rats. Neurosci Lett 12:69–74

D'Amato CJ, Hicks SP (1980) Development of the motor system: effects of radiation on developing corticospinal neurons and locomotor functions. Exp Neurol 70:1–23

Finger S, Simmonds D, Posner R (1978) Anatomical, physiological and behavioural effects of neonatal sensorimotor cortex ablation in the rat. Exp Neurol 60:364–373

Gramsbergen A, IJkema-Paassen J (1982) CNS plasticity after hemicerebellectomy in the young rat. Quantitative relations between aberrant and normal cerebello-rubral projections. Neurosci Lett 33:129–134

Gramsbergen A, IJkema-Paassen J (1984) The effects of early cerebellar hemispherectomy in the rat: behavioural, neuroanatomical and electrophysiological sequelae. In: Finger S, Almli CR (eds) Early brain damage. Academic Press, London, pp 155–177 (Neurobiology and Behaviour, vol 2)

Leonard CT, Goldberger ME (1987) Consequences of damage to the sensorimotor cortex in neonatal and adult cats. I. Sparing and recovery of function. Dev Brain Res 32:1–14

Leong SK (1977) Plasticity of cerebellar efferent after neonatal lesions in albino rats. Neurosci Lett 52:281–289

Manni E, Dow RS (1963) Some observations of the effects of cerebellectomy in the rat. J Comp Neurol 121:189–194

Molinari M, Bentivoglio M, Granato A, Minciacchi D (1986) Increased collateralization of the cerebellothalamic pathway following neonatal hemicerebellectomy. Brain Res 372:1–10

Naus CG, Flumerfelt BA, Hrycyshyn AW (1984) Topographic specificity of aberrant cerebellorubral projections following neonatal hemicerebellectomy in the rat. Brain Res 309:1–15

O'Donoghue DL, Kartje-Tillotson G, Castro AJ (1987) Forelimb motor cortical projections in normal rats and after neonatal hemicerebellectomy: an anatomical study based upon the axonal transport of WGA/HRP. J Comp Neurol 256:274–283

Petrosini L (1984) Task-dependent rate of recovery from hemilabyrinthectomy: an analysis of swimming and locomotor performances. Physiol Behav 33:799–804

Schneider GE (1979) Is it really better to have your brain lesion early? A revision of the Kennard Principle. Neuropsychologia 17:557–583

III Lesions to Specific CNS Subsystems: Spinal Cord

Does Collateral Sprouting from Corticospinal Fibers Participate in Motor Recovery After Spinal Hemisection in Monkeys?

Mamoru Aoki[1], Yutaka Fujito[1], Isao Kosaka[1], and Hajime Satomi[2]
Departments of Physiology[1] and Anatomy[2], Sapporo Medical College, Sapporo 060, Japan

1 Introduction

Hemisection of the spinal cord at the lower thoracic level produces an ipsilateral flaccid paralysis of the hindlimb in the monkey. In the course of several weeks after the operation, the animal shows gradual recovery of some useful motor functions such as standing and locomotion, in parallel with exaggeration of tendon reflexes (Aoki et al. 1976; Aoki and Mori 1978). In the following recovery period several months after the hemisection, the animal appears to have no motor deficits on casual observation. These findings suggest the possibility that the supraspinal descending fibers through the remaining pathways make new functional connections with lumbosacral motoneurons on the hemisected side in the recovery period (Nathan and Smith 1973). We previously demonstrated parallel recovery from the effects of motor cortical stimulation on the hindlimb muscles and the capability of voluntary movement of the affected limb (Aoki and Mori 1978, 1979, 1980). Recent histological study has suggested the possibility that collateral sprouting (McCouch et al. 1958) from the descending fibers in the remaining cord could be correlated with motor recovery in the affected limb (Aoki et al. 1985, 1986).

In the present study, we first attempted to evaluate the role of corticospinal fibers in motor recovery in chronically spinal-hemisected monkeys. Behavioral studies and motor cortical stimulations were performed before and after section of the lateral corticospinal tract in the remaining spinal cord just below the previously hemisected site. Furthermore, to obtain more direct histological evidence for collateral sprouting from the corticospinal fibers, we injected WGA-HRP (wheat germ agglutinin conjugated with horseradish peroxidase) into the precentral motor cortex and traced the distribution of labeled fibers in the lumbosacral segments of the chronically spinal-hemisected monkeys.

2 Methods

A total of 6 adult monkeys (5 Macaca fascicularis, weighing 4.0–5.0 Kg each, and 1 Macaca fuscata, weighing 5.5 Kg) were used.

H. Flohr (Ed.)
Post-Lesion Neural Plasticity

2.1 Behavioral Study

Two chronically spinal-hemisected monkeys were used. The spinal cord of the monkeys had been hemisected at the lower thoracic segments (T_9 and T_{10}) for 6 and 20 months and the animals were carefully kept in the primate cages. The animals were moved from the cage to a primate chair and reflex movements or "voluntary" movements of the legs and the foot digits were observed. Also, the animal was placed on the floor and allowed to stand or make locomotion movements freely. In addition, the animal was trained to climb up a climbing apparatus consisting of two vertical bars (natural tree branches) joined by a horizontal bar 2 m high.

After these observations, the animal was anesthetized with pentobarbital sodium (Nembutal, 20 mg/kg, i.v.) and the lateral corticospinal tract in the remaining cord was aseptically sectioned just below the previously hemisected site with a fine corneal knife and a pair of scissors under a dissection microscope. From the next day of the operation onwards, the above behavioral observations were repeated at 1–3-day intervals and the degree of motor deficit was examined for several weeks. These behavioral observations were supplemented by video recordings for later analysis.

2.2 Motor Cortical Stimulation

Experiments were performed on three monkeys (one normal, two chronically spinal-hemisected). Under Nembutal anesthesia, small craniotomies were made overlying the precentral leg motor areas and the animal was placed in a primate chair. In a lightly anesthetized state, short cortical tetani (5–6 pulses, 0.3 ms duration, 500 Hz) were delivered to the leg areas of the monkey using a movable spring-mounted ball-tipped electrode.

Electromyograms (EMG's) were recorded from selected hindlimb muscles such as the quadriceps femoris, biceps femoris, tibialis anterior and gastrocnemiusisoleus on both sides by means of bipolar hooked wire electrodes (inserted into the muscles 2–3 cm apart, stainless steel, diameter 50 μm).

In one chronically spinal-hemisected monkey, the lateral corticospinal tract was acutely sectioned on the intact side and motor cortical stimulations were repeated. The effects of motor cortical stimulation obtained before and after the operation were compared.

2.3 Histological Study

Three monkeys (two *Macaca fascicularis* and one *Macaca fuscata*), including two monkeys used for motor cortical stimulation, were used for the WGA-HRP study. WGA-HRP (Toyobo) was injected into the leg motor area on one side of the cerebral cortex at seven to eight sites about 3 mm apart. A microsyringe was used to deliver 1 μl of a 10% concentration of WGA-HRP solution to each site.

After a 5-day survival period, the animals were deeply anesthetized with Nembutal and perfused transcardially with normal saline followed by a fixative of 2% paraformaldehyde and 0.5% glutaraldehyde in 0.1 M phosphate buffer (pH 7.4). The brain and spinal cord were dissected and immersed overnight in the same fixative followed by soaking in phosphate buffer with 20% sucrose for 2 days. Frozen sections (60 and 80 μm thick) were made in the frontal plane of the pre-central motor areas of both sides. Serial cross-sections (100 μm thick) of the spinal cord were made using a microslicer (DTK-3000W, Dohsaka EM, Kyoto). The sections of the spinal cord were processed using the tetramethylbenzidine (TMB) method to develop HRP reaction products (Mesulam 1978). The WGA-HRP injection sites of the motor area were estimated by the TMB or benzidine dihydrochloride (BDHC) methods (Mesulam 1976).

3 Results

3.1 Behavioral Study

The two monkeys with thoracic cords chronically hemisected moved their hindlimbs spontaneously while in the primate chair and in an awake state. There were no appreciable differences in the gross flexor and extensor movements between the intact and impaired hindlimbs. However, when the monkey reached for and grasped a wooden stick or a finger of the examiner, the decrease of force in grasping was noted in the affected limb. Tendon reflexes such as knee and ankle jerks were brisk and exaggerated in the affected limb as previously described (Aoki et al. 1976). The monkeys could support themselves with both hindlimbs and had little difficulty in maintaining a stable posture. Hindlimb stepping during four-legged locomotion appeared intact.

The monkeys were successful in climbing the climbing apparatus. However, there were some apparent motor deficiencies in the affected limb in arboreal progression. The affected limb occasionally slipped off from the pole due to incorrect foot placing and weak grasping of the tree with the foot digits.

After these behavioral observations, the ipsilateral corticospinal tract of the remaining cord contralateral to the hemisection was sectioned. From the next day onwards, the monkeys were still able to stand and were capable of locomotion. However, there was a marked reduction of force in the distal movements of both hindlimbs, such as grasping a stick or climbing the climbing apparatus. For the initial few days, diminished general usage of both hindlimbs occurred, and the animal was reluctant to climb the tree-branch pole. When urged to cling on to the pole, the monkey could neither fixate nor firmly grasp the pole with the foot digits, and easily slipped down. This motor impairment was much more marked in the affected limb (contralateral to the corticospinal lesion), which showed almost complete paralysis of the foot digits. In the subsequent several weeks there was a slow improvement in the distal movements. However, the clumsy tree climbing persisted; the affected limb frequently slipped and extended into the air due to an apparent spacial disorientation of the limb during arboreal progression.

3.2 Motor Cortical Stimulation

In a normal monkey, cortical motor stimulation elicited "flick" movements and associated EMG activities of the contralateral hindlimb at an intensity of 1.0–2.0 mA. EMG's from flexor muscles (biceps femoris and tibialis anterior) and exten-

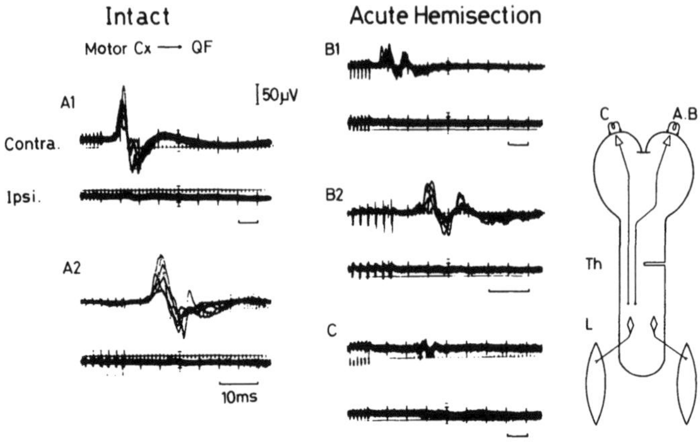

Fig. 1. EMG responses in the hindlimb muscles elicited by trains of stimuli applied to the cortical leg motor area before and after spinal hemisection. In each set of records (**A, B**), upper and lower traces indicate EMG's from the quadriceps femoris muscles contralateral and ipsilateral to the cortical stimulation sides respectively. **A 1** and **A 2** EMG responses before hemisection at different sweep speeds, stimulus intensity 1.5 mA. **B 1** and **B 2** Immediately after hemisection at T_{10}, stimulus intensity 1.8 mA. **C** Cortical stimuli were applied contralateral to the hemisection side, stimulus intensity 2.0 mA. All records are from five superimposed individual recordings. *Inset* drawing illustrates the stimulation an recording setup and the thoracic cord hemisection

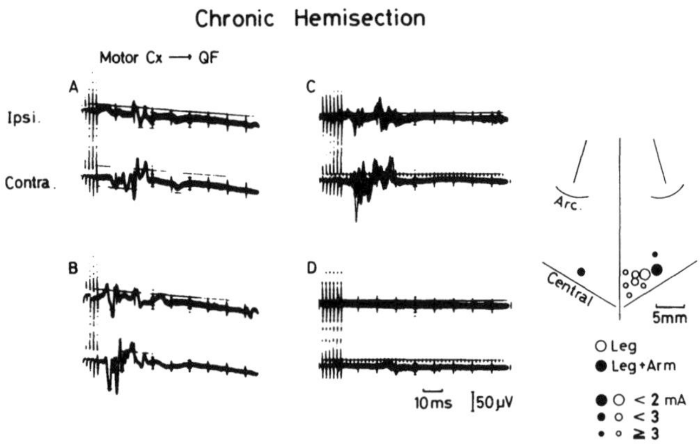

Fig. 2. EMG responses in a chronically spinal-hemisected monkey. In each set of records (**A–C**) upper and lower traces indicate EMG's from the quadriceps femoris muscles ipsilateral and contralateral to the hemisected side; stimulus intensity 1.0 mA (**A**) and 1.8 mA (**B**). **C** and **D** Five superimposed records. **D** Cortical stimuli were applied to the opposite side; stimulus intensity 2.0 mA. *Inset* drawing indicates the motor cortical thresholds for eliciting "flick" movements of the leg or arm on the contralateral side

sor muscles (quadriceps femoris and gastrocnemius-soleus) could be elicited with latencies of 12–20 ms by stimulating the cortical leg area (Fig. 1). Stronger stimuli above 2.5–3.0 mA elicited flick movements of the bilateral hindlimbs, often accompanied by forelimb movements and trunk muscle contractions. This normal monkey was used to observe the effect of spinal hemisection on motor cortical stimulation results. The spinal cord was acutely hemisected at the T_{10} segment. After hemisection, cortical motor stimulation on the hemisected side produced EMG activity of the contralateral hindlimb with similar latencies to the normal state, although the thresholds were slightly elevated and the EMG amplitudes became smaller. When cortical stimuli were applied to the opposite side, small EMG's were observed in the ipsilateral hindlimb.

In the two monkeys with chronic spinal hemisection, cortical stimulation of the hemisected side was found to produce simultaneous movements of the contralateral intact limb and the ipsilateral affected limb. The thresholds for evoking flick movements and EMG's of both hindlimbs were similar and ranged from 0.7 to 1.0 mA. There was a general tendency for the EMG amplitudes to be smaller, and the latencies were slightly longer in the affected limb, with the stimulus intensity being below 2.0 mA (Fig. 2). Cortical stimulation contralateral to the hemisection, with a stimulus intensity of up to 2.0 mA, produced neither flick movements nor EMG's in either hindlimb. With stronger stimuli above 2.5 mA, small EMG's of the quadriceps femoris could be occasionally observed in the hindlimbs bilaterally.

After acute section of the lateral corticospinal tract in the remaining cord of the chronically hemisected monkeys, motor cortical stimuli of intensities below 2.0 mA produced neither EMG activity nor flick movements in either hindlimb.

3.3 WGA-HRP Study

In an intact control monkey, WGA-HRP was injected into the unilateral cortical leg area. Dense anterograde labeling occurred in the lateral corticospinal tract in the white matter of the lumbosacral segments contralateral to the side of the cortical injection. Also, sparse labeling was noted in the lateral and ventral corticospinal tracts ipsilateral to the cortical injection side (Fig. 3). In the gray matter contralateral to the injection side there were diffuse, anterogradely labeled fibers distributed from the dorsal horn to the ventral horn. On the other hand, anterograde labeling was very scanty and only a few labeled fibers were noted in the ventral horn of the gray matter ipsilateral to the injection side.

There were few labeled fibers crossing the commissural region; no such crossing fibers were found in 32 sections of the L5 segment examined, and only 4 crossing fibers were found in 54 sections of the L6 segment.

In one chronically spinal-hemisected monkey, in which the spinal cord had been hemisected at the T_{11} segment for 15 months, WGA-HRP was injected into the cortical leg area ipsilateral to the hemisected side. The lateral corticospinal tract of the lumbar segments on the contralateral side (intact limb side) was densely labeled. There were also diffuse and densely labeled fibers in the gray matter distributed from the base of the dorsal horn to the ventral horn (Fig. 4). On the other hand, neither the lateral nor the ventral corticospinal tract was labeled on the injected side (hemisected side). However, there was a marked increase in labeling in the gray matter on the hemisected side compared with the findings in the intact control. Labeled fibers were distributed densely in the ventromedial and intermediate parts, and a few fibers reached the motor nuclei in the lateral part of the ventral horn (lamina IX). Labeled terminal fibers were observed occasion-

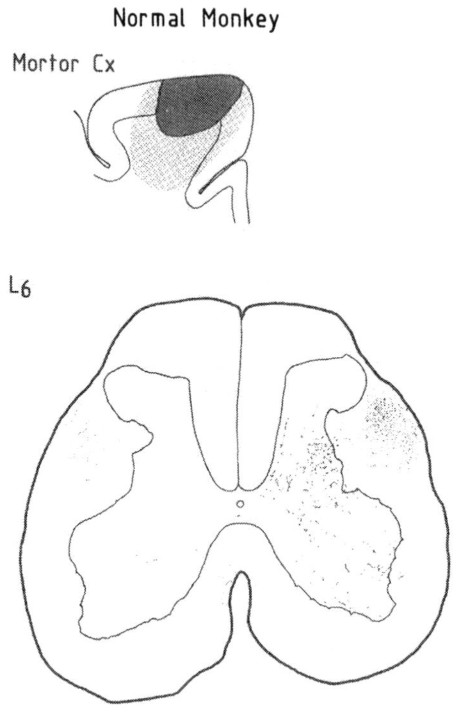

Fig. 3. The WGA-HRP injection site in the motor cortex and the distribution of anterogradely labeled fibers in the lumbar segment in an intact control. *Thickly* and *lightly shaded* areas indicate the virtual injection site and probable extent of diffusion respectively. Distribution of labeled fibers was reconstructed from a representative section of the L6 segment

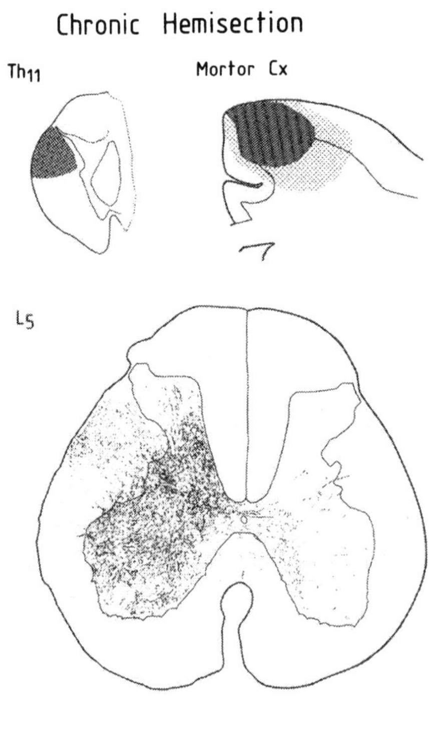

Fig. 4. Diagrams illustrating the cortical WGA-HRP injection site and the distribution of anterogradely labeled corticospinal fibers at T11 hemisected segment and L5 segment. Note that the spinal cord is reduced in size on the hemisected side and labeled fibers are found in the gray matter on the hemisected side

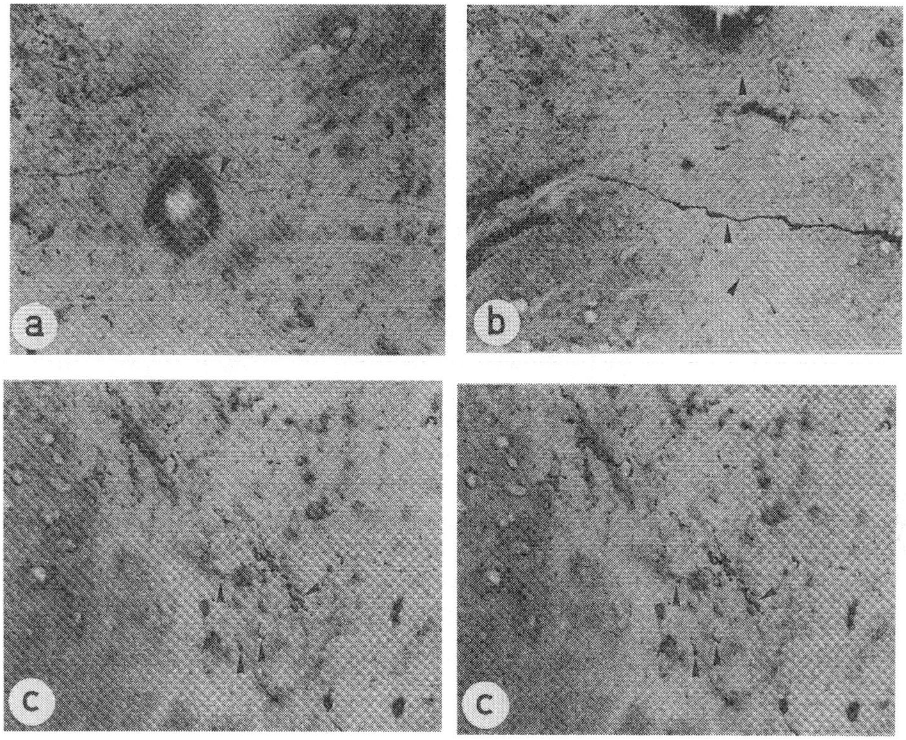

Fig. 5. Photomicrographs showing labeled commissural fibers and terminal fibers in the gray matter on the hemisected side. **a, b** labeled fibers in the dorsal and ventral commissures respectively. **c, d** preterminal fibers in the ventromedial and lateral parts of motoneuron pools (lamina IX) respectively. *Arrows* indicate labeled fibers. Bar = 100 μm

ally along the basal dendrites and perikaryon of the large cells, probably representing motoneurons (Fig. 5).

It was noted that there was a remarkable increase in labeled crossing fibers in the chronically spinal-hemisected monkeys. The number of crossing fibers per section was 6.2 ± 2.4 (mean $\pm$S.D.; range 3–12, N = 33) in the L4 segment. In the L5 and L6 segments, the number of crossing fibers was 11.1 ± 4.6 (5–23, N = 29) and 9.2 ± 2.9 (5–15, N = 20), respectively. Thus, the total number of crossing fibers from the L4 to L6 segments was 8.7 ± 4.0 (3–23, N = 82). Crossing fibers passed through both the ventral and dorsal commissural regions and occasionaly were found to run across the ventral white commissure.

4 Discussion

The initial motor deficits and the subsequent recovery of movement in animals with their spinal cords hemisected has been repeatedly described by us and other investigators (Aoki and Mori 1978, 1979, 1980; Aoki et al. 1985; Murray and

Goldberger 1973). In contrast to the considerable recovery of hindlimb movements engaging proximal and trunk muscles, impairment of distal movements involving the foot digits persisted over many months. We confirmed this by observing the imperfect arboreal progression in the chronically spinal-hemisected monkeys in the present study.

Selective section of the lateral corticospinal tract of the remaining cord produced motor deficits in the distal movements of the hindlimbs bilaterally as previously described (Tower 1940; Goldberger 1974; Woolsey et al. 1972). Impairment of hindlimb digit movement was found to be much more severe on the hemisected side. This observation (together with the results of motor cortical stimulation), that the EMG's evoked in the affected limb by cortical stimuli were abolished by section of the lateral cortical tract, strongly suggests that the corticospinal fibers participate in motor recovery, at least of the distal movements.

In our previous study, in which HRP was injected into the lumbosacral cord unilaterally on the hemisected side in the chronically spinal-hemisected monkeys, the number of retrogradely labeled neurons in the motor cortex was significantly increased on the hemisected side compared with that in an acute control. In the present study, WGA-HRP was injected into the leg motor area ipsilateral to the hemisection in these monkeys. There were densely labeled fibers anterogradely in the gray matter of the hemisected side and a marked increase in the number of labeled fibers crossing the spinal commissure compared with findings in an intact control.

We must interpret the results carefully, since there is considerable quantitative variation in HRP labeling and the number of observed preparations was limited. However, the appearance of the labeled corticospinal fibers in the lateral part of the ventral horn (lamina IX) in the chronically spinal-hemisected monkeys, which has not been reported previously in normal animals, may have resulted from the formation of new connections and may not be due to the preexisting fibers. The previous and the present results together strengthen the view that the crossed lateral corticospinal fibers increase their connections by recrossing to the spinal gray matter on the hemisected side. These recrossing fibers may be due to collateral sprouting of descending corticospinal fibers on the intact side (Tsukahara et al. 1975). The increase of corticospinal fibers in the chronically spinal-hemisected monkeys is in accord with our previous study (Aoki and Mori 1978), in which corticomotoneuronal EPSP's were recorded by motor cortical stimuli ipsilateral to the hemisection. These sprouting fibers may connect via the spinal interneurons or directly to the motoneurons on the hemisected side, and lead to the recovery of movements in the affected limb (Kuypers and Brinkman 1970; Liu and Chambers 1964; Porter 1985).

The present results do not exclude the possibility that the sprouting fibers occur from other supraspinal descending systems such as reticulospinal, rubrospinal, or vestibulospinal fibers. Also, ipsilateral corticospinal fibers may sprout and participate, in part, in motor recovery (Kucera and Wiesendanger 1985). It remains to be determined how the sprouting fibers make functional synaptic connections with the spinal interneurons and motoneurons on the hemisected side, eventually giving rise to motor recovery.

Acknowledgements. The authors thank Dr. Daniel A. Lupin for his helpful advice on the English usage and Miss Yumiko Koshiishi for typing the manuscript.

References

Aoki M, Mori S (1978) Changes in monosynaptic EPSPs of quadriceps motoneurons in monkeys with the spinal cord chronically hemisected at the thoracic level. In: Ito M et al. (eds) Integrative control functions of the brain, vol I. Kodansha, Tokyo/Elsevier, Amsterdam, pp 170–171

Aoki M, Mori S (1979) Recovery of hindlimb movement elicited by motor cortical stimulation after spinal hemisection in monkeys. In: Ito M et al. (eds) Integrative control functions of the brain, vol II. Kodansha, Tokyo/Elsevier, Amsterdam, pp 152–154

Aoki M, Mori S (1980) Corticomotoneuronal EPSPs evoked in hindlimb motoneurons in monkeys with the spinal cord chronically hemisected at the thoracic level. In: Ito M et al. (eds) Integrative control functions of the brain, vol III. Kodansha Tokyo/Elsevier, Amsterdam, pp 156–158

Aoki M, Mori S, Fujimori B (1976) Exaggeration of knee-jerk following spinal hemisection in monkeys. Brain Res 107:471–485

Aoki M, Fujito Y, Kurosawa Y, Satomi H (1985) Functional restitution of corticospinal connections after spinal hemisection. Electroencephalogr Clin Neurophysiol 61:44

Aoki M, Fujito Y, Satomi H, Kurosawa Y, Kasaba T (1986) The possible role of collateral sprouting in the functional restitution of corticospinal connections after spinal hemisection. Neurosci Res 3:617–627

Goldberger ME (1974) Recovery of movement after CNS lesions in monkeys. In: Stein DG et al. (eds) Plasticity and recovery of function in the central nervous system. Academic Press, London, pp 265–338

Kucera P, Wiesendanger M (1985) Do ipsilateral corticospinal fibers participate in the functional recovery following unilateral pyramidal lesions in monkeys? Brain Res 348:297–303

Kuypers HGJM, Brinkman J (1970) Precentral projections to different parts of the spinal intermediate zone in the rhesus monkey. Brain Res 24:29–48

Liu CN, Chambers WW (1964) An experimental study of the cortico-spinal system in the monkey (*Macaca mulatta*) J Comp Neurol 123:257–284

McCouch GP, Austin GM, Liu CN, Liu CY (1958) Sprouting as a cause of spasticity. J Neurophysiol 21:205–216

Mesulam M-M (1976) The blue reaction product in horseradish peroxidase neurohistochemistry. Incubation parameter and visibility. J Histochem Cytochem 24:1273–1280

Mesulam M-M (1978) Tetramethylbenzidine for horseradish peroxidase neurohistochemistry. A noncarcinogenic blue reaction product with superior sensitivity for visualizing neural afferents and efferents. J Histochem Cytochem 26:106–117

Murray M, Goldberger ME (1973) Restitution of function and collateral sprouting in the cat spinal cord: The partially hemisected animal. J Comp Neurol 158:19–36

Nathan PW, Smith MC (1973) Effects of two unilateral cordotomies on the motility of the lower limbs. Brain 96:471–494

Porter R (1985) The corticomotoneuronal component of the pyramidal tract: corticomotoneuronal connections and functions in primates. Brain Res Rev 10:1–26

Tower SS (1940) Pyramidal lesion in the monkey. Brain 63:36–90

Tsukahara N (1981) Synaptic plasticity in the mammalian central nervous system. Annu Rev Neurosci 4:351–379

Tsukahara N, Hultborn H, Murakami F, Fujito Y (1975) Electrophysiological study of formation of new synapses and collateral sprouting in red nucleus neurons after partial denervation. J Neurophysiol 38:1359–1372

Woolsey CN, Gorska T, Wetzel A, Erickson TC, Earls FJ, Allman JM (1972) Complete unilateral section of the pyramidal tract at the medullary level in *Macaca mulatta*. Brain Res 40:119–123

Pathfinding and Synaptic Specificity of Regenerating Spinal Axons in the Lamprey

M. E. Selzer [1], D. Lurie [2], and Scott S. A. Mackler [3]

[1] Department of Neurology and the David Mahoney Institute of Neurological Sciences, University of Pennsylvania School of Medicine, Philadelphia, PA 19104-4283, USA
[2] Department of Neuroscience and the David Mahoney Institute of Neurological Sciences, University of Pennsylvania School of Medicine, Philadelphia, PA 19104, USA
[3] Department of Anatomy, University of Pennsylvania School of Medicine, Philadelphia, PA 19104, USA

1 Introduction

Recent developments in fetal CNS implantation and use of peripheral nerve bridges and grafts to circumvent a scar have led to optimism that regeneration of injured axons in the CNS may some day be used as a strategy in the treatment of CNS injuries (Björkland and Stenevi 1984; David and Aguayo 1981; Reier 1985; Reier et al. 1983 b). Spinal cord injuries are of particular concern because so much function passes through such a restricted cross-section of CNS and even relatively minor contusions often lead, via vascular reactions, to complete functional transection. Despite evidence of anatomical regeneration of fibers into the distal segments of spinal cord (David and Aguayo 1981; Bregman 1987), bridging and grafting experiments have not yet resulted in functional restoration in spinal cord-injured mammals. Unfortunately, it is difficult in such preparations to answer some fundamental questions concerning the fate of regenerating fibers. Do they conduct normal electrical impulses? Do they regenerate in a specific direction? Can they form physiologically functioning synapses with target neurons distal to the lesion? Are such synapses random or are they specific in some way which might lead to functional recovery? And finally, what guides the regenerating axon as it grows?

In order to address these and other questions one must first have a preparation that does regenerate. We have used the large larval sea lamprey *(Petromyzon marinus)*, 4–5 years old and 10–15 cm long. This is a primitive vertebrate in a stable phase of neurological development which recovers coordinated swimming within 6–10 weeks after complete spinal transection. It has large identifiable neurons visible in the transilluminated brainstem (Müller and Mauthner cells), whose giant axons course unbranched through most of the length of the spinal cord. The spinal cord contains large neurons of recognizable type which can also be seen in vivo. This preparation has substantially met the five criteria set forth by an NIH advisory panel (Guth et al. 1980) for functional neuritic regeneration in the spinal cord:

1. *The experimental lesion must cause disconnection of nerve processes.* The spinal cord is completely transected under the dissecting microscope and completeness of the transection has been verified by histological examination of the scar (Rovainen 1976; Selzer 1978).

H. Flohr (Ed.)
Post-Lesion Neural Plasticity

2. *Processes of central nervous system neurons must bridge the level of injury.* Regenerated neurites of identified types of neurons have been followed across a transection in serial light microscopic sections (Rovainen 1976; Selzer 1978) and in spinal wholemounts with intracellularly labeled neurons (Wood and Cohen 1979, 1981; Yin and Selzer 1983).
3. *The regenerated fibers must make junctional contacts.* Horseradish peroxidase (HRP) injected into brainstem neurons has labeled vesicle-containing synaptic profiles caudal to a scar following recovery from high spinal transection (Wood and Cohen 1979, 1981).
4. *The regenerated fibers must generate postjunctional responses.* Intracellular stimulation of neurons on one side of a healed transection has elicited monosynaptic EPSP's in neurons distal to the scar (Mackler and Selzer 1985, 1987).
5. *Changes in function must derive from regenerated connections.* Stimulation of the head elicits tail movements across a long length of isolated cord containing the transection scar (Selzer 1978). Electrical stimulation of the floor of the fourth ventricle elicits rhythmic undulations of the body and tail caudal to the healed transection (Curry and Ayers 1983). In the isolated spinal cord, ventral root discharges from segments rostral and caudal to a healed transection are time locked during "fictive locomotion" (Cohen et al. 1986). Retransection eliminates recovered locomotor behavior (Borgens et al. 1981; Park et al. 1986).

This work has laid the foundation for the use of the lamprey to answer some basic questions about the potential of the vertebrate central nervous system to recapitulate the original developmental blueprint and thus derive function from regeneration.

2 Directional Specificity in Axonal Regeneration

It was noticed early in our investigations that giant reticulospinal axons regenerated preferentially on their original side, and even in the same cross-sectional quadrant of the spinal cord, as the original axon (Selzer 1978). This proved not to be a trivial result of a tendency for neurites to grow in an already established direction. Observations on the response of giant interneurons (GI's) to axotomy provided the most convincing illustration of the point. GI's are large second order mechanosensory neurons located in the caudal half of the lamprey spinal cord. They normally have a single unbranched axon which arises from the anteromedial part of the cell body (Fig. 1 A). The axon decussates and projects to the brainstem in the contralateral dorsolateral and lateral columns. Axotomy within a few mm of the cell body of GI's frequently gave rise to the sprouting of long axon-like neurites from anomalous locations in the dendritic tree (Fig. 1 B). These neurites were thus initially misoriented, but the majority crossed the midline and turned rostralward to grow in the normal trajectory for axons of GI's (Yin et al. 1984).

Four possible explanations for this specificity occurred to us. First, the axons could have been seeking the scar. This would not explain their correct decussation, but it could explain the looping of axons toward the brain. In order to rule

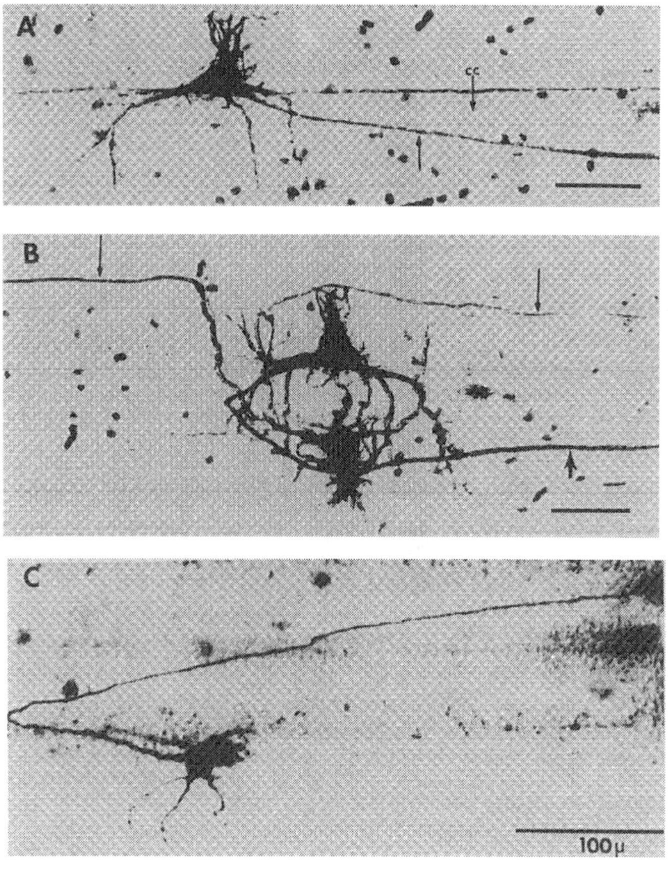

Fig. 1. Projection patterns of axotomized HRP-injected giant interneurons (GI's) in spinal cord wholemounts. **A** Normal axon origin from the anterior end of the cell body. The axon (*arrow*) projects rostrally and contralaterally; *cc* central canal. **B** Two GI's located approximately 1.2 mm caudal to a 20-day-old transection. Their original axons have undergone complete retrograde degeneration. Two neurites (*small arrows*) grow from a caudal dendrite of the cell on the bottom. One neurite (*large arrow*) extends from a caudal dendrite of the top cell. Two of the three long neurites that initially project caudally reverse direction and grow rostrally and contralaterally in their normal projection path. **C** A GI in an island formed by double transection 60 days earlier. This cell was located 6.5 mm below the rostral cut and 1.5 mm above the caudal cut. Despite its proximity to the caudal scar, a long neurite that initially grew posteriorly loops to project rostrally on the contralateral side. A second neurite (not in plane of focus) originating medially also projects rostrally. (Yin et al. 1984)

out this hypothesis, simultaneous transections were made 10 mm apart and the orientation of regenerating neurites determined for GI's in the island between the lesions. If the scar were releasing a trophic substance, the presence of a second scar caudal to the cell body should have exerted a countervailing effect on neurite orientation and the regenerating fibers should have grown randomly. Instead, most axon-like neurites were oriented rostrally (and contralaterally) suggesting that the location of the scar does not dictate the orientation of axonal regeneration (Fig. 1 C).

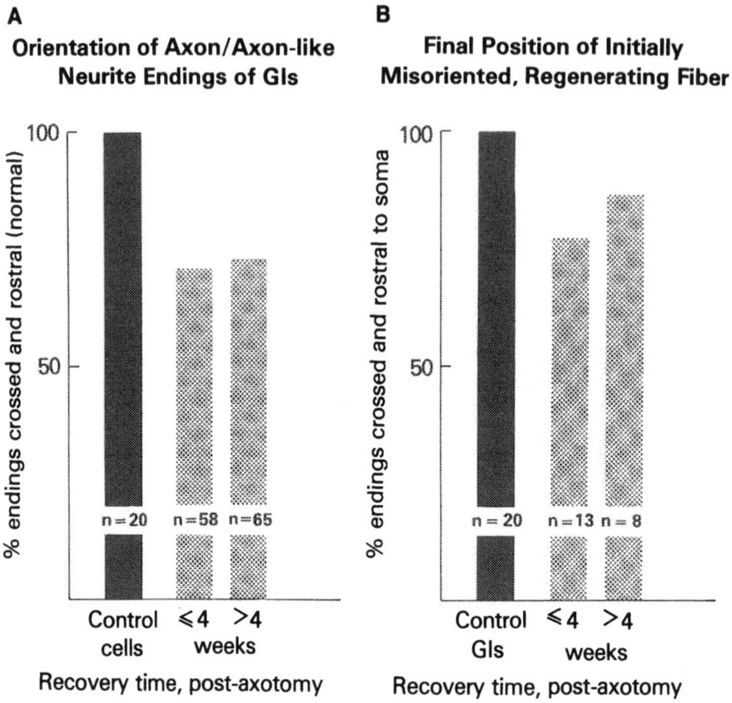

Fig. 2. Early manifestation of directional specificity by regenerating neurites of GI's. **A** The majority of axons or axon-like neurites are already correctly oriented during the first 2–4 weeks, i.e., before the time at which most axonal regeneration proceeds beyond the scar. **B** Even neurites originating from anomalous locations on the dendritic tree are correctly oriented within the first 2–4 weeks

A second possible explanation for directional specificity was that the initial regeneration of neurites was indeed random, but that specificity was imposed secondarily by retraction of incorrectly oriented neurites, i.e., those that had not found appropriate postsynaptic targets. This was tested by studying the orientations of neurites early in their growth (Mackler et al. 1986). The earliest time at which neurites could be detected which were longer than 400 μm (a somewhat arbitrary length which differentiated them from ordinary dendrites) was at 2 weeks post-transection. At 2 to 4 weeks the majority of long axon-like neurites were already oriented rostrally and contralateral to the cell body (Fig. 2).

Thus neurites appear to be growing correctly from the earliest times at which they can be identified. This leaves open a third and fourth explanation for directional specificity, that regenerating fibers are following specific cues, either a trophic influence emanating from their potential targets distal to the transection, or local cues in their immediate environment (such as adhesion molecules on the surfaces of glial cells or neurons). To distinguish between these two possibilities, a 10-mm gap was produced in the spinal cord and the orientation of neurites determined for GI's caudal to the gap (Mackler et al. 1986). Once again, most neurites were oriented correctly, even though they did not grow beyond the caudal stump. In the case of a simple transection, most neurites regenerate less than 10 mm beyond the scar. Thus the 10 mm gap eliminated the majority of the potential postsynaptic target neurons for regenerating fibers. Since this did not alter their projection patterns, it seems unlikely that distant targets are the source of a trophic influence.

These results favor the hypothesis that local guidance mechanisms strongly influence axonal regeneration in the lamprey spinal cord.

3 Specificity of Synaptic Regeneration

It has been shown that regenerating reticulospinal axons form vesicle-containing junctions with processes of unidentified neurons caudal to a lesion in the rostral part of the lamprey spinal cord (Wood and Cohen 1979, 1981). In the caudal half of the cord, simultaneous intracellular recordings have demonstrated monosynaptic connections formed by the regenerating axons of GI's caudal to the scar onto postsynaptic GI's located rostral to the scar (Mackler et al. 1986). Since, in unlesioned adult sea lampreys, approximately 70% of GI-GI pairs are synaptically linked (Rovainen 1974b), it was of interest to determine whether the formation of synapses between pairs of GI's across a scar reflected specificity in the selection of postsynaptic targets by regenerating axons.

Maps were made of the synaptic connections among recognizable neurons in the CNS of large larval sea lampreys (Mackler and Selzer 1987; Figs. 3A, 4A). A total of 162 cell pairs were tested in untransected control animals. In general the types of connections were similar to those previously described in adults (Rovainen 1974a, b), although the frequencies with which particular synaptic connections were found in the larvae were generally less. For example, in the larvae monosynaptic linkage was found between only 44% of GI-GI pairs.

Maps were then made of the synaptic interactions between recognizable neurons separated by a scar in animals which had received spinal transection 7–40 weeks previously. In animals with spinal lesions at the level of the cloaca, GI's located within 10 mm of the scar were studied as postsynaptic neurons, while various other neurons and large axons were stimulated intracellularly as potential presynaptic elements (Fig. 3B). In animals with spinal lesions at the level of the 7th gill, lateral cells (LC's) located within 10 mm of the scar were studied as the postsynaptic neuron (Fig. 4B). A total of 188 cell pairs were tested in previously transected animals. In general, cell types which were synaptically linked infrequently or not at all in control animals were never found to be connected after recovery from transection, while those which were frequently linked in controls did reestablish connections across the healed transection. Thus, GI's synapsed with other GI's 44% of the time in control animals and 17% of the time across a transection scar. Müller and Mauthner axons, which never showed synaptic linkages with GI's in control animals, failed to do so in animals recovered from transection. Similarly, dorsal cells (DC's; intraspinal primary sensory neurons), which synapsed with GI's in only 16% of cell pairs in control animals, were never synaptically linked across a transection (Fig. 3).

In the gill region of the cord, LC's received monosynaptic input from ipsilateral Müller cells B_{2-4} in 58% of cell pairs in control animals. None of the other

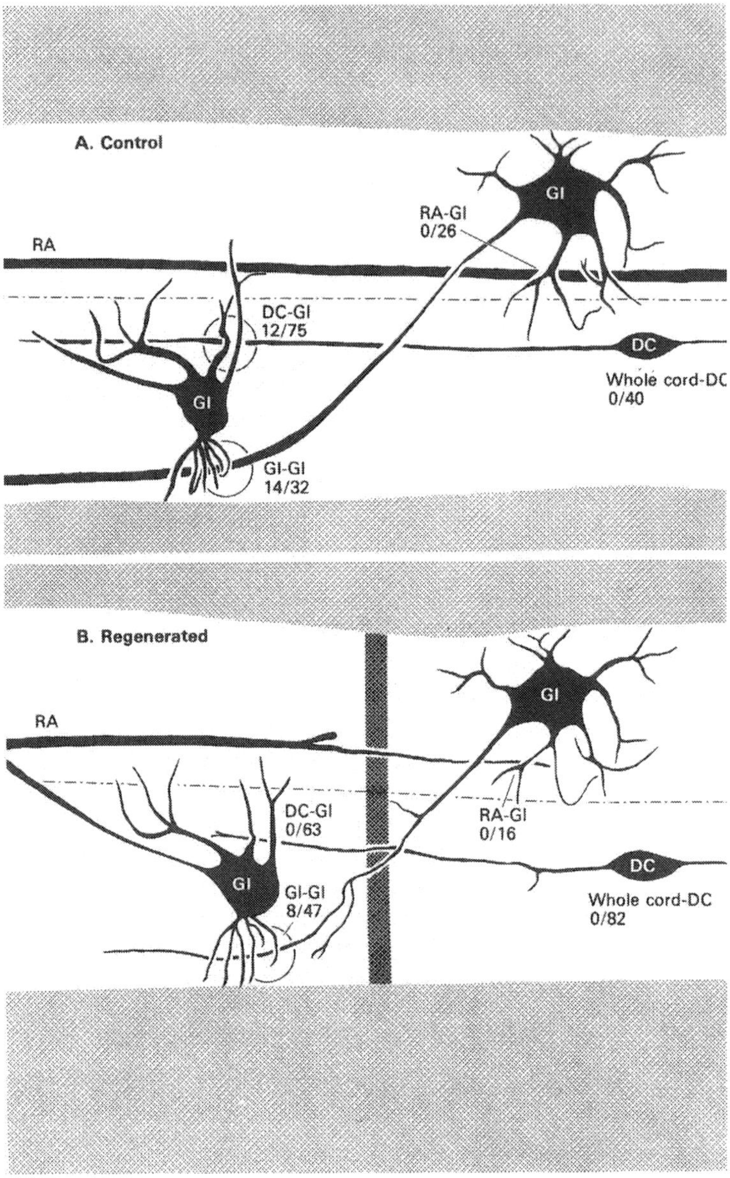

Fig. 3. Specificity of synaptic input to GI's. **A** In control larvae, GI's were monosynaptically excited by other GI's in 14 of 32 cell pairs (44%). Synapses between dorsal cells (*DC's*) and GI's were observed in only 16% of cell pairs. Giant reticulospinal axons (*RA's*) never synapsed with GI's. In addition, DC's never showed synaptic responses to stimulation of the spinal cord. **B** Following recovery from spinal transection, GI's rostral to the scar received monosynaptic EPSP's from GI's caudal to the scar in 8 of 47 cell pairs (17%). No synaptic input to rostral GI's was observed from caudal DC's or to caudal GI's from rostral RA's. The DC's remained without afferent input. (Mackler and Selzer 1987)

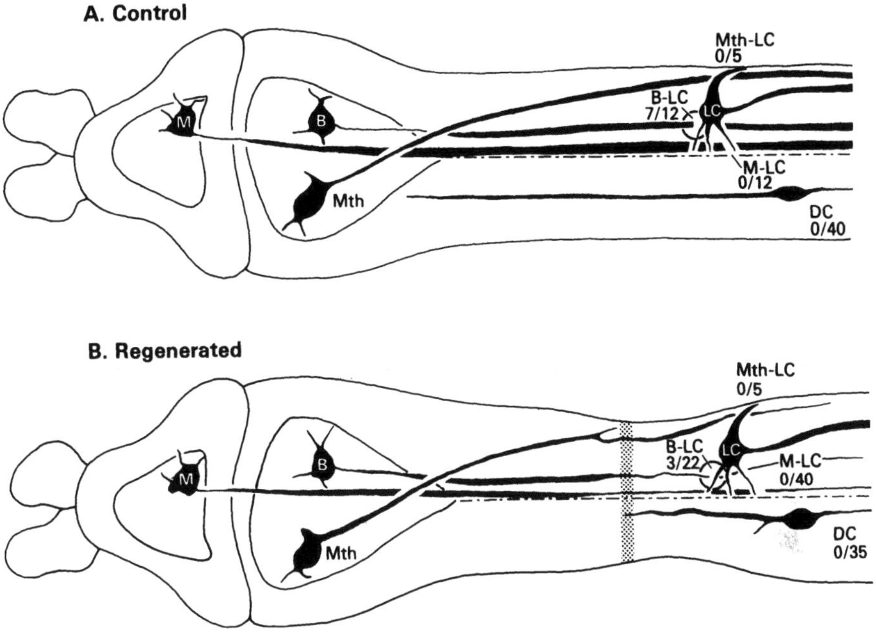

Fig. 4. Specificity of regenerated synaptic input to lateral cells (LC's). **A** In control larvae, LC's were monosynaptically excited by bulbar Müller cells 2–4 (*B*) in 7 of 12 cell pairs (58%). Contralateral Mauthner cells (*Mth*) and ipsilateral Müller cells other than B_{2-4} (*M*) did not synapse with LC's. The DC's of the rostral cord showed no synaptic activity upon cord surface stimulation. **B** Following recovery from spinal transection, LC's caudal to the scar were monosynaptically activated by B cells in 3 of 33 cell paris (14%). These LC's remained without monosynaptic input from Mth and M cells. The DC's remained without afferent input. (Mackler and Selzer 1987)

ipsilateral Müller cells or contralateral Mauthner cells (i.e., those whose axons project to the region of the LC dendritic tree) synapsed with LC's. In previously transected animals, B_{2-4} had synapses onto ipsilateral LC's caudal to the transection scar in 14% of cases, while none of the other Müller or Mauthner cells did (Fig. 4). In addition DC's received no synaptic input from whole cord stimulation, either before or after transection (Figs. 3, 4).

These and other findings have suggested that regenerating axons in the lamprey spinal cord reestablish synaptic linkages selectively with those types of neurons with which they normally form synapses during development. A similar specificity was suggested in the reinnervation of frog motoneurons by regenerating axons of dorsal root ganglion cells (Sah and Frank 1984). It may be that the molecular mechanisms responsible for the specificity of synaptogenesis during development persist or can be reexpressed years after the initial formation of neuronal connections. This would also seem to be true for directional specificity demonstrated by the regenerating axons in lamprey.

4 Role of the Scar in Directional Specificity of Regeneration

It is generally agreed that the glial scar in injured mammalian CNS is the major impediment to axonal regeneration (Kiernan 1979; Puchala and Windle 1977; Reier et al. 1983a). On the other hand, immature glial cells may actually promote axonal elongation when they are transplanted into the adult CNS (Smith et al. 1986). Similarly, the ability of fetal CNS implants to innervate adult host CNS matrix and vice versa may be due to the presence of immature glial cells in the graft.

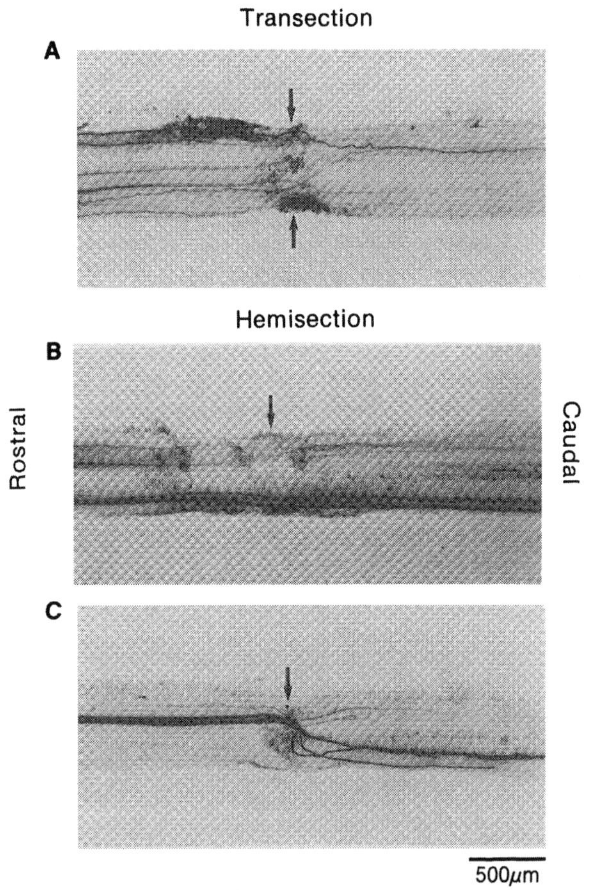

Fig. 5. Appearance of regenerating neurites of HRP-injected giant reticulospinal axons (RA's) in cord wholemounts. **A** Ten weeks after complete transection at the level of the third gill (*arrows*), most neurites have regenerated across the scar and continued on the same side of the cord. One axon decussated anomalously and another looped rostralward before crossing the scar. **B** Ten-week hemisection at the level of the third gill (*arrow*). Neurites of axons cut by the lesion are attenuated in caliber but regenerate through the scar and remain ipsilateral to their parent axon. For comparison, untransected RA's have also been injected with HRP (*bottom*). **C** In only one hemisected animal, several regenerating RA neurites decussated at the level of the scar (*arrow*) and remained contralateral to their parent axons. Some neurites in this cord grew through the scar (*top*)

In the lamprey, spinal transection gives rise to a scar composed of glial and ependymal elements. While the sources of these cells are not known definitively, the histological appearance of the scar during successive stages of its development suggests that ependymal cells are multiplying and differentiating into glial cells (Rovainen 1976; Selzer 1978). If so, this is quite a different origin for the glial scar than that proposed in mammals, where reactive astrocytes are thought to multiply and migrate to the region of injury (Reier et al. 1983a). The role of the glial scar in the regeneration of lamprey spinal axons is still not known. Does it actively support axonal growth or is it an impediment which is not absolute? In order to shed light on this question it is necessary to offer regenerating axons a choice of scar vs. normal CNS matrix through which to grow and to determine which is the preferred substrate. We have done this in the large larval sea lamprey by hemisecting the spinal cord and observing the paths taken by regenerating axons.

Spinal cords were hemisected at the level of the third gill and the animals allowed to recover for 10 weeks, by which time the majority of extending neurites had grown through the scar. The brain and spinal cord were dissected out and the giant reticulospinal cell bodies and axons impaled and injected with HRP. The tissue was reacted with the Hanker-Yates reagent and the CNS examined in wholemount (Figs. 5, 6). Similar to the behavior of regenerating axons in animals with complete transection (Yin and Selzer 1983), 55% of the 74 labeled neurites of transected axons regenerated beyond the level of the hemisection. Of these neurites, 80% grew through the scar and remained on the appropriate side of the cord (Figs. 5B, 6). Only 20% decussated proximal to the scar and grew around it (Fig. 5C). None of these recrossed to the ipsilateral side. In fact, all the neurites which grew around the scar (n=8) came from the same animal, while in every other animal, neurites which regenerated beyond the lesion site (n=74) grew through the scar. Once in the distal segment of the cord, axons invariably remained on the same side. The results of all hemisection experiments are summarized in Fig. 7.

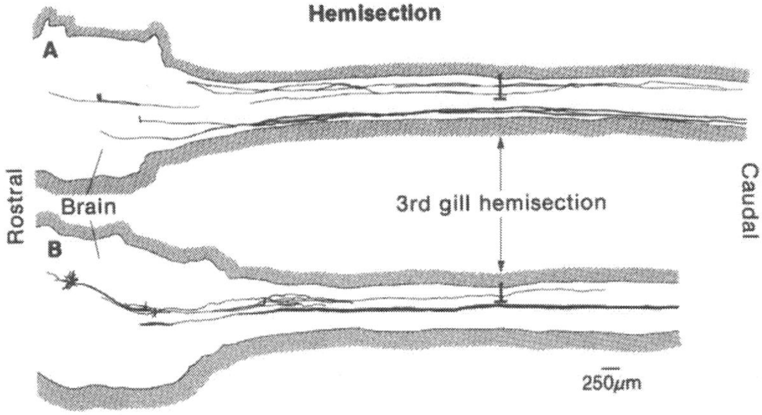

Fig. 6. Camera lucida drawings of HRP-injected RA's in spinal cords hemisected at the level of the third gill (*arrows*). **A** The majority of regenerating neurites grew through the scar. **B** In this preparation, only one neurite regenerated past the level of the scar while a number of neurites terminated proximal to the scar. Even these neurites remained ipsilateral

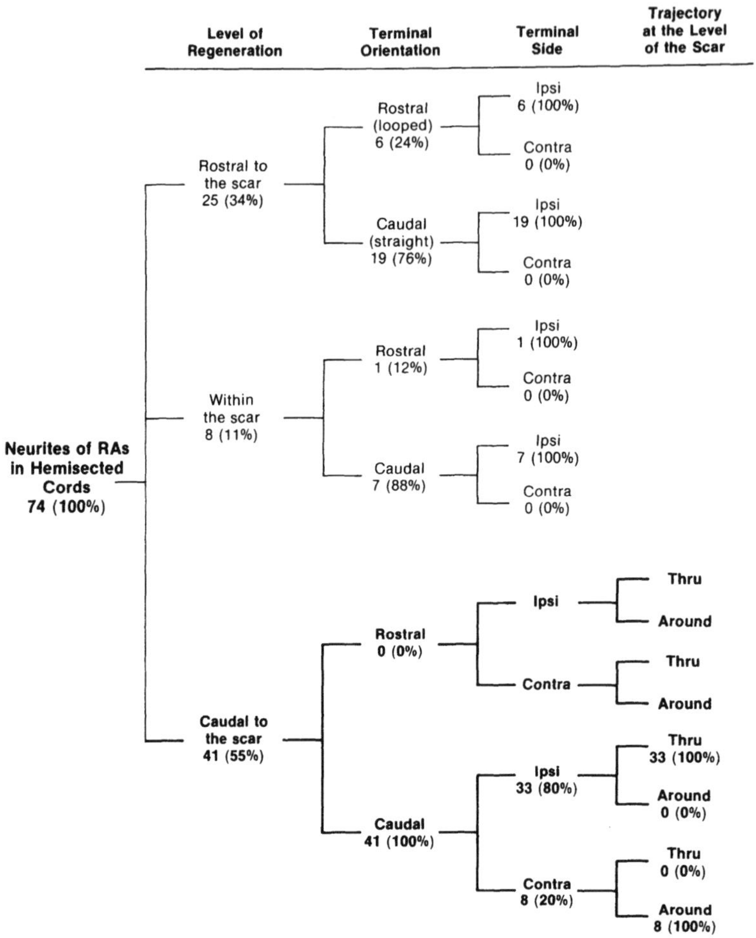

Fig. 7. Summary of neurite projection paths in 10-week hemisected spinal cords. Note that 55% of labeled neurites regenerated beyond the level of the scar (bold type). All of these remained oriented caudally (i.e., once past the scar they did not loop rostrally). Eighty percent remained on the ipsilateral side of the cord and thus grew through the scar

Thus, regenerating neurites grow preferentially through the scar, rather than around it, in order to remain correctly oriented. This could reflect either a supportive attribute of the scar or a reluctance of neurites to grow into a region of spinal cord devoid of the axonal debris and empty spaces produced by axotomy. To rule out the latter possibility, contralateral simultaneous hemisections were performed approximately 10 mm apart. This created empty spaces on both sides of the cord, allowing regenerating neurites to select between paths of similar physical characteristics. Once again, regenerating neurites grew preferentially through their respective scars rather than around them (Figs. 8, 9). Of 132 neurites, 60% regenerated beyond their respective lesion sites. Of these, 100% grew through the scar (Figs. 8, 9). The results of all double hemisection experiments are summarized in Fig. 10.

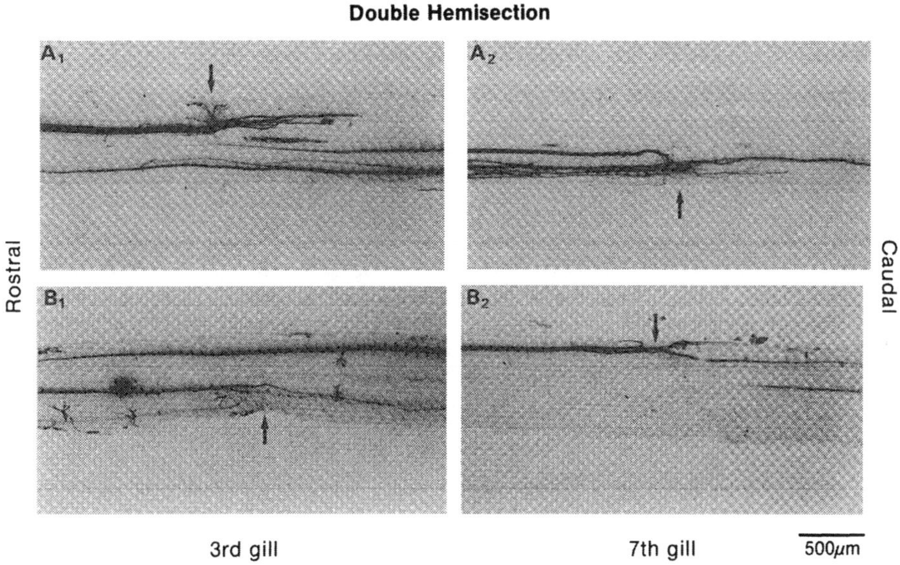

Fig. 8. Spinal wholemounts of 10-week double hemisected animals. A_1 and B_1, third gill (rostral) hemisections (*arrows*). Even though some axons terminated at the level of the scar and begin to loop rostrally, those neurites which regenerated past the level of the hemisection remained on the correct side of the cord. Axons at the bottom of $\mathbf{A_1}$ and the top of $\mathbf{B_1}$ were not cut at this level but were lesioned by the caudal hemisection (7th gill) shown in $\mathbf{A_2}$ and $\mathbf{B_2}$ (*arrows*). The behavior of neurites at the caudal hemisection was similar

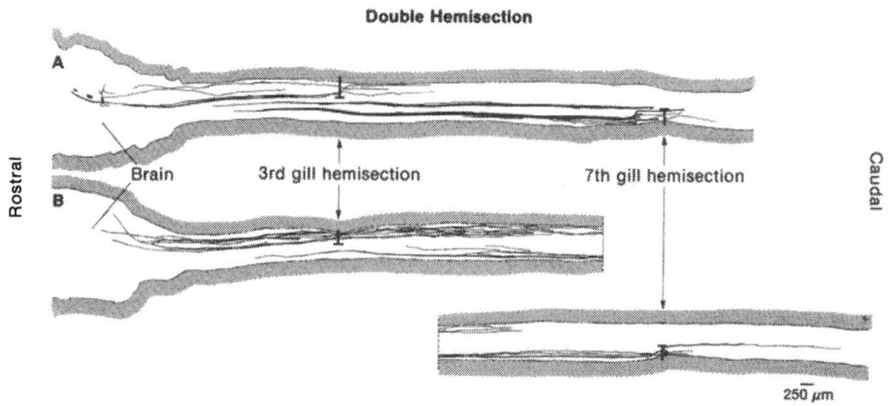

Fig. 9. Camera lucida drawings of HRP-injected RA's in spinal cords doubly hemisected at the levels of the third and seventh gills (*arrows*). All regenerating neurites remained ipsilateral

The results of these experiments strongly suggest that regenerating fibers in the lamprey grow selectively on the correct side of the spinal cord and will regenerate through the glial-ependymal scar in preference to spinal cord matrix, even if the latter contains cellular debris and/or empty spaces resulting from axonal degeneration. This contrasts sharply with data obtained from studies in neonatal mammals. In these systems, although axonal outgrowth occurs after spinal hemi-

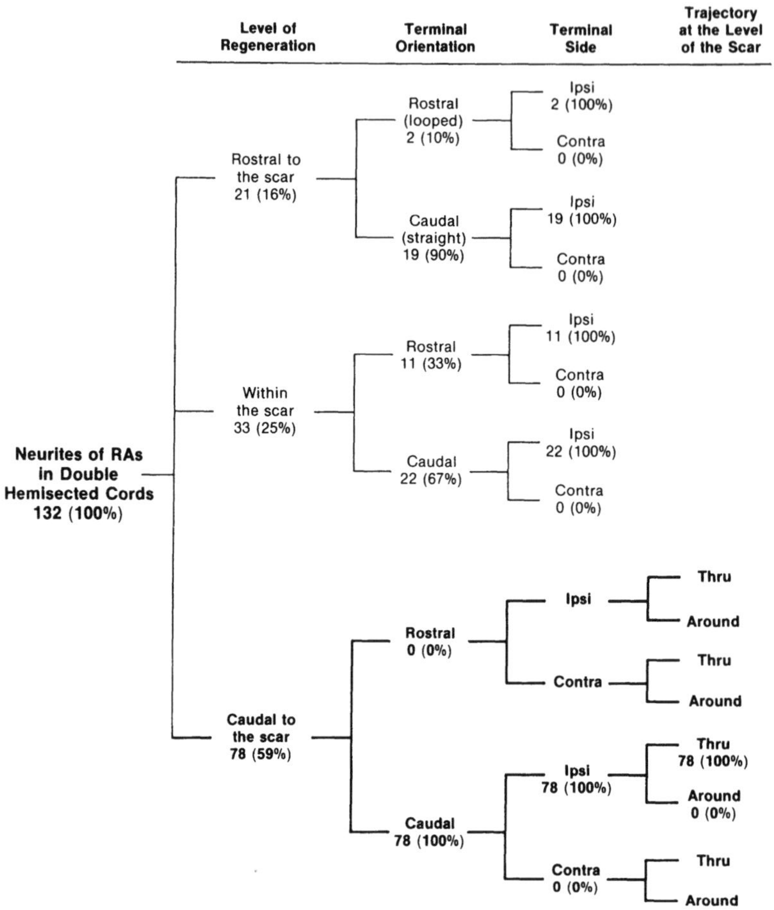

Fig. 10. Summary of neurite projection paths in 10-week double hemisected spinal cords

section in the kitten (Bregman and Goldberger 1982, 1983) or unilateral medullary pyramid section in the hamster (Kalil and Reh 1979, 1982; Reh and Kalil 1982), the extending axons never grow through the scar. Instead, these neurites grow around the lesion site in order to reach their correct terminal fields and establish correct functional pathways. Moreover, it is not clear whether the elongating axons are regenerating or represent late axonal development. In the hamster, although experiments did not employ double labeling techniques, it was argued that pyramidotomies were performed after the medullary pyramids had acquired their full complement of fibers and therefore the elongating axons must have been regenerating (Kalil and Reh 1982). On the other hand, after spinal hemisection in the kitten, only the latest developing pathways displayed axonal outgrowth and associated sparing of function. These pathways had not yet elongated to the level of the hemisection and were undamaged by the lesion (Bregman and Goldberger 1982, 1983). Thus, normal development, and not regeneration, was responsible for the preservation of function and for anatomical plasticity.

In a combined cortical ablation/spinal hemisection experiment, analogous to our double hemisection, neonatal rat corticolspinal tract (CST) axons grew only around the spinal injury, with some fibers redecussating in order to reach normal CST sites (Bernstein and Stelzner 1983). In these experiments, the cortical ablation was performed in order to provide empty spaces for axonal growth in the contralateral hemicord because the number of fibers elongating around the lesion was small following simple hemisection.

Regrowth following hemisection in mammals is limited with respect to the number of fibers and the distance they grow (Bernstein and Stelzner 1983; Kalil and Reh 1982), as is the case for sea lamprey. However, regenerating neurites in the lamprey maintain their appropriate pathways preferentially through the scar. It is now important to determine more precisely the nature and origin of the cells comprising the scar and to characterize the molecular factors which make these cells supportive of regeneration.

5 Guidance of Growth Cones in the Lamprey Spinal Cord

If we are to discover the mechanisms which guide regenerating axons through the spinal cord to their appropriate targets, it will first be necessary to determine where the guidance cues are. As we have seen, the evidence suggests that neurites follow cues in the local environment of their growth. In the unlesioned animal, Müller and Mauthner axons are surrounded primarily by glial processes (Fig. 11 A). This does not necessarily mean that during development the axon grew along such processes. The latter might have grown into the ultrastructural field subsequent to axon elongation. Even if glial processes did form the substrate for the elongation of giant reticulospinal axons during development, the guidance mechanisms during regeneration might be different. Therefore, we have begun to examine the ultrastructural environment of the growth cone and, in particular, to determine which structures are contacted by the filopodia and more proximal growth cone and axon during regeneration.

Animals transected at the level of the third gill were allowed to recover 3, 4, or 7 weeks. At that time Müller and Mauthner cells or their giant axons were injected intracellularly with HRP rostral to the scar. The tissue was reacted with the Hanker-Yates reagent and processed for electron microscopy. Serial thick sections were examined until a labeled growth cone was identified. The block was then serially thin sectioned and the labeled profiles identified with the electron microscope. Although our observations are still very preliminary, it appears that the growth cone is contacted primarily by glial processes (Fig. 11 B). However, the filopodia appear to contact both neuronal and glial processes. Thus, a picture is emerging which is consistent with the view that filopodia sample the environment randomly and that the growth cone extends selectively along preferred surfaces.

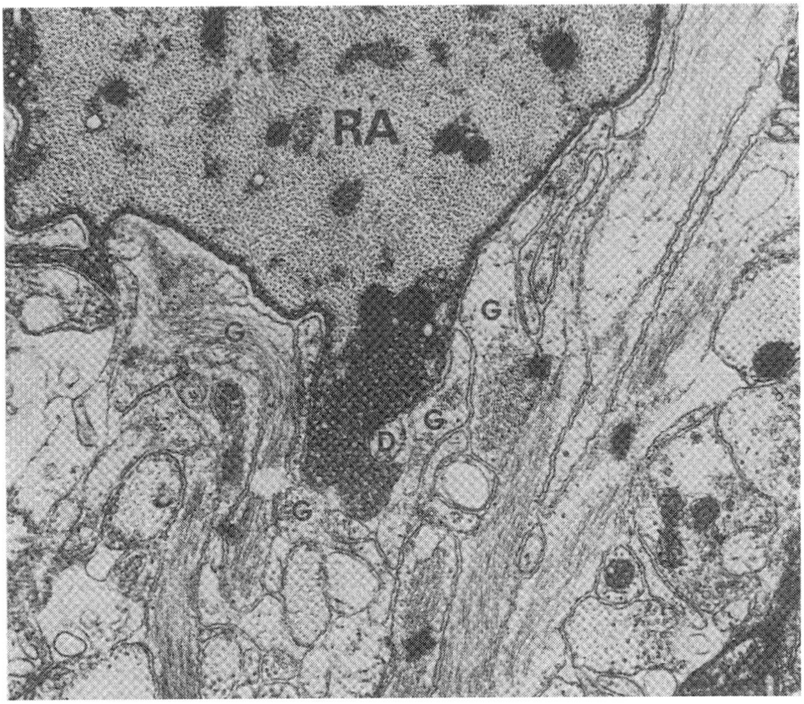

A

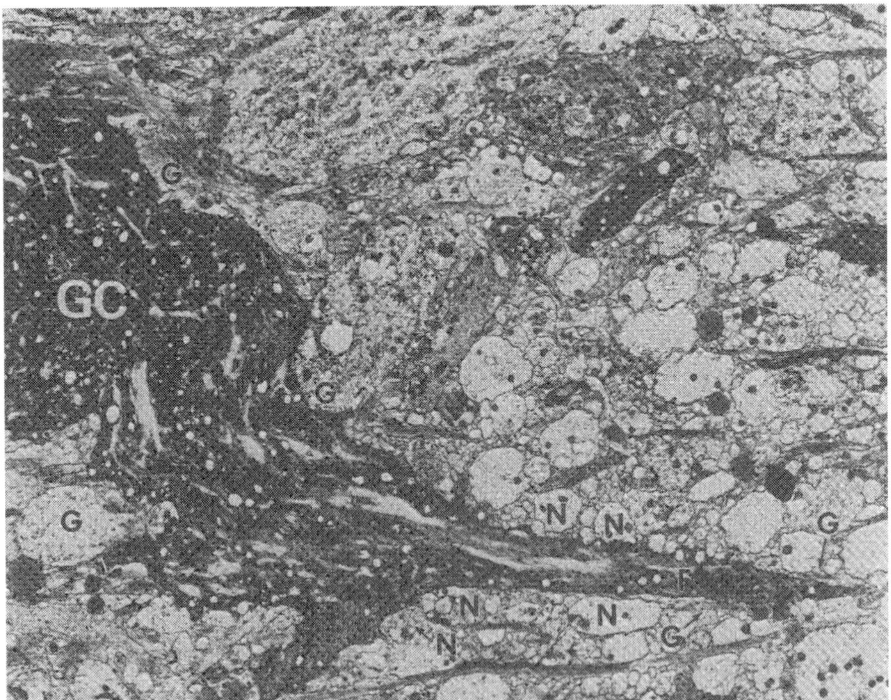

B

6 Summary and Conclusions

The spinal transected lamprey is a favorable preparation in which to study cellular mechanisms of CNS regeneration. It has met stringent criteria for functional axonal regeneration. Moreover, the lamprey CNS is translucent and individual neurons and axon bundles can be seen and impaled in vitro. These advantages have allowed us to determine that axons regenerate selectively in their normal directions and establish functioning synapses specifically with appropriate target neurons distal to a lesion. The fact that the glial/ependymal scar in the lamprey actively supports regeneration, offers a clue to the difference between the regenerative capacities of lampreys and mammals. This finding can now form the basis for studies of the cellular and molecular basis for axon guidance during regeneration.

References

Bernstein DR, Stelzner DJ (1983) Plasticity of the corticospinal tract following midthoracic spinal injury in the postnatal rat. J Comp Neurol 221:382–400
Björklund A, Stenevi V (1984) Intracerebral neural implants: neuronal replacement and reconstruction of damaged circuitries. Annu Rev Neurosci 7:279–308
Borgens RB, Roederer E, Cohen MJ (1981) Enhanced spinal cord regeneration in lamprey by applied electric fields. Science 213:611–617
Bregman BS (1987) Spinal cord transplants permit the growth of serotonergic axons across the site of neonatal spinal cord transection. Dev Brain Res 34:265–279
Bregman BS, Goldberger ME (1982) Anatomical plasticity and sparing of function after spinal cord damage in neonatal cats. Science 217:553–555
Bregman BS, Goldberger ME (1983) Infant lesion effect. III. Anatomical correlates of sparing and recovery of function after spinal cord damage in newborn and adult cats. Dev Brain Res 9:137–154
Cohen AH, Mackler SA, Selzer ME (1986) Functional regeneration following spinal transection demonstrated in the isolated spinal cord of the larval sea lamprey. Proc Natl Acad Sci USA 83:2763–2766
Curry SN, Ayers J (1983) Regeneration of locomotor command systems in the sea lamprey. Brain Res 279:23–240
David S, Aguayo AJ (1981) Axonal elongation into peripheral nervous system "bridges" after central nervous system injury in adult rats. Science 214:931–933
Guth L, Brewer CR, Collins WF, Goldberger ME, Perl ER (1980) Criteria for evaluating spinal cord regeneration experiments. Exp Neurol 69:1–3
Kalil K, Reh T (1979) Regrowth of severed axons in the neonatal central nervous system: establishment of normal connections. Science 205:1158–1161
Kalil K, Reh T (1982) A light and electron microscopic study of regrowing pyramidal tract fibers. J Comp Neurol 211:265–275
Kiernan JA (1979) Hypotheses concerned with axonal regeneration in the mammalian nervous system. Biol Rev 54:153–197

Fig. 11. Ultrastructural environment of the RA growth cone. **A** Electron micrograph of a Müller axon (*RA*) in cross-section lightly labeled with HRP in an unlesioned control spinal cord. The fiber is surrounded by glial processes (*G*) except at points of synaptic contact with dendrites (*D*). Note accumulations of round, clear vesicles. **B** The darkly labeled growth cone (*GC*) of a Müller axon. The growth cone is contacted primarily by glial processes, while its filopodia (*F*) are in contact with both glial and neuronal (*N*) elements

Mackler SA, Selzer ME (1985) Regeneration of functional synapses between individual recognizable neurons in the lamprey spinal cord. Science 229:774–776
Mackler SA, Selzer ME (1987) Specificity of synaptic regeneration in the spinal cord of the larval sea lamprey. J Physiol (Lond) 388:183–198
Mackler SA, Yin HS, Selzer ME (1986) Determinants of directional specificity in the regeneration of lamprey spinal axons. J Neurosci 6:1814–1821
Park S, Snedeker JA, Selzer ME (1986) Behavioral recovery in spinal transected lamprey does not require specific behavioral feedback. Soc Neurosci Abstr 12:425–426
Puchala E, Windle WF (1977) The possibility of structural and functional restitution after spinal cord injury. A Review. Exp Neurol 55:1–42
Reh T, Kalil K (1982) Functional role of regrowing pyramidal tract fibers. J Comp Neurol 211:276–283
Reier P (1985) Neural tissue grafts and repair of the injured spinal cord. Neuropath Appl Neurobiol 11:81–104
Reier P, Stensaas LJ, Guth L (1983 a) The astrocytic scar as an impediment to regeneration in the central nervous system. Spinal Cord Reconstruction, Raven, New York, pp 163–195
Reier PJ, Perlow MJ, Guth L (1983 b) Development of embryonic spinal cord transplants in the rat. Dev Brain Res 10:201–219
Rovainen CM (1974 a) Synaptic interactions of identified nerve cells in the spinal cord of the sea lamprey. J Comp Neurol 154:184–206
Rovainen CM (1974 b) Synaptic interaction of reticulospinal neurons and nerve cells in the spinal cord of the sea lamprey. J Comp Neurol 154:207–224
Rovainen CM (1976) Regeneration of Müller and Mauthner axons after spinal transection in larval lampreys. J Comp Neurol 168:545–554
Sah DWY, Frank E (1984) Regeneration of sensory-motor synapses in the spinal cord of the bullfrog. J Neurosci 4:2784–2791
Selzer ME (1978) Mechanisms of functional recovery and regeneration after spinal cord transection in larval sea lamprey. J Physiol 277:395–408
Smith GM, Miller RH, Silver J (1986) Changing role of forebrain astrocytes during development, regenerative failure, and induced regeneration upon transplantation. J Comp Neurol 251:23–43
Wood MR, Cohen MJ (1979) Synaptic regeneration in identified neurons of the lamprey spinal cord. Science 206:344–347
Wood MR, Cohen MJ (1981) Synaptic regeneration and glial reactions in the transected spinal cord of the lamprey. J Neurocytol 10:57–79
Yin HS, Selzer ME (1983) Axonal regeneration in the lamprey spinal cord. J Neurosci 3:1135–1144
Yin HS, Mackler SA, Selzer ME (1984) Directional specificity in the regeneration of lamprey spinal axons. Science 224:894–896

Altered Kinematic Patterns and Synapses in Recovery of Motor Behavior After Hindlimb Deafferentation in Cats

M. E. Goldberger

Department of Anatomy, The Medical College of Pennsylvania/EPPI Division, 3200 Henry Avenue, Philadelphia, PA 19129, USA

1 Introduction

Recovery of motor behavior following damage to the adult mammalian CNS could occur in a number of different ways. One possibility is that the initially lost or impaired physiological mechanisms underlying normal motor behavior are restored due to plasticity of the remaining systems. For example, when a pathway or system is only partially damaged, the spared parts of the pathway may increase the control they exert. One way in which this might occur is if cells post-synaptic to the partially spared pathway increase their responsiveness to the reduced input by changes in regulation of their receptors or if the reduced input increases by sprouting. The prediction that follows from that model is that the behavior that is lost immediately following the lesion and the behavior that recovers at a later time will be similar in many respects. Another way in which motor behavior may recover involves the production of movement by physiological mechanisms which, because of the conditions imposed by CNS injury, differ from normal ones. This mode of recovery may be anticipated when an entire system has been eliminated. In this case, if movement does indeed recover, the underlying physiological mechanisms, no longer available, must be replaced by others. Since the anatomical pathways mediating the relevant motor functions will then differ to some extent from the normal ones, it seems reasonable to expect alterations in the behavior as well.

There are examples of both types of recovery. When the pyramidal tract has been severed in the monkey, but a small portion is spared, considerable recovery of movements similar to normal ones occurs (Beck and Chambers 1970; Chapman and Wiesendanger 1982), but when the whole tract has been cut, the movement that recovers has a character that is quite different from normal (Goldberger 1972). Some systems may be amendable to recovery of either type. For example, when one eighth nerve or labyrinth has been destroyed, alterations in central pathways controlled by the remaining labyrinthine input can mediate considerable recovery (see Schaefer and Meyer 1981 for review). If, however, both labyrinths are destroyed, compensation is still possible but is mediated by systems that are qualitatively different from the one destroyed, e.g., by visual and somatosensory systems (Schaefer and Meyer 1981). The way in which motor function might differ in the latter paradigm (bilateral labyrinthectomy) has not been analyzed, but clearly the sensory guidance for movement has been markedly altered. Fur-

H. Flohr (Ed.)
Post-Lesion Neural Plasticity

thermore, since the destruction of the second labyrinth results in a reappearance of vestibular symptoms before compensation again recurs, some plasticity must take place among the participating (spinal, visual) systems.

One reason to propose models which rely on alterations in behavior as a result of recovery is that the extent of neuronal plasticity in the adult nervous system is quite limited. For example, although the level of receptor maintenance can be up- or down-regulated, it seems unlikely that receptors can change the transmitter to which they are sensitive in vivo. Similarly, although undamaged pathways can sprout new terminals, the new connections made are not aberrant, i.e., qualitatively different from those found normally. the implication of these observations for known mechanisms of plasticity is that a pathway that contributes to recovery of motor behavior is unlikely to change qualitatively the functions it mediates. If this is so, then the functions mediated by that pathway must be very similar to those subserved by the destroyed pathway if the recovered behavior is to resemble normal behavior. Otherwise, it seems worthwhile to examine closely the recovered behavior and attempt to determine if it is different, perhaps in the way it is executed, when compared with its execution in normal animals.

We have chosen two models in which to compare pre- and post-operative behavior and also to document changes that occur during the recovery period. They are cats with complete unilateral lumbosacral dorsal rhizotomy/ganglionectomy (L_1–S_2) and those with one dorsal root, the sixth lumbar, spared. In the first (complete deafferentation), a whole system that subserves hindlimb topographic somatosensory feedback has been eliminated. Thus, the compensation has to be mediated by systems that are qualitatively different from the primary afferent system. In the second (spared root deafferentation), input from the limb is sharply reduced but not entirely eliminated. In this case, the likelihood that recovery will be mediated by systems qualitatively different from the primary afferent system is diminished since part of it is still present. The possibility that a competition between the spared root and other intact systems, e.g., descending pathways, might be observed, is also raised in this preparation.

2 Methods of Behavioral Examination in Normal Animals and After Lesions

2.1 Somatosensory Postural Reflexes

Several of the classical limb postural reflexes (Rademaker 1931) have been examined and some aspects of their execution have been quantitated. The somatosensory postural reflexes include placing, monopedal hopping, and positive supporting responses. The placing response consists of a flexor-extensor sequence made in response to a somatosensory stimulation of the animal's unsupported foot. The stimulus can be hair contact, skin contact or joint bend (including muscle stretch). The last is called proprioceptive placing. The response is normally executed in all directions depending on which aspect of the cat's foot has been stimulated. The stimulus strength required for placing is measured in the present studies by using a force transducer as the placing stimulus; force is measured as deflections on an attached inkwriter. Normal cats respond in most trials to forces that only bend the hairs (Fig. 1). Relatively few trials require additional skin contact or joint displacement (see, however, Bradley et al. 1983). The force of support is measured by allowing the animal to place on a horizontal force plate. Although this resulting measurement of the positive supporting response is variable from one animal to the next, we have found that the right and left hindlimbs of the same animal behave similarly (Fig. 2). Monopedal hopping is tested on a moving treadmill. The animals are trained pre-operatively to stand with their forelimbs on a small platform raised over the treadmill belt. One hindlimb stands on the belt for testing while the other stands on the nonmoving frame adjacent to the belt. The belt displaces the limb either posteriorly to elicit forward hopping or anteriorly to elicit backward hopping. For lateral and medial hopping, the animal is hand held with one limb in contact with the treadmill belt. The displacement required to elicit hopping can be measured (see Robinson and Goldberger 1986) and the response analyzed kinematically (see Sect. 2.2).

2.2 Locomotion

There are several ways in which locomotion in the cat can be elicited. The simplest form is bipedal stepping in which the hindlimbs are isolated either by lower thoracic spinal transection or by training the animals to stand on a platform with the forelimbs while the treadmill displaces the hindlimbs. Quadrupedal reflex stepping is also elicited by movement of the treadmill belt, but all four limbs participate. In both types of treadmill locomotion, a stationary food reward, placed at the front of the belt helps to make the animal's movements more uniform. For overground locomotion, a runway is used and the animals may be conditioned by food reward or a conditioned stimulus to cross in either direction. A requirement for accurate foot placement is added by using a narrow beam instead of a wide runway (Goldberger 1977), or by placing obstacles or holes in the animal's path. Measurements of joint angles during the sub-phases of the step cycle are made by placing markers over hip, knee, ankle, and toe joints. The measurements are then made from films (Goslow et al. 1973). The normal animal uses a characteristic kinematic pattern which is similar between the two sides. The kinematic pattern is not identical from animal to animal. The pattern used for ordinary overground locomotion is similar to that used by the same animal on a narrower (5 cm) beam.

3 Spared Root Preparation

3.1 Postural Reflexes

Cutting all of the lumbosacral dorsal roots except the sixth lumbar results initially in a severe impairment in somatosensory postural reflexes, in locomotion requiring accurate limb placement, and in the kinematic pattern used for locomotion. The threshold for placing is raised ipsilaterally so that proprioceptive placing predominates. This deficit does not depend on which surface of the animal's limb is stimulated and is not, therefore, related to the peripheral distribution of the spared root. The responses are slow and often hypermetric. Recovery of low threshold placing begins around the sixth day post-operative and soon thereafter

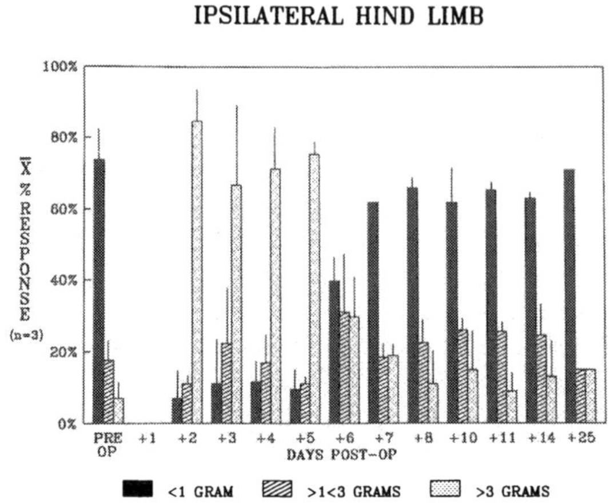

Fig. 1. Loss and recovery of hair-contact placing ipsilateral to spared-root deafferentation. Values pooled for 40 trials (10 in each direction per day for three cats). The values, measured from oscillograph readings are placed in three bins corresponding to hair-contact (*black* < 1 g), skin contact (*diagonal lines* > 1 < 3 g), and proprioceptive (*dots* > 3 g) stimuli as described in Section 2. These are expressed as grams (1 g is equivalent to 10^3 dynes). Hair-contact placing is the most frequent response of the normal animal with very few responses requiring greater stimulation. The frequency of low threshold placing responses is greatly reduced post-operatively thus increasing the frequency of proprioceptive placing. Demonstrable recovery begins on the 6th post-operative day and values approach normal by the 7th

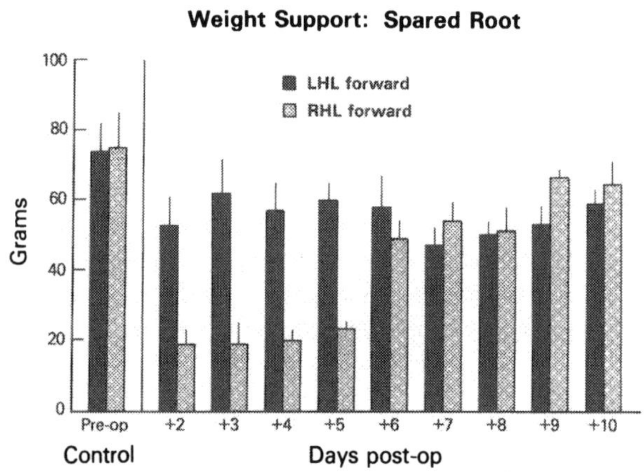

Fig. 2. Weight support. *RHL* right hindlimb, the one ipsilateral to the lesion. *LHL* left hindlimb. Preoperatively, the right and left hindlimbs provide comparable weight support. The supporting response of the right hindlimb is considerably weakened by ipsilateral spared-root deafferentation. Recovery begins on the 6th post-operative day and reaches a plateau beginning day 9. The contralateral hindlimb shows a small deficit. There is a persistent but slight decrease in the strength of the positive supporting responses

most of the responses are elicited by hair-bend (Fig. 1). The proportion of such low threshold placing responses does not recover to the pre-operative level, however. Some parts of the foot remain responsive only to larger stimuli. The distribution of such loci is variable from animal to animal and this variability is due, presumably, to pre- and post-fixation of the lumbosacral plexus (Prendergast et al. 1980, Ungar-Sargon and Goldberger 1987a, b). The loss of lowest threshold responses from small restricted regions of the foot seems to account for the failure of complete recovery, since the majority of responses are low threshold responses and are similar to normal. The execution of placing responses also becomes faster and less hypermetric during the second post-operative week. Positive supporting responses undergo comparable changes. Initially, the force generated by the partially deafferented hindlimb is decreased compared to pre-operative measurements and to the contralateral limb. Recovery begins about the sixth post-operative day and soon stabilizes (Fig. 2). Again, recovery is not complete.

3.2 Locomotion

Locomotion, overground or on the treadmill is present on the second, and sometimes on the first post-operative day. It is not normal however. The lower vertebral column and pelvis are curved toward the side of the lesion. This is not present in sham-operates (Rasmussen et al. 1986). In addition, or as a result of the curvature, the hindlimb ipsilateral to the lesion is abducted and externally rotated. The step-cycle appears somewhat simplified into extensor and flexor coactivations, and the E_2 phase is usually eliminated. On the difficult runways, accurate limb placement is impaired in that the limb reaches the required target or misses it with equal frequency. On both wide and narrow runways, the pattern of joint excursions is different from the pre-operative pattern (Fig. 3). The major recovery begins around day 7 post-operatively. The number of misplacements made on the difficult runways decreases and the kinematic pattern begins to resemble a normal

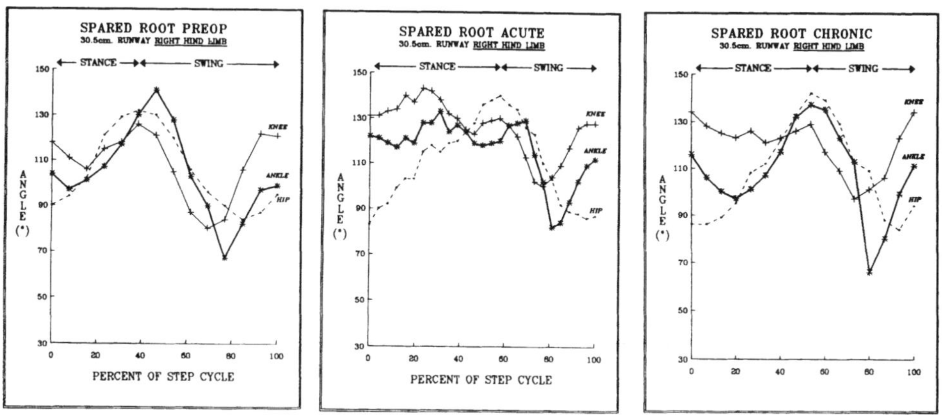

Fig. 3. Single frame representation of step cycles for preoperative, acute and chronic periods for spared root hindlimbs. In the acute period, the kinematic pattern is altered. In the chronic period, after spared root deafferentation the pattern is comparable to the preoperative pattern

one on all runways although significant differences remain indefinitely. The curvature of vertebral column and pelvis also does not recover.

4 Complete Lumbosacral Dorsal Rhizotomy/Ganglionectomy

4.1 Postural Reflexes

All somatosensory postural reflexes are abolished on the side of the lesion. Crossed segmental reflexes elicited by stimulation of the other limb are weak, permanently. Crossed placing is not elicited when placing is tested on the intact side. Vestibular placing is initially weak and becomes stronger and even hyperactive around the 7th or 8th day post-operative, as does the scratch reflex elicited by stimulation of the ipsilateral external auditory meatus.

4.2 Locomotion

The three forms of locomotion tested display different patterns of recovery. Overground locomotion, although very impaired, recovers on the 2nd post-operative day. Participation of the deafferented limb in quadripedal treadmill locomotion begins to recover on the 8th or 9th post-operative day. Prior to that time, the treadmill elicits locomotion of the three intact limbs while the motionless deafferented limb drags behind. Even after quadripedal locomotion returns, the participation of the deafferented limb is inconsistent. Bipedal treadmill locomotion never recovers; the intact hindlimb hops and the deafferented limb drags behind. Overground locomotion on the narrow runway cannot be performed during the first post-operative week. Accurate limb placement begins to recover during the second week and although its behavior is somewhat unpredictable, the impaired

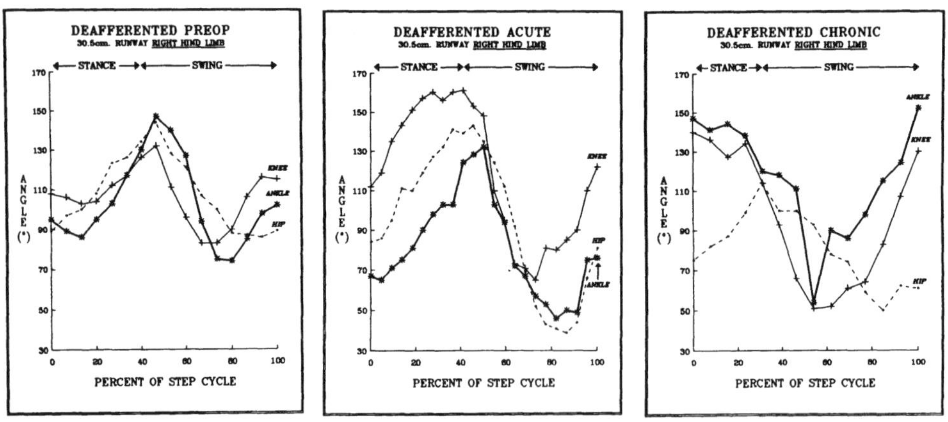

Fig. 4. Single frame representation of step cycles for pre-operative acute and chronic periods for completely deafferented hindlimb. In the acute period after complete deafferentation, the kinematic pattern is altered, moreso than after spared root deafferentation. In the chronic period, after recovery of accurate movement, the kinematic pattern is still significantly different from normal. Thus, the behavior recovers but the execution does not

limb can place accurately during locomotion and bear weight. Even long after recovery of accurate limb placement however, the kinematic pattern of the deafferented limb remains abnormal (Fig. 4). This impairment persists indefinitely (1 year).

5 The Effect of Hemisection

As previously reported (Goldberger 1977), hemisection at T_{13}–L_1 on the side of the previous deafferentation, made long after the recovery of locomotion and accurate limb placement, abolishes all overground locomotion of the deafferented limb. In contrast, hemisection in spared root animals abolishes locomotion and placing only temporarily. Both overground locomotion and accurate limb placement recover during the second post-operative week. This recovery is associated with the recovery of proprioceptive, not contact, placing and also monopedal hopping. Thus, although recovery occurs after complete and incomplete deafferentation, the effect of the additional lesion is qualitatively different for the two preparations.

6 Anatomical and Physiological Correlates

Following a massive denervation such as that resulting from complete lumbosacral deafferentation, one would like to know whether post-synaptic structures in the spinal cord remain denervated or if, instead, some replacement of synapses occurs. This question has been addressed by quantitative electron microscopic studies (Murray and Goldberger 1986; Murray et al. 1987). The stereological method permits identification in the tissue of the area/fraction of that tissue occupied by its components, e.g., dendrites, terminals, etc. If shrinkage has occurred, then this can be taken into account and the area examined must therefore be a measurable one. Two regions in the cat spinal cord have boundaries that are readily discernible; lamina II of the dorsal horn and Clarke's nucleus. Lamina II is the recipient of flexor reflex afferents and is probably an important reflex center, while Clarke's nucleus may be important for cerebellar control of hindlimb movement. Therefore, these areas may both be relevant to recovery of function after deafferentation. Area measurements show that neither region is significantly changed in size by acute or chronic deafferentation, while stereological analysis shows that the maintenance of normal dimensions is probably achieved by the glia. More importantly, measurements and counts of terminals reveals a maintenance of normal terminal number after chronic deafferentation. In Clarke's nucleus, a small decrease in terminal number is seen in the acute period (2–6 days) which recovers in the chronic. In Lamina II a fall in terminal number is never ever seen suggesting that reinnervation is very rapid indeed. The synaptology of Clarke's nucleus is unchanged by the replacement terminals since in cross-sections, which do not demonstrate the "giant" terminals (see Tracey and Walmsley 1984), they are indistinguishable from normal terminals. In lamina II, however, the large central or scalloped (Coimbra et al. 1974) terminals that carry several synaptic specializations and contact several different postsynaptic structures degenerate and are replaced by a proliferation of small terminals that each make a single synapse. The resulting synaptic organization differs from the normal and the physiology of the dorsal horn might be expected to be altered accordingly. Having documented the recovery of synapses, the systems responsible for the lesion-induced synaptogenesis should be identified. Descending supra- and propriospinal input to Clarke's nucleus increases after chronic deafferentation (Goldberger and Murray 1978) and immunocytochemical methods reveal a dramatic increase in the descending serotonergic input to Clarke's nucleus from the brainstem (Goldberger and Paige 1986). This input also increases in lamina II after chronic deafferentation as does the interneuronal, substance P containing projection to the dorsal horn (Tessler et al. 1980, 1981). The reinnervation is therefore partly heterotypic in that one of the reacting systems, the serotonergic one, is qualitatively different from the one that was removed (dorsal roots) which does not contain serotonin. Reinnervation by substance-P-containing interneurons is homotypic in that regard, since substance P is first depleted by the rhizotomy and partially recovers.

When one dorsal root is spared, the morphological changes are quite different. Synapse number is not maintained by a proliferation of small terminals in this preparation but rather by the large spared dorsal root terminals. Although it is unclear whether the number of large terminals increases, this is probably the case, since in chronic spared root preparations, a single root supplies 60% of the large central terminals for the dorsal horn from L_3 to S_1. Only an increase in number of large terminals could account for that observation and for light microscope observations demonstrating an increased projection of the spared root (Liu and Chambers 1958; Goldberger and Murray 1982; Polistina et al. 1987). Furthermore, there is a significant increase in the size of these terminals and in the number of post-synaptic structures with which they make synaptic contact (Murray et al. 1987). Finally, this plasticity of spared root terminals appears to prevent the increase in serotonergic (Goldberger and Paige 1986) and other descending projections (Goldberger and Murray 1982) in the dorsal horn and in most parts of Clarke's nucleus that is seen after complete deafferentation. The increase in dorsal root input has interesting physiological correlates. The organization of dorsal roots in cat dorsal horn indicates a major dense projection to the medial part in the segment of entry and a more diffuse projection to the lateral part where the input of one root overlaps that of ascending and descending collaterals of adjacent roots. This weak projection which increases anatomically after spared root deafferentation also appears to become physiologically stronger when the overlapping input from adjacent roots is eliminated (Pubols and Goldberger 1980; Wilson et al. 1987; Mendell et al. 1986; Mendell and Sedivec 1985). This strengthening of the normally weak input is presumably due to the increase in synaptic input from the spared root although other mechanisms may also contribute to the strengthening of the previously weak input (Devor 1987). The local increase in the spared root's control over spinal networks is consistent with the recovery of somatosensory postural reflexes post-operatively which, since it does not occur after complete deafferentation, depends on the presence of the spared root.

7 Conclusion

Recovery of complex, goal-directed movement may be achieved in more than one way. When part of a system is left intact, that spared part can increase its influence and mediate the recovery of reflex patterns rather similar to normal. Evidence for this type of recovery, which is based on sparing, can be seen in several systems (Finger et al. 1982; Chapman and Wiesendanger 1982; Beck and Chambers 1970). In the deafferentation studies discussed here, when one dorsal root remains, most of the recovery can be attributed to changes in that root's input. The recovery does not, in this situation, depend on descending pathways. A kinematic analysis shows that the recovered movements are quite similar (but not identical) to those initially impaired after the lesion, i.e., to the animal's normal manner of execution. In contrast, when all the dorsal roots have been eliminated, recovery of useful, accurate locomotion depends entirely on those motor pathways that descend on the same side of the spinal cord that was deafferented. The mechanisms underlying the recovery appear to be different, in that recovery in spared root animals is associated with enhancement of somatosensory postural reflexes, whereas recovery after complete deafferentation is associated with enhanced descending (e.g., vestibular) reflexes. Furthermore, the deafferented limb, although it recovers accurate limb placement during locomotion, uses new kinematic patterns, i.e., patterns not seen prior to deafferentation. These differences in motor function could be related to the anatomical correlates of partial and complete deafferentation. Although either lesion results in some permanent alteration in synaptic organization, the organization after complete deafferentation is more abnormal; large terminals are replaced by small ones and some of the reinnervation is heterotypic, since the increased serotonergic input is not related to the removal of a serotonergic projection. The presence of the spared root appears to protect the spinal cord from the development of these abnormal projections and to prevent the appearance of abnormal, albeit useful, motor patterns.

This research was supported by grants from NIH: NS 16629 and NS 24707 and from NSF: BN 586-05441.

References

Beck CH, Chambers WW (1970) Speed, accuracy and strength of forelimb movement after unilateral pyramidotomy in rhesus monkeys. J Comp Physiol Psycol Part 2 70:1–22

Bradley NS, Smith JL, Villablanca JR (1983) Absence of hindlimb tactile placing in spinal cats and kittens. Exp Neurol 82:1–16

Chapman CE, Wiesendanger M (1982) Recovery of function following unilateral lesions of the bulbar pyramid in the monkey. Electroencephalogr Clin Neurophysiol 53:374–387

Coimbra A, Sodre-Borges BP, Magalhaes MM (1974) The substantia gelatinosa Rolandi of the rat. Fine structure, cytochemistry (acid phosphatase) and changes after dorsal root section. J Neurocytol 3:199–217

Devor M (1987) On mechanisms of somatotopic plasticity. In: Pubols LM, Sessle BJ (eds) Effects of injury on trigeminal and spinal somatosensory systems. Liss, New York, pp 215–225

Finger S, Hart T, Jones E (1982) Recovery time and sensorimotor cortex lesion effects. Physiol Behav 29:73–78

Goldberger ME (1972) Restitution of function in the CNS: The pathologic grasp reflex. Exp Brain Res 15:79–96

Goldberger ME (1977) Locomotor recovery after unilateral hindlimb deafferentation in cats. Brain Res 123:59–74
Goldberger ME, Murray M (1978) Axonal sprouting and recovery of function may obey some of the same laws. In: Cotman C (ed) Neuronal plasticity. Raven, New York, pp 73–96
Goldberger ME, Murray M (1982) Lack of sprouting and its presence after lesions of the cat spinal cord. Brain Res 241:227–239
Goldberger ME, Paige E (1986) Anatomical plasticity of serotonin (5HT) projections after deafferentation of the cat spinal cord. Neurosci Abstr 12:512
Goslow GE Jr, Reinking RM, Stuart DG (1973) The cat step cycle: hindlimb joint angles and muscle lengths during unrestrained. J Morphol 141:1–41
Liu CN, Chambers WW (1958) Intraspinal sprouting of dorsal root axons. Arch Neurol Psychiatr 79:46–91
Mendell LM, Sedivec MJ (1985) Physiological consequences of dorsal rhizotomy on the receptive field organization of dorsal horn neurons. In: Dacey RG (ed) Trauma of the central nervous system. Raven, New York, pp 235–244
Mendell LM, Sedivec MJ, Traub RJ (1986) Response of an identified population of spinal cord neurons to partial deafferentation. In: Goldberger ME, Gorio A, Murray M (eds) Development and plasticity of the mammalian spinal cord. Liviana, Padova and Springer, New York, pp 111–121
Murray M, Goldberger ME (1986) Replacement of synaptic terminals in lamina II and Clarke's nucleus after unilateral dorsal rhizotomy in adult cats. J Neurosci 6:3205–3217
Murray M, Wu L-F, Goldberger ME (1987) Spared root deafferentation of cat spinal cord: anatomical recovery. In: Pubols LM, Sessle BJ (eds) Effects of injury on trigeminal and spinal somatosensory systems. Liss, New York, pp 261–271
Polistina D, Murray M, Goldberger ME (1987) Myelinated fibers contribute to dorsal root sprouting after partial deafferentation in rats. Neurosci Abstr 13:166
Prendergast J, Murray M, Goldberger ME (1980) Sprouting and reflex recovery after spinal nerve lesions in cats. Exp Neurol 73:732–749
Pubols LM, Goldberger ME (1980) Recovery of function in the dorsal horn following partial deafferentation. J Neurophysiol 43:102–117
Rademaker VGGJ (1931) Standing: Static reactions, equilibrium and muscle tonus, with special consideration of their retention in animals without a cerebellum. Translated by A. Musen. Arch Neurol 28:141–163
Rasmussen SA, Goslow GE Jr, Hannon R (1986) Kinematics of locomotion in cats with partially deafferented spinal cords: the spared-root preparation. Neurosci Lett 65:183–188
Robinson GA, Goldberger ME (1986) Development and recovery of function in spinal kittens. Brain Res 346:400–413
Schaefer KP, Meyer DL (1981) Aspects of vestibular compensation in guinea pigs. In: Flohr H, Precht W (eds) Lesion-inducted plasticity in sensorimotor systems. Springer, Berlin Heidelberg New York, pp 261–280
Tessler A, Glazer E, Artymyshyn R, Murray M, Goldberger ME (1980) Recovery of substance P in the cat spinal cord after unilateral lumbosacral deafferentation of cat spinal cord. Brain Res 191:459–470
Tessler A, Himes BT, Artymyshyn R, Murray M, Goldberger ME (1981) Spinal neurons mediate return of substance P following deafferentation of cat spinal cord. Brain Res 230:263–281
Tracey DJ, Walmsley B (1984) Synaptic input from identified muscle afferents to neurones of the dorsal spinocerebellar tract in cat. J Physiol (Lond) 350:599–604
Wilson P, Meyers DER, Snow PJ (1987) Changes in spinal circuitry in response to changes in input: an evaluation of the development and basis of contemporary theories. In: Pubols LM, Sessle BJ (eds) Effects of injury on trigeminal and spinal somatosensory systems. Liss, New York, pp 227–238
Ungar-Sargon J, Goldberger ME (1987a) Maintenance of specificity by sprouting and regenerating peripheral nerves. I. Normal variability. Brain Res 407:117–123
Ungar-Sargon J, Goldberger ME (1987b) Maintenance of specificity by sprouting and regenerating nerves. II. Variability after lesions. Brain Res 407:124

Influence of the Noradrenergic Coeruleospinal System on Recurrent Inhibition in the Spinal Cord and its Role During Postural Reflexes

O. Pompeiano
Dipartimento di Fisiologia e Biochimica, Universita' di Pisa, Via S. Zeno 31, I-56100 Pisa, Italy

1 Introduction

Renshaw (R)-cells and recurrent inhibition of α-motoneurons can be influenced in the cat by a variety of excitatory and inhibitory inputs arising from supraspinal structures, such as the cerebral cortex (MacLean and Leffman 1967), the internal capsule (Koehler et al. 1978), the thalamus (MacLean and Leffman 1967), the red nucleus (Henatsch et al. 1986), the reticular formation (Koizumi et al. 1959; Haase and Van der Meulen 1961; MacLean and Leffman 1967; Haase and Vogel 1971) and the cerebellum (Granit et al. 1960; Haase and Van der Meulen 1961; see for ref. Pompeiano 1984). These excitatory or inhibitory volleys are of physiological importance, as they serve to increase or decrease the functional coupling between R-cells and their input motoneurons, thus modifying the input-output relation of α-motoneurons to a given excitatory input (Hultborn et al. 1979).

Experiments performed in decerebrate cats have shown that the postural activity depends in part at least upon the tonic discharge of norepinephrine (NE)-containing locus coeruleus (LC) neurons, the higher the firing rate of these neurons in the animal at rest, the greater the extensor rigidity (Pompeiano and Hoshino 1976a, b). In connection with this finding is the demonstration that activation of noradrenergic coeruleospinal (CS) synapses increases the amplitude of lumbar monosynaptic reflexes (Chan et al. 1986; cf. Fung and Barnes 1984), which has been attributed in part at least to the heightened motoneuron membrane excitability (Fung and Barnes 1981, 1986, 1987). An additional possibility, however, is that the LC exerts a depressive effect on inhibitory R-cells, thus leading to disinhibition of α-motoneurons. This hypothesis is indirectly supported by the fact that in spinal cats systemic injection of DOPA (a NE-precursor for the descending aminergic system) causes a flexion reflex afferent depression in recurrent inhibitory postsynaptic potential in motoneurons and R-cell discharge (Bergmans et al. 1969), thus increasing the efficacy of the Ia afferent volleys to motoneurons.

The experiments summarized in the present report were performed to find out: (1) whether stimulation of the LC depressed the activity in the inhibitory R-cells, thus leading to disinhibition of α-motoneurons; (2) whether the LC neurons, particularly those projecting to the spinal cord, responded to sinusoidal stimulation of macular labyrinth receptors, thus being involved in the control of posture during the vestibulospinal reflexes; (3) whether R-cells linked with limb extensor mo-

H. Flohr (Ed.)
Post-Lesion Neural Plasticity

toneurons also responded to this labyrinth stimulation as if they were under the control of CS neurons; finally, (4) whether the NE-containing LC neurons intervened in the gain regulation of the vestibulospinal reflexes.

2 Locus Coeruleus Influence on Recurrent Inhibition in the Lumbar Spinal Cord

The effects of LC stimulation on recurrent inhibition in the lumbar spinal cord were investigated in precollicular decerebrate cats (Fung et al. 1987, 1988). Both hindlimbs were denervated except for the left gastrocnemius-soleus (GS) and common peroneal (CP) nerves, which were dissected free for stimulation, while the ventral roots (VR) L6–S1 were sectioned intradurally. The animals were then paralyzed with gallamine triethiodide and ventilated artificially. Stable control monosynaptic reflexes (MSR's), which were adjusted to half-maximal amplitudes, were recorded from the proximal end of the split VR bundle of L7 in the left side by stimulating the central cut end of either the GS or the CP nerve with single or double shocks (0.1–0.2 ms pulses), while recurrent inhibition was achieved by conditioning the test MSR with a single volley delivered to the remaining bundle of the L7 VR or the adiacent VR (0.05–0.1 ms pulse at different intensity, VR-to-GS or CP interval = 6.4–12.1 ms). Preconditioning stimuli (3–4 cathodal current pulses at 770/s, 0.7 ms pulse width) were delivered to the ipsilateral LC (LC-to-VR interval = 1.2–5.6 ms) via a stainless steel microelectrode placed stereotaxically at P2 to P3, L2 to L3, H-1.5 to 2.5 (AC impedance, 5 MΩ). Several stimulation sites were studied in each animal by moving the LC electrode systematically. The final site was then marked and processed according to the Prussian blue reaction technique. In some experiments the effects of LC stimulation were tested on individual R-cells recorded with intracellular electrodes filled with 3 M potassium chloride.

With a given set of VR and GS (or CP) stimulus strength, the most conspicuous effect of recurrent inhibition occurred at a conditioning test (C-T) interval of 6.1–18.3 ms; beyond that point the inhibitory effect declined progressively. The time course of the recurrent inhibition curves resembled that reported originally by Renshaw (1941) in his monosynaptic reflexes studies and by Eccles et al. (1954) for the inhibitory postsynaptic potentials recorded in cat motoneurons.

The effectiveness of the LC stimulation on recurrent inhibition and thus on R-cell activity was investigated by changing the following parameters:

a) C–T Interval. At a fixed VR-to-GS interval of 6.1 ms (i.e., at a constant level of recurrent inhibition, open circles, Fig. 1), stimulation of the LC produced marked disinhibition of the MSR, with its peak at 3.4 ms (closed circles, Fig. 1); this finding occurred in the absence of any significant changes in the facilitation of the GS MSR (triangles, Fig. 1). The most pronounced effect of the LC-induced disinhibition as calculated in these experiments (i.e., LC-to-GS interval = 6.1 + 3.4 or 9.5 ms) correlated well with the LC-induced inhibition of VR-evoked Renshaw cell discharges as illustrated in Fig. 2 (LC-to-VR interval = 10 ms). Upon varying the VR-to-GS interval while fixing the LC-to-VR interval at 5.2 ms, the disinhibitory effect was clearly linked to that of the recurrent inhibition, the

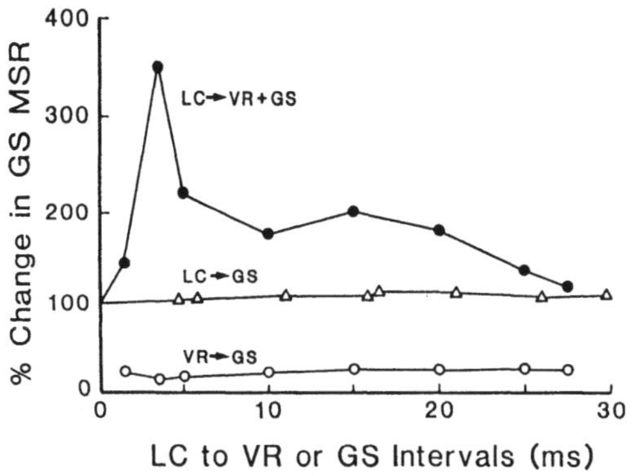

Fig. 1. Time course of the LC influence on recurrent inhibition at fixed VR-to-GS interval. At the fixed VR-to-GS interval of 6.1 ms, i.e., at a constant level of recurrent inhibition (*open circles*), stimulation of the LC (100 μA × 4 cathodal pulses at 770/sec delivered at P2, L 2.5, H-2) performed at various LC-to-VR intervals produced marked disinhibition, expressed as percent recurrent inhibition, with its peak at 3.4 ms interval (*closed circles*). These changes in LC-induced disinhibitory effect occurred in the absence of any significant facilitation of the GS MSR (*triangles*). The most pronounced effect of the LC-induced disinhibition as calculated in this experiment (6.1 + 3.4 or 9.5 ms) correlates well with the LC-induced inhibition of VR-evoked R-cell discharge, as reported in Fig. 2 (LC-to-VR interval = 10 ms)

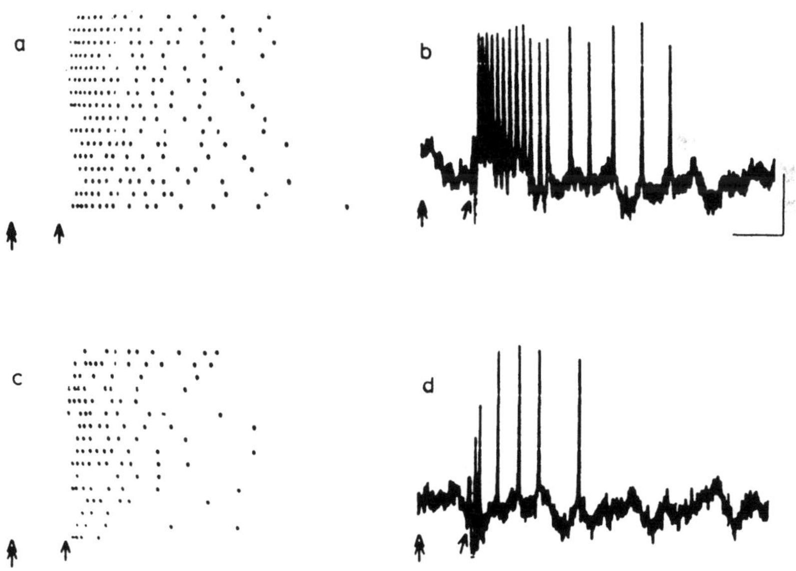

Fig. 2. Inhibitory LC influence on Renshaw cell firing. **a, c** Raster-dot displays of 16 consecutive sweeps (from top downward) of control and conditioned Renshaw cell discharges evoked by single VR volleys, respectively. Corresponding intracellular records were shown in single sweeps in **b** and **d.** *Single arrows* VR shocks: *double arrows* LC stimuli (100 μ A × 4 pulses at 770/s). Calibrations: 10 ms, 5 mV

greater the recurrent inhibition, the larger the disinhibition. Again, with the optimal VR-to-GS interval of 3.9 ms, the time of maximal disinhibitory effect (i.e., LC-to-VR interval = 5.2+3.9 or 9.1 ms) closely resembled the time during which R-cells were being depressed by LC stimulation.

b) Strength of Orthodromic Stimulation. The magnitude of the disinhibitory influence was also dependent on the size of the test MSR. In particular, once maximal recurrent inhibition was attained at a given Ia intensity, it became weakened with further increase in Ia intensity, due to the enhanced firing of motoneurons which overwhelmed the VR-induced recurrent inhibition. In these instances the disinhibitory changes decreased in parallel with the changes in recurrent inhibition.

c) Depth of Stimulation. Besides the variable such as C–T intervals (i.e., LC-to-VR and VR-to-GS or CP) and Ia stimulus intensity, another factor relevant to the parametric study is the location of the LC stimulation point. Its importance was emphasized by tracking with a stimulating electrode systematically while keeping the above-mentioned variables constant. For a given electrode penetration, the disinhibition effect was generally seen to be markedly dependent upon the depth of LC stimulation. Thus maximal disinhibition was clearly related to the depth of stimulation (e.g., at H-1.75 or -2.0), despite the fact that no significant change was noticed in the facilitation curve at this depth.

d) Strength of LC Stimulation. The amount of disinhibition was shown to increase with the LC stimulus intensity, due to graded recruitment of various neuronal aggregates within the LC. This finding was apparently unrelated to that of facilitation, as the LC stimulus was raised from 40 to 150 μA, thus implying that differential sources underlying disinhibition and facilitation phenomena may exist among the LC neuropil.

e) Spatial Relevance. If the LC stimulus intensities remained constant as well as the C–T intervals and the afferent nerve stimulus intensities, the disinhibition affected both GS and CP MSR's. Subsequent qualitative analysis of the disinhibition pattern of the GS compared with the CP MSR's following LC stimulation at different intensities and depths, as well as quantitative analysis, showed that the LC exerted a differential inhibitory control of R-cells coupled to these antagonistic motonuclei.

Histological controls taken at the end of these experiments indicated that the placement of the effective stimulating electrodes at the stereotaxic coordinates of P2 to P3, L2 to 3, H-1.5 to -2.5 corresponded to the LCd and LCα, as delimited by Sakai et al. (1979) and Leger and Hernandez-Nicaise (1980).

In conclusion, the experiments reported above indicated that stimulation of the LC induced removal of recurrent inhibition, an effect which could in part at least be attributed to an inhibitory LC control of R-cell activity, uncoupling it from the corresponding input motoneurons.

3 Responses of Locus Coeruleus Neurons to Sinusoidal Stimulation of Labyrinth Receptors

Experiments were performed in precollicular decerebrate cats, paralyzed with pancuronium bromide and artificially ventilated, to find out whether LC neurons responded to natural stimulations of macular labyrinth receptors (Barnes et al. 1985; Pompeiano et al. 1988). In particular, the unit activity was recorded extracellularly with glass microelectrodes (5–10 MΩ impedance), filled with 0.5 M sodium acetate saturated with pontamine sky blue dye for iontophoretic marking of the recording site. The microelectrode was lowered at the stereotaxic coordinates of P2 to 3, L2.5 to 2.8, H-2 to -3, i.e., in that region where the NE-containing LC and subcoeruleus (SC) neurons are located. Stimulating electrodes made of insulated tungsten wires were implanted into the lateral funiculi of both sides at T_{12}–L_1 for antidromic activation of CS neurons.

Each unit was tested during sinusoidal rotation about the longitudinal axis of the whole animal at the standard parameters of 0.15 Hz, $\pm 10°$, leading to stimulation of labyrinth receptors. The unit activity was averaged over several cycles and processed on-line with a computer system equipped with a fast Fourier analyzer. A spectral analysis of the angular input (table rotation) and of the output (unit activity) was performed and the gain (imp./s/deg) and phase angle of the first harmonic of response were evaluated, the latter being the difference in degrees between the peak of the fundamental component of response and the peak of the side-down animal displacement. At the end of penetrations a mark was made by passing a cathodal current through the tip of the microelectrode, thus allowing identification of the location of the recorded units (cf. Boyle and Pompeiano 1980).

Among the 141 units recorded from the dorsolateral pontine tegmentum, 41 units were histologically located into the LCd, 67 in the LC_α and 33 in the SC. In agreement with the results of previous studies (Chu and Bloom 1974; Cedarbaum and Aghajanian 1976, 1978a; Aghajanian et al. 1977; Watabe and Satoh 1979; Watabe et al. 1982; cf. Foote et al. 1983), the LC neurons recorded above had some of the following physiological characteristics attributed to the NE-containing units: (1) a prolonged extracellular spike wave duration ($\geqq 2$ ms); (2) a very slow and regular resting discharge; and (3) a response to a pinch stimulus applied to the ipsilateral paws characterized by a transient excitatory response followed by a prolonged decrease in impulse activity, due to recurrent and/or lateral inhibition of the NE-containing neurons. In addition, 16 out of the 141 units were CS neurons antidromically activated as projecting to the lumbosacral segments of the spinal cord.

Among the 141 neurons located in the LC complex and tested during roll tilt of the animal at the standard parameters, 80 (56.7%) showed a periodic modulation of their firing rate in response to the sinusoidal input; the average gain of these unit responses corresponded to 0.19 ± 0.26, S.D. imp./s/deg. However, the proportion of units responsive to standard parameters of tilt as well as the response gain were on the average higher in the LCd (25/41, i.e., 61.0% and 0.19 ± 0.23, S.D. imp./s/deg) and the SC area (24/33, i.e., 72.7% and 0.32 ± 0.36, S.D. imp./s/deg) than in the LC_α (31/67, i.e., 46.3% and 0.09 ± 0.13, S.D. imp/s/

deg). Moreover, while the LCα units did not show any predominant phase angle distribution of responses, a large proportion of LCd and SC neurons responded to the extreme animal displacements (Fig. 3 A). In particular, 31 units (63.3%) were excited during side-up tilt of the animal (β-responses), while 13 units (26.5%) by side-down tilt (α-responses); both populations of neurons fired with an average phase lead of the first harmonic of responses of $+19.3 \pm 26.1$ °, S.D. with respect to the extreme animal displacements. The remaining 5 units (10.2%) showed intermediate phase angle of the responses. It should be noticed that the gain of the 31 LCd and SC units that showed β-responses (0.28 ± 0.26, S.D. imp./s/deg) was on the average higher than that of the 13 units displaying α-responses (0.12 ± 0.17, S.D. imp./s/deg) (Fig. 3 A).

Similar response properties were also found for 11 out of the 16 antidromically identified CS neurons which responded to animal tilt. In fact, the majority of these neurons were located in the LCd and the SC area (n = 10 units) and showed large-amplitude β-responses (n = 8 units).

It is of interest that the phase angle of response of the LC-complex neurons to animal tilt remained unmodified by increasing frequency of tilt from 0.008 to 0.32 Hz (at ± 10 °), thus indicating that the responses were due to stimulation of macular receptors; however, the response gain either remained unmodified or slightly increased by increasing frequency of tilt, due to static and dynamic properties of macular receptors (cf. for ref. Boyle and Pompeiano 1980).

The observation that most of the LCd and the SC neurons, including those projecting to the lumbosacral segments of the spinal cord, were excited during side-up tilt (β-responses) differs from that obtained from the vestibulospinal (VS) neurons histologically identified as being located in the lateral vestibular nucleus (LVN) and projecting to the same segments of the spinal cord (cf. Marchand et al. 1987), the majority of which were excited during side-down tilt (α-responses) (Fig. 3 B). Similar result was also obtained at the level of the primary afferents originating from the utricular macula, thus in apparent agreement with the morphological polarization of the majority of the corresponding receptors. The β-responses of the LC-complex neurons can, at least, in part be attributed to direct projection of vestibular nuclear neurons to the LC complex (Cedarbaum and Aghajanian 1978 b; cf. however Aston-Jones et al. 1986). An alternative possibility, however, is that the macular input of one side is transmitted not only to the ipsilateral LVN, but also, via a crossed pathway, to the ventral aspect of the con-

→

Fig. 3. Polar diagrams showing the gain and the phase angle of responses of LC-SC neurons (**A**) and vestibulospinal neurons (**B**) to slow sinusoidal tilt of the animal at 0.026–0.15 Hz, ± 10 °. The response gain of each unit is indicated by the distance of the corresponding symbol to the center of the diagram (see the scale along the vertical meridian); 6 units in **A** had a gain higher than 0.5 and 6 units in **B** had a gain higher than 1.0. The relative position of the symbol with respect to 0 ° meridian indicates in degrees the phase led (positive values) or the phase lag (negative values) of responses with respect to the extreme side-down position of the animal. The *dashed lines* outline the standard deviation of the phase angle of response of the main populations of units responsive to animal tilt in **A** and **B**. In particular, the mean phase angle of the LC-SC units excited during side-down and side-up animal tilt in **A** corresponded to $+27.0 \pm 27.6$, S.D. deg (n = 13) and -163.9 ± 25.2 S.D. deg (n = 31), respectively, while that of the vestibulospinal units displaying the same response patterns in **B** corresponded to $+25.0 \pm 23.8$, S.D. deg (n = 51) and -172.6 ± 33.9, S.D. deg (n = 15), respectively

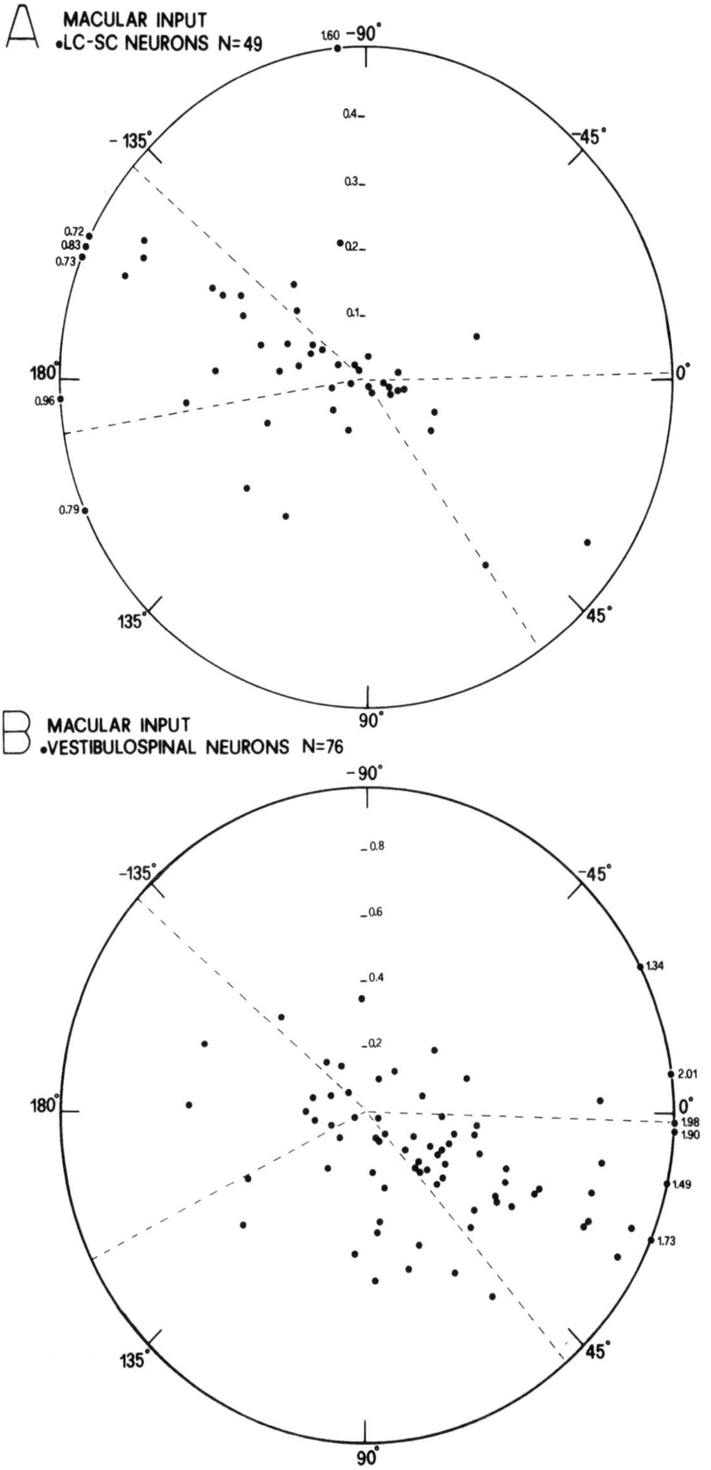
A
MACULAR INPUT
•LC-SC NEURONS N=49
1.60
-90°
-135°
-45°
0.4
0.3
0.2
0.1
0.72
0.83
0.73
180°
0°
0.96
0.79
135°
45°
90°
B
MACULAR INPUT
•VESTIBULOSPINAL NEURONS N=76
-90°
0.8
0.6
0.4
0.2
-135°
-45°
1.34
2.01
180°
0°
1.98
1.90
1.49
1.73
135°
45°
90°

tralateral medullary reticular formation (Manzoni et al. 1983 b), whose neurons may at least in part project to the LC complex (Aston-Jones et al. 1986), thus exerting an excitatory influence on the corresponding NE-containing neurons (Ennis and Aston-Jones 1986).

In order to understand the functional significance of the reciprocal pattern of response of VS and CS neurons during the vestibulospinal reflexes, we should consider that the lateral VS neurons exert a direct excitatory influence on ipsilateral limb extensor motoneurons (cf. Pompeiano 1975) and through their recurrent collaterals on the related R-cells. On the other hand the CS neurons, whose aminergic terminals are in close contact with ventral horn neurons in both the cervical and lumbar enlargements of the cat spinal cord (cf. Mizukawa 1980), exert an excitatory influence on ipsilateral limb extensor (and flexor) motoneurons (cf. Fung and Barnes 1984), an effect which has been attributed in part at least to inhibition of the corresponding R-cells (see Sect. 2).

Since during the vestibulospinal reflexes there is a contraction of ipsilateral limb extensors during side-down tilt of the animal and extensor relaxation during side-up tilt (Schor and Miller 1981), it appears that the motoneurons innervating the ipsilateral limb extensors are excited by an increased discharge of VS neurons during side-down tilt. However, the R-cells anatomically linked with these extensor motoneurons are disinhibited by a reduced discharge of CS neurons for the same direction of animal orientation. In this instance the R-cells would be more prominently linked with the ipsilateral limb extensor motoneurons, a finding which should reduce the response gain of the corresponding limb extensors to labyrinth stimulation. This hypothesis is supported by the results of the following experiments, showing that indeed: (a) R-cells linked with hindlimb extensor motoneurons were excited during ipsilateral tilt, and (b) functional inactivation of the LC complex increased the response gain of limb extensors to labyrinth stimulation.

4 Responses of Renshaw Cells to Sinusoidal Stimulation of Labyrinth Receptors

The responses of R-cells anatomically linked with GS motoneurons during labyrinth stimulation were recorded in precollicular decerebrate cats submitted to bilateral neck deafferentation, immobilized with pancuronium bromide and artificially ventilated (Pompeiano et al. 1985 a). Moreover, both hindlimbs were denervated except for the nerves supplying the left GS muscle, which was deefferented by section of the ventral roots L6–S2.

The electrical activity of 47 R-cells was monitored from L_7 and upper S_1 spinal cord segments, both in the animal at rest as well as during bidirectional rotation of the head on the coronal plane (at 0.026–0.15 Hz, $\pm 10°$) while the body remained fixed horizontally; due to neck deafferentation, this head displacement led to selective stimulation of labyrinth receptors. Groups of several sweeps, each containing the responses of single units to two successive cycles, were averaged. Among these R-cells which responded monosynaptically to antidromic ventral root stimulation, 22 units were disynaptically excited by the orthodromic group I volleys induced by single shock stimulation of the ipsilateral GS nerve, thus indicating that they were under the direct control of the recurrent collaterals of GS motoneurons monosynaptically driven by the orthodromic group Ia volleys elicited by stimulation of the GS nerve (cf. Ross et al. 1972; Pompeiano et al. 1975). Moreover, the same units responded to static stretch of the corresponding muscle.

Among the 47 R-cells tested, 9 units (2 of which coupled with GS motoneurons) were silent at rest and did not respond to head rotation, while the remaining 38 units (20 of which coupled with GS motoneurons) showed a spontaneous discharge at rest; moreover, among these spontaneously firing R-cells, 31 units (16 of which coupled with GS motoneurons) responded to head rotation at either one or both the parameters indicated above. In resting animals with the ankle joint fixed at 90 ° dorsiflexion, all spontaneously active R-cells fired at low rate and their discharge rate closely corresponded to the mean firing rate (base frequency) evaluated during head rotation at 0.026 Hz, ± 10 °. In particular, the base frequency evaluated for 24 spontaneously active units responsive to this stimulus (15 of which coupled with the GS motoneurons) corresponded on the average to 5.90 ± 5.35, S.D. imp./s, while the gain of the first harmonic responses to head rotation at 0.026 Hz, ± 10 ° corresponded on the average to 0.075 ± 0.059, S.D. imp./s/deg (35 groups of averaged responses, i.e., 35 AR).

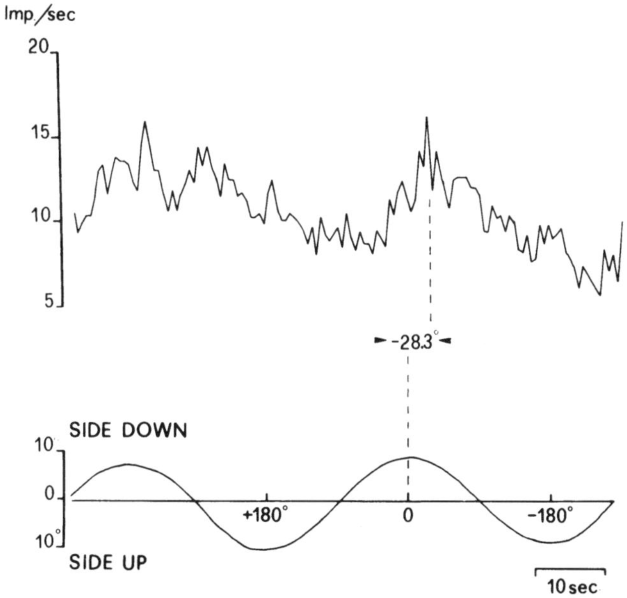

Fig. 4. Response to head rotation, after bilateral neck deafferentation, of a single Renshaw cell coupled with the pool of GS motoneurons. *Upper record* sequential pulse density histogram (SPDH) showing the response of an individual R-cell disynaptically excited by single shock stimulation of the GS nerve during head rotation at the frequency of 0.026 Hz, ± 10 ° peak displacement (average of 8 sweeps, using 128 bins with 0.6 s bin width). *Lower record* trace indicating head position. Ipsilateral side-down displacement of the head is monitored by an upward deflection in the sine wave. The firing rate of this unit increased during side-down and decreased during side-up head rotation (base frequency, 11.0 imp./s). The first harmonic response had a gain of 0.24 imp./s/deg, and a phase lag of -28.3 ° with respect to the extreme side-down displacement of the head

Moreover, the base frequency, as well as the gain of the responses, were on the average slightly higher for the 15 R-cells coupled with the GS motoneurons (25 AR) than for the 9 unidentified R-cells (10 AR).

It is of interest that most of the averaged responses (27 out of 35 AR, i.e. 77.1%) recorded from 17 out of the 24 R-cells were characterized by an increased discharge during side-down head rotation and a decreased discharge during side-up rotation and showed an average phase lead of $+3.1 \pm 54.3$ °, S.D. with respect to the extreme side-down displacement (α-responses). On the other hand the responses recorded from the remaining 7 R-cells (8 AR, i.e., 22.9%) showed an average phase lag of -157.7 ± 69.6 °, S.D. (β-responses). Similar results were also obtained from the averaged responses of R-cells coupled with hindlimb extensor motoneurons, the majority of which (22 out of 25 AR, i.e., 88.0%) recorded from 13 out of 15 R-cells showed α-responses (Fig. 4).

The low firing rate of the R-cells linked with limb extensor motoneurons as observed in decerebrate animals at rest can be attributed, at least, in part to the tonic inhibitory influence exerted on them by the spontaneously firing LC and SC neurons, leading to disinhibition of the corresponding extensor motoneurons. As to the vestibulospinal influences, the observation that most of the R-cells coupled with GS motoneurons increased their discharge rate during side-down head rotation can only in part be attributed to the related discharge of VS neurons acting on ipsilateral limb extensor motoneurons and then, through their recurrent collaterals, on the corresponding R-cells, since this modulation of R-cells activity occurred even in preparations in which the GS muscle did not show any EMG modulation prior to the ventral radicotomy. Therefore, the increased discharge of R-cells during side-down head rotation (α-response) can be attributed, at least, in part to the reduced discharge of the CS neurons (β-response), leading to disinhibition of the R-cells.

5 Effects of Functional Inactivation of Noradrenergic Locus Coeruleus Neurons on Vestibulospinal Reflexes

The demonstration that in decerebrate cats the majority of CS neurons undergo a reduced discharge during side-down tilt of the animal (Pompeiano et al. 1988), thus leading to disinhibition of the corresponding R-cells, may explain why these inhibitory interneurons fire in phase with the corresponding extensor motoneurons (Pompeiano et al. 1985a). This finding also explains why in spite of the increased discharge of excitatory VS neurons during ipsilateral tilt (Boyle and Pompeiano 1980; Marchand et al. 1987), the related contraction of the extensor muscles (Schor and Miller 1981) was quite small in amplitude in the forelimbs or absent in the hindlimbs (Manzoni et al. 1983a, 1984; Boyle and Pompeiano 1984; Pompeiano et al. 1985b).

If one the main roles of the CS neurons is to modify the functional coupling between R-cells and the related extensor motoneurons, one might expect that for a given amount of labyrinth signal, the amplitude of the EMG modulation of limb extensors depends on the background discharge of the CS neurons. In particular, the higher the firing rate of these units in the animal at rest, the greater the disinhibition which would affect the R-cells related to limb extensor motoneurons during side-down animal tilt. In this instance the resulting increase in firing rate of the R-cells would reduce the response of limb extensor motoneurons to the excitatory VS volleys, thus leading to a reduced gain of the vestibulospinal reflex. Just the opposite result would occur following inactivation of the LC neurons.

It is of interest that the NE-containing LC neurons are self-inhibitory, due to the fact that NE, which is released by the LC neurons as a consequence of impulse activity, acts on α_2-adrenoceptors of the same or neighboring neurons, thus leading to recurrent and/or lateral suppression of their activity (cf. Foote et al. 1983). The experiments summarized in the present section were performed to investigate the effects of functional inactivation of LC neurons, produced by local injection into the LC of the α_2-adrenergic agonist clonidine, both on posture as well as on vestibulospinal reflexes elicited by dynamic stimulation of labyrinth receptors (Pompeiano et al. 1987a, b).

In precollicular decerebrate cats with a good symmetric rigidity in the limbs, the multiunit EMG activity of the medial head of the triceps brachii was recorded bilaterally in the animal at rest, as well as during roll tilt of the animal at the standard parameters of 0.15 Hz, $\pm 10^\circ$. In particular, the data of 6–10 sweeps, each containing the responses to two successive cycles, were averaged and repeated at regular intervals of 3–5 min for several hours. These AR were then processed online with a computer system which performed the Fourier and correlation analyses as well as the transfer function of the system. In particular, the base frequency (mean firing rate of the multiunit activity evaluated during each test in imp./s), the gain (imp./s/deg) and phase angle of the first harmonic of the output with respect to the peak of the side-down stimulus displacement ipsilateral to the side of the recording were evaluated before and after microinjection of the α_2-adrenergic agonist clonidine (Sigma) into the LC of one side at the stereotaxic coordinates of P2.5 or 3.0, L2.5 or 2.8, H-2.0. The drug was dissolved in sterile saline with pontamine 5%. The localization of the tip of the needle, as well as the diffusion of the dye mark in the dorsolateral pontine structures, were identified on histological serial sections of the brain stem stained with the neutral red method.

In normal decerebrate cats, the EMG responses of the forelimb extensors to labyrinth stimulation were characterized by an increase activity during side-down tilt of the animal and a decreased activity during side-up tilt. The peak of the responses was related to extreme animal displacement and not to the velocity of animal tilt, thus being attributed to stimulation of position-sensitive utricular receptors. Moreover, the amplitude of modulation and thus the response gain of the triceps brachii to animal tilt were quite small (Fig. 5A). Microinjection of 0.25 μl of the α_2-adrenergic agonist clonidine at the concentration of 0.012–0.15 μg/μl of saline into the LC of one side reduced the extensor tonus in the ipsilateral limbs. Correspondingly, the amplitude of modulation and thus the response gain of the ipsilateral triceps brachii to the same parameters of labyrinth stimulation greatly increased; however, no change in the phase angle of the response was observed (Fig. 5B). In particular, in four experiments, the response gain increased from the mean value of 0.66 ± 0.26, S.D. imp./s/deg (42 groups of averaged responses, i.e., 42 AR) to 1.73 ± 1.07, S.D. imp./s/deg (60 AR) (t-test, $p < 0.001$), while the phase

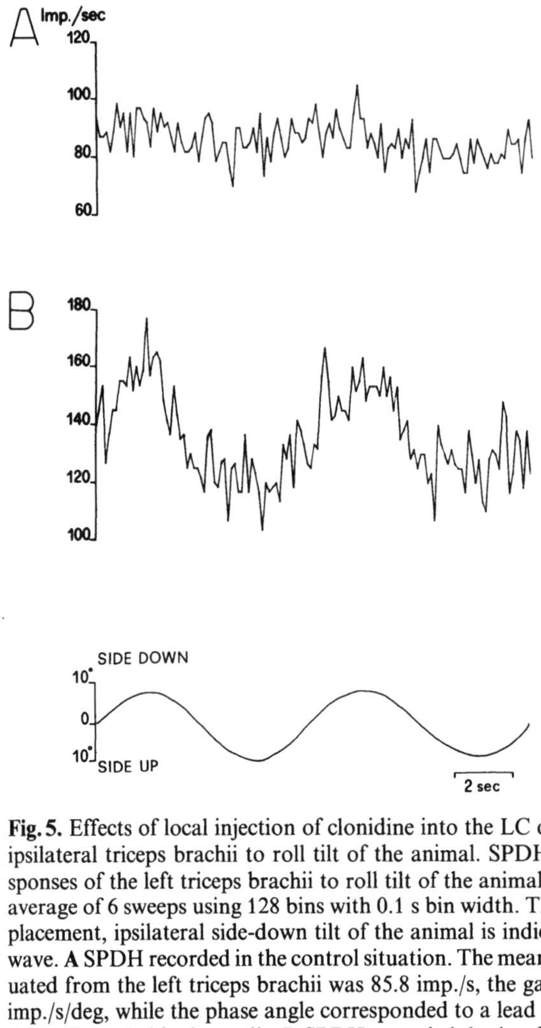

Fig. 5. Effects of local injection of clonidine into the LC of one side on the response gain of the ipsilateral triceps brachii to roll tilt of the animal. SPDHs showing the averaged multiunit responses of the left triceps brachii to roll tilt of the animal at 0.15 Hz, ± 10 °. Each record is the average of 6 sweeps using 128 bins with 0.1 s bin width. The *lower trace* indicates the animal displacement, ipsilateral side-down tilt of the animal is indicated by upward deflection in the sine wave. **A** SPDH recorded in the control situation. The mean discharge rate or base frequency evaluated from the left triceps brachii was 85.8 imp./s, the gain of the first harmonic response 0.41 imp./s/deg, while the phase angle corresponded to a lead of $+45.2$ ° with respect to the peak of the ipsilateral side-down tilt. **B** SPDH recorded 6 min after injection into the ipsilateral LC at P 3, L2.8, H-2 of 0.25 μl of a clonidine solution at the concentration of 0.012 μg/ μl of sterile saline. In this instance the mean base frequency evaluated from the same muscle was 136.1 imp./s, the gain 1.71 imp./s/deg, while the phase angle corresponded to a lead of $+22.7$ ° with respect to the peak of the ipsilateral side-down tilt. In this instance the reflex discharge following an increased static stretch of the muscle compensated for the decrease in EMG activity of the triceps brachii after injection of clonidine

angle, evaluated with respect to the peak of the ipsilateral side-down tilt of the animal, showed an average lead of $+3.5 \pm 14.8$, S.D. deg before injection and $+1.4 \pm 9.8$, S.D. deg after injection of clonidine.

The increased gain of the vestibulospinal reflexes did not depend on the decrease in postural activity of the extensor muscle following injection of the α_2-adrenergic agonist, since it was still observed when the reduced EMG activity was compensated for by the reflex contraction following an increased static stretch of that muscle. In addition to the prominent postural and reflex changes involving the triceps brachii ipsilateral to the side of the injection, there was also a slight decrease in postural activity in the contralateral limbs, which was associated with an increase in the response gain of the triceps brachii of that side.

The postural and reflex changes described above were not due to irritative phenomena following the injection, since neither changes in posture nor in response gain were observed in control experiments when 0.25 μl of saline were injected within the LC. Moreover, the changes in gain produced by clonidine were dose-dependent. Histological controls indicated that the area of injection, which was critically involved in the effects described above, included the LC and the SC region.

The hypothesis that the effects produced in the present study by local injection of the α_2-adrenergic agonist clonidine into the LC were due to reduction or suppression of the spontaneous discharge of the corresponding neurons is supported by the results of previous experiments showing that postural and reflex changes similar to those induced by clonidine were obtained following electrolytic lesion of the LC of one side (d'Ascanio et al. 1985).

We postulate, therefore, that functional inactivation of the LC and the SC neurons, produced by local injection of clonidine, releases the R-cells from the suppressive influence exerted by the noradrenergic terminals. The resulting increase in firing rate of these inhibitory interneurons would reduce the postural activity in the ipsilateral (and the contralateral) limbs in the animal at rest. Moreover, the lack of vestibular modulation of the CS neurons would not allow the functional coupling between these inhibitory interneurons and the corresponding limb extensor motoneurons to increase during side-down animal tilt. These extensor motoneurons would then respond more efficiently to the same excitatory VS volleys elicited by this direction of animal orientation, thus increasing the response gain of limb extensors to labyrinth stimulation.

6 Discussion

Experiments performed in precollicular decerebrate cats have clearly shown that stimulation of the LC induced a removal of recurrent inhibition (Fung et al. 1987, 1988), Although several factors contributed to this effect, a common feature emerged from these experiments, i.e., that the greater the recurrent inhibition, the larger was the LC-induced disinhibitory effect.

The disinhibition process was distinctly independent of the motoneuronal facilitation, as shown by changing the LC-to-VR interval (at fixed VR-to-GS interval, Fig. 1), the depth and the intensity of LC stimulation. These findings implied that the populations of LC neurons responsible for the two phenomena occupied close but unidentical spatial locations. The LC-induced suppression of recurrent inhibition was, at least in part, due to an inhibitory LC control on R-cell activity, as shown by experiments of unit recording (Fung et al. 1987); it appeared, therefore, that stimulation of the LC decoupled the R-cells from the corresponding input motoneurons.

The present results did not indicate the specific transmitter(s) underlying the LC-disinhibitory influence. In our experiments, however, the placement of the stimulating electrodes corresponded to the LCd and LC_α, as delimited by Sakai et al. (1979) and Leger and Hernandez-Nicaise (1980). These LC sites, as well as the SC area in the cat, contain scattered aggregates of fluorescent catecholaminergic neurons (cf. for ref. Jones and Friedman 1983). In particular, with their positive immunoreactivity toward the dopamine-β-hydroxylase enzyme, these cells were regarded as noradrenergic (Miachon et al. 1984), since phenylethanolamine N-methyltransferase containing cell bodies were absent in the LC region. Moreover, by using the retrograde labeling technique in combination with immunocytochemical staining, it was found that these areas contained tyrosine hydroxylase-immunoreactive coeruleospinally projecting cells (Reddy et al. 1987). Indeed, experiments of unit recordings have shown that some of the units located within the confines of the LC and the SC (Fung and Barnes 1981, 1986, 1987; Pompeiano et al. 1988) could be antidromically activated from the spinal cord, thus belonging to coeruleospinal neurons (Fung and Barnes 1981; Pompeiano et al. 1988). Therefore, the descending influences on recurrent inhibition and on R-cells described in our study probably originated from the noradrenergic LC and SC neurons. This is in accord with early findings that R-cell activities were inhibited by iontophoretically applied NE (Biscoe and Curtis 1966; Engberg and Ryall 1966; Weight and Salmoiraghi 1966), a putative transmitter of coeruleospinal synapses (cf. Fung and Barnes 1981, 1984; Chan et al. 1986).

Detailed considerations of possible inclusion of other descending sources originating from the LC having other amines (like serotonin) or peptides (like substance P) as putative transmitters exerting a similar inhibitory action on R-cells (Engberg and Ryall 1966; Belcher and Ryall 1977; Krnjevic and Lekic 1977, Ryall and Belcher 1977) have been reported elsewhere (Fung et al. 1988).

The demonstration that the LC exerts an inhibitory action on R-cells coupled to both extensor and flexor motoneurons, thus releasing these motoneurons from their related recurrent inhibition, suggested that this structure could play a role in regulating the gain setting at the output stage of the motor system (Hultborn et al. 1979). In order to test this hypothesis, the activity of LC and SC neurons, particularly those projecting to the spinal cord, was recorded in decerebrate cats to find out whether in addition to lateral VS neurons even the CS neurons responded to natural stimulation of macular labyrinth receptors, thus being involved in the dynamic control of posture during the vestibulospinal reflexes (Pompeiano et al. 1988). These experiments clearly indicated that the increased contraction of limb extensors which occurs during side-down tilt depends not only on the increased discharge of excitatory VS neurons (α-responses: Boyle and Pompeiano 1980; Marchand et al. 1987), but also on the reduced discharge of CS neurons (β-responses: Barnes et al. 1985; Pompeiano et al. 1988) projecting to the corresponding segments of the spinal cord.

This finding was surprising since, while the lateral VS neurons exert a direct excitatory influence on ipsilateral limb extensor motoneurons and then through their recurrent collaterals on the related R-cells, the CS neurons also exert an excitatory influence on ipsilateral extensor (and flexor) motoneurons, an effect which can in part at least be attributed to an inhibitory influence on R-cells. Thus, VS as well as CS neurons exert an antagonistic influence on R-cells linked with the limb extensor motoneurons.

In order to understand how the descending volleys originating from the LVN and the LC interact at spinal cord level, experiments were performed in decerebrate cats, in which there is a continuous discharge of LC neurons (Pompeiano and Hoshino 1976a, b). In this instance, the R-cells anatomically linked with the GS motoneurons either fired at a low rate or were silent at rest (Pompeiano et al. 1985a), probably due to the tonic inhibitory influence exerted by the CS neurons on R-cells activity (Fung et al. 1987, 1988). Moreover, the same R-cells either did not respond to standard parameters of head rotation, or showed only a small amplitude α-response (Pompeiano et al. 1985a). This response pattern can be attributed to the increased discharge of VS neurons during side-down rotation, which exert an excitatory influence on GS motoneurons and through their recurrent collaterals on the related R-cells. However, the simultaneous decrease in firing rate of the CS neurons for the same direction of head rotation (Barnes et al. 1985; Pompeiano et al. 1988) would reduce the inhibitory influence that these neurons exert on R-cells (Fung et al. 1987, 1988), thus enhancing the functional coupling of the R-cells with their own extensor motoneurons. This finding explains why the response gain of hindlimb extensors was negligible or absent in these preparations (Manzoni et al. 1983a, 1984; Boyle and Pompeiano 1984; Pompeiano et al. 1985b).

This interpretation of the experimental findings is supported by the results of experiments performed in decerebrate cats, leading to suppression of the spontaneous discharge of LC and SC neurons. In particular, inactivation of the LC and SC neurons produced by local injection of minute doses of the α_2-adrenergic agonist clonidine, which acts on the somatodendritic α_2-adrenoceptors by enhancing recurrent and/or lateral inhibition of the corresponding NE-containing neurons (cf. Foote et al. 1983), decreased the postural activity but greatly increased the gain of the vestibulospinal reflexes (Pompeiano et al. 1987a, b). The suppressed discharge of the CS neurons following inactivation of the LC-complex neurons would reduce the postural activity in the ipsilateral limbs in the animal at rest, due to disinhibition of the R-cells; however, due to lack of labyrinthine modulation of LC activity, the functional coupling between R-cells and limb extensor motoneurons would not increase during side-down tilt of the animal. These motoneurons would then respond more efficiently to the same excitatory VS volleys elicited by this direction of animal orientation, thus increasing the response gain of limb extensors to labyrinth stimulation.

In summary, while in normal decerebrate cats the reduced discharge of LC and SC neurons during side-down animal tilt would lead to disinhibition of R-cells, thus enhancing the recruitment of these inhibitory interneurons during the motoneuronal discharge elicited by the VS volleys, after functional inactivation of the LC complex produced the clonidine injection a derecruitment of the R-cells would occur for the same direction of animal tilt. Obviously in the first instance the increased coupling of R-cells with the extensor motoneurons during side-down tilt would *limit the response gain* of the GS muscle to labyrinth stimulation, while in the second instance the reduced coupling of R-cells with the extensor motoneurons for the same direction of animal orientation would *enhance the response gain* of the GS muscle to labyrinth stimulation. In particular, the small tonic motoneurons, which are more powerfully affected by recurrent inhibition (cf. Pompeiano 1984), would be more easily recruited for that direction of tilt. As a result of these findings, the NE-containing LC-complex neurons, which exert an inhibitory influence on R-cells, could play a *permissive* role in the control of posture as well as in the gain regulation of the vestibulospinal reflexes. Since the LC and SC neurons undergo spontaneous fluctuations in their firing rate (cf. Hobson and Steriade 1986) leading to changes in postural activity (cf. Pompeiano 1967) during the sleep-waking cycle, they may well intervene in order to adapt to the animal state the response gain of limb extensors to labyrinth stimulation.

It is worth mentioning that the LC could act in more than one way to modify the response gain of extensor muscles to labyrinth stimulation. In addition to the reported LC-induced inhibition of R-cell activity (Fung et al. 1987, 1988), another means involves the LC enhancement of firing efficacy of extensor motoneurons (Fung and Barnes 1981, 1984, 1986, 1987), modifying the excitatory vestibulospinal drive to extensors. Another means depends on the LC inhibition of the pontine reticular neurons, that provide a tonic excitatory drive to the medullary inhibitory reticulospinal (RS) neurons (d'Ascanio et al. 1985; Pompeiano et al. 1987 a, b; cf. d'Ascanio et al. 1988). The latter neurons were thought to exert an excitatory influence on R-cells, causing them to inhibit the coupled extensor motoneurons (Pompeiano et al. 1985 b). Through this interplay among VS, CS, and RS systems, it becomes possible for the animals to execute balanced postural adjustments in response to labyrinthine signals.

7 Summary

In precollicular decerebrate cats, selective stimulation of the locus coeruleus (LC), in which norepinephrine (NE)-containing neurons are located, produced a consistent reduction of ventral root-induced recurrent inhibition of monosynaptic reflexes in both extensor (gastrocnemius-soleus) and flexor (common peroneal) motonuclei. The magnitudes of the LC-induced decrease in recurrent inhibition, functionally disinhibition, and of recurrent inhibition, were significantly correlated. Direct recordings from Renshaw (R)-cells indicated that LC stimulation decreased the number of spikes and the duration of firing in response to single antidromic (ventral root) volleys. These results suggested that the LC depression of the recurrent inhibitory pathway depended, in part at least, upon coeruleospinal inhibitory influences on R-cell activity.

The influence that this coeruleospinal (CS) system exerts on postural adjustments induced by labyrinthine stimulation was also investigated. In particular, the activity of a large number of neurons located in the LC complex was recorded during roll tilt of the animal leading to sinusoidal stimulation of labyrinth receptors. Among 141 LC-complex neurons, some of which identified antidromically as CS neurons by stimulation of the spinal cord at T_{12}–L_1, 80 responded to animal tilt, the responses being attributed to stimulation of macular utricular receptors. The majority of the units showed β-responses (excitation during side-up tilt), which contrasted with the predominant α-responses (excitation during side-down tilt) displayed by previously recorded vestibulospinal (VS) neurons projecting to the same segments of the spinal cord. Since during the vestibulospinal reflexes there is a contraction of ipsilateral limb extensors during side-down tilt of the animal, it appears that the motoneurons innervating the ipsilateral limb extensors are excited by an increased discharge of VS neurons; however, the R-cells anatomically linked with these extensor motoneurons should be disinhibited by a reduced discharge of the CS neurons for the same direction of animal orientation. In this instance the R-cells would be more prominently linked with the ipsilateral limb extensor motoneurons, a finding which should limit the response gain of the corresponding limb extensors to labyrinth stimulation. This hypothesis was supported by the results of experiments showing that indeed in decerebrate cats R-cells linked with hindlimb extensor motoneurons increased their discharge during side-down head rotation. On the other hand, functional inactivation of the LC following local injection into this structure of clonidine, which acts on the somatodendritic α_2-adrenoceptors by enhancing recurrent and/or lateral inhibition of the corresponding NE-containing neurons, increased the response gain of limb extensors to labyrinth stimulation. The LC may thus play an important role in the gain regulation of the vestibulospinal reflexes.

Acknowledgements. The experiments reviewed in the present report were performed by the author in collaboration with Drs. P. d'Ascanio, C. D. Barnes (USA), S. J. Fung (USA), E. Horn (FRG), D. Manzoni, U. C. Srivastava (India), G. Stampacchia, P. Wand (FRG). This work was supported by the N.I.H. grants NS 07685-15 to 20 and by grants from the Ministero della Pubblica Istruzione, Italy.

References

Aghajanian GK, Cedarbaum JM, Wang RY (1977) Evidence for norepinephrine-mediated collateral inhibition of locus coeruleus neurons. Brain Res 136:570–577

Aston-Jones G, Ennis M, Pieribone VA, Nickell WT, Shipley MT (1986) The brain nucleus locus coeruleus: restricted afferent control of a broad efferent network. Science 234:734–737

Barnes CD, D'Ascanio P, Manzoni D, Pompeiano O, Stampacchia G (1985) Responses of locus coeruleus neurons to sinusoidal stimulation of vestibular receptors. Atti Soc Ital Biol, Soc Ital Fisiol, Soc Ital Nutr Um, Pisa, Abstr n 164

Belcher G, Ryall RW (1977) Substance P and Renshaw cells: a new concept of inhibitory synaptic interactions. J Physiol (Lond) 272:105–119

Bergmans J, Burkes R, Lundberg A (1969) Inhibition of transmission in the recurrent inhibitory pathway to motoneurones. Brain Res 13:600–602

Biscoe TJ, Curtis DR (1966) Noradrenaline and inhibition of Renshaw cells. Science 151:1230–1231

Boyle R, Pompeiano O (1980) Reciprocal responses to sinusoidal tilt of neurons in Deiters' nucleus and their dynamic characteristics. Arch Ital Biol 118:1–32

Boyle R, Pompeiano O (1984) Discharge activity of spindle afferents from the gastrocnemius-soleus muscle during head rotation in the decerebrate cat. Pflügers Arch 400:140–150

Cedarbaum JM, Aghajanian GK (1976) Noradrenergic neurons of the locus coeruleus: inhibition by epinephrine and activation by the alpha-antagonist piperoxane. Brain Res 112:413–419

Cedarbaum JM, Aghajanian GK (1978a) Activation of locus coeruleus neurons by peripheral stimuli: modulation by a collateral inhibitory mechanism. Life Sci 23:1383–1392

Cedarbaum JM, Aghajanian GK (1978b) Afferent projections to the rat locus coeruleus as determined by a retrograde tracing technique. J Comp Neurol 178:1–16

Chan JYH, Fung SJ, Chan SHH, Barnes CD (1986) Facilitation of lumbar monosynaptic reflexes by locus coeruleus in the rat. Brain Res 369:103–109

Chu NS, Bloom FE (1974) Activity patterns of catecholamine-containing pontine neurons in the dorso-lateral tegmentum of unrestrained cats. J Neurobiol 5:527–544

d'Ascanio P, Bettini E, Pompeiano O (1985) Tonic inhibitory influences of locus coeruleus on the response gain of limb extensors to sinusoidal labyrinth and neck stimulations. Arch Ital Biol 123:69–100

d'Ascanio P, Pompeiano O, Stampacchia G (1988) Noradrenergic and cholinergic mechanisms responsible for the gain regulation of vestibulospinal reflexes. In: Pompeiano O, Allum JHJ (eds) Vestibulospinal control of posture and locomotion. Elsevier, Amsterdam, pp 361–374 (Progress in Brain Research, vol 76)

Eccles JC, Fatt P, Koketsu K (1954) Cholinergic and inhibitory synapses in a pathway from motor-axon collaterals to motoneurones. J Physiol (Lond) 126:524–562

Engberg I, Ryall RW (1966) The inhibitory action of noradrenaline and other monoamines on spinal neurones. J Physiol (Lond) 185:298–322

Ennis M, Aston-Jones G (1986) A potent excitatory input to the nucleus locus coeruleus from the ventrolateral medulla. Neurosci Lett 71:299–305

Foote SL, Bloom FE, Aston-Jones G (1983) Nucleus locus coeruleus: new evidence of anatomical and physiological specificity. Physiol Rev 63:844–914

Fung SJ, Barnes CD (1981) Evidence of facilitatory coerulospinal action in lumbar motoneurons of cats. Brain Res 216:299–311

Fung SJ, Barnes CD (1984) Locus coeruleus control of spinal cord activity. In: Barnes CD (ed) Brainstem control of spinal cord function. Academic Press, London, pp 215–255 (Research Topics in Physiology, vol 6)

Fung SJ, Barnes CD (1986) Increased efficacy of antidromic and orthodromic activation of cat alpha motoneurons upon arrival of coerulospinal volleys. Arch Ital Biol 124:229–243

Fung SJ, Barnes CD (1987) Membrane excitability changes in hindlimb motoneurons induced by stimulation of the locus coeruleus in cats. Brain Res 402:230–242

Fung SJ, Pompeiano O, Barnes CD (1987) Suppression of the recurrent inhibitory pathway in lumbar cord segments during locus coeruleus stimulation in cats. Brain Res 402:351–354

Fung SJ, Pompeiano O, Barnes CD (1988) Coeruleospinal influence on recurrent inhibition of spinal motonuclei innervating antagonistic hindleg muscles. Pflügers Arch (in press)

Granit R, Haase J, Rutledge LT (1960) Recurrent inhibition in relation to frequency of firing and limitation of discharge rate of extensor motoneurones. J Physiol (Lond) 154:308–328

Haase J, Meulen JD van der (1961) Effects of supraspinal stimulation on Renshaw cells belonging to extensor motoneurones. J Neurophysiol 24:510–520

Haase J, Vogel B (1971) Direkte und indirekte Wirkungen supraspinaler Reizungen auf Renshaw-Zellen. Pflügers Arch 325:334–346

Henatsch H-D, Meyer-Lohmann J, Windhorst U, Schmidt J (1986) Differential effects of stimulation of the cat's red nucleus on lumbar alpha motoneurones and their Renshaw cells. Exp Brain Res 62:161–174

Hobson JA, Steriade M (1986) Neuronal basis of behavioral state control. In: Bloom FE (ed) Intrinsic regulatory systems of the brain. American Physiological Society, Bethesda, Maryland, pp 701–823 (Handbook of physiology, section 1, The nervous system, vol IV)

Hultborn H, Lindström S, Wigström H (1979) On the function of recurrent inhibition in the spinal cord. Exp Brain Res 37:399–403

Jones BE, Friedman L (1983) Atlas of catecholamine perikarya, varicosities and pathways in the brainstem of the cat. J Comp Neurol 215:382–396

Koehler W, Windhorst U, Schmidt J, Meyer-Lohmann J, Henatsch H-D (1978) Diverging influences on Renshaw cell responses and monosynaptic reflexes from stimulation of capsula interna. Neurosci Lett 8:35–39

Koizumi K, Ushiyama J, Brooks C McC (1959) A study of reticular formation action on spinal interneurons and motoneurons. Jpn J Physiol 9:282–303

Krnjevic K, Lekic D (1977) Substance P selectively blocks excitation of Renshaw cell by acetylcholine. Can J Physiol Pharmacol 55:958–961

Leger L, Hernandez-Nicaise M-L (1980) The cat locus coeruleus: light and electron microscopic study of the neural somata. Anat Embryol 159:181–198

MacLean JB, Leffman H (1967) Supraspinal control of Renshaw cells. Exp Neurol 18:94–104

Manzoni D, Pompeiano O, Srivastava UC, Stampacchia G (1983a) Responses of forelimb extensors to sinusoidal stimulation of macular labyrinth and neck receptors. Arch Ital Biol 121:205–214

Manzoni D, Pompeiano O, Stampacchia G, Srivastava UC (1983b) Responses of medullary reticulospinal neurons to sinusoidal stimulation of labyrinth receptors in decerebrate cat. J Neurophysiol 50:1059–1079

Manzoni D, Pompeiano O, Srivastava UC, Stampacchia G (1984) Gain regulation of vestibular reflexes in fore- and hindlimb muscles evoked by roll tilt. Bull Soc Ital Biol Sper 60, Suppl 3:9–10

Marchand AR, Manzoni D, Pompeiano O, Stampacchia G (1987) Effects of stimulation of vestibular and neck receptors on Deiters neurons projecting to the lumbosacral cord. Pflügers Arch 409:13–23

Miachon S, Berod A, Leger L, Chat M, Hartman B, Pujol JF (1984) Identification of catecholamine cell bodies in the pons and pons-mesencephalon junction of the cat brain, using tyrosine hydroxylase and dopamine-β-hydroxylase immunohistochemistry. Brain Res 305:369–374

Mizukawa K (1980) The segmental detailed topographical distribution of monoaminergic terminals and their pathways in the spinal cord of the cat. Anat Anz 147:125–144

Pompeiano O (1967) The neurophysiological mechanisms of the postural and motor events during desynchronized sleep. Proc Assoc Res Nerv Ment Dis 45:351–423

Pompeiano O (1975) Vestibulo-spinal relationships. In: Naunton RF (ed) The vestibular system. Academic Press, London, pp 147–180

Pompeiano O (1984) Recurrent inhibition. In: Davidoff RA (ed) Handbook of the spinal cord, vol 2 and 3. Dekker, New York, pp 461–557

Pompeiano O, Hoshino K (1976a) Tonic inhibition of dorsal pontine neurons during the postural atonia produced by an anticholinesterase in the decerebrate cat. Arch Ital Biol 114:310–340

Pompeiano O, Hoshino K (1976b) Central control of posture: reciprocal discharge by two pontine neuronal groups leading to suppression of decerebrate rigidity. Brain Res 116:131–138

Pompeiano O, Wand P, Sontag K-H (1975) Response of Renshaw cells to sinusoidal stretch of hindlimb extensor muscles. Arch Ital Biol 113:205–237

Pompeiano O, Wand P, Srivastava UC (1985a) Responses of Renshaw cells coupled with hindlimb extensor motoneurons to sinusoidal stimulation of labyrinth receptors in the decerebrate cat. Pflügers Arch 403:245–257

Pompeiano O, Wand P, Srivastava UC (1985b) Influence of Renshaw cells on the gain of hindlimb extensor muscles to sinusoidal labyrinth stimulation. Pflügers Arch 404:107–118

Pompeiano O, d'Ascanio P, Horn E, Stamacchia G (1987a) Effects of local injection of the α_2-adrenergic agonist clonidine into the locus complex on the gain of vestibulospinal and cervicospinal reflexes in decerebrate cats. Arch Ital Biol 125:225–269

Pompeiano O, Stampacchia G, Horn E, d'Ascanio P (1987b) The role of the locus coeruleus in the gain regulation of vestibulospinal reflexes. Acta Otolaryngol (Stockh) 103:404–409

Pompeiano O, Manzoni D, Barnes CD, Stampacchia G, d'Ascanio P (1988) Labyrinthine influences on locus coeruleus neurons. Acta Otolaryngol (Stockh) 105:576–581

Reddy VK, Fung SJ, Bowker RM, Barnes CD (1987) Quantitative composition of tyrosine hydroxylase (TH)-like immunoreactive nurons of the dorsolateral pons projecting to cervical and lumbar enlargements in the cat. Anat Rec 218:111 A

Renshaw B (1941) Influence of discharge of motoneurons upon excitation of neighboring motoneurons. J Neurophysiol 4:167–183

Ross H-G, Cleveland S, Haase J (1972) Quantitative relation of Renshaw cell discharges to monosynaptic reflex height. Pflügers Arch 332:73–79

Ryall RW, Belcher G (1977) Stubstance P selectively blocks nicotinic receptors on Renshaw cells: a possible synaptic inhibitory mechanism. Brain Res 137:376–380

Sakai K, Sastre JP, Salvert D, Touret M, Tohyama M, Jouvet M (1979) Tegmentoreticular projections with special reference to the muscular atonia during paradoxical sleep in the cat: an HRP study. Brain Res 176:233–254

Schor RH, Miller AD (1981) Vestibular reflexes in neck and forelimb muscles evoked by roll tilt. J Neurophysiol 46:167–178

Watabe K, Satoh T (1979) Mechanism underlying prolonged inhibition of rat locus coeruleus neurons following anti- and orthodromic activation. Brain Res 165:343–347
Watabe K, Nakai K, Kasamatsu T (1982) Visual afferents to norepinephrine-containing neurons in cat locus coeruleus. Exp Brain Res 48:66–80
Weight FF, Salmoiraghi GC (1966) Adrenergic responses of Renshaw cells. J Pharmacol Exp Ther 154:391–397

IV Lesions to Specific CNS Subsystems: Vestibular System

Plastic, Adaptive Changes in the Vestibuloocular Reflex and Their Role in Recovery from Labyrinthine Lesions

B. W. Peterson, J. F. Baker, V. Matsuo, S. A. Rude, and X. N. Sahyouni
Northwestern University Medical School, 303 E. Chicago Ave., Chicago IL, USA

1 Introduction

In order to maintain gaze stability in the face of perturbations that disturb head position, the brain must generate movements of the eyes in the head that compensate for movement of the head in space. Three sensorimotor mechanisms can contribute to these compensatory eye movements: the vestibuloocular reflex (VOR), the cervicoocular reflex (COR) and visually guided eye movements produced by the smooth pursuit (SP) and optokinetic (OK) systems. The dynamic and spatial properties of these four systems in normal subjects have been extensively studied (cf. Cohen et al. 1977, 1981; Lisberger et al. 1981; Peterson 1988; Robinson 1980, 1982). These studies form the basis for understanding the sensorimotor rearrangements that occur during recovery from labyrinthine lesions.

Ideal gaze stabilization requires compensatory eye movements that are equal and opposite to movements of the head. The amplitude of a compensatory eye movement can be expressed as a gain, which is the ratio of eye movement to head movement. For ideal stabilization, this gain should be -1.0. The contribution of the four oculomotor mechanisms to this gain depends upon the frequency of the perturbing head movement. At frequencies of 0.1–0.5 Hz, the baseline VOR, measured in the relaxed state in total darkness, has a gain of -0.6 to -0.8. When image stability is required, this gain can be brought to the required level of -1.0 by voluntary enhancement of the VOR (see Sec. 3) and by adding visually elicited eye movements (due to SP or OK systems) to those produced by the VOR. In normal animals the COR makes a significant contribution to stabilizing eye movements only at frequencies below 0.05 Hz where the VOR begins to fail (Barmack et al. 1981).

2 Adaptive Control of the VOR in Normals

At frequencies of 1 Hz and above, SP, OK and voluntary modulation of the VOR become much less effective, leaving only the baseline VOR to maintain gaze stability in this critical frequency band. Correspondingly, measurements of VOR gain in this band have yielded gain values close to 1.0 (Keller 1978; Pulaski et al. 1981). Such accurate performance would be easy to attain in a closed loop system such as the vestibulocollic reflex where the receptors of the labyrinth can sense

H. Flohr (Ed.)
Post-Lesion Neural Plasticity

the output of the reflex (head movement) and the brain needs only to boost motor output until movement of the head with respect to space is reduced to the desired level. The VOR, however, operates open loop; the labyrinth does not sense the eye movement generated by the reflex. Furthermore, visual feedback mechanisms (SP and OK) are too slow to regulate high frequency vestibuloocular performance. As revealed by early studies in humans (Gonshor and Melvill-Jones 1976a, b) and animals (Robinson 1976; Miles and Eighmy 1980), the brain's solution to this control problem is to use visual feedback to make slow adjustments in the properties of the VOR. These studies, which used lenses to magnify, reduce or reverse image motion produced by turning of the head, emphasized the adaptive regulation of VOR *gain*. Over several days of wearing such lenses the gain could be made to increase or decrease by ~0.7. The adaptive process appeared to have a time constant on the order of 1 day but there was also a rapid initial change that appeared to occur much faster (Collewjin et al. 1983).

Reversing prisms reverse the direction of image motion during yaw (side-to-side) and roll head movements but not during pitch (up-down) head movements. Berthoz et al. (1981) found that the adaptive changes produced by wearing such prisms were spatially specific: VOR gain in roll and yaw was reduced or reversed while the gain in pitch remained unchanged. This observation suggested another adaptation paradigm first explored by Schultheis and Robinson (1981) in which changes in VOR *direction* are induced by coupling rotation of the body in one plane with image motion in another plane. This paradigm has the advantage that the altered component of the VOR occurs in a different direction than the baseline VOR evoked by the rotation and can therefore be readily quantified. For the last several years we have been using this paradigm to characterize the adaptive process that regulates the VOR in cats and humans.

Harrison et al. (1986) studied the effects of exposing a cat to sinusoidal horizontal rotation coupled with sinusoidal vertical image motion. Figure 1a shows typical vertical eye movement responses to horizontal rotation before (upper traces) and after (lower traces) 2 h of coupled rotation. Before the cross-axis training, the gain of the vertical eye movement response was negligible. After training it had increased to 0.25, as indicated by the clear sinusoidal modulation of vertical eye velocity. Figure 1c shows the time course of development of the cross-coupled vertical eye movement response both during the vertical image motion (open circles) and in total darkness (filled circles). As indicated by the dashed line, the adaptive change in the VOR plotted by the filled circles could be closely approximated by a single exponential with a time constant of 30 min. We conclude that this represents the rate of acquisition of the initial rapid component of VOR plasticity. The learned response was maintained intact if the animal was kept in darkness for 14 h after training. It decreased with a time constant of ~15 min when the animal was rotated in a stationary visual surround.

Other experiments revealed considerable specificity of the cross axis learning. Figure 1d shows that in experiments involving horizontal rotation and vertical image motion the gain of the learned response is greatest at the frequency of the sinusoidal stimulation used during training, falling off in the fashion of a band-pass filter at higher or lower frequencies. Different behavior was observed when vertical rotation was paired with horizontal image rotation (Baker et al. 1986).

The vertical rotation will activate both semicircular canal and otolith receptors in the vestibular labyrinth. Data like that shown in Fig. 1e indicated that the learned response had two components: one produced by the vertical semicircular canals, which was maximal at the frequency of the training stimulus (0.25 Hz), the other produced by the otoliths, which was greatest at the lowest frequencies of stimulation.

A further interesting finding is that adaptive changes in the VOR appear to be "gated" by static otolith signals so that the cross-axis learned response is specific to the posture that the animal was in during training. This can be observed, for instance, when an animal lying right ear down is rotated in its sagittal plane while viewing a visual field that moves in the animal's horizontal plane. The cross-coupled horizontal VOR response to sagittal plane rotation will be much stronger when the animal is tested right ear down than left ear down, even though the canal stimulus is the same in both cases (Baker et al. 1987). If the animal receives additional training in the left ear down position, but with the direction of the horizontal visual field motion reversed, the system can even be trained to generate different directions of horizontal eye movements to the same canal stimulus when the animal is lying on its left and right sides (Baker et al. 1988). The remarkable specificity revealed by these studies indicates that VOR adaptation should be capable of compensating for a wide range of VOR deficits produced by labyrinthine dysfunction or other causes.

Changes in VOR gain induced by magnifying or reversing optics are similar in humans and animals (Collewijn et al. 1983; Gonshor and Melvill-Jones 1976a b; Miles and Eighmy 1980; Robinson 1976). We therefore predicated that humans should exhibit cross-axis adaptive changes similar to those in the cat. Our initial studies suggest that this is indeed the case. Figure 1 B illustrates the change in the gain and phase of vertical eye movement response to horizontal rotation that was induced when we rotated a volunteer subject sinusoidally (0.25 Hz, 12 °) for 1 h about the earth vertical axis while he viewed a projected optokinetic stripe pattern that moved at the same frequency and amplitude in his sagittal plane. The subject clearly developed a vertical component to his vestibuloocular response to 0.25 Hz horizontal rotation in the dark. The cross-coupled response was detectable after 15 min of adaptation and was very clear at 45 and 60 min. Thus the time course of onset is likely to be similar to that in the cat. Figure 1 b compares the gain and phase of the vertical eye movement produced by horizontal rotation before and after the adaptation. Initially there was a small cross-coupled response with a negative gain (phase near 180 °) that probably reflected a slight deviation of the EOG electrodes from the vertical plane. After adaptation the direction of the vertical eye movement had reversed to correspond to the direction of coupling between the optokinetic projector and the chair. When the mean response before adaptation was subtracted vectorially from the mean response after adaptation, the net change in gain was calculated to be +0.14. As the figure indicates, this change was highly significant. Longer adaptation times or higher velocities would probably increase the adaptive gain.

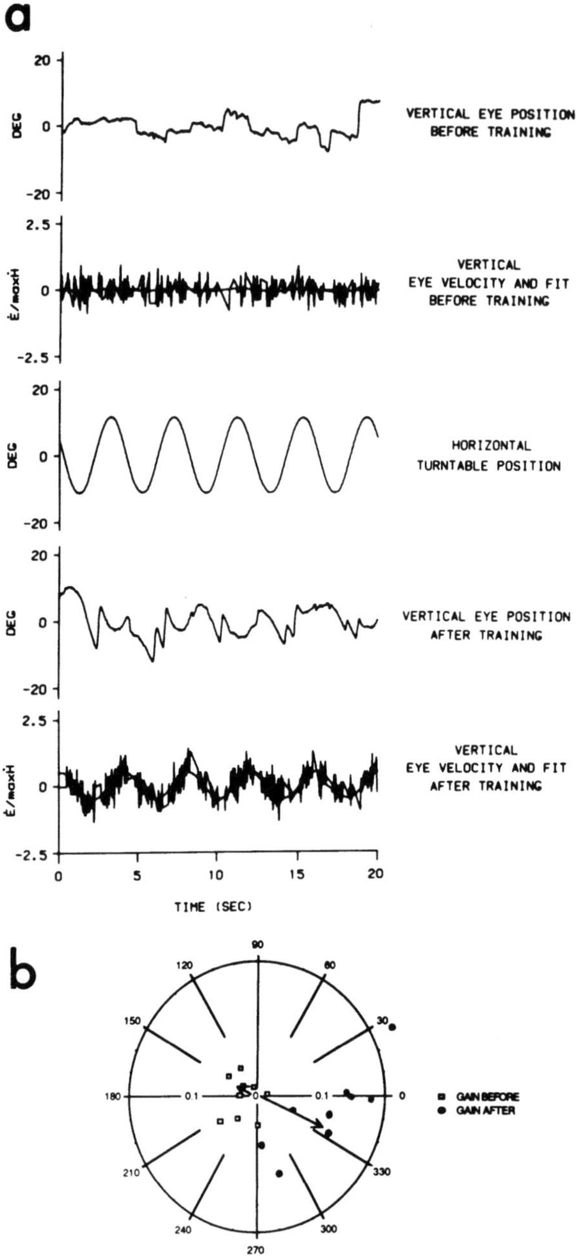

Fig. 1. Adaptive changes in the direction of the VOR in cats and humans. **a** Vertical eye movements produced by 0.25 Hz horizontal rotation in dark before (*top*) and after (*bottom*) 2 h of pairing vertical image motion with horizontal rotation. First and fourth traces show vertical eye position, second and fifth traces, vertical smooth eye velocity after removal of saccades. Middle trace is the rotation stimulus used during training and testing. (Harrison et al. 1986). **b** Polar plot of vertical eye movements elicited by horizontal rotation of a human subject before and after 60 min of pairing vertical image motion with horizontal rotation. Note that after training the response shifted by 180 ° and is about three times the response amplitude before training. The arrows represent vector averages before and after training. The outer radius represents a gain of 0.2.

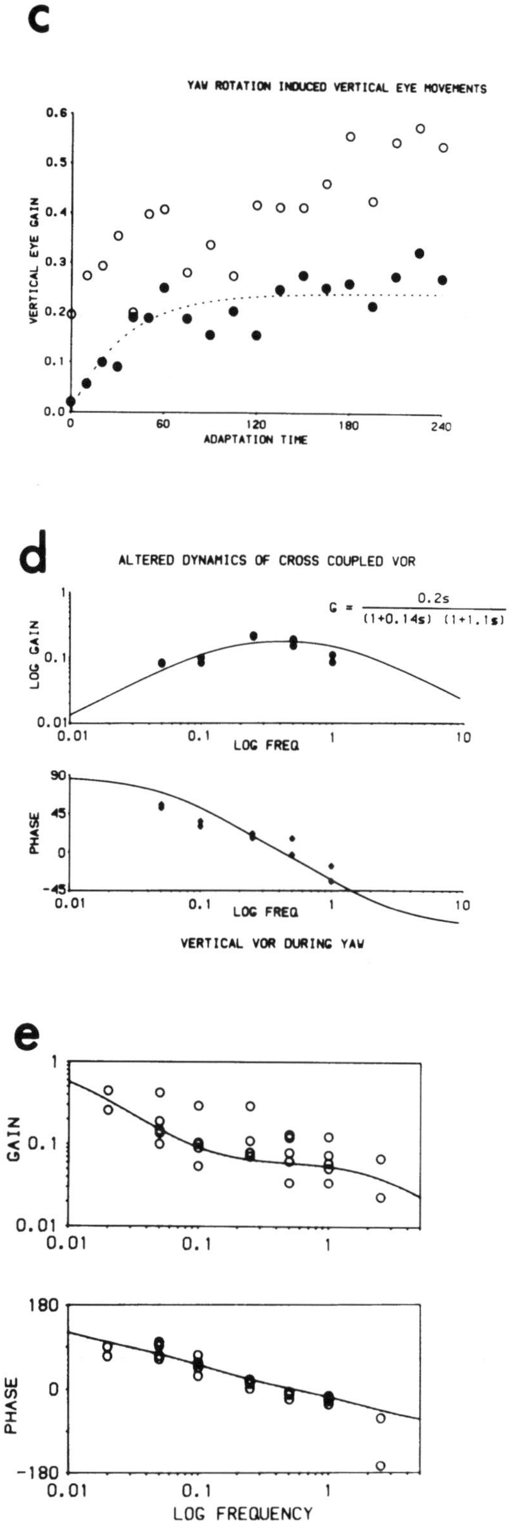

Fig. 1 (continued). **c** Time course of acquisition of vertical eye movement in response to 0.25 Hz horizontal rotation in the cat. *Filled circles* represent gain of vertical eye movements evoked by horizontal rotation in the dark. *Open circles* represent gain of eye movements evoked by the combination of horizontal rotation and vertical image motion during training. *Dashed line* is an exponential curve with time constant of 35 min, which was fitted to the data. (Harrison et al. 1986). **d** Frequency response of vertical vestibuloocular eye movements evoked by horizontal rotations after training at 0.25 Hz. Solid curve is frequency response of a bandpass filter with the transfer function shown in the inset. (Harrison et al. 1986). **e** Similar frequency response for horizontal eye movements evoked by vertical rotations after 2 h of pairing a 0.25 Hz pitch rotation with 0.25 Hz horizontal image rotation. Note that gain peaks at lowest frequency rather than at the frequency of training as in **d**. (Baker et al. 1986)

3 Voluntary Modulation of the VOR

Recent studies (Barr et al. 1976; McKinley and Peterson 1985) have shown that VOR gain can be controlled voluntarily by having subjects attempt to follow imagined targets fixed with respect to their body (voluntary suppression) or to the earth (voluntary enhancement). As shown in Fig. 2, subjects can switch between suppression and enhancement with a reaction time on the order of 0.3 s. The fact that VOR modulation declines for frequencies above ~1.0 Hz, as does SP, has led to the suggestion that the modulation may be produced by the SP system. However, McKinley and Peterson (1985) showed that VOR modulation does not depend on the predictability of target motion as does SP, nor does it exhibit the large phase shifts characteristic of pursuit. Voluntary modulation and SP therefore appear to be independent mechanisms. Nevertheless, their parallel decline above 1 Hz may indicate that they share certain common neural pathways. If this were the case, subjects attempting to track moving targets while rotating would not be able to add SP to suppression of their VOR. Instead, the two systems

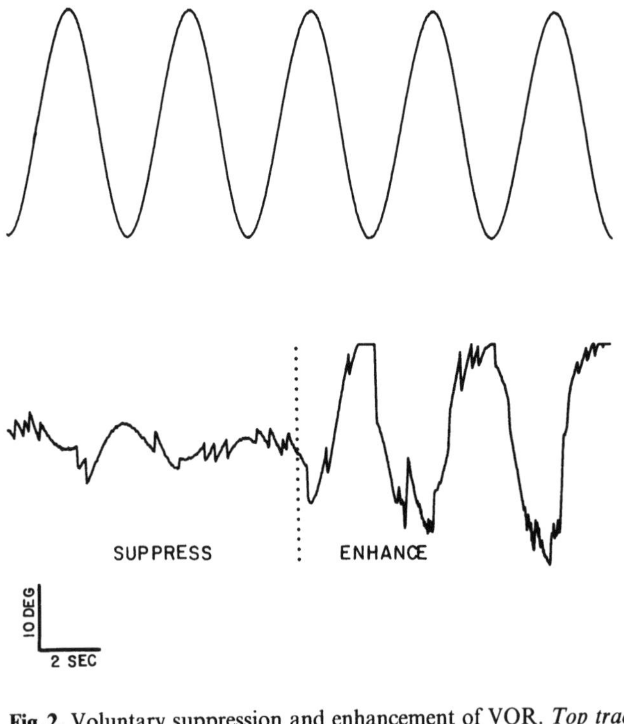

Fig. 2. Voluntary suppression and enhancement of VOR. *Top trace* shows velocity of chair rotation. *Bottom trace* shows horizontal eye position of subject seated in the chair. At beginning of record, subject is suppressing his VOR by attempting to fixate an imagined target rotating with the chair. At time marked *by dashed line,* he was instructed to begin enhancing his VOR by attempting to fixate a target fixed to the wall. In both cases the chair rotation evokes vestibular nystagmus consisting of alternating slow phases, where the eyes move in the opposite direction to the chair and quick phases, which bring the eyes back towards the center of the orbit. Note the large increase in slope (velocity) of the slow phases following the change from suppression to enhancement

would interact to give a combined eye velocity less than the sum of SP and suppressed VOR. We have recently observed such an interaction in subjects seated in a rotating chair, who were attempting to follow a visual target that moved in the same direction as, but faster than the chair (Sahyouni et al., 1987). As velocities were increased into the range in which SP and VOR modulation begin to saturate, none of the subjects could produce a tracking eye velocity equal to that predicted if the two mechanisms were operating independently.

4 Mechanisms Underlying Recovery from Labyrinthine Lesions

The sudden change in vestibular input produced by unilateral labyrinthine loss leads to severe loss of ocular stability due to the unbalancing of tonic vestibular input from the two labyrinths. The imbalance manifests as unidirectional vestibular nystagmus with slow phases toward lesioned side, fast phases toward the normal side. To eliminate these inappropriate responses, the CNS must undergo adaptive changes that restore the balance of tonic activity in the vestibular nuclei on the two sides. In otherwise healthy individuals this adaptation occurs within a few days following the lesion. The underlying mechanisms are poorly understood, but appear to involve both ascending and descending signals from other regions of the CNS since lesions of many structures can lead to decompensation (Schaefer and Meyer 1981).

A peripheral vestibular lesion should also lead to reduction in VOR gain proportional to the loss of receptors. Thus unilateral labyrinthectomy should lead to a 50% reduction in VOR gain. The question then to be answered is how is this loss compensated for in order to restore gaze stability during head movements? On the basis of the material presented above, three possible mechanisms can be suggested: (1) adaptive changes in baseline VOR, (2) voluntary modulation of the VOR and (3) visual tracking mechanisms (SP and OK). Evidence for the involvement of these mechanisms will be discussed below.

4.1 Plastic, Adaptive Increase in Baseline VOR Gain

Studies described above indicate that altered visual feedback during head rotation can lead to recalibration of VOR gain that tends to restore gaze stability. Reports vary concerning whether such a recalibration of VOR gain occurs after unilateral labyrinthine loss. Paige (1983) reported full recovery of baseline VOR gain measured in the dark in squirrel monkeys following unilateral plugging of semicircular canals. On the other hand, unilateral labyrinthectomized rabbits show little gain recovery (Baarsma and Collewijn 1974) and only 50% of the cats followed by Maioli et al. (1983) for several months after a unilateral labyrinthectomy exhibited a significant increase in gain.

Our recent studies of patients following resection of an acoustic neuroma (Hart et al. 1987) revealed a similar variation in amount of recovery of baseline VOR gain that could be attributed to VOR plasticity. When tested in the relaxed state during rotation in the dark, some individuals, such as Patient 1 in Fig. 1, had VOR gains that were half the normal value (filled squares). Patient 3, on the other hand, had a baseline gain in the high normal range indicating that substantial plastic gain recovery had occurred. The typical case, exemplified by patient 2, fell between the two extremes. The question of why plastic readjustment does not occur in all individuals remains a mystery. Maioli and Precht (1985) showed that their low gain cats could plastically increase their gains to normal levels when exposed to passive whole body rotation in a stationary illuminated environment so that the reason does not seem to be a loss of adaptive capacity but rather a change in the VOR gain set point or in gaze strategy.

4.2 Voluntary Enhancement of the VOR

The voluntary control of VOR gain described in Section 3 could play a role in maintaining gaze stability in labyrinthectomized individuals. In fact, Patient 1 in Fig. 3 appears to utilize this mechanism quite effectively. When he attempted to fix his eyes upon an imagined earth-fixed target during rotation in total darkness, he was able to raise his VOR gain from 0.25 to 0.97 (filled diamonds). He thus appears to rely primarily upon voluntary modulation of the VOR to obtain good gaze stabilization. Patient 3 is quite different. While her baseline VOR gain during relaxed rotation in the dark was an above normal 0.8, her ability to override the VOR voluntarily was very poor – attempting to raise the gain by tracking an imagined earth-fixed target actually resulted in a gain reduction! Patient 2 had some ability to improve VOR gain voluntarily but could not attain the fully compensatory gaze performance shown by Patient 1.

4.3 Increased Reliance on Visually Guided Image Stabilization

Smooth pursuit and optokinetic eye movements can help keep the eyes directed toward a target of interest provided that the target does not move too rapidly (Lisberger et al. 1981). The contribution of smooth pursuit to gaze stabilization in the three patients in Fig. 3 was tested by giving them an earth-fixed light to fixate upon during body rotation (open squares in figure). The gain of Patient 2 increased dramatically in this condition, indicating that he was utilizing visual pursuit to help stabilize his gaze. Gain in Patient 1 did not change, since it was already brought to the required value by voluntary enhancement of the VOR. Interestingly, Patient 3 did not do well when given the visual target reference. She appears to rely solely upon a plastic, adaptive increase in VOR gain to approach gaze stability.

Data from four other patients confirmed the patterns illustrated in Fig. 3. Only 2/7 patients had baseline VOR gains in the normal range, indicating effective, plastic recalibration of gain. Two patients exhibited large increases in gain when they attempted to track an imagined target. They appear to rely primarily upon voluntary modulation of the VOR to obtain good gaze stabilization. The remaining three patients were like Patient 2. They had only a modest gain recovery due to plastic, adaptive mechanisms and were not especially adept at increasing this gain by fixating an imagined target. When given a visible target to track, however, they could attain gains approaching 1.0. They therefore appear to rely primarily upon visual tracking mechanisms to stabilize gaze.

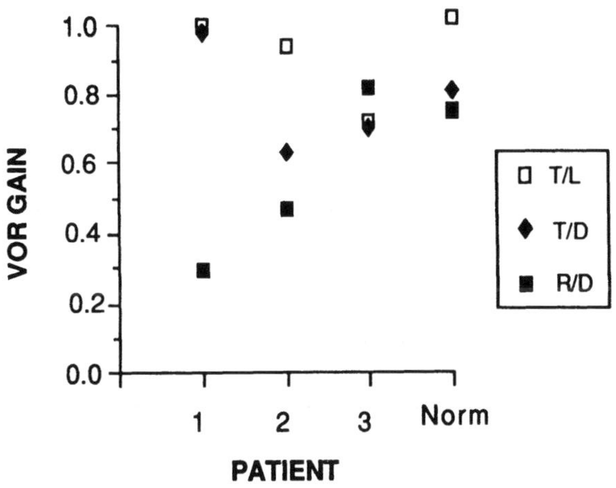

Fig. 3. Vestibuloocular reflex performance in three patients after surgical removal of an acoustic neuroma is compared with the median responses of a group of seven normal subjects. VOR gain was calculated as the ratio of the velocity of smooth eye movements to velocity of chair rotation during 0.25 Hz, 80 °/s sinusoidal chair rotations under three conditions: subject relaxed during rotation in dark (*R*/*D*), subject attempting to track an imagined, earth-fixed target during rotation in dark (*T*/*D*) and subject tracking a visible LED during rotation (*T*/*L*)

These findings raise the question of why there is so much variation in the recovery strategies adopted by different individuals. In particular it is unclear why plastic adaptation does not raise the VOR gain to normal levels in all individuals. Perhaps the answer is that alternative gaze stabilization mechanisms adopted early in the recovery period preempt the recovery process and prevent plastic adaptive changes from occurring.

Individuals might be especially prone to select a specialized gaze stabilizing strategy if it were difficult for them to employ more than one stabilizing mechanism simultaneously. Recent data from our laboratory (Sahyouni et al. 1987) suggest that human subjects may have difficulty employing the VOR and SP together. When subjects seated in a chair that was rotating to the right at 40 °/s were asked to follow a visual target that was moving to the left at 40 °/s, the tracking eye velocity that they produced did not equal the predicted sum of a VOR response to a 40 °/s rotation and a SP response to a 40 °/s moving target. Instead, eye velocity was closer to the slower value observed during pursuit of an 80 °/s moving target. Furthermore if the chair rotation was stopped during the task, no change in eye velocity was seen at the latency expected for the VOR (<30 ms). Rather, the first adjustment of eye velocity occurred at 120–150 ms latency typical of SP. Both of these observations are consistent with the idea that subjects suppress their VOR when using SP to track a moving visual target even when the VOR would aid the tracking. A tendency to suppress the VOR during tracking could lead hemilabyrinthectomized individuals to choose either the VOR or SP as their primary mode of attaining gaze stabilization. Those choosing SP might not develop adaptive changes in VOR gain, since they would tend to suppress their VOR during many activities requiring image stabilization.

5 Conclusions

Studies of plastic, adaptive changes in the VOR have shown that the adaptive system can produce substantial changes in the gain and direction of the VOR, which are remarkably specific to the learning context. Thus the adaptive system is theoretically capable of restoring VOR performance even after major perturbations to the VOR such as those following hemilabyrinthectomy. However, the adaptive system is only one of several mechanisms that can contribute to the restoration of gaze stability. Voluntary modulation of the VOR can produce gain changes rivaling those produced by the adaptive system. Smooth pursuit of visual targets can also make an important contribution to image stabilization. The finding that subjects using SP to follow a target tend to suppress their VOR raises the interesting possibility that these systems may interact in a way that leads individual subjects to adopt a preferred strategy for visual image stabilization.

Acknowledgements. Work in the authors' laboratories was supported by grants EY05049, EY05289 and EY06485.

References

Baarsma EA, Collewijn H (1974) Vestibulo-ocular and optokinetic reactions to rotation and their interaction in the rabbit. J Physiol (Lond)238:603–625

Baker J, Harrison REW, Isu N, Wickland C, Peterson B (1986) Dynamics of adaptive change in vestibulo-ocular reflex direction. II. Sagittal plane rotations. Brain Res 371:166–170

Baker J, Wickland C, Peterson B (1987) Dependence of cat vestibulo-ocular reflex direction adaptation on animal orientation during adaptation and rotation in darkness. Brain Res 408:339–343

Baker JF, Perlmutter SI, Peterson BW, Rude SA, Robinson FR (1988) Simultaneous opposing adaptative changes in cat vestibulo-ocular reflex direction for two body orientations. Brain Res (in press)

Barmack NH, Nastos MA, Pettorossi VE (1981) The horizontal and vertical cervico-ocular reflexes of the rabbit. Brain Res 224:261–278

Barr CC, Schultheis LW, Robinson DA (1976) Voluntary, non-visual control of the human vestibulo-ocular reflex. Acta Oto-Laryngol 81:365–375

Berthoz A, Melvill-Jones G, Begue AE (1981) Differential visual adaptation of vertical canal-dependent vestibulo-ocular reflexes. Exp Brain Res 44:19–26

Cohen B, Matsuo V, Raphan T (1977) Quantitative analysis of the velocity characteristics of optokinetic nystagmus and optokinetic after-nystagmus. J Physiol (Lond) 270:321–344

Cohen B, Henn V, Raphan T, Dennett D (1981) Velocity storage, nystagmus and visual-vestibular interactions in humans. Ann NY Acad Sci 374:421–433

Collewijn H, Martins AJ, Steinmann RM (1983) Compensatory eye movements during active and passive head movements: fast adaptation to changes in visual magnification. J Physiol (Lond) 340:259–286

Gonshor A, Melvill Jones G (1976 a) Short-term adaptive changes in the human vestibulo-ocular reflex arc. J Physiol (Lond) 256:361–379

Gonshor A, Melvill Jones G (1976 b) Extreme vestibulo-ocular adaptation induced by prolonged optical reversal of vision. J Physiol (Lond) 256:381–414

Harrison REW, Baker JF, Isu N, Wickland CR, Peterson BW (1986) Dynamics of adaptive change in vestibulo-ocular reflex direction. I. Rotations in the horizontal plane. Brain Res 371:162–165

Hart CW, McKinley PA, Peterson BW (1987) Compensation following acute unilateral total loss of peripheral vestibular function. In: Graham M (ed) The vestibular system (in press)

Keller EL (1978) Gain of the vestibulo-ocular reflex in monkey at high rotational frequencies. Vision Res 18:311–315

Lisberger SG, Evinger C, Johanson GW, Fuchs AF (1981) Relationship between eye acceleration and retinal image velocity during foveal smooth pursuit in man and monkey. J Neurophysiol 46:229–249

Maioli C, Precht W (1985) On the role of vestibulo-ocular reflex plasticity in recovery after unilateral peripheral vestibular lesions. Exp Brain Res 59:267–272

Maioli C, Precht W, Reid S (1983) Short and long-term modification of vestibuloocular response dynamics following unilateral vestibular nerve lesions in the cat. Exp Brain Res 50:259–274

McKinley PA, Peterson BW (1985) Voluntary modulation of the vestibuloocular reflex in humans and its relation to smooth pursuit. Exp Brain Res 60:454–464

Miles FA, Eighmy BB (1980) Long-term adaptive changes in the primate vestibuloocular reflex. I. Behavioral observations. J Neurophysiol 43:1406–1425

Paige GD (1983) Vestibuloocular reflex and its interaction with visual following mechanisms in the Squirrel monkey. II. Response characteristics and plasiticity following unilateral inactivation of horizontal canal. J Neurophysiol 49:152–168

Peterson BW (1988) Cervico-collic and cervico-ocular reflexes. In: Peterson BW, Richmond FJ (eds) Control of head movement. Oxford Univ Press (in press)

Precht W, Maioli C, Dieringer N, Cochran S (1981) Mechanisms of compensation of the vestibulo-ocular reflex after vestibular neurotomy. In: Flohr H, Precht W (eds) Lesion-induced neuronal plasticity in sensorimotor systems. Springer, Berlin Heidelberg New York, pp 221–230

Pulaski PD, Zee DS, Robinson DA (1981) The behavior of the vestibulo-ocular reflex at high velocities of head rotation. Brain Res 222:159–165

Robinson DA (1976) Adaptive gain control of the vestibuloocular reflex by the cerebellum. J Neurophysiol 39:953–969

Robinson DA (1980) Control of eye movements. In: Brooks VB (ed) Handbook of physiology, the nervous system, III. Motor Control, pp 1275–1320

Robinson DA (1982) The use of matrices in analyzing the three-dimensional behavior of the vestibulo-ocular reflex. Biol Cybern 46:43–57

Sahyouni XN, Matsuo V, Peterson BW (1987) Ocular pursuit during yaw rotation in humans. Soc Neurosci Abstr 13:171

Schaefer K-P, Meyer DL (1981) Aspects of vestibular compensation in guinea pigs. In: Flohr H, Precht W (eds) Lesion-induced neuronal plasticity in sensorimotor systems. Springer, Berlin Heidelberg New York, pp 197–207

Schultheis LW, Robinson DA (1981) Directional plasticity of the vestibulo-ocular reflex in the cat. Ann NY Acad Sci 374:504–512

Vestibulo-Ocular Reflex Compensation in the Vertical and Horizontal Planes Following Unilateral Peripheral Vestibular Deficit in Man

J. H. J. Allum [1], M. Yamane [2], and C. R. Pfaltz [1]

[1] Department of Otorhinolaryngology, University Hospital, CH-4031 Basel, Switzerland
[2] M. Yamane is currently at Dept. of Neurootology, Tokyo, Metropolitan Neurological Hospital, Japan

1 Introduction

A unilateral peripheral deficit of the vestibular system results in a dramatic impairment of oculomotor function which gradually recedes over a period of months. Because the deficit interrupts the flow of vestibular afferent information from the same side the classical symptoms of unstable eye and head movements are observed. The alteration in vestibulo-ocular reflex (VOR) control indicates that the dynamic balance between vestibular signals reaching the vestibular nuclei directly and via commissural pathways (Shimazu and Precht 1966) has been profoundly affected. The CNS is then challenged to restore normal function. The changes in the VOR make vestibular compensation an attractive model to study plastic and adaptive mechanisms of the CNS.

The compensation process undertaken by the CNS, on onset of a unilateral vestibular deficit, is dependent on several neurophysiological factors. Firstly, the removal of the afferent input from one side reduces the dynamic ipsilateral excitatory and contralateral inhibitory input to the vestibular nuclei. Secondly, the levels of resting discharge in the vestibular nuclei are depressed (Ried et al. 1984), causing a reduction in the amount of inhibitory and disinhibitory modulation possible in VOR reflex arcs via type II neurons. Thirdly, the central "storage mechanism" extending the duration of nystagmus (that is, its time constant) beyond response duration of vestibular nerve fibres (Raphan et al. 1979) is affected for horizontal (HOR) rotations (Maioli et al. 1983). The effect of a storage mechanism has not been observed for the vertical (VERT) VOR (Correia et al. 1985) or at least only for downward vestibular nystagmus (Matsuo and Cohen 1984). Thus, relations between durations of nystagmus in different VOR planes are altered by an unilateral deficit. All these factors cause alterations in VOR response gain, asymmetry, and time constant relations for one plane and between canal planes of rotation.

The time elapsed before the static imbalance caused by a unilateral deficit and the resulting HOR spontaneous nystagmus disappears has been extensively studied in man (Brandt et al. 1978; Fisch 1973; Pfaltz and Kamath 1970). However, little attention has been directed in man toward documenting the time course of compensation in the dynamic response of either the HOR or VERT VOR. Although it is known that the gain, but not the time constant, of HOR VOR is restored to near normal values over a period of several months after acoustic neuri-

H. Flohr (Ed.)
Post-Lesion Neural Plasticity

noma surgery (Olson et al. 1981; Olson and Wolfe 1984), the effect of a unilateral vestibular deficit on the VERT VOR is not known. Studies of the compensation process for the deficit in HOR and VERT VOR's might indicate differences in the organization of these two types of reflex arcs.

2 Methods

2.1 Clinical Populations

The VOR responses of three populations were examined in this study: 20 normal subjects without prior history of neurological and otological diseases (11 men and 9 women ranging in age from 18 to 59 years, average 28.4 years); a group of patients with acute unilateral vestibular paralysis unaccompanied by cochlear disorders; and a group of patients with acoustic neurinomas.

The 12 patients with acute unilateral vestibular paralysis (UNILATS) consisted of 6 men and 6 women ranging in age from 25 to 57 (average 37) years. All patients included in this study had vertigo of sudden onset and showed 100% canal paralysis (CP) of the affected inner ear in bithermal caloric tests (30 and 44 °C). Twelve patients with diagnosed and surgically confirmed unilateral acoustic neurinoma (aged 41 to 61, average 49 years) were investigated. Six patients showed saccadic eye movements in eye tracking tests and depressed responses in optokinetic nystagmus tests, indicative of an operatively determined attachment of the tumor to the brain stem. The UNILATS' first VOR examination occurred as soon after clinic entry as their vertigo permitted, whereas the acoustic neurinoma patients (ACNS) were tested first pre-operatively and then again post-operatively. A second (third, for ACN patients) VOR test was repeated 1 month to 1 year later on all patients. As the trend in the results became clear, the second test was regularly performed 2 months after the first for the UNILAT patients.

2.2 Test Procedures

Subjects were rotated about the yaw and pitch axes in total darkness with the eyes open while requested to keep the eyes in the mid-position. Eye movements were recorded with standard electro-oculography techniques (EOG). For rotation about the yaw axis, subjects were seated on a servo-controlled rotating chair (CF10 from Servo-Med, Sweden) with the head erect and braced at the center of rotation. For rotation about the pitch axis, the chair was moved off-center on a cantilever beam system with the back of chair leaning 45 ° toward the center of rotation. The subjects then lay on their side in the chair with the head held by a brace so that the interaural axis could be aligned with the axis of chair rotation.

A modified form of the triangular velocity stimulus waveform introduced by Guedry et al. (1971) was employed in this study (see Fig. 1). It consisted of two segments of oppositely directed constant acceleration; 6 and 12 s in duration, and 6 s of constant deceleration to stop. Two modes of rotation were used at each magnitude of acceleration (10, 15 and 20 deg/s^2), one which started with the clockwise rotation, and another which started with counterclockwise rotation.

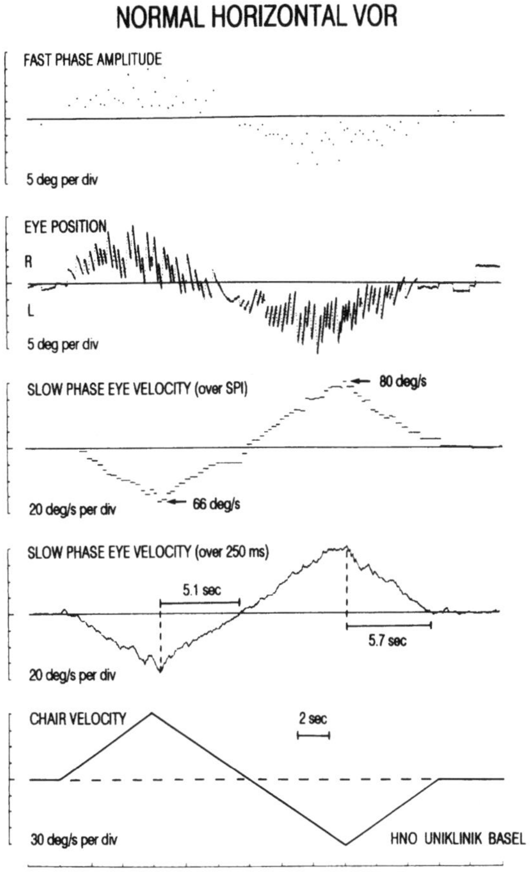

Fig. 1. Normal horizontal (HOR) VOR response. The traces illustrate the manner of data analysis employed throughout this paper. The vestibular nystagmus is in response to a whole body rotation whose velocity profile is shown in the lowest trace (20 deg/s^2 constant acceleration, 120 deg/s maximum velocity). The maximum slow phase eye velocity at 6 and 18 s after rotation onset was measured using the slow phase eye velocity (SPV) averaged across each slow phase interval (SPI); see third trace. Mean values for normals are 54 (s. d. 19) deg/s and 62 (s. d. 17) deg/s at 6 and 18 s after rotation onset respectively. The time for SPV to decay from maximum to zero (ZCT) was determined from a running average of SPV (over 250 ms); see fourth trace. Marked in the SPV (250 msecs) trace are typical ZCT times for normals. Mean ZCT for normals are 4.9 and 5.1 s at 6 and 18 s respectively, standard deviations (s. d.) 0.7 and 0.5 s respectively. (Allum et al. 1988)

For VERT rotations, a half of normal subjects were tested lying on their right side, and half lying on their left side. Patients were always tested lying on the lesion side.

2.3 Data Analysis

Figure 1 illustrates the four measurements taken from each response. The two maximum slow phase velocity (SPV) values at approximately 6 and 18 s after the onset of rotation were located with the aid of an automatic digital nystagmus analyzer (Tönnies FRG) which computed the average SPV across each slow phase (Allum 1985). These provided measures of response amplitude and therefore gain. Two response intervals were defined. These were the times between maximum slow phase velocity and the time when the SPV crossed the pre-stimulus level of SPV, in normals, zero SPV. These intervals are termed, hereafter, zero-crossing times or ZCT. Both times are shown in Fig. 1 for a normal HOR VOR response. Four measurements were taken for each clockwise and counterclockwise starting rotation to yield a total of eight values per acceleration magnitude. The maximum

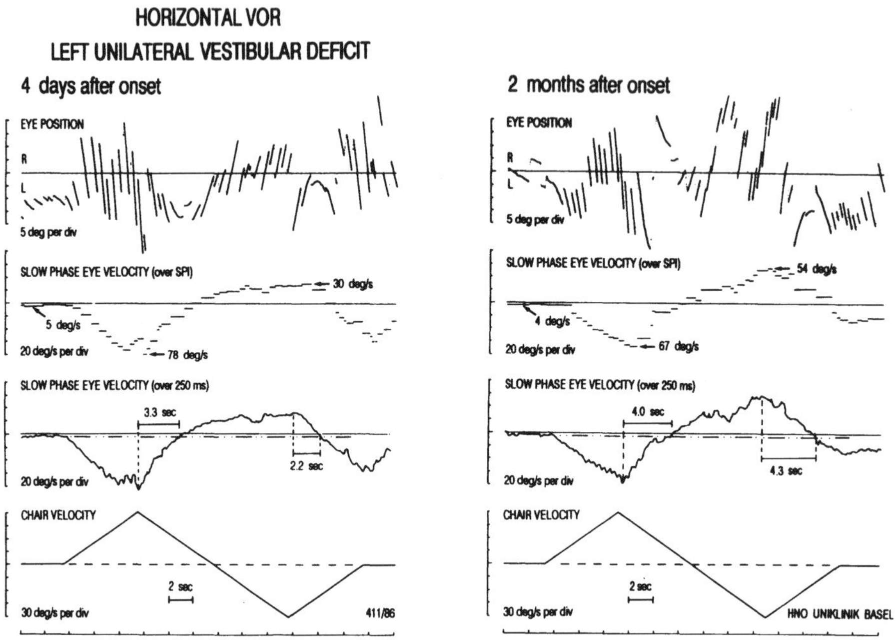

Fig. 2. HOR VOR responses from a patient 4 days (*left*) and 2 months (*right*) after a sudden left unilateral paralysis. Note that the SPV values on the left become 73 and 35 deg/s and those on the right 63 and 58 when corrected for the offset caused by spontaneous nystagmus. The response asymmetry is considerably reduced after 2 months (cf. Fig. 1) but the shortening of ZCT times remains. (Allum et al. 1988)

SPV and the ZCT values were corrected for offsets caused by any spontaneous nystagmus (see Fig. 2).

Each SPV amplitude measurement was first expressed as a percentage of the averaged normal value for the direction of SPV (right, left, up, and down), intensity of stimulation (10, 15, and 20 deg/s^2) and time since stimulus onset (ca. 6 or 18 s). The six values for each SPV direction were condensed into a single average response measure. A similar pooling technique was applied to the time constants. The time constants were calculated by assuming that the nystagmus responses are dominated by the longer time constant τ of the VERT or HOR VOR. Then the first ZCT, T_1, can be expressed as

$$T_1 = \tau \cdot \ln [2\text{-exp} (-6/\tau].$$

Similarly, the second ZCT, T_2, is given by

$$T_2 = \tau \cdot \ln [2\text{-exp} (-12/\tau) + \exp (-18/\tau],$$

where the values of 6, 12, and 18 s for acceleration durations are those corresponding to the acceleration profile employed in this study. The time constant, τ, was calculated recursively from the equations above.

3 Results

3.1 Horizontal VOR

The left set of traces in Fig. 2 shows the well-known reduction of horizontal (HOR) VOR gain, change in response asymmetry and in time constant following an acute UNILAT paralysis. Two less well-known aspects are worth emphasizing. The asymmetry in gain is somewhat less profound but nonetheless significant, when the offset caused by the spontaneous nystagmus is accounted for (see legend to Fig. 2 and cf. Fig. 2 A, Maioli et al. 1983) and the effective time constant of nystagmus is shorter than the normal mean value of 20.4 s [standard deviation (sd), 5.3 s] when rotation was either toward the deficit or toward the intact side.

Although the general characteristics of the response asymmetry remained after 1–3 months of recovery they are by no means so sharp. The right traces of Fig. 2 document the normal SPV response amplitudes for both directions of rotation 2 months after deficit onset in this patient. An asymmetry still exists for some patients but it is not significant once spontaneous nystagmus offsets have been accounted for. Despite the near-normality of response gain, the time constants for both directions of rotation remained abnormally short (compare zero crossing times in Fig. 1 and 2). The distinction between static and dynamic balance recovery is particularly well shown in Fig. 2. Here the level of spontaneous nystagmus SPV was essentially unchanged, while dynamic response gain was reestablished at normal levels. Across both our patient groups no correlation between the level of spontaneous nystagmus (static imbalance) and VOR response gain could be established.

The left half of Fig. 3 and 4 show that the general pattern of the HOR VOR established in Fig. 2 did not depend at all crucially on the pattern of acceleration amplitude, or choice of subject. Here the pattern of recovery, or lack of it, is compared between normals, UNILATS and ACN patients for HOR VOR gain. Over the first days after onset of a unilateral deficit or post-operatively for ACNS the VOR gain is approximately 50% and 75% of normal for rotation toward and away from the lesion side respectively. Within 3 months UNILATS have normal levels of VOR gain which required over 6 months of recovery for ACNS to achieve.

Differences between the recovery of the HOR VOR time constant between the two groups of patients were also observed. Rotations both toward the normal and toward the deficit or lesion side yielded a significant shortening of the time constant with respect to normal values, the greatest shortening occurring for rotation to the deficit side (lower left Fig. 3 and 4). After up to 6 months of recovery in ACNS and 3 months for UNILATS, the time constants were 9.1 and 10.3 s respectively for rotation to the deficit side and 13.4 and 13.7 s respectively for oppositely directed rotation compared to a mean normal value of 20.4 s. Attainment of a time constant within the normal lower limit (mean minus two standard deviations) was only observed in the long term for rotation toward the normal side for UNILATS. Even for ACN patients controlled up to 1.5 years post-operatively HOR VOR time constants were still not in the normal range, although a slightly longer time constant was revealed by rotation toward the normal side.

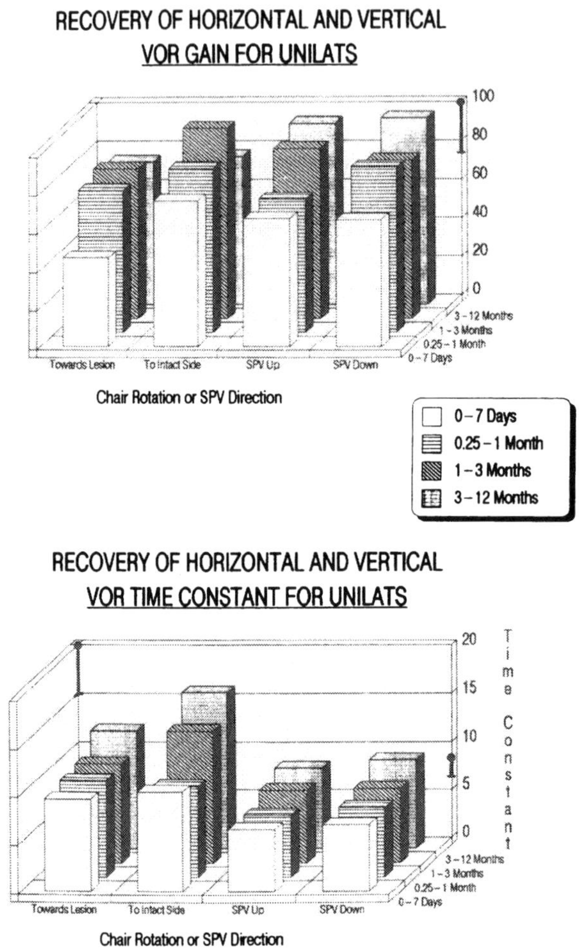

Fig. 3. Recovery course of horizontal (HOR) and vertical (VERT) VOR dynamics for patients with sudden unilateral vestibular paralysis (UNILATS). In the upper part of the figure the recovery course of VOR gain is compared with normal gain at intervals of 0 to 7 days, 0.25 to 1 month, 1–3 months, 3–12 months after deficit onset. HOR VOR gain is plotted separately for HOR rotation toward the lesion and toward the intact side. Vertical VOR gain is plotted for SPV up and down. The standard deviation for normal VOR gain is shown by *thick bars* on the gain axis. In the lower part of the figure the recovery course of VOR time constant is plotted. The normal mean time constant and one standard deviation for HOR (20.4 and 5.3 s) and VERT (8.5 and 2.3 s) rotations is shown by the *filled circle* and bar at the *left* and *right* of the figure respectively. (Allum et al. 1988)

The post-operative compensation for the vestibular deficit induced by ACN surgery is complicated by the existence of pre-operative, tumor-induced deficit for which central compensation was in process. Pre-operatively ACN patients had normal HOR VOR gains for rotation to the deficit side, but surprisingly reduced gain for rotation to the normal side (Fig. 3, upper left). This difference is replicated in the pre-operative time constants. Both are shorter than normal, with the

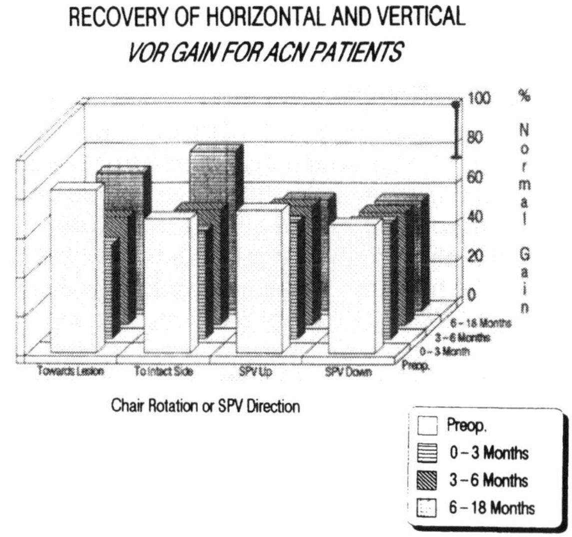

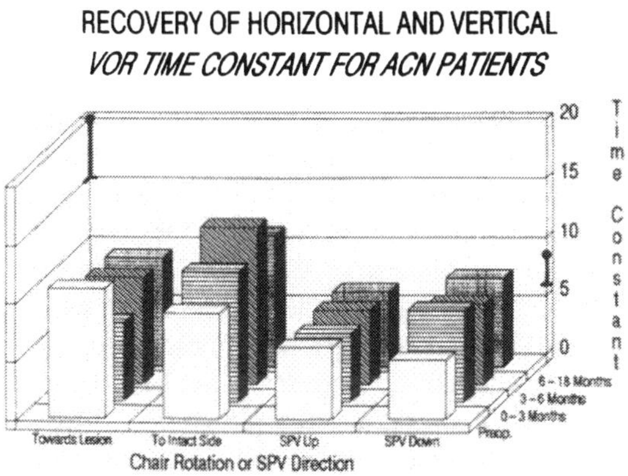

Fig. 4. Recovery course of HOR and VERT VOR dynamics for patients with operated acoustic neurinomas (ACN). The presentation of data is identical to that of Fig. 3, except that the time intervals have been divided into preoperative, 0 to 3, 3 to 6, and 6–18 months post-operative

shorter value for rotation towards the intact side. Although we were not able to test pre-dificit VOR values in the UNILATS it was reasonable to assume from the patients' medical history that these were normal. Presumably the differences in pre-deficit VOR response asymmetry of ACN and UNILATS has a major influence on the recovery of VOR gain following ACN surgery and on the recovery of VOR time constant for rotation to the intact side. Common to both groups of patients is the lack of recovery of VOR time constant for rotation toward the lesion side.

3.2 Vertical VOR

Our data on the vertical (VERT) VOR response in normals were obtained primarily to provide a reference for the assessing the effect of a peripheral vestibular deficit. Because of the considerable differences in VERT VOR findings in primates (Baloh et al. 1983; Correia et al. 1985; Matsuo and Cohen 1984), it is worthwhile to briefly describe our data on normals. Human VERT VOR gain or time constant asymmetry between up and down SPV was not observed across our 40 normals. This finding confirms most findings to data in primates but not those of Baloh et al. (1983), for humans, who found a 17% gain asymmetry in favor of SPV down, nor those of Matsuo and Cohen (1984), for monkeys, who report a 15-s time constant for SPV up. With respect to the time constant, our data are in agreement with those of Correia et al. (1985) and Baloh et al. (1983). The VERT VOR time constant for SPV up (mean 8.3 sd 2.9 s) and down (mean 8.8 sd 1.8 s) was ca. one third to one half shorter than that of the HOR VOR. These time constants are only slightly longer than the average time constant of afferent activity (5–7 s) in the semicircular canal nerves of monkeys (Correia et al. 1985; Fernandez and Goldberg 1971). This result then indicates a neural storage mechanism similar to that influencing the HOR VOR (Raphan et al. 1979) is not a major element in human VERT VOR response dynamics.

Figure 5 illustrates the VERT VOR response from the same patient whose HOR VOR responses are illustrated in Fig. 2. The profound gain reduction and asymmetry observed 4 days after deficit onset in the HOR VOR for rotation to one side, that of the lesion, is surprisingly absent in the VERT VOR responses of Fig. 5. A weaker asymmetry and gain reduction in the SPV UP and SPV DOWN response gain to respectively 45% and 68% of the normal value can be observed at onset of the deficit. The VERT VOR response shown in the compensated state, 2 months later, is identical to a normal response on the basis of gain and time constant (not shown in Fig. 5). At onset of the deficit the response shown in Fig. 5 indicated a slight shortening of the nystagmus time constant with respect to normal response duration (3.5 and 4.0 s are the normal ZCT's) but not of the magnitude observed in Fig. 2.

Considered for themselves, the average patient data in the right hand side of Fig. 3 and 4 replicate the specific findings of the case study in Fig. 5 in not conforming with HOR VOR data of Fig. 2. VERT VOR gain is, on average, symmetrically reduced to 66% of normal on onset of the UNILAT deficit and post-operatively for the ACNS. Within 3 months, gain for the UNILATS increases to within the normal range with no changes in symmetry. No VERT VOR gain recovery at all was traceable in ACN patients even when tested 1.5 years post-surgery. As with the HOR VOR, this lack of recovery may be linked to the gain reduction existing prior to surgery which Fig. 4 indicates is only slightly higher than post-operative values.

4 Discussion

We have presented studies that investigated how the CNS compensates for a unilateral loss of vestibular afferent information. The picture that emerges is clearer

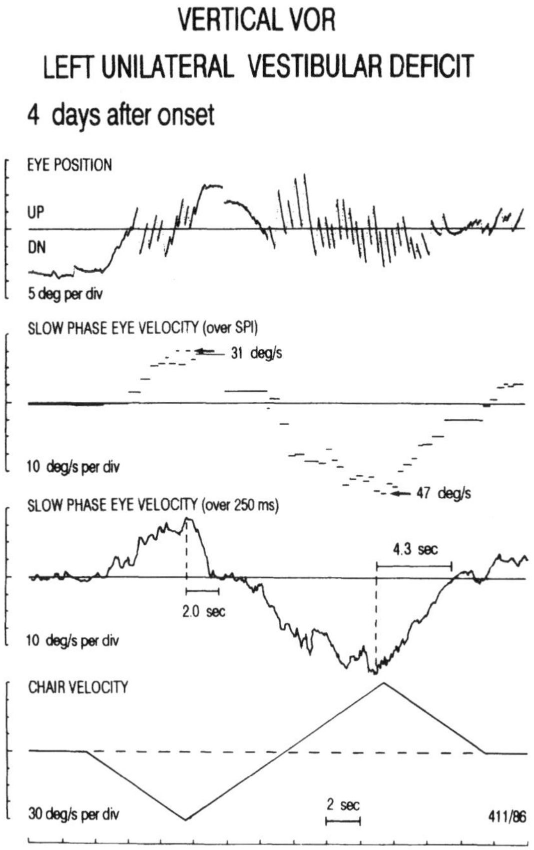

Fig. 5. VERT VOR responses from the patient of Fig. 2, 4 days after onset of unilateral vestibular paralysis. In contrast to marked HOR VOR asymmetry in Fig. 2, note that the maximum SPV values (31 and 47 deg/s at 6 s and 18 s) are more symmetrically reduced with respect to the normal mean values of 64 and 69 deg/s when compared to the reduction in Fig. 2. The layout of the figure is identical with that of Fig. 1 and 2. (Allum et al. 1988)

for patients with a sudden loss (UNILATS) rather for the patients with an acoustic neurinoma (ACN). For this former group HOR VOR gain is reduced to 50% and 75% of normal for rotation toward and away from the deficit side respectively. In contrast VERT VOR gain is symmetrically attenuated to 66% of normal gain. Additionally the horizontal time constant, particularly for rotations toward the lesion, is markedly reduced after the deficit. The first finding on gain suggests that the balance between direct excitatory ipsilateral and crossed disinhibitory effects on second-order vestibular neurons driving ocular motoneurons may not be the same for the HOR and VERT VOR. The second finding confirms previous observations of time constant prolongation solely for the HOR VOR and that this prolongation is uniquely dependent on resting discharges from the HOR canal nerve endings (Cohen et al. 1983). New is our observation of major differences in recovery rates for rotations toward and away from the deficit side. This finding implies that the neural circuitry prolonging the time constant has an ipsilateral and not distributed organization.

Given that both VERT and HOR second-order vestibular neurons send pairs of excitatory and inhibitory signals to ocular motoneurons, the question arises whether differences in post-deficit gains are to be expected. For the HOR system the ipsilateral and contralateral afferent inputs to vestibular nuclei neurons have a roughly equal excitatory effect (Ried et al. 1984). The gain asymmetry on onset of the deficit appears to result from a change in the level about which the contralateral dynamic input can disinhibit the second order neuron once the resting discharge of the nerve ipsilateral to the deficit is reduced (Maioli et al. 1983). Our finding of a symmetrical reduction in VERT VOR gain then indicates that either the balance of dynamic inputs is normally weighted in favour of the contralateral afferent signal or that the deficit has a lesser effect on the resting discharge of VERT second order vestibular neurons and thus on dynamic disinhibitory effects.

4.1 *Differences Between the Recovery Processes of UNILATS and ACN Patients*

Two of the main features of HOR VOR dynamics shown by ACN patients are the lack of recovery of a normal time constant for rotation to the normal side as well as to the deficit side and the slow gain recovery. Although both of these features have been noted before (Olson et al. 1981; Olson and Wolfe 1984), the current work established that recovery is faster, for UNILATS, and that they have practically normal time constants after recovery for rotation to the normal side if not to the deficit side. The cause of these differences in patient population effects are probably twofold: The presence of a deficit prior to surgery which is then often enhanced by the surgery and the greater age of the ACN patients compared to UNIlATS. Both factors presumably affect the capability of the CNS to compensate for the deficit. The surgical reports of our ACN patients were examined to establish whether a correlation between attachment of the tumor to the brain stem (applying a lesion directly on brain stem circuits) and VOR recovery occurred. On average, no significant influence of brain stem attachment of the tumor on VOR dynamics was apparent, an observation which confirms the results of Olson and Wolfe (1984).

The major difference between ACN patients and UNILATS is the lack of recovery of the VERT VOR gain for the former group. The HOR and VERT VOR gains of UNILATS are well within the normal range 1 to 3 months after paralysis onset. However, the VERT VOR of ACN patients remained depressed over our maximum follow-up period of 18 months. The main difference between the deficit of UNILATS and ACN patients is the presumed site of the lesion. Thus the question arises if this is the primary cause of the lack of VERT VOR recovery. Following this line of reasoning, a difference in the nerve's resting discharge, and with it changes in the efficacy of ipsi- and contralateral inputs at second-order neurons, would have to be assumed. For example, for the UNILATS that the lesion was at the receptor rather than on the nerve, as these patients had no accompanying hearing loss (Schuknecht and Kitamura 1981). In the UNILAT patients the resting discharge of vestibular afferents might be maintained whereas for the ACN patients it might be absent as a result of nerve damage or compromised blood supply during surgery. Yet, no differences in the initial shortening of the HOR VOR time constant was observed in these two groups of patients. As this VOR time constant is affected most when the resting discharge is absent (Cohen et al. 1983), it appears unlikely that resting discharges are affected differently in the two populations and therefore underlie the markedly different VERT VOR gain recovery. Alternatively, the lack of VERT VOR gain recovery in ACN patients could be part and parcel of the generalized poorer recovery of VOR dynamics in ACN patients. Because of the greater adaptive capabilities of central vestibular system on the HOR VOR gain and time constant, these dynamcis may simply recover faster than VERT VOR dynamics in ACN patients. In which case, VERT VOR controls after 2 years might show a slow but sure recovery in the VERT VOR dynamics.

The behavior of the VOR after a unilateral deficit either sudden or developing over time, as a tumor on the eighth nerve grows, indicates that separate and different processes of gain and time constant compensation exist for the HOR and VERT VOR . It appears unlikely that the neuronal circuits underlying the compensation process for one VOR plane are distributed over the CNS. Rather neural arcs within restricted regions of the vestibular nuclei, to which vertical and horizontal canal afferent information is sent to be transformed to signals driving motoneurons, are probably responsible for the compensation process. In this case, the most effective catalyst for the compensation process to commence is the long-term reduction in the resting discharge of second order vestibular neurons as a result of the removal of ipsilateral afferent input.

Summary

Horizontal (HOR) and vertical (VERT) vestibulo-ocular reflex (VOR) responses to whole body triangular velocity profiles with constant accelerations of 10, 15, and 20 deg/s were studied in three populations: normales, patients with acute unilateral peripheral vestibular paralysis, and patients with surgically removed acoustic neurinomas. The effect of these two types of unilateral deficit on VOR gain and long time constant were determined, as well as the time course of the compensation processes for HOR and VERT VOR dynamics. In both patient populations HOR VOR gain was asymmetric post-deficit, being, on average, 50 and 75% of normal for rotations toward and away from the deficit respectively. For the VERT VOR, on average, a symmetric 66% reduction occurred. The VERT VOR time constant was marginally affected by the deficit. HOR time constants were reduced for both directions of rotation. HOR and VERT VOR gain was within normal limits 1–3 months following an acute paralysis; time constants required a longer recovery period. For the acoustic neurinoma patients HOR VOR gain required over 6 months to recovery, VERT gain remained below normal values for at least 1 year post-surgery. No recovery in HOR VOR time constants was observed. These studies indicate major differences between the central organization of the compensation process for deficits of the VERT and HOR VOR.

Acknowledgements. This research was supported by Swiss National Foundation grant No. 3.148-0.85. We are grateful for the typographical assistance provided by Ms W. Brunetti.

References

Allum JHJ (1985) Automatic on-line nystagmus analysis: methods and clinical applications. In: Moser M, Ranacher G (eds) Computer-vestibulometrie. Thieme, Stuttgart, pp 65–87
Allum JHJ, Yamane M, Pfaltz CR (1988) Long-term modifications of vertical and horizontal vestibulo-ocular reflex dynamics in man. I. After acute unilateral peripheral vestibular paralysis. Acta Oto-Laryngol 105:328–337
Baloh RW, Richman L, Yee RD, Honrubia V (1983) The dynamics of vertical eye movements in normal human subjects. Aviat Space Environ Med 54:32–38
Brandt Th, Allum JHJ, Dichgons J (1978) Computer analysis of optokinetic nystagmus in patients with spontaneous nystagmus of peripheral vestibular origin. Acta Oto-Laryngol 86:115–122
Cohen B, Suzuki JI, Raphan T (1983) Role of the otolith organs in the generation of horizontal nystagmus: effects of selective labyrinthine lesions. Brain Res 276:159–164
Correia MJ, Perachio AA, Eden AR (1985) The monkey vertical vestibuloocular response: a frequency domain study. J. Neurophysiol 54:532–548
Fernandez C, Goldberg JM (1971) Physiology of peripheral neurons innervating semicircular canals of the squirrel monkey. II. Response to sinusoidal stimulation and dynamics of peripheral vestibular system. J Neurophysiol 39:661–675
Fisch V (1973) The vestibular response following unilateral vestibular neurectomy. Acta Oto-Laryngol 76:229–238
Guedry FE Jr, Stockwell CW, Norman JW, Owens GG (1971) Use of triangular waveforms of angular velocity in the study of vestibular function. Acta Oto-Laryngol 71:439–448
Maioli C, Precht W, Ried S (1983) Short- and long-term modifications of vestibuloocular response dynamics following unilateral vestibular nerve lesions in the cat. Exp Brain Res 50:259–274
Matsuo V, Cohen B (1984) Vertical optokinetic nystagmus and vestibular nystagmus in the monkey: up-down asymmetry and effects of gravity. Exp Brain Res 53:197–216
Olson JE, Wolfe JW (1984) Responses to rotational stimulation of the horizontal canals from patients with acoustic neuromas. Acta Oto-Laryngol Suppl 406:203–208
Olson JE, Wolfe JW, Engelken EJ (1981) Responses to low frequency harmonic acceleration in patients with acoustic neuromas. Laryngoscope 91:1270–1277
Pfaltz CR, Kamath R (1970) Central compensation of vestibular dysfunction. I. Peripheral lesions. Pract Oto-Rhino-Laryngol 32:335–349
Raphan T, Matsuo V, Cohen B (1979) Velocity storage in the vestibulo-ocular reflex arc (VOR). Exp Brain Res 35:229–248
Ried S, Maioli C, Precht W (1984) Vestibular nuclear neuron activity in chronically hemilabyrinthectomized cats. Acta Oto-Laryngol 98:1–13
Schuknecht HF, Kitamura K (1981) Vestibular neuritis. Ann Otol Rhinol Laryngol 90:78
Shimazu H, Precht W (1966) Inhibition of central vestibular neurons from the contralateral labyrinth and its mediating pathway. J Neurophysiol 29:467–492

Recovery from Unilateral Labyrinthectomy in Primates

M. Fetter [1] and D. S. Zee [2]

[1] Neurologische Universitätsklinik, Eberhard-Karls-Universität, Liebermeisterstr. 18–20, D-7400 Tübingen, FR Germany

[2] The Johns Hopkins Hospital, Department of Neurology, Baltimore, MD 21205, USA

1 Introduction

The mechanisms by which the central nervous system assures accurate dynamic vestibular responses when the head is moving and static vestibular balance when the head is still are partially responsible for eye-head coordination and of considerable interest both to clinicians and to basic scientists. Most recent experimental work in primates has focused on artificial manipulation of the visual scene during head rotation with various optical devices (Gonshor and Melvill Jones 1976; Lisberger et al. 1983) as well as plugging of the semicircular canals (Paige 1983) or bilateral labyrinthectomy (Dichgans et al. 1973), respectively, thereby testing the mechanisms of vestibulo-ocular reflex adaptation. Much less attention has been paid to the mechanisms of recovery of the vestibulo-ocular reflex (VOR) after unilateral labyrinthectomy in primate. This is a somewhat more complicated problem for the adaptive mechanisms to solve because unilateral labyrinthectomy creates not only a decrease in the gain of the VOR but also a tonic imbalance that causes both spontaneous nystagmus and directional asymmetry in dynamic vestibular function.

To understand better how primates recover from vestibular lesions we quantified the VOR in rhesus monkeys before and after unilateral labyrinthectomy. We also investigated the importance of visual experience in adaptation to labyrinthectomy by comparing recovery of monkeys kept in the light as opposed to others kept in the dark as well as by studying monkeys that had undergone bilateral occipital lobectomy prior to labyrinthectomy.

2 Methods

Nine juvenile (1 ½–3 years old) rhesus monkeys were subject to labyrinthectomy. Three had undergone a prior bilateral occipital lobectomy (Zee et al. 1987). Eye movements were recorded using the magnetic-field search-coil technique with implanted coils of wire in each eye.

After pre-operative training and testing we performed a complete left-sided labyrinthectomy using a dental drill. The animals were transferred immediately after surgery into a light-proof room and kept in the dark until the first recording session which took place 18–20 h after labyrinthectomy. After recording sponta-

H. Flohr (Ed.)
Post-Lesion Neural Plasticity

neous nystagmus and responses to rotation in darkness the three occipital-lobectomized monkeys and three of the previously intact monkeys (henceforth called light-kept monkeys) were put back into normal light conditions. The other three previously intact monkeys (henceforth called dark-kept monkeys) were kept in complete darkness for a total of 4 days. After first exposure to light each animal was tested 6 h later and then daily for at least 1 week. The dark-kept monkeys were also tested daily prior to their first exposure to light 4 days after labyrinthectomy. Following the initial post-operative period the animals were tested twice a week for 1 week and then at 10–14-day intervals for at least 3 months. Finally, in three monkeys we performed a bilateral occipital lobectomy about 6 months after labyrinthectomy.

Vestibular responses were elicited by rotating with a maximum acceleration of 100 deg/s/s, about the earth-vertical axis, a frame that supported the magnetic-field coils and contained the primate chair. Constant-velocity stimuli were used with different speeds of rotation varying from 30–300 deg/s.

Spontaneous nystagmus in the dark was measured as average slow-phase velocity over 5–10 consecutive beats of nystagmus. Rotational responses were quantified by calculation of the VOR gain as the ratio of peak horizontal slow-phase velocity divided by peak head (chair) velocity. In most instances the gain of the VOR was calculated with a correction for spontaneous nystagmus by subtracting the average slow-phase velocity of the last few beats of spontaneous nystagmus immediately before any change in the rotational stimulus.

3 Results

3.1 Pre-Labyrinthectomy Control Data

Pre-operatively there was a small amount of spontaneous drift, in darkness, with the mean absolute values of individual monkeys ranging from 0–5 deg/s in the horizontal plane. Pre-operatively VOR responses were symmetrical; VOR gain in the dark was about 1.0 for 30–180 deg/s and slightly less than 1.0 for 240 and 300 deg/s.

3.2 Post-Labyrinthectomy Results

Spontaneous Nystagmus: When first measured 18–20 h after labyrinthectomy, the average slow-phase velocity of spontaneous nystagmus in the dark, measured when the eye passed through the primary position in the orbit was similar in all three groups of animals, ranging from 22–62 deg/s in the occipital-lobectomized monkeys, from 24–48 deg/s in the dark-kept monkeys, and from 32–54 deg/s in the light-kept monkeys. Subsequently the level of spontaneous nystagmus, measured in darkness, decreased in a similar fashion in all three groups reaching a level of 4–22 deg/s at day 25 (Fig. 1). A further, slower decrease was observed during the remainder of the observation period (up to 6 months post-labyrinthectomy) to levels between 3 and 10 deg/s.

Recovery of the VOR Gain in the Dark: During the first recording session after labyrinthectomy (before any exposure to light) the value of the gain of the VOR

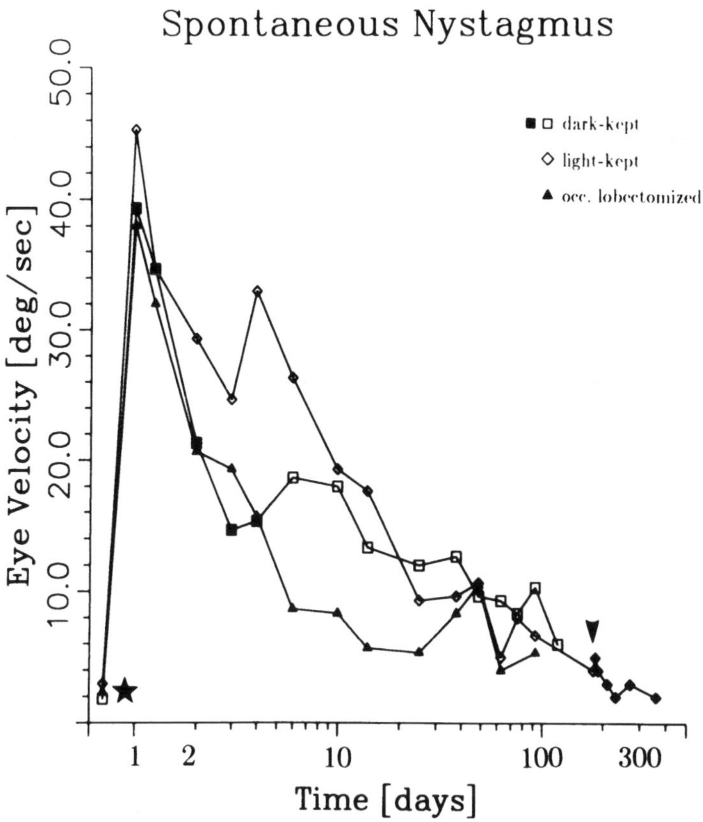

Fig. 1. Mean values of spontaneous nystagmus in the dark pre-(*asterisk*) and post-operatively in the three groups: occipital lobectomized (*triangles*), light-kept (*diamonds*) and dark-kept (squares) monkeys (note: *filled squares* represent dark-kept monkeys while being in the dark for the first 4 days). The time scale is logarithmic. *Day 1* represents one day after labyrinthectomy. The *arrow* indicates the time of occipital lobectomy in three seeing monkeys after they recovered from labyrinthectomy. Note the comparable decrease in spontaneous nystagmus in all groups

in darkness (corrected for spontaneous nystagmus) was low in all animals. For rotations in both directions, at speeds up to 120 deg/s, slow-phase velocity was decreased to about half of the pre-operative value. For higher stimulus velocities slow-phase velocity was about half the pre-operative value for rotations toward the intact side but progressively declined for rotations toward the lesioned side.

Following exposure to light the seeing monkeys showed a characteristic pattern of recovery of VOR gain (Fig. 2). For rotations toward the intact side the increase of the gain was biphasic. In the first phase, lasting about 48–72 h, the gain increased to about 80%–90% of the value that was eventually reached after 3 months. Thereafter the gain increased more slowly to its final value. For rotations toward the lesioned side similar behavior was true for responses to 30 and 60 deg/s stimuli. But for higher velocities the gain increased more gradually throughout the post-operative observation period. This rate of recovery was approximately exponential with a time constant of 3–4 weeks.

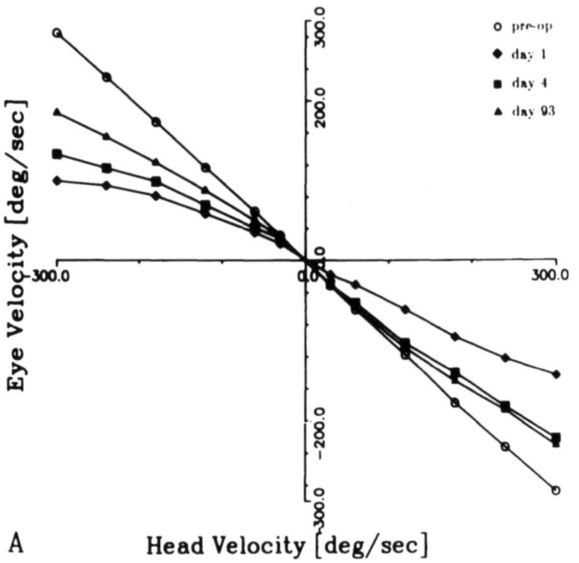

A

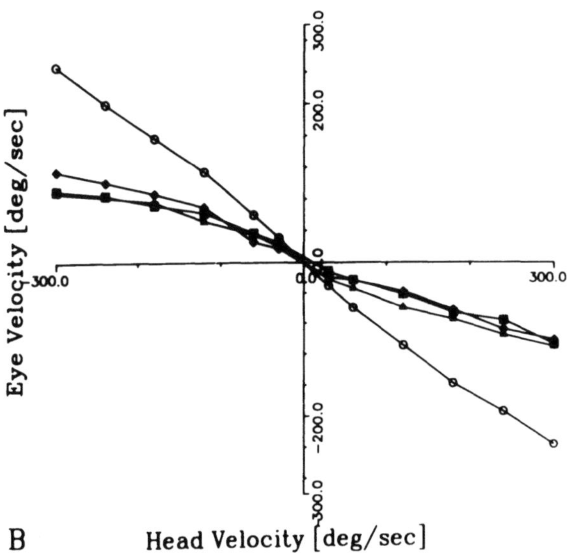

B

Fig. 2. Recovery of slow-phase velocity in response to constant-velocity rotations (30–300 deg/s) in both directions. The mean of six seeing monkeys (**A**) and of three occipital-lobectomized monkeys (**B**) are shown. Data from pre-(*circles*) and post-operative day 1 (*diamonds*), day 4 (*squares*) and day 93 (*triangles*). In seeing monkeys (**A**) note that for rotations toward the intact side (positive head velocities) slow-phase velocity dropped to about 50% of the pre-operative values at all stimulus speeds. For rotations toward the lesioned side (negative head velocities) slow-phase velocity dropped to about 50% for stimulus speeds of 30 to 120 deg/s, and to much less than 50% for higher speeds. In the occipital-lobectomized monkeys, slow-phase velocity initially dropped to about 30–40% of the pre-operative values. Occipital-lobectomized monkeys showed minimal improvement in their VOR, primarily for rotations toward the intact side but only at low velocities.

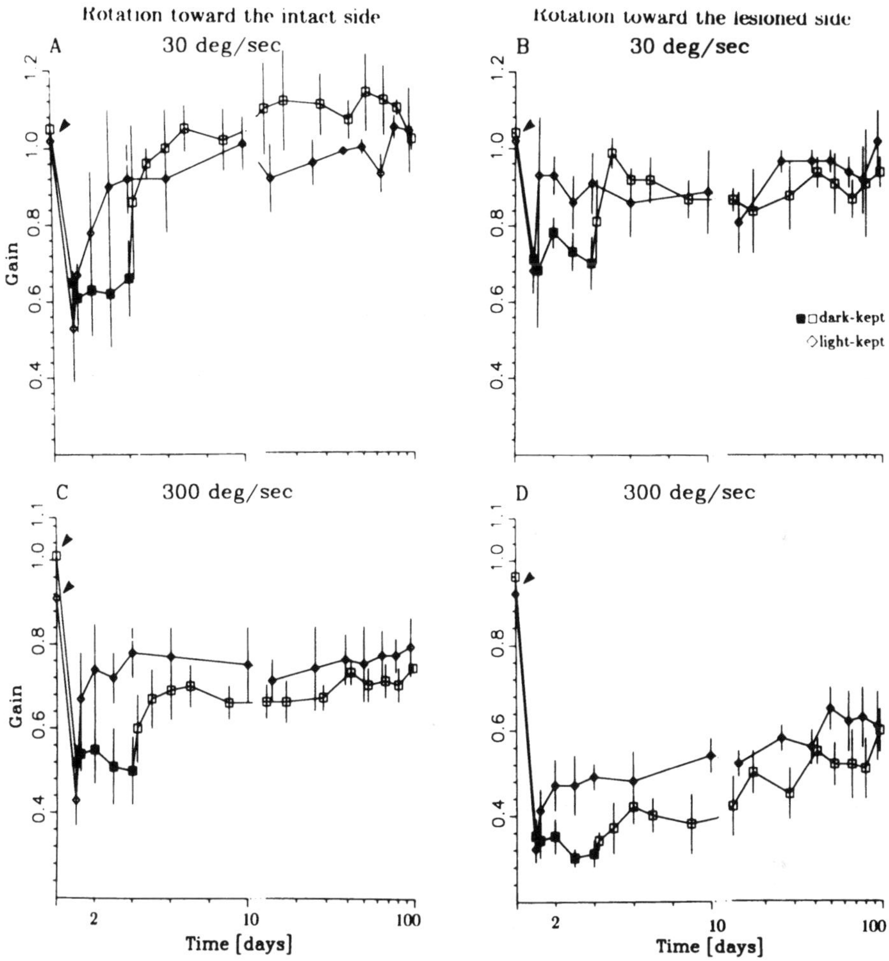

Fig. 3. Mean values +/−1 standard deviation of VOR gain in the dark for the dark-kept (*squares*) and the light-kept (*diamonds*) monkeys. Values for rotations toward the intact side (**A, C**) and toward the lesioned side (**B, D**) are shown. Pre-operative values are nearest the ordinate and denoted by an *arrow*. Note the delay in increase in the VOR gain in the dark-kept monkeys until after light exposure. For rotations toward the lesioned side at the higher velocities, the rate of increase in VOR gain in the dark-kept group was lower and the values smaller. The time scale changes from linear to logarithmic at day 10

In the dark-kept group recovery of the VOR was delayed until the animals were exposed to light (Fig. 3). Then, for rotations toward the intact side the pattern of increase in VOR gain was similar to that of the light-kept group though slightly less complete at high velocities. For rotations toward the lesioned side, at stimulus velocities higher than 60 deg/s, the rate of recovery was considerably lower than that of the light-kept monkeys.

On the other hand, the previously occipital-lobectomized monkeys showed little recovery of the VOR gain. There was only moderate improvement for low-velocity rotations (30–60 deg/s) due to an increase in response to rotations toward the intact side (Fig. 2).

Finally, in three animals an occipital lobectomy was performed about 6 months after labyrinthectomy. There was no recurrence of the spontaneous nystagmus that had been present immediately after labyrinthectomy. However, over the course of 30 days, each animal showed a decrease in VOR gain approaching the value found just after labyrinthectomy and prior to any light exposure.

4 Discussion

Elimination of Spontaneous Nystagmus

The results of the experiments in which we compared resolution of spontaneous nystagmus after unilateral labyrinthectomy in light-kept, in dark-kept and in occipital-lobectomized monkeys were straightforward. Visual experience was not necessary for the acquisition or the maintenance of the compensation that eliminated the spontaneous nystagmus created by unilateral labyrinthectomy. When first measured at 18–20 h post-labyrinthectomy the level of spontaneous nystagmus was much less than that which had been elicited pre-operatively with ice-water caloric stimulus. In other words, considerable rebalancing had already taken place before any exposure to light. Furthermore, some tonic discharge must already have been restored in the vestibular nuclei on the lesioned side since with high velocity rotations a nystagmus with slow phases directed toward the intact side could be elicited. Furthermore, the rate of disappearance of spontaneous nystagmus was not retarded in dark-kept or in occipital-lobectomized animals and spontaneous nystagmus did not reappear when occipital lobectomy was performed after compensation for the labyrinthectomy had already taken place.

These results are largely in accord with those reported from other species (Smith et al. 1986). There is some suggestion that in cats the spontaneous nystagmus may subside more quickly if exposure to light is permitted (Courjon et al. 1977) but even in that species most of the nystagmus disappeared even when the animals were kept in complete darkness after labyrinthectomy.

4.2 Correction of VOR Gain

On the other hand, our results suggest that in monkeys visual experience is critical for the acquisition (and probably the maintenance) of the compensation for the abnormally reduced VOR gain created by unilateral labyrinthectomy. There was no increase in the gain of the VOR until exposure to light and then recovery began promptly whether or not the animals had been previously kept in the dark. These findings are in accord with the results of similar experiments performed in cats (Courjon et al. 1977).

We also found that the compensatory changes in the gain of the VOR that occurred after labyrinthectomy were slowly lost when a subsequent occipital lobectomy was performed. In these animals the occipital lobectomy per se was unlikely to have been responsible for the decrease in the gain of the VOR, since little alteration in the gain or symmetry of the VOR were produced by occipital lobectomy in animals that had undergone no prior vestibular lesion (Fetter et al. 1988). Taken together, these findings suggest that the VOR gain compensation is a dynamic process that requires visual experience for its acquisition and is dependent upon the posterior cerebral hemispheres for its maintenance.

One must ask why visual experience is necessary for compensation for dynamic vestibular performance but not for static imbalance. It may be that deafferentation of the vestibular nucleus initiates alterations in the intrinsic properties of the vestibular neurons themselves, denervation hypersensitivity, or sprouting of extralabyrinthine afferents. Each process might lead to an increase in the level of the spontaneous activity of neurons on the lesioned side. On the other hand, ocular gaze stability is so crucial for normal visual function that nonvisual error signals – perhaps based on proprioception or efference copy – may also stimulate adaptive networks to restore ocular motor balance. This could take place either in or at a level beyond the vestibular nuclei.

Finally, our results indicate that the rate and possibly the ultimate degree of recovery of the gain of the VOR may have been lower in the animals kept in darkness for 4 days prior to light exposure than in the animals allowed immediate exposure to light. Lacour and Xerri (1981) reported an analogous phenomenon for both cats and monkeys in their study of compensation of otolith-induced vestibulo-spinal reflexes after labyrinthectomy. Adaptation was slower and less complete in animals that had been transiently immobilized immediately after unilateral labyrinthectomy. It was suggested that there is a "critical period" immediately after a vestibular lesion during which it is necessary to have the appropriate motor activity and sensory feedback for optimal compensation to take place. If such activity does not take place in this period after the lesion – even if allowed later – the rate and perhaps the degree of recovery will be limited.

Such a phenomenon was suggested by our data but only for rotations at high velocities especially toward the lesioned side. Longer periods of visual deprivation post-labyrinthectomy may be necessary to demonstrate whether or not a "critical period" truly exists for compensation of dynamic VOR performance in monkeys. Of course, if it does, this would have important clinical implications for the treatment of human patients who have suddenly lost vestibular function. Early and vigorous physical activity in lighted circumstances should be encouraged to promote the maximal degree of recovery of function.

Acknowledgements. This research was supported by grant Fe 237/1-1+1-2 (M. F.) from the Deutsche Forschungsgemeinschaft and research grant EYO1849 (D. S. Z.) from the National Eye Institute, N. I. H., U. S. Public Health Service. Core Grant EYO1765.

References

Courjon JH, Jeannerod M, Ossuzio I, Schmid R (1977) The role of vision in compensation of vestibulo ocular reflex after hemilabyrinthectomy in the cat. Exp Brain Res 28:235–248

Dichgans J, Bizzi E, Morasso P, Tagliasco V (1973) Mechanisms underlying recovery of eye-head coordination following bilateral labyrinthectomy in monkeys. Exp Brain Res 18:548–562

Fetter M, Zee DS, Proctor LR (1988) Effect of lack of vision and of occipital lobectomy upon recovery from unilateral labyrinthectomy in rhesus monkey. J Neurophysiol 59:394–407

Gonshor A, Melvill Jones G (1976) Extreme vestibulo-ocular adaptation by prolonged optical reversal of vision. J Physiol (Lond) 256:381–414

Lacour M, Xerri C (1981) Vestibular compensation: new perspectives. In: Flohr H, Precht W (eds) Lesion-induced neuronal plasticity in sensorimotor systems. Springer, Berlin Heidelberg New York, pp 240–253

Lisberger SG, Miles FA, Optican LM (1983) Frequency-selective adaptation: evidence for channels in the vestibulo-ocular reflex? J Neurosci 3:1234–1244

Paige GD (1983) Vestibuloocular reflex and its interactions with visual following mechanisms in the squirrel monkey. II. Response characteristics and plasticity following unilateral inactivation of horizontal canal. J Neurophysiol 49:152–168

Smith PF, Darlington CL, Curthoys IS (1986) The effect of visual deprivation on vestibular compensation in the guinea pig. Brain Res 364:195–198

Zee DS, Tusa RJ, Herdman SJ, Butler PH, Guecer G (1987) Effects of occipital lobectomy upon eye movements in primate. J Neurophysiol 58:883–907

Vestibular Compensation Aspects of Time Course and De-Compensation

K.-P. Schaefer [1] and D. L. Meyer [2]

[1] Dept. of Psychiatry and [2] Dept. of Anatomy, University of Göttingen, D-3400 Göttingen, FR Germany

1 Introduction

Descriptions of labyrinthectomy symptoms date back as far as the 19th century (Flourens 1824; Goltz 1870; Cyon 1874; Bornhardt 1876; Spamer 1880; Bechterew 1883; Ewald 1892). By the turn of the century it was well established that most of the lesion-induced symptoms are gradually compensated (eliminated); comparative studies on various vertebrate species had been carried out, and Bechterew's (1883) ingenious experiments involving subsequent lesions of both labyrinths had been a major breakthrough with regard to an understanding of compensatory mechanisms.

Despite a significant number of investigations carried out since then, our concept of what really occurs in the CNS after a labyrinthine lesion has been inflicted is still relatively vague (Flohr et al. 1985).

Symptoms strikingly similar to the ones caused by unilateral vestibular lesions can, in some species, be induced by unilateral ocular enucleation (Lyon 1909; Greene and Laurens 1923; Main 1928; von Holst 1954; von Holst and Schoen 1954; Jahn 1960). Such (asymmetry) symptoms are compensated in much the same way as the ones resulting from unilateral vestibular lesions.

Considering the well-established interactions between visual and vestibular systems, it is tempting to assume that the circuits involved in generating symptoms caused by either kind of lesion, as well as their compensation, might be identical (see Meyer and Bullock 1977). Many of the papers on compensation of postural abnormalities resulting from unilateral visual and vestibular lesions provide data on the time course of compensation which differs with regard to the different symptoms induced (Schaefer and Meyer 1974), with the age of the specimens studied (Schaefer and Meyer 1973; Horn and Rayer 1978; Horn 1981), due to additional CNS lesions (Schaefer and Meyer 1974; Schaefer et al. 1979), under the influence of drugs (Schaefer and Wehner 1966; Schaefer and Meyer 1973, 1974, 1981; Horn et al. 1979; Flohr and Lüneburg 1982; Flohr et al. 1985), etc. Only a few papers have concerned themselves with the dynamics of the compensatory process and their manipulation by experimental (surgical, pharmacological, etc.) interference. Yet another aspect neglected by many investigators is the one of temporary relapses of labyrinthectomy symptoms for unknown reasons (Magnus and de Kleijn 1913; T'ang and Wu 1936; Schaefer and Meyer 1973; Putkonen et al. 1977; Sirkin et al. 1984), as well as, but to a lesser degree, due to the effect of

H. Flohr (Ed.)
Post-Lesion Neural Plasticity

pharmacological agents (Schaefer et al. 1978; Bienhold and Flohr 1980; Abeln et al. 1981; Bienhold et al. 1981; Flohr et al. 1981; Schaefer and Meyer 1981; Flohr et al. 1985).

It is the purpose of this contribution to present and discuss some data that strongly indicate that a thorough analysis of the phenomena cited may be of relevance for an understanding of vestibular compensation in general.

2 Material and Methods

The experiments reported were carried out on a total of ca. 200 adult guinea pigs. Labyrinthectomy was performed by an injection of chloroform into the middle ear (see Schaefer and Meyer 1973). Unpublished control experiments have demonstrated that symptoms and compensation are identical to those caused by surgical labyrinthectomy.

Drugs administered to specimens were injected i.p., except for ether and paraoxon. Ether was inhaled; it was applied to induce anesthesia during labyrinthectomy and as a test substance (see below). Paraoxon injections were performed s.c.

3 Results and Discussion

3.1 Time Courses of Compensation

It has long been known that different time courses apply to compensation of different labyrinthectomy symptoms (see Schaefer and Meyer 1974); furthermore,

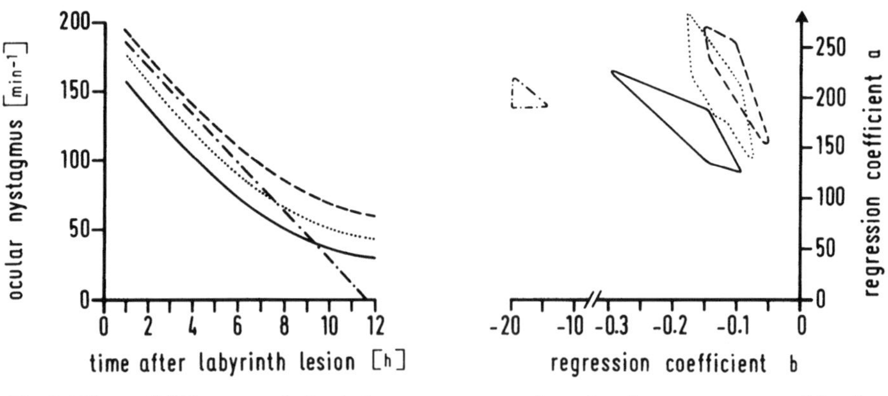

Fig. 1. Effects of different cerebellar lesions on compensation of ocular nystagmus resulting from destruction of right labyrinths in guinea-pigs (N=8 for each kind of lesion and for control group). Time courses of decrease of nystagmus frequency was fitted to either an exponential ($y=ae^{bx}$) or a linear ($y=a+bx$) function for each specimen. Averaged curves are given (*left*). *Right side* shows envelopes of distribution of regression coefficients (*a* initial value; *b* slope).
Note: ablation of right cerebellar hemisphere (---) and combined exstirpation of left flocculus and paraflocculus (·····) do not change "shape" of compensation curve as compared with controls (——). In contrast, combined ablation of right flocculus and paraflocculus result in compensation curve's best fitting function being linear (–·–·–)

it has been established that these are differently affected by surgical and pharmacological interference with the ongoing compensation process (Schaefer and Meyer 1973). Figures 1 and 2 give examples of surgically induced changes of time course of nystagmus compensation. Apart from the length of the time-span until a steady state is reached, the shape of the curves given appears to be of relevance to us. The following arguments are to emphazise this point.

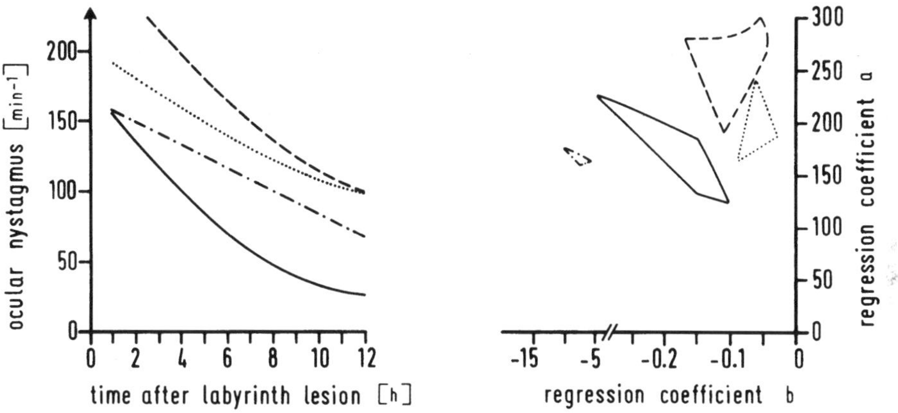

Fig. 2. Plotting procedure as described for Fig. 1 (——) controls; (– – –) ablation of left cerebellar hemisphere; (·····) complete cerebellar exstirpation; (–·–·–) ablation of caudal vermis. Depending on kind of lesion compensation of nystagmus due to right labyrinthectomy follows different time courses. These are best fitted by a linear function in case of caudal vermis lesioning, and exponentially approach values higher than normal in the other two cases given

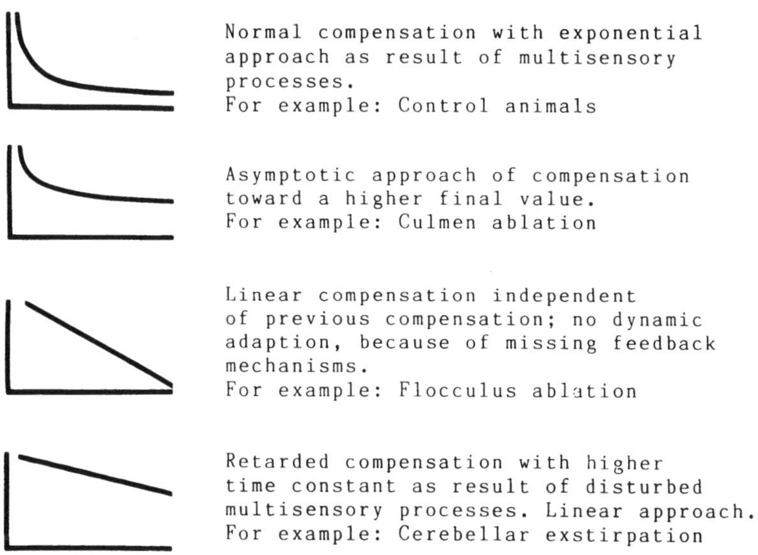

Fig. 3. Different time courses of compensatory processes after labyrinthine lesions (examples given relate to compensation of head deviation)

In our 1973 article we presented results demonstrating that young guinea pigs recover faster than adults from a unilateral labyrinthectomy. Similar findings were later reported by Horn and Rayer (1978), who worked on amphibians. Our data plots clearly show that the reasons for this phenomenon are unrelated to a higher plasticity of the juvenile brain (as concluded by Horn and Rayer). Rather, both "compensation curves" run perfectly parallel to each other; the reason for compensation being completed faster in juveniles is simply due to the fact that the original (shortly after labyrinthectomy) symptoms are more pronounced in adults. (This is likely to result from a higher spontaneous discharge rate in the adult labyrinth).

Hence, the exact features of the time course of compensation can apparently contain useful information. Reviewing plots of compensation of labyrinthectomy symptoms obtained from animals additionally treated with chemical agents or subjected to CNS lesion prior or subsequently to the vestibular lesion, revealed that four basic forms of compensation time courses exist (Fig. 3). At this point we can only speculate about their individual significance. One of the tempting (but not necessarily conclusive) assumptions is that a linear decrease of symptom strength may suggest that feedback loops have been disrupted. The important role of such loops has been previously emphazised by Flohr et al. (1981).

3.2 Chemically Induced De-Compensation

Descriptions of compensated labyrinthectomy symptoms reappearing due to various subsequent CNS lesions (DiGiorgio 1939; Menzio 1949; Carpenter et al. 1959; Azzena 1969; Schaefer and Meyer 1973, 1974; Llinas et al. 1975; Bienhold and Flohr 1978; Llinas and Walton 1979), in response to pharmacological treatment (Schaefer and Meyer 1973, 1974; Schaefer et al. 1978; Bienhold and Flohr 1980; Bienhold et al. 1981; Flohr et al. 1985), or just due to lifting the animals off the ground and thus diminishing proprioceptive afferents (Schaefer and Meyer 1974) have been numerous.

We have studied the effect of 15 chemical agents with regard to their potency to induce de-compensation in guinea pigs that had been subjected to unilateral labyrinthectomy from 1 day to 6 months before. The specimens were investigated for reoccurring nystagmus, head deviation in the horizontal plane, and head turning about the longitudinal axis (see: Schaefer and Meyer 1973, 1974). Results obtained under the influence of Paraoxon (3 mg/kg) shown in Fig. 4 may serve as an example (a detailed description of findings will be given elsewhere).

Two major conclusions were reached on the basis of this set of experiments:

Firstly, there are substances (e.g., alcohol, ether) which can induce significant de-compensation only during the first ca. 10 days. The question arises what changes in the compensating brain at this time. Other substances will also be effective 6 months after labyrinthectomy (e.g., Paraoxon).

Secondly, induced de-compensations may differ significantly with regard to the three labyrinthectomy symptoms considered and mentioned above. For example, chlorpromazine (20–40 mg/kg) causes head turning and head deviation (see above) to reappear, but no nystagmus can be noted. This is in contrast to the effects of Paraoxon, which causes all symptoms, including nystagmus, to be tem-

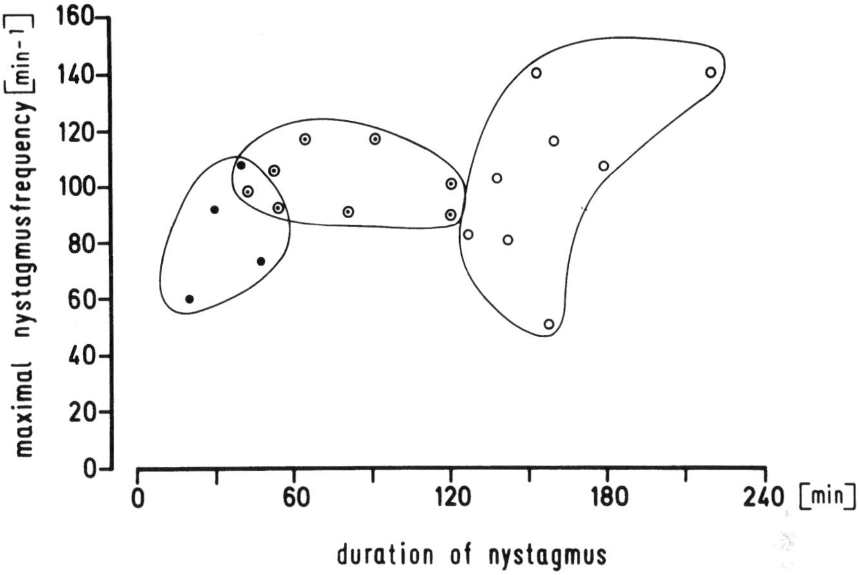

Fig. 4. Ocular nystagmus due to unilateral labyrinthectomy can be re-elicited (released) by cholinesterase inhibitor Paraoxon (E 600, 3 mg/kg) in specimens that have previously completed compensation. Plotted are max. frequency of nystagmus vs. duration of its occurance in animals labyrinthectomized 2 d (○), 8 d (⊙), and 28 days (●) earlier. Data indicate an ongoing stabilization of compensatory mechanisms even after lesions-induced symptoms have ceased to be present

porarily "released". We consider such patterns as indicating where the different substances may mainly act in the CNS. Hence, we think that conclusions on mechanisms/structures involved in the compensation process can be drawn from this kind of evidence. The effects of chlorpromazine on compensated animals are strikingly similar to those observed after depriving the specimens of somatosensory input (see: Schaefer and Meyer 1974), whereas Paraoxon exerts influences that suggest a direct interference with vestibular structures. As proposed by Magnus in 1924, vestibular compensation is unlikely to be achieved by vestibular structures only, rather a significant portion of the brain seems to be involved. Different parts of the CNS may take over different roles and take care of different lesion-induced "subproblems". Thus, selective interference with their function may cause a particular pattern of the various "releasable" labyrinthectomy symptoms to be present in compensated animals.

4 Concluding Remarks

In connection with the data of previous studies (cited above) we consider the findings reported here as suggesting:

1. While investigating vestibular compensation, more attention should be paid to the time course (meaning: analysis of "shape" of compensation curves, rather than just being interested in a possible acceleration or retardation) of compen-

sation induced by various experimental manipulations (surgical, pharmacological, etc.).
2. More thorough investigations involving chemical agents that are capable of eliminating/disturbing a completed compensation might significantly facilitate our understanding of the phenomenon under consideration.

Furthermore, we feel that the lack of comparable studies on compensation of postural asymmetries induced by unilateral enucleations is a significant handicap; this holds, in particular, when trying to determine to what extent vestibular compensation may serve a model for how the CNS handles lesions of itself.

5 Summary

The paper describes results pertinent to the ideas, that the time course of vestibular compensation, as well as the patterns of pharmacologically "released" (decompensated) labyrinthectomy symptoms may contain more information on the nature of compensatory mechanisms than hitherto recognized.

Acknowledgement. This work was supported by grants of the German Science Foundation to both authors.

References

Abeln W, Bienhold H, Flohr H (1981) Influence of cholinomimetics and cholinolytics on vestibular compensation. Brain Res 222:458–462

Azzena GB (1969) Role of the spinal cord in compensating the effects of hemilabyrinthectomy. Arch Ital Biol 107:43–53

Bechterew W von (1883) Ergebnisse der Durchschneidung des N. acusticus, nebst Erörterung der Bedeutung der semicirculären Canäle für das Körpergleichgewicht. Pflügers Arch 30:312–347

Bienhold H, Flohr H (1978) Role of commissural connections between vestibular nuclei in compensation following unilateral labyrinthectomy. J Physiol (Lond) 284:178

Bienhold H, Flohr H (1980) Role of cholinergic synapses in vestibular compensation. Brain Res 195:476–478

Bienhold H, Abeln W, Flohr H (1981) Drug effects on vestibular compensation. In: Flohr H, Precht W (eds) Lesion-induced neuronal plasticity in sensorimotor systems. Springer, Berlin Heidelberg New York, pp 265–273

Bornhardt A (1876) Experimentelle Beiträge zur Physiologie der Bogengänge des Ohrlabyrinthes. Pflügers Arch 12:471–521

Carpenter MB, Fabrega H, Glinsmann W (1959) Physiological deficits occuring with lesions of labyrinth and fastigial nuclei. J Neurophysiol 22:222–234

Cyon E (1874) Über die Funktion der halbzirkelförmigen Kanäle. Pflügers Arch 8:306–340

DiGiorgio AM (1939) Effetti di lesioni unilaterali della corteccia cerebrale sui fenomeni di compenso da hemislabirintazione. Atti Accad Fisiol Fac Med Siena 11:382–384

Ewald JR (1892) Physiologische Untersuchungen über das Endorgan des N. octavus. Bergmann, Wiesbaden

Flohr H, Lüneburg U (1982) Effects of ACTH on vestibular compensation. Brain Res 248:169–173

Flohr H, Bienhold H, Abeln W, Macskovics I (1981) Concepts of vestibular compensation. In: Flohr H, Precht W (eds) Lesion-induced neuronal plasticity in sensorimotor systems. Springer, Berlin Heidelberg New York, pp 153–172

Flohr H, Abeln W, Lüneburg U (1985) Neurotransmitter and neuromodulator systems involved in vestibular compensation. In: Berthoz A, Melvill Jones G (eds) Adaptive mechanisms in gaze control, vol 1. Elsevier, Amsterdam, pp 269–277

Flourens M (1824) Recherches expérimentales sur les propriétés et les fonctions du système nerveux dans les animaux vertèbres. Crevot, Paris

Goltz F (1870) Über die physiologische Bedeutung der Bogengänge des Ohrlabyrinthes. Pflügers Arch 3:172–193

Greene WF, Laurens H (1923) The effect of extirpation of the embryonic ear and eye on equilibrium in *Amblyostoma punctatum*. Am J Physiol 64:120–143

Holst E von (1954) Einfluß der nichtbelichteten Retina auf das Gleichgewichtsverhalten von Fischen. Naturwissenschaften 41:507–508

Holst E von, Schoen L (1954) Der Einfluß mechanisch veränderter Augenstellungen auf die Richtungslokalisation bei Fischen. Z Vgl Physiol 36:433–442

Horn E (1981) An ontogenetic approach to vestibular compensation mechanisms. In: Flohr H, Precht W (eds) Lesion-induced neuronal plasticity in sensorimotor systems. Springer, Berlin Heidelberg New York, pp 173–183

Horn E, Rayer B (1978) Compensation of vestibular lesions in relation to development. Naturwissenschaften 65:441

Horn E, Greiner B, Horn I (1979) The effect of ACTH on habituation of the turning reaction in the toad *Bufo bufo* L. J Comp Physiol 131:129–135

Jahn T (1960) Optische Gleichgewichtsregelung und zentrale Kompensation bei Amphibien, insbesondere bei der Erdkröte (Bufo bufo L.) Z Vgl Physiol 43:119–140

Llinas R, Walton K (1979) Place of the cerebellum in motor learning. In: Brazier MAB (ed) Brain mechanisms in memory and learning: from the single neuron to man, IBRO monograph series 4. Raven, New York, pp 17–36

Llinas, R, Walton K, Hillman DE, Sotelo C (1975) Inferior olive: its role in motor learning. Science 190:1230–1231

Lyon EP (1909) On rheotropism. II. Rheotropism of fish blind in one eye. Am J Physiol 24:244–251

Magnus R, Kleijn A de (1913) Analyse der Folgezustände einseitiger Labyrinthexstirpation mit besonderer Berücksichtigung der Rolle der tonischen Halsreflexe. Pflügers Arch 154:178–306

Main RJ (1928) Phototropism in fishes, and its relation to the results obtained by eye dislocation. Z Vgl Physiol 7:611–616

Menzio P (1949) Rapporti fra la corteccia cerebrale ed i fenomeni di emislabirintazione. Arch Fisiol 49:97–104

Meyer DL, Bullock TH (1977) The hypothesis of sense-organ-dependent tonus mechanisms. History of a concept. Ann N Y Acad Sci 290:3–17

Putkonen PTS, Courjon JH, Jeannerod M (1977) Compensation of postural effects of hemilabyrinthectomy in the cat. A sensory substitution process? Exp Brain Res 28:249–257

Schaefer K-P, Meyer DL (1973) Compensatory mechanisms following labyrinthine lesions in guinea pigs. A simple model of learning. In: Zippel HP (ed) Memory and transfer of information. Plenum, New York, pp 203–232

Schaefer K-P, Meyer DL (1974) Compensation of vestibular lesions. In: Kornhuber HH (ed) Handbook of sensory physiology, vol VI/2. Springer, Berlin Heidelberg New York, pp 463–490

Schaefer K-P, Meyer DL (1981) Aspects of vestibular compensation in guinea pigs. In: Flohr H, Precht W (eds) Lesion-induced neuronal plasticity in sensorimotor systems. Springer, Berlin Heidelberg New York, pp 197–207

Schaefer K-P, Wehner H (1966) Zur pharmakologischen Beeinflussung zentralnervöser Kompensationsvorgänge nach einseitiger Labyrinthausschaltung durch Krampfgifte und andere erregende Substanzen. Naunyn-Schmiedebergs Arch Pharmakol Exp Pathol 254:1–17

Schaefer K-P, Wilhelms G, Meyer DL (1978) Der Einfluß von Alkohol auf die zentralnervösen Ausgleichsvorgänge nach Labyrinthausschaltung. Z Rechtsmed 81:249–260

Schaefer K-P, Meyer DL, Wilhelms G (1979) Somatosensory and cerebellar influences on compensation of labyrinthine lesions. In: Granit R, Pompeiano O (eds) Reflex control of posture and movement. Progress in brain research, vol 50. Elsevier, Amsterdam, pp 591–598

Sirkin DW, Precht W, Courjon J-H (1984) Initial, rapid phase of recovery from unilateral vestibular lesion in rat not dependent on survival of central portion of vestibular nerve. Brain Res 302:245–256
Spamer C (1880) Experimenteller und kritischer Beitrag zur Physiologie der halbkreisförmigen Kanäle. Pflügers Arch 21:479–590
T'ang Y, Wu CF (1936) The effects of unilateral labyrinthectomy in the albino rat. Chin J Physiol 10:571–598

Cervico-Ocular Reflex After Labyrinthine Damage

V. E. Pettorossi [1], P. Errico [2], A. Ferraresi [2], R. Fedeli [2], and R. M. Santarelli [2]

[1] Institute of Human Physiology, Università di Perugia, Via del Giochetto, Perugia, Italy

[2] Institute of Human Physiology, Università Cattolica S. Cuore, Largo F. Vito, Roma, Italy

1 Introduction

The effects of peripheral vestibular damage on eye motility consist of both eye nystagmus and impairment of vestibulo-ocular reflexes (VOR's). Although the compensation of nystagmus is complete in the light (Schaefer and Meyer 1973), the recovery of the VOR was less than complete after unilateral and bilateral labyrinthine lesions (Baarsma and Collewijn 1975; Barmack and Pettorossi 1981; Maioli et al. 1983; Paige 1985). In the rabbit, the gain of the VOR remained much less than half of normal after hemilabyrinthectomy (HL) (Baarsma and Collewijn 1975; Barmack and Pettorossi 1981). Thus, in the presence of a labyrinthine lesion a large deficit would remain in the rabbit's ability to maintain a stable retinal image unless other reflexes may reduce it through a substitution process. Since the optokinetic reflex (OKR) slightly improved the oculomotor reactions (Baarsma and Collewijn 1975) the cervico-ocular reflex (COR) could play a role in the functional recovery. In fact it has been demonstrated that in normal rabbits the COR (1) contributes to maintain gaze stability; (2) sets the range of reflex eye movements toward a more compensatory position depending on the head-body angle (Meiry 1971; Barnes and Forbat 1979; Barmack et al. 1981; Barmack et al. 1985); and (3) undergoes plastic changes in the presence of labyrinthine lesions (Dichgans et al. 1974; Kasai and Zee 1978; Baker et al. 1982). The COR is agonistic to VOR and OKR in compensating for horizontal head displacements by enhancing the gain and decreasing the phase lead of HVOR at low frequencies of stimulation (0.005–0.05 Hz) (Barmack et al. 1981). Cervical input can also maintain the range of the eye movements in a compensatory range by delaying and reducing the anticompensatory vestibular and optokinetic fast phases (AFP's) (Barmack et al. 1985). In addition to these immediate effects, a potentiation of the COR gain has been shown in cats (Baker et al. 1982), in monkeys (Dichgans et al. 1974) and in human beings (Kasai and Zee 1978), following labyrinthine impairment. Furthermore the balancing process concerning the ocular asymmetries following hemilabyrinthectomy is influenced by the cervical input (Pettorossi and Petrosini 1984).

In this study we examined the questions of whether cervical input can substitute for vestibular function in stabilizing gaze and whether it can still control the range of reflex eye movements even in the absence of AFP's. In order to answer these questions, the COR was studied in rabbits which underwent two different

H. Flohr (Ed.)
Post-Lesion Neural Plasticity

labyrinthine lesions: bilateral labyrinthectomy (BL) and bilateral horizontal semicircular canal plugging (BP). The immediate consequences of labyrinthectomy are: (1) the loss of modulation of secondary neurons by vestibular primary afferent signals; (2) the decrease in spontaneous primary vestibular afferent input to secondary vestibular neurons with the consequent reduction of spontaneous secondary vestibular activity; and (3) the abolition of otolithic information. On the other hand, semicircular canal plugging causes deficits in reflex eye movements without altering the spontaneous activity of the vestibular neurons, the vertical VOR's and the otolithic function. The results were that in both groups of impaired animals the COR showed a gain increase at high frequency of stimulation, but the ability to maintain the direction of the responses was definitively lost in BL animals. However, eye position and the range of reflex eye movements were still controlled by means of fast phases of cervical origin.

2 Methods

2.1 Preliminary Surgical Procedure

Six pigmented rabbits (1–2 kg) were anaesthetized with ketamine hydrochloride (Ketalar, Parke-Davis, 50 mg/kg) and diazepam (Valium, Roche, 0.5 mg/kg i.m.). Two head-restraining screws were fixed to the skull by means of dental cement and four smaller screws.

2.2. Labyrinthine Lesions

Under ketamine and diazepam anaesthesia (see above), the middle ear was opened under a dissecting microscope and the tympanic membrane was removed with the malleus to expose the bony surfaces of the horizontal and anterior semicircular canals.

For the plugging operation, a small opening was drilled in the bony wall of the lateral semicircular canal, leaving the membranous canal intact. A small silver spindle was gently inserted into the opening, which was then sealed with bone wax (Barmack and Pettorossi 1981).

Labyrinthectomy was obtained by drilling through the ampullae of the anterior and lateral semicircular canals and extending the opening through the ampullo-utricular duct to destroy the posterior semicircular canal and the otolithic receptors. By this technique Scarpa's ganglion was undamaged.

The adequacy of the surgical procedures was confirmed postoperatively by the VOR tests. A 90% decrease in the HVOR gain, normal vertical VOR (VVOR) and symmetric ocular responses indicated the accuracy of the bilateral horizontal canal plugging, while the disappearance of all VOR's assured the completeness of the bilateral labyrinthectomy.

2:3 Vestibular and Cervical Stimulations

Animals were placed on a servo-controlled three-axial turntable, with the body firmly encased in a plastic tube and the head secured by the previously implanted screws to a stereotaxic frame. Cervical stimulations were given by horizontally rotating the body on the fixed head with the center of rotation passing through the junction of the C1 and C2 vertebrae. Vestibular stimulations were provided by oscillating the whole animal around a vertical axis passing through the head. All tests were performed in the dark. Sinusoidal stimulations at constant amplitude (20 ° peak to peak) over a frequency range of 0.01–0.8 Hz and step stimulations (amplitude: 20 °; velocity: 40 °–60 °/s) were utilized. Table movements were measured by means of servo-potentiometers.

2.4 *Eye Position Recording*

Eye position was monitored with an infrared light projection technique. The eye was topically anaesthetized (Novesina, Wander) and a small suction cup bearing an infrared light emitter diode (LED) was attached to the cornea. The LED projected a small beam of infrared light onto a photosensitive X–Y position detector (SC50, UDT) fixed relative to the head and located 5 mm from the tip of the LED. X and Y signals were sent to a Grass polygraph, a VUKO oscilloscope and stored on magnetic tape.

The system was calibrated by moving the LED on a model of the rabbit eye through known angular displacements and was found to be linear to within 5% for eye movements of ± 15 °.

Gain values were computed by comparing the amplitude of the reflex eye movement with the amplitude of the table movement. The phase of the reflex (eye position:body position) was measured at each cycle of stimulation. Gains and phases were compared by t-tests.

3 Results

3.1 *Pre-Lesion HCOR, HVOR and HVOR+HCOR*

All the animals were tested before the labyrinthine damaging by sinusoidally oscillating in the horizontal plane (0.01–0.8 Hz, 20 ° peak to peak) the body on the fixed head (HCOR), the body and the head as a unit (HVOR), and, finally, the head on the fixed body (HVOR+HCOR). Gains and phases of these reflexes as a function of the frequency are reported in Fig. 1. At lower frequencies the COR contribution to gaze stability was greater than that of HVOR. HCOR gain declined and the phase showed an increasing lag with increasing frequency above 0.05 Hz. At frequencies higher than 0.1 Hz, the COR gain, even if very low, remained almost constant and the phase lag progressively decreased. The orientation of the cervico-ocular responses was found to be spatially fixed in the horizontal plane at various head pitch angles (Figs. 2, 3, pre-BL).

As regards the eye responses to step stimulation, the COR was characterized by an early low gain slow compensatory response followed by a late high gain (0.4–0.5) compensatory fast phase (CFP) (Fig. 4). The VOR consisted in a slow compensatory response interrupted by a large AFP. By combining the cervical and the vestibular stimulation the slow compensatory responses were almost the same as those of VOR, but the AFP's were delayed and shortened (Fig. 4, pre-

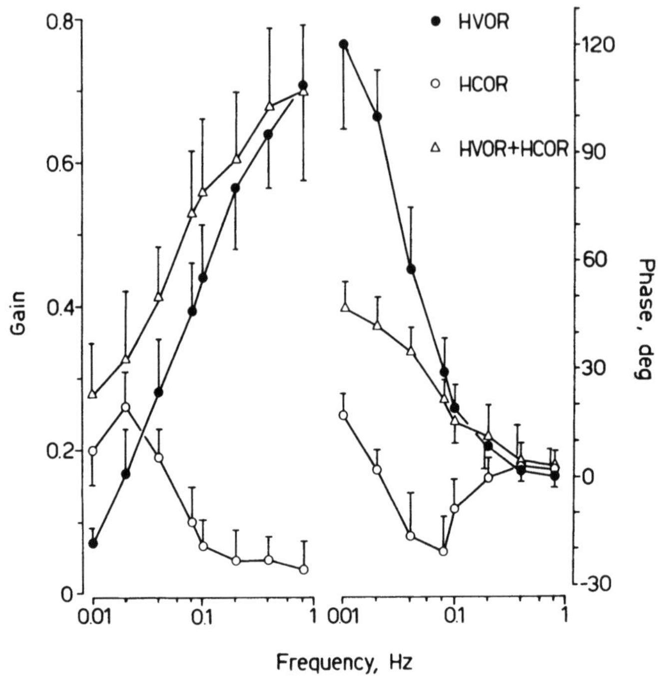

Fig. 1. Comparison of the phase and gain of the HVOR, HCOR and HVOR + HCOR. The mean gain and phase (re180 °) as a function of the frequency is illustrated for six rabbits. *Bars* indicate standard deviations for each data point

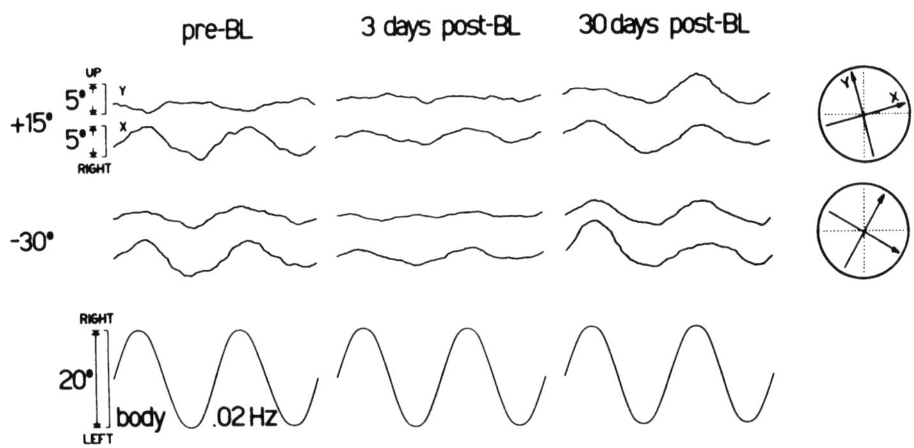

Fig. 2. Effects of bilateral labyrinthectomy (BL) in the HCOR responses at +15 ° and −30 ° static head pitch angles. The eye movements following cervical stimulation before, three days and 30 days after BL are reported. Since the X–Y eye position detector (*right side*) was fixed to the head, the X and Y axes changed their spatial orientation depending upon the head pitch angles

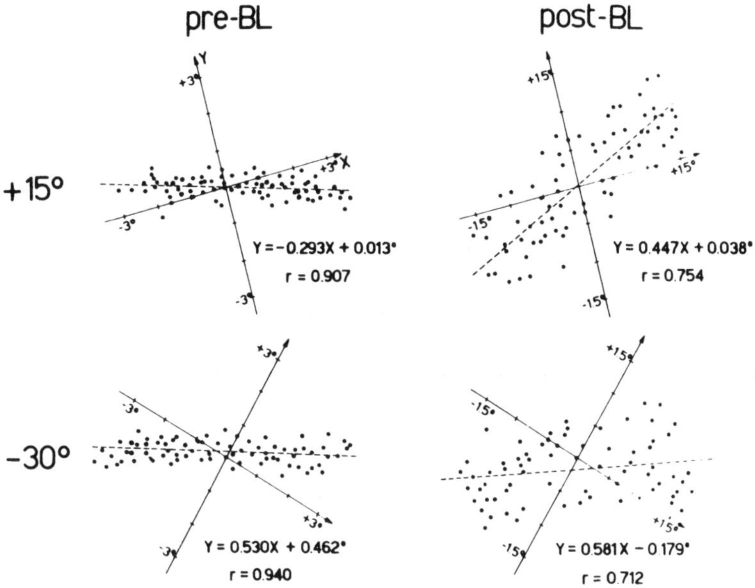

Fig. 3. Orientation of the HCOR responses of one animal before and after BL, at +15° and −30° static head pitch angles (0.02 Hz, 20° peak-to-peak). The regression lines fitting the right eye position points measured every 10 s within 15 cycles are represented. After the lesion (3 days), the average direction of the reflex eye movement is related to the orbit, as indicated by the small difference in the angular coefficient of the regression line at +15° and −30°, showing the lack of spatial reference at different head sagittal positions

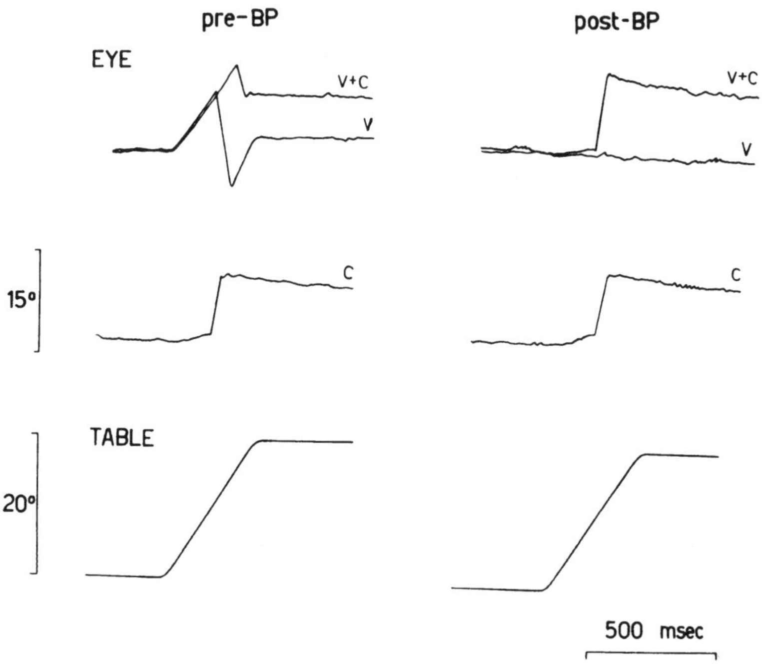

Fig. 4. Ocular responses to step cervical (*C*), vestibular (*V*) and combined (*V+C*) stimulations before (*pre-BP*) and 3 days after (*post-BP*) bilateral plugging. (For explanation see text)

BL). As a result of this interaction, the eye was maintained in a compensatory position. Hence, the range of reflex eye movements remained shifted according to the head-body angle.

3.2 HCOR in BL Rabbits

The HCOR was measured in three BL animals at three different postoperative times. Three days after the lesion the animals showed marked eye and head instability. However, as would be expected in symmetric vestibular lesions, spontaneous nystagmus was absent. The HCOR gain was reduced by about 50%–70% over all the tested frequencies (Fig. 5). Thirty days after the labyrinthectomy the COR gain was enhanced so as to reach values higher than the pre-lesion ones. However, the orientation of the eye responses was mainly oblique with respect to the horizon (Fig. 3, post-BL). Therefore, to analyze the plastic changes, the gain was studied by differentiating vertical and horizontal components of the ocular responses. A statistically significant increase ($p<0.01$) was observed for the horizontal component only at the higher frequencies, while it was significant over the whole range of frequencies for the vertical component. Seventy days after the lesion the COR gain had increased in both the components to 0.16 for the horizontal component and to 0.14 for the vertical one at 0.4 Hz of stimulation.

As regards the orientation of the cervico-ocular responses at various head pitch angles (from 15° nose up to 30° nose down), the reflex eye movements were maintained in a plane defined by orbital reference points instead of spatial ones (Fig. 3, post-BL).

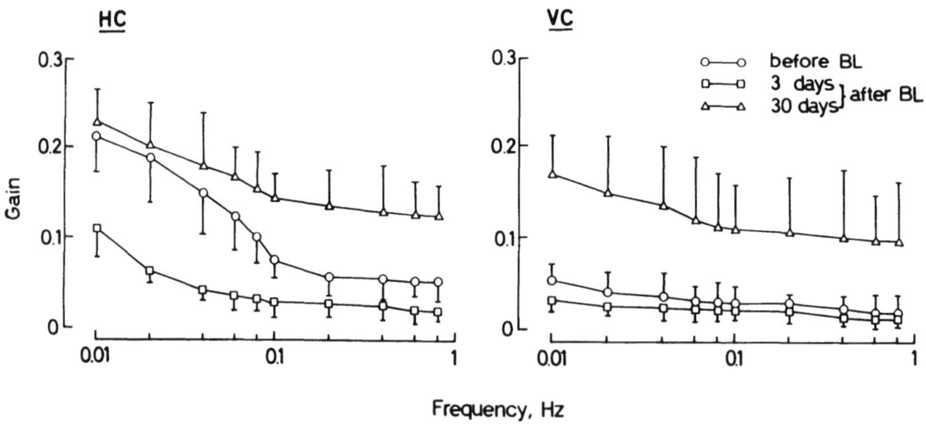

Fig. 5. Comparison between the gains of the horizontal (*HC*) and vertical (*VC*) components of the HCOR before BL, 3 days and 30 days after BL

3.3 HCOR in BP Rabbits

Bilateral plugs were placed in the horizontal semicircular canals of three rabbits. The effect of the plugging consisted of an about 90% reduction of the HVOR gain and a phase lead of the residual response at high frequency stimulation. The re-

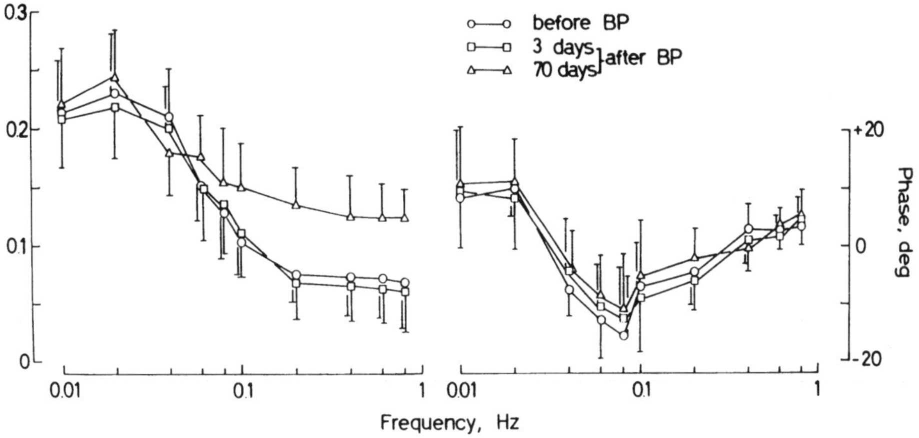

Fig. 6. Gain and phase values of the HCOR before BP, 3 days and 70 days after BP

sidual responses have been attributed to activation of the vertical semicircular canals (Barmack unpublished observation). No change in the HVOR was observed during the post-lesion period at least until 70 days post-lesion. On the contrary, the COR, which in the immediate post-lesion period did not show any change concerning amplitude and phase as well as orientation of the response, underwent a 200% gain increase ($p < 0.01$) in the range of frequencies between 0.1 and 0.8 Hz (Fig. 6).

3.4 Step Cervical (COR), Vestibular (VOR) and Combined (VOR+COR) Stimulations in BL and BP Rabbits

In BL and BP animals the step cervical stimulation elicited ocular responses similar to those previously observed in normal animals. On the contrary, the responses to vestibular and combined stimulations were deeply affected by the labyrinthine damage. In fact the vestibulo-ocular responses were small or absent, while the combined ones were the same as those observed following step cervical stimulation alone (Fig. 4). Instead of large slow compensatory responses with fast resetting phases, the responses consisted of a small compensatory eye movement followed by compensatory fast phases, which allowed the resetting of the eye in compensatory position even if the vestibular slow phases were absent.

4 Discussion

In normal rabbits, COR contributes to horizontal gaze stability by enhancing gain and reducing the phase lead of the ocular responses at low frequencies of vestibular stimulation. At higher frequencies the gain becomes insignificant, but the cervical input can still influence the eye vestibulo-ocular reflex by delaying and reducing the AFP's (Barmack et al. 1985). By this interaction the eye remains in a compensatory position following fast head displacement. Also the range of the reflex eye movements is shifted towards the same direction.

In this study we demonstrated (1) the improvement of gaze stability by enhancement of COR gain and (2) the maintenance of cervical position control after labyrinthine lesions.

As far as the COR gain is concerned, an increase of up to 200% was observed at the higher frequencies during the postoperative period, both in BP and BL (horizontal component) groups. The frequency range within which this increase occurred was the same as that within which the VOR was mostly effective.

The frequency selective gain increase suggests that a signal error may influence the remodeling process (Flohr et al. 1981). Feed-back of the system's output into the system may be part of the mechanism guiding the compensatory process. In fact, the bilateral vestibular lesions induced head instability which was mostly evident following quick external displacement stimuli and caused prolonged head oscillations. As a consequence of this, the retinal slip of the image increased mainly during fast head movements. This head instability can be revealed by the cervical proprioception and could induce gain enhancement of COR at high frequencies to reduce the image retinal slip. The COR is an open-loop system and its gain should be regulated by a visually mediated long term learning process.

The frequency selective gain increase leads one to assume that the central pathways of the VOR and COR are organized into parallel overlapping temporal frequency channels (Miles et al. 1985). It could be suggested that the proprioceptive neck signal undergoes "low-pass" filtering and, to a minor extent, "high-pass" filtering. The high-pass filters could be the same as those to which the semicircular canal input is exposed. The adaptive change could consist of the potentiation of this secondary cervical circuit by increasing the synaptic efficacy of the cervical input on those vestibular secondary neurons which were no longer modulated by the primary vestibular afferents.

The result is a high frequency COR gain increase which represents a partial process of substitution of the impaired vestibular function.

There are, however, differences between the BL and BP animals concerning the time course of the COR changes during the postoperative period (Fig. 7). The BL group, in fact, showed an immediate reduction of the COR gain, and then a rapid gain enhancement overcoming the normal values in about one month. In the BP group there was only a gain increase, but it was slower than that of the BL animals. The differences could be explained by taking into account that, unlike BP, BL induces not only a loss of stimulus-modulated activity, but also the loss of spontaneous primary and secondary vestibular afferent activity. The reduction of COR gain immediately after the lesion could be due to the decreased level of activity of the vestibular neurons on which vestibular and cervical inputs converge (Rubin et. al. 1977; Anastasopoulos and Mergner 1982). The following gain increase, which is faster than that observed in plugged animals, suggests that the lack of activity in the vestibular nuclei could be a factor that potentiates synaptic contact from other sensory systems.

Other differences between BL and BP animals were observed regarding the spatial orientation of the COR responses. In spite of the constant horizontal direction of the ocular responses observed in the BP group, BL caused the eye reflex movements to follow orbital coordinates instead of spatial ones. Thus, COR did not change the extraocular muscle pattern of activation which is dependent upon

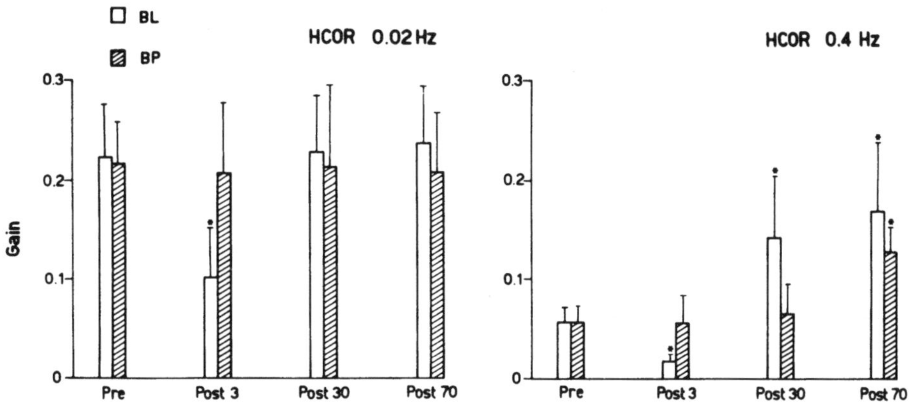

Fig. 7. Time course of HCOR gain in BL and BP animals at 0.02 and 0.4 Hz stimulations. *Bars* represent standard deviation of means, *asterisks* significant differences with respect to the pre-lesion values ($P < 0.01$)

the sagittal head position. This lack of orientation could be related to the otolith destruction, whose role in the COR orientating mechanism we have recently demonstrated (Pettorossi et al. 1987). Furthermore, the proper orientation of COR was progressively altered in BL animals by a plastic increase of the response orthogonal to the normal plane. It can be suggested that the loss of the otolithic and vertical semicircular canal inputs could elicit sprouting of cervical afferents toward the secondary vestibular neurons connected with the vertical reflex pathway. In this case the lack of activity was detrimental to functional recovery.

Concerning the cervical ability to set the eyes in a compensatory position and to readjust the range of ocular movements, we observed that this function was maintained after labyrinthine damage, even if through a different mechanism. This task, normally accomplished by an interaction with AFP's, was achieved in vestibular damaged animals by compensatory fast phases (CFP's) of cervical origin. In order to explain both cervical functions, that is the decrease in AFP's and the triggering of CFP's, a model is proposed based on a center which compares the required with the actual eye position. The output of this center should influence the direction and threshold level of the fast phases. In the presence of vestibular slow compensatory phases, the output of the comparator should be reduced and the cervical input could affect only AFP parameters. The absence of vestibular responses should increase the comparator output so that CFP's could be triggered.

5 Conclusion

In rabbits, neck proprioception can influence eye position by inducing slow compensatory responses to horizontal head displacements and by maintaining the range of the reflex eye movements in a compensatory direction. In bilaterally labyrinthectomized and plugged animals it was demonstrated that COR can partially substitute for the vestibular function in gaze stability by increasing its gain at high frequencies of stimulation. A signal error coming from the head-body angle amplitude could guide this specific gain increase. However, differences between BP and BL groups were observed concerning the time course of the COR gain changes and the spatial orientation of the responses. Also, it was demonstrated that cervical input can maintain the eyes in compensatory position after fast head displacements even in the absence of anticompensatory fast phases by triggering compensatory fast phases.

References

Anastasopoulos D, Mergner T (1982) Canal-neck interaction in vestibular nuclear neurons of the cat. Exp Brain Res 46:269–280

Baarsma EA, Collewijn H (1975) Changes in compensatory eye movements after unilateral labyrinthectomy in the rabbit. Arch Oto-Rhino-Laryngol 211:219–230

Baker J, Goldberg J, Peterson B, Schor R (1982) Oculomotor reflexes after semicircular canals plugging in cats. Brain Res 252:151–155

Barmack NH, Pettorossi VE (1981) The influence of unilateral horizontal semicircular canal plugs on the horizontal vestibulo-ocular reflex of the rabbit. In: Flohr H, Precht W (eds) Lesion-induced neuronal plasticity in sensorimotor systems, symposium proceedings, July 1980, Bremen. Springer, Berlin Heidelberg New York, pp 231–239

Barmack NH, Nastos MA, Pettorossi VE (1981) The horizontal and vertical cervico-ocular reflexes of the rabbit. Brain Res 224:261–278

Barmack NH, Draicchio F, Errico P, Ferraresi A, Pettorossi VE (1985) Characteristics of the combination of the vestibulo- and cervico-ocular reflexes in the control of rabbit eye movements. Proc Physiol Soc J Physiol (Lond) 369:47

Barnes GR, Forbat LN (1979) Cervical and vestibular afferent control of oculomotor response in man. Acta Oto-Laryngol 88:79–87

Dichgans J, Bizzi E, Morasso P, Tagliasco D (1974) The role of vestibular and neck afferents during eye-head coordination in the monkey. Brain Res 71:225–232

Flohr H, Bienhold H, Abeln W, Macskovics I (1981) Concepts of vestibular compensation. In: Flohr H, Precht W (eds) Lesion-induced neuronal plasticity in sensorimotor systems, symposium proceedings, July 1980, Bremen. Springer, Berlin Heidelberg New York, pp 153–172

Kasai T, Zee DS (1978) Eye-head coordination in labyrinthine-defective human beings. Brain Res 144:123–141

Maioli C, Precht W, Ried S (1983) Short- and long-term modifications of vestibulo-ocular response following unilateral vestibular nerve lesions in the cat. Exp Brain Res 50:259–274

Meiry JL (1971) Vestibular and proprioceptive stabilization of eye movements. In: Bach-y-Rita P, Collins CC, Hyde JE (eds) The control of eye movements. Academic Press, London, pp 483–496

Miles FA, Optican LM, Lisberger SG (1985) An adapting equalizer model of the primate vestibulo-ocular reflex. In: Berthoz A, Melvill Jones G (eds) Adaptive mechanisms in gaze control. Elsevier, Amsterdam, pp 313–326

Paige GD (1985) Plasticity in the vestibulo-ocular and optokinetic reflexes following modification of canal input. In: Berthoz A, Melvill Jones G (eds) Adaptive mechanisms in gaze control. Elsevier, Amsterdam, pp 145–153

Pettorossi VE, Petrosini L (1984) Tonic cervical influences on eye nystagmus following hemilabyrinthectomy: immediate and plastic effects. Brain Res 324:11–19

Pettorossi VE, Errico P, Ferraresi A, Fedeli R (1987) Vestibular contribution to the orientation of cervico-ocular reflex in rabbit. Brain Res 403:58–65

Rubin AM, Liedgren SCT, Milne AC, Young JA, Fredrickson JM (1977) Vestibular and somatosensory interaction in the cat vestibular nuclei. Pflügers Arch 371:155–160

Schaefer KP, Meyer DL (1973) Compensatory mechanisms following labyrinthine lesions in the guinea pig. A simple model of learning. In: Zippel HP (ed) Memory and transfer of information. Plenum, New York, pp 203–232

Compensation of Cerebellar-Vestibular Lesions and Microgravity

O. G. Gazenko [1], R. A. Grigorian [2], Yu. V. Kreidich [1], G. S. Aizikov [1], and A. S. Markin [1]

[1] Institute of Biomedical Problems, USSR Ministry of Health, Moscow, USSR
[2] Institute of Evolutionary Physiology and Biochemistry, USSR Academy of Sciences, Leningrad, USSR

1 Introduction

There is evidence that abnormalities in movement coordination and spatial orientation that occur in microgravity are primarily induced by vestibular changes (Shock 1961; Fiorica et al. 1962; Lacour and Xerri 1980; Kozlovskaja et al. 1984). The direct investigation of the labyrinthine function of nonanesthetized cats exposed to short-term microgravity (up to 2 s) showed that both the frequency and amplitude of vestibular biopotentials almost doubled (Fiorica et al. 1962). The study of the electrical activity of the cerebral cortex of cats demonstrated that during short-time exposure to microgravity the most significant and characteristic changes developed in the anterior compartment of the suprasylvian and ectosylvian convolutions, which is the site of localization of the vestibular function in the cortex. On the other hand, it is well known that in microgravity, changes can be seen in the otolith bioelectrical activity and in vestibulo-motor responses (Gualtierotti et al. 1972; Aizikov et al. 1981). In addition there are publications indicating changes in the frequency of electrical discharges of stem neurons of Deiters' nucleus when the anterior lobes of the cerebellum and labyrinths are polarized (De Vito et al. 1956) as well as the convergence of labyrinthine, optic and proprioceptive signals to the cerebellum (Dow and Moruzzi 1958; Ito 1984; Rasumeev and Grigorian 1976), which suggest that the cerebellum is involved in the motor coordination and spatial orientation.

In view of the above observations it was interesting to investigate the effect of the removal of the cerebellum and vestibular apparatus on the mechanisms of motor compensation at 1 g and 0 g.

2 Methods

The effect of short-term microgravity of 20 to 30 s duration was investigated on cats and rhesus monkeys flown on board an aircraft along the Keplerian trajectory. The exposure to microgravity was followed by a 15-s effect of acceleration of 1.8–2.0 g. The experiments were performed on four cats. One of them had the cerebellum removed 4 months before the study. By the time the experiments began, the animal showed no extensor rigidity, which is the main symptom of the "dynamic period" accompanying cerebellectomy. Another cat had the left cere-

H. Flohr (Ed.)
Post-Lesion Neural Plasticity

bellar hemisphere with the adjacent dorsal and ventral paraflocculus removed 5 months before the study. Two remaining cats served as controls. Two months before the experiment, a rhesus monkey was implanted with electrodes to record the neuronal activity of vestibular nuclei (superior and medial), EEG, EOG, and EMG of leg extensor and flexor muscles. Each animal was exposed to microgravity at least 12 times. The effect of long-term microgravity was investigated using Wistar-SPF white rats weighing 200 to 250 g flown in Cosmos-782, Cosmos-936, Cosmos-1129 for as long as 18.5–19.5 days. A detailed description of the flight experimental protocol was published previously (Il'in et al. 1979). Before and after flight the rat posture, motor reactions, balance function, turning reaction in free fall, and lifting reaction were investigated. This was done using specially designed devices in which the animals were allowed to fall freely in the supine position from as high as 1 m to the guard net. These devices were described in detail elsewhere (Aizikov et al. 1971, 1980, 1981).

The experimental fall was filmed at the rate of 40 frames/s, i.e., 1 frame/25 ms. The experiments were carried out on intact and labyrinthectomized rats in which the vestibular apparatus was inactivated by electrocoagulation 2–3 weeks preflight. In addition, intact rats were flown in Cosmos-936: these rats were rotated in a 320-mm arm centrifuge at a velocity of 53.5 ± 3.0 rpm to generate artificial gravity equal to 1 g (Aizikov et al. 1981). The effect of long-duration microgravity was also investigated on two rhesus monkeys flown for 5 days in Cosmos-1514. These animals were implanted with electrodes similar to those used in parabolic flights. The Cosmos-1514 monkeys were examined twice before flight, 4 to 6 h after recovery and at R+1, 3, 6, 11, and 24 days. The same protocol was used in ground-based control studies of two groups of animals: vivarium controls and synchronous controls that were exposed to space flight effects in the biosatellite mock-up (or in the centrifuge in the case of Cosmos-936).

3 Results and Discussion

3.1 *Effect of Short-Term Microgravity*

In conditions of weightlessness, normal cats exhibit a sudden increase of motor activity (running and catching movements, scratching, steadying movements of the tail) in the initial period of adaptation to weightlessness, always with characteristic predominance of extensor tone in the muscles of the back, neck, and extremities. These signs were, however, observed only during the first few periods of weightlessness, but later, with repeated production of the state, the animals gradually became adapted to the altered gravitational conditions and there was significant reduction in their motor activity. Figure 1 A, 1–5 frames shows five very characteristic successive positions assumed by the normal cat in one of the last periods of weightlessness.

The characteristic feature in the decerebellate cat was the very pronounced state of extensor rigidity in the muscles of trunk, posterior and, particularly, the anterior limbs (Fig. 1 B), the rigidity being much more intense than that observed in the control animals in initial periods of weightlessness. These signs remained undiminished, with no sign of adaptation to weightlessness, throughout the 12 ex-

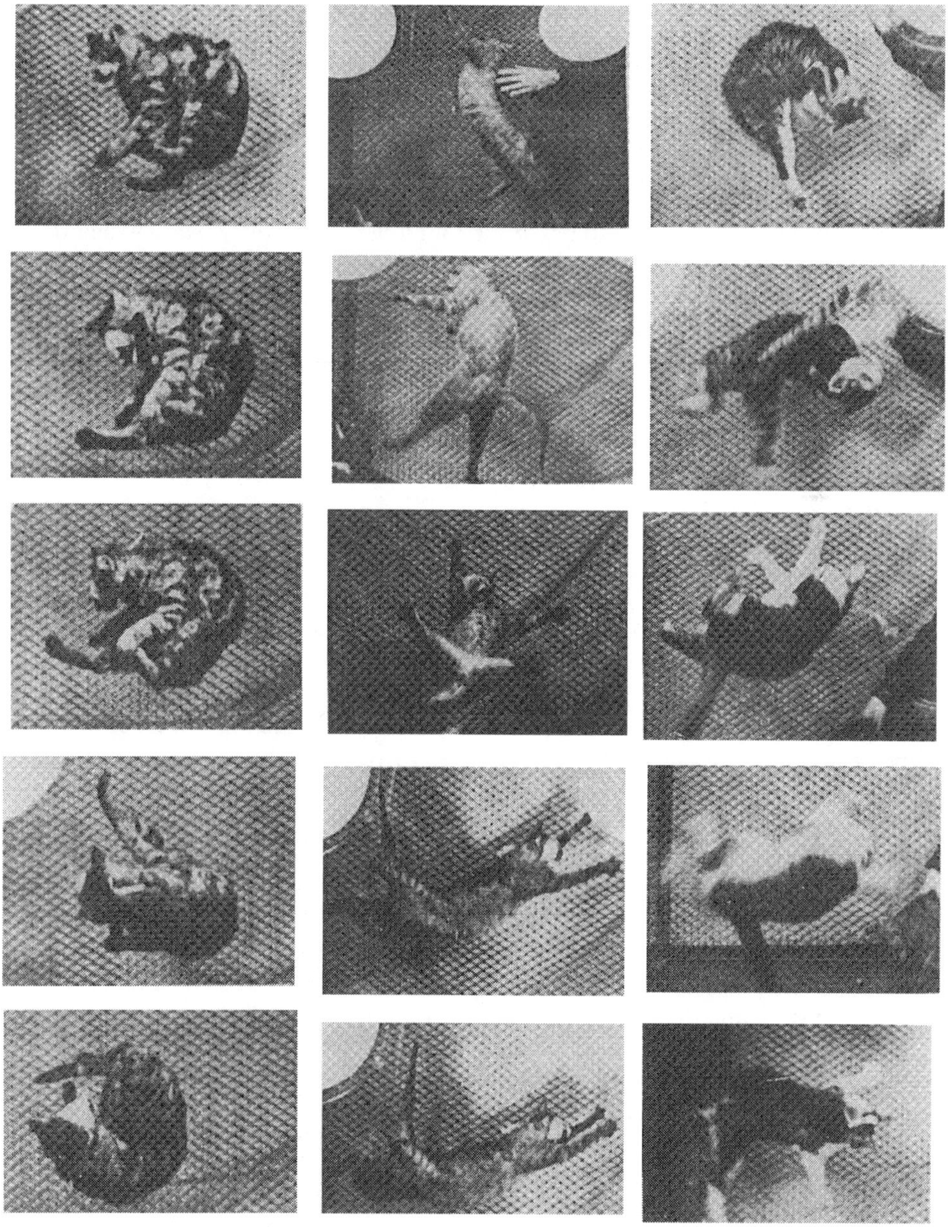

A B C

Fig. 1. Effects of short-term microgravity on motor reactions of cats. **A** intact animal; **B** cerebellectomized cat; **C** cat with the left cerebellar hemisphere removed. *1–5* successive photographs periments. Rhythmic, pendulum-like movements of extremities (1 or 2 per s) were only observed during the first four to six periods of weightlessness, when the animal tried unsuccessfully to grip surrounding objects, but was unable to hold on because of the considerable excess of extensor tone.

Loss of the assistance of vision had practically no effect on the character of the decerebellate cat's motor activity in the state of weightlessness. The animal with only part of the cerebellum removed exhibited a very marked increase of motor activity and made efforts to grasp surrounding objects; there was asymmetrical contraction of the trunk muscles on the side away from the cerebellar lesion, with lateral flexion of the trunk to the right and rotation of the animal to the left (Fig. 1 C, 1–3). No differences could be demonstrated by vestibular tests in the periods preceding and following the experiments with weightlessness. Active vestibular reflexes in the operated animals indicated that the vestibular system was still functioning normally.

3.2 Effect of Prolonged Microgravity

Preflight, the intact rats showed distinct otolith reflexes, motor activity, and orientation reaction. The animals moved along the tunnel quickly. Their steps were normal and smooth. They came out of the tunnel in a careful manner, sometimes waiting there for a while. Having emerged from the tunnel, the rat cleaned their hair, "washed" themselves and lay down. In the lying position they hid their paws beneath them. When placed on a bar, the animals could walk freely and rapidly, changing direction easily; they remained sitting on the bar or went down along the pillars (Fig. 2 A).

The motor function and orientation reaction of labyrinthectomized rats differed drastically from those of the intact animals. The labyrinthectomized rats moved too much, with movements being abrupt, sharp, poorly coordinated. In the enclosure they moved quickly and chaotically, stopped suddenly or changed direction abruptly. Sometimes they moved along one direction only. The animals showed specific neurological symptoms, viz. neck muscle ataxia and extension, torsion spasm, intention tremor. These symptoms attenuated or disappeared when they scratched or "washed" themselves or in lying, and increased when they were walking or maintaining posture. The animals walked through the tunnel quickly and held their head and body slightly bent; normally they did not stop before entering the enclosure. Occasionally they lost balance when taking a turn or stopping abruptly or assuming the "kangaroo" posture. The attempts to take this posture were frequently futile. When sitting on the bar, the rats were continuously equilibrating their tail and body. In this situation their head was shaking, which made it very difficult for them to maintain equilibrium and the animals were hanging with their back downwards. Sometimes rats tried to walk along the bar but lost balance and fell (Fig. 2 A).

After flight (18.5–19.5 days), the hair of the intact rats was as a rule wet, sticky and dirty. The otolith reflexes were well expressed. However, the animals were inert, apathic, slow. These animals could walk with their hindlimbs kept widely apart, rocking from side to side; they often stopped and lay down. They were unable to stay on the bar, lost balance and fell down (Fig. 2 B). Next day the state

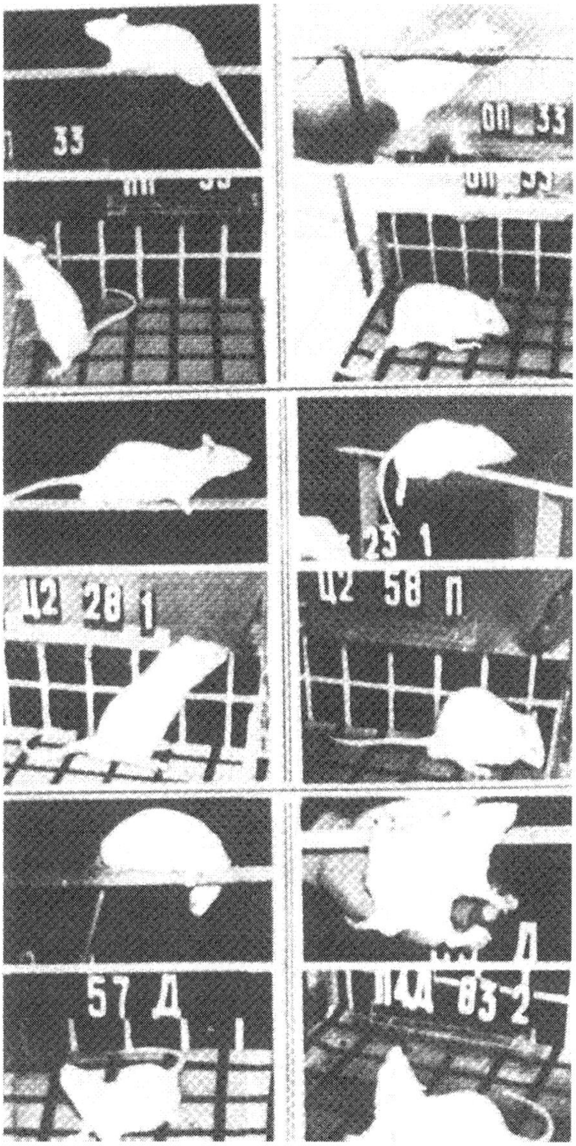

A B

Fig. 2. Posture, motor and equlibrium reactions of rats in pre- (**A**) and postflight (**B**) periods. *1–2* intact animals; *3–4* animals affected by artificial gravity (1 g); *5–6* labyrinthectomized animals

of the flight animals became better. They were moderately active, cleaned their hair, "washed" themselves and slowly moved within the enclosure. They walked with their body kept low and hindlimbs half bent. They did not try to stick their head out of the enclosure or take the "kangaroo" posture. When walking through the tunnel, the rats made two or three stops and sometimes lay down. In the horizontal position they assumed an unusual configuration of one or two hindlimbs, which were thrust out from the body with the fingers pulled apart and the shin unnaturally pronated (Fig. 2 B). When changing their posture or turning aside from the prone position, the rats did not raise their body but drew it together by actively moving the forelimbs. Thus muscles of the forelimbs retained their function, while those of the hindlimbs almost lost it.

On the 3rd postflight day, the motor behavior of the animals improved significantly. The rats were adequately active, they often stood up on hindlimbs, stuck out the enclosure, took the "kangaroo" posture, but infrequently and for a short period of time only. Normally these rats walked through the tunnel rapidly, straightening their limbs; nevertheless their total motor activity remained below the preflight level. All in all the rats were still slow and inactive.

On the 6th postflight day the motor activity of the animals became good and very similar to that seen preflight. They showed insignificant changes in their step and walked along the bar. Later the posture-motor reactions returned to normal, although walking along the bar remained a problem.

After recovery, the appearance of the labyrinthectomized rats was very similar to that of the intact ones. Their motor activity was sufficiently high albeit lower than preflight. When walking through the tunnel, they remained there for a long time. In the enclosure they made a lot of chaotic movements, stumbled, circled, and often lost balance. All movements were accompanied by head tremor and ataxia symptoms that were stronger than in the intact rats. They could not stay on the bar and in most cases fell with the head downward (Fig. 2 B).

By the 11–15th postflight day their motor activity and behavioral pattern were similar to those of controls, but the equilibrium function did not recover and only at R + 24 days did it tend toward normalization.

On the day of recovery the rats exposed to artificial gravity were in good condition with hair clean and dry. They displayed good motor activity and well-expressed orientation reaction (Fig. 2 B). They walked through the tunnel quickly, moved actively in the enclosure, looked out, and rose on their hindlimbs. However, their equilibrium function was impaired: they rocked when walking and failed to assume the "kangaroo" function. They could hardly walk along the bar and soon lost balance, hanging with their back downwards. By the 11–15th postflight day their posture-motor reactions and the equilibrium function returned to normal.

Examination of the intact rats from the vivarium group revealed no significant changes in the posture-motor reactions or the equilibrium function. Only some animals showed a minor decline of these parameters. The synchronous control rats exhibited a slight inertness and change in the equilibrium function on days 0–3. These changes disappeared rapidly and subsequently the motor behavior of these animals did not differ from that of the vivarium rats. The general status and vestibulo-motor reactions of the labyrinthectomized rats from both control groups were identical. Their motor activity improved gradually and ataxia symptoms weakened. By recovery day 24, the changes in the labyrinthectomized rats of the two control groups were minimal. However, the effects of labyrinthectomy were easily identified with the aid of special tests or situations in which the animals had to make very quick and accurate decisions (how to maintain a stable posture or equilibrium on a bar).

The synchronous rats that were exposed to acceleration were slow and inactive. They walked through the tunnel slowly, stopped before they emerged from it, often circled around the enclosure. Their step was ataxic. When trying to assume the "kangaroo" posture, they lost balance. When placed on the bar, they could hardly maintain equilibrium and infrequently hung from it with their back downwards. By recovery day 11 their posture-motor reactions and equilibrium function returned to normal.

3.3 Turning Reaction in Free Fall

The turning reaction of free falling helped them to orient their bodies and land on their limbs. The pattern of the reaction remained unchanged with closed eyes. The labyrinthectomized rats showed a drastic inhibition of the turning reaction. It can be seen from the recordings that the animals fell longer back downward and began to orient the body only immediately before landing, so that they landed on the back or side but not on the limbs (Fig. 3). During postflight experiments, the turning reaction of the intact rats with eyes open or closed was adequate, and allowed them to land on the limbs (Fig. 4 A, B). The turning reaction of the rats exposed to artificial gravity of 1 g was normal when they were falling with eyes open. However, when the eyes were closed, the turning reaction was completely inhibited. After the rats lost support they made no attempts to orient the body and landed on their back (Fig. 5 A, B). Postflight the labyrinthectomized rats displayed an inhibited reaction of turning, and when falling with eyes open, they were unable to orient the body adequately and had to land on their back (Fig. 5 C).

3.4 The Jump Reflex and the Lift Reaction

Before flight all intact rats showed a distinct jump reflex and typical lift reaction movements. The jump reflex was characterized by a rapid rise of the head, forelimb stretching, back muscle straining, and slight extension of the body. When the rat container was allowed to fall, the animals extended their forelimbs and then hindlimbs, rose and curved their back, lifting the lid of the container. The latency of the lift reaction calculated with respect to this complex movement was 26 ± 6.0 ms. The latency of the lift reaction measured in terms of EMG responses of oculomotor muscles was 17.8 ± 0.4 ms. After total labyrinthectomy, the animals showed no lift reaction at all for as long as 3–4 weeks; they remained indifferent to the fall of their container, and only some of them weakly moved the head upon its touchdown. The lift reaction was restored in more than half of them a month after operation. These animals became able to stand on their hindlimbs and open the lid. This reaction was, however, unstable (the lift reaction was recorded in four to six cases out of ten falls) and the latency was extended to 47.0 ± 10 ms. The pattern of the lift reaction depended to a large extent on the original state of labyrinthectomized animals: restless and aggressive rats showed the lift reaction more frequently with the latency being shorter, whereas quiet and inert rats exhibited either no lift reaction or reaction with a longer latent period.

After flight, the latency of the intact rats measured with respect to the time it took them to lift the lid increased significantly. By postflight day 3 the latency was very close to the initial value. The rats exposed to artificial gravity and vi-

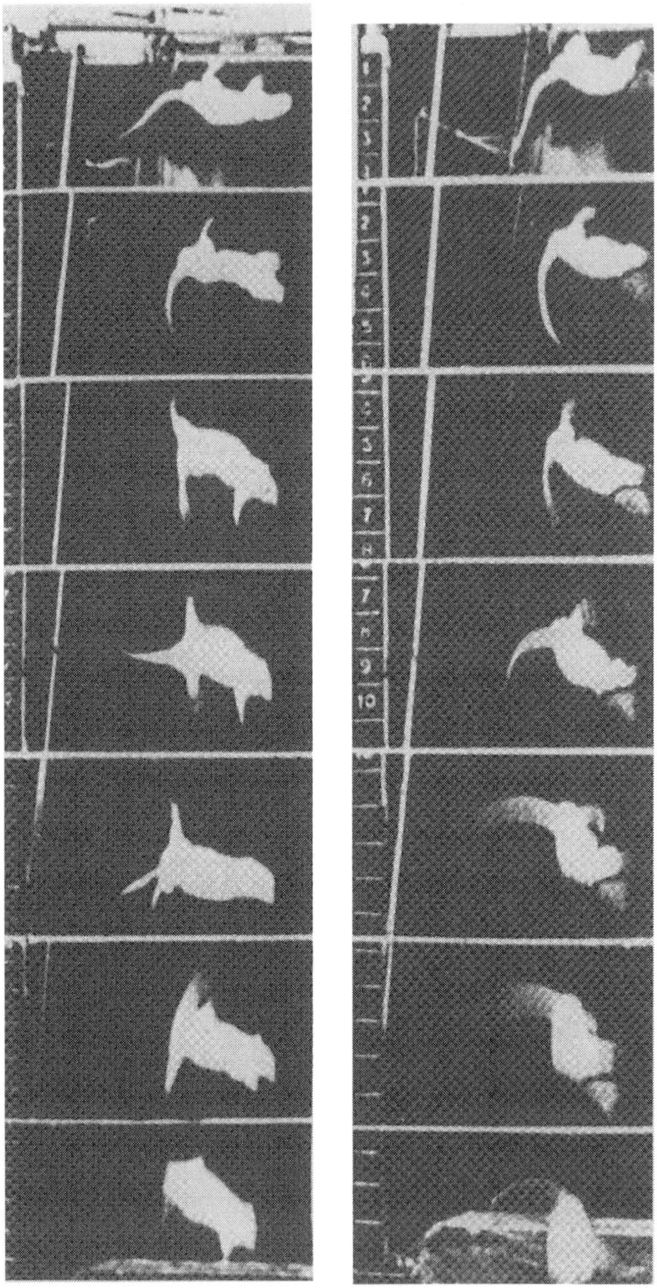

A B

Fig. 3. Turn up reactions of labyrinthectomized rats with open (**A**) and closed (**B**) eyes. Free fall altitude = 1 m

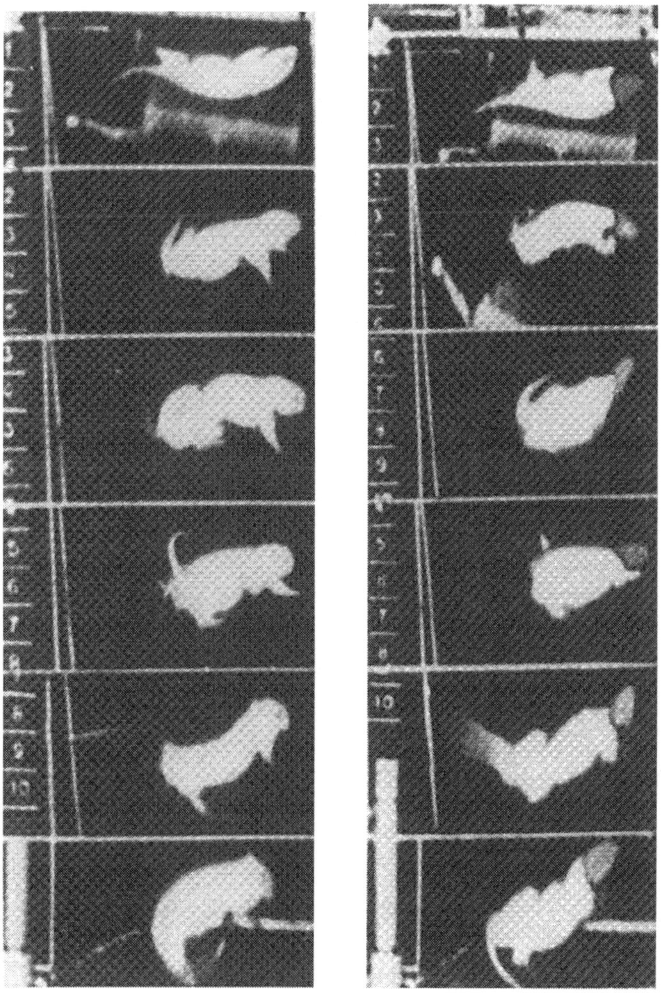

A B

Fig. 4. Turning reactions of intact rats with open (**A**) and closed (**B**) eyes after a stay in microgravity conditions. Free fall altitude = 1 m

varium rats showed no significant changes in latent period. However, the synchronous rats, both resting and rotated, showed an increase in the latent period of response, which was nevertheless smaller than in the rats exposed to real microgravity.

It should be noted that the latent period of the lift reaction, as follows from EMG data, remained unchanged in all intact rats' flight and ground-based controls. Following recovery, the jump reflex was present in the intact animals and absent in the labyrinthectomized rats. The level of the jump reflex was different in various animal groups: it was weak in the animals exposed to artificial gravity and in the intact animals exposed to long-term microgravity.

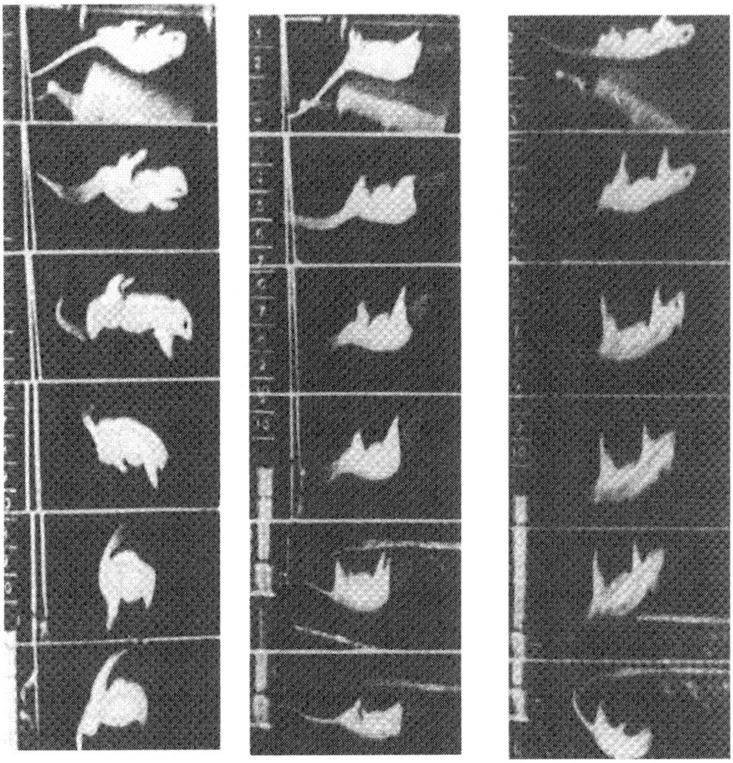

A B C

Fig. 5. Turning reactions of animals with open (**A**) and closed (**B**) eyes affected by artificial gravity (1 g) during space flight. **C** the same for labyrinthectomized rat with open eyes

After flight the latency of the lift reactions of the rhesus monkeys Abrek and Bion, as measured by leg muscle EMG, varied insignificantly and amounted to 72 and 70 ms for the soleus and anterior tibial muscles, respectively (versus 69 and 58 preflight). The latency of the first and second trains of EMG bioelectric activity increased to a greater extent – 133 and 144 ms, respectively (versus 80 and 112 ms preflight) (Fig. 6).

Our investigations in short-term microgravity demonstrated that in cats with the cerebellum totally or partially removed, extensor rigidity increased markedly. A similar change occurred in nonoperated cats but it was much weaker and disappeared as the animals became adapted to microgravity. In the normal gravitational field hypertonicity of extensors manifested itself soon after cerebellectomy (the cerebellar release phenomena) or after decerebration (decerebration rigidity). In the first case it is attributed to the elimination of the cerebellar inhibitory effect on brain stem vestibular nuclei (De Vito et al. 1956). In the second case hypertonicity of extensors is associated with the elimination of cortical and mesencephalic inhibitory effects.

An important part in the regulation of postural tonicity is played by labyrinthine impulsation and skin-muscle reception (Dow and Moruzzi 1958). It can

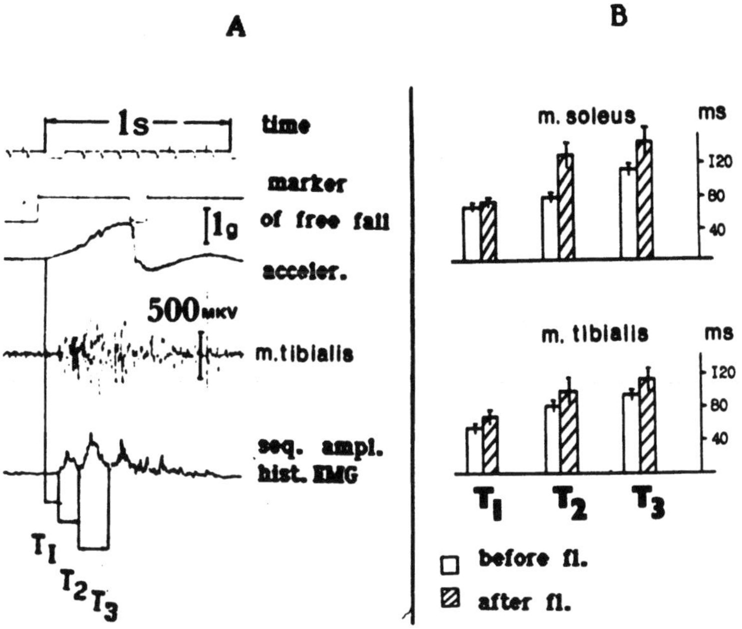

Fig. 6. Effects of microgravity on latency of EMG reactions of monkey. **A** Specimen of test lift reaction. **B** Mean latency of EMG reaction of flexor and extensor muscles before and after a 5-day space flight

therefore be concluded that the cerebellar release phenomenon is compensated by the inhibitory effects mainly of the brain stem and additionally of the brain cortex (Dow and Moruzzi 1958; Gazenko et al. 1965). It appears, however, that this compensation cannot entirely replace cerebellar inhibition; this may be the reason why we failed to reveal adaptation to microgravity in cerebellectomized cats, which normally develops in nonoperated animals.

Our findings have also demonstrated that a prolonged exposure to space flight noticeably influences the general condition of the animals.

3.5 Posture-Motor and Orientation Reactions Change to the Greatest Extent

These changes occur to a high degree in intact rats and to a low degree in labyrinthectomized and centrifugated rats. The insignificant changes in the latter can be attributed to the effect of artificial gravity close to 1 g. This can be viewed as the normalization effect of artificial gravity on the posture-motor activity of animals.

However, rotation in a short-arm centrifuge and at a high angular velocity may cause adverse effects because it produces Coriolis acceleration and artificial gravity gradient. These factors may have a negative effect on movement coordination, vestibular reactions, and precision of movements.

The labyrinthectomized rats initially showed high motor activity which can be explained by the involvement of redundant sensory systems in the analysis of spatial orientation and movement coordination. It is probable that in space flight the response to labyrinthectomy is the same, and has some conditioning effect in microgravity and provides a better state of animals after recovery. It is also known that labyrinthectomized animals can readily survive their transition to microgravity and do not restrict their movements in flight, in contrast to intact animals (Henry et al. 1952; Aizikov et al. 1981, 1986). It is very likely that the absence of the vestibular apparatus is responsible for lower responses of labyrinthectomized animals to afferent gradients in a different gravity field. Similar observations have been made both on the ground and in short-term microgravity (Aizikov et al. 1971; Yuganov et al. 1974). Postflight examination also showed deterioration of such parameters as ataxia and head tremor in the labyrinthectomized rats. In comparison to the ground-based labyrinthectomized controls they displayed a delay in the compensation of the vestibular lesion and enhancement of cerebellar symptoms.

After space flight the equilibrium function of all animals considerably deteriorated. Immediatly after recovery this function of the intact animal was comparable to that of the labyrinthectomized vivarium and synchronous controls. Similar changes were seen in the centrifugated animals in space and on the ground. The equilibrium function tended to normal at R+3 to R+6. However, equilibrium response of flight and control animals became identical only by recovery day 24. The labyrinthectomized animals of the control groups showed a progressive improvement of the equilibrium function, which suggests a stable development of compensatory processes. In contrast, the labyrinthectomized rats of the flight group exhibited similar changes which developed at a far slower rate and were considered as a trend toward improvement. These data can be viewed as further evidence of delayed compensation which may be attributed to the unloading of skeletal muscles. On the ground, similar effects were recorded in experiments with animal suspension or chordotomy of spinal cord dorsal roots (Schaefer and Meyer 1974) accompanied by drastic reduction or elimination of tactile and proprioceptive impulses to the central nervous system and vestibular nuclei.

Summary

1. In the state of microgravity motor disturbances are due to a great extent to the loss of cerebellar control of compensated and/or elaborated motor programs.
2. A stay of normal animals in weightlessness has little effect on the basic mechanisms responsible for vestibular and vestibulomotor reactions.
3. The obtained data provide strong evidence that microgravity leads to disturbances in the compensation processes which are similiar to cerebellar ataxia.
4. Decompensation of vestibulo-cerebellar disorders takes place during adaptation to the state of microgravity as well as in the readaptation period, when the animals return to earth.

References

Aizikov GS, Markin AS (1981) Posture movements and equilibrium functions of rat flown on the biosatellites. Cosm Biol Med N6:33–38 (in russian)

Aizikov GS, Emeljanov MD, Ovechkin VG (1971) Reflex activity of spinal cord in intact and labyrinthectomized animals exposed to radial accelerations. Cosm Biol Med N3:23–27 (in russian)

Aizikov GS, Markin AS, Sarkisov YY (1980) The divice for investigations of turn up reactions in small laboratory animals. Cosm Biol Med N6:76–78 (in russian)

Aizikov GS, Markin AS, Mokrousova AV (1981 a) Lifting reflex of white rats after prolonged space flight. Effect of weightlessness and artificial gravity. Bull Exp Biol Med N12:722–725 (in russian)

Aizikov GS, Markin AS, Mokrousova AV, Sarkisov YY (1981 b) Methods of investigation of lift reaction in white rats. Cosm Biol Med N4:82–84 (in russian)

Aizikov GS, Grigorian RA, Kreidich YV (1986) Posture-vestibular reflexes in rats and monkeys in microgravitational conditions. In: Disorders of posture and gait. 8th Intern Symp on Posturography. Free Univ of Amsterdam, p 85

Batini C, Moruzzi G, Pompeiano O (1957) Cerebellar release phenomena. Arch Ital Biol 95(N1):71–95

Dow RS, Moruzzi G (1958) The physiology and pathology of the cerebellum. University of Minnesota Press, Minneapolis

Fiorica V, Semba T, Steggerda FR (1962) Vestibular response of the unanesthetized cat recorded during free fall. Aerosp Med 33:475–481

Gazenko OG, Grigorian RA, Kitaev-Smyk LA, Klochkov AM (1965) Increase of extensor tone in completely or partially decerebellate cats in a state of weightlessness. Bull Exp Biol Med N7:7–12 (in russian)

Gualtierotti T, Bracchi F, Rocca E (1972) Orbiting frog otolith experiment. Final report of the data reduction and control experimentation. NASA Report

Henry JP et al. (1952) Animal studies of the subgravity state during rocket flight. J Aviation Med 23:421–432

Ito M (1984) The cerebellum and neural control. Raven, New York

Il'in EA, Korolkov VI et al. (1979) Goals and conditions for physiological experiments on rats flown aboard the biosatellite Cosmos-936. Cosm Biol Med N6:18–22 (in russian)

Kozlovskaja IB et al. (1984) The effect of weightlessness on motor and vestibulo-motor reactions. Physilogist 27(N6):111–114

Lacour M, Xerri C (1980) Compensation of postural reactions to free fall in the vestibular neuroectomized monkey. Exp Brain Res 40(N1):103–110

Rasumeev AN, Grigorian RA (1976) The cerebellum and gravitation. Problems of cosmic biology, vol 29. Nauka, Moscow, p 451

Schaefer KR, Meyer DL (1974) Compensation of vestibular lesions. In: Handbook of sensory physiology, vol VI/2. Springer, Berlin, pp 463–490

Shock GD (1961) A study of animal reflexes during exposure to subgravity and weightlessness. Aerosp Med 32(N4):336–340

Vito R de, Brusa A, Arduini A (1956) Cerebellar and vestibular influences on Deitersian units. J Neurophysiol 19(N3):241–253

Yuganov EM, Kasjan II, Asamolov BF (1974) Bioelectrical activity of skeletal muscles as a sequence of overloading and weightlessness. In: Weightlessness. Moscow, pp 213–218 (in russian)

HRP Study on Structural Changes in the Commissural Fiber System of *Rana temporaria* Following Labyrinthectomy

U. Will[1], H. Kortmann[2], and H. Flohr[2]

[1] Dept. of Neurophysiology, Faculty of Biology, University of Bielefeld, D-4800 Bielefeld 1, FR Germany
[2] Dept. of Neurobiology, University of Bremen, D-2800 Bremen 33, FR Germany

1 Introduction

Unilateral labyrinthectomy results in a characteristic disorganization of posture and movement that disappears spontaneously in the course of time. The basic phenomena of this compensation process have been known since the end of the last century (Bechterew 1883; Ewald 1892). The fact that compensation occurs without the regeneration of vestibular end organs implies an extensive reorganization of central structures (Magnus 1924). Several lines of experimental evidence (Bienhold and Flohr 1978; Dieringer and Precht 1979a, b; Flohr et al. 1981) indicate that the commissural fiber (CF) system, connecting the vestibular zones (VZ) of both sides of the medulla, is involved in this reorganization process. Initially, hemilabyrinthectomy leads to an asymmetry in the resting activity of the two VZs. Restoration of symmetrical resting activity seems to be a prerequisite for compensation and may be achieved by changes in the intervestibular CF system that lead to an increased influence on the deafferented VZ by the intact VZ (Galiana et al. 1984). Indeed, Dieringer and Precht (1979a, b) have shown that while the response pattern of commissural neurons in compensated frogs had not changed, their efficacy on vestibular neurons on the lesioned side was increased as compared to that in acutely lesioned animals. However, there are no morphological data about the changes that take place. Accordingly, in the present investigation a qualitative and quantitative study was made of the structural changes in the intervestibular CF system following hemilabyrinthectomy in the frog, *Rana temporaria* (Fig. 1).

The CF system of the alar plate in *Bufo bufo* has been reported to consist of internal and external arcuate fibers (Grofova and Corvaja 1972). Arcuate fibers of the ventral zone were claimed to arise as internal and external fibers. Fuller (1974) could not confirm this latter result. In *Rana catesbeiana* the CF's were found to arise only as internal arcuate fibers and to terminate as internal or external fibers on the contralateral side. A CF system as described by Fuller was also found for other anurans (Will et al. 1985b; Will 1988). The latter study demonstrated the existence of two differently organized parts of the CF system in anurans: Caudal to the level of entrance of the VIIIth nerve the CF's are organized as described by Fuller (1974) and Will et al. (1985b). Rostral to the VIIIth nerve, however, CF's arise as internal and external fibers, and, while decussating, do not change their position, i.e., external fibers remain external and internal

H. Flohr (Ed.)
Post-Lesion Neural Plasticity

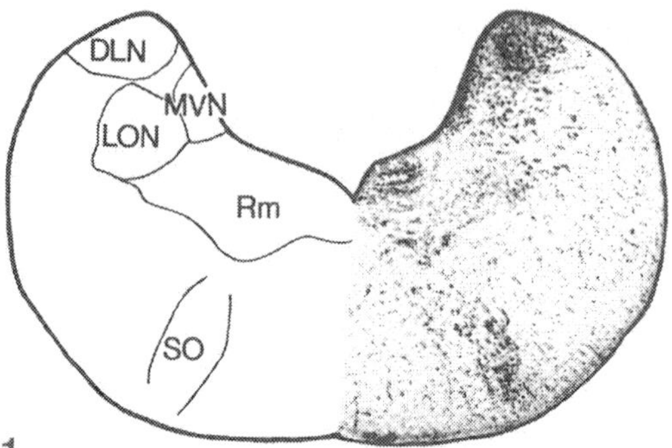

1

Fig. 1. Cross-section through the medulla of *Rana temporaria. Right half* is from a Nissl-stained section, *left half* delineates the main nuclear areas. The ventral (octavus) zone is formed by the lateral octavus nucleus (*LON*) and the medial vestibular nucleus (*MVN*). The level of the section is between the VIIIth and the Xth cranial nerve. *DLN* dorso-lateral nucleus; *Rm* medial reticular formation; *SO* superior olive

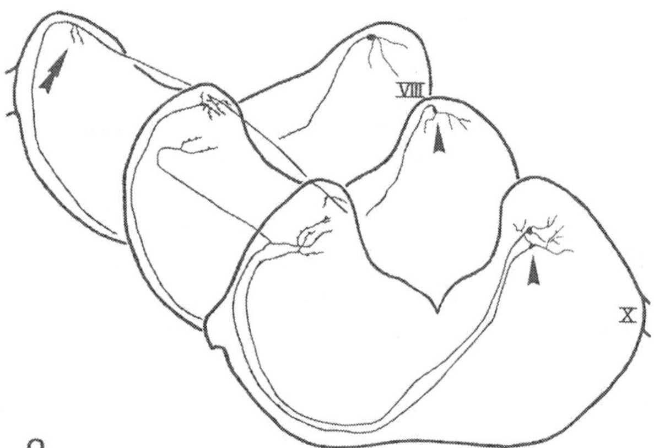

2

Fig. 2. Schematic drawing of the main types of commissural neurons (*arrowheads*) and fibers of the medullary octavus zone in the frog. *Roman numerals* indicate the level of the respective cranial nerves

fibers remain internal on the contralateral side. More detailed information about the commisural system has been revealed in a recent HRP study (Will in preparation). The main results of importance for the present report are summarized in Fig. 2 and are as follows: We can discern three types of commissural neurons with respect to their contralateral termination pattern. (1) At all rostro-caudal levels there are fibers that terminate contralaterally at around a level that corresponds to the position of their ipsilateral cell body (arrowheads in Fig. 2). (2) Many fibers from the anterior nucleus (AN) decussate, ascend to the contralateral AN and descend to the caudal end of the ventral zone within the neuropil dorsally adjacent to the ventral zone, giving off collaterals during their whole course (double arrowhead in Fig. 2). (3) From the midlevel between the VIIIth and the Xth nerve to the caudal end of the Xth nerve, fibers are found that, after decussation, ascend to the contralateral ventral zone and turn rostrally within the ventral zone up to the level of the entrance of the VIIIth nerve, giving rise to collaterals throughout their course. The cell bodies of type 1 and 3 commissural neurons are predominately located at the medial border of the lateral octavus nucleus (LON) and the caudal nucleus (CN).

2 Materials and Methods

Experiments were performed on 45 *Rana temporaria*. Thirty frogs were hemilabyrinthectomized and one bilaterally labyrinthectomized under anesthesia employing a ventral approach as described by Ewald (1892). The deviation of the head position from the horizontal, i.e., the angle of head tilt, was measured in order to quantify the compensation process prior to HRP application.

A first series of experiments was performed to reveal whether any anatomical differences between normal (nonlabyrinthectomized) and chronically deafferented frogs (2–6 months after hemilabyrinthectomy) could be detected by HRP injectons into various parts of the ventral zone. The animals were anaesthetized by immersion in an aqueous solution of Tricaine methanesulfonate (MS 222, 1:1000). The skull was opened by means of a dental drill and the dorsal medullary surface was exposed. Micropipets with tip diameters of 50–150 μm were filled with a concentrated aqueous solution of HRP, connected to a syringe via a flexible tube, driven into the brain by means of a micromanipulator and left in place for about 5 min. Two days after HRP application the animals were killed and the brains were treated for the HRP-DAB reaction as described elsewhere (Will et al. 1985a).

Differences in collateral and bouton density in the LON on the lesioned side were found between normal and chronic preparations which had received HRP injections into the contralateral ventral zone at the level of the Xth nerve. Therefore, this type of injection was performed on a second series of acutely hemilabyrinthectomized frogs to reveal the onset of this changes. This series consisted of animals 2, 4, 8, and 12 days after hemilabyrinthectomy and one frog with HRP injection 2 months after bilateral labyrinthectomy. Following HRP injection, the animals were treated as described above for the first series.

To allow for comparison and quantitative analysis, all preparations were examined according to the following criteria: (1) general appearance of labeling, (2) position of the injection site with respect to alar plate nuclei and fiber tracts, (3) size of the injection tract (diameter and length), (4) size of the HRP diffusion area, (5) labeling pattern rostral and (6) caudal to the injection site. Those preparations with injections identical according to these criteria were grouped together and analyzed quantitatively. From our two experimental series we obtained the following four groups: (a) 1 normal (N), 1 acute (4 D/days), 1 chronic 6 M/months); (b) 1 normal, 1 acute (6 D), 1 chronic (6 M); (c) 1 normal, 1 chronic (2.5 M), 1 chronic (6 M); (d) 1 normal, 1 chronic (Bi 2 M, i.e., bilateral labyrinthectomy), 1 chronic (ipsi 6 M, i.e., HRP injection ipsilateral to the labyrinthectomized side). To account for possible remaining differences in tissue damage and HRP uptake, the number of labeled CF's in each preparation was counted at the level of the tractus descendens trigemini contralateral to the injection in four sections rostral and four sections caudal to the level of the injection center. Within each group the number of CF's in every preparation was then compared with the lowest number of CF's in that group. The ratio of these two numbers, ranging from 1.1 to 2.2, reflects the difference in the amount of HRP labeling and was therefore used as a normalization factor for the comparison of collateral and bouton counts within each group.

In each preparation the number of labeled boutons and fibers were counted within the LON in three successive sections 150 μm caudal to the VIIIth nerve and in three consecutive sections rostral to the Xth nerve. Boutons and fibers were counted in each of the 20 columns of a 20 × 20 ocular grid which covered an area of 5760 μm^2 in the center of the LON. The cumulative sum of each section was then normalized to that preparation with the lowest number of CF within the respective group and divided by 10.

The problem of shrinkage due to hemilabyrinthectomy was addressed in an additional series of preparations. Brains from four compensated frogs (2.5 months after hemilabyrinthectomy) were dehydrated, embedded in paraplast, sectioned transversely at 24 μm, mounted onto slides and stained with cresyl violet. In all four preparations the alar plate on both sides of the medulla was measured in ten successive sections between the VIIIth and the Xth nerve in the following way: The width (W) of the alar plate was measured at its base, right above the spino-cerebellar tract, which was easily delineable with darkfield illumination. The height (H) was measured from the spino-cerebellar tract to the dorsal tip of the alar plate where it connects to the chorioid plexus (Fig. 3). The area of the alar plate was approximated by calculating the area of a semi-ellipse according to the formula $W/2 \times H \times \pi/2$. In all four frogs the alar plate on the lesioned side was smaller than on the intact side. For the frog with the largest reduction the mean differences between the normal and the lesioned side were dW = 69.75 μm (N = 10;s = ±10.96 μm); dH = 66.22 μm (N = 10;s = ±12.61 μm). The alar plate areas were calculated to 145 267 μm^2 on the intact and 112 740 μm^2 on the lesioned side, i.e., the area on the lesioned side was reduced to 77.6% of that of the intact side. The mean reduction was 17% for all four frogs. Under the assumption that the whole alar plate is homogeneously affected by this shrinkage, we applied a mean correction factor of 0.83 for the fiber and bouton counts in our chronic preparations.

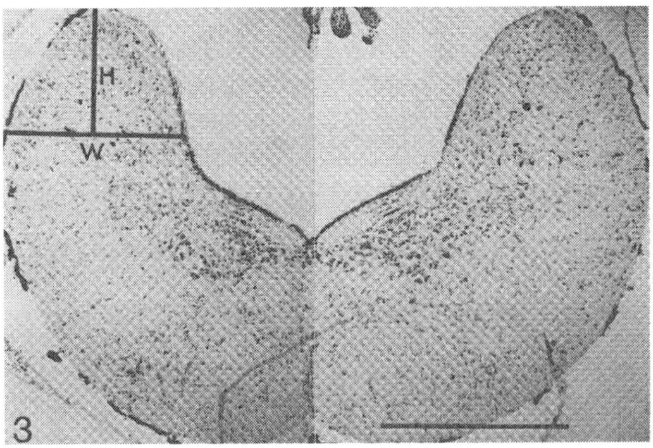

Fig. 3. Cross-section through the medulla of *R. temporaria* 2 months after hemilabyrinthectomy (*left side*). Note the difference in area of the two sides. *Bars* in the *top left* of the section indicate where height (*H*) and width (*W*) were measured. *Bar* at the *bottom* = 1 mm

3 Results

We did not detect any reliable difference between the CF system of normal and labyrinthectomized frogs following HRP injections into the rostral part of the ventral zone, i.e., from the AN down to the midlevel of the LON/MVN. HRP applications to the caudal VZ (Fig. 4), however, revealed the following: If the injections were performed contralateral to the side of the labyrinthectomy, the density of labeled CF's on the lesioned side was considerably increased compared to that in normal and acute frogs (Fig. 5). The increased density was found in an area stretching from the caudal end of the VIIIth nerve to the caudal root of the Xth nerve (Fig. 5a–d) and was due to an increase in the number of CF collaterals as well as in the number of boutons (Fig. 5e, f).

Comparing the number of boutons in normal and chronic animals there is an average increase in chronic animals by about 79% with no significant differences between chronic animals at 2 and 6 months postoperatively, Figure 6a, b display the results for the six animals of group a and d, respectively. Each column represents the mean value for three successive rostral (darkly hatched) or caudal (lightly hatched) sections. The values for the chronic animals (>2 months) have been corrected by a factor of 0.83 (see methods). The onset of the increase cannot unequivocally be determined from our material. However, group a (Fig. 6a) indicates an onset at about the fourth day after labyrinthectomy in the rostral part of the terminal field.

An increase of boutons may be established by two processes, the formation of new boutons on existing or on new collaterals. Comparing the number of collaterals per unit area in the different states (Fig. 6c, d) clearly shows an increase in collaterals. On the average there was an increase by 65% for chronic animals with no significant differences between chronic specimens 2 and 6 months postoperatively. The increase in fibers/unit apparently had the same time course as the increase in boutons/unit.

Group a (Fig. 6a, c) and d (Fig. 6b, d) are two groups with slightly different injection sites. It we compare the number of fibers/unit in the rostral and caudal sections of these groups, the ratio is about 1:2 for group a and 1:3 for group d. This indicates that the commissural fiber system consists of two subpopulations with slightly different sites of origin. Most likely, these two subpopulations cor-

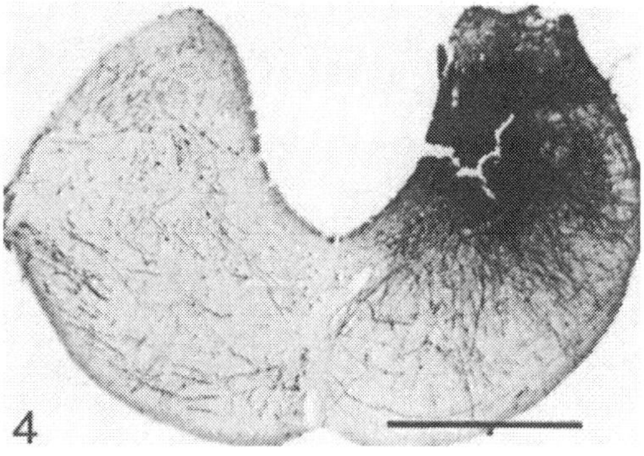

Fig. 4. Cross-section through the caudal medulla showing a large HRP injection site in the VZ and labeled commissural fibers (the rupture on the right appeared during mounting of the section). *Bar* = 1 mm

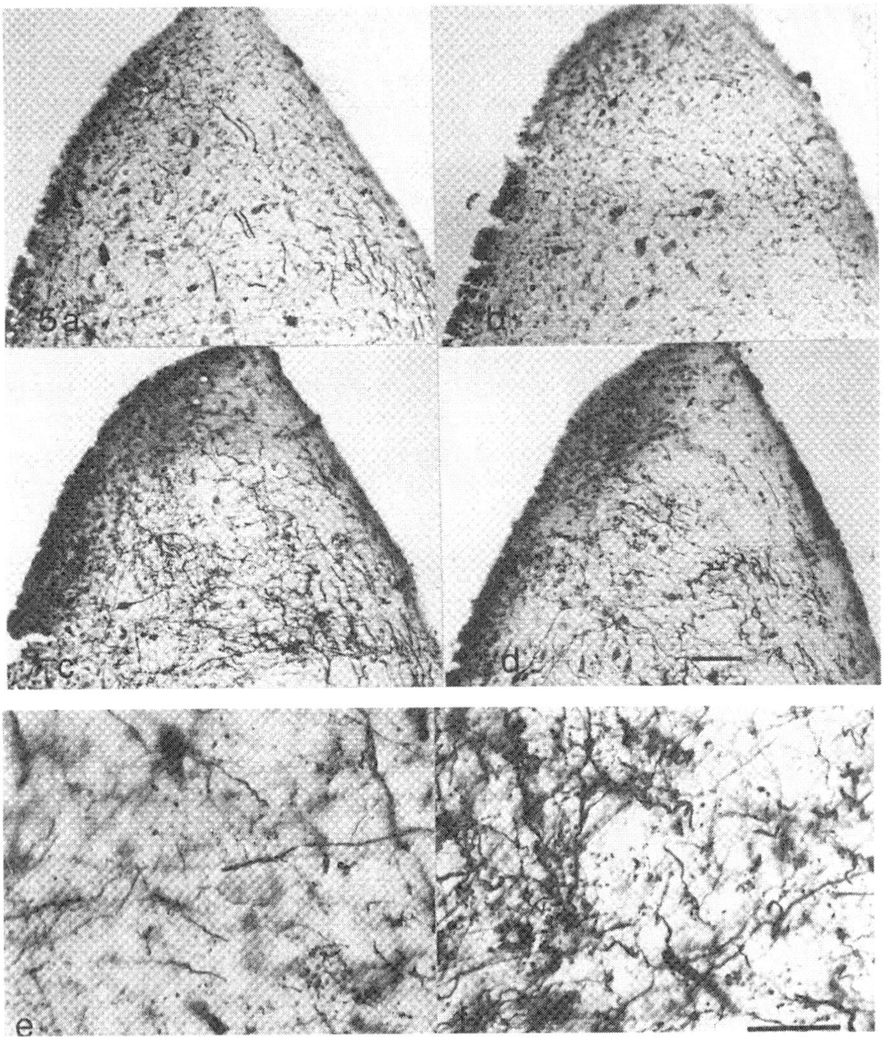

Fig. 5. a–d Sections through the LON on the lesioned side in an acute (**a, b**) and a chronic (**c, d**) hemilabyrinthectomized frog. HRP was injected into the caudal VZ on the contralateral side. **a** and **c** are at the level of the Xth nerve; **b** and **d** are 200 μm caudal to the VIIIth nerve. **Bar** in **d** (for **a–d**) is 100 μm. **e, f** CF terminal fields in the LON on the lesioned side at higher magnification. **e** Is from a normal and **f** is from a chronic frog. Both sections are 100 μm rostral to the Xth nerve. *Bar* in **f** (for **e, f**) = 50 μm

respond to the two types of commissural fibers shown in the lower sections of Fig. 2: One type projects directly opposite to their site of origin, the other type decussates and ascends within the LON up to the level of the VIIIth nerve.

Figure 6b and d display the results of group d containing two special cases. In one animal (ipsi 6 M) HRP was injected into the VZ as in the previous cases except that the injection was ipsilateral to the lesioned side. There were no signif-

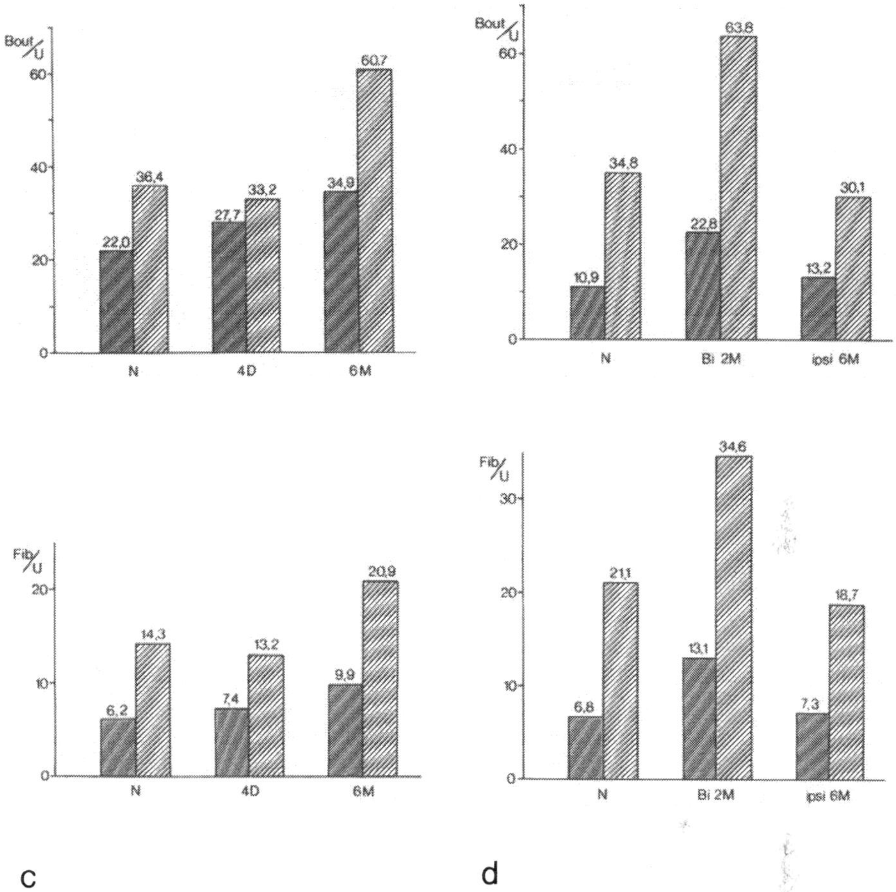

Fig. 6. Results of bouton (**a, b**) and fiber counts (**c, d**) from two groups. Each *column* represents the mean value of three successive sections. *Darkly hatched* columns are from rostral, *lightly hatched* columns from caudal sections. Except for ipsi 6M, all HRP injections were contralateral to the lesion. *N* non-labyrinthectomized; *D* days, *M* months after labyrinthectomy; *Bi* a frog with bilateral labyrinthectomy. Unit (*U*) is an area of 0.000576 mm^2

icant changes, neither in the number of boutons nor of collaterals, as compared to a nonlabyrinthectomized animal. Another animal (Bi 2 M) had been bilaterally labyrinthectomized 2 months before HRP injection. The results are essentially the same as those from hemilabyrinthectomized animals.

The identification of growth cones was somewhat hampered by our method because we aimed at relatively large HRP injections to label a large portion of CF's. It was thus not possible to trace individual fibers to their site of termination and, as the terminal fields were quite densely occupied by labeled fibers and boutons, it was impossible to distinguish between normal boutons and growth cones of similar shape. However, two types of structures were found that may be considered as growth cones. Both are flattened and less densely labeled than boutons and differ considerably from both boutons and axonal swellings like Hering

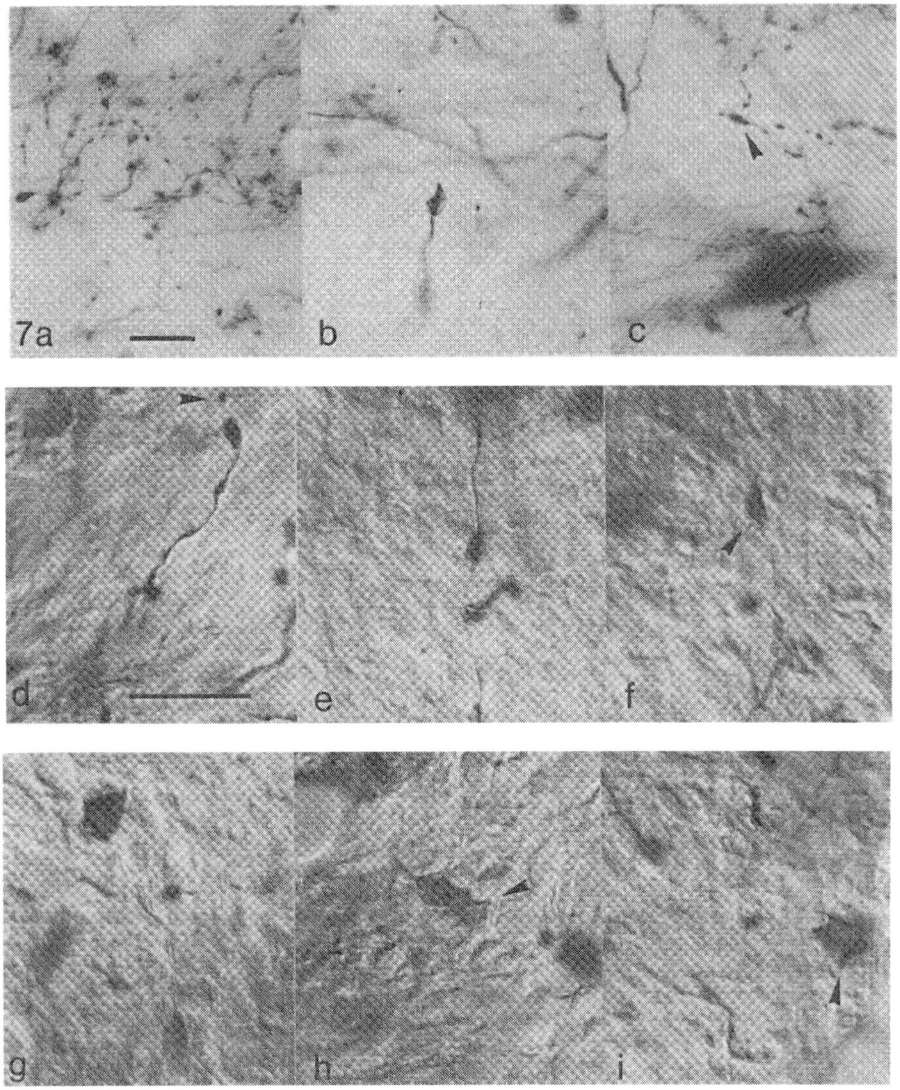

Fig. 7. Microphotographs of HRP-filled boutons, axonal swellings and presumed growth cones from the LON on the lesioned side of acute frogs. **a** A set of "normal" en passant and terminal boutons. **b** A typical axonal swelling. The dorsal continuation of the fiber is found in the next section. **c** Spatulate type 1 growth cone (*arrowhead*) with a short apical process. **d** Typical en passant and terminal boutons under Nomarski optics. The connection between the terminal bouton (*arrowhead*) and the rest of the fiber is out of focus. **e, f** Type 1 growth cones with long apical processes. In **e** this process terminates in a small thickening. **g, h, i** Type 2 growth cones. Some of the apical indentations (*arrowheads*) appear as small spines. **d–i** are Nomarski micrographs. Bars in **a** (for **a–c**) and **d** (for **d–i**) are 20 μm

bodies or transport swellings (Fig. 7b). The first type is lance-shaped with a long (up to 18 μm), filopodium-like process on the distal tip (Fig. 7c, e, f). The second type is larger (diameters of about 10–12 μm) and less elongated than the first. Its shape is ragged or notched and occasionally shows small spiny processes (Fig. 7–i). The first type resembles those growth cones described in the literature as spindle- or lance-shaped and ending in a long, fine, tapering filament which may itself terminate in a small expansion (Cajal 1959; Tennyson 1970; Scalia and Matsumoto 1985). The second type has, to our knowledge, not been described before. However, both types of structures were never found in nonlabyrinthectomized nor in chronic (>2 months) frogs. They were identified only 4, 10, and 12 days following labyrinthectomy. Therefore both types might represent growth cones.

4 Discussion

The measurement of shrinkage due to deafferentation was done without consideration of intraindividual variations of the right and left alar plate for which there were no data available. The methods applied, however, had the advantage that both the lesioned and the intact alar plate of each specimen were subjected to exactly the same procedure. The shrinkage due to the embedding process would influence both alar plates in the same way but would not affect the relative differences caused by labyrinthectomy. Therefore, the relative reduction calculated from the paraplast-embedded material may also be applied to the HRP material even though the latter was treated differently. As mentioned above, the use of this reduction coefficient implies that the whole alar plate is uniformly affected. This can be assumed because the major factor causing shrinkage is the degeneration of octavus afferents and these are more or less equally distributed throughout the alar plate (Matesz 1979; Will 1988). To date it is not known whether some of the secondary neurons do also degenerate.

Under the assumption of a uniform reduction in area of 17% of the lesioned alar plate 2.5 months after hemilabyrinthectomy, our HRP preparations show an increase in fiber and bouton density in the LON on the lesioned side. The onset of this process is detectable at about the fourth day after labyrinthectomy and concurs with the appearance of degenerating primary afferents in the octavus nuclei (Gregory 1972; Hillman 1972). The process appears to be completed after 2 months at the latest and results in an increase of collaterals/unit by about 65%. Likewise, the number of boutons/unit increases by 79%. The increase seems to be due to a collateral sprouting as defined by Davis (1985) because (1) with our type of HRP injection no fibers other than CFs were labeled in the deafferented VZ, and (2) the number of collaterals/unit in different animals were normalized to the same number of CF's.

A sprouting process is also indicated by the finding of two types of structures resembling growth cones. One of these types has, to our knowledge, not been described before and such a structure may represent growth cones in a specific state, such as retraction or they may be specific for the vestibular system. However, as we applied only light microscope criteria, the presence of growth cones needs to be confirmed by other methods. Furthermore, the relatively small number of structures identified as growth cones (17 in 3 preparations) suggests that other types of growth cones may have been masked by the methods employed (e.g., large HRP injections). On the other hand, the time course of the sprouting process indicated by our preparations, is in good agreement with behavioral studies showing that compensation, as measured by the restoration of head position, is completed about 2 months after labyrinthectomy (Flohr et al. 1981).

Galiana et al. (1984) have recently proposed a closed loop positive feedback model for the interconnected VZ. Their analytic study showed that the organized control over commissural gains in a closed loop feedback system is fully compatible with behavioral and electrophysiological findings on vestibular compensation processes. The present study supports the proposed model in so far as collateral sprouting and the generation of new boutons can be considered as structural correlates for gain changes. Furthermore, a previous study (Will 1988) has revealed morphological correlates of a closed loop feedback system in anurans: Neurons of the LON/MVN complex project to the ipsilateral CN, whilst neurons of the CN project to the contralateral LON/MVN complex and vice versa. The present study indicates that another fiber system is probably also involved, namely, those neurons of the caudal VZ that project to a level directly opposite to their cell bodies. At present it is not clear whether these neurons form a morphologically separated closed loop system or whether they are part of the feedback system just described. Moreover, further physiological studies are needed to demonstrate whether these two CF systems also perform different functions.

References

Bechterew W von (1883) Ergebnisse der Durchschneidung des N. acusticus, nebst Erörterung der Bedeutung der semicirculären Canäle für das Körpergleichgewicht. Pfluegers Arch 30:312–347

Bienhold H, Flohr H (1978) Role of commissural connections between the vestibular nuclei in compensation following unilateral labyrinthectomy. J Physiol (Lond) 284:178

Cajal S, Ramon Y (1959) Degeneration and regeneration of the nervous system (May RM trans). Hafner, New York

Davis JN (1985) Neuronal rearrangements after brain injury: a proposed classification. In: Becker DP, Povlishock J (eds) Central nervous system trauma status report. NIH, USA, pp 491–501

Dieringer N, Precht W (1979a) Mechanisms of compensation for vestibular deficits in the frog. I. Modification of the excitatory commissural system. Exp Brain Res 36:311–328

Dieringer N, Precht W (1979b) Mechanisms of compensation for vestibular deficits in the frog. II. Modification of inhibitory pathways. Exp Brain Res 36:329–341

Ewald JR (1882) Physiologische Untersuchungen über das Endorgan des N. octavus. Bergmann, Wiesbaden

Flohr H, Bienhold H, Abeln W, Macskovics I (1981) Concepts of vestibular compensation. In: Flohr H, Precht W (eds) Lesion-induced neuronal plasticity in sensorimotor systems. Springer, Berlin Heidelberg New York, pp 153–172

Fuller PM (1974) Projection of the vestibular nuclear complex in the bullfrog. Brain Behav Evol 10:157–169

Galiana HL, Flohr H, Melvill Jones G (1984) A reevaluation of intervestibular nuclear coupling: its role in vestibular compensation. J Neurophysiol 51:242–259

Gregory KM (1972) Central projections of the eighth nerve in frogs. Brain Behav Evol 5:70–88

Grofova I, Corvaja N (1972) Commissural projections from the nuclei of termination of the VIIIth cranial nerve in the toad. Brain Res 42:189–195

Hillman DE (1972) Vestibulocerebellar input in the frog: anatomy. In: Brodal A, Pompeiano O (eds) Progress in brain research, vol 37. Elsevier, Amsterdam, pp 329–339

Magnus R (1924) Körperstellung. Springer, Berlin

Matesz C (1979) Central projection of the VIIIth cranial nerve in the frog. Neurosci 4:2061–2071

Scalia F, Matsumoto DE (1985) The morphology of growth cones of regenerating optic nerve axons. J Comp Neurol 231:323–338

Tennyson VM (1970) The fine structure of the axon and growth cones of the dorsal root neuroblast of the rabbit embryo. J Cell Biol 44:62–79

Will U (1988) Organization and projections of the area octavolateralis in amphibians. In: Fritzsch B et al. (eds) Evolution of the amphibian auditory system. Wiley, New York, pp 185–208

Will U, Luhede G, Goerner P (1985a) The area octavolateralis in *Xenopus laevis*. I. Primary afferent projections. Cell Tissue Res 239:147–161

Will U, Luhede G, Goerner P (1985b) The area octavolateralis in *Xenopus laevis*. II. Second order projections and cytoarchitecture. Cell Tissue Res 239:163–175

Multimodal Sensory Substitution Process in Vestibular Compensation

C. Xerri, M. Lacour, and L. Borel

Université de Provence, Département de Psychophysiologie, U.A., C.N.R.S., N° 372, Centre de St Jérôme, F-13397 Marseille Cedex 13, France

1 Introduction

The post-lesion plasticity which contributes to functional recovery has long been recognized as a basic property of the nervous system. Substitution and/or restitution of neuronal response characteristics in deafferented nervous structures are thought to occur through a remodelling of the neural circuits (structural plasticity) and/or a modification of the neural transmission (synaptic plasticity). Moreover, the adaptation process involving an extensive reorganization of the concerted activity in functionally dependent neural networks might consist of readjusting the sensorimotor programs and redefining appropriate motor strategies.

Although recent studies have provided convincing evidence that axonal growth as well as collateral sprouting occur in many central nervous structures and may form new functional synapses (Raisman 1969; Moore et al. 1971; Tsukahara et al. 1974), the regenerative capacity of the Central Nervous System (CNS) seems to be more limited in adults than in the young. It is thus generally agreed that structural plasticity plays a complementary rather than a leading role in functional recovery in the adult CNS. Sensory substitution processes which do not necessarily involve structural changes may mainly depend, at the cellular level, on specific modifications in the synaptic transmission, leading to increased efficiency in the spared inputs.

It seems to be clearly established that post-lesion recovery involves several nervous structures and sensory modalities which co-operate in the same function. Integrated functions are controlled by various sensorimotor systems which can be regarded as complementary subsystems, the working ranges of which partly overlap. In this respect, sensorimotor subsystems mediating a common function can potentially be substituted – at least to some extent – for the injured subsystem.

The vestibular system can be said to be a "subsystem" co-operating with the visual and somatosensory "subsystems" in oculomotor and postural control as well as in space orientation. Remarkable functional recovery, referred to as vestibular compensation, occurs after unilateral vestibular lesion in many species (cf. Schaefer and Meyer 1974). In the present chapter we provide evidence that vestibular compensation can be regarded as a multimodal sensory substitution process and constitutes a suitable model for studying the plasticity mechanisms underlying post-lesion recovery.

H. Flohr (Ed.)
Post-Lesion Neural Plasticity

2 Vestibular Compensation

Acute unilateral lesions of the peripheral vestibular system lead to dramatic impairments of the vestibulo-spinal and vestibulo-ocular regulations, which result in severe head and limb postural asymmetries, associated with spontaneous eye nystagmus and asymmetrical vestibulo-ocular reflexes. It is generally admitted that all these deficits are basically determined by the strong imbalance between the tonic and phasic neuronal responses in opposite vestibular nuclei. Indeed, single-unit recordings in the vestibular nuclei have shown that a strong depression of the spontaneous firing rate occurs on the side of the lesion (Mc Cabe and Ryu 1969; Shimazu and Precht 1966; Xerri et al. 1983) which is associated with an increase in the activity on the intact side, at least in canal-related neurons (Shimazu and Precht 1966). Moreover, canal-related (Precht 1974) as well as otolith-related neurons (Xerri et al. 1983) in the deafferented vestibular nuclei lose their dynamic sensitivity to rotation. In line with the electrophysiological findings, asymmetries in metabolic activity not only between the bilateral vestibular nuclei but also within many other structures have been demonstrated by means of the 2-deoxy-D-glucose technique (Llinas and Walton 1979; Flohr et al. 1981).

The symptoms induced by peripheral vestibular lesion disappear with a time course which varies from one species to another (cf. Schaefer and Meyer 1974). In the cat and the monkey most of the motor disturbances become greatly attenuated with the first post-lesion week, which suggests that the adaptive mechanisms intervene rapidly. However, complete or near-complete compensation for the deficits is not achieved before 4 or 5 weeks in these species. In the course of time after the lesion, symmetrical resting rates have been recorded in both the deafferented and intact vestibular nuclei but the firing levels remained below normal on both sides (Precht et al. 1966; Pompeiano et al. 1984; Ried et al. 1984). Owing to contralateral labyrinth influences, the dynamic response properties were partially recreated in deafferented vestibular neurons (Precht et al. 1966; Xerri et al. 1983). Interestingly, Llinas and Walton (1979) and Flohr et al. (1981) have reported a recovery of near-symmetrical metabolic activity in the vestibular nuclei and other structures such as the cerebellum, the reticular formation, and the inferior olive which may be crucially involved in functional recovery. The recovery process occurring after unilateral vestibular lesion attenuates the static and dynamic asymmetries which severely impair posture, locomotion, and oculomotor control. For this purpose the CNS has not only to rebalance resting activity in the nervous structures affected by the lesion but also to recalibrate vestibulo-spinal and vestibulo-ocular influences. This implies that remaining (contralateral) vestibular afferents, and visual and proprioceptive inputs involved in the perception of space, body posture and body motion in space contribute to recreating dynamic neuronal responses which are functionally equivalent to the missing vestibular signals.

3 Contribution of Remaining Labyrinthine Influences

The fact that the intact labyrinth plays a part in the compensation for postural deficits after hemilabyrinthectomy has long been recognized by clinicians. Recent

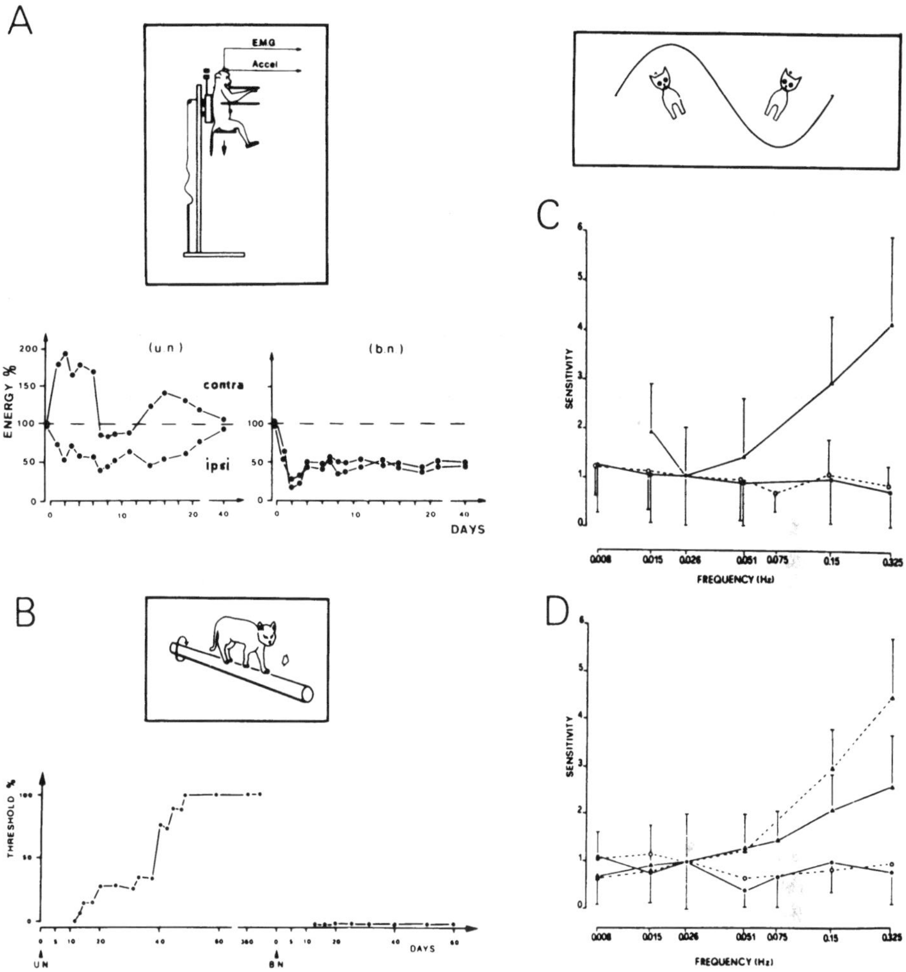

Fig. 1. Role of remaining labyrinthine afferents in vestibular compensation. **A** Mean time course of the muscular energy (surface of integrated EMG) developed in soleus muscles during vertical falling (8.8 m/s^2) after unilateral (*u.n.*) and bilateral (*b.n.*) vestibular neurectomy in the baboon. Post-lesion measurements of energy are plotted as a function of time in days and expressed as a percentage of control values. Note that initial asymmetries between ipsilateral and contralateral muscle responses decreased considerably within 6–8 days and normal values were recovered in u.n. baboons, while recovery remained partial in b.n. baboons. **B** Recovery of locomotor equilibrium performances on a rotating beam in the cat. Maximum performance (highest speed of rotation which did not cause the animal to fall) evaluated after u.n. and b.n. is expressed as a percentage of maximum control performance. Note that recovery was complete at about 50 days in u.n. cats, while the second neurectomy prevented animals from recovering. **C, D** Changes in the dynamic response characteristics of Deiters' neurons after unilateral vestibular deafferentation in the cat. The response sensitivity of each unit to sinusoidal tilt (peak amplitude: 10 °) was tested by increasing the rotation frequency from 0.008 Hz to 0.32 Hz. The sensitivity of unit responses was normalized at 0.026 Hz. Recordings were performed ipsilaterally (*broken line*) and contralaterally (*solid line*) to the lesion in the acute state (**C**). Note the presence of two populations of neurons on the intact side: neurons which displayed a stable response sensitivity and neurons exhibiting an increased response sensitivity over the frequency range. The dynamic unit responses disappeared after acute lesion but recovered significantly (*solid line*) within about 1 month post-lesion (**D**) although the values were infranormal (*broken line*)

studies have precisely quantified the post-lesion course of motor disorders. These studies have clearly shown that remaining (contralateral) vestibular afferents are necessary to the recovery of phasic muscular responses to falling (Lacour et al. 1979) and rapid locomotor adjustments in the monkey (Igarashi et al. 1970) and the cat (Xerri and Lacour 1980) (Fig. 1 A, B).

Interestingly, Dieringer and Precht (1979 a, b) have demonstrated that changes in the synaptic efficiency of the commissural system, linking canal-related neurons in opposite vestibular nuclei, develop during compensation in the frog: the synaptic transmission of excitatory semi-circular canal influences increased significantly in the deafferented vestibular nuclei. Furthermore, the efficiency of inhibitory pathways involving cerebellar and brain stem neurons was also increased. In the cat the picture is not so clear. Indeed, in their original study, Shimazu and Precht (1966) reported that after an acute lesion, very few type I canal neurons were to be found in the deafferented vestibular nuclei, in the cat; moreover, these neurons responded weakly to angular acceleration. These neuronal events, together with the removal of commissural vestibular influences, were assumed to be responsible for asymmetry in the vestibulo-ocular reflex (VOR) gain. At the chronic stage, the potency of commissurally mediated inhibition was reported to increase on the side of the lesion (Precht et al. 1966). Functionally the commissural pathways may have modulated the activity of deafferented neurons, by inhibition and disinhibition, thereby producing responses which were qualitatively equivalent to that depending on ipsilateral canal inputs. Recent findings do not support this assumption: using polarizing stimulation of the labyrinth, Ried et al. (1984) showed that contralateral vestibular influences do not become more effective at the chronic lesion stage in the cat. Methodological differences (electrical versus natural stimulation) may explain the discrepancy between the results of these two studies. Yet the findings reported by Precht et al. (1966) that commissural efficiency was not significantly changed and that the poor sensitivity to angular acceleration persisted in the deafferented nuclei agree with the permanent deficits observed in VOR symmetry (Maioli et al. 1983).

Concerning otolith-related neurons, acute vestibular neurectomy results, on the side of the lesion, in selective suppression of unit responses characterized by a frequency-dependent sensitivity during animal tilt (Xerri et al. 1983) (Fig. 1 C). This loss of dynamic neuronal properties creates a considerable asymmetry in the output of vestibulospinal units during motion, as confirmed by Lacour et al. (1985). We observed that dynamic response characteristics were partially but significantly recovered in the chronic state. This suggests that remaining otolith signals can to some extent replace the missing dynamic otolith inputs on the deafferented side and rebalance neuronal activities (Fig. 1 D). At present it is not known whether these vicarious otolith signals are mediated through the fastigial (excitatory influences) and/or the spino-cerebellar pathways (inhibitory influences). Whatever it may be the recovery of unit response characteristics necessarily implies that an extensive reorganization of the contralateral labyrinthine influences has taken place. This reorganization probably contributes to the recovery of dynamic motor functions.

4 Role of Visual Cues

Ever since the early research by Ewald (1892), Magnus (1924), and Kolb (1955), it has been known that vision plays an important role in vestibular compensation. More recent findings indicate that static visual cues provide necessary information in order to compensate for tonic head tilt following hemilabyrinthectomy in the cat (Putkonen et al. 1977). The same conclusion was drawn by Courjon et al. (1977) concerning spontaneous vestibular nystagmus. Other experiments have demonstrated that visual motion cues contribute to the recovery of dynamic reflexes. The poor VOR gain and strong phase shift measured in the dark show little or no improvement with time, but in the light near-compensatory eye movements are observed (Precht et al. 1981). Furthermore we have shown that vision plays a crucial role in the recovery of reflex responses evoked in various extensor muscles during vertical falling in the monkey (Lacour and Xerri 1980; Lacour et al. 1981): in the acute stage after the lesion, visual motion cues are selected as the most relevant sensory information for eliciting appropriate muscle responses, whereas in the chronic state the remaining labyrinthine afferents appear to be sufficient (Fig. 2A, B). Our data strongly suggest that visual information fulfills a sensory substitution function in that it compensates for the lack of dynamic vestibular input in deafferented central nervous structures.

Electrophysiological investigations on various species have yielded convincing evidence that visual and vestibular inputs converge and interact in numerous central nervous structures, particularly in the vestibular nuclei complex in which motion velocity coding improves when both types of signal are combined (cf. Xerri et al. 1987). This important result suggests that visual afferents can potentially be substituted for missing vestibular afferents in individual neurons. It is well known that the visual system is "tuned" to low frequency motion whereas the vestibular system is "tuned" to higher frequency motion. The sensory substitution hypothesis would therefore require plastic changes in the working range of the visual system compensating for the loss of neuronal coding in the high frequency range of motion. Surprisingly no neuronal data supporting the existence of this kind of visual substitution process had been provided as yet. We recently investigated the post-lesion influence of visual motion cues on vestibular nuclei units in the awake cat. In intact animals, an optokinetic stimulus (pseudo-random visual pattern) moving sinusoidally in the vertical plane induced a sinusoidal modulation of the unit firing rate which was maximum at the 0.10 Hz stimulus frequency level, averaging the neurons. Above 0.25 Hz the visually-induced response was poorly developed, while a strong discharge modulation was elicited by vestibular signals during animal displacement in the dark (Fig. 2C). Interestingly, during the first weeks of functional recovery, recordings in the deafferented vestibular nuclei showed that optokinetic stimuli evoke a powerful discharge modulation above 0.25 Hz. The unit responses increased almost linearly within the 0.025 Hz–1 Hz frequency bandwidth, indicating that the working range of the visual system was extended towards higher frequencies. However, visually induced neuronal modulation remained below that recorded in the intact animal during vestibular stimulation in the high frequency range. It might well be that a combination of visual and contralateral vestibular inputs fully compensates for

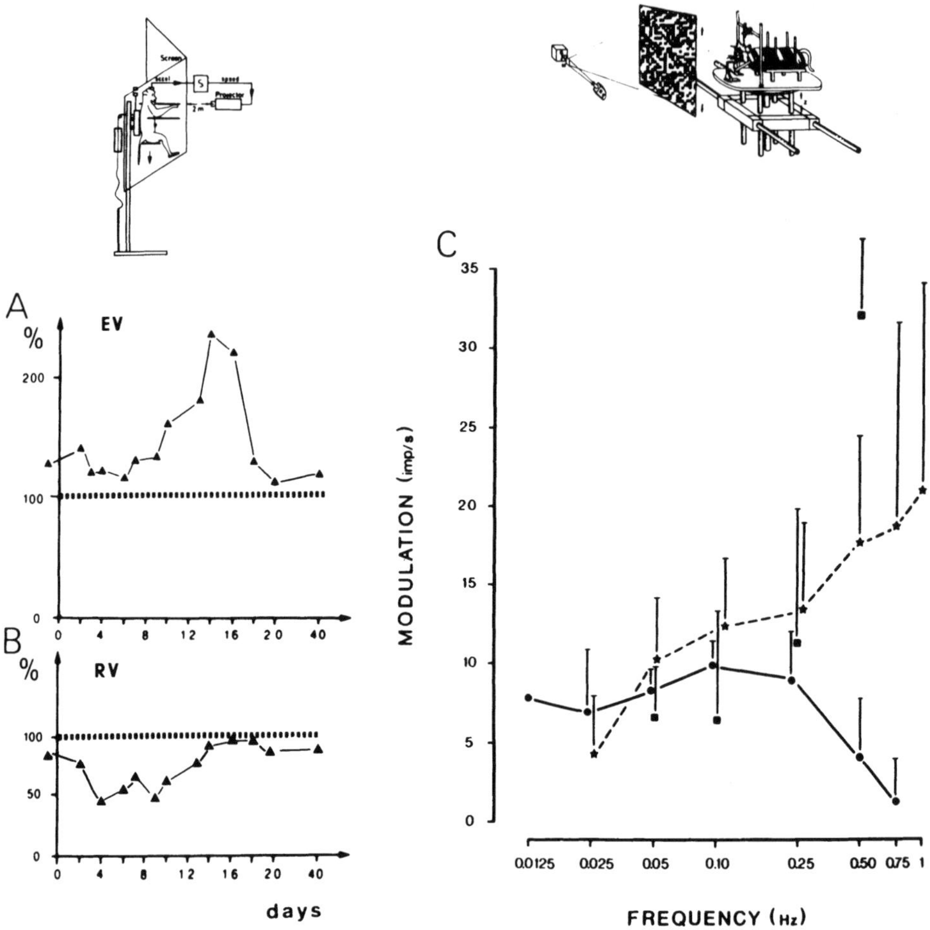

Fig. 2. Role of visual motion cues in vestibular compensation. **A, B** Mean time course of the muscular energy developed in the splenius capitis muscle during vertical falling at 2.2 m/s^2, after contralateral vestibular neurectomy, in the baboon. The visual surround (structured black and white pattern) projected onto a screen placed in front of the animal remained stationary (normal vision, NV) or moved either upward (enhanced vision, EV) or downward (reduced vision, RV) during the animal fall. The relative velocity of the visual surround was 1.6 (EV) or 0.4 (RV) times the velocity (0.55 m/s) of the falling monkey. The energy recorded under conditions EV (**A**) and RV (**B**) are expressed as a percentage of the value obtained under NV conditions and plotted as a function of the post-operative time. Note that the visual-dependent variation in muscular energy was stronger during the two first weeks after the lesion. **C** Change in firing rate modulation of vestibular nuclei neurons after ipsilateral vestibular deafferentation in the cat. Unit responses were evaluated in the 0.0125 Hz-1 Hz frequency range (corresponding peak velocities: 0.02 m/s–1.82 m/s) during otolith stimulation (animal motion in the dark; *black squares*) in intact cats and during visual stimulation (visual surround motion) in intact animals (*filled circles*) and animals with lesions (*stars*). Unit responses recorded from the 4th to the 15th days post-lesion were averaged. Note that during this period, the efficacy of visual signals was potentiated within the upper frequency range

the loss of ipsilateral vestibular signals. The effects of vision seem to be confined to the magnitude of discharge modulation, since no change in the input-output phase relationship was observed after the lesion. These data provide new arguments supporting the notion that visual signals constitute a functional sensory substitution source in the vestibular compensation process.

5 Influence of Somatosensory Afferents

Numerous studies have focused on the contribution of somatosensory inputs to compensation for static (Kolb 1955; Azzena 1969; Schaefer and Meyer 1973; Lacour et al. 1976; Jensen 1979) and dynamic (Xerri and Lacour 1980; Dieringer and Precht 1981) vestibular deficits. These authors have suggested that the potency of ascending proprioceptive inputs projecting onto vestibular neurons may increase with time and these inputs might thereby be substituted for part of the defective labyrinthine input. However, very little is known about the nature and origin of these proprioceptive afferents. Evidence has been provided that neck afferents intervene in the compensation for the postural syndrome in the guinea pig (De Jong et al. 1977; Jensen 1979; Schaefer et al. 1979). Moreover, after bilateral labyrinthectomy in the monkey (Dichgans et al. 1973) as well as in man (Kasai and Zee 1978) the cervico-ocular reflex elicited during head rotation is potentiated with time. These findings suggest that neck afferent inputs can indeed compensate for the loss of vestibular input. It is striking that little attention had been paid to the underlying neurophysiological mechanisms.

Vestibulospinal neurons integrate not only visual and vestibular signals but also information originating from neck receptors (cf. Boyle and Pompeiano 1980) which contribute to the modulation of limb reflexes (Kim and Partridge 1969) as well as to the postural adjustments (Lindsay et al. 1976; Ezure and Wilson 1983; Manzoni et al. 1983) that occur during head motion. Analysis of vestibulo-spinal unit responses to neck proprioceptor input during the course of the vestibular compensation is therefore a valuable means of better understanding the neural substrates involved in sensory substitution. We performed this analysis to investigate how neck afferent signals are able to compensate for the loss of vestibular afferents and/or participate in the elaboration of adaptive postural reactions (Xerri et al. 1985, 1988). In the acute state a significant increase in the proportion as well as the sensitivity of neuronal responses to neck rotation was encountered in the deafferented Deiters' nucleus (Fig. 3 A, B). This strongly suggests that the synaptic efficiency of the proprioceptive neck input increases after vestibular deafferentation. In line with this hypothesis is the finding that transection of the spinal cord in compensated hemilabyrinthectomized guinea pigs induces a strong depression of the tonic activity of vestibulospinal neurons ipsilaterally to the vestibular lesion (Azzena et al. 1976; Tolu et al. 1980) thus resulting in a decompensation, i.e., the reappearance of vestibular deficits (Jensen 1979). The above results are also in good agreement with the finding that postural compensation is delayed in the guinea pig when the animal is prevented from contacting the ground (Schaefer and Meyer 1974). Likewise it has been found that hindlimb restraint prevents and delays locomotor ataxia compensation in the monkey (La-

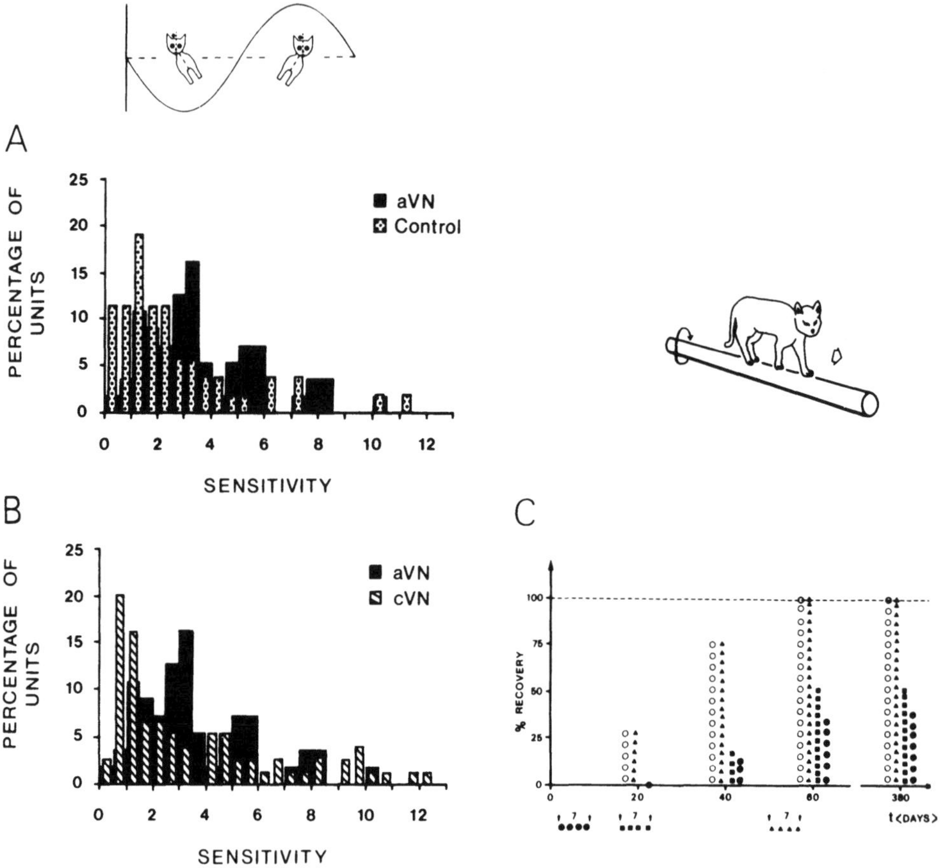

Fig. 3. Role of somatosensory afferents in vestibular compensation. **A, B** Histograms illustrating modifications in the distribution of the response sensitivity (%/deg) of vestibular nuclei units to neck receptor stimulation after acute (*aVN*) and chronic (*cVN*) ipsilateral vestibular deafferentation in the cat. Mean increase in sensitivity after aVN was significant at $P<0.05$. **C** Effect of sensorimotor restriction (SMR) on the recovery of locomotor performance after unilateral vestibular neurectomy in the cat. Post-operative time course of the maximum performance (see legend to Fig. 1) in control animals (*empty circles*) and animals subjected to SMR (duration 7 days) during the 1st (*filled circles*), the 3rd (*squares*), and the 8th (*triangles*) post-lesion week. The degree of recovery is expressed as a percentage of the performance level evaluated prior to lesion in each animal. Note that SMR applied in the 1st and 3rd weeks prevented and delayed locomotor equilibrium recovery

cour et al. 1976) and that locomotor restriction applied in the early stages of compensation also delays recovery of the dynamic equilibrium function in the cat (Xerri and Lacour 1980).

6 Concluding Remarks

Behavioral, neurophysiological and biochemical investigations have clearly demonstrated that vestibular compensation is achieved by means of a multisensory substitution process. Functional recovery depends moreover upon the concerted activity of many nervous structures (reticular formation, cerebellum, inferior olive, vestibular nuclei) which integrate labyrinthine, visual, and somatosensory afferents and contribute to the regulation of sensorimotor programs. This is attested by the decompensation phenomenon occurring after destruction of these various CNS structures (see Schaefer and Meyer 1974) and by the post-lesion changes in their metabolic activity (Llinas and Walton 1979).

Post-lesion plasticity is an ongoing process which has to meet two separate requirements, i.e., the achievement and the maintenance of the functional recovery. Data in the literature show that the nervous structures and sensory modalities involved in the process do not serve these two purposes in an identical manner. Moreover, it seems relevant to consider that recovery means dealing with two distinct although not completely independent tasks. The first, which relies on a "*short-term process*", is to restore rapidly a new balance in the tonic and phasic properties of sensorimotor systems so as to reduce the asymmetry in oculomotor and posturo-kinetic functions. The second task, which relies on a "*long-term process*", is to handle dynamic performance in order to recalibrate the gain of vestibulo-ocular and vestibulo-spinal reflexes by producing functionally adaptive changes in sensorimotor systems cooperating in gaze and postural stabilization. Ultimately in order to understand the recovery process, the nature of the plasticity mechanisms involved must also be taken into consideration.

Functional Rebalance. The significant reduction in static and dynamic functional asymmetry, which develops rapidly after unilateral vestibular lesion has been well documented. For instance, a fairly clear symmetry was found to exist in spinal motoneuron excitability at rest (Lacour et al. 1976: monkey) and postural symmetry was regained (Xerri and Lacour 1980: cat) within about 1 week post-lesion. The spontaneous nystagmus observed in the light was found to disappear within the same period of time. The rapid reduction of static deficits is determined to a large extent by recovery of the symmetrical although infranormal resting discharge rate in opposite vestibular nuclei (VN). Electrophysiological investigations have shown that balance is regained through an increased and decreased spontaneous unit discharge in the deafferented and intact VN, respectively (see Sect. 2). Although the sensory inputs involved in the increase in spontaneous activity has not been identified, it is reasonable to assume that all the spared afferents converging onto partially deafferented neurons will contribute to this restoration. The Bechterew phenomenon (decompensation occurring after removal of the remaining labyrinth about 1 week after the first lesion) indicates that nonlabyrinthine afferents contribute to maintaining the newly established resting discharge.

Dynamic rebalance is also achieved: the gain in vestibulospinal reflexes becomes symmetrical but at an infranormal level (Lacour et al. 1979: monkey). It seems to be well established that vision is an important factor in dynamic rebalance: visual motion cues exert a greater influence on these muscles which are mainly in relation with the deafferented vestibular nuclei (Lacour and Xerri 1980: monkey). Studies dealing with the VOR reflex have also shown that a significant increase in the right-left symmetry occurs within the first 10 days post-lesion (Maioli et al. 1982: cat).

Functional Recalibration. Complete functional recovery requires several weeks to be achieved. Obviously, a long-term process is needed to recalibrate the gain in dynamic output of the "vestibular subsystem" and other subsystems acting jointly. In chronic animals, the gain in phasic vestibulospinal influences onto extensor muscles is comparable to that recorded in intact animals (Lacour et al. 1979: monkey). Locomotor adjustments are fully recovered only if one labyrinth is left intact (Igarashi et al. 1970: monkey; Xerri and Lacour 1980: cat). The fact that the remaining labyrinth is particularly involved in the recovery of dynamic sensorimotor performances has been attested by single-unit studies. In the deafferented vestibular nuclei, contralateral labyrinth influences recreated dynamic neuronal properties which were previously determined by ipsilateral labyrinth inputs (Precht 1974; Xerri et al. 1983: cat). Commissural (Bienhold and Flohr 1978: frog; Galiana et al. 1984) and it is very likely that non-commissural pathways mediate appropriate changes in contralateral labyrinthine inputs. Visual cues also have a recalibrating function in the recovery of the energy developed in neck and hindlimb extensor muscles (Lacour and Xerri 1980: monkey), as well as in the recovery of a stable head posture (Putkonen et al. 1977; Courjon and Jeannerod 1979: cat). Furthermore, visual input is needed to improve the gain in both horizontal VOR and OKR (optokinetic reflex) (Precht et al. 1981: cat). A neural substrate for the vicarious visual function has been found in vestibular nuclei units: visual motion cues can supply missing vestibular information during rapid displacement and thereby contribute to recalibrating the dynamic output of these neurons (see Sect. 4).

As far as gaze stabilization is concerned, post-lesion potentiation of the cervico-ocular reflex gain (Dichgans et al. 1973: monkey; Kasai and Zee 1978: man) helps to compensate for remaining deficits in VOR and OKR in the chronic state. These post-lesion modified reflexes can act synergistically to accomplish a proper stabilizing function. Somatosensory afferents have also been suggested to contribute to the recalibration of locomotor adjustments, although continuous availability is not necessary to maintain compensation (Xerri and Lacour 1980: cat). We observed that, however, somatosensory inputs are not involved in the recovery of static posture in the cat, contrary to what was reported in the guinea pig (Schaefer and Meyer 1974). Besides the potentiation of synergistic reflexes spared by vestibular lesion, adaptive changes can also involve antagonistic reflexes. In intact animals, symmetrical posture is achieved owing to continuous cancellation of vestibular and neck reflexes acting on neck and limb extensor muscles (see Manzoni et al. 1979 for review). Postural asymmetries induced by unilateral vestibular lesion are due to a large extent to the action of neck reflexes producing contraction of neck extensors and relaxation of limb extensors on the side of the lesion. Interestingly, in the chronic stage, we observed drastic changes in the response phase of vestibulospinal neurons to neck rotation which probably contribute to attenuating postural asymmetries (Xerri et al. 1985: cat). It can be postulated that vestibular compensation relies upon adaptive modification of the strength of covariant sensory modalities devoted to the same function. An appropriate adjustment of the balance between these inputs may be necessary for sensorimotor control to be properly achieved.

Plasticity Mechanisms. Although the nature of the plasticity mechanisms underlying vestibular compensation is still a controversial issue, synaptic as well as non-synaptic processes are generally thought to be involved. The rapid increase (within a few hours of the lesion) in the response sensitivity of deafferented vestibular nuclei neurons to spared inputs suggests that activation of previously non-functional synapses and/or potentiation of synaptic efficacy have occurred (Nelson and Mendell 1979). If they do not simply reflect reaction to injury, these rapid changes may to some extent prevent deafferented neurons from becoming silent. As a result of denervation supersensitivity deafferented cells might also become more sensitive to the remaining inputs. This mechanism would facilitate the recovery of spontaneous activity and presumably contribute to re-establishing balance in the tonic and dynamic properties of neuronal responses. Recent findings suggest that as far as deafferented vestibular structures are concerned, true denervation supersensitivity does not involve cholinergic neurons (Bienhold et al. 1981: frog) even though evidence has been provided on the other hand that cholinergic transmitter systems play an important role in the maintenance and presumably in the acquisition of compensation (Bienhold and Flohr 1980: frog). Other brain stem transmitter (adrenergic, GABA-ergic) systems seem to be involved in vestibular compensation, suggesting that synaptic changes take place (see Flohr et al. 1985). The intervention of biochemical agents could account for the transitory substitution role played by visual motion cues in motor control (Lacour and Xerri 1980), which is not really compatible with the possibility that morphological rearrangements may occur. Biochemical changes in neural transmission leading to a potentiation of synaptic efficiency provide a plausible mechanism for sensory substitution at the neuronal level.

The reversibility of the compensated state, i.e., reappearance of the vestibular syndrome reported after pharmacological intervention, suggests that functional recovery depends on asymmetrical effects of neurotransmitters in brain structures. A long-term reconstruction mechanism has been postulated to account for vestibular compensation. Dieringer and Precht (1977a, b) have presented convincing evidence that modifications in crossed intervestibular connections take place during compensation in the hemilabyrinthectomized frog. The efficacy of excitatory commissural and inhibitory non-commissural fibers ending on vestibular neurons on the deafferented side was found to increase with time. This may account for part of the restored discharge of deafferented neurons and help to counterbalance the asymmetry in the tonic and dynamic outputs of vestibular nuclei. It has been proposed that new synapses formed within the deafferented nuclei might permanently change the efficiency of these pathways. These changes were reported to begin 4–5 days after the lesion. This interval is sufficiently long for vestibular afferent terminals to degenerate (2–3 days: cf. Hillman 1972) and fits the minimal time required for Bechterew's phenomenon to occur in the frog. The hypothesis that new functional synapses have been formed to replace degenerated axosomatic vestibular inputs is compatible with the morphological changes observed by electron microscopy within vestibular nuclei in the compensated hemilabyrinthectomized cat (Korte and Friedrich 1979). Collateral sprouting in the vestibular nuclei might also involve other sensory modalities such as somatosensory afferents, as directly proved by autoradiographic data on compensated frogs (Dieringer et al. 1984).

In conclusion, whatever the nature (morphological, biochemical) of the plasticity mechanisms involved in sensory substitution, we believe that, although directly initiated by the lesion, these adaptive capabilities are dependent upon subject-environment interactions. In other words, predetermined plasticity mechanisms operating at the cellular level might be influenced by post-lesion sensorimotor experience. If we assume, as previously stated (Xerri and Lacour 1980; Lacour et al. 1981), that vestibular compensation is a relearning process, experience will cause reafferent signals evoked during sensorimotor activities to be validated as the error signals needed to correct deficits and recalibrate motor outputs. Biochemical factors such as neuromodulators acting diffusely in the brain, might contribute to the relearning process by converting significant experience into concrete traces in neuronal networks and by inducing appropriate changes in neural parameters (cf. Flohr 1983).

Acknowledgements. The authors wish to thank Mrs M. R. Ibrahim for typing the manuscript, L. Agate for preparing the photographs and Dr. Jessica Blanc for revising the English.

References

Azzena GB (1969) Role of the spinal cord in compensating the effects of hemilabyrinthectomy. Arch Ital Biol 107:43–53

Azzena GB, Mameli O, Tolu R (1976) Vestibular nuclei of hemilabyrinthectomized guinea pigs during decompensation. Arch Ital Biol 114:389–398

Bienhold H, Flohr H (1978) Role of commissural connections between vestibular nuclei in compensation following unilateral labyrinthectomy. J Physiol (Lond) 284:178

Bienhold H, Flohr H (1980) Role of cholinergic synapses in vestibular compensation. Brain Res 195:476–478

Bienhold H, Abeln W, Flohr H (1981) Drug effects on vestibular compensation. In: Flohr H, Precht W (eds) Lesion-induced neuronal plasticity in sensorimotor systems. Springer, Berlin Heidelberg New York, pp 265–273

Boyle R, Pompeiano O (1980) Responses of vestibulospinal neurons to sinusoidal rotation of neck. J Neurophysiol 44:633–649

Courjon JH, Jeannerod M (1979) Visual substitution of labyrinthine defects. In: Granit R, Pompeiano O (eds) Reflex control of posture and movement. Progress in brain research 50. Elsevier, Amsterdam, pp 783–791

Courjon JH, Jeannerod M, Ossuzio I, Schmidt R (1977) The role of vision in compensation of vestibulo-ocular reflex after hemilabyrinthectomy in the cat. Exp Brain Res 28:235–248

Dichgans J, Bizzi E, Morasso P, Tagliasco V (1973) Mechanisms underlying recovery of eye-head coordination following bilateral labyrinthectomy in monkeys. Exp Brain Res 18:548–562

Dieringer N, Precht W (1979a) Mechanism of compensation for vestibular deficits in the frog. I. Modification of the excitatory commissural system. Exp Brain Res 36:311–328

Dieringer N, Precht W (1979b) Mechanism of compensation for vestibular deficits in the frog. II. Modifications of the inhibitory pathways. Exp Brain Res 26:329–341

Dieringer N, Precht W (1981) Functional restitution of static and dynamic reflexes in the frog after hemilabyrinthectomy. In: Flohr H, Precht W (eds) Lesion-induced neuronal plasticity in sensorimotor systems. Springer, Berlin Heidelberg New York, pp 185–196

Dieringer N, Künzle H, Precht W (1984) Increased projection of ascending dorsal root fibers to vestibular nuclei after hemilabyrinthectomy in the frog. Exp Brain Res 55:574–578

Ewald JR (1892) Physiologische Untersuchungen über das Endorgan der Nervus Octavus. Bergmann, Wiesbaden

Ezure K, Wilson VJ (1983) Dynamics of neck-to-forelimb reflexes in the cecerebrate cat. J Neurophysiol 50:688-695

Flohr H (1983) Control of plastic processes. In. Basar E, Flohr H, Haken H, Mandell AJ (eds) Synergetics of the brain. Springer, Berlin Heidelberg New York, pp 60–74
Flohr H, Bienhold H, Abeln W, Macskovics I (1981) Concepts of vestibular compensation. In: Flohr H, Precht W (eds) Lesion-induced neuronal plasticity in sensorimotor systems. Springer, Berlin Heidelberg New York, pp 153–172
Flohr H, Abeln W, Lüneburg U (1985) Neurotransmitter and neuromodulator systems involved in vestibular compensation. In: Berthoz A, Melvill Jones G (eds) Adaptative mechanisms in gaze control. Reviews of Oculomotor Control V. I. Elsevier, Amsterdam, pp 269–277
Galiana HL, Flohr H, Melvill Jones G (1984) A reevaluation of intervestibular nuclear coupling: its role in vestibular compensation. J Neurophysiol 51:242–259
Hillman DE (1972) Vestibulocerebellar input in the frog: anatomy. In: Brodal A, Pompeiano O (eds) Basic aspects of central vestibular mechanisms. Progress in brain research 37. Elsevier/North-Holland, Amsterdam, pp 329–339
Igarashi M, Watanabe T, Maxian PM (1970) Dynamic equilibrium in squirrel monkeys after unilateral and bilateral labyrinthectomy. Acta Oto-Laryngol 69:247–253
Jensen DW (1979) Vestibular compensation: tonic spinal influence upon spontaneous descending vestibular nuclear activity. Neuroscience 4:1075–1084
Jong PTVM de, Jong JMBV de, Cohen B, Jongkess LBW (1977) Ataxia and nystagmus induced by injection of local anesthetics in the neck. Ann Neurol 1:240–246
Kasai T, Zee DS (1978) Eye-head coordination in labyrinthine-defective human beings. Brain Res 144:123–141
Kim JH, Partridge LD (1969) Observations on types of response to combination of neck, vestibular, and muscle stretch signals. J Neurophysiol 32:239–250
Kolb E (1955) Untersuchungen über zentrale Kompensation und Kompensationbewegungen einseitig enstateter Frösche. Z Vgl Physiol 37:136–160
Korte GE, Friedrich VL (1979) The fine structure of the feline superior vestibular nucleus: identification and synaptology of the primary vestibular afferents. Brain Res 176:3–32
Lacour M, Xerri C (1980) Compensation of postural reactions to fall in the vestibular neurectomized monkey. Role of the visual motion cues. Exp Brain Res 40:103–110
Lacour M, Roll JP, Appaix M (1976) Modifications and development of spinal reflexes in the alert baboon (Papio papio) following an unilateral vestibular neurotomy. Brain Res 113:255–269
Lacour M, Xerri C, Hugon M (1979) Compensation of postural reactions to fall in the vestibular neurectomized monkey. Role of the remaining labyrinthine afferences. Exp Brain Res 37:563–580
Lacour M, Vidal PP, Xerri C (1981) Visual influence on vestibulospinal reflexes during linear vertical motion in normal and hemilabyrinthectomized monkeys. Exp Brain Res 43:383–394
Lacour M, Manzoni D, Pompeiano O, Xerri C (1985) Central compensation of vestibular deficits. III. Response characteristics of lateral vestibular neurons to roll tilt after contralateral labyrinth deafferentation. J Neurophysiol 54:988–1005
Lindsay KW, Roberts TDM, Rosenberg JR (1976) Asymmetric tonic reflexes and their interaction with neck reflexes in the decerebrate cat. J Physiol (Lond) 261:583–601
Llinas R, Walton K (1979) Vestibular compensation: a distributed property of the central nervous system. In: Asanuma H, Wilson VJ (eds) Integration in the nervous system. Igaku Shoin Tokyo, pp 145–166
Magnus R (1924) Körperstellung. Springer, Berlin
Maioli C, Precht W, Ried S (1982) Vestibulo-ocular and optokinetic reflex compensation following hemilabyrinthectomy in the cat. In: Roucoux A, Crommelinck M (eds) Physiological and pathological aspects of eye movements. Junk, The Hague, pp 200–208
Maioli C, Precht W, Ried S (1983) Short- and long-term modification of vestibulo-ocular response dynamics following unilateral vestibular nerve lesions in the cat. Exp Brain Res 50:259–274
Manzoni D, Pompeiano O, Stampacchia G (1979) Tonic cervical influences on posture and reflex movements. Arch Ital Biol 117:81–110
Manzoni D, Pompeiano O, Srivastava UC, Stampacchia G (1983) Responses of forelimb extensors to sinusoidal stimulation of macular labyrinth and neck receptors. Arch Ital Biol 121:205–214

McCabe BF, Ryu JH (1969) Experiments on vestibular compensation. Laryngoscope 79:1728–1736
McCabe BF, Ryu JH, Sekitani T (1972) Further experiments on vestibular compensation. Laryngoscope 82:381–396
Moore RY, Bjorklund A, Stenevi U (1971) Plastic changes in the adrenergic innervation of the rat septal area in response to denervation. Brain Res 33:13–35
Nelson SG, Mendell LM (1979) Enhancement in Ia-motoneuron synaptic transmission caudal to chronic spinal cord transection. J Neurophysiol 42:642–654
Pompeiano O, Xerri C, Gianni S, Manzoni D (1984) Central compensation of vestibular deficits. II. Influences of roll tilt on different size lateral vestibular neurons after ipsilateral labyrinth deafferentation. J Neurophysiol 52:18–38
Precht W (1974) Characteristics of vestibular neurons after acute and chronic labyrinthine destruction. In: Kornhuber HH (ed) Handbook of sensory physiology, vol VI/2. Springer, Berlin Heidelberg New York, pp 451–462
Precht W, Shimazu H, Markham CH (1966) A mechanism of central compensation of vestibular function following hemilabyrinthectomy. J Neurophysiol 19:996–1010
Precht W, Maioli C, Dieringer N, Cochran S (1981) Mechanisms of compensation of the vestibulo-ocular reflex after vestibular neurotomy. In: Flohr H, Precht W (eds) Lesion-induced neuronal plasticity in sensorimotor systems. Springer, Berlin Heidelberg New York, pp 222–230
Putkonen PTS, Courjon JH, Jeannerod M (1977) Compensation of postural effects of hemilabyrinthectomy in the cat. A sensory substitution process? Exp Brain Res 28:249–257
Raisman G (1969) Neuronal plasticity in the septal nuclei of the adult rat. Brain Res 14:25–48
Ried S, Maioli C, Precht W (1984) Vestibular nuclear neuron activity in chronically hemilabyrinthectomized cats. Acta Oto-Laryngol 98:1–13
Schaefer KP, Meyer DL (1973) Compensatory mechanisms following labyrinthine lesions in the guinea pig. A simple model of learning. In: Zippel HP (ed) Memory and transfer of information. Plenum, New York, pp 203–232
Schaefer KP, Meyer DL (1974) Compensation of vestibular lesions. In: Kornhuber HH (ed) Handbook of sensory physiology, vol VI/2. Springer, Berlin Heidelberg New York, pp 463–490
Schaefer KP, Meyer DL, Wilhems G (1979) Somatosensory and cerebellar influences on compensation of labyrinthine lesions. In: Granit R, Pompeiano O (eds) Reflex control of posture and movement. Progress in brain research, vol 50. Elsevier, Amsterdam, pp 591–598
Shimazu M, Precht W (1966) Inhibition of central vestibular neurons from the contralateral labyrinth: its mediating pathway. J Neurophysiol 29:467–492
Tolu E, Mameli O, Azzena MT, Azzena GB (1980) Dynamic responses of vestibular cell during spinal decompensation. Physiol Behav 25:637–640
Tsukahara N, Hultborn H, Murakami F (1974) Sprouting of cortico-rubral synapses in red nucleus neurons after destruction of the nucleus interpositus of the cerebellum. Experientia (Basel) 30:57–58
Xerri C, Lacour M (1980) Compensation des déficits posturaux et cinétiques après neurectomie vestibulaire unilatérale chez le Chat. Rôle de l'activité sensorimotrice. Acta Oto-Laryngol 90:414–424
Xerri C, Gianni S, Manzoni D, Pompeiano O (1983) Central compensation of vestibular deficits. I. Response characteristics of lateral vestibular neurons to roll tilt after ipsilateral labyrinth deafferentation. J Neurophysiol 50:428–448
Xerri C, Gianni S, Manzoni D, Pompeiano O (1985) Central compensation of vestibular deficits. IV. Response characteristics of lateral vestibular neurons to neck rotation after labyrinth deafferentation. J Neurophysiol 54:1006–1025
Xerri C, Barthélémy J, Borel L, Lacour M (1987) Neuronal coding of linear motion in the vestibular nuclei of the alert cat. III. Dynamic characteristics of visual-otolith interactions. Exp Brain Res 70:299–309
Xerri C, Pompeiano O, Manzoni D (1988) Central compensation of vestibular deficits. Convergence and interaction fo neck and contralateral macular inputs on lateral vestibular neurons after ipsilateral labyrinth deafferentation. Arch Ital Biol (submitted)

Electrophysiological Changes in Vestibular Cortex After Labyrinthine Input Deprivation

G. B. Azzena, M. A. Caria, and F. Melis
Institute of Human Physiology, Sassari University Medical School,
Viale San Pietro 43/B, 07100 Sassari, Italy

1 Introduction

Vestibular compensation is the typical substitution process that corrects the effects of unilateral vestibular deprivation. This phenomenon has been deeply investigated, not only to discover the intrinsic mechanism of the process but also as an experimental tool for studying neuronal plasticity. Several nervous structures have been proposed as responsible for readjusting the vestibular balance deranged by vestibular deafferentation (Azzena 1969; Precht 1974; Azzena et al. 1976, 1977, 1979, 1981; Putkonen et al. 1977; Dieringer and Precht 1979; Flohr et al. 1981; Courjon et al. 1982; Dieringer et al. 1984; Ried et al. 1984; Mameli and Tolu 1986; Smith et al. 1986). However, up to now no reports have been presented on the relationship between unilateral vestibular lesion and consequent modifications in the cortical vestibular function. Several papers indicated that the cerebral vestibular area could be considered as the center for vestibular perception (Penfield 1957, Kornhuber and da Fonseca 1964). Therefore, alterations in vestibular nuclear activity might find a corresponding disturbance at cortical level. In man, this could be identified as vertigo. During compensation, also the vestibular derangement sensation gradually disappears, which could be related to changes in the electrical activity of the cortical vestibular area. The experiments in this paper were aimed at verifying this possibility. Since spinal transection causes the reappearance of symptoms elicited by a previous unilateral vestibular deprivation (Azzena 1969), the electrical activity of the same cortical zone was also studied during spinal decompensation.

2 Methods

The experiments were performed on 72 rabbits weighing 2–2.5 kg, subdivided into two groups:

a) in the first group (27 rabbits), the functional characteristics of the vestibular cortex were studied during the acute period of vestibular deafferentation;

b) in the second group (45 rabbits), the vestibular cortex was studied following the transection of the spinal cord. This group of animals was further subdivided into two subgroups. One, formed of 30 hemilabyrinthectomized and compensated animals and the second of 15 normal animals. The vestibular deafferen-

H. Flohr (Ed.)
Post-Lesion Neural Plasticity

tation was performed during barbiturate anesthesia and in asepsis by section of the left VIII nerve, which was reached by a translabyrinthine approach (Azzena et al. 1986). The animals were allowed to compensate the labyrinthine syndrome for at least 40 days.

First Group. The animals were anesthetized with alpha-chloralose (80 mg/kg), tracheotomized, cannulated and fixed to the stereotaxic frame. During surgical preparation, additional doses of barbiturate were given every 45 min. ECG, EEG, and blood pressure were monitored during the experiment. The exposition of both VIII nerves was performed by a retrocochlear approach along the course of the facial nerve; the preparation was completed by removing the VII nerve, while the VIII nerves and their branches were kept intact (Azzena et al. 1986). The surgical procedure was completed by exposing both cerebral hemispheres which were protected by a pool of warm mineral oil. A pair of stainless steel electrodes, insulated except at the tips and 1 mm apart, was applied to the right VIII nerve, which was protected, as was the left nerve, by a mixture of low melting point paraffin and warm mineral oil. Electric stimulation was performed by rectangular pulses of 0.1 ms duration; 0.5–20 Hz; 1.5–3 V. This intensity was able to produce ocular jerks and was kept constant throughout the experiment. Furthermore, this intensity elicited in control animals a clear sequence of P1–N1–N2 waves in the complex of the vestibular nuclei (Azzena et al. 1976). The animals were then paralyzed and artificially ventilated by controlling the composition of the respiratory gases.

In each experiment localization of the vestibular cortex of both sides, according to previous results (Boisacq-Schepens and Rocoux-Hanus 1972; Ödkvist et al. 1973a, 1973b, 1974, 1975; Fredrickson et al. 1974; Rocoux-Hanus and Boisacq-Schepens 1974; Büttner and Lang 1979; Mergner 1979; Azzena et al. 1986), was performed by analyzing field potentials generated during stimulation of the right vestibular nerve. The following recordings were made before and after section of the left VIII nerve:

1. extracellular recordings of the spontaneous discharge rate of single cortical units at a depth between 1.6–1.8 mm;
2. extracellular recordings of the responses of the same cells to electrical stimulation of the right vestibular nerve.

Tungsten and glass microelectrodes (0.8–1.5 MOhm) were employed. They were advanced through the gray matter by an electronic microdrive. Spikes and field potentials were recorded on magnetic tape for later computerized analysis (Tecfen Computerscope ISC-16 program) based on the average of 16 traces for evoked potentials, instantaneous frequency and post-stimulus time histograms (PSTH) constructed following 100 stimuli applied to the right vestibular nerve.

Second Group. The experiment procedures were the same as for the first group. In this group of animals, however, the study was performed in rabbits with previous left section of the VIII nerve and fully compensated. In addition, the spinal cord at T3–T4 level was exposed. The electrophysiological analyses made in the first group were also performed in this second group of animals, before and after spinal cord transection.

3 Results

The first phase of each experiment consisted in the localization of the cortical vestibular area, according to the results by Ödkvist et al. (1973 a, 1973 b, 1974, 1975). This zone of the cerebral cortex, 2 mm^2, is 4.5–5 mm from the midline, 22–24 mm from the intercollicular sulcus and is bilaterally represented. However, each vestibular nerve projects its stronger input to the contralateral vestibular area. The main electrophysiological characteristic is the inversion of the polarity of the potential generated by the stimulation of the vestibular nerve as the microelectrode goes deeper through the gray matter (Ödkvist et al. 1973 a). After identification of the vestibular zone, the study was based on the analysis of single unit activity.

First Group. The results are based on 27 cortical neurons. Sixteen were localized in the left and 11 in the right vestibular area. These neurons were analyzed follow-

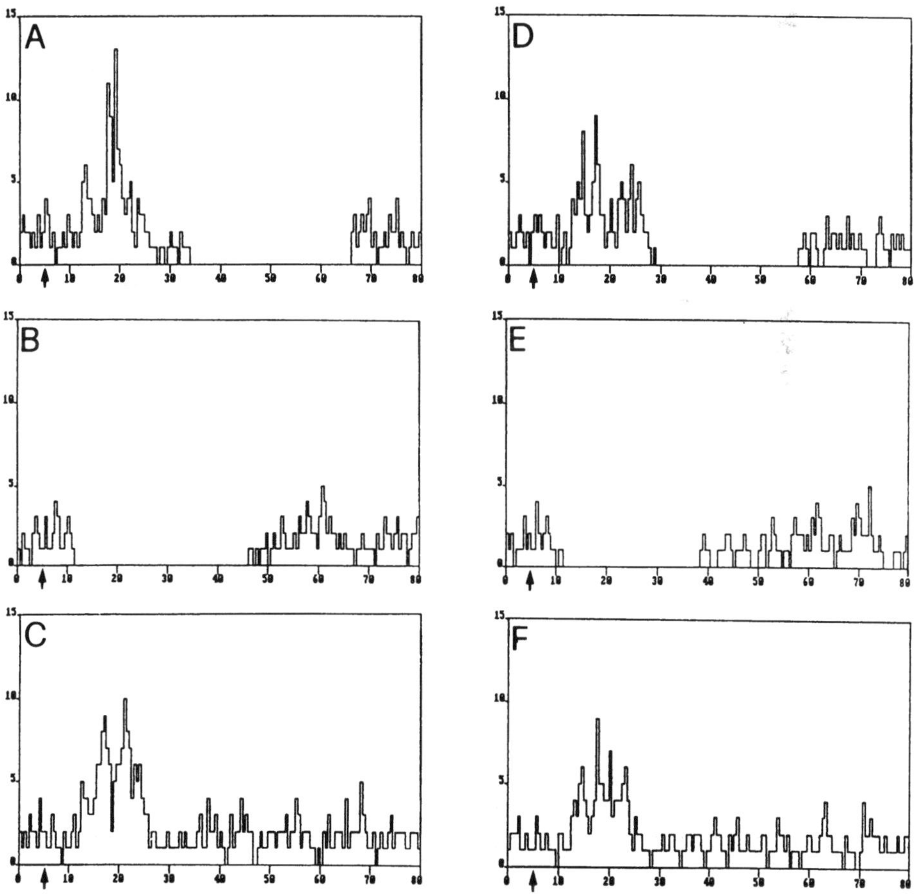

Fig. 1. Patterns of single cortical vestibular neuron response reported as PSTHs, during stimulation of the right vestibular nerve in normal animals. **A, B,** and **C** from left vestibular area; **D, E** and **F** from right vestibular area. *Arrows* indicate stimulation. Horizontal calibrations in ms; vertical calibrations in number of counts/address

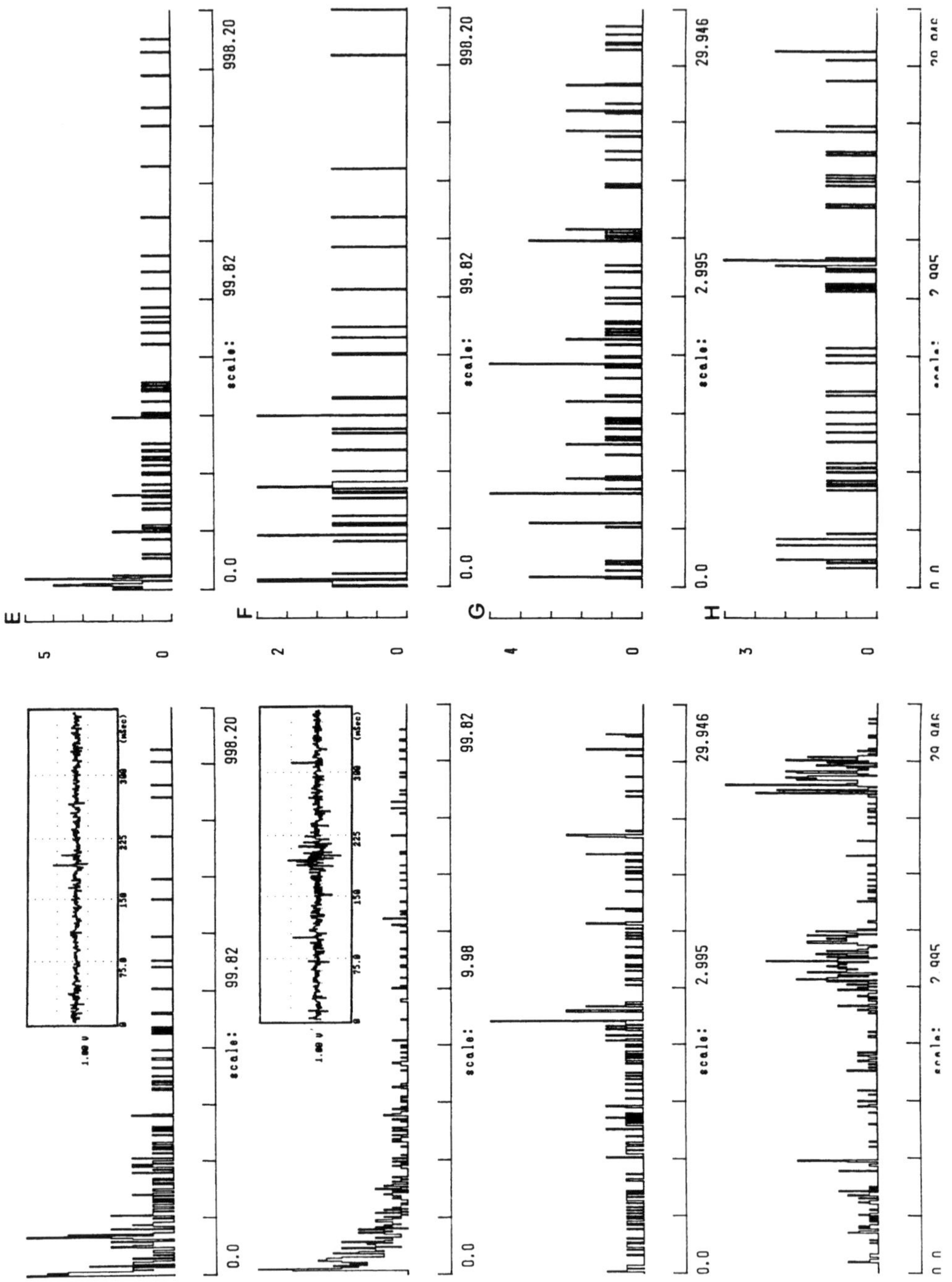
A
B
C
D
E
F
G
H
scale: 99.82
scale: 9.98
scale: 2.995
0.0
99.82
998.20
9.98
2.995
29.946
1.00 V
(mSec)

Fig. 2

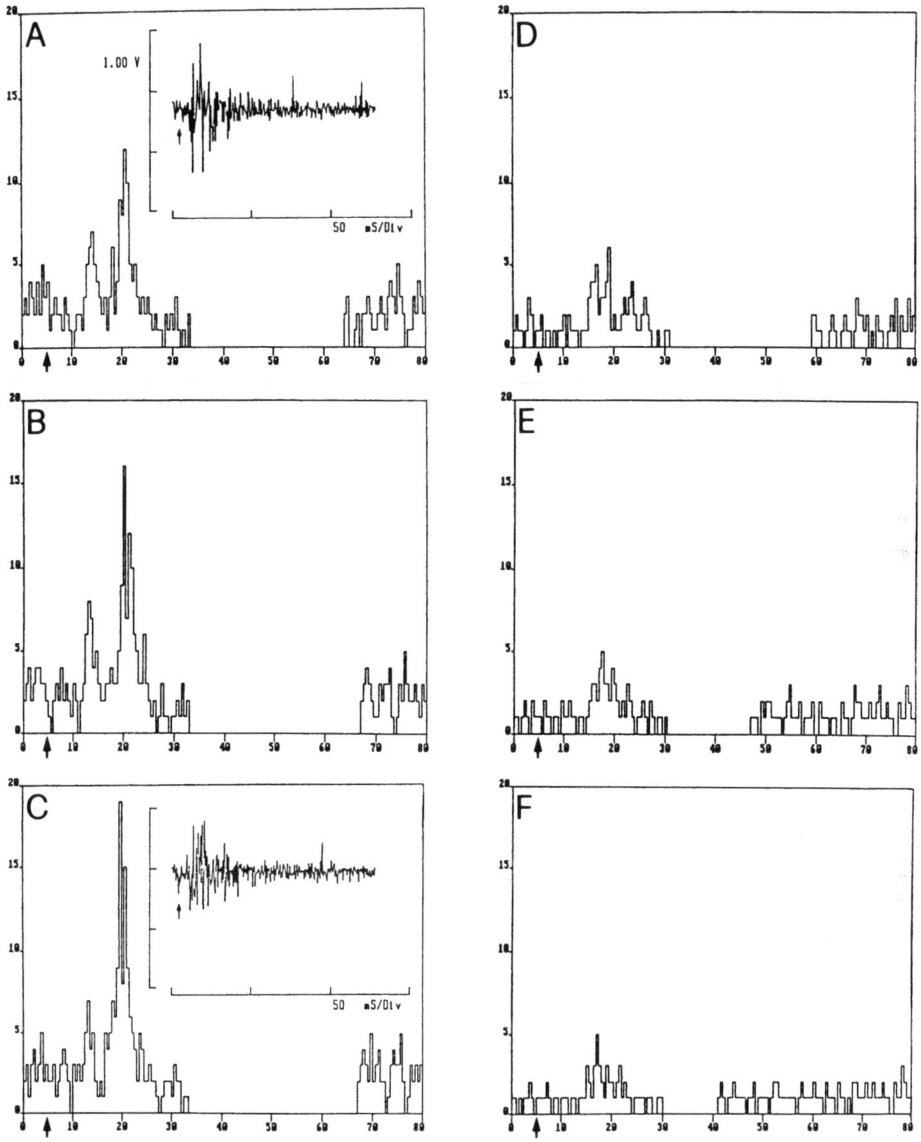

Fig. 3. Responses of two cortical vestibular neurons to stimulation of the right vestibular nerve in two normal rabbits. **A, B** and **C** from left vestibular area; **D, E,** and **F** from right vestibular area. **A** and **D** in basal conditions; **B** and **E** 10 min after vestibular deprivation; **C** and **F** 30 min after. *Arrows* indicate stimulation. Horizontal calibrations: in ms; vertical calibrations: number of counts/address. *Inserts* in **A** and **C** show the filtered specimen records. Vertical calibration of 1 V/div in *insert* **A** applies also to *insert* **C**

◄

Fig. 2. Effects provoked by acute left vestibular deafferentation on the discharge rate of two neurons, one in the left cortical vestibular area (**A, B, C, D**) and one in the right (**E, F, G, H**). **A, C, E** and **G** in normal conditions; **B, D, F** and **H** following left vestibular deafferentation. **A, B, E** and **F** are the rate histograms (the reciprocal of the rate of event occurrences). Horizontal calibrations are in ms, but for **B** with a different scale. **C, D, G** and **H** are the rate meter histograms. Horizontal calibrations are in s. For all traces, vertical calibrations are number of counts/address. *Inserts* in **A** and **B** show the specimen records ing the electrical stimulation of the right vestibular nerve. Several patterns of response were observed, formed by different combinations of excitation-inhibition sequences. Fig. 1 A–C reports the patterns recorded from the left vestibular area contralateral to the right stimulated vestibular nerve. In particular, Fig. 1 A shows excitation followed by inhibition, Fig. 1 B reports pure inhibition and Fig. 1 C illustrates pure excitation. Similar patterns of response were recorded from the right vestibular area ipsilateral to the stimulated vestibular nerve (Fig. 1 D–F). Latencies of contralateral excitation were 7.5 ± 0.4 ms ($\bar{x} \pm$ S.D.) and inhibition 6.2 ± 0.3 ms. Longer latencies were recorded from the ipsilateral vestibular area. Excitation started at 10.1 ± 0.15 ms, while no pure inhibition was observed. The latencies of inhibition following excitation were variable depending on the duration of the excitation, which fluctuated from 12 to 35 ms. The cortical vestibular representation was stronger on the contralateral cerebral area. The amplitude of excitatory responses ranged between 0.8 to 1.4 spikes/stimulus for the contralateral and varied from 0.35 to 0.65 spikes/stimulus for the ipsilateral responses. By analyzing the latency values it is evident that the excitatory peak of unit responses was superimposed on the negative limb of the respective field potential. The normal discharge rate of these neurons ranged from 9 to 13 spikes/s (Fig. 2 A and C).

The vestibular deafferentation provoked profound modifications in the spontaneous and evoked electrical activity of the cortical vestibular neurons in both hemispheres. The spontaneous discharge of neurons localized in the cerebral vestibular area ipsilateral to the vestibular lesion increased to 177% (19–25 spikes/s) (Fig. 2 A–D), and the characteristics of the evoked responses were enhanced in both excitation and inhibition (Fig. 3 A–C). In particular, the amplitude of excitatory responses increased to 90% (0.95–1.52 spikes/stimulus) with longer response duration. On the contrary, in the vestibular cortex contralateral to the vestibular lesion the excitability became reduced, as shown by lowering of the spontaneous activity to 85% (2 to 8 spikes/s) and of the excitatory responses to 0.1–0.38 spikes/stimulus (Figs. 2 E–H; 3 D–F). No changes in latency of the evoked responses were observed.

Second Group. The animals of this group were subdivided into two groups: 15 normal and 30 left hemilabyrinthectomized and fully compensated animals. Evoked potentials were recorded during stimulation of the right vestibular nerve in five normal and five deafferentated animals. The recordings were performed before and after the transection of the spinal cord at T3–T4 level. The morphological characteristics of the field potentials recorded in both groups of animals were virtually the same. Greatly different were the effects provoked on the responses evoked during stimulation of the right intact nerve by the spinal transection. In fact, no changes were observed in the normal animals, while evident and consistent modifications were provoked in the animals with the previous lesion of one labyrinth. Again an increase in the left evoked potentials (ipsilateral to vestibular lesion) and a decrease in the right waves (contralateral to vestibular lesion) were recorded. In this group also the latency values remained unchanged. These effects started immediately after the spinal decompensation and persisted throughout the experiment, reaching a maximum after 15–30 min. The stability

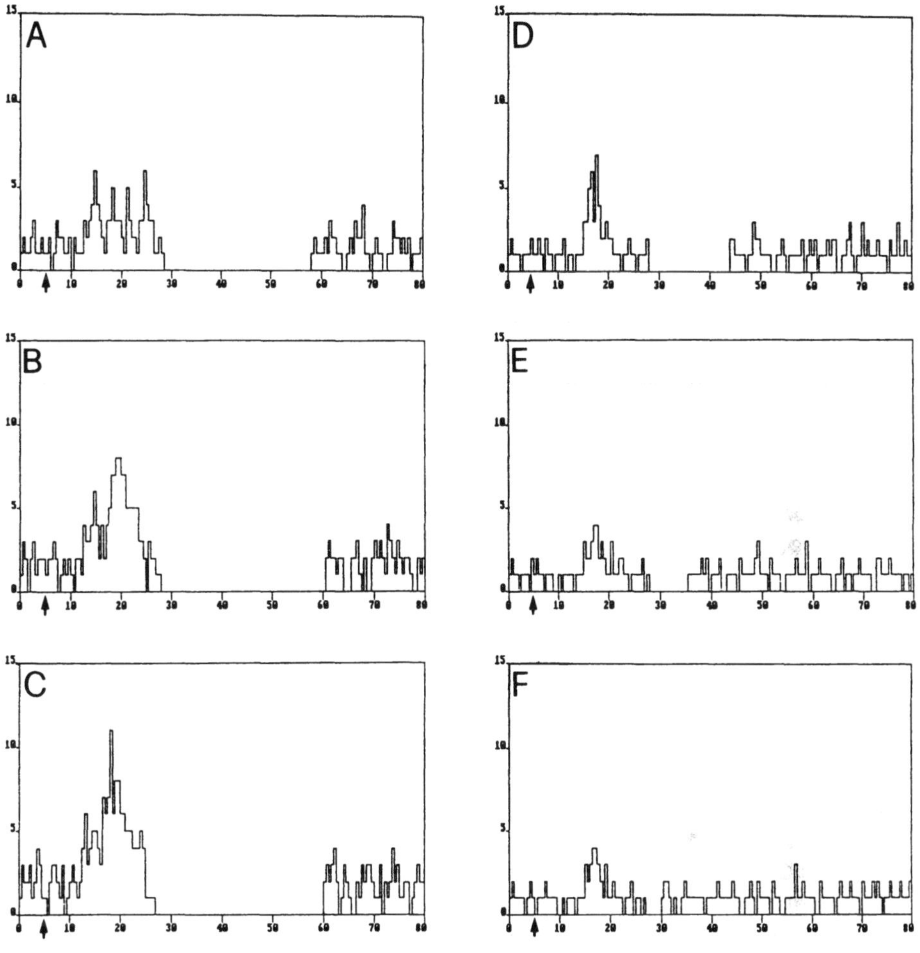

Fig. 4. Responses of two cortical vestibular neurons to stimulation of the right vestibular nerve in two left hemilabyrinthectomized and compensated rabbits. **A, B**, and **C** from left and **D, E** and **F** from right vestibular area. **A** and **D** in basal conditions; **B** and **E** 10 min after spinal cord transection; **C** and **F** 30 min after. Horizontal calibrations in ms; vertical calibrations: number of counts/address

of blood pressure, which was monitored throughout the experiment, showed that the modifications in cortical excitability were independent of the cerebral blood flow variations, possibly provoked by the spinal lesion.

Parallel modifications were induced by the spinal decompensation in the responses of single units located in the same cortical areas. Ten cortical vestibular neurons in the control animals were studied (four in the right and six in the left zone), and 25 in the hemilabyrinthectomized and compensated animals (10 in the right and 15 in the left zone). The response patterns were similar to those described for the first group, except for a lower amplitude. Left responses ranged from 0.54 and 1.2 spikes/stimulus, while right responses were between 0.25 and 0.45 spikes/stimulus. Spinal decompensation provoked a stronger excitation or a more evident inhibition. The excitatory responses recorded from the left vestibular area following stimulation of the right vestibular nerve increased to 150%, being between 0.62 and 1.35 spikes/stimulus (Fig. 4 A–C). On the contrary, the effects induced in the right vestibular area by stimulation of the right vestibular nerve were characterized by a reduction to 67% of the excitatory burst with 0.15–0.33 spikes/stimulus (Fig. 4 D–F). By recording the variations of single unit discharge rates, the results were of the same nature.

4 Discussion

The present research was undertaken in order to analyze the neuronal activity of the cerebral vestibular area during two different, but highly correlated functional stages: first, during the acute stage of the unilateral vestibular deafferentation and, second, during the resulting compensatory processes revealed by transection of the spinal cord. Previous researches showed that the symptoms resulting from the unilateral vestibular deafferentation are the consequence of the asymmetrical electrical activity in the vestibular nuclei of both sides, and that the recovery of their balance is the basis for compensation (Azzena et al. 1977; Dieringer and Precht 1979; Ried et al. 1984). This effect arises from the labyrinth of the intact side, the retinae and the spinal cord (Precht 1974; Azzena et al. 1977; Putkonen et al. 1977; Dieringer and Precht 1979; Ried et al. 1984; Smith et al. 1986). In particular, it has been shown that the influence of signals ascending through the spinal afferents takes place through precerebellar structures such as the inferior olive and the lateral reticular nucleus (Azzena et al. 1981; Mameli and Tolu 1986). The results of the present research showed that neuronal excitability of the cortical vestibular zone is modified by both acute unilateral vestibular deafferentation and by spinal decompensation performed in animals previously hemilabyrinthectomized and compensated. The main consequence was an asymmetrical distribution of the responses. Increase of the excitability was observed in the cortical vestibular area ipsilateral to vestibular deafferentation. An opposite response was recorded from the cortical vestibular zone of the opposite side. This condition recalls similar events which characterize the vestibular nuclei of both sides during the same experimental conditions (Azzena et al. 1977). These experiments offer additional support. In fact, the increase of the cortical excitability is present in the vestibular area which receives the input from the vestibular nuclei activated by the vestibular deafferentation, i.e., those of the opposite side. On the contrary, the excitability decrease of the contralateral vestibular area is due to the reduced activity in the projecting vestibular nuclei of the deafferentated side. Moreover, these experiments demonstrated that electrophysiological changes observed in both groups of animals are comparable, thus demonstrating that at cortical level, on the analogy of the vestibular nuclei, the spinalization also reproduces the same conditions elicited by the acute deprivation of the vestibular signals from one side. The specificity of the effects was strengthened by recordings in control animals where the spinalization did not provoke significant variations of cortical excitability. These results suggest that the activity of the vestibular cortex is deeply influenced by the vestibular deafferentation and that the spinal afferents also influence the cortical vestibular units during the compensation stage. The question is whether these changes are the expression of the electrophysiological alterations at subcortical level or the consequence of a process directly involving the vestibular cortex. At level of subcortical somatosensory structures partial deafferentation provokes a clear reorganization (Basbaum and Wall 1976; Brenowitz and Pubols 1981). On the other hand, restricted areas of somatosensory cortex rapidly become reactivated by additional inputs from close peripheral fields. In monkeys, Merzenich et al. (1978, 1983) showed that following section of median nerve the areas 3 b and 1 develop the appearance of new duplications of organized representations of the body (Goldberger and Murray 1978; Kaas et al. 1983). Similar results were also observed in cats (Franck 1980) and rats (Wall et al. 1981). From the experiments described in this chapter, it appears that the vestibular area recovery is sustained by the afferents ascending from the vestibular nuclei until the electrical activity within the vestibular complexes of both sides returns to its previous state of balance.

Acknowledgement. This research was supported by grants from Italian CNR and MPI, Rome, Italy.

References

Azzena GB (1969) Role of the spinal cord in compensating the effects of hemilabyrinthectomy. Arch Ital Biol 65:43–53

Azzena GB, Mameli O, Tolu E (1976) Vestibular nuclei of hemilabyrinthectomized guinea pigs during decompensation. Arch Ital Biol 114:389–398

Azzena GB, Mameli O, Tolu E (1977) Vestibular units during decompensation. Experientia (Basel) 33:234–235

Azzena GB, Mameli O, Tolu E (1979) Cerebellar contribution in compensating the vestibular function. In: Granit R, Pompeiano O (eds) Reflex control of posture and movement. Elsevier/North Holland Biomed Press, Amsterdam, pp 599–606 (Progress in Brain Research, vol 50)

Azzena GB, Tolu E, Mameli O (1981) The lateral reticular nucleus. Role in vestibular compensation. In: Flohr H, Precht W (eds) Lesion induced neuronal plasticity in sensorimotor systems. Springer, Berlin Heidelberg New York, pp 254–264

Azzena GB, Melis F, Caria MA, Salis S, Teatini GP, Ghiselli F, Blotta P (1986) The vestibular cortical projection during spinal decompensation. Arch Ital Biol 124:15–26

Basbaum AI, Wall PD (1976) Chronic changes in the response of cells in adult cat dorsal horn following partial deafferentation: the appearance of responding cells in a previously nonresponsive region. Brain Res 116:181–204

Boisacq-Schepens N, Rocoux-Hanus M (1972) Motor cortex vestibular responses in the chloralosed cat. Exp Brain Res 14:539–549

Brenowitz GL, Pubols LM (1981) Increased receptive field sizes of dorsal horn neurons following chronic spinal cord hemisections in cats. Brain Res 216:45–49

Büttner U, Lang W (1979) The vestibulocortical pathway: neurophysiological and anatomical studies in the monkey. In: Granit R, Pompeiano O (eds) Reflex control of posture and movement. Elsevier/North-Holland Biomed Press, Amsterdam, pp 581–588 (Progress in Brain Research, vol 50)

Courjon JH, Flandrin JM, Jeannerod M, Schmid R (1982) The role of the flocculus in vestibular compensation after hemilabyrinthectomy. Brain Res 239:251–257

Dieringer N, Precht W (1979) Synaptic mechanism involved in compensation of vestibular function following hemilabyrinthectomy. In. Granit R, Pompeiano O (eds) Reflex control of posture and movement. Elsevier/North-Holland Biomed Press, Amsterdam, pp 607–615 (Progress in Brain Research, vol 50)

Dieringer N, Künzle H, Precht W (1984) Increased projection of ascending dorsal root fibers to vestibular nuclei after hemilabyrinthectomy in the frog. Exp Brain Res 55:574–578

Flohr H, Bienhold H, Abeln W, Macksovics I (1981) Concepts of vestibular compensation. In: Flohr H, Precht W (eds) Lesion-induced neuronal plasticity in sensorimotor systems. Springer, Berlin Heidelberg New York, pp 153–172

Franck JI (1980) Functional reorganization of cat somatic sensorymotor cortex (SmI) after selective dorsal root rhizotomies. Brain Res 186:458–462

Fredrickson JM, Kornhuber HH, Schwarz DWF (1974) Cortical projections of the vestibular nerve. In: Kornhuber HH (ed) Vestibular system, Part 1. Springer, Berlin Heidelberg New York, pp 565–582 (Handbook of sensory physiology, vol VI/1)

Goldberger ME, Murra M (1978) Recovery of movement and axonal sprouting may obey some of the same laws. In: Cotman CW (ed) Neuronal plasticity. Raven, New York, pp 73–96

Kaas JH, Merzenich MM, Killackey HP (1983) The reorganization of somatosensory cortex following peripheral nerve damage in adult and developing mammals. Annu Rev Neurosci 6:325–356

Kornhuber HH, Fonseca JS da (1964) Optovestibular integration in the cat's cortex: a study of sensory convergence on cortical neurons. In: Bender MB (ed) The oculomotor system. Harper & Row, New York, pp 239–279

Mameli O, Tolu E (1986) Involvement of the inferior olivary neurons in vestibular compensation. Brain Res Bull 16:739–743

Mergner T (1979) Vestibular influences on the cat's cerebral cortex. In: Granit R, Pompeiano O (eds) Reflex control of posture and movement. Elsevier/North-Holland Biomed Press, Amsterdam, pp 567–579 (Progress in Brain Research, vol 50)

Merzenich MM, Kaas JH, Sur M, Lin C-S (1978) Double representation of the body surface within cytoarchitectonic areas 3b and 1 in "S-1" in the owl monkey (*Aotus trivirgatus*). J Comp Neurol 181:41–74

Merzenich MM, Kaas JH, Wall J, Nelson RJ, Sur M, Felleman D (1983) Topographic reorganization of somatosensory cortical areas 3b and 1 in adult monkeys following restricted deafferentation. Neuroscience 8:33–55

Ödkvist LM, Rubin AM, Schwarz DWF, Fredrickson JM (1973a) Vestibular cortical projection in the rabbit. J Comp Neurol 149:117–120

Ödkvist LM, Rubin AM, Schwarz DWF, Fredrickson JM (1973b) Vestibular and auditory cortical projection in the guinea pig (Cavia Porcellum) Exp Brain Res 18:279–286

Ödkvist LM, Schwarz DWF, Fredrickson JM, Hassler R (1974) Projection of the vestibular nerve to the area 3a arm field in the squirrel monkey (Saimiri sciureus) Exp Brain Res 21:97–105

Ödkvist LM, Liedgren SRC, Larsby B, Jervall L (1975) Vestibular and somatosensory inflow to the vestibular projection area in the postcruciate dimple region of the cat cerebral cortex. Exp Brain Res 22:185–196

Penfield W (1957) Vestibular sensation and the cerebral cortex. Ann Otol (St Louis) 66:691–698

Precht W (1974) Characteristics of vestibular neurons after acute and chronic labyrinthine destruction. In: Kornhuber HH (ed) Vestibular system, Part 2. Springer, Berlin Heidelberg New York, pp 451–462 (Handbook of sensory physiology, vol VI/2)

Putkonen PTS, Courjon JH, Jeannerod M (1977) Compensation of postural effects of hemilabyrinthectomy in the cat. A sensory substitution process? Exp Brain Res 28:249–257

Ried S, Maioli C, Precht W (1984) Vestibular nuclear neuron activity in chronically hemilabyrinthectomized cats. Acta Otolaryngol 98:1–13

Roucoux-Hanus M, Boisacq-Schepens N (1974) Projections vestibulaires au niveau des aires corticales suprasylvienne et postcruciée chez le chat anesthesié au chloralose. Arch Ital Biol 112:60–76

Smith PF, Darlington CL, Curthoys IS (1986) The effect of visual deprivation on vestibular compensation in the guinea pig. Brain Res 364:195–198

Wall JT, Cusick CG, Kaas JH (1981) Evidence for incomplete reorganization of the S-1 foot representation following sciatic nerve section in the adult rat. Neurosci Abstr 7:758

Optokinetic and Vestibular Oculomotor Responses and their Interaction Following Hemilabyrinthectomy in the Cat

C. Maioli

Istituto di Fisiologia dei Centri Nervosi – CNR, Via Mario Bianco 9, 20131 Milano, Italy

1 Introduction

One of the most striking examples of plasticity in the central nervous system is the compensation of the highly disabling deficits which result from unilateral peripheral vestibular lesions. In fact, many studies in different species (cf. Schaefer and Meyer 1974) have shown that the vestibular compensation processes are remarkable both in their speed and in the degree of functional recovery. In particular, ocular and postural imbalance subside almost completely within the first post-operative week, and behavioral tests in cats and monkeys have shown that 30–40 days after hemilabyrinthectomy (HL) the recovery of postural and locomotor abnormalities is so extensive that it is often difficult to distinguish between operated and non-operated animals (Lacour et al. 1976; Xerri and Lacour 1980).

However, an intriguing aspect of vestibular compensation is the persistence of considerable deficiencies in the vestibulo-ocular reflex (VOR) (Baarsma and Collewijn 1975; Wolfe and Kos 1977; Maioli et al. 1983). In the cat, the VOR remains asymmetrical even several months after the lesion, the phase lead being larger than normal and the gain being approximately half of control values (Maioli et al. 1983). This finding is somewhat surprising, since a dysmetric VOR is known to induce blurred vision during head movements as well as an impairement of motor performances (Gauthier and Robinson 1975; Melvill-Jones 1977). While in normal conditions the vestibulo-ocular system responds to such a situation by promptly adapting its gain to match the functional demands (cf. Miles and Lisberger 1981) in chronic HL cats this does not occur, even though the plastic properties of the VOR have been shown to be unaffected by HL (Maioli and Precht 1985).

One can then surmise that a satisfactory functional restoration of the impaired VOR is achieved through a partial sensory substitution process, that is, through an increase of the relative contribution of other sensory inputs (e.g., vision and neck proprioception) to the stabilization of the visual scene on the retina during head movements. Since VOR and the optokinetic reflex (OKR) are normally synergistic in inducing compensatory eye movements during head motion, the hypothesis that visually guided eye movements can compensate for the deficiencies of the VOR is quite appealing. However, it has to be pointed out that HL affects not only VOR but OKR as well (Cohen et al. 1973; Precht et al. 1981). Since the same vestibulo-ocular pathways convey both optokinetic signals and VOR, and

H. Flohr (Ed.)
Post-Lesion Neural Plasticity

the latter can remain deficient for a long time, an independent mechanism for OKR recovery would then seem necessary for a sensory substitution process to be possible.

In this chapter I shall consider the problem of whether OKR can indeed recover after HL independently of VOR, and present some speculations about possible mechanisms underlying the compensation of the responses to combined visual-vestibular stimulation in chronic HL cats. The data used come from a set of experiments initiated in collaboration with W. Precht, which will be reported *in extenso* elsewhere (Maioli and Precht in preparation). In this study 11 adult cats underwent HL on the right side. Eye movements were recorded by chronically implanted EOG electrodes. For details on the eye movement recording and calibration and on the surgical procedures see Maioli et al. (1983).

2 OKR Deficiencies Following HL

Figure 1 shows the changes in optokinetic response to constant velocity stimuli after HL, in two cats, and how these deficits recover within 20–30 days after the lesion. Shortly after HL, OKR is strongly asymmetrical, being basically unaffected for stimuli directed toward the lesioned side and almost absent for stimuli toward the intact side. Actually, when responses are measured 2–3 days after HL (as in the case represented by dots in Fig. 1), stimuli toward the intact side usually induce a paradoxical nystagmus with slow phases going in the opposite direction. This phenomenon can be explained by the fact that, at this stage, a spontaneous

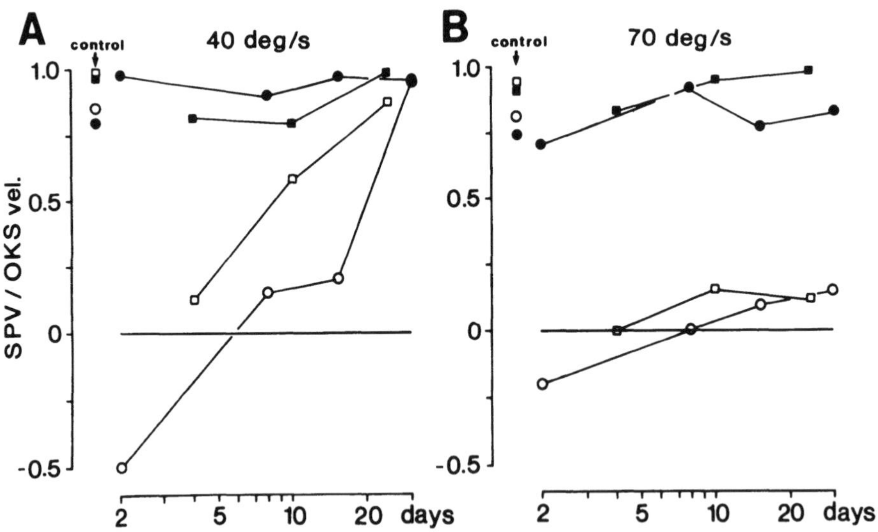

Fig. 1. Changes of OKR, during constant velocity stimulation, induced by HL in two cats (different symbols). The ratio between steady-state *SPV* and the velocity of the optokinetic stimulus (*OKS*) is plotted versus the time elapsed after the lesion. *Empty* and *filled symbols* correspond to a stimulation toward the intact and lesioned side, respectively. OKS velocity was of 40 (**A**) and 70 °/s (**B**)

nystagmus is still present in the dark, which is effectively suppressed in the presence of a stationary visual background. When the visual background moves in the direction in which OKR is deficient, vision can no longer be used to stabilize the eye in the orbit and a release of the spontaneous nystagmus is produced. Once the spontaneous nystagmus in the dark is completely compensated, responses with slow phases directed in the same direction of the stimulus (which saturate at a velocity of few degrees per second) are always observed, also for stimuli moving toward the intact side. However, at longer times after the lesion, some compensation of OKR can be observed. In most animals a good symmetry of the responses is restored within one month for low-velocity stimuli (Fig. 1 A), while a strong asymmetry always persists for higher velocities (Fig. 1 B). Moreover, marked deficiencies are present also in the responses to sinusoidal stimuli. Figure 2 A shows the optokinetic response at 0.17 Hz (±16 °/s) recorded one month after HL. Besides the large response asymmetry, a remarkable saturation of the slow phase velocity (SPV) at a few degrees per second is present when the stimulus is directed toward the intact side. Within the range of values tested in these experiments the SPV at which saturation occurs is independent of the frequency of stimulation. In the case illustrated in Fig. 2 B saturation occurs at about 5 °/s. By contrast, when the stimulus is directed toward the lesioned side, the responses are not impaired by HL, as in the case for constant velocity stimuli. Actually, as in the example of Fig. 2, they are often slightly larger than the control (not shown). Furthermore, OKR gain (maximum SPV/maximum stimulus velocity) varies with the stimulation frequency as in the intact animal (Maioli and Precht 1984).

These findings are in keeping with the notion that the vestibular nuclei (VN's) are an important link in the optokinetic pathways (Dichgans et al. 1973; Waespe and Henn 1977). In fact, OKR impairement can be readily explained by the dramatic depression of the resting discharge of secondary vestibular neurons ipsilateral to the lesioned labyrinth (Gernandt and Thulin 1952; Shimazu and Precht 1966). Indeed, it has been demonstrated that canal plugging (a procedure that

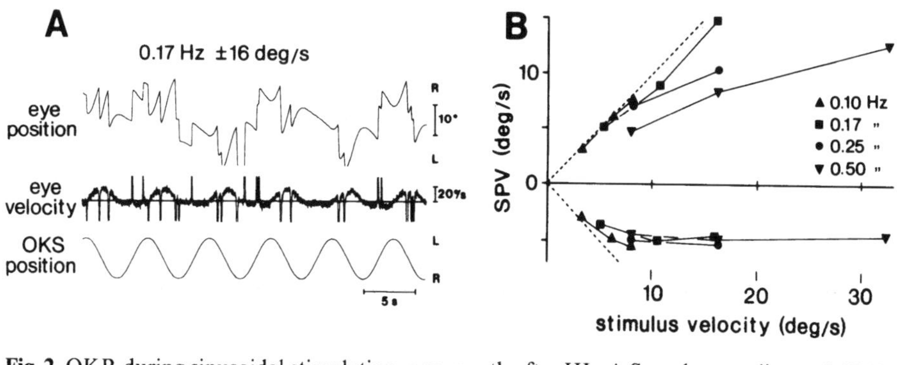

Fig. 2. OKR during sinusoidal stimulation, one month after HL. **A** Sample recording at 0.17 Hz (±16 °/s). **B** Plot of maximum SPV of the response versus stimulus velocity for various frequencies (different symbols). Negative SPV values refer to stimuli directed toward the intact side. *Dashed lines* with a slope equal to one are shown for reference

abolishes VOR without altering the tonic discharge of primary afferents and, consequently, also of secondary vestibular neurons) does not affect OKR (Cohen et al. 1983). Furthermore, the findings are also in agreement with the notion that labyrinthectomy affects only the "slow" component of the OKR by disabling the velocity storage mechanism while sparing the "fast" component (Cohen et al. 1973; Collewijn 1976; Zee et al. 1976). In the cat, the "fast" component has a low gain and saturates early (Maioli and Precht 1984) so that OKR is almost completely abolished when the "slow" component is not operating. After HL the "fast" component is dominating the responses especially during sinusoidal stimuli of medium-high frequency. This explains the unidirectional frequency-independent [1] saturation observed in Fig. 2 B. By contrast, for stimuli directed toward the lesioned side the dependency of gain on stimulus frequency can be accounted for by the fact that, because of the low-pass filter properties of the velocity storage, the contribution of the "slow" OKR component decreases as the frequency of stimulation increases.

A more complete description and discussion of these data will appear in a subsequent paper (Maioli and Precht in preparation).

3 Comparison Between VOR and OKR Compensation

3.1 OKR Can Recover Independently of VOR

We come now to address the question of whether there is a correlation between VOR and OKR recoveries. Such a correlation would be expected if the neuronal plastic changes were to involve exclusively the vestibulo-ocular pathways (which are known to be shared by both reflexes). Before tackling this issue, however, it is worthwhile to notice that the OKR changes described in the previous section are qualitatively different from those reported in the literature for the VOR. In fact, HL is known to induce a decrease of the VOR gain for rotations in both directions (Baarsma and Collewijn 1975; Maioli et al. 1983). This is due to a 50% decrease in sensitivity of secondary vestibular neurons consequent to the lack of the normal interaction between the VN's of the two sides through the vestibular commissure (Markham et al. 1977). Instead, OKR for stimuli toward the lesioned side appears to be very little affected by the lesion, as revealed by the fact that in all cases also the responses to high-velocity stimuli are very similar to the control [2]. By contrast, for stimuli toward the intact side the "slow" OKR component is completely absent in the acute stage and only the small "fast" component can be observed as soon as the spontaneous nystagmus is compensated. The VOR, instead, can always be evoked to either side in HL cats (Maioli et al. 1983), even

[1] The "fast" OKR component can be considered with a very good approximation to result from a pure gain element with no dynamics, placed in parallel with the velocity storage path.

[2] If G is the open-loop gain in a negative feed-back system, the closed-loop gain is $G/(1+G)$. Thus, if G is sufficiently large, even a reduction by half of its value would induce only small changes in the closed-loop gain. As a consequence, a drop in the open-loop gain of OKR could not result in a decrease of the response amplitude for low-medium velocity stimuli, but the closed-loop gain would be reduced with high-velocity stimuli, since in this case G has a low value (Maioli and Precht 1984).

Table 1. Changes in OKR and VOR symmetry after HL

Cat no.	Days after HL (Δ)	OKR symmetry (stim. = 40°/s) (Δ)	VOR symmetry (Δ)	Δ OKR / Δ VOR
1	8–79 (71)	0.49–0.40 (−0.09)	0.82–0.73 (−0.09)	1.00
2	5–8 (3)	0.00–0.14 (0.14)	0.27–0.40 (0.13)	1.08
3	4–22 (18)	0.13–0.71 (0.58)	0.32–0.85 (0.53)	1.09
4	8–95 (87)	0.40–0.45 (0.05)	0.60–0.64 (0.04)	1.25
5	6–10 (4)	0.52–0.70 (0.18)	0.50–0.58 (0.08)	2.25
6	4–8 (4)	0.03–0.23 (0.23)	0.43–0.35 (−0.08)	−2.87
7	4–23 (19)	0.20–0.63 (0.43)	0.51–0.65 (0.14)	3.07
8	5–10 (5)	0.15–0.78 (0.63)	0.65–0.75 (0.10)	6.30
9	15–32 (17)	0.20–1.00 (0.80)	0.51–0.52 (0.01)	80.0

though the responses are strongly asymmetrical. The reasons for such differences are at the moment unclear, but it is evident that unilateral OKR is much less dependent than VOR on the reciprocal interaction between VN's of the two sides via the vestibular commissure.

Since OKR in one direction is basically not affected by HL, the variation in response symmetry is a good way to measure the amount of OKR recovery. Thus, the variations of OKR symmetry (Δ OKR) were compared to those of the VOR (Δ VOR). The results obtained in nine cats at different times after HL are shown in Table 1. The ratios between Δ OKR and Δ VOR are also reported (last column). The data show that the increase in symmetry was considerably larger for OKR than for VOR in five out of nine animals, the most striking example being cat No. 9, in which OKR symmetry increases from 0.2 to 1.0, while VOR symmetry is virtually unchanged.

However, some caution is required in interpreting these data. In fact, VOR operates in open-loop conditions, while the optokinetic system has to be considered as a closed-loop velocity servo. Thus, a symmetrical increase of the forward gain in the vestibulo-ocular pathway may induce an improvement in OKR symmetry with no change in the symmetry of the VOR. This is due to the fact that almost no OKR improvement is possible with stimuli toward the lesioned side since the closed-loop gain cannot be more than one (see also footnote 2). However, within the time period considered, only in four of the nine cats was the increment of VOR gain (with rotation to the intact side) larger than 3%. This increase was of 12, 34, 65, and 26% in cats No. 1, 3, 6, and 8, respectively. It can be shown by simple computation that the increase in VOR gain was never large enough to account for the fact that in cats No. 5 to 9 the recovery of symmetry

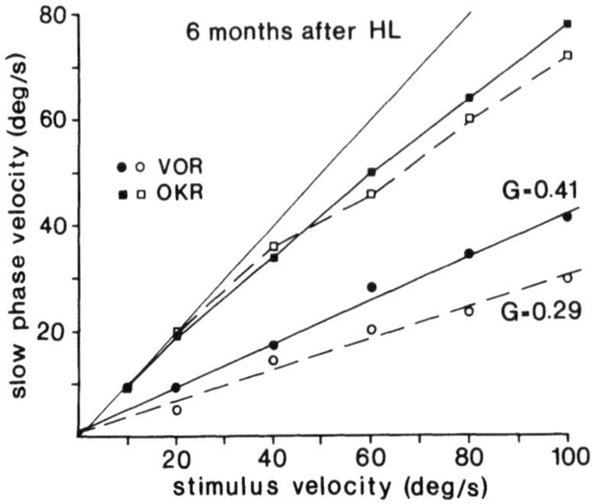

Fig. 3. Measurements of the response amplitude of VOR and OKR at different velocities of stimulation, performed 6 months after HL. *Filled* and *open symbols* correspond to responses with SPV directed toward the lesioned and intact side, respectively

was greater for OKR than for VOR. Consequently, one has to conclude that in these cats OKR did recover independently of VOR. It should also be noticed that the opposite possibility, that is, a recovery of the VOR not accompanied by a corresponding recovery of the OKR, was never observed in this sample.

The fact that OKR can compensate independently of VOR can be even more strikingly demonstrated in some cats at a very late stage of recovery. A comparison of the response amplitude of VOR and OKR recorded from one cat 6 months after HL is shown in Fig. 3. One can see that the compensation of the OKR is exceedingly good in all ranges of stimulation velocity studied. Nevertheless, the VOR has a low gain and is asymmetrical, the responses being quite similar to those recorded acutely after the lesion.

3.2 Are the Vestibular Nuclei a Necessary Link to the Oculomotor Nuclei for the "Slow" OKR Component?

The obvious implication of the fact that OKR can recover independently of VOR is that the classical vestibulo-ocular pathways are not the only site in which plastic changes occur during compensation following HL. There are two possibilities: (1) plastic changes in the optokinetic pathway occur upstream the VN's, so that the VOR is not affected; (2) the "slow" optokinetic component is also mediated by a pathway which does not involve the VN's.

Although the first possibility cannot be excluded, it should be noticed that the imbalance of VN activity induced by HL seems to persist in the cat for a long time after the full compensation of the spontaneous nystagmus (Ried et al. 1984). In fact, one month after HL, it is still much more difficult to find VN units responding to vestibular canal stimuli on the lesioned side than on the intact side. Since no neuronal loss can be demonstrated in VN as a consequence of the labyrinthine lesion (Ried et al. 1984), it can be assumed that a large number of secondary vestibular neurons deprived of the potent exitatory drive from the ipsilateral VIIIth nerve remain silent for a long time after HL, and that the recovery of ocular balance is obtained also with the help of other structures besides VN's. The persistance of a VN imbalance is also supported by the fact that a VOR asymmetry is usually observed for many months after HL (see Fig. 3 and Maioli et al. 1982). Thus, it can be inferred that the unilateral depression of secondary vestibular neurons should greatly limit the effectiveness of a gain increase in the optokinetic pathway before VN's.

Conversely, data present in the literature support the hypothesis that the "slow" OKR component is conveyed by a pathway which bypasses VN's. Specifically, several observations indicate that the nucleus prepositus hypoglossi (PHN) subserves a crucial role in the optokinetic control of eye movements. Anatomical (Cazin et al. 1982; Magnin et al. 1983) and electrophysiological evidence (Cazin et al. 1984) have shown that PHN conveys optokinetic signals to VN's. Accordingly, the firing of PHN units is modulated during optokinetic stimulation (Lopez-Barneo et al. 1982; Precht et al. 1982) and lesions of the rostral PHN abolish the "slow" OKR component (Uemura and Cohen 1973; Cheron et al. 1986). However, there is also overwhelming evidence that the role of PHN in oculomotor control is not exerted exclusively through its projection to VN's. In fact, it has been anatomically and electrophysiologically demonstrated that PHN projects directly to all oculomotor nuclei (Graybiel and Hartwieg 1974; Baker and Berthoz 1975; Maciewicz et al. 1977; Baker et al. 1977; McCrea et al. 1979; Steiger and Büttner-Ennever 1979) as well as to areas of the reticular formation which are known to be involved in oculomotor functions (McCrea et al. 1979; Hikosaka and Igusa 1980). The role of PHN in the control of gaze which has been mostly emphasized is that of providing a "corollary discharge" to other areas of the brain (Lopez-Barneo et al. 1982). However, despite the lack of conclusive data concerning the function of the monosynaptic projection from PHN to ocular motoneurons, a possible involvement of PHN as a center carrying motor commands in parallel with the classical vestibulo-ocular pathways has to be considered. This viewpoint is also suggested by the extensive similarities between the efferent anatomical organization of PHN and VN.

It can then be proposed that the differences described above between the time courses of VOR and OKR recovery can be accounted for by assuming that the PHN-reticular-oculomotor pathway is plastically modifiable and participates to the processes of OKR compensation after HL.

4 Combined Visual-Vestibular Stimulation

The results presented up to this point support the view that OKR can in principle compensate for VOR deficits. More specifically, they show that a recovery of OKR deficits is possible and that OKR compensation can occur also in absence of a significant recovery of VOR. However, a complete recovery of OKR (that is, a restoration of the responses to a level comparable to the control values as in Fig. 3) was observed only in two out of the four cats which survived for at least 3 months after HL. Thus, a sensory substitution process whereby OKR could overcome (at least partially) the VOR deficits would not become functionally useful before several months after HL, if at al. Consequently, a discrepancy remains between the large and fast recovery of postural and locomotor symptoms and the presence of considerable deficits of both VOR and OKR.

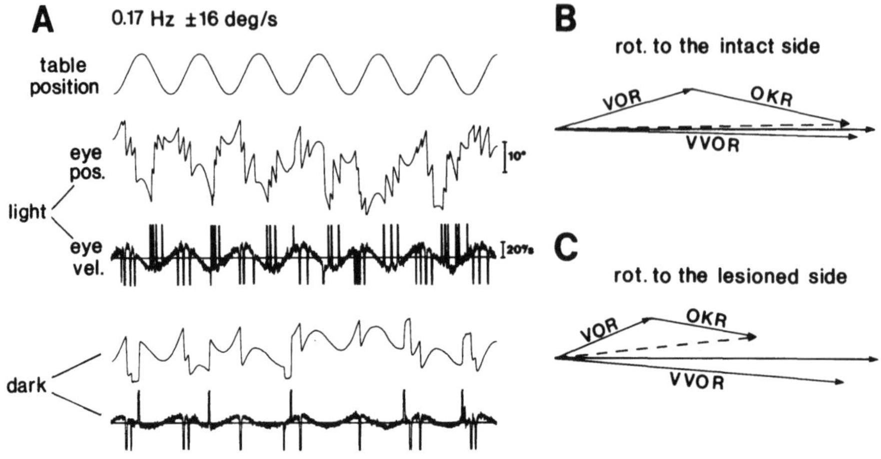

Fig. 4. Responses to a combined visual-vestibular stimulation, one month after HL. **A** Eye movements during sinusoidal oscillation of the animal placed in front of a stationary visual background (*light*) and in the dark. **B–C** Vectorial analysis of the responses shown in **A**, for rotation toward the intact (**B**) and lesioned side (**C**). The stimulus is represented by the thick horizontal vector. The theoretical responses computed by vectorial summation of the measured VOR and OKR (*dashed lines*) are compared with the recorded visual-vestibular ocular reflex (*VVOR*)

It becomes therefore relevant to ascertain the performance of the system in conditions of combined visual-vestibular stimulation, i.e., when both VOR and OKR are activated together. Figure 4A shows the responses recorded from the same cat of Fig. 2 one month after HL. The stimulus consisted of an horizontal oscillation of the animal at 0.17 Hz (± 16 °/s) in the presence of a stationary visual background (middle traces) and in the dark (bottom traces). It can be seen that in the dark VOR is reduced in amplitude and is asymmetrical, the gains for rotation toward the intact and lesioned side being 0.46 and 0.32, respectively. As for the responses to pure optokinetic stimulation, they were shown in Fig. 2. For stimuli toward the lesioned side (same frequency and amplitude as in Fig. 4A) the OKR gain was about 0.9, while for stimuli in the opposite direction the responses saturate at 5 °/s. Surprisingly, the amplitude and symmetry of the responses obtained during combined visual-vestibular stimulation (VVOR) are remarkably good, the SPV during rotation to the lesioned side being considerably better than what expected on the basis of the responses obtained with either optokinetic or vestibular stimuli alone. In fact, by vectorial analysis (Fig. 4B and C) one finds that the response amplitude predicted by the simple sum of VOR and OKR (dashed lines) corresponds quite well to the measured value during rotation to the intact side (Fig. 4B), but diverges markedly from the data during rotation to the lesioned side (Fig. 4C). Typically, VVOR was found to be symmetrical with a gain close to one when low-medium frequencies of stimulation were used, while the gain decreased toward the VOR values at higher frequencies.

The reasons of this nonlinear interaction of VOR and OKR during combined visual-vestibular stimulation cannot be identified with certainty. Drowsiness during VOR cannot be invoked to explain the data because of the small variability of the responses, the sharpness of the quick phases and the fact that cats were under amphetamine during the experiments. It is possible that a mechanism of "visual attentiveness" [such as that suggested by Fuller (1981) to account for the change of the gain of the vestibulo-collic reflex induced by vision in the rabbit] could also cause an increase of VOR gain in presence of a textured visual field, by changing the "internal state" of the system. Although this possibility cannot be discarded, another speculative explanation could be found in the fact that after HL a large number of VN units on the lesioned side are silent (see Sect. 3.2 above). Since during rotation to the ipsilateral side these units are both disinhibited by the contralateral VN and facilitated by the optokinetic input, one can argue that during combined stimulation a larger number of neurons are driven above threshold than during either vestibular or optokinetic stimuli alone. This larger recruitement could explain the high symmetry of the responses observed during VVOR.

5 Conclusion

The data have shown the complexity of the compensation processes for the reflex control of the eye movements following HL. Plastic changes develop over a very long period of time after the lesion, and the compensation can differ quantitatively from animal to animal, as suggested by the variable degree of recovery shown by OKR and VOR (see also Maioli and Precht 1985). Interestingly, OKR seems to have a larger potentiality for compensation than VOR, since adaptive processes may take place also at sites which do not lie along the vestibulo-ocular pathway. Nevertheless, such changes progress very slowly and a satisfactory restoration of function is meanwhile achieved through a nonlinear summation of the responses to visual and vestibular stimuli. However, it should be stressed that, as in normal animals, vision induces effective compensatory eye movements only with head movements in the low-medium frequency range. For fast movements of the head, instead, other sensory inputs (e.g., the proprioceptive afferents from the neck) may play an important role in the compensation processes following HL.

References

Baarsma EA, Collewijn H (1975) Changes in compensatory eye movements after unilateral labyrinthectomy in the rabbit. Arch Oto-Rhino-Laryngol 211:219–230

Baker R, Berthoz A (1975) Is the prepositus hypoglossi nucleus the source of another vestibulo-ocular pathway? Brain Res 86:121–127

Baker R, Berthoz A, Delgado-Garcia J (1977) Monosynaptic excitation of trochlear motoneurons following electrical stimulation of the prepositus hypoglossi nucleus. Brain Res 121:157–161

Cazin L, Magnin M, Lannou J (1982) Non-cerebellar visual afferents to the vestibular nuclei involving the prepositus hypoglossi complex: an autoradiographic study in the rat. Exp Brain Res 48:309–313

Cazin L, Lannou J, Precht W (1984) An electrophysiological study of pathways mediating optokinetic responses to the vestibular nucleus in the rat. Exp Brain Res 54:337–348

Cheron G, Gillis P, Godaux E (1986) Lesions in the cat prepositus complex: effects on the optokinetic system. J Physiol (Lond) 372:95–111
Cohen B, Uemura T, Takemori S (1973) Effects of labyrinthectomy on optokinetic nystagmus and optokinetic afternystagmus. Equilibrium Res 3:88–93
Cohen B, Suzuki J, Raphan T (1983) Role of the otolith organs in generation of horizontal nystagmus: effects of selective labyrinthine lesions. Brain Res 276:159–164
Collewijn H (1976) Impairment of optokinetic (after-)nystagmus by labyrinthectomy in the rabbit. Exp Neurol 52:146–156
Dichgans J, Schmidt CL, Graf W (1973) Visual input improves the speedometer function of the vestibular nuclei in the goldfish. Exp Brain Res 18:319–322
Fuller JH (1981) Eye and head movements during vestibular stimulation in the alert rabbit. Brain Res 205:363–381
Gauthier GM, Robinson DA (1975) Adaptation of the human vestibulo-ocular reflex to magnifying lenses. Brain Res 92:331–335
Gernandt BE, Thulin CA (1952) Vestibular connections of the brain stem. Am J Physiol 171:121–127
Graybiel AM, Hartwieg EA (1974) Some afferent connections of the oculomotor complex in the cat: an experimental study with tracer techniques. Brain Res 81:543–551
Hikosaka O, Igusa Y (1980) Axonal projection of prepositus hypoglossi and reticular neurons in the brainstem of the cat. Exp Brain Res 39:441–451
Lacour M, Roll JP, Appaix M (1976) Modifications and development of spinal reflexes in the alert baboon (*Papio papio*) following an unilateral vestibular neurotomy. Brain Res 113:255–269
Lopez-Barneo J, Darlot C, Berthoz A, Baker R (1982) Neuronal activity in prepositus nucleus correlated with eye movement in the alert cat. J Neurophysiol 47:329–352
Maciewicz RJ, Eagen K, Kaneko CRS, Highstein SM (1977) Vestibular and medullary brain stem afferents to the abducens nucleus in the cat. Brain Res 123:229–240
Magnin M, Courjon JH, Flandrin JM (1983) Possible visual pathways to the cat vestibular nuclei involving the nucleus prepositus hypoglossi. Exp Brain Res 51:298–303
Maioli C, Precht W (1984) The horizontal optokinetic nystagmus in the cat. Exp Brain Res 55:494–506
Maioli C, Precht W (1985) On the role of vestibulo-ocular reflex plasticity in recovery after unilateral peripheral vestibular lesions. Exp Brain Res 59:267–272
Maioli C, Precht W, Ried S (1982) Vestibuloocular and optokinetic reflex compensation following hemilabyrinthectomy in the cat. In: Roucoux A, Crommelinck M (eds) Physiological and pathological aspects of eye movements. Junk, The Hague, pp 202–208
Maioli C, Precht W, Ried S (1983) Short- and long-term modifications of vestibulo-ocular response dynamics following unilateral vestibular nerve lesions in the cat. Exp Brain Res 50:259–274
Markham CH, Yagi T, Curthoys IS (1977) The contribution of the contralateral labyrinth to the second order vestibular neuronal activity in the cat. Brain Res 138:99–109
McCrea R, Baker R, Delgado-Garcia J (1979) Afferent and efferent organization of the prepositus hypoglossi nucleus. In: Granit R, Pompeiano O (eds) Reflex control of posture and movements. Elsevier/North-Holland, Amsterdam, pp 653–665
Melvill-Jones G (1977) Plasticity in the adult vestibulo-ocular reflex arc. Phil Trans R Soc Lond 3 Biol Sci 278:319–334
Miles FA, Lisberger SG (1981) Plasticity in the vestibulo-ocular reflex: a new hypothesis. Annu Rev Neurosci 4:273–299
Precht W, Maioli C, Dieringer N, Cochran S (1981) Mechanisms of compensation of the vestibulo-ocular reflex after vestibular neurotomy. In: Flohr H, Precht W (eds) Lesion-induced neuronal plasticity in sensorimotor systems. Springer, Berlin Heidelberg New York, pp 221–230
Precht W, Cazin L, Blanks R, Lannou J (1982) Anatomy and physiology of the optokinetic pathways to the vestibular nuclei in the rat. In: Roucoux A, Crommelinck M (eds) Physiological and pathological aspects of eye movements. Junk, The Hague, pp 153–172
Ried S, Maioli C, Precht W (1984) Vestibular nuclear neuron activity in chronically hemilabyrinthectomized cats. Acta Oto-Laryngol 98:1–13

Schaefer KP, Meyer DL (1974) Compensation of vestibular lesions. In: Kornhuber HH (ed) Handbook of sensory physiology, vol VI/2. Springer, Berlin Heidelberg New York, pp 463–490

Shimazu H, Precht W (1966) Inhibition of central vestibular neurons from the contralateral labyrinth and its mediating pathway. J Neurophysiol 29:467–492

Steiger HJ, Büttner-Ennever JA (1979) Oculomotor nucleus afferents in the monkey demonstrated with horseradish peroxidase. Brain Res 160:1–15

Uemura T, Cohen B (1973) Effects of vestibular nuclei lesions on vestibulo-ocular reflexes and posture in monkeys. Acta Oto-Laryngol 315:1–71

Waespe W, Henn V (1977) Neuronal activity in the vestibular nuclei of the alert monkey during vestibular and optokinetic stimulation. Exp Brain Res 27:523–538

Wolfe JW, Kos CM (1977) Nystagmic responses of the rhesus monkey to rotational stimulation following unilateral labyrinthectomy: final report. Trans Am Acad Ophthalmol Oto-Laryngol 84:38–45

Xerri C, Lacour M (1980) Compensation des déficits posturaux et cinétiques après neurectomie vestibulaire unilatérale chez le chat. Rôle de l'activité sensorimotrice. Acta Oto-Laryngol 90:414–424

Zee DS, Yee RD, Robinson DA (1976) Optokinetic responses in labyrinthine-defective human beings. Brain Res 113:423–428

"Acute" Vestibular Compensation in the Goldfish: A Visual Substitution Process?

A. Burt and H. Flohr

Department of Neurobiology, University of Bremen, D-2800 Bremen 33, FR Germany

1 Introduction

Unilateral lesion of a labyrinth causes impairment of posture and locomotion. The subsequent recovery of function with time is usually termed vestibular compensation. The behavioral deficits manifested after hemilabyrinthectomy (HL) have been used to divide the course of compensation into different stages or phases. Most authors (see, e.g., Schaefer and Meyer 1974) distinguish between an acute and a chronic phase, whereby the acute phase begins directly after HL and lasts from a few hours to days depending on species. Another classification (McCabe and Ryu 1969), which is usually applied to mammals, includes an additional, immediately post-operative "critical" phase, which is not always considered to be part of the compensation process (Lacour et al. 1979).

Although the lesion-induced deficiencies in posture and locomotion are basically similar in all vertebrates (for reviews see Precht 1974; Schaefer and Meyer 1974), there are a number of interesting differences, particularly in the time course of compensation. In general, functional recovery appears to take place more rapidly in mammals than in submammalian species; compensation of head deviation, for example, occurs within hours in the rat (Llinás and Walton 1979a) but requires several weeks in the frog (Kolb 1955). Typically, within the same animal, the different deficits are compensated at different rates, but, here too, characteristic differences are apparent between species, e.g., in the cat, ocular balance is restored more quickly than a normal head position, whilst the opposite is true of the rat (Schaefer and Meyer 1974). In short, vestibular compensation appears to be effected by species-specific recovery mechanisms.

Since the lesioned substrate does not regenerate, recovery of function must be based on a reorganization of intact central structures. A fundamental step in this process is generally considered to be the restoration of the lost tonic activity in the vestibular nuclear complex of the deafferented side (see, e.g., Llinás and Walton 1979a; Flohr et al. 1981; Luyten et al. 1986). Evidence exists that this newly established afferent input can be derived from several sources, including various brain stem and cerebellar structures (for references see Llinás and Walton 1979a, b) and a number of sensory systems. The involvement of somatosensory (Lacour et al. 1976; Petrosini 1983) and visual (Courjon et al. 1977; Putkonen et al. 1977; Courjon and Jeannerod 1979) afferents as well as commissural vestibular afferents from the intact contralateral labyrinth (Bienhold and Flohr 1978;

H. Flohr (Ed.)
Post-Lesion Neural Plasticity

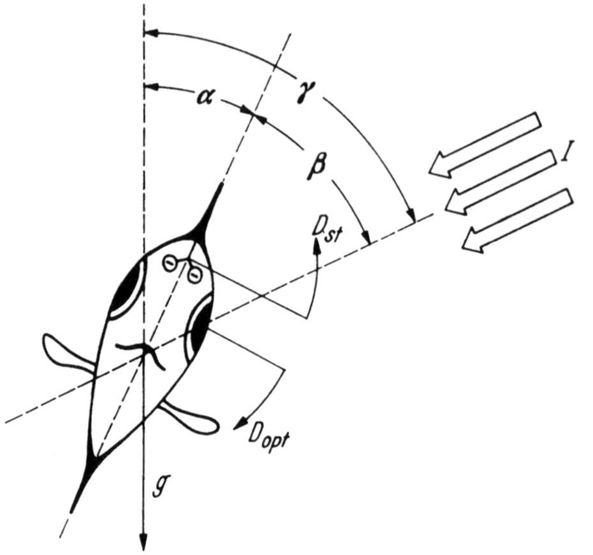

Fig. 1. Schematic representation of the interaction between vestibular and visual inputs in the so-called „dorsal light response" of a teleost fish. When the two eyes of a teleost fish are asymmetrically illuminated, it tilts its dorsal surface away from the vertical (*g*) towards the region of highest light intensity (*I*). According to v. Holst (1935a, b, 1950), who was the first to describe this behavior in teleost fish, the position adopted by the animal is an equilibrium position, in which the optically-induced tilting tendency (D_{opt}) is equal and opposite to the statically-induced righting tendency (D_{st}). (Illustration from Bogenschütz 1961.)

Dieringer and Precht 1979a, b, c; Lacour et al. 1979) in the recovery process has led some authors (Courjon et al. 1977; Putkonen et al. 1977; Lacour et al. 1983; Pfaltz 1983) to propose that vestibular compensation constitutes a multisensory substitution process.

Teleost fish would appear to be ideal experimental subjects for investigating the extent to which visual and vestibular information might contribute to the recovery of function following HL, since the role of the somatosensory system is probably less important than in other vertebrates. Muscle spindles are not found in teleost fish and, with few exceptions, the proprioceptive system is poorly developed (Bone and Marshall 1985, p. 154). Furthermore, the motor performance of a fish swimming freely in water is not subserved by tactile information as in terrestrial vertebrates. Finally, relative to other animals, the degree of somatosensory-vestibular convergence is likely to be much reduced in fish, because movement of head and body are not independent. Equilibrium orientation in the intact animal may thus be considered to be under the bimodal control of the vestibular and visual systems (v. Holst 1935a, b, 1950) (see Fig. 1).

The only previously reported study of the immediately post-operative effects of HL in teleost fish was conducted by Löwenstein (1932) on the minnow, a close relative of the goldfish used in the present investigation. Löwenstein provides a most detailed qualitative description of the postural and locomotory impairment following HL. Although he did not deal with compensation per se, his work suggests that the post-operative symptoms disappeared remarkably quickly, "in some animals after only a few minutes" (p. 833). By contrast, he found that blinded animals required many weeks to months in order to overcome the lesion-induced deficits.

These findings of Löwenstein form the basis of the present investigation, which was designed to examine two questions:

1. Does the rapid disappearance of the immediately post-operative deficits represent a distinct stage in the compensation process?
2. Does the visual input play a role in the compensation of these early deficits?

2 Methods

2.1 Animals

The experimental subjects were 10–12 cm-long goldfish *(Carassius auratus)* of the Comet variety. They were obtained from a local supplier and acclimated to laboratory conditions in a large (360-l) stock tank for at least 6 weeks prior to use. One week before the operation, groups of four to six fish were transferred to small (20-l) plastic tanks, in which they remained throughout the experimental period. The aquarium room was illuminated from 6.00 to 20.00 h and maintained at a temperature of 18 ± 1 °C.

2.2 Hemilabyrinthectomy

Topographically and anatomically, the inner ear of the goldfish (cf. Klinke 1970) closely resembles that of another cyprinid fish, the minnow *(Phoxinus laevis)*, detailed anatomical studies of which have been published by v. Frisch and his coworkers (v. Frisch and Stetter 1932; Wohlfahrt 1932). The surgical technique outlined below is based largely on that employed by v. Frisch and Stetter (1932; cf. Löwenstein 1932) in their ablation experiments on the minnow.

The teleost membranous labyrinth is divided into two distinct halves, the pars superior (utricle and semicircular canals) and the pars inferior (saccule and lagena), joined by the relatively narrow utriculo-saccular duct. The physiological relevance of this structural dichotomy remains the subject of debate (e.g., Platt 1983). Although there would appear to be general agreement that the pars superior is exclusively involved in the control of posture and locomotion, it is still not clear whether the pars inferior responds only to vibrational and acoustic stimuli, or whether it also has some gravistatic sensitivity. As far as the goldfish is concerned, destruction of the saccular and lagenar otoliths in situ has been reported to have no effect on equilibrium orientation (Manning 1924), a finding which is supported by more recent electrophysiological investigations suggesting that the pars inferior is purely auditive in function (Furukawa and Ishii 1967; Fay and Olsho 1979).

In short, even if the pars superior is not solely responsible for the maintenance of equilibrium, it is without doubt the major gravistatic organ. For this reason, the pars inferior was not extirpated in the present investigation and the inflicted vestibular lesion thus represents a partial labyrinthectomy: the utricular pouch, containing the otolith and sensory macula, was removed together with the ampullae of the anterior vertical and horizontal canals and innervating branches from the anterior root of the eighth cranial nerve. The ampulla of the posterior vertical canal is embedded in bone and its removal requires deeper surgery, a technically difficult procedure which has been reported to involve a high mortality rate in the minnow (v. Frisch and Stetter 1932; Löwenstein 1932). Since this additional operation was found to cause no further deficits in gravistatic sensitivity (Löwenstein 1932), it was not attempted here.

Each animal was unilaterally labyrinthectomized on the left side. Following anesthesia in (0.3 g/l) tricaine methane sulfonate (Sigma Chemical Co., St. Louis), the fish was placed in a sponge-lined container immersed in water such that only the cranium remained exposed. Since the operation lasted maximally 10–15 min, the fish was not perfused over the gills.

When the head of a goldfish is viewed from above, a triangular area is usually readily distinguishable medial to the upper margin of the operculum. The base of this triangular area runs along a line between the cerebellum, the outlines of which are visible through the cranium, and the posterior edge of the operculum, whilst the apex points towards the eye. Immediately below the apical region is a piece of cartilage which permits easy access to the pars superior in the angle between the anterior vertical and horizontal canals.

The cranium was cut along the two sides of the triangular area with a small, thin circular saw mounted on a dentist's drill. With the aid of a scalpel, the V-shaped piece of bone was then hinged back caudally at the base and the cartilage in the apical region removed. Under a dissecting microscope, the utricular pouch was grasped with fine, curved forceps (#7) and gently pulled out of the hole. Bleeding was usually negligible. Fluid and any blood were cleared by aspiration before the V-shaped piece of bone was returned to its original position and sealed with a drop of tissue glue (Histoacryl blau, Braun-Dexon, Spangenberg). The animal was then placed in the observation tank.

The completeness of the lesion was subsequently controlled by examining the ablated parts under a microscope. In all cases, the utricle, its otolith and macula, the two anterior semi-circular canal ampullae with their sensory cristae and innervating nerve branches were successfully removed. The examination revealed that the utricular pouch was transected between the utriculo-saccular duct and the ampulla of the posterior canal. The anterior branch of the eighth cranial nerve was severed immediately caudal to the plexus supplying the utricular macula and the two anterior cristae.

2.3 Observation Tank and Lighting Conditions

The observations were conducted in a quiet, darkened room, the temperature of which was maintained at 18 ± 1 °C. The observation tank was mounted on a narrow frame 20 cm above the top of a table. It was constructed completely of glass (5 mm thick) with the internal dimensions $30 \times 27 \times 30$ cm ($L \times B \times H$) and was filled to a depth of 27 cm with aged tap water.

The animals were observed through one end of the tank. The four lighting conditions were: overhead illumination (OI), unilateral illumination (UI), illumination from below (IFB) and darkness (i.e., infra-red illumination, IRI). A sheet of clear Plexiglas (2 mm thick) covered with white silk was affixed to the inner surface of the area to be illuminated such that the silk faced the inside of the tank. Sheets of Plexiglas covered with black velvet were affixed in a similar manner to all other surfaces. The surface of the water was treated accordingly, i.e., a sheet of Plexiglas covered with either silk or velvet was affixed just below the water surface. The animals in the different treatment groups were thus presented with exactly the same conditions with the exception of the direction of incident light.

The light source was equipped with two white neon tubes (40 W, 5000 K, Osram) and a pane of opalescent glass (2 mm thick). It covered the whole of the surface to be illuminated and was positioned parallel to the surface at a distance of 10 cm. The infra-red light source ($\lambda > 850$ nm, DVF 102, Philips) was mounted 100 cm above the surface of the water.

2.4 Recording and Analysis of Behavior

Recordings were made either with a color video camera (C 550, Nordmende) or an infra-red camera (VC 33005, Sanyo), mounted at a distance of 160 and 30 cm respectively from the front wall of the observation tank. A stopwatch display could be superimposed onto the recordings, enabling an accurate (± 0.1 s) temporal analysis of the course of acute compensation. Recorder and monitor were housed in a separate room. Since the camera could be operated from the recorder, it was not necessary to enter the observation room after the animal had been placed in the tank.

Recovery from anesthesia took place in the observation tank. Initially, each animal was placed on the bottom of the tank, lying on its intact side. The observation period began with the animal's first bodily movement, which invariably consisted in turning to lie on the operated side. The animal's post-operative behavior was videotaped for 2 h and subsequently analyzed.

Following HL, four static and dynamic symptoms were observable (for details see Results, Sect. 3.1). The term "acute vestibular syndrome" (AVS) is used to describe the presence of one or more symptoms; the animal's posture or locomotion is referred to as "impaired". Conversely, in the absence of symptoms, the animal's posture or locomotion is said to be "unimpaired". Impaired and unimpaired behavior could be manifested by both active, i.e., freely moving, and inactive, i.e., resting (on the bottom), animals.

The following parameters were used to characterize the course of acute compensation:

1. The time which elapsed until the first occurrence of unimpaired activity. This is referred to as stage 1. It is a stage of continuous impairment.
2. The time which elapsed until the last appearance of symptoms of the AVS. This is referred to as stage 2. It is a stage of interpolated impairment and unimpairment.
3. The time which elapsed until the completion of 5-min continuous and unimpaired swimming. This is referred to as stage 3. It is a stage of continuous unimpairment and animals which reached this stage showed no further impairment unless disturbed. In other words, it may be regarded as the criterion of successful compensation (free-swimming criterion, FSC).

In all cases, zero time was the start of the observation period, i.e., the animal's first bodily movement, and all times were rounded to the nearest quarter of a minute. For diagrammatic and statistical purposes, animals failing to reach a particular stage within the 2-h observation period were attributed a time of 120 min.

The possible influence of general activity on the course of acute compensation was evaluated by analyzing the animals' behavior in terms of impaired and unimpaired activity. In each case, measurements were made of:

1. The total amount of activity (expressed as a percentage of the total compensation or observation time),
2. the frequency of activity (number of bouts of activity per minute), and
3. the mean duration (s) of each bout of activity.

In order to evaluate possible confounding effects of anesthesia or surgery on post-operative behavior, records were made of:

1. The duration of exposure to anesthetic, i.e., the time spent in tricaine methane sulfonate solution,
2. the duration of HL, i.e., the time spent out of water,
3. the duration of recovery from anesthesia, i.e., the time which elapsed until spontaneous respiratory movements recommenced, and
4. the latency (from zero time) to the first bout of activity.

Where appropriate, the results were analyzed statistically with the Wilcoxon-Mann-Whitney U-Test (Sachs 1968, pp. 293–300) and the Spearman Rank Correlation Coefficient (Sachs 1968, pp. 390–394).

3 Results

3.1 Acute Effects of Hemilabyrinthectomy

HL caused impairment of both posture and locomotion. The static symptoms (Fig. 2) were:

Fig. 2. Static symptoms of a goldfish labyrinthectomized on the left side. Note the flexion of the longitudinal body axis towards the operated side and the tonic ventral deviation of the ipsilateral eye. The abnormal displacement of some of the fins can also be seen (for details see text)

1. Flexion of the body towards the operated side;
2. tonic deviation of the eyes towards the operated side, i.e., the ipsilateral eye was deviated ventrally and the contralateral eye dorsally; and
3. abnormal displacement of the fins. The dorsal fin and the dorsal margin of the caudal fin were tilted contralaterally, whilst the anal fin and the ventral margin of the caudal fin were tilted ipsilaterally. The displacement of the paired fins was less consistent; on the whole, the ipsilateral pectoral fin was displaced towards the intact side, whereas the contralateral pectoral fin and both pelvic fins were displaced towards the lesioned side.

The dynamic deficit manifested itself in abnormal swimming with rolling around the longitudinal axis towards the ipsilateral side. During such bouts of impaired activity, the postural asymmetries listed above remained evident.

The symptoms developed almost immediately following recovery from anesthesia. In stage 1 (continuous impairment), resting animals always lay on the operated side. Typically, relatively long periods of inactivity were interpolated with brief bouts of impaired activity. The latter were often followed, after resumption of the resting position, by a pronounced but short-lasting nystagmus.

The first unimpaired activity, which marked the start of stage 2 (interpolated impairment and unimpairment), could by no means be described as "normal" swimming. Locomotion was slow and "jerky", often accompanied by slight pendulating movements about the longitudinal axis. The animals seemed to swim only with the pectoral fins. Initially, such bouts of unimpaired activity lasted only a few seconds and were terminated by the sudden reappearance of the AVS. Disappearance of symptoms prior to bouts of unimpaired activity was likewise abrupt and all symptoms disappeared more or less simultaneously. Periods of inactivity were less frequent and of shorter duration than in stage 1. Indeed, many animals were continuously active from the first bout of unimpaired activity onwards.

Eventually, the animals succeeded in swimming continuously and without impairment for a period of 5 min, whereby the FSC (stage 3) was reached and acute compensation was considered complete. Even though the animals showed no further impairment after this time, providing they were left undisturbed, they still swam less quickly and smoothly than intact fish. Some 2 h post-criterion, however, the only anomaly indicative of an operated animal was a slight tilt (1–3 °) of the body towards the lesioned side. The AVS reappeared briefly when the animals were transferred to their home tanks and could also be re-elicited up to 5–6 days after HL by startling or feeding the animals. Thereafter, under "normal" lighting conditions, it was impossible to distinguish between operated and intact animals.

3.2 Acute Compensation Under Different Lighting Conditions

The different lighting conditions had no observable effect on the nature of the symptoms, but did exert a marked influence on the rate at which the AVS disappeared. Relative to OI, all other conditions significantly increased the time required to reach each stage of compensation ($p \leqq 0.001$ in all cases; see Fig. 3). Similarly, UI had a significantly less deleterious effect on the rate of compensation than either IRI ($p \leqq 0.025$ at each stage) or IFB ($p \leqq 0.01$ at each stage). The effects of IRI and IFB did not differ significantly.

The duration of acute compensation under OI was extremely short, the longest time to criterion of any animal being 18.5 min. By contrast, animals exposed to IFB remained impaired throughout the whole of the 2-h observation period. The results under IRI were similar. However, one of the six animals tested did actually reach criterion within 2 h (after 55 min). Of the remaining five animals, three showed no signs of compensation, whilst the other two reached stage 1 (after 35 and 75 min respectively) but failed to satisfy the criterion of 5-min free-swimming. (It is perhaps significant that the unimpaired activity of these two animals took place near the bottom or the walls of the tank; as soon as they moved into free water, their locomotion became impaired).

Animals maintained under UI exhibited a form of behavior not seen in other groups. Under the conditions of unilaterally incident light, the fish often assumed a position in which the two eyes were asymmetrically illuminated. They then performed a so-called "dorsal light response" (see Fig. 1). During stage 2, illumination of the ipsilateral eye induced a sudden and extreme increase in the degree of tilt, which invariably led to a loss of equilibrium with rolling around the longitudinal axis. With time, the animals seemed to learn to keep the ipsilateral eye away from the light or to adopt a position in which the eyes were not asymmetrically illuminated, i.e., with the longitudinal body axis parallel to the direction of light

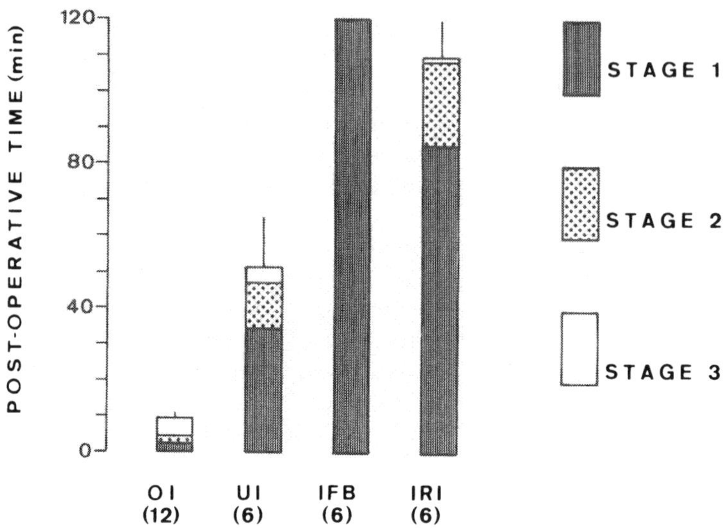

Fig. 3. Course of acute compensation under the different lighting conditions (*OI* overhead illumination; *UI* unilateral illumination; *IFB* illumination from below; *IRI* infra-red illumination; N in brackets). The histograms show the mean time (min) required to reach each stage of compensation. The bars indicate s.e. values of the mean times to the 5-min FSC. For further details see Methods Section 2.4

and the head pointing towards or away from the light source. Under these conditions, bouts of unimpaired activity of 2–3 min duration were not exceptional.

Once the FSC had been reached, the dorsal light response was no longer accompanied by a loss of equilibrium. When illuminated from the lesioned side, extreme degrees of tilt were assumed and maintained until the animals changed their position relative to the light. The angle of tilt varied slightly with the distance from the light source and from the background wall but it was rarely less than 80 ° and some animals tilted more than 90 °. A dorsal light response was also performed on the intact side when the contralateral eye was illuminated. The angles of tilt were less pronounced, however, lying between 10 and 40 °.

3.3 Activity Measures

Measures of impaired and unimpaired activity are presented in the form of histograms in Fig. 4. The four treatment groups differed little in terms of impaired activity. The total amount of impaired activity was similar in each group and significant differences were found only in the frequency and mean duration of activity. As can be seen from Table 1, bouts of impaired activity were less frequent but longer lasting in IRI-animals than in other groups.

In terms of unimpaired activity, however, the different groups were less readily comparable. IFB-animals showed no unimpaired activity and IRI-animals executed less unimpaired activity than either of the other two groups, significantly so in comparison with UI (see Table 2). It is noticeable that bouts of unim-

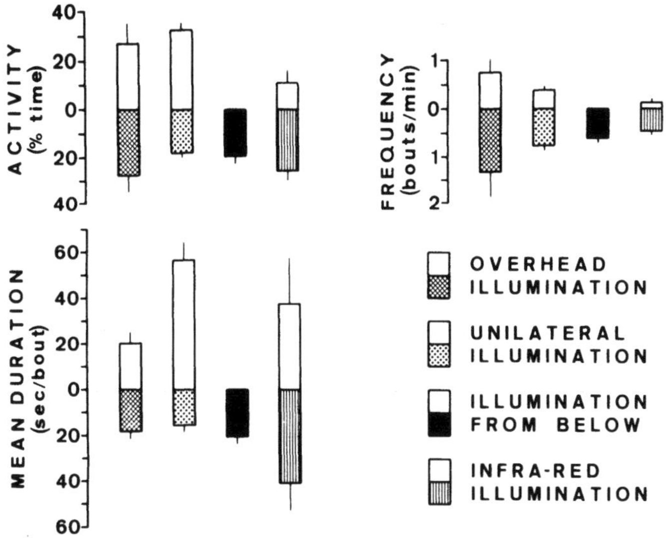

Fig. 4. Measures of impaired (*hatched columns*) and unimpaired (*open columns*) activity in each treatment group (see *inset legend*). The histograms show mean values (±s.e.) of activity over total time, i.e., either compensation time (AVS-duration) or observation time (2 h) in animals which failed to compensate. For further details see Methods, Section 2.4

Table 1. Statistical comparison (U-Test) of impaired activity measures (cf. Fig. 4) between all treatment groups

Comparison	% Activity	Frequency	Mean duration
OI vs. UI	ns (26)	ns (27)	ns (28)
IFB	ns (36)	ns (26)	ns (25)
IRI	ns (30)	ns (21)	0.05 (15)
UI vs. IFB	ns (13)	ns (9)	ns (8)
IRI	ns (9)	0.05 (6)	0.025 (4)
IFB vs. IRI	ns (11)	ns (10.5)	ns (11)

OI, overhead illumination; UI, unilateral illumination; IFB, illumination from below; IRI, infrared illumination; probability values for a one-tailed test (Milton 1964); ns, not significant; U-values in brackets.

Table 2. Statistical comparison (U-Test) of unimpaired activity measures (cf. Fig. 4) between all treatment groups

Comparison	% Activity	Frequency	Mean duration
OI vs. UI	ns (23)	ns (25)	0.005 (6)
IFB	0.005 (6)	0.005 (6)	0.005 (6)
IRI	ns (21)	0.025 (14)	ns (34)
UI vs. IFB	0.005 (0)	0.005 (0)	0.005 (0)
IRI	0.005 (1)	0.025 (4)	ns (9)
IFB vs. IRI	ns (9)	ns (9)	ns (9)

For details see legend to Table 1.

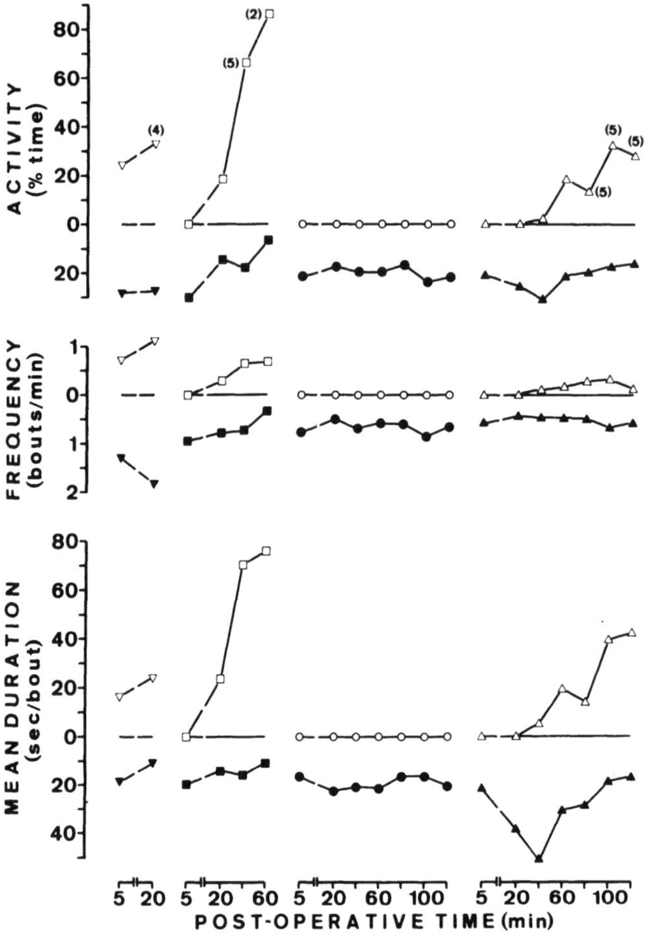

Fig. 5. Measures of impaired (*closed symbols*) and unimpaired (*open symbols*) activity as a function of post-operative time in each treatment group (inverted triangles *OI*; squares *UI*; circles *IFB*; triangles *IRI*). The *symbols* show mean values of activity over the stipulated time periods: 0–5 min, 5–20 min and thereafter at intervals of 20 min until all the animals in a particular group had compensated or the observation time (2 h) had elapsed. (Animals were excluded from the analysis when they reached the end of stage 2; the *figures in brackets in the top row* indicate where N differs from the values given in Fig. 3.) For further details see Methods, Section 2.4, and legend to Fig. 3

paired activity were of lower frequency and longer duration in animals exposed to either UI or IRI than in those maintained under OI.

Changes in the two forms of activity with time are presented in Fig. 5. During the first 5 min, unimpaired activity was exhibited only by OI-animals. In the UI- and IRI-groups, there was a general tendency for the unimpaired activity measures to increase with time, whilst the impaired activity measures underwent a corresponding decrease. In IFB-animals, the level of impaired activity remained fairly constant throughout the 2-h observation period.

Differences in activity were apparent not only between groups but also within groups. Such differences were not reflected in compensation time, however, as il-

Table 3. Mean duration (min ± s.e.) of exposure to anesthetic, hemilabyrinthectomy and recovery from anesthesia and mean latency (sec ± s.e.) to the first bout of activity for each of the four treatment groups

Treatment	Anesthetic	Hemilab.	Recovery	Latency
OI (N=12)	4.0±0.3	11.8±1.0	3.5±1.4	66.3±28.5
UI (N=6)	4.5±0.4	11.5±1.0	4.7±2.8	62.8±19.2
IFB (N=6)	4.5±0.4	13.5±1.5	2.5±1.6	62.0±18.2
IRI (N=6)	4.2±0.3	12.3±1.7	4.0±1.4	66.8±30.9

OI, overhead illumination; UI, unilateral illumination; IFB, illumination from below; IRI, infra-red illumination.

lustrated by the following representative example of two OI-animals with an almost identical AVS-duration of 2 and 2.25 min respectively. The first animal showed 5 s impaired activity (three complete rolls about the longitudinal axis) and 5 s unimpaired activity, whilst the corresponding values for the second animal were 113 s (97 complete rolls) and 17 s. An analysis of combined data from all animals which compensated during the observation period (N=19) revealed no significant correlation between AVS-duration and the total amount of impaired or unimpaired activity ($r_s = -0.148$ and 0.084 respectively).

3.4 Effects of Anesthesia and Surgery

Mean values of the duration of exposure to anesthetic, HL and recovery from anesthesia and of the latency to the first bout of activity are compiled in Table 3. There were no significant differences between treatment groups.

4 Discussion

Electrophysiological recordings from the vestibular nuclei of mammals have demonstrated that, immediately following HL, neuronal activity undergoes a decrease on the deafferented side and a corresponding increase on the intact side as a result of reduced contralateral inhibition (for review see Precht and Dieringer 1985). This marked asymmetry in resting activity between the bilateral vestibular nuclei results in an unbalanced output to the oculomotor and spinal motor systems. The asymmetries in motor innervation and the attendant lack of coordinated interaction between the two labyrinths may be held responsible for the static and dynamic symptoms constituting the "acute vestibular syndrome" observed in the present investigation. The postural and locomotory deficits described here for the goldfish are very similar to those reported previously by Löwenstein (1932) in another cyprinid fish, the minnow.

Qualitatively, the acute effects of HL in the goldfish are directly comparable with those observed in other vertebrates. Flexion of and rolling movements around the longitudinal body axis are symptoms common to almost all investigated species (Schaefer and Meyer 1974). Although there was no persistent nystagmus, as in birds and mammals (Schaefer and Meyer 1974), the tonic deviation of the eyes resembles that seen in frogs (Kolb 1955; Jahn 1960). The abnormal displacement of the fins may be considered analogous to the ab- and adduction of the limbs in tetrapods.

A quantitative comparison of the course of initial recovery in the goldfish with that in other vertebrates, however, reveals two important differences: firstly, the various static and dynamic deficits do not seem to be compensated at different rates but disappear more or less simultaneously and with unusual abruptness; secondly, and more importantly, the immediately post-operative behavioral deficiencies in the goldfish are compensated with a rapidity not encountered in other vertebrates. These differences are probably attributable to species-specific recovery mechanisms, which have been related to qualitative and quantitative differences in synaptic circuitry (Dieringer and Precht 1979 a) and to different modes of locomotion or organization of motor programs (Llinás and Walton 1979 a) in the various species. In addition to those differences in postural and motor control between fish and terrestrial vertebrates already mentioned in the Introduction, it is perhaps significant that the majority of fish are continually active, i.e., unlike other vertebrates, they do not normally rest on supporting substrates but are either actively moving or actively maintaining a stable position in free water. (The relationship between general activity and vestibular compensation will be taken up again below).

We regard the initial phase of compensation in the goldfish to be a separate stage in the recovery process. Within this stage, which we term the "acute" phase, all substantial post-operative deficits are compensated to such an extent that, following attainment of the FSC, the posture and locomotion of lesioned animals is difficult to distinguish from that of intact animals even for an experienced observer. As will be discussed below, visual information determines the rate of acquisition of the acutely compensated state. Elsewhere we will be presenting evidence that the maintenance of this state is likewise dependent on light. For a certain length of time after completion of the acute stage, symptoms of the AVS can be re-elicited by changing the lighting conditions; the susceptibility to such treatment gradually decreases over a number of weeks and eventually disappears. On the basis of these behavioral findings, the course of vestibular compensation in the goldfish may be divided into two distinct stages: an immediately post-operative acute phase and a subsequent chronic phase. It is assumed that the neuronal changes underlying these two stages of behavioral recovery are effected by different mechanisms.

The short duration of acute compensation under optimal lighting conditions and the relatively abrupt transition from severe impairment to "normal" motor performance are probably the reasons why earlier workers failed to recognize this phase as a separate stage in the recovery process. It is possible that the impairment was attributed to the effects of anesthesia and/or surgical trauma following severance of the vestibular nerve. There are a number of arguments which can be brought to bear against such an interpretation but the most convincing is the light-dependency of the recovery process, which will now be discussed.

The present results show that the rate of acute compensation in the goldfish is dependent not only on the presence of light but also on its direction of incidence. Under OI, acute compensation was completed in less than 10 min, whilst under UI the duration of compensation was significantly increased. There was no compensation up to 2 h when the fish were maintained under IFB and a similar, significantly deleterious effect on the course of acute compensation was also observed in the absence of light, i.e., under IRI.

This pronounced influence of light on the course of acute compensation is consistent with the finding of Löwenstein (1932) that blinded minnows required many weeks to months in order to overcome the immediately post-operative effects of HL, whereas visually intact animals recovered in a much shorter time. Blinding also delayed compensation of head deviation in frogs (Kolb 1955), although, puzzlingly, these animals are reported to have compensated just as quickly in the dark as in the light. Putkonen et al. (1977) and Courjon and Jeannerod (1979) have conducted similar experiments on cats, which were placed in the dark immediately after HL. There was no compensation of head deviation as long as visual deprivation was maintained. On the contrary, the severity of the postural asymmetry actually increased during the course of dark exposure. Comparable results have been observed in the development of spontaneous nystagmus in kittens (Berthoz et al. 1975) and cats (Courjon et al. 1977). Similarly, Jensen et al. (1979) studied both spontaneous nystagmus and head deviation in hemilabyrinthectomized guinea pigs held either in the dark or blindfolded in the light: both treatments slowed the rate of acute compensation.

Two alternative explanations of the present findings need to be discussed:

1. The various lighting conditions differentially influence the level of motor activity, which, in turn, might exert an effect on the rate of acute compensation.

It may be assumed that information about deficiencies in posture is available at rest, whereas information about deficits in motor performance can only be obtained during active movement. Indeed, according to Lacour and Xerri (1981), active sensorimotor exploration is a prerequisite for the development of vestibular compensation. This conclusion is supported by experimental evidence. Sensorimotor activation, usually in the form of enforced physical exercise, has been shown to enhance the rate of compensation in a variety of animals (Igarashi et al. 1975, 1979, 1981; Petrosini 1983) including man (Cawthorne 1946; Pfaltz 1983). Conversely, sensorimotor restriction has been reported to delay the disappearance of symptoms (Lacour et al. 1976; Xerri and Lacour 1980; Schaefer and Meyer 1981; Pettorossi and Petrosini 1984).

In the present investigation, an analysis of the general activity manifested by the animals during the observation period revealed no consistent differences between treatment groups in terms of impaired or unimpaired activity and only negligible changes in the amount of total activity with time. Furthermore, in those animals which compensated during the observation period, no significant correlation could be found between AVS-duration and the total amount of impaired or unimpaired activity. In short, comparable levels of activity were observed under all treatment conditions and an explanation of the present findings in terms of light-related differences in motor activity must thus be considered inadequate.

2. Acute vestibular compensation is based on a visual substitution process, in which the missing labyrinthine afferents are functionally replaced by an increase in the relative weight of the visual input to the deafferented side.

A fundamental event in the recovery of function following HL is generally assumed to be the restitution of a balanced resting activity in the vestibular nuclei of the intact and deafferented sides. As demonstrated by the pronounced asymmetry in the dorsal light behavior of animals maintained under UI, unilateral removal of the vestibular input results in an increase in the relative weight of the visual input to the ipsilateral side. This asymmetrical strengthening of the visual input could help to reduce the imbalance in output from the bilateral vestibular nuclei by supplementing the decreased activity on the deafferented side.

Experiments conducted in v. Holst's laboratory (Schoen 1950) on other teleost fish also provide evidence of a change in the strength of the visual input following unilateral labyrinthine lesion. Measurements of the dorsal light behavior of acutely compensated animals exposed to UI revealed an extreme asymmetry in tilt to the ipsilateral and contralateral sides similar to that described for the goldfish in the present investigation.

A supplementation of activity in the deafferented vestibular nucleus by an increase in the relative weight of the visual input would afford an adequate explanation for the occurrence of compensation under OI and the lack of compensation in the absence of light, i.e., under IRI. The present findings suggest, however, that the substitution process is dependent not only on the presence of light but also on its direction of incidence: in the visual-vestibular conflict situations, UI and IFB, where the direction of incident light did not correspond to the direction of gravity, acute compensation was prolonged (relative to OI) or failed to take place within the 2-h observation period. The visual substitution process would thus appear to consist of a supplementation of the lost or reduced tonic activity on the deafferented side in conjunction with a recalibration of coherent spatial coordinates using the available sensory information, viz. visual cues and vestibular cues from the intact labyrinth.

A similar interpretation of the role played by the visual system in the restoration of central symmetry following HL has been developed by Lacour and his coworkers on the basis of their studies in monkeys and cats (Lacour et al. 1979, 1981; Lacour and Xerri 1980, 1981; cf. Courjon et al. 1977). These workers, however, consider that visual information is not the only vicariant input conditioning compensation. Rather, all sensory inputs providing information about the animal's position and movement in space are assumed to intervene in the recovery process. We have restricted our interpretation to the role of the visual input, because its participation is both necessary and sufficient to explain the observed differences in post-operative behavior. Further experiments may well indicate that information from other sensory systems is also involved in the substitution process. In this respect, our deliberations in the Introduction concerning the presumed negligible contribution of the somatosensory system in fish require some qualification. Although tactile information is not perceived by the freely swimming fish, such information is obviously available to a fish which is resting on a solid substrate, such as the bottom of an aquarium. Tactile cues might thus be a further source of information in the recalibration process discussed above.

At present, we can only speculate about where the postulated changes might be localized and how they might occur. The present interpretation presupposes the existence of a central structure (or structures), the output of which goes directly or indirectly via the vestibular nuclei to the ocular and spinal motoneurons and into which both vestibular and visual afferents converge. In the vestibular nuclei of the goldfish, neurons have been found which respond to both visual and vestibular stimulation (Dichgans et al. 1973; Allum et al. 1976). Such neurons, or those in other centers also receiving multimodal information, such as the cerebellum or the reticular formation, could be the site of the proposed modulation in visual-vestibular interaction. Post-synaptic supersensitivity and pre-synaptic hyperactivity, as defined by Gage et al. (1982), are both mechanisms by which changes in the relative strength of neuronal input can be effected following partial denervation. Whether such changes would occur quickly enough to account for the rapid rate of acute compensation under OI is open to question. In this respect, it is perhaps significant that the visual-vestibular interaction responsible for the maintenance of equilibrium in intact teleost fish (see Fig. 1) is subject to both stimulus-dependent and spontaneous changes. Such changes in the central evaluation of peripheral inputs have been termed "Umstimmungen" (v. Holst 1967); they have often been described in teleost fish and, more importantly, they take place rapidly (v. Holst 1935b, 1948, 1967; Schoen 1950; Traill and Mark 1970; Vinyard and O'Brien 1975; Pfeiffer and Riegelbauer 1978). It may therefore be conjectured that the asymmetrical strengthening of visual afferents following unilateral withdrawal of the vestibular input might be effected by a similar mechanism to that underlying visual-vestibular interaction in the intact animal.

Acknowledgement. This work was supported by the Deutsche Forschungsgemeinschaft (Fl 139/2-1).

References

Allum JHJ, Graf W, Dichgans J, Schmidt CL (1976) Visual-vestibular interactions in the vestibular nuclei of the goldfish. Exp Brain Res 26:463–485

Berthoz A, Jeannerod M, Vital Durand F, Oliveras J (1975) Development of vestibulo-ocular responses in visually deprived kittens. Exp Brain Res 23:425–442

Bienhold H, Flohr H (1978) Role of commissural connections between vestibular nuclei in compensation following unilateral labyrinthectomy. J Physiol (Lond) 284:178

Bogenschütz H (1961) Vergleichende Untersuchungen über die optische Komponente der Gleichgewichtshaltung bei Fischen. Z Vgl Physiol 44:626–655

Bone Q, Marshall NB (1985) Biologie der Fische. Fischer, Stuttgart

Cawthorne T (1946) Vestibular injuries. Proc R Soc Med 39:270–273

Courjon JH, Jeannerod M (1979) Visual substitution of labyrinthine defects. In: Granit R, Pompeiano O (eds) Reflex control of posture and movement. Elsevier, Amsterdam, pp 783–792 (Progress in brain research, vol 50)

Courjon JH, Jeannerod M, Ossuzio I, Schmid R (1977) The role of vision in compensation of vestibulo ocular reflex after hemilabyrinthectomy in the cat. Exp Brain Res 28:235–248

Dichgans J, Schmidt CL, Graf W (1973) Visual input improves the speedometer function of the vestibular nuclei in the goldfish. Exp Brain Res 18:319–322
Dieringer N, Precht W (1979a) Synaptic mechanisms involved in compensation of vestibular function following hemilabyrinthectomy. In: Granit R, Pompeiano O (eds) Reflex control of posture and movement. Elsevier, Amsterdam, pp 607–615 (Progress in brain research, vol 50)
Dieringer N, Precht W (1979b) Mechanisms of compensation for vestibular deficits in the frog. I. Modification of the excitatory commissural system. Exp Brain Res 36:311–328
Dieringer N, Precht W (1979c) Mechanisms of compensation for vestibular deficits in the frog. II. Modification of inhibitory pathways. Exp Brain Res 36:329–341
Fay RR, Olsho L (1979) Discharge patterns of lagenar and saccular neurones of the goldfish eighth nerve: displacement sensitivity and directional characteristics. Comp Biochem Physiol 62A:377–386
Flohr H, Bienhold H, Abeln W, Macskovics I (1981) Concepts of vestibular compensation. In: Flohr H, Precht W (eds) Lesion-induced neuronal plasticity in sensorimotor systems. Springer, Berlin Heidelberg New York, pp 153–172
Frisch K von, Stetter H (1932) Untersuchungen über den Sitz des Gehörsinnes bei der Elritze. Z Vgl Physiol 17:686–801
Furukawa T, Ishii Y (1967) Neurophysiological studies on hearing in goldfish. J Neurophysiol 30:1377–1403
Gage FH, Dunnett SB, Bjorklund A, Stenevi U (1982) Functional recovery following brain damage: conceptual frameworks and biological mechanisms. Scand J Psychol Suppl 1:112–120
Holst E von (1935a) Die Gleichgewichtssinne der Fische. Verh Dtsch Zool Ges 37:108–114
Holst E von (1935b) Über den Lichtrückenreflex bei Fischen. Pubbl Stn Zool Napoli 15:143–158
Holst E von (1948) Quantitative Untersuchungen über Umstimmungsvorgänge im Zentralnervensystem. I. Der Einfluß des „Appetits" auf das Gleichgewichtsverhalten bei Pterophyllum. Z Vgl Physiol 31:134–148
Holst E von (1950) Die Arbeitsweise des Statolithenapparates bei Fischen. Z Vgl Physiol 32:60–120
Holst E von (1967) Quantitative Messungen von Stimmungen im Verhalten der Fische. Symp Soc Exp Biol 4:143–173
Igarashi M, Alford BR, Kato Y, Levy JK (1975) Effect of physical exercise upon nystagmus and locomotor dysequilibrium after labyrinthectomy in experimental primates. Acta Oto-Laryngol 79:214–220
Igarashi M, Levy JK, Takahashi M, Alford BR, Homick JL (1979) Effect of exercise upon locomotor balance modification after peripheral vestibular lesions (unilateral utricular neurotomy) in squirrel monkeys. Adv Oto-Rhino-Laryngol 25:82–87
Igarashi M, Levy JK, O-Uchi T, Reschke MF (1981) Further study of physical exercise and locomotor balance compensation after unilateral labyrinthectomy in squirrel monkeys. Acta Oto-Laryngol 92:101–105
Jahn T (1960) Optische Gleichgewichtsregelung und zentrale Kompensation bei Amphibien, insbesondere bei der Erdkröte (*Bufo bufo* L.) Z Vgl Physiol 43:119–140
Jensen DW, Kubo T, Igarashi M (1979) Visual influences on vestibular compensation. Soc Neurosci Abstr 5:691
Klinke R (1970) Efferent influence on the vestibular organ during active movements of the body. Pflügers Arch 318:325–332
Kolb E (1955) Untersuchungen über zentrale Kompensation und Kompensationsbewegungen einseitig entstateter Frösche. Z Vgl Physiol 37:136–160
Lacour M, Xerri C (1980) Compensation of postural reactions to free-fall in the vestibular neurectomized monkey: role of the visual motion cues. Exp Brain Res 40:103–110
Lacour M, Xerri C (1981) Vestibular compensation: new perspectives. In: Flohr H, Precht W (eds) Lesion-induced neuronal plasticity in sensorimotor systems. Springer, Berlin Heidelberg New York, pp 240–253
Lacour M, Roll JP, Appaix M (1976) Modifications and development of spinal reflexes in the alert baboon (Papio papio) following a unilateral vestibular neurotomy. Brain Res 113:255–269

Lacour M, Xerri C, Hugon M (1979) Compensation of postural reactions to fall in the vestibular neurectomized monkey. Role of the remaining labyrinthine afferences. Exp Brain Res 37:563–580
Lacour M, Vidal PP, Xerri C (1981) Visual influences on vestibulospinal reflexes during vertical linear motion in normal and hemilabyrinthectomized monkeys. Exp Brain Res 43:383–394
Lacour M, Vidal PP, Xerri C (1983) Dynamic characteristics of vestibular and visual control of rapid postural adjustments. In: Desmedt JE (ed) Motor control mechanisms in health and disease. Raven, New York, pp 589–605
Llinás R, Walton K (1979a) Vestibular compensation: a distributed property of the central nervous system. In: Asanuma H, Wilson VJ (eds) Integration in the nervous system. Igaku-Shoin, Tokyo New York, pp 145–166
Llinás R, Walton K (1979b) Place of the cerebellum in motor learning. In: Brazier MAB (ed) Brain mechanisms in memory and learning: from the single neuron to man. Raven, New York, pp 17–36 (IBRO monograph series 4)
Löwenstein O (1932) Experimentelle Untersuchungen über den Gleichgewichtssinn der Elritze (Phoxinus laevis L.) Z Vgl Physiol 17:806–854
Luyten WHML, Sharp FR, Ryan AF (1986) Regional differences of brain glucose metabolic compensation after unilateral labyrinthectomy in rats: a [^{14}C]2-deoxyglucose study. Brain Res 373:68–80
Manning FB (1924) Hearing in the goldfish in relation to the structure of its ear. J Exp Zool 41:5–20
McCabe BF, Ryu JH (1969) Experiments on vestibular compensation. Laryngoscope 79:1728–1736
Milton RC (1964) An extended table of critical values for the Mann-Whitney (Wilcoxon) two-sample statistic. J Am Statist Ass 59:925–934
Petrosini L (1983) Compensation of vestibular symptoms in hemilabyrinthectomized guinea pigs. Role of the sensorimotor activation. Behave Brain Res 8:335–342
Pettorossi VE, Petrosini L (1984) Tonic cervical influences on eye nystagmus following hemilabyrinthectomy: immediate and plastic effects. Brain Res 324:11–19
Pfaltz CR (1983) Vestibular compensation: physiological and clinical aspects. Acta Oto-Laryngol 95:402–406
Pfeiffer W, Riegelbauer G (1978) The effect of the alarm substance on the central nervous excitation of the black tetra *Gymnocorymbus ternetzi* (Characidae, Ostariophysi, Pisces) indicated by dorsal light response. J Comp Physiol 123:281–288
Platt C (1983) The peripheral vestibular system of fishes. In: Northcutt RG, Davis RE (eds) Fish neurobiology, vol 1: brain stem and sense organs. University of Michigan Press, Ann Arbor, pp 89–123
Precht W (1974) Characteristics of vestibular neurons after acute and chronic labyrinthine destruction. In: Kornhuber HH (ed) Vestibular system, part II. Springer, Berlin Heidelberg New York, pp 451–462 (Handbook of sensory physiology, vol VI/2)
Precht W, Dieringer N (1985) Neuronal events paralleling functional recovery (compensation) following peripheral vestibular lesions. In: Berthoz A, Melvill Jones G (eds) Adaptive mechanisms in gaze control: facts and theories. Elsevier, Amsterdam, pp 251–268 (Reviews of oculomotor research, vol 1)
Putkonen PTS, Courjon JH, Jeannerod M (1977) Compensation of postural effects of hemilabyrinthectomy in the cat. A sensory substitution process? Exp Brain Res 28:249–257
Sachs L (1968) Statistische Auswertungsmethoden. Springer, Berlin Heidelberg New York
Schaefer KP, Meyer DL (1974) Compensation of vestibular lesions. In: Kornhuber HH (ed) Vestibular system, part II. Springer, Berlin Heidelberg New York, pp 463–490 (Handbook of sensory physiology, vol VI/2)
Schaefer KP, Meyer DL (1981) Aspects of vestibular compensation in guinea-pigs. In: Flohr H, Precht W (eds) Lesion-induced neural plasticity in sensorimotor systems. Springer, Berlin Heidelberg New York, pp 197–207
Schoen L (1950) Quantitative Untersuchungen über die zentrale Kompensation nach einseitiger Utriculusausschaltung bei Fischen. Z Vgl Physiol 32:121–150
Traill AB, Mark RF (1970) Optic and static contributions to ocular counter-rotation in carp. J Exp Biol 52:109–124

Vinyard GL, O'Brien WJ (1975) Dorsal light response as an index of prey preference in bluegill (*Lepomis macrochirus*) J Fish Res Board Can 32:1860–1863
Wohlfahrt TA (1932) Anatomische Untersuchungen über das Labyrinth der Elritze (Phoxinus laevis L.) Z Vgl Physiol 17:659–685
Xerri C, Lacour M (1980) Compensation des déficits posturaux et cinétiques après neurectomie vestibulaire unilatérale chez le chat. Rôle de l'activité sensorimotrice. Acta Oto-Laryngol 90:414–424

V Lesions to Specific CNS Subsystems: Cerebellum and Oculomotor System

Recovery from Smooth Pursuit Impairments After Successive Unilateral and Bilateral Chemical Lesions in the Dorsolateral Pontine Nucleus of the Monkey

J. G. May III and E. L. Keller
The Smith-Kettlewell Eye Research Foundation, San Francisco, Ca. 94115 USA

1 Introduction

Primates have a well-developed ability to move their eyes accurately in order to track the motion of small moving objects. These tracking eye movements, which consist of both saccadic (fast) and smooth pursuit (slow) responses, are driven by target image motion across the retina. The resulting retinal image slip velocity or retinal error velocity (REV) has been shown to be an important visual signal used by the nervous system for the control of smooth pursuit eye movement (Lisberger and Westbrook 1985).

Directionally selective units responsive to visual motion are found in a variety of visual cortical regions, including primary visual cortex (area 17) and a group of cortical areas within the caudal superior temporal sulcus and inferior parietal lobule (parieto-occipital cortex). These areas have also been implicated in the control of smooth pursuit by lesion studies. Damage to these cortical regions can result in specific deficits in the generation and maintenance of smooth pursuit eye movements (Dürsteler et al. 1987; Newsome et al. 1985; Lynch and McClaren 1982).

Anatomical studies suggest that visual information originating in these areas may reach the flocculus and posterior vermis, cerebellar regions important in smooth pursuit control (Zee et al. 1986), via the pontine nuclei (Brodal 1978, 1979; Glickstein et al. 1980; Langer et al. 1985; May and Andersen 1986). Our laboratory has been studying the role of the dorsolateral pontine region in smooth pursuit control by examining the response characteristics of single units and by looking at the effects of pharmacological lesions and microstimulation in this area (May et al. 1986, 1988; Suzuki and Keller 1984).

Unilateral injections of ibotenic acid (an excitotoxin specific to cell somata) into the dorsolateral pontine nucleus produced dramatic but transient impairments in smooth pursuit tracking similar to the impairments seen after cortical lesions (Dürsteler et al. 1987; Newsome et al. 1985). We were surprised at the speed of recovery after pontine lesions, since we viewed the pons as funneling information from a variety of cortical regions onto the cerebellum. In this report we present recent results related to how this pursuit recovery may be mediated.

H. Flohr (Ed.)
Post-Lesion Neural Plasticity

2 Methods

Animals were prepared for chronic electrophysiological recording by implanting a scleral eye coil (for precise measurement of eye movements), a stainless steel recording chamber, and a device for atramatic head restraint. Conventional electrophysiological recording procedures were used to locate and chart visually responsive regions of the dorsolateral pontine nuclei. After the area had been identified and baseline pursuit performance was collected, microinjections of the excitotoxins ibotenic acid (IBO) or N-methyl-DL-aspartic acid (NMDA) were made through micropipets using an air pressure delivery system.

Eye movement responses were collected primarily with the step-ramp paradigm of Rashbass (1961). The animal was first required to fixate a stationary target spot. After a variable period, the target would step out into either the ipsilateral or contralateral visual field and simultaneously begin to ramp either toward or away from the fovea. The animal was then required to track this target for a variable interval. The animal's eye position was monitored by computer and compared with a signal representing target position. If at the end of the tracking period the animal's eye postion fell within a 2–5 ° electronic window around the target position he received a liquid reinforcement. This paradigm was chosen since it allowed us to examine the initial response of the pursuit system to retinal image motion confined to a known retinal location. This initial pursuit response (over the first 100 ms) can be viewed as the open loop performance of the system (Lisberger and Westbrook 1985), since there is not enough time for visual feedback to make a contribution to the response.

3 Effects of Unilateral Pontine Lesions

An example of the effects of a unilateral ibotenate lesion on smooth pursuit is presented in Fig. 1. Pre- and post-lesion eye position traces from individual trials are superimposed in the upper portion of the figure with averaged eye velocity for these same trials presented below. The pursuit impairments are easiest to see on these averaged velocity traces. Eye acceleration was markedly reduced when target motion (and pursuit tracking) was directed ipsilateral to the lesion (see Fig. 1, panel A). Significant deficits were seen in the initial pursuit response as well during steady state or maintained pursuit tracking. There was a smaller impairment present for the initiation of pursuit to target motion in the opposite direction (contralateral to the lesion), most apparent during the late portion of this "open loop" response (30–100 ms). All five unilaterally lesioned animals exhibited asymmetrical horizontal impairments (greater in the ipsilateral direction) as well as deficits in at least one of the vertical directions. The degree of impairment seen for pursuit initiation in response to target motion in the four cardinal directions is presented in the right panel of Fig. 2 (the same animal presented in Fig. 1).

There appeared to be little dependence of the retinal location of target image motion on the magnitude of the pursuit impairments seen after DLPN ibotenate injections. By selecting appropriate steps on the step-ramp paradigm, we could control the retinal locus of the target image motion which generated the initial

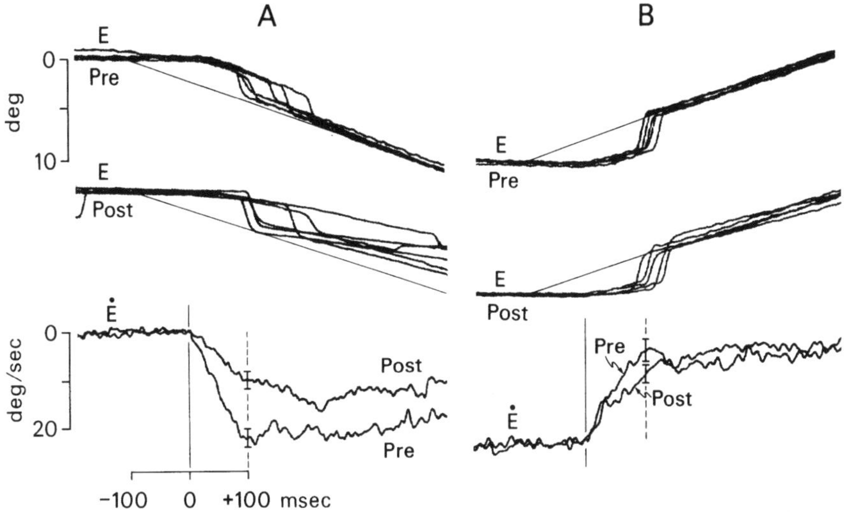

Fig. 1. Smooth pursuit eye movement responses during constant velocity ramp tracking for monkey P. Each set of records show a representative collection of eye position responses (*E*) just prior to toxin injection (*pre*) and one day after a 1.5 µl injection of ibotenate in the left pontine nuclei (*post*). The averaged eye velocity traces presented in the bottom represent the average of the same trials presented in the top position traces. **A** The eye movement response in the leftward (ipsilateral) direction. **B** Similar responses in the rightward direction. The pre and post eye responses were constructed by averaging individual trial velocity responses which were desaccaded prior to averaging. Position traces are aligned on the onset of target motion. Velocity traces are shown aligned on the pursuit response onset (*vertical line*). The *vertical dashed line* to the *right* on the velocity averages shows the time epoch of eye velocity measurements (100 ms after response onset) used to quantify the responses in Fig. 2. The *vertical bars* on the averaged eye velocity responses show the standard deviations of the mean eye velocity values at the 100-ms measurement point

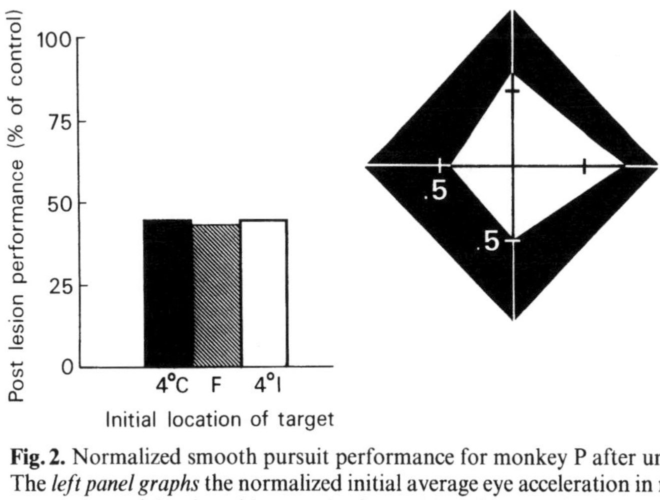

Fig. 2. Normalized smooth pursuit performance for monkey P after unilateral ibotenate lesions. The *left panel graphs* the normalized initial average eye acceleration in response to ipsilateral target motion originating either on the fovea (*F*) or from 4 ° in the ipsilateral (*I*) or contralateral (*C*) visual field. The degree of impairment produced by this lesion was not dependent on the retinal locus of initial target motion for any of the step sizes tested (4, 6, and 10 degrees). The plot on the *right* shows the relative eye acceleration generated in response to target motion for the four cardinal directions. Pre-lesion control performance is represented as the *dark outer rectangle*. The post-lesion pursuit performance was normalized by dividing post-lesion measurements by control values from the same animal and is depicted by the *internal white rectangle*

100-ms pursuit response. This paradigm allowed us to compare the effects of retinal image motion in the foveal region or from the ipsilateral and contralateral visual field on the initial pursuit response (see Fig. 2, left panel). There was little difference in the pursuit impairments seen either for foveal or parafoveal (either ipsilateral or contralateral field) target motion. Thus, pursuit impairments created by unilateral DLPN injections are related to the direction of visual motion and are not dependent upon the hemifield in which the visual motion stimulus occurs.

4 Recovery of Pursuit

All of the unilateral lesioned animals showed complete, or near complete recovery of pursuit performance, as measured by average eye acceleration over the first 100 ms. The greatest deficits were always seen on the day immediately following the injection. A rapid recovery then ensued, such that most animals regained eye accelerations close to pre-lesion levels within 7 days following the lesion (see Fig. 3). Several explanations may be proposed to account for this recovery including: (1) recovery of transiently impaired pontine neurons spared by the injections; (2) functional compensation by units in the contralateral pontine nucleus; and (3) mediation of pursuit by alternate nonpontine pathways.

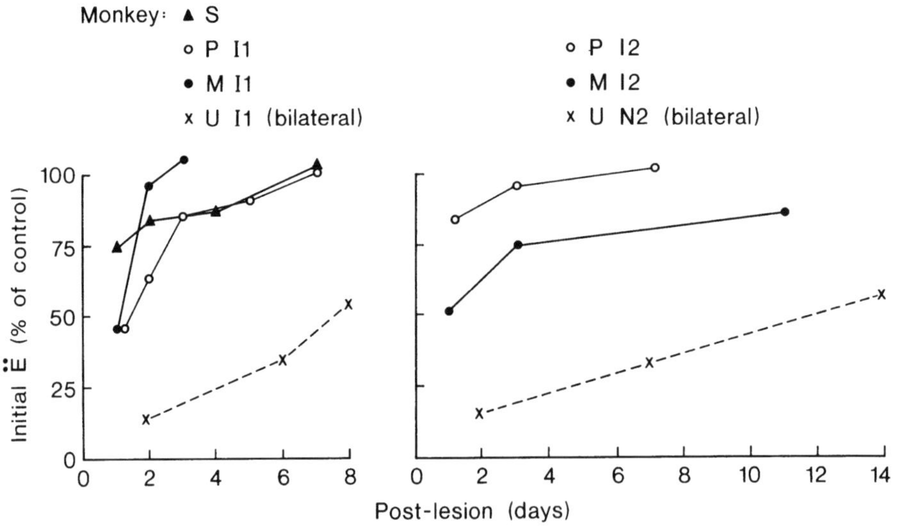

Fig. 3. Recovery of initial eye acceleration responses after pontine lesions. The symbols plot relative post lesion performance (initial eye acceleration after the lesion divided by the initial acceleration prior to the lesion) for ramp velocities of 20 deg/s. Each *data point* shows an average obtained from at least six individual trials. The effects of a second series of repeated injections is presented in the *right panel* for three animals. Note the similar degree of impairment and the somewhat slower recovery curve produced by these repeated injections. The *dashed line* illustrates the more severe impairments produced by bilateral injections in the dorsolateral pontine nucleus

The functional extent of the inactivated region produced by the ibotenic acid injections is hard to assess. Although an extensive region of cell loss was seen in each animal, we suspect that a much larger population of cells was suppressed by the toxin in the days immediately following the injection. Recovery of these cells might account for much of the rapid behavioral recovery exhibited by these animals. This is similar to a suggestion put forth by Newsome et al. (1985) to explain the recovery seen in their cortically lesioned animals.

Occasional neurons were found within the region of extensive cell loss and gliosis, although these neurons often had a rather pale staining cytoplasm and were relatively small. It is possible that recovery of these cells, presumably rendered nonfunctional initially, may also contribute to the observed recovery in pursuit performance.

5 Repeated Injections

Observations from animals receiving repeat injections subsequent to the recovery from initial injections support the notion that the recovery of a surrounding halo of cells, initially depressed by the action of the toxin but not killed, is responsible for much of the observed recovery in pursuit initiation. The effects of this second injection on ipsilaterally directed pursuit and its subsequent recovery is presented in Fig. 3 B for two unilateral animals. The fact that the second injection created a pursuit deficit of almost identical severity to that produced by the first injection weighs against the suggestion that recovery is mediated through functional compensation by units in the contralateral pontine nuclei. If this were the major mechanism for recovery, then a second injection into the same area should not have much effect on the animal's recovered pursuit performance.

6 Bilateral Lesions

Simultaneous bilateral injections were made in one animal, first with ibotenic acid followed one week later by bilateral NMDA injections. The effects on pursuit of these bilateral lesions were more severe, though more directionally symmetric, than those seen after unilateral injections. There was still a rather rapid recovery, despite the severity of the lesions, as shown by the dotted curves in Fig. 3 and trace C in Fig. 4. There was significant damage to the dorsolateral regions on both sides in this case, however the lesion on the right side was much more extensive. Average eye velocity traces are presented in Fig. 4 for pursuit directed toward the side of the largest lesion. Trace B in the left panel shows the impairment seen one day after the first set of ibotenic injections. Note the lowered eye velocity seen throughout pursuit initiation and the failure to match eye velocity to target velocity during pursuit maintenance. After one week, pursuit maintenance appeared normal although pursuit initiation was still slowed (trace C). Repeated bilateral injections (now with NMDA) were similar to the repeated unilateral cases where the effects produced by the second set of repeat injections were very similar to the impairments seen after the original injections (righthand panel Fig. 4). Note the

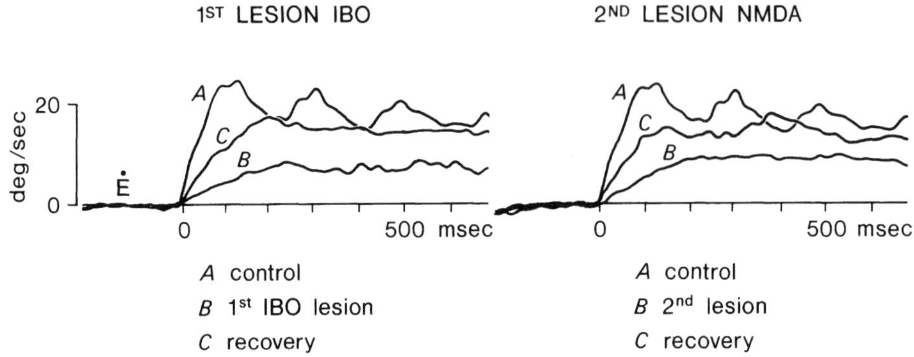

Fig. 4. Bilateral ibotenate (*IBO*) lesion in animal U (*left panel*) followed one week later by bilateral NMDA injections (*right panel*). Trace *A* represents pre-lesion performance and is presented in both panels for comparison purposes. Trace *B* shows the impairments seen in eye velocity 1 day after each injection. Trace *C in the right panel* depicts the recovery seen one week after the first set of injections. Trace *C in the left panel* represents pursuit recovery seen 2 weeks after the second set of injections. Note the similarities in the impairments produced and degree of recovery seen after each set of injections

similarities both of the initial impairments seen one day post-injection (Fig. 4 traces B) and the degree of recovery after these two sets of injections (Fig. 4 traces C). These results support the interpretation of significant nonlethal suppression of surrounding pontine regions.

Two weeks after the NMDA injections there was still a residual impairment in steady-state pursuit tracking. In addition, whereas all of the unilateral animals showed complete recovery of the first 30 ms component of pursuit initiation, this response remained impaired over the entire 2-week test period after the bilateral lesions. This initial 30-ms response has been described as an early mechanism which is relatively independent of both the speed and the retinal locus of stimulus motion (Lisberger and Westbrook 1985).

7 Summary and Conclusions

Ibotenate lesions of the dorsolateral pontine region produce transient impairments in the initiation and maintenance of smooth pursuit tracking eye movements. Unlike cortical lesions in area MT, which produce omnidirectional retinotopic impairments in pursuit initiation, pontine lesions result in directionally asymmetric impairments in pursuit initiation independent of the location of initial target motion in the visual field. These pontine lesions usually produce directional impairments in pursuit maintenance similar to the deficits seen after cortical lesions of foveal MT and MST.

The occurrence of nonretinotopic deficits in pursuit initiation as well as problems in pursuit maintenance may be explained by the prominate convergence of cortical inputs from widespread parts of MT, foveal MT and MST within the ibotenate lesioned area. The origin of the directional asymmetry seen for pursuit initiation, however, is not clear. Similarly, although the directional nature of the steady-state impairment is consistent with the effects of cortical lesions, the mechanism producing these directional effects has not been adaquately explained.

Ibotenate injections made into areas which had received previous injections produced a similar pattern of deficits. The similarity of the deficits produced by the repeated injection argues that the observed recovery cannot be fully explained due to a functional compensation by the contralateral dorsolateral pons or even by alternate pathways carrying similar visual motion information to pursuit structures.

These observations provide evidence that areas surrounding the central locus of the injection may be responsible for many of the initial behavioral deficits seen. The dramatic recovery may reflect, in large part, the recovery of cells in those areas surrounding the central lethal zone. This idea is consistent with observations of astrocytic proliferation within the area surrounding this central zone of cell loss. Similar arguments have been proposed by Newsome et al. (1985) to explain the rapid recovery seen after cortical MT lesions.

Severe impairments for both directions of horizontal pursuit were produced by bilateral ibotenate injections. Unlike the unilateral lesions seen in five animals, this bilateral case exhibited a persistent impairment in the first 30 ms of the initial pursuit response throughout the 14-day post-lesion observation period. There was also a persistent impairment in steady-state pursuit maintenance for tracking target motion directed ipsilateral to the side exhibiting the largest extent of damage. These results suggest that the contralateral pontine nuclei may indeed be involved in some degree of compensation after unilateral damage (of the DLPN region). This may be especially true for the first 30 ms component of pursuit initiation.

These results do not rule out the possibility of significant contributions to the recovery by additional nonpontine pathways. Such contributions are hard to evaluate in the present experiments since all cases revealed significant intact portions of DLPN and/or surrounding pontine regions known to receive visual signals.

References

Brodal P (1978) The cortico-pontine projection in the rhesus monkey. Origin and principles of organization. Brain 101:251–283

Brodal P (1979) The pontocerebellar projection in the rhesus monkey: an experimental study with retrograde axonal transport of horseradish peroxidase. Neuroscience 4:193–208

Dürsteler MR, Wurtz RH, Newsome WT (1987) Directional pursuit deficits following lesions of the foveal representation within the superior temporal sulcus of the macaque monkey. J Neurophysiol 57:1262–1287

Glickstein M, Cohen J, Dixon B, Gibson A, Hollins M, LaBossiere E, Robinson F (1980) Corticopontine visual projections in macaque monkey. J Comp Neurol 190:209–229

Langer T, Fuchs A, Scudder C, Chubb M (1985) Afferents to the flocculus of the cerebellum in the rhesus macaque as revealed by retrograde transport of horseradish peroxidase. J Comp Neurol 235:1–25

Lisberger SG, Westbrook LE (1985) Properties of visual inputs that initiate horizontal smooth pursuit eye movements in monkeys. J Neurosci 5:1662–1673

Lynch JC, McClaren JW (1982) The contribution of parieto-occipital association cortex to the control of slow eye movements. In: Lennerstrand G, Keller E (eds) Functional basis of ocular motility disorders. Pergamon, Oxford, pp 501–510

May JG, Andersen RA (1986) Different patterns of cortico-pontine projections from seperate cortical fields within the inferior parietal lobule and dorsal prelunate gyrus of the macaque. Exp Brain Res 63:265–278

May JG, Keller EK, Crandall WF (1986) Changes in eye velocity during smooth pursuit tracking induced by microstimulation in the dorsolateral pontine nucleus of the macaque. Soc Neurosci Abst 11:79

May JG, Keller EK, Suzuki DA (1988) Smooth pursuit eye movement deficits with chemical lesions in the dorsolateral pontine nucleus of the monkey. J Neurophysiol 59:952–977

Newsome WT, Wurtz RH, Dürsteler MR, Mikami A (1985) Deficits in visual motion processing following ibotenic acid lesions of the middle temporal visual area of the macaque monkey. J Neurosci 5:825–840

Rashbass C (1961) The relatonship between saccadic and smooth pursuit tracking eye movements. J Physiol (Lond) 294:335–348

Suzuki DA, Keller EK (1984) Visual signals in the dorsolateral pontine nucleus of the alert monkey: their relationship to smooth-pursuit eye movements. Exp Brain Res 53:473–478

Zee DS, Tusa RJ, Herdman SJ, Butler PH, Gucer G (1986) In: Keller EK, Zee DS (eds) Adaptive processes in visual and oculomotor systems. Pergamon, Oxford, pp 267–276

Regeneration of Cerebellofugal Projection in Kittens

S. Kawaguchi

Department of Physiology, Institute for Brain Research, Faculty of Medicine, Kyoto University, Kyoto 606, Japan

1 Introduction

Occurrence of marked, functionally active, regeneration of the cerebellofugal projection was proved in kittens after complete transection of the decussation of the brachium conjunctivum (BCX)[1]. The origin, course, and destination of the regenerated fibers were morphologically identified unambiguously; functional reconnection of the regenerated fibers was confirmed electrophysiologically; time-course of axonal regeneration was clarified (Kawaguchi et al. 1981, 1986). Most of the regenerated axons took a course similar to that of the normal projection and terminated in the normal projection areas, whereas a small proportion of fibers showed an aberrant course and termination. In all animals in which marked axonal regeneration occurred, cerebello-cerebral responses, as in intact animals, were evoked in the frontal motor and parietal association cortices. This indicates the occurrence of functional reconnection of the regenerated projection because transection of the BCX in control animals completely abolished the cerebellocerebral responses. The cut ends of axons began to swell as early as 15 min after transection, produced terminals tipped by growth cones in 14–24 h, grew to cross the lesion in 3 days, and distributed dense terminals in the thalamus by 19 days. The majority of such terminals were fine, as in intact animals, but some were much coarser; the latter was presumed to be in various stages of maturation.

2 Experimental Design

It is self-evident that axonal regeneration can be accepted only when axons under consideration are proved to originate from cut axons but not from neighboring spared axons nor from immature axons arriving late in ontogenesis. Proving it, however, is not easy. Complete transection of a central fiber bundle is rather difficult even when a razor-sharp knife is used, and spared fibers tend to be dislocated remarkably. Collaterals sprouting from such fibers can be seen near the lesion. Thus, neither the presence of fibers coursing in aberrant pathways nor the presence of newly formed fibers tipped by growth cones is convincing evidence for axonal regeneration.

---

[1] List of Abbreviations see page 429

H. Flohr (Ed.)
Post-Lesion Neural Plasticity

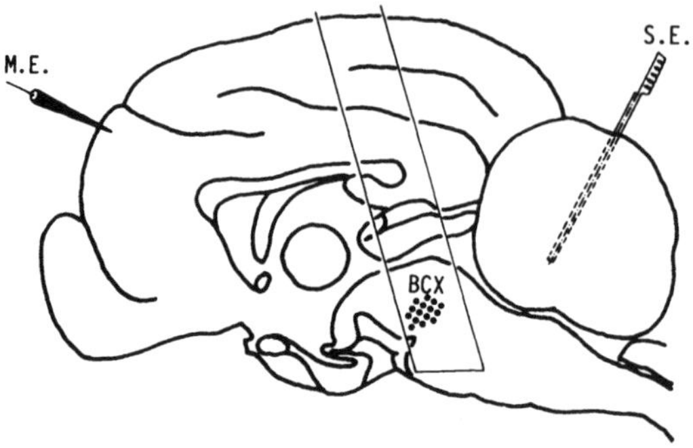

Fig. 1. Experimental design to prove axonal regeneration. A U-shaped trilateral tungsten wire of 200 μm in diameter, a razor-sharp edged bottom side connected with arms of two parallel lateral sides, was used for BCX transection. Complete transection of the BCX was accomplished by implantation of the cutting device which was pushed down at the midline to reach the brainstem base and left in situ to mark the lesion. Later, HRP was injected into the cerebellar nuclei to label anterogradely the cerebellofugal projection. Potentials were recorded from the cerebral cortex on stimulation of the cerebellum to test the functional connectivity of the regenerated fibers in the same animals for the histological examination

The cerebellofugal projection arising from the lateral and interpositus nuclei appears an appropriate material for proving axonal regeneration because the projection is a compact fiber bundle and crosses completely at the BCX where the fiber bundle can be transected completely. The BCX was transected in kittens; to mark lesion, it was performed by implantation of an edged U-shaped trilateral tungsten wire (200 μm in diameter) in the brainstem at the midline (Fig. 1). Transection was complete when the bottom side of the cutting device had been pushed down to the brainstem base. Later, horseradish peroxidase (HRP) was injected unilaterally into the cerebellar lateral and interpositus nuclei to label the cerebellofugal projection by anterograde axonal transport. Axonal regeneration can be proved provided fibers labeled with HRP are demonstrated to cross the area enclosed with the trilateral tungsten wire because such fibers can be ascribed neither to collateral sprouting from spared axons nor to late arrival of developing axons. The reason for the latter is that growth and termination of this projection in the cat are considered a prenatal occurrence (Kawaguchi et al. 1979a, 1979b, 1983, 1986). Functional connectivity of the regenerated projection was tested by recording potentials from the cerebral cortex on stimulation of the cerebellar nuclei in the same animals examined morphologically. Transection of the BCX interrupts completely the cerebellothalamic projection arising from the lateral and interpositus nuclei, and thus abolishes completely the cerebellocerebral response induced by stimulation of these nuclei. Functional reconnection of the regenerated projection, therefore, can be proved provided the cerebellocerebral response is evoked in the animals in which the BCX had been transected. To reveal morphological changes in axons occurring very early after transection, HRP was injected immediately after or prior to transection.

3 Results

3.1 Morphological Evidence for Regeneration of the Cerebellofugal Projection

The entire course of the cerebellofugal projection arising from the lateral and interpositus nuclei can be visualized up to the terminals by labeling of horseradish peroxidase (HRP) that was injected into these nuclei. As shown in Fig. 2 in a horizontal (A) and a frontal section (B) of the brainstem, the superior cerebellar peduncle in intact animals decussates completely at the BCX. A large proportion

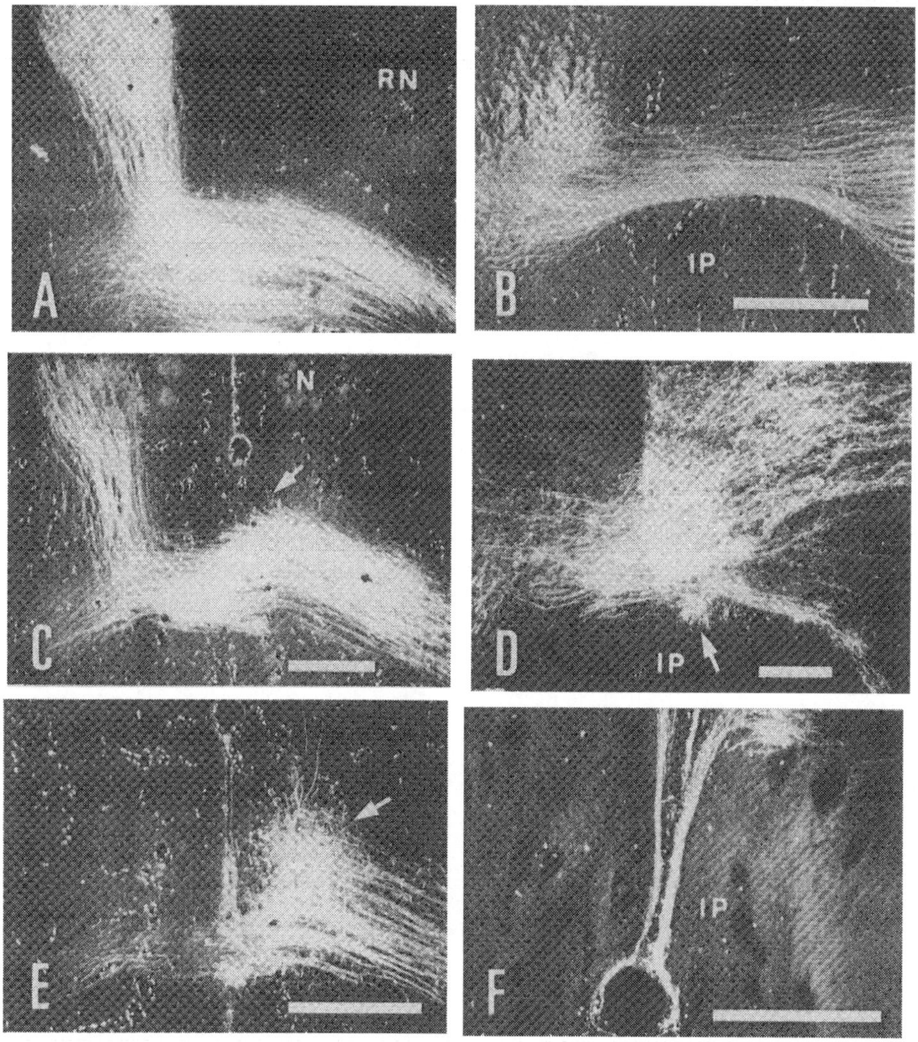

Fig. 2. Darkfield photomicrographs showing the intact and regenerated superior cerebellar peduncle. Horizontal (*left panel*) and frontal sections (*right panel*) of the brainstem, taken from a 6-day-old intact kitten (**A**), a 1-day-old intact kitten (**B**), 6 PO 19 (**C**), 14 PO 33 (**D**), 6 PO 19 (**E**), and 20 PO 57 (**F**). *Arrows* indicate aberrant terminations. Bars = 500 μm. Bar in **B** is applicable to **A** and **B**

of the crossed fibers turns rostralward almost rectangularly to form the ascending limb that traverses the brainstem to enter and surround the contralateral red nucleus (Fig. 2 A); and a relatively small proportion of the crossed fibers turns caudalward to form the descending limb. While fibers at the decussation course closely around the interpeduncular nucleus, none of them enter the nucleus (Fig. 2 B).

After injection of HRP into the cerebellar lateral and interpositus nuclei, a large number of labeled fibers crossing the area enclosed with the trilateral tungsten wire were demonstrated in two out of 24 (Kawaguchi et al. 1981) and in eight out of 82 kittens whose BCX had been transected (Kawaguchi et al. 1986). In the following descriptions, animal's age is given as A (days), the post-lesion time is given as B (days), and the animal is identified as A PO B. An example of the marked regeneration of the cerebellofugal projection is shown in Fig. 2 C which is a horizontal section of the brainstem taken from a kitten (6 PO 19). Many fibers labeled with HRP crossed the previously transected sagittal plane and turned rostralward to ascend the brainstem to enter the red nucleus, just like normal projection fibers. However, there were some aberrant projectios in the tegmental field rostral to the BCX (arrow). Another example of the marked regeneration is shown in Fig. 2 D which is a frontal section of the brainstem taken from a kitten (14 PO 33). In this case, a portion of the labeled fibers crossed the midline to enter the contralateral brainstem but some did not cross and concentrated at the midline where the tissue was initially divided. The crossing fibers were much more dense and twisted as compared with those in control animals. Aberrant projections were seen in the interpeduncular nucleus (arrow). In these animals, dense glial scars surrounded the area of fiber crossing, whereas gliosis in the area of fiber crossing was diffuse and not dense.

Occassionally, regenerated fibers showed aberrant course and termination. A large proportion of such fibers formed the uncrossed ipsilateral cerebellofugal projection. An example is shown in a horizontal section of the brainstem (Fig. 2 E) that was taken from the same kitten in Fig. 2 C at a more dorsal level. In other animals, labeled fibers were observed to make "detours", i.e., the fibers bent ventrally and coursed around the bottom side of the cutting device, and then, bent dorsally. Figure 2 F illustrates an example of the detours which were disrupted in the bottom of the brainstem at the time when the implanted wire was removed from the ventral side for the histological examination. These fibers appeared most likely due to axonal regeneration and very unlikely due to persistence of initially deflected fibers because a razor-sharp wire had been pushed down to reach the brainstem base and a similar procedure in acute experiments transected the BCX completely.

The consequence of BCX transection in the vast majority of animals was failure of marked axonal regeneration. In a large proportion of such animals, however, there were trace amounts of the uncrossed ipsilateral cerebellofugal projection which was never observed in intact animals.

3.2 Course and Termination of the Regenerated Projection

Figure 3 illustrates schematically the course and termination of the regenerated fibers which were largely similar to those of normal projection fibers (filled lines

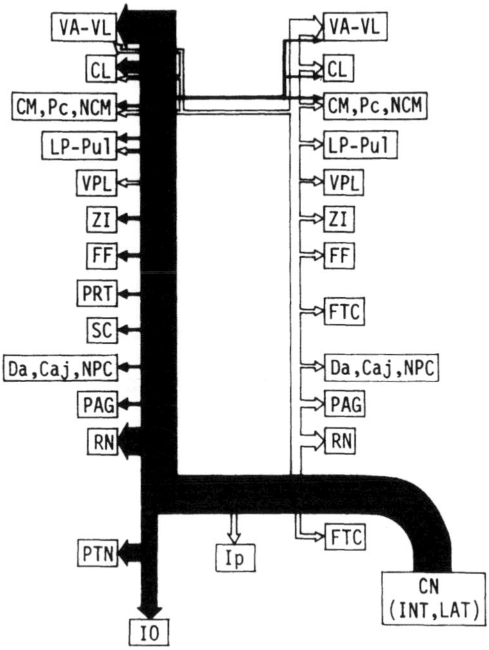

Fig. 3. Schematic illustration of the course of destination of the regenerated cerebellofugal projection. They were mostly similar to those in the normal projection (*filled lines* and *arrows*) but partly aberrant (*open lines* and *arrows*). The majority of the aberrant fibers formed the uncrossed ipsilateral projection that terminated in the homologous structures of the normal projection

and arrows) but partly aberrant (open lines and arrows). Fibers either crossing the lesion or making detours separated into two bundles in a manner similar to the normal projection fibers: one consisted of a large proportion of fibers which turned rostralward to form the ascending limb and the other consisted of a small proportion of fibers which turned caudalward to form the descending limb. The ascending limb, as in control animals, gave the bulk of its terminals to the contralateral red nucleus and thalamic VA-VL nuclear complex, a relatively small number of terminals to the thalamic CL nucleus, and much smaller number of terminals to the periaqueductal gray, interstitial nucleus of Cajal, nucleus of Darkschewitsch, nucleus of posterior commissure, superior colliculus, pretectum, Forel's field, zona incerta, and several thalamic nuclei (LP-Pul, CM, Pc, and NCM). There were a small number of fibers crossing within the thalamus to form the recrossed ipsilateral cerebellothalamic projection terminating in the ipsilateral thalamic VA-VL nuclear complex, CL, CM, and Pc nuclei.

The destinations of the uncrossed ipsilateral cerebellofugal projection as shown in Fig. 2E were mostly homologous structures of the destinations of the crossed contralateral cerebellofugal projection. Besides these fibers projecting onto homologous structures of the normal projection, there were other fibers terminating in entirely anomalous destinations such as the VPL, caudal part of the LP-Pul, central tegmental field surrounding the BCX (arrows in Fig. 2C and E), and interpeduncular nucleus (arrow in Fig. 2D). The proportion of fibers terminating in the normal destinations was much larger than that of fibers terminating in the aberrant destinations; and for the latter, the proportion of fibers terminating in the homologous structures of the normal projection was much larger than that of fibers terminating in the entirely anomalous destinations.

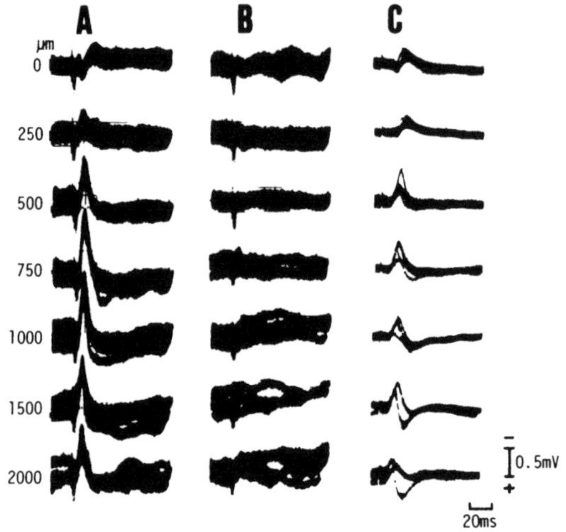

Fig. 4. Cerebellocerebral response in intact and BCX-transected animals. **A** Potentials in the frontal motor cortex evoked by stimulation of the cerebellar nucleus, before (**A**) and after transection (**B**) in a control animal, and in 6 PO 19 (**C**). Numerals to the left of specimen records indicate depths from the cortical surface. Calibrations are applicable to all records

3.3 Functional Connectivity of Regenerated Projection

Stimulation of the interpositus or the lateral nucleus of the cerebellum in intact animals induces contralaterally two distinct types of responses in the frontal motor (areas 4 and 6) and parietal association cortices (areas 5 and 7) respectively (Kawaguchi et al. 1983). The response in the frontal motor cortex is a diphasic positive-negative wave in the superficial cortical layers and a diphasic negative-positive wave in the deep cortical layers (Fig. 4 A). The response in the parietal association cortex is a negative wave in the superficial cortical layers and a positive wave in the deep cortical layers. Transection of the BCX abolished these responses completely (Fig. 4 B). It is natural because all fibers arising from the lateral and interpositus nuclei cross completely at the BCX. The cerebellocerebral response could not be evoked in any of the animals in which labeled terminals were virtually absent in the thalamus at the subsequent histological examination. By contrast, as shown in Fig. 4 C, marked cerebellocerebral responses were evoked in all kittens in which labeled terminals arising from regenerated axons were distributed densely in the thalamic VA-VL nuclear complex. When the aberrant ipsilateral cerebellofugal projection was virtually absent, the responses were evoked only contralaterally with the reponsive area and the configuration similar to those in intact animals. When a considerable amount of the aberrant ipsilateral cerebellofugal projection was present (Fig. 2 E), weak responses of normal configuration were also evoked ipsilaterally in the areas homologous to the contralateral cortex.

3.4 Time-Course of Regeneration

As early as 15 min after transection, some proximal and distal stumps of cut axons began to swell, and the swelling became increasingly remarkable over a distance of 200 µm by 30 min. In 14 h all fiber endings showed polymorphic swelling near the lesion: some were presumably growth cones and some were presumably retraction bulbs. The typical endings which were presumed growth cones were large, ramified, and very densely labeled, whereas the typical endings which were

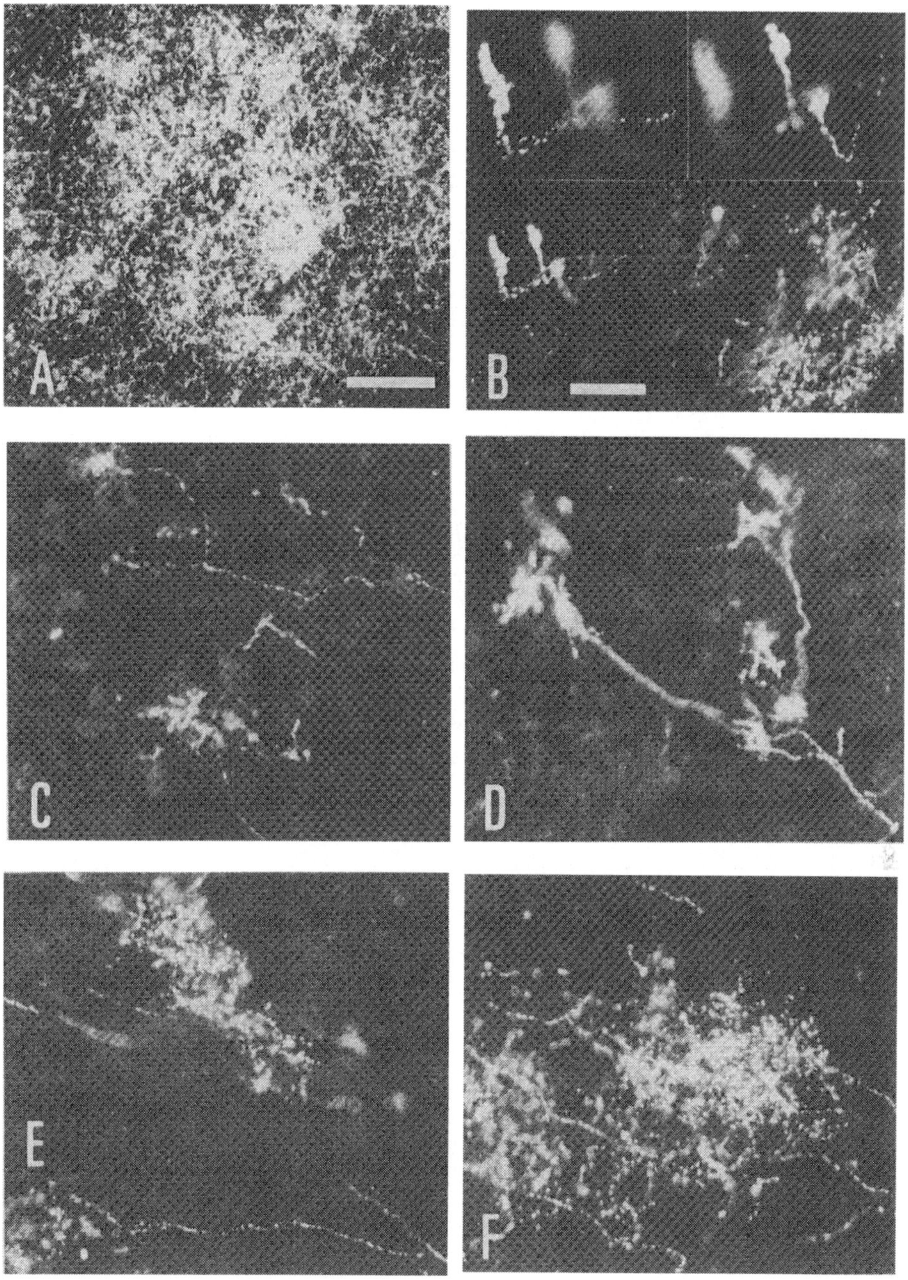

Fig. 5. Photomicrographs of polymorphic terminals of regenerated cerebellothalamic projection fibers in the thalamus. Fine terminals shown in **A** are similar to those in intact animals. Among them, however, much coarser terminals were coexistent as shown in **B**; the three coarse terminals on the *left* in which two were focused and illustrated on the top were branches of a single axon. Metamorphosis of terminals during development presumably proceeds from the most immature stage as shown in the coarse terminals in **B**, through the intermediate stages in **C–E**, to the final form as shown in **F**. Bar = 200 μm in **A**, and 50 μm in **B** applicable to **B–F**

presumed retraction bulbs were small, round, and faintly labeled. Besides these typical ones, there were numerous endings which could not be placed under the two categories; they were smooth or spinous club-shaped endings with or without dense labeling. In 24 h after transection the presumed growth cones and club-shaped endings were scattered not only near the lesion but also far behind the lesion, and the presumed retraction bulbs faintly labeled with HRP were found 500 to 1000 μm behind the lesion. Swelling of distal stumps changed into retraction bulbs which were labeled faintly. In 3 days a small number of axons tipped by a swollen ending crossed the midline where the BCX had been transected. After 7 days, such axons increased in number and ascended through the contralateral brainstem to reach the red nucleus. Swollen endings, presumably growth cones, were observed sporadically not only along the way to the red nucleus but also near the lesion and behind the lesion of more than 1 mm. In one and the same kitten (6 PO 19) shown in Fig. 3, dense terminals were distributed in the contralateral red nucleus and thalamic VA-VL nuclear complex (Fig. 5 A), and cerebellocerebral responses were evoked (Fig. 4 C). This indicates that a large number of regenerated axons can re-establish their terminal distribution within 3 weeks, but does not necessarily mean that the regenerative process ceases. Actually, it continues for a much longer period of time since swollen endings, presumably growth cones, were observed in various portions of the brainstem and in the thalamic VA-VL nuclear complex more than 4 months after the operation.

Most of the labeled terminals in the thalamus 19 days after transection were very fine (Fig. 5 A); they were exactly similar to those in intact kittens. However, various forms of much coarser terminals unseen in intact animals, very likely the developing terminals, were also present. Polymorphism of these terminals appears to indicate that terminals in various stages of maturation are coexistent. Because of their gradual changes in the grade of metamorphosis, it is easy to arrange them in the presumptive order of development, e.g., from the most immature stage as shown in the three club-shaped endings in Fig. 5 B to the most matured stage as shown in the fine terminals of a large sphere in Fig. 5 F.

4 Comments

It has been widely accepted that axonal regeneration in the mammalian central nervous system is deficient (Clemente 1964), since Ramón y Cajal reached the conclusion of its abortive regeneration. Nevertheless, Cajal himself knew the latent regrowth potential of mammalian central axons and wrote in his book (Cajal 1928) as "It is for the science of the future to change, if possible, this harsh decree. Inspired with high ideals, it must work to impede or moderate the gradual decay of the neurons, to overcome the almost invincible rigidity of their connections, and to re-establish normal nerve paths, when disease has severed centers that were intimately associated." Presently we have convincing evidence for the occurrence of marked regeneration of mammalian central axons (Foerster 1982; Kalil and Reh 1982; So and Aguayo 1985), which can be functionally active (Kawaguchi et al. 1981, 1986; Reh and Kalil 1982).

Regarding the neoformation of axon terminal, Cajal proposed the concept of "direct or distal sprouting" and "indirect or proximal sprouting": the former occurs at the end of axons, whereas the latter occurs at a certain proximal segment of axon when the distal portion is necrotic (Cajal 1928). This distinction is not absolute, because all axons lose some protoplasm as a result of direct injury. The formation of uncrossed ipsilateral cerebellofugal projections shown in Fig. 2 E are accounted for by axonal neoformation due to "indirect or proximal sprouting". Axons tipped by retraction bulbs, therefore, do not necessarily mean that they are under the down-hill degenerative process; but rather, they may be turned into regenerating axons. On the other hand, axons tipped by regenerating endings must be finally turned into degenerating ones unless they reach target neurons. Thus, it appears very likely that the regenerative and degenerative processes are concomitant in each injured axon and that whether the axon regenerates or degenerates depends on which process overwhelmed the other as in a tug of war. Leaving years of pessimism about the failure of mammalian central axons to regenerate, we must endeavor to find out factors which bring about successful regeneration. Repairing of neural tracts after brain injury may not be an insubstantial dream.

List of Abbreviations

BCX	Decussation of brachium conjunctivum
Caj	Interstitial nucleus of Cajal
CL	Central lateral nucleus of thalamus
CM	Centromedian nucleus of thalamus
CN	Cerebellar nuclei
Da	Nucleus of Darkschewitsch
FF	Forel's field
FTC	Central tegmental field of the brainstem
INT	Interpositus nucleus of cerebellum
IO	Inferior olivary nucleus
IP	Interpeduncular nucleus
LAT	Lateral nucleus of cerebellum
LP	Lateral posterior nucleus of thalamus
M.E.	Microelectrode
NCM	Central medial nucleus of thalamus
NPC	Nucleus of posterior commissure
PAG	Periaqueductal gray
Pc	Paracentral nucleus of thalamus
PN	Pontine nucleus
PRT	Pretectum
PTN	Pontine tegmental nucleus
Pul	Pulvinar nucleus of thalamus
RN	Red nucleus
SC	Superior colliculus
S.E.	Stimulating electrode
VA	Ventral anterior nucleus of thalamus

VL Ventral lateral nucleus of thalamus
VPL Ventral posterolateral nucleus of thalamus
ZI Zona incerta

References

Cajal S, Ramón y (1928) Degeneration and regeneration in the nervous system. May RM (transl 1959) Hafner, New York

Clemente CD (1964) Regeneration in the vertebrate central nervous system. Int Rev Biol 6:257–301

Foerster AP (1982) Spontaneous regeneration of cut axons in adult rat brain. J Comp Neurol 210:335–356

Kalil K, Reh T (1982) A light and electron microscopic study of regrowing pyramidal tract fibers. J Comp Neurol 211:265–275

Kawaguchi S, Yamamoto T, Samejima A (1979a) Electrophysiological evidence for axonal sprouting of cerebellothalamic neurons in kittens after neonatal hemicerebellectomy. Exp Brain Res 36:21–39

Kawaguchi S, Yamamoto T, Samejima A, Itoh K, Mizuno N (1979b) Morphological evidence for axonal sprouting of cerebellothalamic neurons in kittens after neonatal hemicerebellectomy. Exp Brain Res 35:511–518

Kawaguchi S, Miyata H, Kawamura M, Harada Y (1981) Morphological and electrophysiological evidence for axonal regeneration of axotomized cerebellothalamic neurons in kittens. Neurosci Lett 25:13–18

Kawaguchi S, Samejima A, Yamamoto T (1983) Post-natal development of the cerebello-cerebral projection in kittens. J Physiol (Lond) 343:215–232

Kawaguchi S, Miyata H, Kato N (1986) Regeneration of the cerebellofugal projection after transection of the superior cerebellar peduncle in kittens: morphological and electrophysiological studies. J Comp Neurol 245:258–273

Reh T, Kalil K (1982) Functional role of regrowing pyramidal tract fibers. J Comp Neurol 211:276–283

So K-F, Aguayo AJ (1985) Lengthy regrowth of cut axons from ganglion cells after peripheral nerve transplantation into the retina of adult rats. Brain Res 328:349–354

Compensatory Mechanisms at the Level of the Vestibular Nuclei Following Post-Natal Degeneration of Specific Cerebellar Cell Classes and Ablation of the Cerebellum in Mutant Mice

U. Grüsser-Cornehls

Department of Physiology, Freie Universität Berlin, Arnimallee 22, D-1000 Berlin (West) 33

1 Introduction

Neurological diseases with genetic cerebellar defects play a not inconsiderable role in human medicine (Gilman et al. 1981). Differentiation and treatment and the exploration of the dynamic pathophysiological processes, e.g., the degenerative processes and the resulting compensation mechanisms, are to date still very unsatisfactory. Mutant mice with cerebellar degenerative defects (Sidman et al. 1965; Herrup and Mullen 1979) may advance our understanding in such diseases by providing a tool for investigating the neurophysiological and neuroanatomical bases of behavioral deficiencies and compensation mechanisms resulting from a particular lesion, the loss of a single or a restricted number of nerve cell classes. For this reason, neuroanatomical (Grover and Grüsser-Cornehls 1976), electrophysiological (Grüsser-Cornehls 1983; Helmchen et al. 1985), synaptic transmitter (Kahle and Grüsser-Cornehls 1987) and immunohistochemical investigations were undertaken in two "extreme" specimens of these mutant mice: the Weaver mice and the Purkinje cell degeneration (pcd) mice.

In Weaver mice almost all granule cells die within the first postnatal weeks during their migration from the external germinal layer (Rakic and Sidman 1973a, b) mossy fibers forming heterologous synapses with Purkyně cell spines (Sotelo 1982) and in pcd mice almost all Purkyně cells (P-cells) degenerate during the first postnatal months (Mullen et al. 1976). Despite the fact that in the pcd the main class of cells in the cerebellum, the Purkyně cells, is missing, cerebellar symptoms, like ataxia, tremor, and instability of gait, are much less prominent in these pcd than in Weaver mice. This raises the question whether the fate of these mutant mice depends on their "specific" lesion defects, i.e., whether the function of the residual cerebellar brain structure determines which compensation mechanisms are involved, leading to the final functional performance of the system and its motor expression.

In this chapter some electrophysiological and transmitter results shall be presented and will then be correlated with immunocytochemical results.

2 Material and Methods

10 Weaver (strain B6CBA) and 30 controls and 7 pcd mice (strain C57BL6J) and 6 controls were investigated electrophysiologically. Briefly, the animals were pre-

H. Flohr (Ed.)
Post-Lesion Neural Plasticity

pared for microelectrode recordings (glass capillaries) from the flocculus and vestibular nuclei, fixed on a specially constructed animal holder, placed on a turntable, artificially respirated and restrained (0.04 ml Flaxedil). They were then subjected to horizontal sinusoidal rotation in the dark (amplitude $\pm 35°$, frequency 0.05–0.5 Hz) in order to stimulate the vestibular input. The axis of horizontal rotation was located in the middle of the ear-to-ear axis. The head was tilted 45° forwards into the plane of the horizontal canals. Signals were amplified by a high impedance amplifier and a 7A22 plug-in-unit of a Tectronix 7603 oscilloscope and stored together with the potentiometer recorded position signal of the turntable on a 7-channel-analog magnetic tape.

In 34 animals (Weaver, pcd, and controls) the P-cell transmitter GABA (Gamma-aminobutyric acid) was injected iontophoretically through multibarrelled microelectrodes onto the vestibular nuclear cells (20 nA to 60 nA; "hold back" current -10 nA to -30 nA) during vestibular stimulation and during recording of spontaneous activity.

In a further series of experiments, immunohistochemical procedures were carried out on Weaver (9), pcd (5) and control (9) animals. In three Weaver the cerebellum was removed uni- or bilaterally between postnatal day 35 and day 100 and the animals allowed to survive for 2 to 5 months. In four controls (3 to 4 months old) the cerebellum was removed unilaterally and the animals allowed to survive 3 to 4 days.

For immunohistochemistry the animals were anaesthetized with a lethal dose of chloral hydrate (1.75 g/kg body weight) and perfused with 1% paraformaldehyde and 1% glutaraldehyde in 0.1 M phosphate buffer (pH 7.4) at 4 °C. The brains were blocked and immediately cut on a vibratome at 30 μ. Serial sections were collected in 0.1 M phosphate buffer and processed freefloating for GABA immunocytochemistry using the unlabeled antibody peroxidase-antiperoxidase (PAP) method (Sternberger 1979). To begin with, sections were incubated in 10% methanol/3% H_2O_2 to prevent labeling of endogenous peroxidase. Sections were then incubated for 30 minutes at room temperture in 10% normal swine serum (NSS) and 1% Triton X-100 in PBS and then for 12 to 16 h at 4 °C in rabbit anti-GABA serum (Immuno-Nuclear Corp.) diluted 1:3000 in PBS containing 5% NSS and 0.3% Triton X-100. Control sections were incubated in diluted normal rabbit serum. To label the specific antigen, sections were then incubated in swine anti-rabbit IgG (Dakopatts) diluted 1:1000 in the same solution as the first antibody for 1 h at room temperature and secondly in rabbit PAP (Dakopatts) diluted 1:50 in the same solution for 1 h at room temperature. For visualizing the PAP complexes the Diaminobenzidine (DAB) reaction was used by incubating the sections in 0.05% DAB in 0.05 M Tris-buffered saline (TBS), pH 7.6 and 1% H_2O_2 for 25 to 30 min. Rinsing, mounting, drying, dehydrating, and coverslipping as usual.

Sections were examined with the light microscope (phase contrast or brightfield) and camera lucida drawings prepared.

3 Results

3.1 Electrophysiological Experiments

Stimulus Frequency-Response Relationship. Subjecting the animals to horizontal sinusoidal rotation in the dark and recording either from the flocculus or the vestibular nucleus, two main classes of neurons could be distinguished: *class 1* acti-

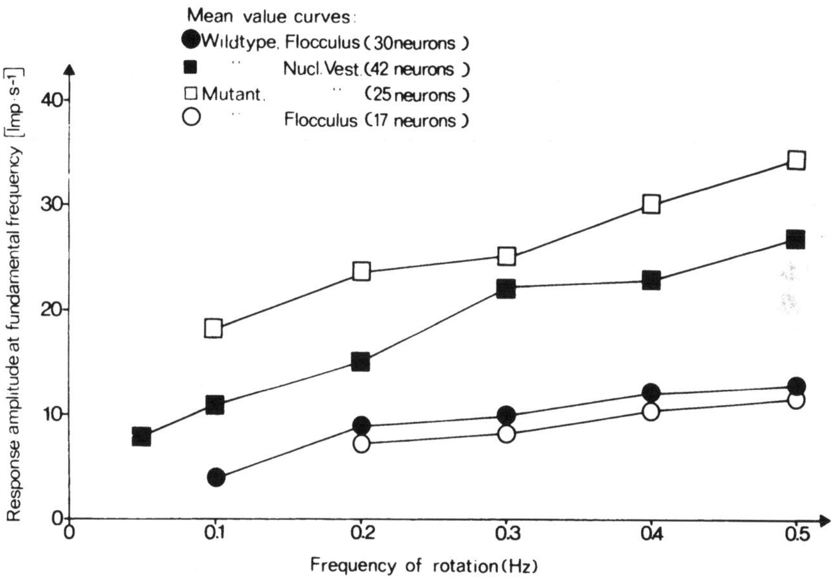

Fig. 1. Mean value curves of the response vs. frequency of rotation for flocculus (*circles*) and Nucleus vestibular neurons (*squares*) for wild type (B6CBA) (*closed symbols*) and Weaver mutant mice (*open symbols*)

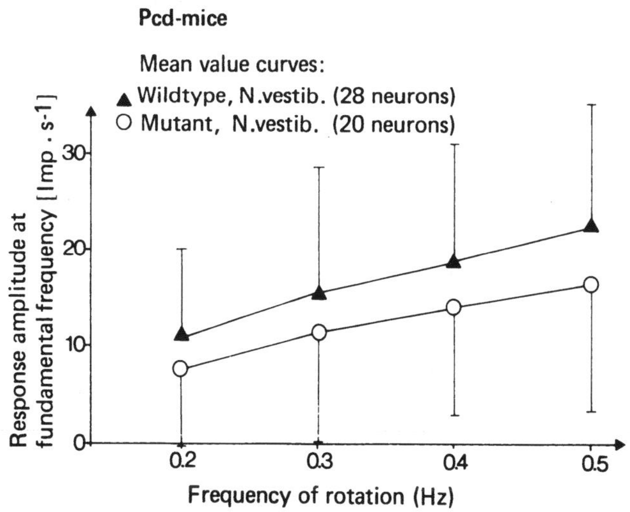

Fig. 2. Mean value curves with standard deviations of Nucleus vestibular neurons for wild type (C57BL6J) and pcd mutant mice for the relationship between neuronal response and frequency of rotation

vated by movement to the *ipsilateral side* and *class 2* activated by movement to the *contralateral side* (Duensing and Schaefer 1958). Very seldom *class 3* neurons, activated in *both directions*, could be encountered as well. When the impulse rate at fundamental frequency was plotted against rotation frequency or stimulus amplitude and the mean value curves calculated, response amplitude increased with both stimulus amplitude and frequency. The latter case is shown in recordings from the flocculus or the vestibular nuclei (nucleus vestibularis superior) for Weaver mutants and their controls and for pcd mutants and their controls (Figs. 1 and 2). The open symbols indicate the recordings from wild type and the closed symbols from mutant mice. In Fig. 1 both upper curves represent the results of the recordings of the vestibularis neurons and the lower two curves those from the flocculus neurons. The curve obtained from the responses of vestibularis neurons in Weaver mutants is clearly higher than the corresponding curve from the wild type. Cerebellar P-cells normally exert an *inhibitory influence* on their target cells in the vestibular nuclei. This influence seems to be diminished in mutants, resulting in a disinhibition of the target cell response. Although the standard de-

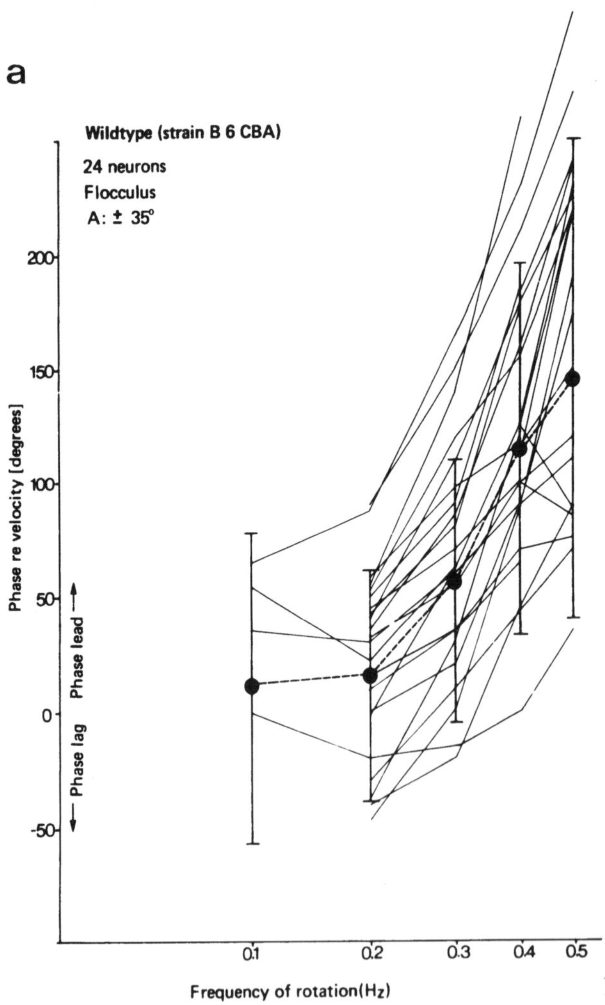

Fig. 3 a. Phase relationship for wild type (B6CBA) floccular neurons

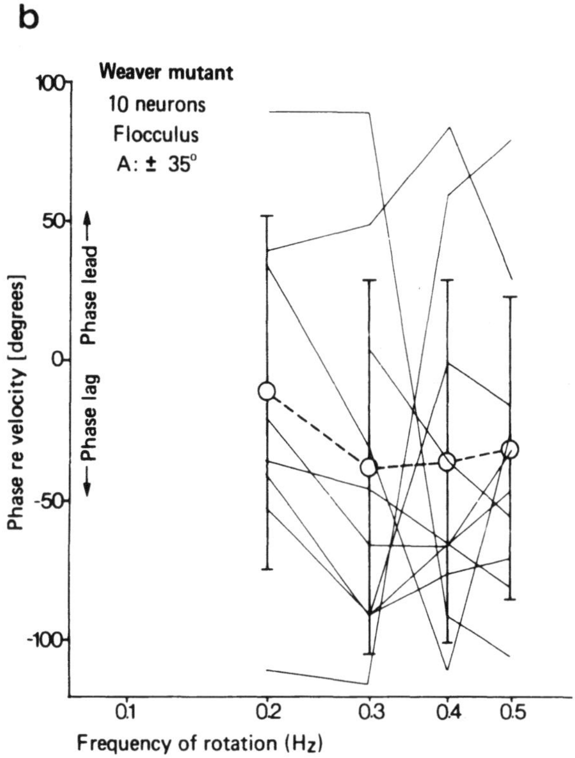

Fig. 3b. Same for Weaver mutant mice. Note the irregular course of the curves

viations (omitted in Fig. 1 for clarity) of the mean values are fairly large, the difference between the two upper curves is highly significant ($p<0.001$). The same is true for the curves shown in Fig. 2. Here, the curve obtained from the responses of vestibularis neurons in pcd mutants is shifted downwards in relation to the corresponding curve obtained from the normals. This indicates an increased inhibition. In consequence, when calculating gain re velocity curves, the gain for Weaver mutant mice was higher and the gain for pcd mutant mice lower than that for the wild type controls.

Phase Responses. Considering the phase relationships of vestibular neuron responses (best fitting sinus function) for wild type mice, maximum response in the range of 0.2 to 0.5 Hz appears to be nearly in phase with maximum angular velocity. In Weaver mutants the phase angle increased from 0 to +30° in this range and in pcd mutants it was +30° at 0.2 Hz and −30° at 0.5 Hz. Quite different were the results for floccular units, which, of course, could only be obtained in Weaver mutants and in the controls, since pcd mutants lack P-cells. Figure 3a demonstrates the results for flocculus neurons in wild type controls. The phase relationship (response maximum re stimulus angular velocity) displays a strong increase in phase lead with increasing stimulus frequency (0.2 to 0.5 Hz). Despite considerable interneuronal variability the single curves are very regular and show identical tendencies. Figure 3b demonstrates the corresponding phase relationship for the Weaver mutant mice. One can note a marked difference from the phase relationship of controls shown in Fig. 3a: the phase curves for the mutant

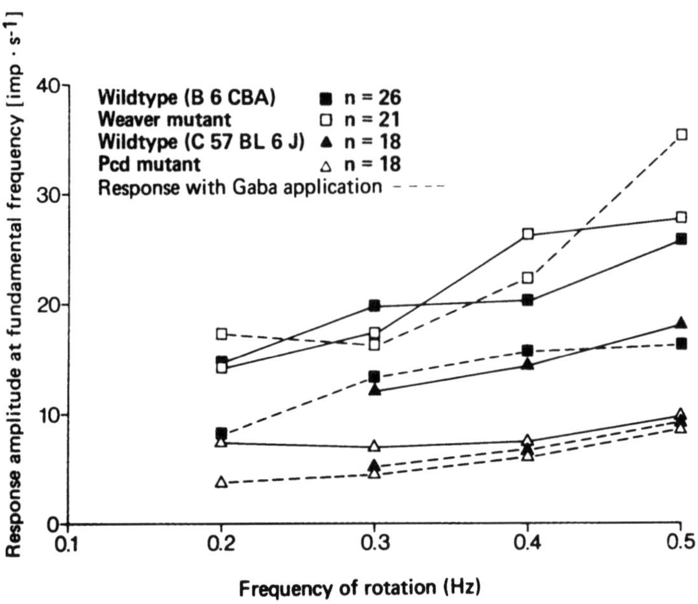

Fig. 4. Mean value curves for Nucleus vestibularis neuron activation before and during GABA application. For explanation see symbols on figure. The *dashed lines* always indicate the response during GABA application

mice are very irregular, show no uniform tendency, and display a greater standard deviation than the responses from the wild type mice, indicating a disruption in timing in cerebello-vestibular action (Grüsser-Cornehls 1983; Grüsser-Cornehls et al. 1986).

3.2 Transmitter-Studies

When the P-cell transmitter GABA was applied iontophoretically to the vestibularis cells, a different response appeared between neurons of mutants and controls. In controls, spontaneous activity and the neuronal response to vestibular stimulation was greatly suppressed. In Weaver mutant mice, on the other hand, GABA hardly affected spontaneous activity or vestibular stimulation responses, if at all. In pcd mice spontaneous activity was not reduced and response amplitude to vestibular stimulation was not as severely affected as in normals. Figure 4 demonstrates this relationship, in displaying the mean value curves for the different animals. The dashed lines always indicate the results with GABA application (Kahle and Grüsser-Cornehls 1987).

3.3 Immunohistochemical Investigations

The immunohistochemical investigations, which were started in collaboration with K.P. Hoffmann and A. Horn in Ulm, again revealed differences between mutants and controls and mutants among themselves.

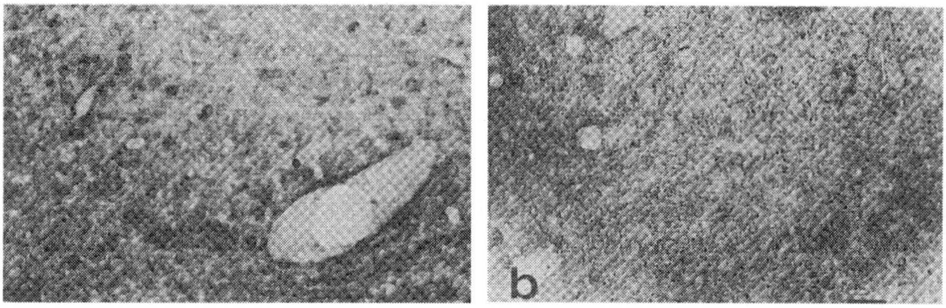

Fig. 5. Photomicrographs of GABA immunostained sections. **a** Normal mouse. Medial vestibular nucleus at the level of Deiters' nucleus. Immunopositive (*dark*) immunonegative (*light*) cells are seen as well as immunopositive terminals. **b** Weaver mutant mouse. Medial vestibular nucleus at level comparable to **a**. Only a few cells appear weakly immunopositive. Immunopositive terminals are easily seen as black puncta surrounding negative cells [25 × PLAN, N.A. = 0.46; photographed with a green filter; bar (see **b**) = 40 μ]

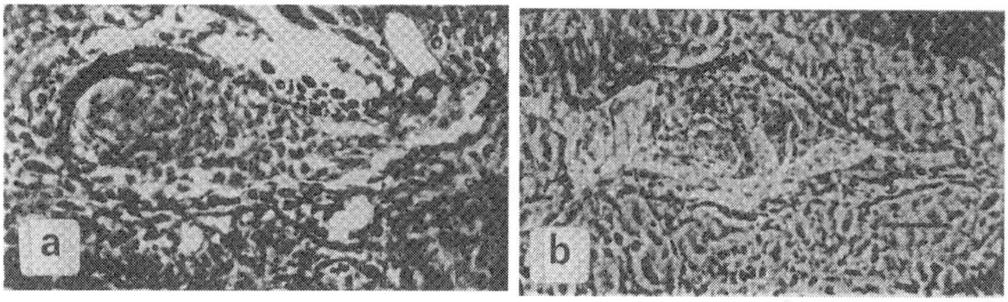

Fig. 6 a. A photographic reconstruction (photo collage) of the soma membrane in serial "optical" sections through a Deiters' nucleus cell of a Weaver mutant mouse. Phase contrast, 100 × objective (N.A. = 1.32, calculated resolution = 0.21 μ; depth of field 0.8 μ. Bar = 10 μ). **b** As in **a** for a normal control. Note that there are more small-sized terminals on the normal cell

In the material of controls immunoreacted for GABA, positive somata may be identified in several brain regions including the cerebellum and vestibular nuclei (Fig. 5a). Stained punctate structures determined by electron microscopic studies to be axon terminals or collections of such boutons (Katsumaru et al. 1986) could be identified by light microscopy in apposition to somata and dendrites. In normal mice these structures in the superior and lateral vestibular nuclei ranged in size from 0.3 to 3 μ and may be roughly categorized as small, medium and large. In Weaver mutants, these structures tended to be larger. Figure 6 demonstrates photo collages of the membrane associated boutons of two giant Deiters' cells, one of a mutant (left) and one of a normal (right). It can be clearly seen that the percentage of larger structures is higher in the mutant mouse than in the control. The same phenomenon is shown in Fig. 7. Inspection of these large structures with higher resolving power reveals that the largest ones consist of collections of two to four boutons. On the other hand, cytoplasmic staining of GABA-positive cells, in the vestibular nuclei only, is consistently weaker in the

Fig. 8 a. Sketch of a Weaver mutant
tion of the same mouse brain from
Upper row for comparison Deiters' r
mouse (*right*); *lower row left* Deiters
operated side of a Weaver mutant m
apochromatic immersion objective (
4 × 10 μ

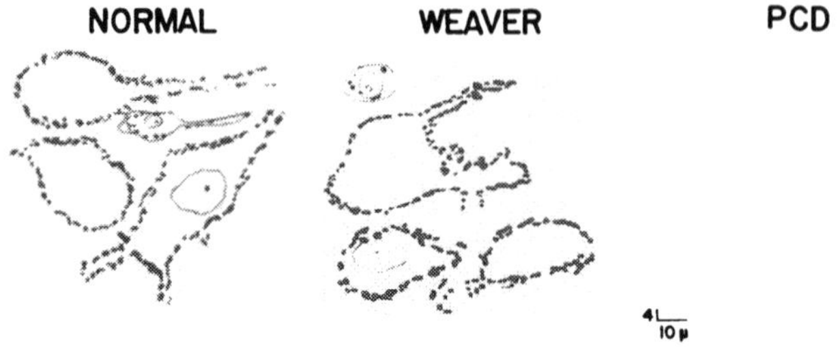

Fig. 7. Camera lucida drawings of cells and GABA-like terminals in the Deiters' nucleus of normal (*left*), Weaver (*middle*) and pcd mutant mice (*right*). Drawings made using phase contrast optics with a 100 × objective and a 10 × wide-field ocular. Bar = 4 × 20 μ

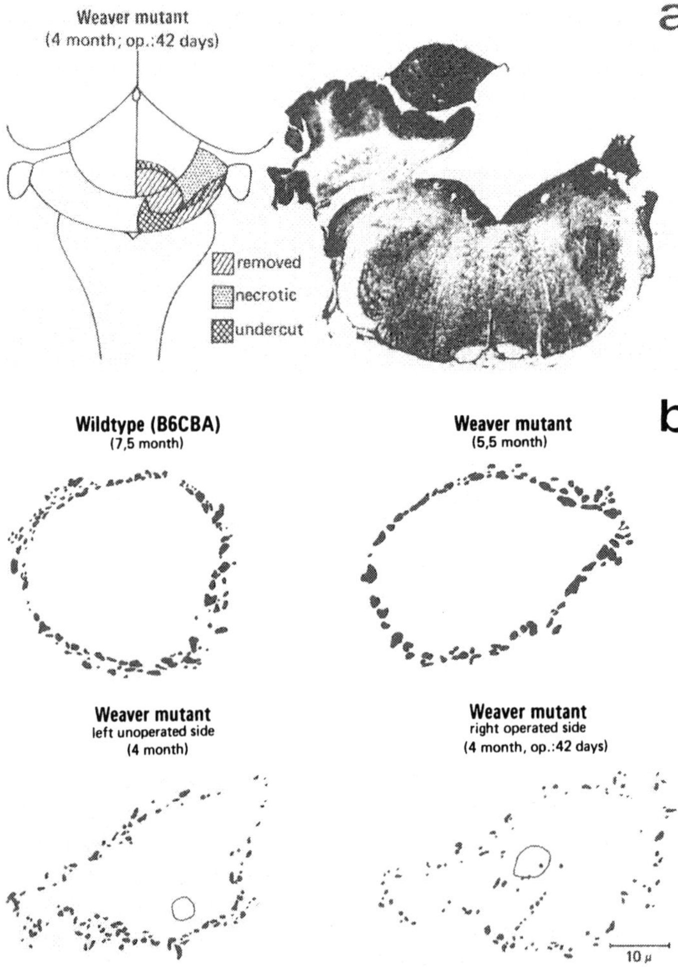

Fig. 8 a. Sketch of a Weaver mutant mouse brain with operated parts as indicated (*left*) and section of the same mouse brain from which the camera lucida drawings in **b** are taken (*right*). **b** *Upper row* for comparison Deiters' nucleus cell of control (*left*) and unoperated Weaver mutant mouse (*right*); *lower row left* Deiters' nucleus cell from the unoperated side and *right* from the operated side of a Weaver mutant mouse. Drawings were made using a 63 × phase contrast plan apochromatic immersion objective (N.A. 1.4) at a total optical magnification of 2125 ×. Bar = 4 × 10 μ

Weaver mutants than in controls (Fig. 5 b) (Grüsser-Cornehls and Grover 1987; Grüsser-Cornehls, Grover, Horn and Hoffmann in preparation).

In contrast, GABA-positive terminal structures in the superior and lateral vestibular nuclei of pcd mice were only of the small and medium size (Fig. 7). Cytoplasmic staining in these nuclei was not different from controls (Grüsser-Cornehls, Grover and Bäuerle in preparation).

When removing both halves of the cerebellum or part of it 4 to 5 weeks postnatally, consistent plastic changes occurred. First of all, motor performance in Weaver mutant mice was considerably improved when both halves of the cerebellum were removed. Even mutants deprived of one half of the cerebellum performed better than unoperated animals. Figure 8 a presents a coronal section and a schematic diagram of the hemi-cerebellectomized brain of a Weaver mutant operated at 42 days of age which was allowed to survive for 2.5 months. The camera lucida drawings of Fig. 8 b are taken from this section. In the upper row of Fig. 8 b, for comparison, cells from comparable regions of the ventral Deiters' nucleus from a wild type and an unoperated mutant are shown. In the bottom row are the camera lucida drawings of a Deiters' cell of the unoperated side (left) and one of the operated side (right) of the Weaver mutant. In the operated animal the *large* GABA-positive structures on the Deiters' cells were absent and the total number of GABA-positive terminal structures was reduced. The same results were obtained from a bilaterally cerebellectomized mutant, in which the flocculi and paraflocculi, however, remained intact. No data are available to date for comparable normal animals, but investigating Deiters' cells of 3-month-old wild types at the third and fourth day after operation revealed no difference in size and density of the GABA terminals between the operated and unoperated side or the unoperated normal controls (Grüsser-Cornehls, Grover and Bäuerle in preparation).

4 Discussion

The neurophysiological, transmitter, and immunohistochemical results support the hypothesis that in Weaver mutants the disturbance of cerebellar function (as seen, for example, in the phase irregularities in the responses of neurons in the Weaver flocculus) is accompanied by plastic changes outside the cerebellum. The desensitization of vestibular neurons to GABA inhibition may represent an attempt to reduce the disturbing influence of the physiologically disorganized cerebellar input in this mutant (Grüsser-Cornehls et al. 1986). This results in a disinhibition of vestibular neurons in Weaver, and in turn, worse motor performance. In contrast, in pcd mice where P-cell input to vestibular nuclei neurons is absent, other more effective kinds of compensation mechanisms can occur. For example, the sensitivity of the vestibular cells to inhibition seems not to be affected or even to be enhanced and motor performance is better. From this hypothesis we concluded that removal of the total cerebellum at a very early stage of life in the Weaver mutant mice would lead to a better motor performance of the animals. This was indeed observed.

How can the immunocytochemical results be related to these facts? Let us first consider the weaker soma labeling in the Weaver mutant mice throughout the vestibular nuclei. Does this reflect a depletion of intracellular GABA stores due to higher release rate or a turning off of GABA production in vestibular nuclei GABAergic neurons? (Grüsser-Cornehls and Grover 1987). We cannot answer this question at this point. It is also difficult to interpret the change in size distribution in GABA-positive terminal structures. Are the boutons comprising the large structure sprouts of a single axon branch or do they come from different axons?

Changes similar to those observed by us have been seen by Katsumaru et al. (1986) at the light microscope level in the Nucleus ruber of cats 20 days after lesion of the rubro-projecting Nucleus interpositus. Their EM sections show that the newly appearing larger terminal structures (which appeared to be strings of GAD-immuno reactive puncta with Nomarski optics) are composed of *several* discrete boutons. The phenomenon of terminal sprouting in our material may differ from the classic axonal sprouting first described by Lin and Chambers (1958) in that the number of end terminals of an axon branch may increase without an increase in axonal length.

What determines whether such collections are formed and what is the origin of the large terminal structures? Are they terminals of P-cells, commissural, inter- or intranuclear neurons? The absence in pcd of those large structures found in the normal argue that the large structures are of P-cell origin and that the increase in Weaver represents P-cell terminal sprouting. However, the decrease in large structures on the unoperated side in Weaver suggests that terminal sprouting may or may also be under secondary control of vestibular commissural signals and that the terminals which sprout do not necessarily originate from the P-cells. In any case, the increase in large GABAergic terminal structures seen immunohistologically in Weaver does not appear to be adaptive. Both cerebellectomized Weavers and pcd mutants show fewer large structures than unoperated Weavers, and their motor performance is also better. Further, the increase in GABA terminals in the Weaver is in apparent conflict with the electrophysiologically determined insensitivity to iontophoretic GABA application. Studies of GABA receptor density in normals and mutants are underway to determine if plastic changes occur here also.

In conclusion, the degeneration of a specific cerebellar nerve cell class in Weaver and pcd mice leads to a cascade of functional and anatomical changes inside and outside of the cerebellum. Further, the changes in GABA sensitivity of vestibular neurons and the number of GABA terminals on vestibuar nucleus neurons in the Weaver mutant mice are not primarily induced by a loss of cerebellar efferents (the integrity of the Purkinje cell axons is preserved) but by changes in the signal flow in these efferents (Grüsser-Cornehls and Grover 1987). Finally, although these physiologically induced extracerebellar changes may not be adaptive in Weaver, such mechanisms may underly the normal adaptive changes present in motor learning.

5 Summary

Electrophysiological, transmitter and immunocytochemical investigations were undertaken in Weaver (postnatal degeneration of granular cells) and in Purkinje-cell-degeneration (pcd) mutant mice (degeneration of Purkyně cells) and their respective normal controls. Extracellular responses were recorded with micropipets in the flocculus and/or the Nucleus vestibularis during horizontal sinusoidal rotation in the dark. When plotting the maximal impulse rate at fundamental frequency against rotation frequency, flocculus neuron responses in Weaver mutants were not greatly altered, but Nucleus vestibularis neuron responses were stronger (disinhibited) when compared to wild type responses. The contrary was found for pcd mice: Nucleus vestibularis neurons were less strongly activated when compared to wild type responses, indicating an increased inhibition. Considering the values for the phase shift (phase re velocity), a consistent difference between Weaver mutant and wild type appeared: The phase shift for flocculus neuron responses in the mutant was very irregular and displayed a large deviation.

When injecting the Purkinje cell transmitter GABA iontophoretically onto the Nucleus vestibularis cells of normals and mutants, the neuronal response to vestibular stimulation in normals was greatly reduced, while in pcd neuronal responses were only slightly affected and in Weaver not at all.

Recent immunocytochemical investigations revealed that in material immuno-reacted for GABA, positive terminal structures surrounding the cell soma profiles of Nucleus vestibular neurons in normals can be divided into small, medium and large ones. In Weaver mutants, stained terminal structures in Nucleus superior and Deiters' tend to be of the larger size, while in pcd only small and medium-sized are present. The large terminal structures could be identified as collections of sprouts comprising 2 to 4 boutons. Cytoplasmic staining of GABA positive cells in the vestibular nuclei only is consistently weaker in Weaver mutants than in normals and pcd mutants. In uni- or bilateral cerebellectomized Weaver mutants (survival time 2.5 months), large GABA-positive terminal structures are lacking and the total number of stained terminals is reduced. In hemicerebellectomized normal mice (survival time 3 to 4 days) no consistent changes could be found in GABA-positive terminal size or number.

A comparison with cerebellar symptoms and motor performance was established.

Acknowledgements. This work was supported by grants of the Deutsche Forschungsgemeinschaft (Gr 276/19,1–5) and by the Maria Sonnenfeld-Gedächtnis-Stiftung. I wish to thank Dr. B.G. Grover and Prof. O.-J. Grüsser for critical reading of the manuscript, Mrs. A. Thiele and Mrs. H. Wolynski for excellent technical assistance and Mrs. J. Dames for help with the English translation. Mr. L. Weiss wrote the computer programs.

References

Duensing F, Schaefer JP (1958) Die Aktivität einzelner Neurone im Bereich der Vestibulariskerne bei Horizontalbeschleunigung unter besonderer Berücksichtigung des vestibulären Nystagmus. Arch Psychiatr 198:225–252

Gilman S, Bloedel JR, Lechtenberg R (1981) Disorders of the cerebellum. FA Davis Company, Philadelphia

Grover BG, Grüsser-Cornehls U (1986) Cerebellar afferents in normal and Weaver mutant mice. Brain Behav Evol 29:162–175

Grüsser-Cornehls U (1983) Responses of vestibular neurons from the nucleus vestibularis and the flocculus in wild type mice and mutants. Neurosci Abstr 9:524
Grüsser-Cornehls U, Grover BG (1987) Is the change in GABA sensitivity of vestibular neurons in the mutant Weaver an example of physiologically induced plasticity? In: Elsner N, Creutzfeldt O (eds) New frontiers in brain research. Thieme, Stuttgart, p 36
Grüsser-Cornehls U, Plassmann W, Helmchen C, Kahle G (1986) Single unit activity in the vestibular nuclei of normal mice and mutant mice with genetic cerebellar defects. In: Abstr Developments in oculomotor research, IUPS satellite meeting. Gleneden Beach, Oregon, July 20–24, p 11
Helmchen C, Thiele A, Grüsser-Cornehls U (1985) Vestibular responses of nucleus vestibularis neurons in wild type mice and pcd-mutant mice. Pflügers Arch 405: suppl 2, R56
Herrup K, Mullen RJ (1979) Regional variation and absence of large neurons in the cerebellum of the staggerer mouse. Brain Res 172:1–12
Kahle G, Grüsser-Cornehls U (1987) Transmitter investigations in the vestibular nuclei of normal mice and mutant mice with genetic cerebellar defects. In: Elsner N, Creutzfeldt O (eds) New functions in brain research. Thieme, Stuttgart, p 215
Katsumaru H, Murakami F, Wu JY, Tsukahara N (1986) Sprouting of GABAergic synapses in the red nucleus after lesions of the nucleus interpositus in the cat. J Neurosci 6:2864–2874
Lin CN, Chambers WW (1958) Intraspinal sprouting of dorsal root axons. Arch Neurol Psychiatry 79:46–61
Mullen RJ, Eicher EM, Sidman RL (1976) Purkinje cell degeneration, a new neurological mutation in the mouse. Proc Natl Acad Sci USA 73:208–212
Rakic P, Sidman RL (1973a) Sequence of developmental abnormalities leading to granule cell deficit in cerebellar cortex of Weaver mutant mice. J Comp Neurol 152:103–132
Rakic P, Sidman RL (1973b) Organization of cerebellar cortex secondary to deficit of granule cells in Weaver mutant mice. J Comp Neurol 152:133–162
Sidman RL, Green MC, Appel SH (1965) Catalogue of the neurological mutants of the mouse. Harvard University Press, Cambridge
Sotelo C (1982) Synaptic remodeling in agranular cerebella. In: Palay SL, Chan-Palay V (eds) The cerebellum – New Vistas. Springer, Berlin Heidelberg New York, pp 50–68
Sternberger LA (1979) Immunocytochemistry, 2nd edn. Wiley, New York, p 354

Functional Recovery of the Primate Oculomotor Pursuit System Following Cerebellar and Cerebral Cortical Lesions

R. Eckmiller

Department of Biophysics, University of Düsseldorf, Universitätsstr. 1, D-4000 Düsseldorf 1, FR Germany

1 Introduction

The vertebrate central nervous system (CNS) as exemplified here by the primate oculomotor pursuit system is not based on rigid hardware and functions, but exhibits a surprising extent of adjustability and neural plasticity. Throughout life there are numerous occasions requiring functional adjustments to new sensory input combinations (e.g., visual-vestibular rivalry), to eye muscle dysfunctions, or to various functional deficits of circumscribed brain regions. In this respect the pursuit system can be regarded as a particularly sensitive monitor of sensory-motor brain functions and demonstrates the CNS as a highly fault-tolerant information-processing system, which is based on "sloppy workmanship but superb architecture".

The following paragraphs describe first the entire primate pursuit eye movement (PEM) system and then give examples for functional recovery from pursuit deficits following lesions in the cerebellar and posterior parietal cortex in macaque monkeys. Special attention is payed to the neural network architecture rather than subcellular neural mechanisms, which allows for fault-tolerance and recovery from local defects.

2 Circuit Diagram of the Primate Pursuit System

Pursuit eye movements (PEM) in primates serve continuous maintenance of vision of a small moving target in the photopic range as occurs during fixation. One distinguishes two different oculomotor functions for initiation of PEM and maintenance of PEM. The neural evaluation of the target projection movement relative to the foveal center (position error and slip velocity) during PEM by means of the afferent visual system is not completely understood. Further details on the neural control of PEM and a complete circuit diagram based on more than 400 primate studies have been compiled in a recent review (Eckmiller 1987).

Figure 1 shows a reduced circuit diagram of the primate pursuit system. The visual input reaches the various brain regions via parallel pathways (interrupted lines from the visual input box). Afferent visual signals reach the medial superior temporal area (MST) from the middle temporal visual area (MT, within area 19). Visual inputs also arrive at area 7a in the inferior parietal cortex from pulvinar

H. Flohr (Ed.)
Post-Lesion Neural Plasticity

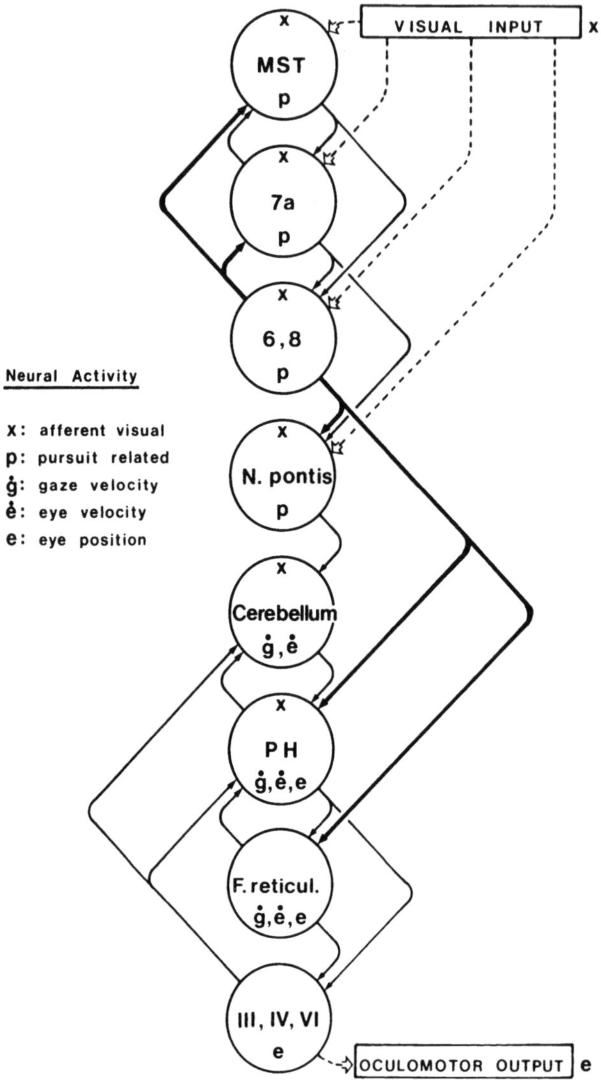

Fig. 1. Circuit diagram of primate pursuit system

(PUL), internal medullary lamina (IML), superior colliculus (SC), and areas V1 (area 17) and V2 (within area 18). In parallel, visual afferents reach the precentral cortex (area 8 or frontal eye field and area 6a beta or supplementary eye field) from PUL, IML, nucleus of the optic tract (NOT), V2, and MT. Furthermore, visual signals reach various pontine nuclei (N. pontis) directly from NOT, SC, and MT.

The layout of this circuit diagram gives connections between any two brain regions always in clockwise direction to distinguish down-stream connections on the right from up-stream connections on the left like for example from the oculomotor nuclear complex (III) to the nucleus prepositus hypoglossi (PH).

The different types of neural activity that were recorded in the corresponding brain regions in alert macaques are indicated by symbols: x, p, ġ, ė, and e as explained in Fig. 1.

Pursuit-related activity p occurs in MST, 7a, the precentral cortex (6, 8), as well as in some pontine nuclei. Pursuit-related neurons are specifically activated during PEM in a given direction, even during temporary target disappearance.

I propose that p activity represents a first draft of the required pursuit motor program. Since only the precentral cortex has direct projections to all other brain regions (including NOT) where p activity has been found, it is the most likely source of its origin. This postulate, that the motor program for PEM arises from a first draft signal in the oculomotor related regions of the motor cortex (6, 8), needs further experimental verification.

Figure 1 indicates that p activity (traveling along the thick lines which originate in the precentral cortex) reach the pontine nuclei as input stage for the cerebellar cortex, and in parallel both PH and reticular formation (F. reticul.). Interestingly, the refined pursuit motor program occurs as gaze velocity ($\dot{g}$) and pursuit eye velocity ($\dot{e}$) signal in the cerebellar cortex as well as in PH and the reticular formation. According to my hypothesis (for reasoning see: Eckmiller 1987) the refinement happens in PH and F. reticul. and is subsequently sent to the cerebellar cortex via one of the up-stream connections, thus assigning a monitoring and modifying role to the cerebellar cortex with regard to pursuit motor control.

On the basis of this circuit diagram, various lesion experiments involving parts of cerebellar or cerebral cortex can be interpreted.

3 Recovery from Cerebral Cortical Lesions

Bilateral surgical lesions in area V1 (striate cortex) were found to cause irreversible impairment of PEM (Denny-Brown and Chambers, 1976; Miller et al. 1980), whereas unilateral V1 lesions only impaired PEM of targets in the affected hemifield (Goldberg et al. 1982). In contrast, extensive lesions in areas 18 and 19, however, sparing area V1 (Denny-Brown and Chambers 1976) left PEM intact. One is forced to conclude that area V1 is an essential (in the sense of irreplacable) part of the primate pursuit system as is the retina and the dorsal lateral geniculate nucleus (LGN).

Chemical lesions using Ibotenic acid in area MT (within area 19) caused a reduction of smooth eye velocity for initiation of PEM toward the side of the lesion (Newsome et al. 1986). However, maintenance of PEM appeared normal as soon as the target had reached the fovea. These pursuit deficits disappeared completely within several days following the lesion in adult macaques, although the neurotoxin had irreversibly destroyed the cortical cells in a circumscribed region about the injection site. This finding of an initial severe pursuit deficit with quite rapid subsequent recovery could mean that (a) MT is no essential link between V1 and the motor part of the pursuit system and (b) the initial deficit was caused by a temporary imbalance of signals within the neural network, until related brain regions managed to compensate for this imbalance.

Effects of frontal eye field (area 8) lesions on the human pursuit system have not been described (Milner 1982). However, persisting pursuit deficits were found in macaques following surgical lesions of area 8 (Lynch 1986). Interestingly, combined bilateral surgical lesions of both frontal eye fields and inferior parietal lobes in macaques (Lynch et al. 1986) led to severe and permanent pursuit deficits, although some improvement could be observed. These findings were taken by the authors as indication that these two cortical regions have an overlapping oculomotor function and that deficits in one can be compensated by the intact other region. Figure 1 indicates the reciprocal connections between both cortical regions, which should, however, not be interpreted as simple feedback loops.

A further example for two brain regions supporting the same function in parallel was provided by another study (Schiller et al. 1980): combined bilateral surgical lesions in both frontal eye fields and superior colliculi led to permanent oculomotor deficits, which did not occur following lesions in one region alone. In this case, PEM were not studied specifically.

Malfunction of the parietal cortex in human patients (Baloh et al. 1980) or surgical lesions in macaque parietal cortex (Lynch and McLaren 1982; Lynch et al. 1986) have been associated with pursuit deficits as well as optokinetic nystagmus deficits toward the side of the lesion. These deficits, however, could not be linked to a specific subregion within the inferior parietal cortex.

In collaboration with R. Andersen and O. Bock, we studied specific pursuit deficits in a monkey *(Macaca fascicularis)* following small chemical microlesions

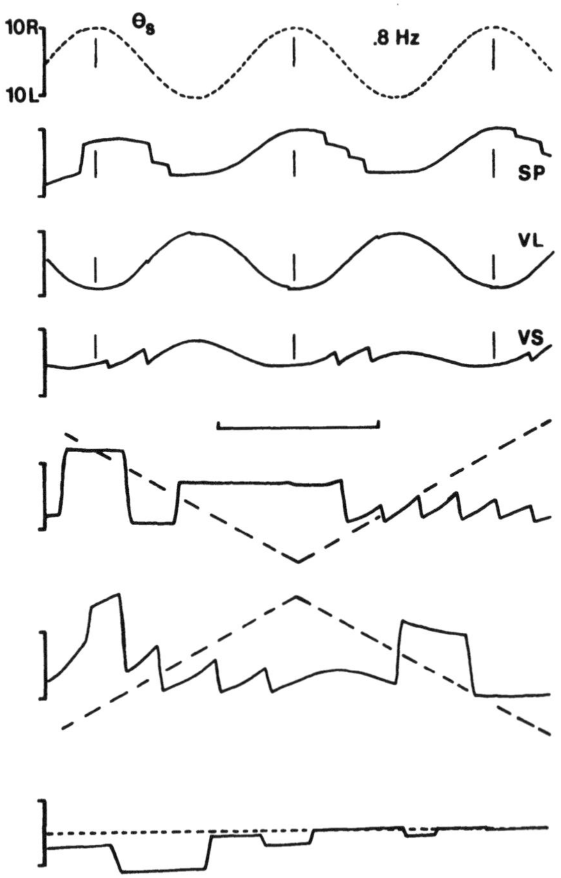

Fig. 2. Evidence for defect of pursuit and OKN to the left following ibotenic acid lesion in left area 7a

in the inferior parietal cortex. The monkey had been trained to pursue a sinusoidally moving small target (8 min of arc) in the horizontal plane in the lower photopic range against a homogenous background. Eye movements were recorded with an infrared oculometer (Bouis). Three intracortical microinjections of 1 µl 15% ibotenic acid solution were administered about 1 mm apart in area 7a including the border region between 7a and 7b on the left side during Ketanest anesthesia.

Figure 2 gives several horizontal eye movement recording episodes of the right eye during binocular vision 3 days following the lesion. The horizontal bar, which separates PEM recordings above from OKN recordings below, indicates 1 s.

The eye movement traces SP for smooth pursuit (target moves, head stationary), VL for VOR-Light (target stationary, monkey with chair rotates), and VS for VOR-Suppression (target moves together with rotating chair) correspond to a common top trace for sinusoidal stimulus movement at 0.8 Hz.

The SP trace indicates that pursuit is severely impaired for PEM toward the side of the lesion (left). In contrast, VOR-Light appears normal (VL trace). During VOR-Suppression (VS trace) those phases, which require an intact pursuit system for PEM to the left in order to suppress the vestibular signal during leftward rotations, are severely impaired.

The bottom trace indicates a slight movement drift to the right during fixation of the stationary target straight ahead.

The upper OKN recording shows that OKN to the left (stripe movement indicated by interrupted line beginning with a negative slope) is evidently not possible. The monkey performs saccades and fixations instead. Upon reversal of the OKN direction to the right (in the middle of the recording) OKN occurs with slow phase velocity in the normal range. The lower OKN recording gives another example, however, with OKN reversal in the opposite direction. Again OKN to the left is not possible.

Two points have to be made:

1. A similar lesion in a second trained monkey, however, located more in area 7b including the border region between 7a and 7b did not yield any measurable pursuit deficits.
2. The effected monkey (Fig. 2) did not recover from this severe unilateral pursuit defect, but rather developed epileptic seizures with increasing intensity.

The presented findings can be interpreted as preliminary evidence that chemical microlesions in small (about 10%) portions of area 7a can lead to permanent pursuit defects towards the side of the lesion. This interpretation could be reconciled with another study applying larger surgical lesions (Lynch and McLaren 1982).

4 Recovery from Cerebellar Cortical Lesions

Cerebellar cortical lesions involving the flocculus and paraflocculus (Westheimer and Blair 1974; Zee et al. 1981) or even vermis lobule V and VII (Ritchie 1976) in macaques have been shown to disturb the pursuit system. These deficits, however, are only transient in young monkeys (Westheimer and Blair 1974), and recovery is complete in adult macaques who received cerebellar lesions as neonates as long as the vestibular and deep cerebellar nuclei were spared (Eckmiller and Westheimer 1983; Eckmiller et al. 1984).

The clinical literature contains a number of case reports of cerebellar agenesis or substantial hypoplasia that remained undetected throughout life due to apparently complete compensation (Hitzig 1884; Macchi and Bentivoglio 1977).

Figure 3 gives a circuit diagram of the cerebellum with its major inputs and outputs in order to evaluate the functional damage caused by cerebellar lesions as well as the corresponding secondary degeneration sites. Please, note that the bilateral (mostly contralateral) input from the pontine nuclei (N. pontis), the ipsilateral input from the lateral reticular nucleus (N. ret. lat.), and the contralateral input from the inferior olive (O. inf.) reach not only the cerebellar cortex (rectangular block) but in parallel also the corresponding cerebellar nuclei (D for N. dentatus, I for N. interpositus, F for N. fastigii, V for vestibular nuclear complex). This parallel input to cerebellar cortex and corresponding nucleus allows the following predictions:

1. A circumscribed cortical lesion will induce secondary degeneration in extracerebellar input and output structures according to Fig. 3.
2. Only lesions in the cerebellar nuclei interrupt the information flow through the cerebellum, whereas pure cortical lesions "only" destroy a "bypass loop" via cortex to the nuclei.

The extent of extracerebellar secondary degeneration following neonatal cerebellar cortical lesions in a macaque is shown in Fig. 4. This monkey (monkey F in: Eckmiller and Westheimer 1983; Eckmiller et al. 1984) had been examined 4 years later. The surgical lesion in the second week after birth was restricted to lobulus ansiformis, lobulus paramedianus, flocculus, and paraflocculus of the cerebellar cortex on the right side. The final histological examination (see: Eckmiller et al. 1984) revealed a clear reduction in volume of N. interpositus and N. dentatus on the right side.

Extensive examination of the pursuit system following a normal training period in the adult stage, revealed no residual pursuit deficits (see: Eckmiller and Westheimer 1983), although the pursuit system had been considerably impaired for several weeks after the neonatal lesion.

Figure 4 gives semi-schematic drawings made from representative Nissl sections at the levels of lateral reticular nuclei (RL) in the upper left, inferior olives (upper right), cerebellar nuclei (lower left), and pons (lower right; SN for substantia nigra, RTP for N. reticul. tegmenti pontis, GPO for pontine gray). At each level the top section is at the caudal end. Dotted areas represent regions with normal or relatively normal cellularity. Hatched areas represent sites with clear indications of cell degeneration and loss. Histology based on serial sections confirmed that the principle olive as well as parts of the pontine nuclei (especially N. peduncularis, N. ventralis, and N. lateralis) on the left side as cerebellar input structures, and pars parvocellularis of the red nucleus on the left side as cerebellar output were almost completely degenerated.

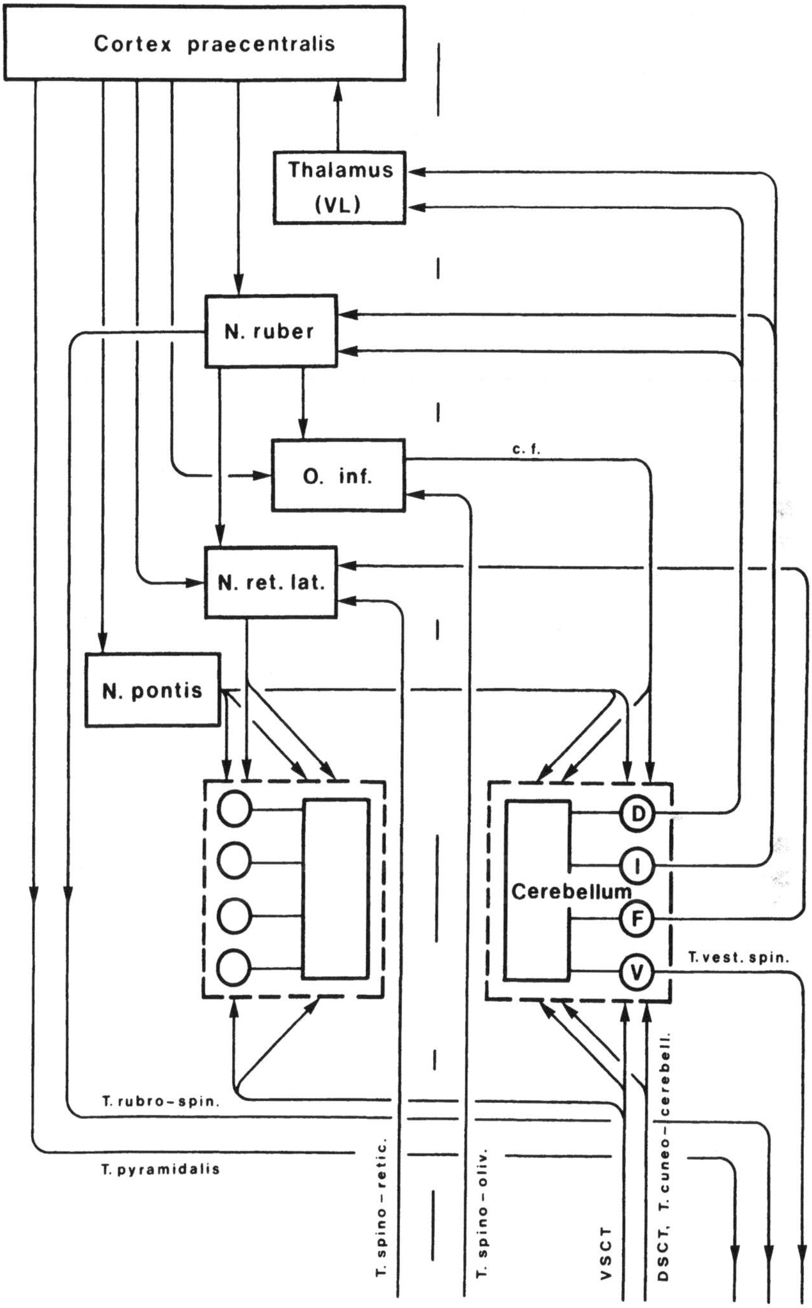

Fig. 3. The cerebellum with its major inputs and outputs

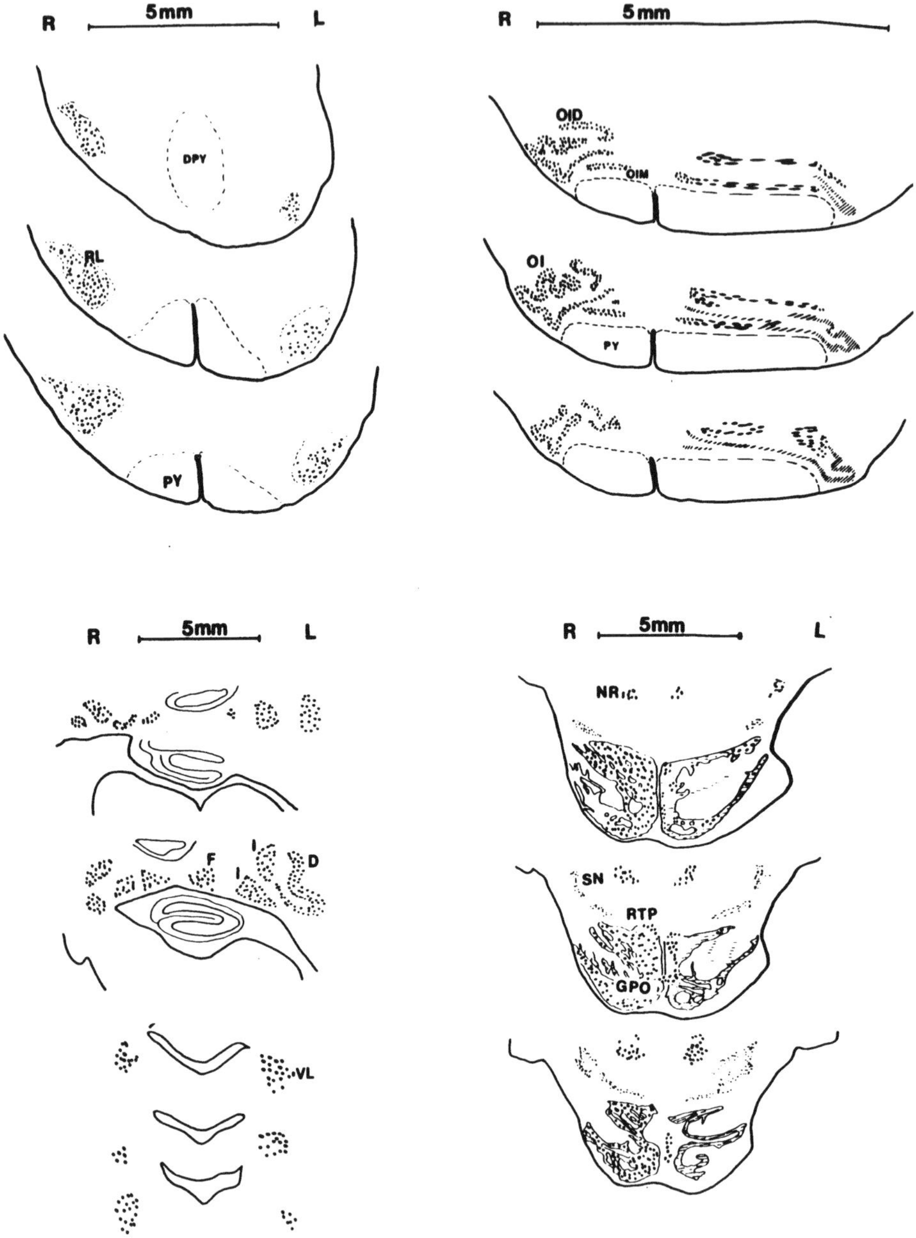

Fig. 4. Extracerebellar degeneration sites in adult macaque following complete recovery from neonatal lesions of right cerebellar cortex

In the light of complete recovery from these substantial cerebellar cortical and extracerebellar destructions, the question arises, which brain regions take over the function and whether there may be morphological manifestations of this functional compensation. Careful planimetry of motor cortex and pyramidal tract (Eckmiller et al. 1984) yielded no evidence at the light microscopic level in support of the notion in the old clinical literature (Anton 1922; Hitzig 1884) that unilateral cerebellar defects lead to an enlargement of the contralateral cerebral cortex (particularly areas 4 and 6) and the pyramidal tract due to its compensatory additional function.

Acknowledgement: This work was supported by the Deutsche Forschungsgemeinschaft SFB 200, B10.

References

Anton G (1922) Über Ersatz der Bewegungsleistungen beim Menschen und Entwicklungsstörungen des Kleinhirns. Z Ges Neurol Psychiatr 30:372–374

Baloh RW, Yee RD, Honrubia V (1980) Optokinetic nystagmus and parietal lobe lesions. Ann Neurol 7:269–276

Denny-Brown D, Chambers RA (1976) Physiological aspects of visual perception. I. Functional aspects of visual cortex. Arch Neurol 33:219–227

Eckmiller R (1987) The neural control of pursuit eye movements. Physiol Rev 67:797–857

Eckmiller R, Westheimer G (1983) Compensation of oculomotor deficits in monkeys with neonatal cerebellar ablations. Exp Brain Res 49:315–326

Eckmiller R, Meisami E, Westheimer G (1984) Neuroanatomical status of monkeys showing functional compensation following neonatal cerebellar lesions. Exp Brain Res 56:59–71

Goldberg ME, Bruce CJ, Ungerleider L, Mishkin M (1982) Role of the striate cortex in generation of smooth pursuit eye movements. Ann Neurol 12:113

Hitzig E (1984) Über einen Fall von halbseitigem Defect des Kleinhirns. Arch Psychiatr Nervenkr 15:266–268

Lynch JC (1986) The interaction of prefrontal and parieto-occipital cortex in the control of purposive eye movements in rhesus monkeys. In: IUPS Satellite Meeting. Developments in oculomotor research, Gleneden Beach, p 37

Lynch JC, McLaren JW (1982) The contribution of parieto-occipital association cortex to the control of slow eye movements. In: Lennerstrand G, Zee DS, Keller EL (eds) Functional basis of ocular motility disorders. Pergamon, Oxford, pp 501–510

Lynch JC, Allison JC, Hines RS, Roark RL (1986) Oculomotor impairment following combined lesions of parieto-occipital cortex and frontal eye fields in rhesus monkeys. Soc Neurosci Abstr 12:1086

Macchi G, Bentivoglio M (1977) Agenesis or hypoplasia of cerebellar structures. In: Vinken PJ, Bruyn GW (eds) Handbook of clinical neurology, vol 30. Congenital malformations of the brain and skull, part I. Elsevier, New York, pp 367–393

Miller M, Pasik P, Pasik T (1980) Extrageniculostriate vision in the monkey. VII. Contrast and sensitivity functions. J Neurophysiol 43:1510–1526

Milner B (1982) Some cognictive effects of frontal-lobe lesions in man. Philos Trans R Soc Lond Biol Sci 298:211–226

Newsome WT, Dürsteler MR, Wurtz RH (1986) The middle temporal visual area and the control of smooth pursuit eye movements. In: Keller EL, Zee DS (eds) Adaptive processes in visual and oculomotor systems. Pergamon, Oxford, pp 223–230

Ritchie L (1976) The effects of cerebellar lesions on saccadic eye movements. J Neurophysiol 39:1246–1256

Schiller PH, True SD, Conway JL (1980) Deficits in eye movements following frontal eye-field and superior colliculus ablations. J Neurophysiol 44:1175–1189

Tusa RJ, Zee DS, Herdman (1986) Recovery of oculomotor function in monkeys with large unilateral cortical lesions. In: Keller EL, Zee DS (eds) Adaptive processes in visual and oculomotor systems. Pergamon, Oxford, pp 209–216

Westheimer G, Blair SM (1974) Functional organization of primate oculomotor system revealed by cerebellectomy. Exp Brain Res 21:463–472

Zee DS, Yamazaki A, Butler PH, Gücer G (1981) Effects of ablation of flocculus and paraflocculus on eye movements in primate. J Neurophysiol 46:878–899

VI Lesions to Specific CNS Subsystems: Visual System

The Effect of Monocular Pattern Deprivation and Open-Loop Stimulation on Optokinetic Nystagmus in Squirrel Monkeys (*Saimiri sciureus*)

F. Behrens and O.-J. Grüsser
Department of Physiology, Freie Universität, Arnimallee 22, 1000 Berlin (West) 33

1 Introduction

Under normal visual conditions the optokinetic reflex (OKR) mainly serves the purpose of minimizing retinal image movement during head or body movements. The gaze-stabilizing OKR of primates is evoked by the movement-induced visual flow of foveal and extrafoveal stimuli (including parallax movement). It supports the foveal fixation mechanism (saccades and attentive gaze pursuit system) and the compensatory eye movements evoked by the vestibulo-ocular reflex. Optokinetic stimulation by means of a large continuously or sinusoidally rotating visual pattern (e.g., an optokinetic drum) activates in untrained monkeys the extrafoveal OKR predominantly, thus evoking binocularly conjugate optokinetic stare-nystagmus (OKN). Barring special instructions to the human observer, his OKN under similar conditions also belongs to this class of stare-nystagmus (ter Braak 1936).

Monocular pattern deprivation during sensitive periods is known to lead, if at all, to only minor changes in signal transmission of the retina and the lateral geniculate nucleus (LGN), but does induce cell size changes in monkeys in the magnocellular and parvocellular layers of the LGN connected with the deprived eye (e.g., Headon and Powell 1973; von Noorden 1973; Vital-Durand et al. 1978; Garey et al. 1979: Grüsser 1979). As in cats, lid suture during the neonatal "sensitive period" leads in monkeys to a dramatic impairment of signal processing in the primary visual cortex (V1) and to amblyopia (Wiesel and Hubel 1963; Baker et al. 1974; Hubel et al. 1977; Blakemore et al. 1978; Freeman 1979; review).

Studies performed during the last few years (Collewijn 1976, 1981; Hoffmann und Schoppmann 1975; Hoffmann 1979, 1982; Grasse and Cynader 1984; Hoffmann und Distler 1986; Kato et al. 1986) provided good evidence that the optokinetic pathway in rabbits, cats, and monkeys is characterized by a specialized retinal input. Movement-sensitive and direction-selective ganglion cells, having rather large receptive fields, project from the retina to the pretectal nuclei of the accessory optic tract which control optokinetic eye movements (review: Simpson 1984; Hutchins and Weber 1985; Weber 1985). For the horizontal OKR-system, the nucleus of the optic tract (NOT) and the dorsal terminal nucleus (DTN) are the main target areas of ganglion cells originating in the nasal retina of the contralateral eye and responding preferentially to horizontal pattern movement through the visual field in the temporo-nasal direction (Hoffmann 1987a; Fig. 1).

H. Flohr (Ed.)
Post-Lesion Neural Plasticity

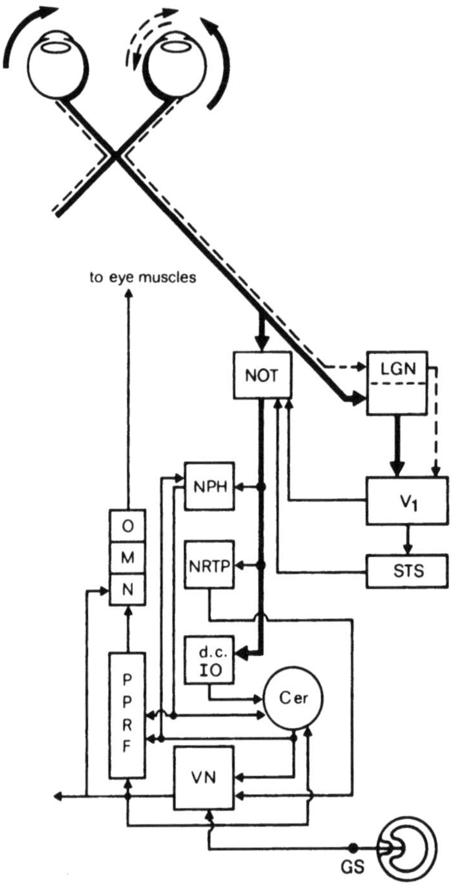

Fig. 1. Scheme of the optokinetic reflex system of the right half of the brain. The connections of the nasal retina of the left eye and the temporal retina of the right eye are shown. *NOT* nucleus of the optic tract; *LGN* lateral geniculate nucleus; *V1* primary visual cortex; *STS* movement-sensitive areas in the cortex around the supratemporal sulcus; *NPH* nucleus praepositus hypoglossi; *NRTP* nucleus reticularis tegmenti pontis; *dcIO* dorsal cap of inferior olive; *Cer* cerebellum; *VN* complex of the brainstem vestibular nuclei; *GS* ganglion Scarpae; *PPRF* praepontine reticular formation; *OMN* oculomotor nuclei

In addition to this direct input from the contralateral retina, NOT and DTN receive corticofugal fibers originating in the striate visual cortex (VI) and, in monkeys, also from movement-sensitive neurons of the STS-region (supratemporal sulcus region). Through this pathway, NOT neurons can be activated by the contralateral and the ipsilateral eye, whereby most of the corticofugal axons rise from binocularly driven pyramidal cells of layer 3. The corticofugal input to the NOT conveys information on naso-temporal and temporo-nasal movement from both eyes and is derived from a cortical neuronal system especially sensitive to border contrast and its orientation.

The nerve cells of the NOT and DTN have large receptive fields (> 30 degrees) which always include the fovea and respond best to ipsilaterally moving visual stimuli in a velocity range of 10–80 degrees/s (Hoffmann and Distler 1986). They send their axons to the dorsal cap of the inferior olive. Via climbing fibers the neurons of this structure form one class of visual movement input to the "oculomotor" and "vestibular" cerebellum (Fig. 1). Other connections of the NOT and DTN reach the nucleus praepositus hypoglossi (NPH), which also has connections via mossy fibers to the "oculomotor" part of the cerebellum (Baker and Berthoz 1975). The nucleus reticularis tegmenti pontis (NRTP) seems to be another target area of NOT and DTN axon collaterals (Keller and Crandall 1983). NRTP and perhaps NPH as well transmit visual movement signals to the brainstem vestibular nuclei, which in turn are connected with the praepontine horizontal gaze centers (PPRF) and the oculomotor nuclei (OMN, Fig. 1). These structures receive additional inputs from the cerebellum via the brainstem vestibular nuclei (Boyle et al. 1985). All the structures mentioned (Fig. 1) form the brainstem horizontal OKR-network (review: Waespe and Henn 1987).

Since monocular pattern deprivation impairs signal processing in area V1 and the latter projects either directly or via the STS-region to the pretectal complex of NOT and the nuclei of the accessory optic system, one can predict that monocular pattern deprivation should have a strong impact on OKN in monkeys. The present study presents data on the effect of long-lasting monocular pattern deprivation on Squirrel monkey horizontal OKN. The deprivation through monocular lid suture began 6 to 10 days after birth and lasted several years. Since the movement of the visual stimulus across the retina ("retinal slip velocity") is the essential driving stimulus for OKN, it became important to know the open-loop gain of the horizontal OKR-system. In pursuing this question we carried out measurements of horizontal OKN evoked by monocular stimulation of an immobilized eye.

2 Methods

The experiments were performed during the last 3 years in three normal and four monocularly deprived adult Squirrel monkeys (*Saimiri sciureus*), which had undergone monocular lid suture between day 6 and day 10 after birth (performed under general Ketamine anesthesia). The lid-closure was maintained for 1.5 to 9 years. A few weeks prior to the experiments, the eyelids were reopened under general anesthesia. The monkeys had been raised in two Squirrel monkey colonies. The monocularly deprived animals showed no behavioral abnormalities compared to others in the colony.

Implantation of the Scleral Search Coil and the Head Holder. After the monkeys were adapted to the experimental chair, a scleral search coil was implanted in one eye outside the corneal limbus behind the conjunctiva (Robinson 1963; Collewijn 1981). No postoperative complications were observed and the coil and the head implants (plug and fixation device) were well tolerated by all animals for up to 28 months. All seven animals are still alive and will be used for further studies (for details see Behrens et al. 1988).

Eye Movement Recordings. The animals which were accustomed to the monkey chair tolerated mechanical immobilization of the head without difficulty for recording periods of 1 to 2 h per day. They received 1 mg DL-Amphetamine i.m. before recording and were rewarded for cooperation during the experiments by small amounts of juice. The eye with the scleral coil was placed in the center of two pairs (one vertical, one horizontal) of Helmholtz electromagnetic field coils (58 cm in diameter, 10 and 20 kHz inducing field), which were fixed on four sides of a cube of 64 cm side length. Eye position could be measured by this method with an error of less than 0.5%, provided one took into consideration that a deviation in the position of the scleral search coil from the parallel to the vertical or horizontal plane determined by the Helmholtz coils led to an error in the position signals. This could be easily compensated by taking into account the signals induced from the other pair of Helmholtz coils (cf. Behrens 1988).

Monocular Immobilization. In one normal monkey the left eye was immobilized by retrobulbar injection of 11 units Botulinum toxin (BoTx), kindly provided by Prof. G. Kommerell from the Freiburg University Eye Clinic, which was dissolved in 0.1 cm^3 Ringer solution. BoTx-uptake in the external eye muscles and the m. levator palpebrae was complete after 72 h. Neither vertical nor horizontal eye movements could be seen under the 24 × binocular microscope. The pupil diameter of the left eye was about 30% larger than that of the right eye. This state lasted about 14 days and recovered slowly within the following 3 weeks. During the experiments the upper eyelid was held open by a small piece of tape. No spillover of BoTx to other muscles was observed. Thus the BoTx eyeball immobilization method provides a simple and efficient tool for further research on open-loop oculomotor responses.

Optokinetic Stimulation. The monkey, placed in a plastic monkey chair, was surrounded by a vertical striped drum of 58 cm diameter and 68 cm height. The striped cylinder moved inside the Helmholtz-coils and was constructed exclusively of plastic and fiberglass. A thin piece of cardboard with silk-screen-printed, vertical, black and white stripes of equal widths (2.37 or 15 degree periods) covered the inside of the cylinder (white-black modulation about 0.7). Continuous rotation of the striped drum at different speeds by means of a servomotor to the left or right (0.5 to 400 degrees/s) was applied. The cylinder was continuously illuminated by a light source placed above the head of the monkey; its shape guaranteed a homogeneous illumination of the whole striped cylinder (average illumination was 15 Lux).

Data Analysis. Horizontal and vertical eye position, speed of the drum and the drum illumination were continuously monitored and simultaneously stored after AD-conversion on the digital computer memory. By computer programs the average speed of individual slow nystagmus periods was measured and printed out numerically or graphically. Algebraic means of these values were usually taken for ten successive slow OKN-periods. By means of digital differentiation the eye angular velocity could be continuously monitored. The computer programs are described in Behrens et al. (1988).

3 Results

3.1 Optokinetic Nystagmus in Normal Squirrel Monkeys (N-Animals)

Figure 2a and b illustrates recording examples of OKN at different stimulus velocities. The high regularity of OKN is evident. To obtain gain vs. stimulus-velocity curves, the velocity of the striped drum was altered in steps upward or downward. A fairly regular horizontal OKN was recorded at stimulus angular velocities up

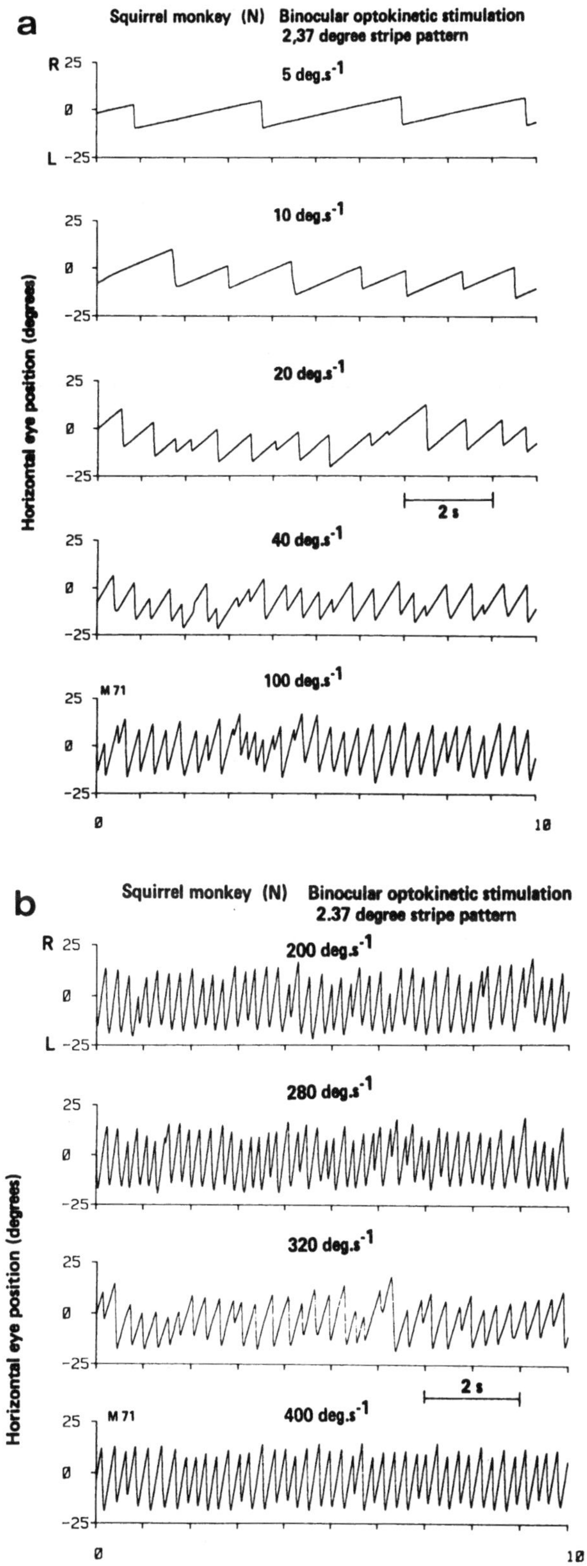

Fig. 2. Horizontal optokinetic nystagmus of normal Squirrel monkey evoked by a rotating cylinder covered with vertical black and white stripes of 2.37 degrees period. (**a, b**): Stimulus angular velocities as indicated

to 350 degrees/s. The average gain of at least ten consecutive slow nystagmus phases was measured. It varied between 0.85 and 0.97 in the velocity range between 1 and 150 deg/s. At angular velocities above 160 deg/s, the gain of OKN slow-phase velocity decreased with increasing stimulus velocity. At stimulus velocities between 350 and 400 deg/s the gain had dropped to 0.1–0.3 (Fig. 3a–c). The following general properties of horizontal OKN were found in all three normal monkeys and also in the monocularly pattern-deprived monkeys for monocular stimulation of the normal eye on which the coil was also implanted:

a) Maximum speed of horizontal OKN slow phases varied between 160 and 180 deg/s (Fig. 3c).
b) No systematic directional preference was obtained with monocular closed-loop stimulation, i.e., naso-temporal and temporo-nasal pattern movement had about the same effect on OKN. Animal 71, from which the data shown in Figs. 2 and 3 were obtained, exhibited a slight left-right difference for both eyes at drum velocities above 150 deg/s.
c) No significant differences in OKN gain were found for monocular OKN evoked by stimulation of the right or the left eye.

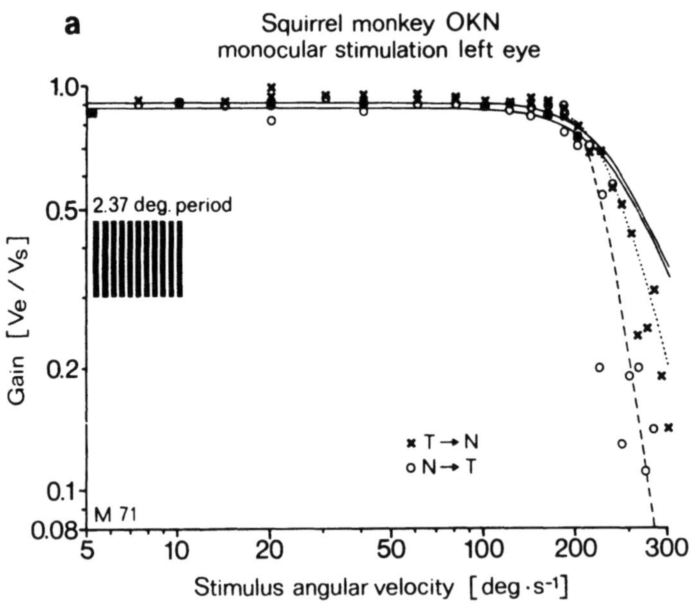

Fig. 3. a Monocular horizontal OKN gain (*ordinate*) vs. stimulus angular velocity (*abscissa*). Monocular stimulation of the left eye, optokinetic stimulation towards the right (*crosses*) and the left (*circles*). The *solid lines* are curves computed by fitting optimally to the data a Butterworth filter of second order (−3dB point at 258 and 256 deg/s). The *punctuated line (crosses)* are computations of a Butterworth filter of third order and the *dashed line (circles)* one of fifth order. **b** Same relatioship as in **a** but binocular stimulation and stripe pattern of 15 degrees period. **c** Relationship between the OKN slow-phase angular velocity V_e (*ordinate*) and the stimulus angular velocity V_s (*abscissa*) for horizontal OKN evoked by monocular stimulation of the right eye. Note that maximum V_e values reach about 160 deg/s. Stripe pattern of 15 degrees period

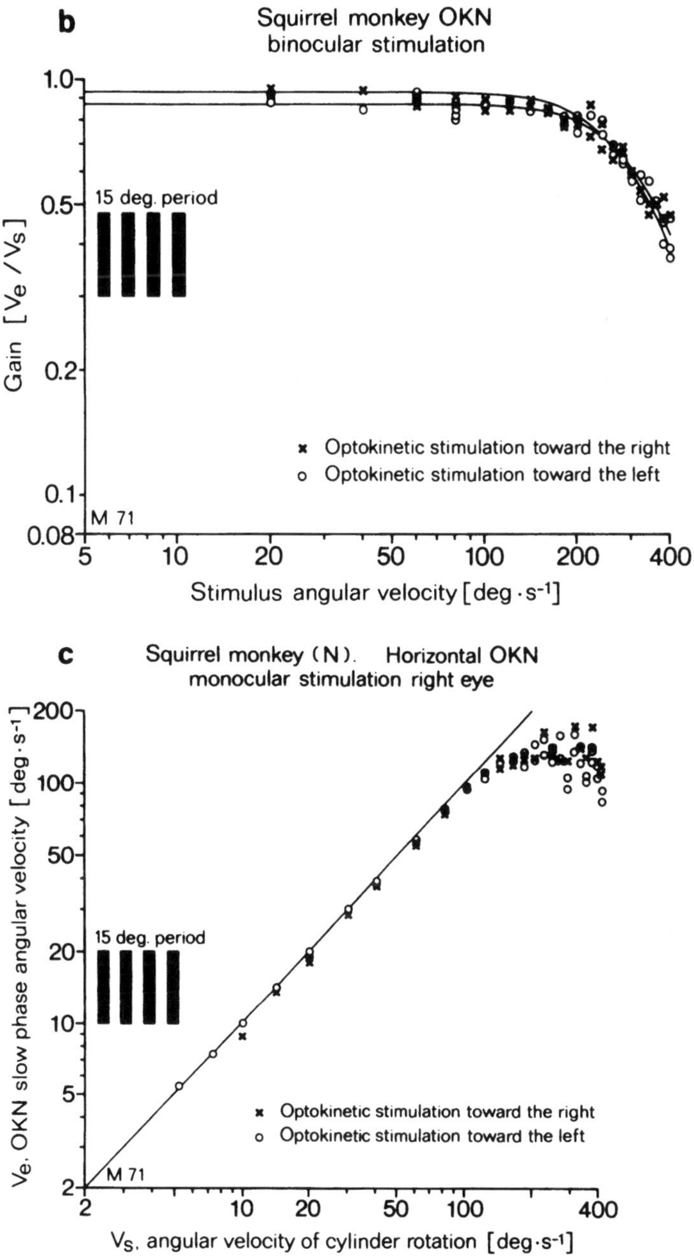

Fig. 3b and c

d) Gain of OKN at stimulus velocities above 150 deg/s was consistently higher for binocular stimulation than for monocular stimulation of the left or the right eye. At lower speeds these gain differences were marginal.
e) When an OKN, evoked by the 2.37-degree-period stripe pattern, was compared with that evoked by the 15-degree-period pattern, the upper limiting velocity (−3dB value) was not always shifted to higher values with the larger pattern period (Fig. 3).

3.2 Retinal Slip Velocity and Open-Loop OKN

The ongoing OKN is controlled by three mechanisms: The retinal slip velocity caused by the difference between the angular velocities of the stimulus and the eye, the internally generated efference copy signals of the gaze movement and a slow storage mechanism responsible for OKAN I and most likely "fed" by the efference copy signals (Raphan et al. 1977). The "OKN storage complex", which can be characterized as a neuronal leaky integrator, reached its maximum load 3–5 s after the closed-loop OKN began. When light was suddenly turned off during OKN, it discharged relatively slowly with a time constant of about 12–15 s, as seen in the first phase of optokinetic afternystagmus (OKAN I, Fig. 4a, b). The presentation of stationary visual patterns after OKN inhibited the integrator mechanism or facilitated the discharge of this integrator: OKAN I, II and III were reduced.

In Squirrel monkeys (as in man and other primates), the "driving" retinal stimulus, i.e., retinal slip velocity V_r, depended on stimulus angular velocity V_s (Fig. 5a). For steady-state stimulus conditions, it was meaningful to plot the angular velocity of the slow nystagmus phase V_e not only as a function of drum velocity V_s but also of retinal slip velocity V_r (Fig. 5b).

Stimulating the immobilized eye and recording horizontal OKN from the other covered mobile eye, the relationship between OKN slow-nystagmus phase and retinal slip velocity can be easily measured in "open loop" conditions. As Fig. 6 illustrates, this relationship depended on the duration of the movement stimulus. Therefore, when one plots the open-loop gain of OKN as a function of stimulus angular velocity (Koerner and Schiller 1972), one has to regard the time after the onset of stimulation as a parameter (examples are shown in Fig. 6). It

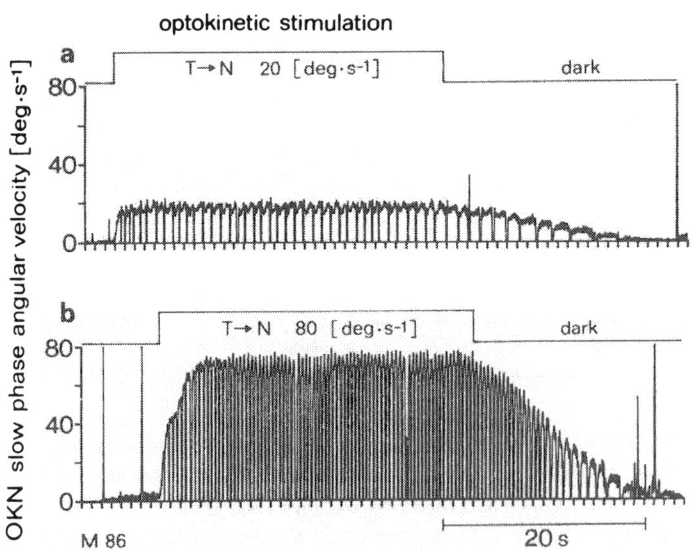

Fig. 4. OKN slow phase angular velocity (ordinate) of horizontal OKN of normal Squirrel monkey evoked by monocular stimulation of the left eye. Data obtained by differentiation of eye position signals. 20 deg/s (**a**) and 80 deg/s (**b**) optokinetic stimulation; 2.37-degree-period stripe pattern

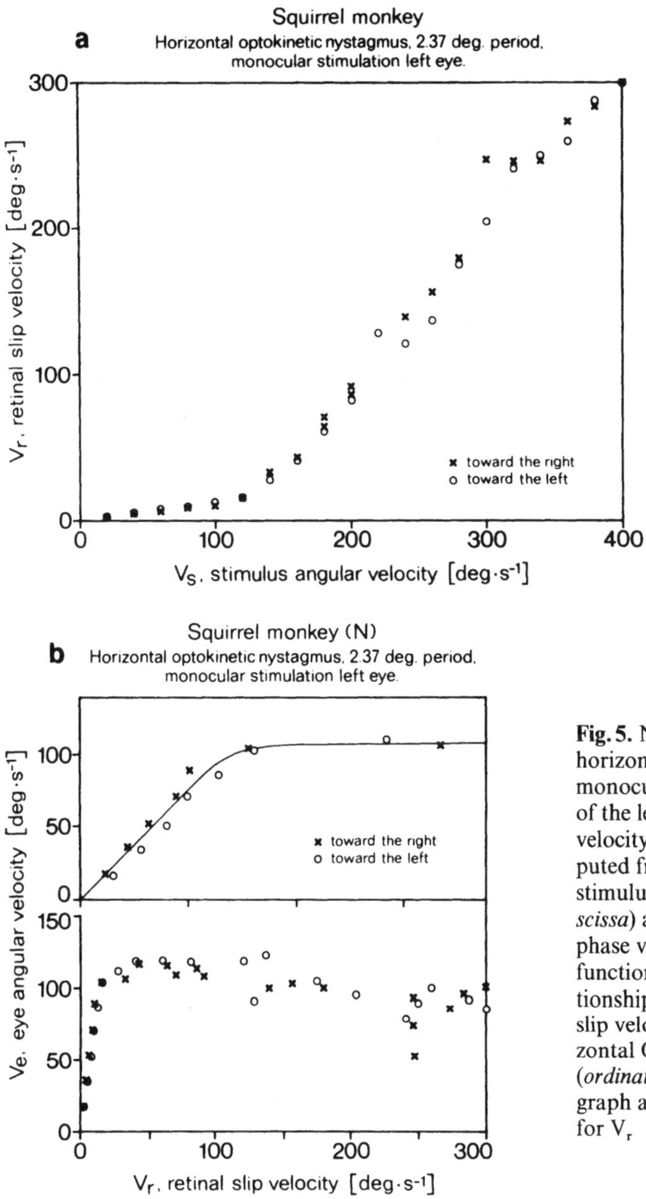

Fig. 5. Normal Squirrel monkey horizontal OKN. **a** Horizontal monocular optokinetic stimulation of the left eye. The retinal slip velocity V_r (*ordinate*) was computed from the difference between stimulus angular velocity V_s (*abscissa*) and horizontal OKN slow-phase velocity V_e. V_r is plotted as a function of V_s (*abscissa*). **b** Relationship between horizontal retinal slip velocity V_r (*abscissa*) and horizontal OKN angular velcoity V_e (*ordinate*). In the upper part of the graph a different scale is applied for V_r

became evident that the time constant of increasing V_e with duration of open-loop stimulation was considerably longer with retinal slip velocities below 2 deg/s than with higher V_r values. This non-linear dichotomy might be due to the relative involvement of the subcortical and the cortical loop in driving OKN, whereby the higher V_r, the more dominant becomes the cortical loop.

For open-loop and closed-loop OKN, the efficacy of retinal pattern movement reached a maximum at rather low V_r-values. The relationship between OKN-gain and V_r was very similar for closed-loop and open-loop stimulus conditions (Fig. 7a, b).

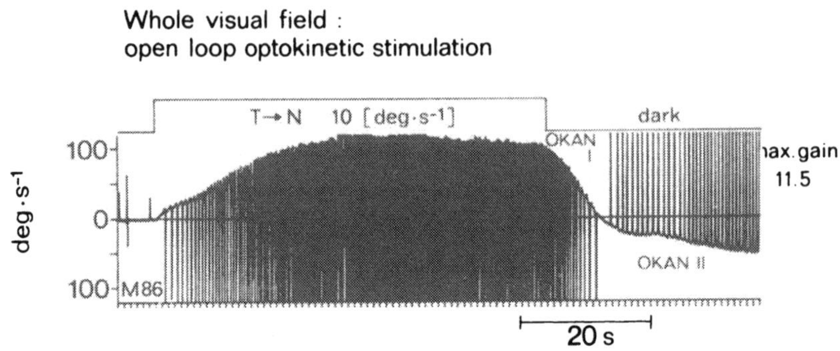

Fig. 6. Squirrel monkey open-loop OKN evoked by monocular stimulation of the immobilized left eye and recorded from the mobile right eye. The slow-phase angular velocity (*abscissa*) plotted as a function of stimulus duration. Retinal slip velocity ($V_r = V_s = 10$ deg/s).

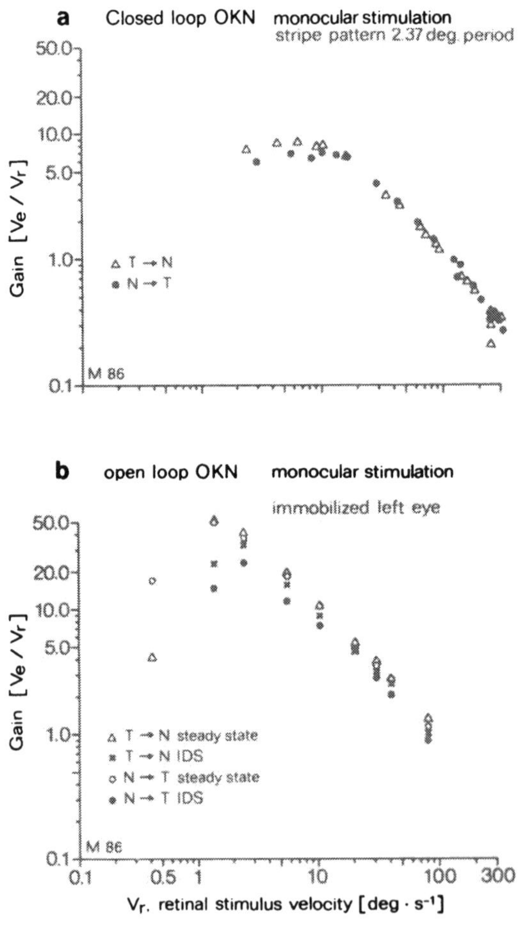

Fig. 7. a Horizontal OKN. Closed-loop gain (*ordinate*) plotted as a function of retinal slip velocity V_r (*abscissa*). **b** Open-loop gain (*ordinate*) as measured for plateau values 45 s after onset of stimulation and 10 s IDS after onset. Horizontal OKN evoked by optokinetic stimulation of the immobilized left eye. Retinal stimulus angular velocity is plotted on the *abscissa;* 2.37 degree stripe pattern

3.3 Differences in Stimulation Efficacy of the Nasal and the Temporal Visual Hemifield in Open-Loop OKN

Figure 8 a and b demonstrates the time course of the slow-phase velocity of open-loop horizontal OKN, obtained by striped pattern movements at one selected angular velocity (10 deg/s) in the nasal or the temporal visual hemifield. It is evi-

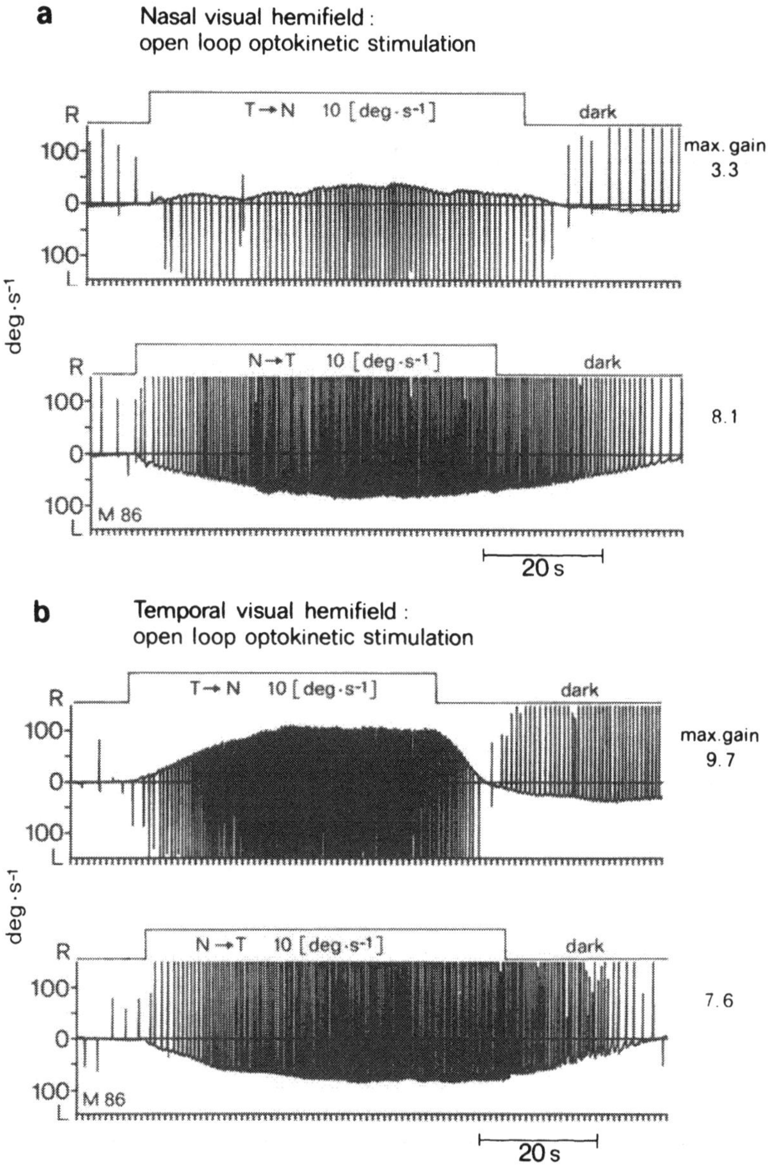

Fig. 8. Time course of open-loop horizontal OKN slow-phase angular velocity. Stimulation of the nasal (**a**) or the temporal (**b**) visual hemifield of the immobilized left eye. OKN was recorded from the mobile right eye. Stimulus angular velocities ($V_s = V_r$) 10 deg/s. Note the strong directional difference in OKN-gain for temporo-nasal and naso-temporal stimulation in the nasal visual hemifield (**a**) but not in the temporal hemifield (**b**). 2.37 degree stripe pattern

dent from these data that a distinct directional selectivity is present for the efficacy of optokinetic stimuli moving through the nasal visual hemifield. Principally the same directional sensitivities were also found with higher or lower optokinetic velocities.

3.4 The Effect of Long-Lasting Monocular Deprivation on Horizontal OKN (D-Animals)

When the eyelid of the deprived eye was reopened a couple of weeks before the first experimental session, it became immediately evident that the deprived eye was squinting preferentially toward the nasal side (adducting strabismus). This strabismus did not improve in the course of the following months. For technical reasons we could only implant one coil (on the nondeprived eye) to date. Therefore, the precise conditions of this squinting in the "D-animals" have not been explored so far quantitatively.

Horizontal OKN evoked monocularly by the normal eye revealed the same properties in the D-animals as described above for the N-animals. Since the deprived eye showed considerable deficiencies in its responses to optokinetic stimuli,

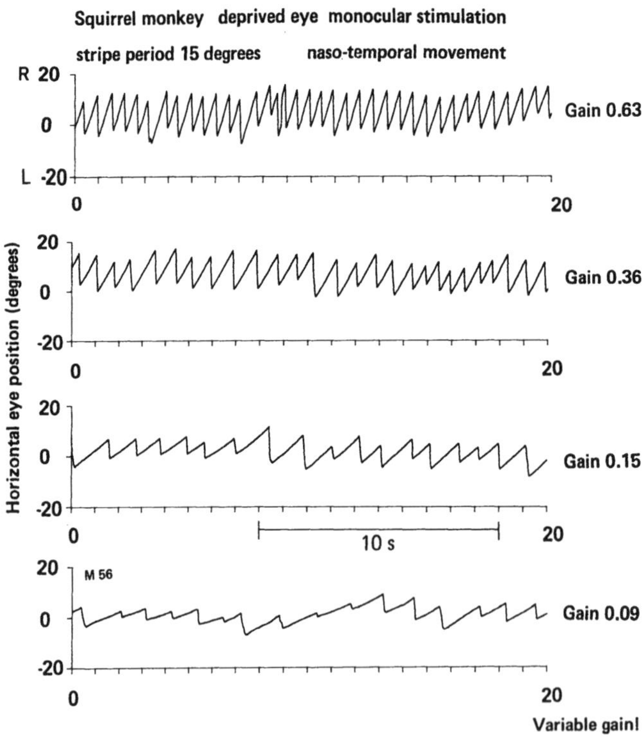

Fig. 9. Monocular optokinetic stimulation of a pattern-deprived eye. Squirrel monkey, 8 years monocular pattern deprivation. 15 degrees stripe period, stimulus angular velocity 40 deg/s. Note the very changeable OKN leading to a gain varying between 0.09 and 0.63 as indicated

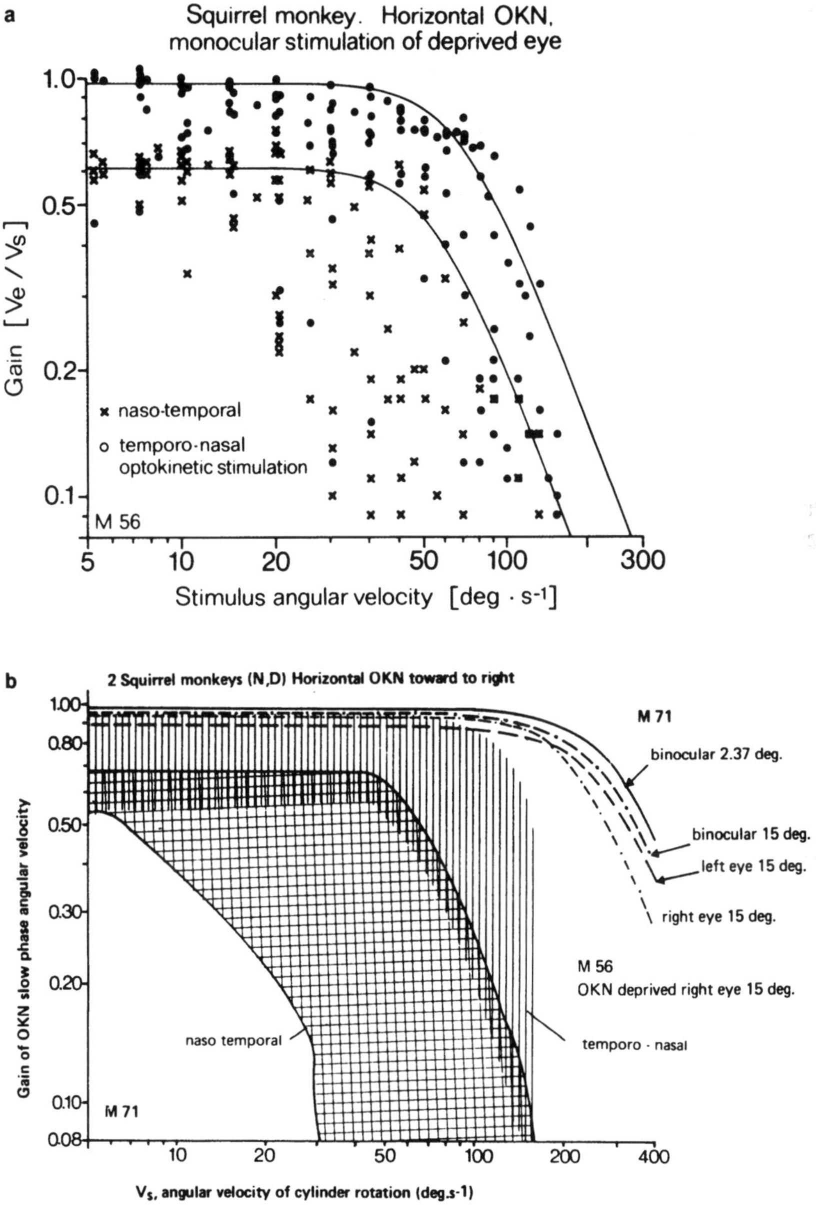

Fig. 10. a Monocular stimulation of the pattern-deprived right eye of a Squirrel monkey. Gain (*ordinate*) vs. stimulus angular velocity V_s (*abscissa*). Optokinetic stimulation towards the right (naso-temporal direction) led to a considerably higher variability and lower gain of OKN than monocular stimulation towards the left (temporo-nasal direction). **b** Normal monkey and monocular pattern deprived monkey. Comparison of the distribution of gain values for monocular horizontal optokinetic stimulation at different stimulus angular velocities V_s (abscissa). In the normal monkey the values varied minimally along the drawn lines (monocular and binocular stimulation with a stimulus pattern of 15 degrees period and binocular stimulation with a stimulus pattern of 2.37 degrees period). In the pattern-deprived animal (Fig. 9a), naso-temporal and temporo-nasal optokinetic stimulation evoked rather different responses from those in the normal eye. The data are distributed in the *hatched regions*

the binocular OKN was predominantly determined by the normal eye. Monocular stimulation of the deprived eye evoked a rather variable OKN even in highly alert monkeys under constant stimulus conditions. An example is shown in Fig. 9. Maximum V_e reached values of about 80 deg/s, i.e., at maximum about half of the speed evoked by stimulation of the the normal eye. When average V_e-values were computed for short periods of ten successive OKN slow phases and plotted as a function of stimulus angular velocity, the scatter of the data was very high (Fig. 10 a). In addition, horizontal OKN evoked by naso-temporal movement of the stripes had a higher irregularity and lower maximum V_e-values than when temporo-nasal stimulation was applied (Fig. 10 a). Further it became evident that at angular velocities between 1 and 50 deg/s, OKN evoked by naso-temporal movement reached lower gain values on the average than temporo-nasal movements. Under optimal conditions the latter could reach a gain at about 0.95, while naso-temporal OKN-gain did not exceed 0.75. In Fig. 10 b these findings are contrasted to the gain V_s velocity curves of a normal monkey.

4 Discussion

The horizontal OKN-system of Squirrel monkeys operates in untrained monkeys with a precision which exceeds that of the OKN-system in man at least with longer optokinetic stimulation and at higher angular velocities. When the stripe pattern moving at constant angular velocities was suddenly visible, however, the OKN-gain of Squirrel monkeys reached its maximum considerably later than that of an attentive human subject. The Squirrel monkey evidently did not employ the foveal pursuit system to accelerate the OKN, as his attentive human counterpart does. A neuronal leaky intergrator seems to sum up the retinal movement signals with a time constant of about 15 s (open-loop conditions), which leads to a V_e "build-up" with a time constant of about 4 s (for theoretical discussion see Behrens and Grüsser 1988). This slow component of the optokinetic reflex is presumably also responsible for the fact that OKAN I, II and III are more pronounced and persist longer in Squirrel monkeys than in man.

The open-loop experiment revealed another striking difference between the Squirrel monkey OKN-system and that in man. When one records gaze-nystagmus of man under open-loop conditions (Grüsser et al. 1981), the fast phases of OKN (backward saccades) lead to a reset of the open-loop gain (Fig. 11). This reset does not operate in open-loop OKN of Squirrel monkeys (Fig. 6 a), at least not under our experimental conditions.

The data described in the present report provided clear evidence that the deteriorating effect of monocular pattern deprivation on cortical signal processing also affected OKN evoked by stimulation of the pattern-deprived eye: OKN exhibited a reduced precision and had an upper cut-off angular velocity at much lower values than the OKN evoked from the normal eye. At low angular stimulus velocities the gain of OKN evoked by patterns moving towards the nasal side of the deprived eye's visual field was on the average lower and varied considerably more than OKN evoked by naso-temporal movement. As in monkey infants *(Macaca mulatta)* and in human babies of less than 2 months of age (Atkinson

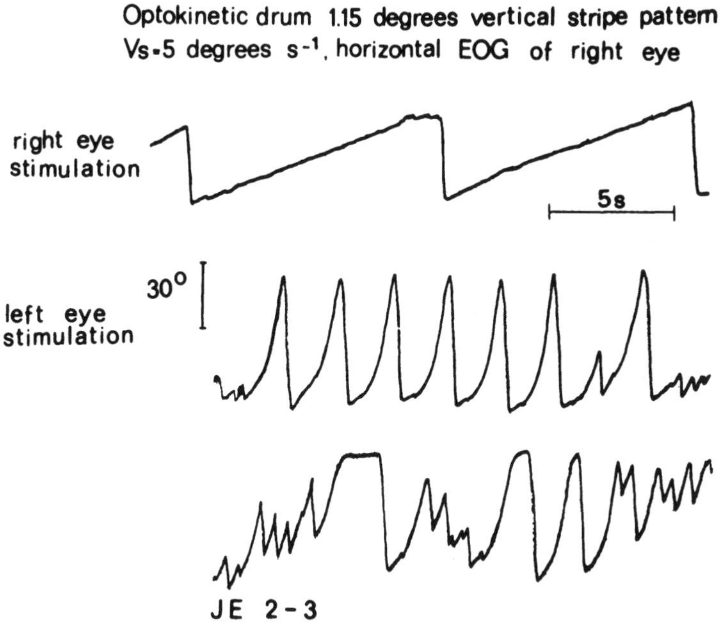

Fig. 11. Horizontal OKN in man. Monocular stimulation of the mobile right eye at 5 deg/s by a 1.15 degree stripe pattern moving towards the right; attentive gaze-nystagmus (*upper row*). In the *lower row* open-loop horizontal OKN. Monocular stimulation of the immobilized left eye. The time course of the slow phases of the open-loop optokinetic gaze nystagmus demonstrates a velocity function accelerating in time. Backwards saccades led to a reset of eye angular velocity recorded from the covered right eye. This reset was correlated with the reduction in the perceived speed of the continuously moving stripe pattern. From unpublished data of Grüsser, Kulikowski, Pause and Wollensak (1981). Eye immobilization by retrobulbar injection of 5 ml. 1.5 percent Scandicaine (Subject O.-J. G)

1979), naso-temporal stimulation was less effective in the pattern-deprived eye than temporo-nasal movement of the optokinetic stimulus across the visual field. It seems that pattern deprivation maintains the connections between the deprived eye and the OKN- or OKR-system at a functional level already attained very early in life (Rakic 1979): temporo-nasal pattern movement is processed more efficiently via the nasal retina and the contralateral NOT/DTN-complex, while naso-temporal pattern movement reaches the ipsilateral NOT predominantly via the insufficient cortical loop.

Some Thoughts on the Development of Squint by Monocular Pattern Deprivation. From the scheme of Fig. 1 and the data presented in the preceding report, one can deduce that monocular pattern deprivation reduces the efficiency of the corticofugal input to the contralateral NOT but does not disturb the direct connections from the retina to the NOT. The connections of the pattern-deprived retina to the ipsilateral NOT via LGN, area V1 and movement-sensitive STS areas of the brain are less effective than those of the normal retina. Furthermore, the contralateral connections from the retina to the NOT (direct or via the cortical loop) are dominant over the ipsilateral connections. Activation of the NOT leads to horizontal eye movements in the ipsiversive direction (relative to the side of NOT). Pattern deprivation presumably changes the activity of NOT-neurons contralateral to the deprived eye differently than that of the NOT-neurons ipsilateral to the deprived eye. In addition, OKN evoked by naso-temporal optokinetic stimulation of the pattern-deprived eye was more impaired than that evoked by temporo-nasal optokinetic stimulation. Pattern deprivation led to a left-right imbalance of the NOT-inputs to the gaze motor system with a reduced activity for ipsiversive eye movements (relative to the deprived eye side). The adducting squint observed in the monocularly pattern-deprived monkeys might have been the consequence of this left-right imbalance. The movement of the nondeprived eye, however, is impaired very little, while the movement of the deprived eye is reduced whenever eye movements are attempted in the naso-temporal direction of that eye.

5 Summary

1. Horizontal optokinetic nystagmus (OKN) was measured in three normal and four monocularly pattern-deprived Squirrel monkeys by means of the electromagnetic search coil technique. Binocular stimulation and monocular stimulation of each eye towards the left and the right by moving vertical stripe patterns of 2.37 or 15 degrees were applied at angular velocities of 0.5–400 deg/s. In one animal, open-loop OKN gain was measured by monocular stimulation of an eyeball immobilized by means of retrobulbar injections of botulinum toxin (BoTx).
2. In normal Squirrel monkeys gain of optokinetic nystagmus reached values between 0.8–0.97 at angular velocities below 150 deg/s. A slightly higher gain was found for binocular than for monocular stimulation. No significant differences were found between monocular stimulation in the naso-temporal and in the temporo-nasal direction. The upper cut-off angular velocity (−3dB-point) reached values (180–250 deg/s) significantly above those observed in man.
3. Monocular optokinetic stimulation of an immobilized eye led to a vigorous optokinetic nystagmus and OKAN of the other eye, whereby maximum gain was found at low retinal stimulus velocities (2–5 deg/s, gain above 30). Increasing retinal angular velocity led to a decrease in gain with a slope of about 20 dB per decade. When a striped pattern suddenly appeared under open-loop stimulus conditions, rotating at a constant angular velocity, the open-loop OKN slow phase reached its maximum angular velocity only 15–40 s after stimulus onset. The time course of the slow "build-up" of V_e expressed the function of two different, slow, central integrating mechanisms operating with a time constant of about 12–15 s at V_r-values below 2 deg/s and a time constant of about 7–9 s at higher V_r-values.
4. Monocular deprivation led to an impairment of horizontal OKN evoked by the deprived eye, but to no change in the OKN evoked by the nondeprived eye. The reduction in OKN precision, average gain and upper cut-off angular velocity was stronger for naso-temporal optokinetic stimulation of the deprived eye than for temporo-nasal stimultion. This observation can be explained by the deterioration of the cortical loop for the signals reaching NOT

from the deprived eye via LGN, the visual cortex V_1 and STS-areas. Direct connections between retina and NOT exist predominantly for the optic tract fibers originating in the nasal retina of the contralateral eye. This connection is not impaired by pattern deprivation.

5. Monocular deprivation led to left-right imbalance of activity in the OKR-system responsible for horizontal OKN. This imbalance presumably is responsible for the adducting squint observed in the deprived eye.

Acknowledgements. This work was supported by a grant of the Deutsche Forschungsgemeinschaft (Gr 161/32). We thank Mr. J. Lerch for his excellent work in the construction of the monkey chair and other mechanical components used in the present study, Dipl.-Ing. L.-R. Weiss for his assistance with the computer programs, Mrs. Ch. Dreykluft for technical assistance, Mrs. B. Hauschild and Mrs. I. Knierim for typing and Mrs. J. Dames for her help in the English translation of the manuscript.

References

Atkinson J (1979) Development of optokinetic nystagmus in the human infant and monkey infant: An analogue to development in kittens. In: Freeman RD (ed) Developmental neurobiology of vision. Plenum, New York, pp 277–287

Baker FH, Grigg P, Noorden GK von (1974) Effects of visual deprivation and strabismus on the responses of neurons in the visual cortex of the monkey, including studies on the striate and pre-striate cortex in the normal animal. Brain Res 66:185–208

Baker R, Berthoz A (1975) Is the prepositus hypoglossi nucleus the source of another vestibular ocular pathway? Brain Res 86:121–127

Behrens F (1988) Some calibration problems in using the electromagnetic scleral search coil technique for eye movement recordings in animals. J Neurosc Methods (submitted)

Behrens F, Grüsser O-J (1987) Open-loop optokinetic nystagmus in Squirrel monkeys (in preparation)

Behrens F, Grüsser O-J, Weiss L-R (1988) Optokinetic nystagmus of Squirrel monkeys (in preparation)

Blakemore C, Garey LJ, Vital-Durand F (1978) The physiological effects of monocular deprivation and their reversal in the monkey's visual cortex. J Physiol (Lond) 283–223–262

Boyle R, Büttner U, Markert G (1985) Vestibular nuclei activity and eye movements in the alert monkey during sinusoidal optokinetic stimulation. Exp Brain Res 57:362–369

Braak JWG Ter (1936) Untersuchungen über optokinetischen Nystagmus (translated "Investigations on optokinetic nystagmus") Arch Neerl Physiol 21:309–376. In: Collewijn H (1981) The oculomotor system of the rabbit and its plasticity. Studies in brain function, vol 5. Springer, Berlin Heidelberg New York

Brodal A (1984) The vestibular nuclei in the macaque monkey. J Comp Neurol 227:252–266

Brodal P (1978) The corticopontine projection in the rhesus monkey: origin and principles of organisation. Brain 101:251–283

Collewijn H (1976) Direction-selective units in the rabbit's nucleus of the optic tract. Brain Res 100:489–508

Collewijn H (1981) The oculomotor system of the rabbit and its plasticity. In: Braitenberg et al. (eds) Studies in brain function. Springer, Berlin Heidelberg New York

Freeman RD (1979) (ed) Developmental neurobiology of vision. Plenum, New York

Garey LJ, Vital-Durand F, Blakemore C (1979) Reversal of effects of visual deprivation in monkeys. In: Freeman RD (ed) Developmental neurobiology of vision. Plenum, New York, pp 261–265

Grasse KL, Cynader MS (1984) Electrophysiology of lateral and dorsal terminal nuclei of the cat accessory optic system. J Neurophysiol 51:276–293

Grüsser O-J (1979) Neuronal activity in the afferent visual system and monocular pattern deprivation. In: Freeman RD (ed) Developmental neurobiology of vision. Plenum, New York, pp 73–77

Grüsser O-J, Kulikowski J, Pause M, Wollensak M (1981) Optokinetic nystagmus, sigmaoptokinetic nystagmus and eye pursuit movements elicited by stimulation of an immobilized human eye. J Physiol (Lond) 320:21–22
Headon MB, Powell TPS (1973) Cellular changes in the lateral geniculate nucleus of infant monkey after suture of the eyelids. J Anat (Lond) 116:135–145
Hoffmann K-P (1979) Optokinetic nystagmus and single-cell responses in the nucleus tractus opticus after early monocular deprivation in the cat. In: Freeman RD (ed) Developmental neurobiology of vision. Plenum, New York, pp 63–72
Hoffmann K-P (1982) Cortical versus subcortical contributions to the optokinetic reflex in the cat. In: Lennerstrand G (ed) Functional basis of ocular motility disorders. Pergamon, Oxford, pp 303–310
Hoffmann K-P (1986) Visual inputs relevant for the optokinetic nystagmus in mammals. In: Freund H-J, Büttner U, Cohen B, Noth J (eds) The oculomotor and skeletalmotor system: differences and similarities. Prog Brain Res 64, Elsevier, Amsterdam pp 75–84
Hoffmann K-P (1987a) The influence of visual experience on the otogony of the optokinetic reflex in mammals. In: Rauschecker J, Marler P (eds) Imprinting and cortical plasticity. Wiley, New York (in press)
Hoffmann K-P (1987b) Neural basis for changes of the optokinetic reflex in animals and man with strabismus and amblyopia. In: Lennerstrand O (ed) Strabismus and amblyopia – experimental basis for advances in clinical management. MacMillan, London (in press)
Hoffmann K-P, Distler DS (1986) The role of direction selective cells in the nucleus of the optic tract of cat and monkey during optokinetic nystagmus. In: Keller EL, Zee DS (eds) Adaptive processes in visual and oculomotor system. Pergamon, Oxford, pp 261–266
Hoffmann K-P, Schoppmann A (1975) Retinal input to direction selective cells in the nucleus tractus opticus of the cat. Brain Res 99:359–366
Hubel DH, Wiesel TN, S LeVay S (1977) Plasticity of ocular dominance in monkey striate cortex. Philos Trans R Soc Lond Biol Sci 278:377–402
Hutchins B, Weber JT (1985) The pretectal complex of the monkey: a reinvestigation of the morphology and retinal terminations. J Comp Neurol 232:425–442
Kato I, Harada K, Hasegawa T, Igarashi T, Koike Y, Kawasaki T (1986) Role of the nucleus of the optic tract in monkeys in relation to optokinetic nystagmus. Brain Res 364:12–22
Keller EL, Crandall WF (1983) Neuronal responses to optokinetic stimuli in pontine nuclei of behaving monkey. J Neurophysiol 49:169–187
Koerner F, Schiller PH (1972) The optokinetic response under open and closed loop conditions in the monkey. Exp Brain Res 14:318–330
Noorden GK von (1973) Histological studies of the visual system in monkeys with experimental amblyopia. Invest Ophthalmol 12:727–738
Rakic P (1979) Genesis of visual connections in the Rhesus monkey. In: Freeman RD (ed) Developmental neurobiology of vision. Plenum, New York, pp 249–260
Raphan T, Cohen B, Matsuo V (1977) A velocity-storage mechanism responsible for optokinetic nystagmus (OKN), optokinetic afternystagmus (OKAN) and vestibular nystagmus. In: Baker R, Berthoz A (eds) Control of gaze by brainstem neurons. Development Neurosci, vol I. Elsevier, Amsterdam, pp 37–47
Robinson DA (1963) A method of measuring eye movements using a scleral search coil in a magnetic field. IEEE Trans Biomed Eng 10:137–145
Simpson JI (1984) The accessory optic system. Annu Rev Neurosci 7:13–41
Vital-Durand F, Garey LJ, Blakemore C (1978) Monocular and binocular deprivation in the monkey: Morphological effects and reversibility. Brain Res 158:45–64
Waespe W, Henn V (1987) Gaze stabilization in the primate. The interaction of the vestibulo-ocular reflex, optokinetic nystagmus, and smooth pursuit. Rev Physiol Biochem Pharmacol 106:37–125
Weber JT (1985) Pretectal complex and accessory optic system of primates. Brain Behav Evol 26:117–140
Wiesel TN, Hubel DH (1963) Single-cell responses in striate cortex of kittens deprived of vision in one eye. J Neurophysiol 26:1003–1017

Return of Function After Optic Tract Lesions in Adult Rats: Spontaneous Axonal Regeneration?

A. P. Foerster

Department of Neurosciences, McMaster University, 1200 Main St. W., Hamilton, Ontario, L8N 3Z5, Canada

1 Introduction

It is unusual to see a spontaneous regeneration of severed axons suggested as one of the many possible mechanisms which may contribute to the often-observed improvement in functions which were compromised immediately after injury to the adult mammalian central nervous system (CNS). Severed axons in the CNS do not generally show the vigorous, albeit often randomly directed, regrowth of peripheral axons (Ramón y Cajal 1928), and it is rare to find that they have crossed a cut unless induced to do so by adjacent implanted tissue which they enter, but within which they tend to remain (Kromer et al. 1981; So and Aguayo 1985; Berry et al. 1986; Vidal-Sanz et al. 1987).

In earlier anatomical studies (Marks [née Foerster] 1972; Foerster 1982) I attempted a new strategy in the study of CNS regeneration: that of using a special wire cutting device to make a precisely defined lesion, and implanting it to provide a clear and permanent anatomical marker of the location of the axons which it had cut. This allowed the identification of new axonal growth and the confirmation that it originated from within the zone of severed axons. I now have evidence consistent with a spontaneous regeneration of severed optic axons in the rat, and their mediation of a long-term, progressive return of visual function in their target, the superior colliculus (SC). A short-term return of function which stabilizes within hours has also been characterized, and seems likely to be the result of recovery from pressure-block in spared optic axons next to the lesion.

2 Anatomical Evidence for Spontaneous Regeneration

2.1 The Lesion

This was made by a device consisting of a fine (90 μm diameter) horizontal cutting crosswire, 1–2.5 mm long, supported between two vertical wires (also 90 μm diameter) which projected slightly beneath the cutting wire; these projections helped secure underlying tissue from being pushed aside as the device was lowered, under micromanipulator control, from the top of the brain to the floor of the skull. After the descent was completed the protruding vertical support wires were cemented to the calvarium. The brain was fixed at times ranging from min-

H. Flohr (Ed.)
Post-Lesion Neural Plasticity

utes to months later. The device was removed via the floor of the now-solid brain, leaving two vertical channels which had contained the support wires; thus the tissue that had been cut by the crosswire was located between these channels.

2.2 Immediate and Later Appearances Postlesion

Measurements on *horizontal* sections from brains fixed immediately postlesion showed that no more than 200 μm of tissue had been passively deflected around each of the vertical wires, whose channels in cross-section provided the two holes that marked each end of the cut; the severed axons between the two marker holes abutted the line of the lesion. These measurements were important in interpreting the unexpected later appearance of populations of axons which originated from between the support wires, and had deviated *en masse* to curve around them, from as far away as 800 μm. Clearly, these represented a genuine new development, rather than the persistence of a mechanically produced deformation. Nor could these later "detours" be attributed to brain shrinkage, since this was rarely observed; all the illustrations in Foerster (1982) were made from brains in which measurements revealed that there had been no shrinkage of the injured side. The occurrence of anterograde degeneration within tracts cut by the device, the progressive increase in number and size of the fiber detours which curved around the lesion, and the swollen axon terminals found within them at 3–8 days postlesion, were consistent with the interpretation that the detours were produced by a spontaneous regeneration of the severed axons which grew *around,* rather than across, the lesion.

Perhaps the most remarkable feature of this hypothesized regeneration was the invariable finding that the rerouted axons appeared to have selectively regrown toward, eventually to rejoin, the distal portions of their appropriate pathways. Even a tract which had been totally severed could do this. All tested pathways could form detours, and at 3 or more weeks postlesion it was unusual to find that the population of axons nearest to each end of the cut had not done so. Classical signs of regeneration failure, consisting of severed axons facing the lesion and perhaps somewhat withdrawn from it, were also seen but were generally restricted to the central zone of the population which had been severed; the distal pathways which corresponded to those which had failed to regrow would then show a loss of axons. In these instances the immediately adjacent severed regions on each side of the "arrested" zone would show the typical signs of having regrown. Spatial factors seemed to be involved in the success or failure of regrowth to occur, and in particular no intercepted tract has formed a detour around the end of a cut which had extended more than about 400 μm beyond the anatomical boundary of that tract.

Clearly, if these findings do indicate the occurrence of spontaneous regeneration, then they also pose intriguing questions concerning the location, effectiveness, and mode of action of the growth cues that must have evoked it.

3 Return of Function in the Retinocollicular Preparation

3.1 The Experiment

In order to investigate whether these suggestions of a spontaneous regeneration of severed axons in the CNS might be correlated with an eventual return of physiological function, I turned to the retinocollicular system of the adult rat (Diamond et al. 1983; Foerster et al. 1983; Foerster 1985a, b). The pathway to be investigated, the optic tract (OT), is compact, about 1 mm wide, and carries the axons of virtually all contralateral retinal ganglion cells; these terminate in a linear visuotopic pattern in the upper layers of the superior colliculus (SC) (Siminoff et al. 1966; Linden and Perry 1983), which are rather flat with a horizontal area of about 6 mm^2. This projection was mapped in the anesthetized rat by making a systematic set of microelectrode penetrations and determining, for each site, the location of the (small, generally 5 °) area of the contralateral visual field within which a moving stimulus evoked the characteristic burst of impulses from many units (Siminoff et al. 1966). An important feature of the normal map is that there is a sequential representation of the ventral-to-dorsal retina on the medial-to-lateral SC (Siminoff et al. 1966).

My strategy was to map the retinal projection across the SC, make a lesion of the OT, and re-map immediately to measure the loss (if any) of response. Following this, in some instances the animals were allowed to recover and a third (perhaps a fourth, and even a sixth) mapping was done hours to months later, to test for a recovery of visual response. (Rats tolerated the post-surgical remappings very well). In other animals, the anatomical correlates of the initial deficit were obtained by using an immediate intraocular injection of HRP to permit tracing of the retinal projection. In the long-term animals similar injections were done in terminal experiments to obtain the analogous correlates of recovery (or of its failure).

Fortunately, the response of visually driven multiunits in the SC was robust, and was not significantly affected either by the mapping and remapping procedures or by a lesion which, though very close, actually spared the OT. Only a lesion of optic axons produced a "silencing" of visual response in the SC. These studies differed from the earlier purely anatomical ones in that the lesion was made with a razor-ended blade which was lowered vertically through the overlying brain, the OT anterior to the lateral geniculate nuclei, and finally the tough tissue (which included cranial nerves) underlying the OT until it reached the ventral skull, after which it was withdrawn from the brain. The latter procedure meant that this method did not provide a definition at later times of the precise extent of the original lesion, as had the earlier technique. It was adopted specifically to ensure the total and reliable cutting of all the OT which lay beneath the blade. The new feature introduced in these studies to offset this disadvantage was the obtaining of a physiological correlate of the extent of the initial lesion (see below). The size and orientation of the blade was such that if the entire tract were not severed, then it was always either a medial, or a lateral, part of it which was spared. This (predictable) item turns out to be of importance.

3.2 The Immediate Effects of a Lesion of the OT

After lesions of the OT, there was sometimes a silencing of all the recording sites in the SC. Immediate intraocular HRP then has always revealed a total incision of the OT, and loss of the contralateral optic innervation (the appearance of the small ipsilateral projection provided a convenient technical control). When only part of the OT was severed, there was a loss of response in a discrete area (small to near-total) of the SC. This silenced area always included either the medial (after the lateral OT was severed) or lateral (after lesions of the medial OT) boundary of the SC. The still-responsive SC adjacent to a silenced area showed a reduced response, but there was never a patchy diminution or scattered silencing of mapped sites. HRP-filled severed axons faced the rostral part of such partial lesions, and HRP-filled axons in continuity were observed in the spared remnant of the OT, even immediately adjacent to the lesion, indicating that their axoplasmic transport persisted. These spared axons were distributed mainly within the still-responsive area of SC, but a very sparse innervation was also seen in the region of the silenced area immediately bordering the responsive one.

The results of partial lesions thus support the findings of other workers that in the rat there is a medial and lateral segregation in the OT of axons respectively from the dorsal and ventral parts of retina (Yamadori et al. 1975; Bunt and Lund 1982). They also suggest, together with the nature of the short-term recoveries mentioned below, that there may be a somewhat more precise ordering of these axons, perhaps resembling those in the OT of the cat, where more extensive studies have revealed some degree of medial-to-lateral ordering within the OT of axons from the dorsal-to-ventral retina (Torrealba et al. 1982; Mastronarde 1984).

3.3 Return of Visual Function

When examined at post-lesion intervals of 3 weeks to 6 months, a return of visual responsiveness has been observed in the previously silenced SC. The recovered responses are often unusual in many respects. They nearly always involved fewer units recorded at each microelectrode penetration. They sometimes required a stronger stimulus to be evoked; e.g., the normally effective contrast (1 log unit) between the hand-held moving black card stimulus and its backilluminated white background might require a tenfold increase (to 2 log units). When even this stimulus was ineffective, the motion of a 5 ° focused bright light in darkness might evoke a response, but from only one or a few units, and not at all of the recording sites. Additionally, weak responses were often associated with the absence of the characteristic pentobarbital-dependent background of a repetitive (10–20/s) and synchronous firing of units. (This bursting is retina-derived, and disappears upon enucleation).

These long-term recoveries could occur even after a total silencing of the SC by the lesion. They could develop progressively in time in the same animal, as measured both by the area of SC which had recovered and by the vigor of the responses within it. When the responses were faint, they would often fatigue to repeated stimulation. However, when responses were vigorous enough to allow the multiunit receptive field centers to be localized and their boundaries determined, they were either small, or alternatively might subtend as much as 30 °. Interestingly, the topography of a recovered map was sometimes relatively normal, but in other animals there were some bizarre representations of the retina.

There was another, very early, type of recovery which was quite different from the later-developing one just described. It was observed after partial lesions of the OT, and was always detected at recording sites in the sparsely innervated zone of the initially silenced part of the SC, just adjacent to the still-responsive area. The number of sites which showed these particular returned responses usually reached a maximum by 2 hours postlesion. When these rapidly developing recoveries were vigorous enough to be characterized, their quite normal, small, multiunit receptive fields appeared to be activated by their formerly appropriate areas of retina. A likely explanation is that spared axons located near to the cut suffered a reversible pressure-block when the lesion was made. The topographic localization of the recoveries in the SC again suggests that these spared axons were organized somewhat topograpically in the OT.

4 Possible Mechanisms of Long-Term Recovery

The most important feature of the long-term, later appearing, recoveries, which have ocurred even after the total silencing of the SC initially, was that they could encompass both the most dorsal and ventral visual field. The recovered responses were always associated with a direct, anatomically demonstrable, retinocollicular connection from the contralateral eye. This connection, as it passed by the medial or lateral side of the lesion, occupied a small fraction of the normal area of the OT; the rest of the OT faced the site of the lesion. Electron microscope examination of the SC following intraocular HRP has revealed apparently normal optic terminals, some making synapses in areas that also contained profiles of degenerating axons, and occasional unusual retina-derived profiles that may have been growing endings.

Since a direct retinocollicular connection is implicated in the return of function, the question arises as to whether the OT was totally severed by the lesion and, if not, whether there is such a random arrangement of axons in the normal OT that a spared remnant of it could carry information from a set of axons representing the entire visual field. Even an apparently complete physiological silencing immediately postlesion does not in itself constitute *proof* that there had been a total incision of the OT; there is always the possibility that a small remnant (necessarily a medial or lateral one) of the tract was spared, but not detected by the remapping.

The evidence from deliberately made *partial* lesions of the OT indicates that if a remnant of medial or lateral OT were not detected physiologically, it must have been very small indeed. The HRP tracing after partial lesions supports the findings of others (Yamadori et al. 1975; Bunt and Lund 1982; Torrealba et al. 1982; Mastronarde 1984) that such a remnant would carry axons from the dorsal retina or the ventral retina, but not both. Yet the long-term recoveries (in contrast to the short-term ones) clearly involved a restoration of response to both these parts of the retina. At this point, one seems forced to consider the possibility that a spontaneous regeneration of optic axons has originated in severed axons deriving from a widespread population of retinal ganglion cells, which then grew within the degenerating OT to the deafferented SC. Here it would appear that they were able to restore a less effective, but somewhat orderly, physiological representation of the retina.

5 Conclusion

Anatomical studies, in which the location and extent of a lesion in the adult brain are identified by the implantation of the cutting device which made the lesion, have produced evidence consistent with a spontaneous axonal regeneration. The regenerating populations of axons appear to have selected their appropriate pathways for further regrowth, after elongating within the intact tissue around the lesion. The physiological correlate of an initial incision of the OT has now been established by microelectrode mapping of the retinal projection across the SC before and after the lesion is made; the robust character of the responses, and the rat's toleration of repeated mappings, have allowed the measurement of visual recovery in the SC at any time postlesion.

A short-term recovery has been characterized after partial lesions, and is likely to represent a reversal of pressure block. However, a long-term type of recovery with rather different characteristics has also been found. A key item in its interpretation is the evidence that the medial-to-lateral OT contains at least a rough retinotopic ordering of the optic axons, which precludes the possibility that a spared medial, or lateral, remnant of it could be responsible for the observed long-term recovery to stimulation of points representing the *entire* retina. It seems likely that the explanation of this recovery may very well be a spontaneous regeneration of severed optic axons.

References

Berry M, Rees L, Sievers J (1986) Regeneration of axons in the mammalian visual system. Exp Brain Res Suppl 13:18–33

Bunt SM, Lund RD (1982) Optic fiber arrangements in the visual pathways of Long Evans hooded rats. Proc Soc Neurosci 8:451

Diamond J, Foerster A, Visheau B (1983) Optic tract regeneration in the adult rat: anatomy. Proc Soc Neurosci 9:694

Foerster AP (1982) Spontaneous regeneration of cut axons in the adult rat brain. J Comp Neurol 210:335–356

Foerster AP (1985a) Return of function in rat superior colliculus after optic tract lesions. Can J Physiol Pharmacol 63:Axii

Foerster AP (1985b) Superior collicular correlates of partial optic tract lesions in rats. Proc Soc Neurosci 11:1007

Foerster AP, Holmes M, Diamond J (1983) Optic tract regeneration in the adult rat: physiology. Proc Soc Neurosci 9:694

Kromer LF, Björklund A, Stenevi U (1981) Regeneration of the septohippocampal pathways in adult rats is promoted by using embryonic hippocampal implants as bridges. Brain Res 210:173–200

Linden R, Perry VH (1983) Massive retinotectal projection in rats. Brain Res 272:145–149

Marks A (née Foerster AP) (1972) Regenerative reconstruction of a tract in a rat's brain. Exp Neurol 34:455–464
Mastronarde DN (1984) Organization of the cat's tract as assessed by single-axon recordings. J Comp Neurol 227:14–22
Mesulam M-M, Mufson EJ (1980) The rapid anterograde transport of horseradish peroxidase. Neurosci 5:1277–1286
Ramón y Cajal S (1928) Degeneration and regeneration in the nervous system, (May RM, Transl 1959). Hafner, New York
Siminoff R, Schwassman HO, Kruger L (1966) An electrophysiological study of the visual projection to the superior colliculus of the rat. J Comp Neurol 127:435–444
So K-F, Aguayo AJ (1985) Lengthy regrowth of cut axons from ganglion cells after peripheral nerve transplantation into the retina of aculd rats. Brain Res 328:349–354
So K-F, Xiao Y-M, Diao Y-C (1986) Effects on the growth of damaged ganglion cell axons after peripheral nerve transplantation in adult hamsters. Brain Res 377:168–172
Torrealba F, Guillery RW, Eysel U, Polley EH, Mason CA (1982) Studies of retinal representation within the cat's optic tract. J Comp Neurol 211:377–396
Vidal-Sanz M, Bray GM, Villega-Perez MP, Thanos S, Aguayo AJ (1987) Axonal regeneration and synapse formation in the superior colliculus by retinal ganglion cells in the adult rat. J Neurosci 7:2894–2909
Yamadori T, Inamura Y, Minra M (1975) The termination and pattern of localized distribution of the nerve fields in rats. Hirosaki Med J 271:48–71

Lesion-Induced Relative Supersensitivity to an Excitatory Amino Acid Agonist in the Cat Dorsal Lateral Geniculate Nucleus

U. Th. Eysel [1], H.-Chr. Pape [1], T. Salt [2], and A. Sillito [2]

[1] Institute of Physiology, University of Essen, D-4300 Essen, FR Germany
[2] Department of Physiology, University College, Cardiff, United Kingdom

1 Introduction

Chronic monocular deafferentation has been shown to produce an increased excitability in cells of the dorsal lateral geniculate nucleus (dLGN) (Eysel 1979). Furthermore, a recent quantitative morphological study (Somogyi et al. 1987) has shown that chronic monocular deafferentation results in a relative increase in density of the excitatory cortical input to the dLGN neuron. This effect and the changes related to transneural atrophy of the deafferented cells might explain part of the functional changes observed after chronic deafferentation. Nevertheless, denervation supersensitivity might also be involved. Accumulating evidence indicates that the neural transmitter mediating the visual input to dLGN cells acts at an excitatory amino acid receptor (Kemp and Sillito 1982). Natural excitatory amino acids such as L-glutamate and L-aspartate can act at a number of receptors which have been characterized in terms of the specific agonists N-methyl-D-aspartate, quisqualate and kainate (Watkins and Evans 1981). In the present experiments we have addressed the question of possible denervation supersensitivity to excitatory amino acid agonists as a consequence of chronic monocular deafferentation. With the iontophoretic technique the responses of normal and denervated cells in the A-laminae of the dLGN to acetylcholine, N-methyl-D-aspartate and quisqualate were compared. The results indicate that chronic monocular deafferentation increases the relative sensitivity of the denervated cells to quisqualate.

2 Methods

Seven adult cats with body weight between 2.5 and 4.0 kg were used. The experiments were performed in Cardiff (4) and in Essen (3). Three animals had been monocularly deafferented by photocoagulation three months prior to electrophysiological recordings, three cats were acutely deafferented during the experiments and two normal cats served as controls. All electrophysiological experiments were performed under halothane, nitrous oxide and oxygen anaesthesia. The end expiratory CO_2 was set to 3.8%, the mean arterial blood pressure remained above 90 mm of mercury. The animals were artificially respirated after paralysis with 0.3 mg/kg/h d-tubocurarine and 4.0 mg/kg/h gallamine triethio-

H. Flohr (Ed.)
Post-Lesion Neural Plasticity

dide in a glucose and Ringer solution. Atropine sulfate (1%) and phenylephrine hydrochloride (5%) were applied to the eyes for mydriasis and retraction of the nictitating membranes. Clear contact lenses of 0 power were used to prevent dehydration of the corneae.

Recordings were made with five-barrel glass-micropipettes (tip diameter 7–9 μm, resistances 15–35 MΩ). The recording barrel was filled with a solution of 0.5 M sodium acetate for recording and 2% pontamine sky blue for marking of the recording site. Acetylcholine chloride (ACh), N-methyl-D-aspartate (NMDA), quisqualic acid (Quis) and gamma-amino-n-butyric acid (GABA) could be iontophoretically injected from the other barrels with a Neurophore-2 system (Medical Systems Corp., Great Neck, USA). Visual stimuli were presented on a large screen oscilloscope in front of the cat's eyes to subdivide the majority of cells into X and Y types.

Acute and chronic deafferentations were produced by photocoagulation. While recording from a given cell the receptive field area was determined and photocoagulated with high intensity xenon light producing lesions of 5 °–10 ° diameter. The acutely deafferented cells immediately lost their light excitability and most of the maintained activity. Whole geniculate layers were chronically deafferented by photocoagulation of the papilla and the central retinal vessels of one eye. In these animals, penetrations through layers A and A1 allowed direct comparisons of innervated and denervated cells in the same experiment. In addition to excluding inter-individual differences, this also minimized influences introduced by different electrodes. Peristimulus time histograms were computed from single cell activities. To quantify the relative sensitivities of the cells to the different excitatory amino acid agonists we adjusted ejection currents in two ways: firstly, to elicit threshold responses, and secondly, to produce excitatory responses of equal amplitude.

Complete analyses were made from 136 neurons (33 normal controls, 46 from cats with acute deafferentation and 57 from animals with chronic deafferentation).

3 Results

The synaptic connections in the cat lateral geniculate nucleus are quite well understood, and the chemical transmitters mediating the inputs to dLGN cells are partially known (Fig. 1). Gamma-amino butyric acid is the main inhibitory transmitter, and excitation is mediated by acetylcholine (Sillito et al. 1983; Eysel et al. 1986; Pape and Eysel 1988) as well as excitatory amino acids (Kemp and Sillito 1982). It has been suggested that excitatory amino acid transmitters are involved in afferent visual and corticofugal visual transmission, while acetylcholine appears related to inputs from the mesencephalic reticular formation (de Lima and Singer 1987).

Following chronic deafferentation in the adult cat the relay cells undergo transneuronal atrophy but no cell death occurs and all synaptic contacts except for the optic tract terminals remain present (Eysel and Wolfhard 1984; Somogyi et al. 1987; Fig. 1). Under this condition, the responses to electrical stimulation

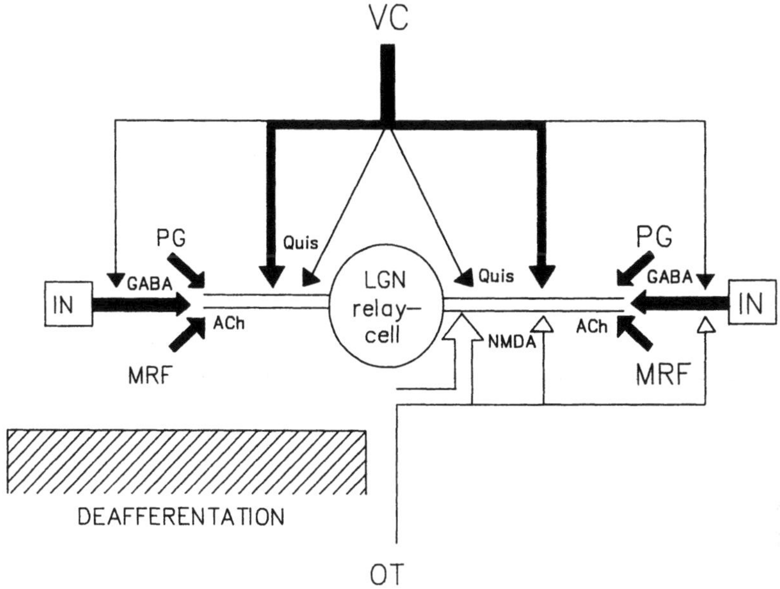

Fig. 1. Schematic drawing of input systems at the innervated (*right half*) and deafferented (*left half*) dLGN relay cell. Optic tract (*OT*) terminals are situated mainly at the dendrites close to the soma, while corticofugal inputs (from visual cortex, *VC*) are more frequently found in the periphery of the dendrites (*thick arrows*). N-methyl-D-aspartate (*NMDA*) is indicated as the excitatory amino acid receptor type most involved in retinogeniculate transmission, while the transmitter of the corticofugal fibers presumably operates mainly at quisqualate (*Quis*) receptors. Acetylcholine (*ACh*) is shown as one of the transmitters of the input systems from the mesencephalic reticular formatin (*MRF*). Intrageniculate interneurons (*IN*) and recurrent inhibitory interneurons from the perigeniculate nucleus (*PG*) mediate GABAergic inhibition. Following chronic loss of the optic input, cells and dendrites show transneuronal atrophy (indicated by the shorter dendritic process) which leads to a relative increase of corticofugal (and presumably also the other) synapses

of optic tract and visual cortex display clear changes in excitability (Fig. 2), indicating stronger inhibition as well as increased excitability by corticofugal inputs (Eysel 1979).

The present study was aimed at investigating possible changes in transmitter sensitivity that might be involved in the above described functional effects of chronic deafferentation at relay cells in the dLGN. Single cells were tested with iontophoretic pulses of acetylcholine (ACh), NMDA, and quisqualate.

Peristimulus time histograms averaged from four trials (Fig. 3) show the relative sensitivities to the different drugs in terms of ejection currents necessary to evoke equal response rates. The normal cell exhibits very similar relative sensitivities to NMDA, quisqualate, and acetylcholine (Fig. 3 A). To quantify the results, we computed ratios of the ejection currents necessary to evoke equal response rates with NMDA and ACh, and quisqualate and ACh. In normal cells the ejection currents to meet this criterion only seldom varied by more than a factor of 2. The mean current ratios for NMDA to ACh and quisqualate to ACh were computed from 33 normal cells from layers A and A1 and are shown in Table 1. There

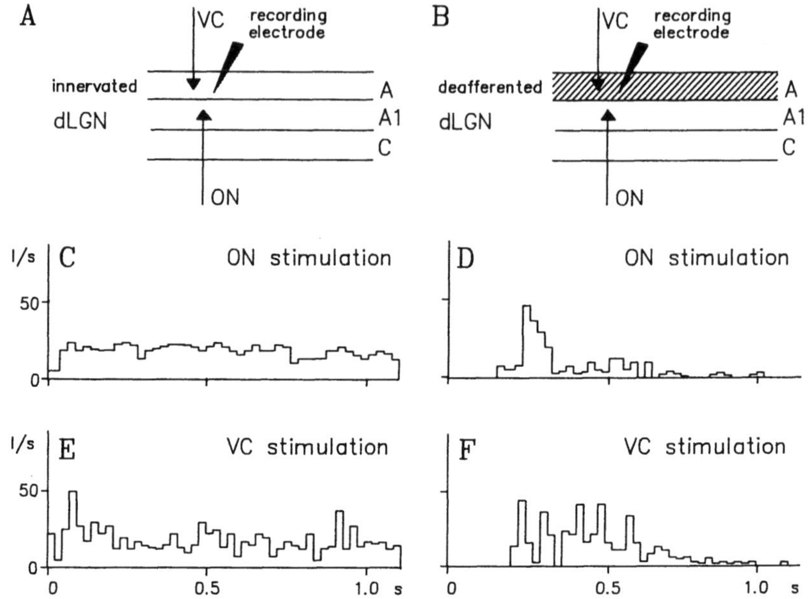

Fig. 2. Responses of electrical stimulation of the ipsilateral optic nerve in normal animals and in cats with chronic contralateral enucleation. **A** Recordings were made in layer *A* of normal animals with electrical stimulation of the ipsilateral optic nerve (innervating layer *A1*) and with stimulation of the visual cortex. **B** The same approach was chosen in cats with chronically deafferented layer A. **C** Peri-stimulus time (PST) histogram averaged from 255 responses to electrical stimulation of the ipsilateral optic nerve (note initial "interocular inhibition"). **D** In cells with chronic deafferentation (10 weeks duration) the inhibition appears increased and followed by an excitatory rebound (averaged from 41 responses). **E** PST histogram obtained from a normal dLGN relay cell with VC stimulation (31 responses averaged) shows the typical initial sequence of inhibition and excitation. **F** The same cell as in D but with electrical stimulation of the VC (42 responses averaged). Note the earlier onset and longer duration of the excitatory period

is no significant difference between the relative sensitivities for both drug pairs between layers A and A1.

Our first step in investigating possible changes introduced by deafferentation was to compare normal cells with cells deprived of their visual inputs by photocoagulation during an acute experiment. Approximately equal numbers of innervated and acutely denervated cells were sampled. The example shown in Fig. 3 B, C demonstrates equal sensitivity in a continuously recorded cell before and after deafferentation. Compared to the control group of normally innervated cells from the same experiment (Table 1), the quantitative statistic of the acutely deafferented group shows results similar to those obtained from the completely normal animals. In the cases of quisqualate versus ACh a first indication of increased relative sensitivity to QUIS was present ($p < 0.01$), while the NMDA to ACh current ratio remained unchanged in deafferented cells.

In the chronically deafferented animals we found a clear difference in the relative sensitivity to NMDA and quisqualate as compared to acetylcholine. An extreme example of a cell with a very high relative sensitivity to quisqualate is shown

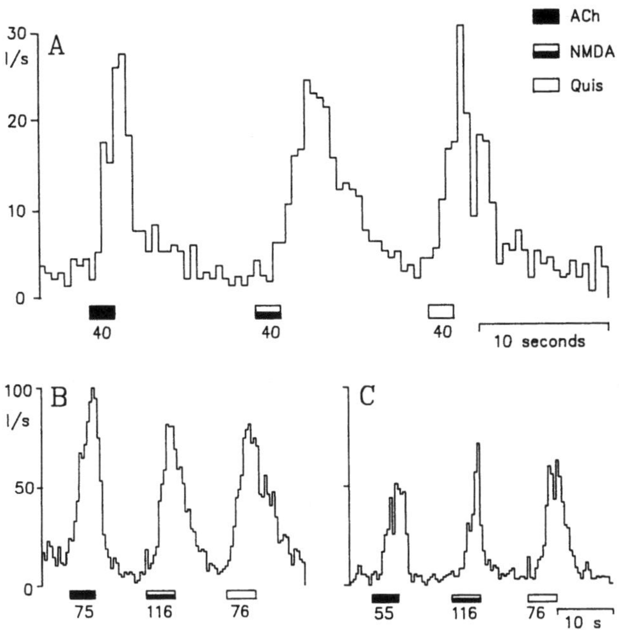

Fig. 3. Relative sensitivities to ACh, NMDA, and Quis in normal cells and after acute deafferentation. **A** Normal dLGN cell tested with 2 s ejection pulses in a control experiment. **B, C** Single cell recorded before (**B**) and after (**C**) acute deafferentation by retinal photocoagulation tested with 5 s ejection pulses. All histograms averaged from four trials. Different drugs are marked by individual symbols with ejection current in nA indicated below each symbol

Table 1. Current ratios for NMDA versus ACh and quisqualate versus ACh were computed for cases where each substance elicited equal peak response rates. For three groups: Normal (layer A), acutely deafferented (DA) and chronically deafferented (DC) controls were taken from the same experiments (layer A1, NA, NC). Mean ± standard deviation and number (n) of cells as well as t-test results are given (significance levels indicated below)

		NMDA/ACh		QUIS/ACh	
Normal	A	1.33±0.49	n=18	1.22±0.84	n=18
	A1	1.43±0.56	n=15	1.35±0.93	n=15
Acute	DA	1.30±0.83	n=22	1.05±0.43	n=22[a]
	NA	1.56±0.74	n=24	1.63±0.88	n=24
Chronic	DC	2.79±2.30	n=17	0.74±0.65	n=24[b]
	NC	3.09±1.91	n=14	2.31±1.64	n=29

[a] $p<0.01$.
[b] $p<0.0001$.

in Fig. 4. Very low injection currents in the range normally necessary for threshold excitation in control cells produced response rates often well above 100 impulses per second. In several cases even lower ejection currents in the range between 2–10 nA elicited strong responses. Differences in relative sensitivity by a factor of more than 10 were found in the comparison between quisqualate and acetylcholine in some cases. Table 1 shows no significant difference for the NMDA to ACh

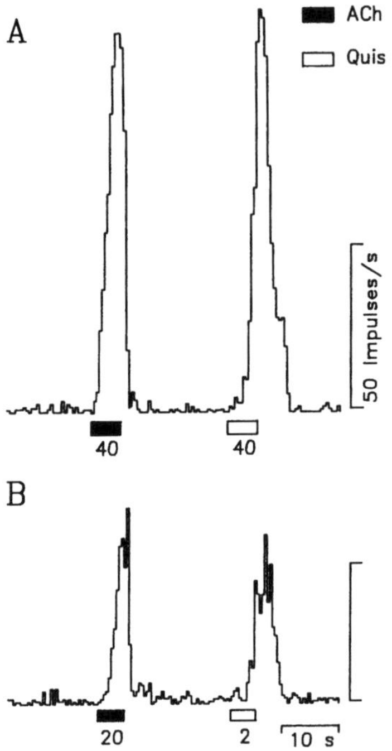

Fig. 4. Relative supersensitivity to quisqualate in a chronically deafferented cell. **A** Four times higher relative sensitivity to Quis as compared to ACh at peak response rates of approximately 125 impulses/s. **B** A response to ACh with 20 nA reaches about 70 impulses/s which is matched by a response to Quis with 2 nA (corresponding to a tenfold sensitivity in terms of necessary currents). All histograms averaged from four trials, 5 s ejection pulses, with currents indicated below each drug symbol as in Fig. 3.

current ratios in 17 chronically deafferented cells as compared to 14 normally innervated cells from the same sample. In marked contrast, the quisqualate to ACh current ratios show a highly significant difference between chronically deafferented and control cells. The mean current ratio, which in all other cases was similar and above one for cells tested with the same electrodes in the same experiment, now shifted from a value larger than one in 29 normal cells to a significantly different value smaller than one ($p < 0.0001$). This means that in chronically deafferented cells there is a much higher relative sensitivity to quisqualate and even a reversal of the relative sensitivities before and after chronic deafferentation.

4 Discussion

For a number of years we have tried to explain the functional changes originally observed after chronic monocular deafferentation in the cat (Eysel 1979). These changes basically indicated an increased efficiency of remaining input systems after loss of visual innervation of dLGN cells. In a recent paper (Somogyi et al. 1987) we were able to show that in the chronically deafferented dLGN there is neither a reduction of GABA-ergic inhibitory interneurons nor an indication of sprouting of remaining inputs such as the corticofugal pathway. The morphological changes observed in this quantitative electron microscope study were: (1) reduction of cell size as a consequence of transneuronal atrophy, and (2) a relative increase in density of the excitatory cortical input per neuron. The following picture emerges: the deafferented cells have lost their visual input as a consequence of which they undergo a reduction in size and thereby increase the synaptic density of remaining inputs which stay equal in absolute numbers. This was shown for inhibitory interneurons as well as for corticofugal synapses (Somogyi et al. 1987).

The present results suggest that there might be, in addition, a relative supersensitivity to quisqualate, one of the excitatory amino acid agonists. Denervation supersensitivity has been functionally observed in muscle (Cannon and Haimovici 1939), in the spinal cord (McCough et al. 1958), and in the cuneate nucleus (Kjerulf and Loeser 1973). Several authors observed increases in sensitivity to different excitatory transmitters (Bird and Aghajanian 1975; McCall and Aghajanian 1979; Lamour et al. 1982; Menkes et al. 1983). According to recent observations the NMDA receptor seems to be related to visual excitation of dLGN cells (Kemp and Sillito 1982). Excitatory amino acids seem to be involved in the corticofugal projection from layer 6 (Baughman and Gilbert 1981), and the quisqualate receptor might be activated by the corticofugal transmitter.

The anatomically observed effects of transneuronal atrophy might contribute to an overall increase of excitability due to higher density of synapses per dendrite and higher input resistances at the peripheral dendrites (leading to higher synaptic voltages in response to equal synaptic currents), while the observations made in this study require a specific shift in the influence of one group of excitatory amino acid receptors.

According to our anatomical observations, the reduction in dendritic volume implies a higher density of corticofugal terminals which at the same time come closer to the cell body. Thus, if the quisqualate receptors are located in the subsynaptic density of these connections at the dLGN cells, after chronic deafferentation they might be present in greater density and closer to the cell body than before.

Since corticofugal terminals are predominantly distributed at the peripheral dendrites while optic synapses are mainly found at the proximal dendrites (Guillery 1971), shunting effects exerted by the optic synapses will be reduced. In fact, the trend to a higher relative sensitivity of the quisqualate receptor already in acutely deafferented cells might indicate such reduced shunting. An additional mechanism seems to act at chronically deafferented cells which increases the sensitivity to quisqualate and leads to the impressive absolute changes and even to the reversal of the relative sensitivities tested in this study. After deafferentation, the NMDA receptors related to the optic inputs appear to remain unchanged (or reduced in parallel to the ACh sensitivity), while the quisqualate receptors might come into the zone of microiontophoretic influence due to dendritic shrinkage.

More information about the specific distributions of NMDA and QUIS receptors on the surface of dLGN cells is necessary to resolve the open questions and confirm the speculations made in this report. The present results, however, indicate that the monocularly deafferented dLGN of the cat might be among those central nervous structures that develop denervation supersensitivity in response to the loss of excitatory inputs.

Acknowledgements. This study was supported by the Deutsche Forschungsgemeinschaft (SFB 200/A4) and the Ministerium für Wissenschaft und Forschung des Landes Nordrhein-Westfalen.

References

Baughman RW, Gilbert CD (1981) Aspartate and glutamate as possible neurotransmitters in the visual cortex. J Neurosci 1:427–439

Bird SJ, Aghajanian GJ (1975) Denervation supersensitivity in the cholinergic septohippocampal pathway: a microiontophoretic study. Brain Res 100:355–370

Cannon WB, Haimovici H (1939) The sensitization of motoneurones by partial "denervation". Am J Physiol 126:731–740

Eysel UT (1979) Maintained activity, excitation and inhibition of lateral geniculate neurons after monocular deafferentation in the adult cat. Brain Res 166:259–271

Eysel UT, Pape H (1987) Lateral excitation in the cat lateral geniculate nucleus. Exp Brain Res 67:291–298

Eysel UT, Wolfhard U (1984) The effects of partial retinal lesions on activity and size of cells in the dorsal lateral geniculate nucleus. J Comp Neurol 229:301–309

Eysel UT, Pape H-C, Schayck R van (1986) Excitatory and differential disinhibitory actions of acetylcholine in the lateral geniculate nucleus of the cat. J Physiol 370:233–254

Guillery RW (1971) Patterns of synaptic interconnections in the dorsal lateral geniculate nucleus of the cat and monkey: a brief review. Vision Res Suppl 3:211–227

Kemp JA, Sillito AM (1982) The nature of the excitatory transmitter mediating X and Y cell inputs to the cat dorsal lateral geniculate nucleus. J Physiol 323:377–391

Kjerulf TD, Loeser JD (1973) Neuronal hyperactivity following deafferentation of the lateral cuneate nucleus. Exp Neurol 39:70–85

Lamour Y, Dutar P, Jobert A (1982) Excitatory effect of acetylcholine on different types of neurones in the first somatosensory neocortex of the rat: laminar distribution and pharmacological characteristics. Neuroscience 10:107–117

Lima AD de, Singer W (1987) The brain stem projection to the lateral geniculate nucleus in the cat. Identification of cholinergic and monaminergic terminals. J Comp Neurol 259:92–121

McCall RB, Aghajanian GK (1979) Denervation supersensitivity to serotonin in the facial nucleus. Neuroscience 4:1501–1510

McCough GP, Austin GM, Liu CN, Liu CY (1958) Sprouting as a cause of spasticity. J Neurophysiol 21:205–216

Menkes DB, Gallagher DW, Reinhard JF, Aghajanian GK (1983) α_1-Adrenoceptor supersensitivity in brain: physiological and receptor binding studies. Brain Res 272:1–12

Pape H-C, Eysel UT (1988) Cholinergic excitation and inhibition in the visual thalamus of the cat – influences of cortical inactivation and barbiturate anesthesia. Brain Res 440:79–86

Salt TE (1986) Mediation of thalamic sensory input by both NMDA receptors and non-NMDA receptors. Nature 322:263–265

Sillito AM, Kemp JA, Berardi N (1983) The cholinergic influence on the function of the cat dorsal lateral geniculate nucleus (dLGN) Brain Res 280:299–307

Somogyi J, Eysel UT, Hamori J (1987) A quantitative study of morphological reorganization following chronic optic deafferentation in the adult cat dorsal lateral geniculate nucleus. J Comp Neurol 255:341–350

Watkins JC, Evans RH (1981) Excitatory amino acid transmitters. Annu Rev Pharmacol 21:165–204

Path- and Homefinding of Regenerating Retinal Axons in Goldfish

C. A. O. Stuermer

Friedrich-Miescher-Laboratorium der Max-Planck-Gesellschaft,
Spemannstr. 37–39, D-7400 Tübingen, FR Germany

1 Introduction

The reformation of the retinotopic map by regenerating goldfish retinal axons has been well documented with behavioral and electrophysiological techniques (reviews: by Jacobson 1978; Gaze 1978). To account for the reliability with which regenerating axons return to their retinotopic target sites in tectum, Sperry (1963) proposed that "cells and fibers must carry some kind of individual identification tags, presumably cytochemical in nature, by which they are distinguished one from another almost to the level of the single neuron". In their classical anatomical paper, Attardi and Sperry (1963) conveyed the view that regenerating axons follow their old pathways and proceed directly to their prespecified loci on tectum by 4 weeks after nerve section.

Inconsistent with their claims are more recent data, which provide evidence that regenerating axons do not recapture their normal paths (Cook 1983; Stuermer and Easter 1984 a). Anatomical mapping experiments showed further that the tectal projection is initially diffuse and lacks its normal retinotopic organization for up to 3 months (Meyer et al. 1985; Rankin and Cook 1986).

These findings led to support the hypothesis that regenerating axons might lack any goal-directedness (Rankin and Cook 1986) and course instead in random routes through tectum. Lack of goal-directedness might imply either that regenerating axons are predestined for specific target sites but lack global information in which direction the target might be located. Or alternatively, the tectum does not provide any cues and activity mediated axon-axon and axon-tectum interactions mediate the reassociation with retinotopically related neighbors at the target sites.

One approach to clarify which mechanisms govern axonal navigation is to observe the axonal trajectories.

To decide whether regenerating axons course through tectum at random or whether they grow in target-directed routes, we visualized the axonal pathways. We examined tecta at early regeneration stages, that is at stages when the retinotectal projection is still diffuse (Meyer et al. 1985; Rankin and Cook 1986), as well as at late stages when the projection has become refined to near-normal retinotopic precision (Meyer et al. 1985; Rankin and Cook 1986). Further we compared the paths of axons with normal impulse activity and axons whose impulse activity was blocked by intraocular injections of TTX.

H. Flohr (Ed.)
Post-Lesion Neural Plasticity

In fish of survival periods between 10 to 30 days and at 3, 9, and 12 months after optic nerve section, small groups of regenerating axons were labeled with horseradish peroxidase (HRP) by intraretinal tracer applications to either the dorsotemporal or dorsonasal retina. The axonal paths were pursued in DAB reacted tectal whole mounts (Stuermer 1988) and recorded in camera lucida tracings.

2 Preference of Regenerating Retinal Axons for Retinotopically Appropriate Tectal Regions and the Achievement of Target-Directed Routes

To illustrate the difference between the paths of normal and regenerating axons, the routes of dorsotemporal axons in normal and in regenerating fish are demonstrated in Figs. 1 and 2. In normal fish all dorsotemporal axons course through tectum in orderly routes toward their retinotopic target sites in the rostroventral hemitectum (Stuermer and Easter 1984b; Easter and Stuermer 1984; Stuermer 1984).

Regenerating dorsotemporal axons, illustrated in a fish at 9 months after optic nerve section (Fig. 2), terminate at retinotopic sites. However, they cross in abnormal routes through both the dorsal and ventral hemitectum, as well as through the inappropriate caudal tectal half.

Despite these pathway aberrations, the dorsotemporal axons course preferentially through their appropriate rostral tectum, whereas nasal axons proceed

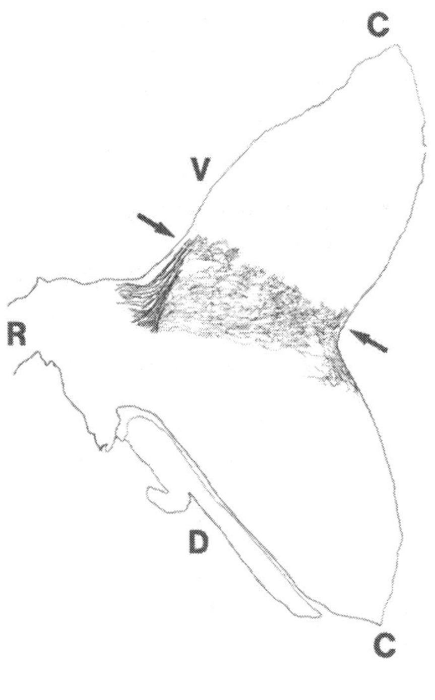

Fig. 1. Tectal whole mount with labeled dorsotemporal axons in a normal fish. Axons run in orderly routes through the ventrorostral tectum toward their termination sites (*arrows*) *R* rostral; *C* caudal; *D* dorsal; *V* ventral

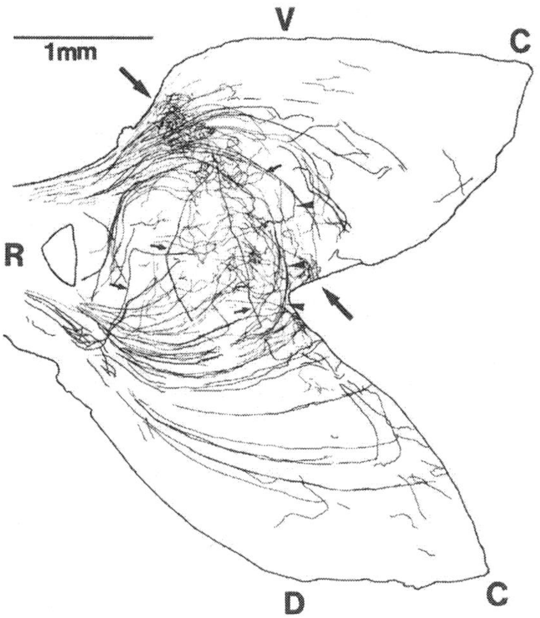

Fig. 2. Tectal whole mount with regenerated dorsotemporal axons, 9 months after optic nerve section. Axons travel through both the dorsal and ventral hemitectum towards their retinotopic target sites (*large arrows*). Misrouted axons correct their course abruptly (*small arrows*) or more gradually (*arrowheads*). Abbreviations *R, C, D, V* as in Fig. 1

into their appropriate caudal tectum. This differential distribution of temporal and nasal axons was quantified by counting the labeled axons at three representative sites in the dorsal and ventral hemitectum, at rostral, midtectal and caudal levels, in 12 fish. The graph in Fig. 3 shows that 43% and 23% of the dorsotemporal axons are represented in the ventro- and dorsorostral tectum respectively, 12% and 19% in midtectum and only 1% and 3% in caudal tectum. In contrast, dorsonasal axons pass in equal numbers through rostral (20% and 22%) and midtectal regions (24% and 19%) and decline to 7% and 8% in regions caudal to their retinotopic sites (Stuermer 1988).

In addition to their affinity for the appropriate rostral or caudal hemitectum, respectively, the regenerating axons also exhibit a specificity for their appropriate target sites. For example, dorsotemporal axons which erred into inappropriate dorsal and caudal tectal regions undergo course corrections and redirect their routes toward their target sites. Dorsotemporal axons misrouted in the dorsal hemitectum cease to grow caudally, bend and cross the tectal equator, and course toward their termination region in the rostroventral tectum. Axons which had passed beyond their target region bend and return into the direction of the target (Fig. 2). Comparable course corrections are established by misrouted dorsonasal axons (Stuermer 1988).

Thus, the paths of regenerating axons through tectum are far from being random; instead, axons reveal a preference for appropriate tectal territories and establish target oriented routes, i.e., they exhibit a high degree of goal-directedness.

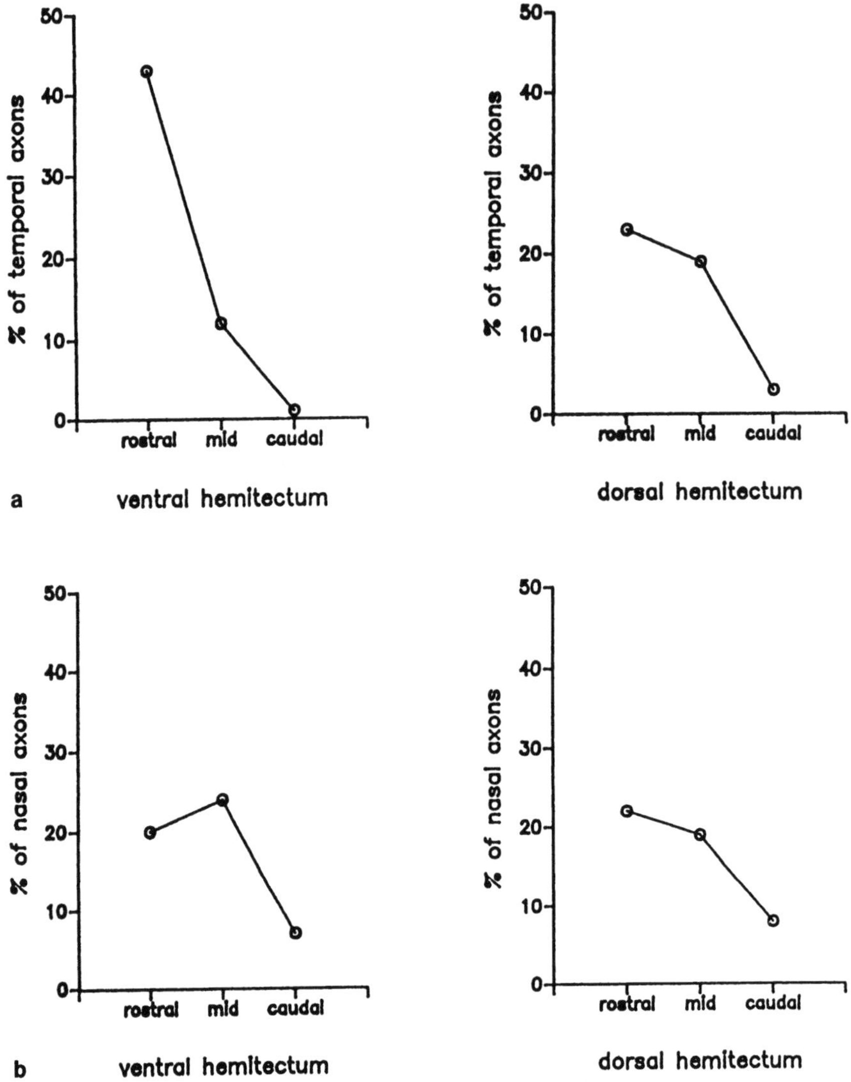

Fig. 3. Relative distribution of regenerating dorsotemporal (**a**) and dorsonasal (**b**) axons over the ventral and dorsal hemitectum

3 Exploratory Growth of Regenerating Axons in Early Regeneration Periods

To elucidate the strategies that are employed by the regenerating axons to gain access to their target sites, we visualized the paths of regenerating axons and their morphology at early regeneration stages between 10 and 30 days after optic nerve section (i.e. at times when the projection is still diffuse).

Most axons have not reached their target sites by 4 weeks, but have their growing endings in various regions in tectum. The growing axons show a variety of growth related morphological characteristics (Fig. 4). The simplest morphol-

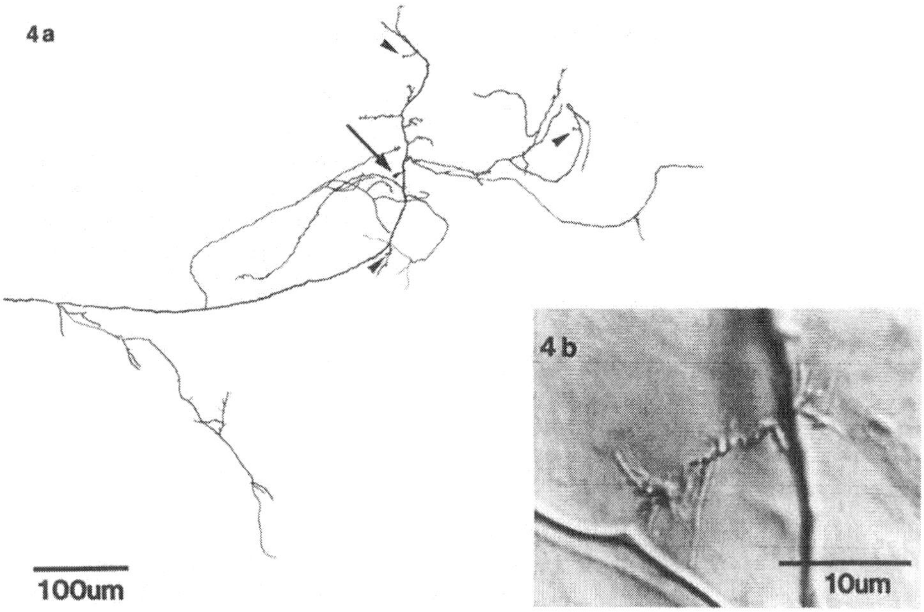

Fig. 4. a Ramified regenerating axon at 4 weeks after optic nerve section at an ectopic site in tectum. The long branches carry growth cones (*arrowheads*) and filopodia. **b** Photomicrograph of one of the growth cones in a (*arrow*)

ogy have all axons in the superficial fascicle or fiber layer (SO = Stratum opticum) of tectum and a few axons in the synaptic layer SFGS (Stratum fibrosum et griseum superficiale). They are unbranched and tipped with a leading growth cone. Such axons begin to or have already achieved target directed orientations. Other axons in SFGS are more complex. They are equipped with a large number of side branches, located at more than 800 µm of the axon's extent, and numerous (up to 15) growth cones on their branches and the axonal shaft. Yet another group residing in foreign tectal territories emit long side branches of up to 500 µm in length into various directions. Each one of these side branches carries one to several growth cones and filopodia (Fig. 4) (Stuermer 1988).

Nearly all of the branches are transient extensions that disappear when the axons have arrived at their target sites. The branches appear to be sent out like spies to explore the tectal surface and to seek out the direction of the target. Most likely, those branches that have target-directed orientations are preferentially elongated, whereas ectopically oriented ones are withdrawn or eliminated.

Thus the growing axons have either a simple morphology and establish target-oriented routes directly, or they have numerous sidebranches and go through a phase of exploratory growth, which is followed by a preferential selection of appropriately oriented branches (Fujisawa et al. 1982).

These findings lead to the conclusion that axons do not course at random through tectum. Instead, the exploration of tectal territories and the achievement of target-directed routes strongly suggests that regenerating axons are able to read directional cues even when they are far away from the target.

4 Path- and Homefinding of Regenerating TTX-Silenced Axons

Regenerating axons that were deprived of their normal impulse activity by repeated intraocular injections of tetrodotoxin (TTX), exhibited preferences for their appropriate tectal halves, underwent course corrections and had their terminal arbors clustered at retinotopic regions similar to active axons from control eyes (Hartlieb and Stuermer 1987). Thus activity-mediated interactions of axons and between axons and tectum are not necessary for successful path- and homefinding.

5 Target-Directed Navigation of Regenerating Axons in Long-Term Denervated Tecta

Earlier publications proposed that guidance cues could be provided by axonal debris or transient cues left behind by the old axons (Murray 1976; Schmidt 1978). Whether path- and homefinding relies on the presence of such debris or transient cues was tested by pursuing the growth of axons into tecta that had been deprived of retinal afferents for 1–12 months prior to the axons' arrival (Busse and Stuermer 1987).

The paths of the axons regenerating into these tecta, denervated for 1–12 months, showed no signs of greater disorder. Instead, axons had an even stronger preference for their appropriate hemitectum than axons in tecta which had not been deprived of retinal afferents for extended time periods. Course corrections and target directed orientations of misrouted axons occurred in similar ways as in nondeprived tecta (Busse and Stuermer 1987).

Thus, the preference of axons for appropriate tectal regions and the goal-directed growth of regenerating retinal axons does not depend on the presence of axonal debris or transient cues. This suggests that long-lasting, tectum-inherent markers serve as guidance cues.

6 Axonal Preference for Target-Derived Membranes in Vitro

In an elegant in vitro assay, Bonhoeffer and his colleagues (Walter et al. 1987) discovered that tectal positional markers are membrane associated components that are recognized by embryonic retinal axons of the chick. We (Vielmetter and Stuermer unpublished) employed their in vitro assay to test the preference of goldfish regenerating retinal axons in vitro. Membranes from rostral and caudal tectum were prepared according to Walter et al. (1987) and attached to a filter in alternating stripes of 100 μm width. Segments from temporal or nasal retina were then explanted onto the filter perpendicular to the membrane stripes.

Similarly to chick embryonic axons, the regenerating goldfish axons from temporal retina accumulate preferentially on membranes of their retinotopically appropriate rostral tectum (Fig. 5). Nasal axons of both, chick (Walter et al. 1987) and fish, do not exhibit a preference but instead grow randomly over both membranes. It is beyond the scope of the present paper to discuss the different behav-

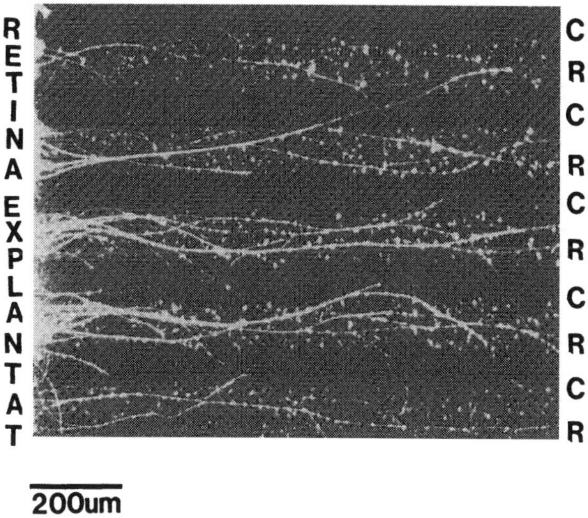

Fig. 5. Photomicrograph of regenerating temporal retinal axons in vitro, growing on two types of tectal membranes, rostral (*R*) and caudal (*C*) in alternating stripes. Temporal axons prefer rostral over caudal membranes

ior of the nasal axons (see Walter et al. 1987). However, the selectivity of temporal axons for membranes of their appropriate target region strengthens the point that the surfaces of tectal cells carry the markers that are recognized by the growing axons (Walter et al. 1987).

7 Conclusion

Our findings that regenerating retinal axons return to their retinotopic target sites in tectum confirm Sperry's (Sperry 1963) statement. However, the mode in which axons reach their targets is inconsistent with Attardi and Sperry's claims (Attardi and Sperry 1963) and the rigid chemospecificity concept (Sperry 1963).

Regenerating axons travel in abnormal but nonrandom routes over the surface of tectum. They exhibit a strong preference for their appropriate tectal regions and achieve from any ectopic site in tectum target-oriented routes. Target-directed routes can be established by the axons either immediately upon their ingrowth into tectum, or after extension of exploratory sidebranches followed by a preferential selection of appropriately oriented processes. Routes of axons deprived of their normal impulse activity are similar to those of axons with normal activity. These findings foster the view that regenerating axons navigate under the influence of tectal markers which give directional cues to the growing axons over considerable distances.

Axonal debris or cues left behind by the old projection are irrelevant for the path- and homefinding of regenerating axons.

Tectal positional markers are most probably (at least in rostral tectum) associated with tectal cell surfaces.

The mode in which axons course through tectum let us to evaluate how such positional markers should be distributed over tectum. Since axons from any ectopic site in tectum can "perceive" their actual position, reorient themselves, and achieve target directed routes, the directing cues must be accessible to axons even when they are far away from their target. This rules out the idea that positional markers are localized to specific and discrete regions. However, the observed axonal behavior is in agreement with the gradient theory (Gierer 1981; Bonhoeffer and Gierer 1984). This theory implies that marker molecules (and two different molecules suffice for each direction) have a graded distribution over the rostrocaudal and dorso-ventral axis of tectum. In such a coordinate system of graded markers the growth cone can perceive its actual position and the direction of the target.

References

Attardi DG, Sperry RW (1963) Preferential selection of central pathways by regenerating optic fibers. Exp Neurol 7:46–64
Bonhoeffer F, Gierer A (1984) How do retinal axons find their targets on tectum? TINS, pp 378-381
Busse U, Stuermer CAO (1987) Navigation and target recognition of regenerating retinal axons in long-term denervated tecta in goldfish. Neurosci Abstr 13
Cook JE (1983) Tectal paths of regenerated optic axons in the goldfish: evidence from retrograde labelling with horseradish peroxidase. Exp Brain Res 51:433–442
Cook JE, Horder TJ (1977) The multiple factors determining retinotopic order in the growth of optic fibres into the optic tectum. Philos Trans R Soc Lond B Biol Sci 278:261–276
Cook JE, Rankin ECC (1986) Impaired refinement of the regenerated retinotectal projection of the goldfish in stroboscopic light: a quantitative study. Exp Brain Res 63:421–430
Easter SS, Stuermer CAO (1984) An evaluation of the hypothesis of shifting terminals in goldfish optic tectum. J Neurosci 4:1052–1063
Fujisawa H, Tani N, Watanabe K, Ibata Y (1982) Branching of regenerating retinal axons and preferential selection of appropriate branches for specific neuronal connection in the newt. Dev Biol 90:43–57
Gaze RM (1970) The formation of nerve connections. Academic Press, London
Gierer A (1981) Development of projections between areas of the nervous system. Biol Cybern 42:69–78
Hartlieb E, Stuermer CAO (1987) Path and target finding of retinal axons under the influence of TTX in the retinotectal system of goldfish. New Frontiers in Brain Res. Thieme, Stuttgart New York
Jacobson M (1978) Developmental neurobiology. Plenum, New York
Meyer RL, Sakurai K, Schauwecker E (1985) Topography of regenerating optic fibers in goldfish traced with local wheat germ injections into retina: evidence for discontinious microtopography in the retinotectal projection. J Comp Neurol 239:27–43
Murray M (1976) Regeneration of retinal axons in the goldfish optic tectum. J Comp Neurol 168:175–196
Rankin ECC, Cook JE (1986) Topographic refinement of the regenerating retinotectal projection of the goldfish in standard laboratory conditions. A quantitative WGA-HRP study. Exp Brain Res 63:409–420
Schmidt JT (1978) Retinal fibers alter tectal positional markers during the expansion of the half retinal projection in goldfish. J Comp Neurol 177:279–300
Sperry RW (1963) Chemoaffinity in the orderly growth of nerve fiber patterns and connections. Proc Natl Acad Sci USA 50:703–709
Stuermer CAO (1984) Rules for retinotectal terminal arborizations in the goldfish optic tectum. A whole-mount study. J Comp Neurol 229:214–232
Stuermer CAO (1988) The trajectories of regenerating retinal axons in the goldfish tectum I. J Comp Neurol 267:55–68
Stuermer CAO (1988) The trajectories of regenerating retinal axons in the goldfish tectum II. J Comp Neurol 267:69–91

Stuermer CAO, Easter SS (1984a) A comparison of the normal and regenerated retinotectal pathways of goldfish. J Comp Neurol 223:57–76
Stuermer CAO, Easter SS (1984b) Rules of order in the retinotectal fascicles of goldfish. J Neurosci 4:1045–1051
Walter J, Kern-Veits B, Huf J, Stolze B, Bonhoeffer F (1987) Recognition of position-specific properties of tectal cell membranes by retinal axons invitro. Development 101:685–696

Regeneration of the Retinotectal Projection in Goldfish: Selective Stabilization of Retinotopic Synapses by Correlated Activity

J.Schmidt

Department of Biological Sciences, State University of New York at Albany, 1400 Washington Avenue, Albany, N.Y. 12222, USA

1 Introduction

In most studies of postlesion neural plasticity, the question centers around the rearrangement of connections of the remaining unlesioned neurons. However since the retinotectal projection of fish and frogs readily regenerates, the question centers around the mechanisms by which the regenerating axons select appropriate termination sites to reestablish the precise retinotopic map on the tectum.

The mechanisms thought to be involved can be divided into three categories: (1) chemoaffinity mechanisms for selection of pathways and target regions (Sperry 1963; Bonhoeffer and Huf 1982), (2) Competition between fibers for terminal space (Schmidt 1982; Murray et al. 1982) and (3) an activity dependent sharpening of the crudely retinotopic initial projection (Willshaw and von der Malsburg 1976; Meyer 1983; Schmidt and Edwards 1983). The first two have been thoroughly reviewed several times (Easter 1985; Fraser 1985; Schmidt 1982, 1985). It should suffice to say that the chemoaffinity mechanisms have limited resolution as demonstrated firstly by numerous competition experiments that lead to systematic retinotopic reorganizations (half retinal removal leads to expansion of the remaining half projection over the tectum and half tectal removal leads to a compression of the full projection onto the remaining half) and secondly by the natural experiment of continued growth of retina and tectum in geometrically dissimilar patterns that causes a continuous shifting of the retinotopic map. In addition, the results reviewed herein show that the initial retinotopic map regenerated to the tectum is only crudely retinotopic and remains so when activity is either blocked or disrupted from its normal pattern.

2 Crude Retinotopy During Early Regeneration

2.1 Anatomical Studies

Meyer (1980) used radioautography combined with acute partial retinal lesions to trace the projection from the remaining retina. Although very small retinal lesions normally produce a corresponding denervated zone in tectum indicating precise retinotopy, at least half of the retina had to be removed to create a denervated zone during the first 2 months after the fibers grew back to the tectum. Both Stuermer and Easter (1984) and Rankin and Cook (1986) assessed retinotopy via

H. Flohr (Ed.)
Post-Lesion Neural Plasticity

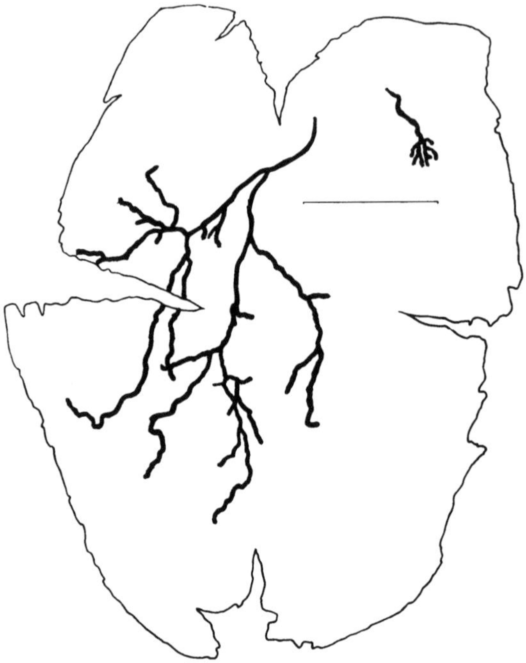

Fig. 1. Camera lucida drawing of a regenerated optic arbor, stained with HRP at 21 days after nerve crush, showing the widespread branching. The *outside lines* mark the edge of the tectum which was split to allow it to lie flat. A normal optic arbor is shown for comparison. Bar = 1 mm

retrograde transport of HRP from punctate injections made into the tectum. Normally the labeled ganglion cells are confined to a tight cluster at the retinotopic site. For several months after nerve crush, however, the labeled cells were scattered over large areas of the retina. The cluster sharpened considerably by 70 days, but was still larger than in normals.

The diffuse projections could result either because the arbors make errors in targeting or because the arbors are much larger than normal although centered on the right spot. This question has been addressed by anterograde HRP staining of individual optic axons in whole mounts of the tectum where the entire arbor can be seen. In the newt, Fujisawa et al. (1982) reported that the regenerating fibers often took aberrant paths into and through the tectum and made many branches in inappropriate places along the way. Schmidt et al. (1988) used this technique in goldfish and quantified the extents of the regenerating arbors. Whereas normal arbors ranged in size from about 100 to 400 μm, the arbors at 2 to 3 weeks were on average 5 times larger, the largest being more than 2 mm across (Fig. 1). At 4 to 5 weeks the average size dropped markedly towards normal although a few large arbors were seen for several more weeks. (See also Stuermer, this volume) In similar studies of development (*Xenopus:* Sakaguchi and Murphey 1985; Fujisawa 1987; mammals: Schneider et al. 1981; Sachs et al. 1986), the trajectory of the axons tends to be more stereotyped, as they achieve the correct mediolateral level upon entry, but give off extra branches along the way as they travel rostrocaudally within the tectum. Thus, there may be an elimination of excess branches during development as well.

2.2 *Electrophysiological Studies*

In the frog, the sharpening of the map can be followed by electrophysiological mapping (Adamson et al. 1984; Humphrey and Beazley 1982). From 3 to 4 weeks postcrush, recording at each tectal point yielded many units (arbors of retinal ganglion cells) responding to stimulation of a wide area of the visual field. This area is called the multiunit receptive field and its large size indicates a crude level of retinotopic organization. Over the next 3 to 4 weeks, these large responsive areas shrunk to the normal size. Adamson et al. (1984) also used the intertectal relay to assess the postsynaptic effects of the crude retinotopic map. This relay from tectum to N. Isthmi to the opposite tectum provides an indirect ipsilateral retinal projection in addition to the direct contralateral one. Early in regeneration, the receptive fields of single relay fibers in the normal tectum were enlarged to the same degree as the multiunit receptive fields recorded from the opposite tectum. Thus, the misdirected arbors and branches of arbors within the crude map establish effective synaptic connections.

In the goldfish, electrophysiological recordings during the early diffuse phase are difficult because the responses are very small and fatigue extremely rapidly. The earliest recordings at 34 days postcrush show maps that are already normal in organization and size of multiunit receptive fields. This suggests that the electrodes select for the small arbors with their greater density of processes over the large sparsely branched arbors. The small ones are abundant by this time although the large ones are still occasionally present. This selectivity for small arbors may help to explain why the sharpening is seen earlier with recording than with retrograde labeling, even though both techniques theoretically look at the distribution of ganglion cells that project to a given point in tectum. On the other hand, the retrograde labeling techniques may underestimate the sharpness of the map due to labeling through axons of passage, which are increased due to the tortuous trajectories of regenerated axons, or through very small ectopic branches that constitute a tiny fraction of the arbor.

3 Evidence that Activity Sharpens Retinotopy

3.1 *Intraocular Tetrodotoxin Studies*

In adult goldfish, both anatomical and electrophysiological studies have demonstrated that the sharpening of the initially diffuse regenerated projection requires activity. Meyer (1983) used radioautographic tracing and retinal lesions to show that the regenerated projection does not sharpen but remains diffuse when activity is blocked with intraocular injections of tetrodotoxin (TTX). Blocking from 32 to 80 days postcrush prevented the return to normal retinotopy. However, the map could still sharpen if the fish were allowed a further 24 days with activity after the TTX was discontinued, indicating that the block of sharpening was readily reversible in these small fish.

Schmidt and Edwards (1983) used electrophysiological recording to assess the sharpening of retinotectal maps in large 12-cm goldfish. In control fish, the map is sharp by 35 days postcrush with normal 11 to 12° multiunit receptive fields.

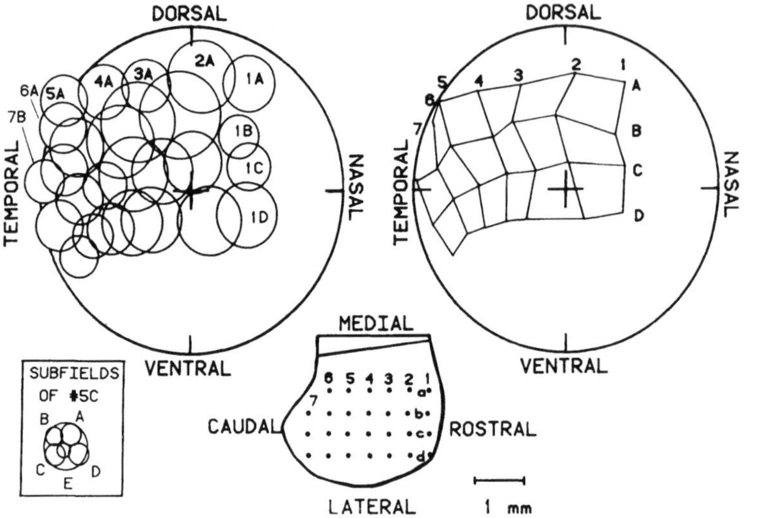

Fig. 2. Retinotectal map recorded 63 days postcrush in a fish blocked intraocularly with TTX from 0 to 28 days. The *two large circles* are two representations of the visual field that contain the outlines of the receptive fields (*left*) and the positions of their centers (*right*). Below, electrode penetrations are marked on the drawing of the tectal surface. The receptive fields recorded at this array of tectal points fall into an orderly array in the visual field, and are correspondingly numbered. The *inset* at the lower left diagrams the relationship between a multiunit receptive field and several of its single unit components mapped separately with a spike height discriminator. (Schmidt and Edwards 1983)

In projections that were blocked for the first 28 days, however, the multiunit receptive fields were substantially enlarged, averaging around 30 ° (Fig. 2). The centers of these enlarged fields were in the retinotopically appropriate region of the visual field, indicating that the gross organization of the map was correct. The enlarged multiunit fields reflect the convergence onto each tectal point of retinal arbors from a wide area of retina. Several lines of evidence support this interpretation. First, the units persist following the application of alpha Bungarotoxin (which acts presynaptically to decrease retinotectal synaptic transmission) thereby demonstrating the presynaptic origin of these recordings. Second, the receptive fields of single ganglion cells, either recorded from the retina or isolated from the tectal recordings by a spike height discriminator, were found to be normal in size and in other characteristics. Therefore the enlarged multiunit receptive fields appear to reflect errors in targeting of the regenerated arbors, errors that are usually corrected when the fibers have normal activity. This interpretation is consistent with the finding of many enlarged arbors early in regeneration, and with the presence of enlarged multiunit receptive fields in the frog as a normal phase in regeneration. If the lack of activity interferes with the elimination of nonretinotopic branches from these initial arbors, we should expect to see abnormal morphologies of optic arbors regenerated during a block of activity. Preliminary results using anterograde HRP staining show that these arbors fail to concentrate their branches, which remain scattered over an area much larger than normal, and this undoubtedly contributes to the unsharpened maps recorded electrophysiologically.

Although the above studies show that neural activity is necessary for refining the pattern of retinotopic connections, they do not distinguish between a permissive role and an instructive role for activity. In the former case, the lack of activity may merely make the neuronal metabolism abnormal so that it is unable to form an exact set of connections. In the latter case, there must be information encoded within the pattern of activity. Several studies (Arnett 1978; Ginsberg et al. 1984) have shown that there is correlated activity in the retina between neighboring ganglion cells of the same type, both under dark and light conditions. In addition, computer models show how this information could be used to establish retinotopic maps (Willshaw and von der Malsburg 1976). The strobe rearing experiments discussed below demonstrate an instructive role for activity in sharpening.

3.2 Stroboscopic Rearing Studies

The instructive model predicts that the sharpening would still be disrupted even if the ganglion cells were allowed activity, as long as the strictly local correlations in firing of neighboring ganglion cells are disrupted. The strobe light causes correlated firing in all ganglion cells: both "on" and "off" cells respond to the strobe by firing one or two spikes at a constant latency (Schmidt and Eisele 1985). When all cells (not just near neighbors) fire in synchrony, correlated activity can no longer be a cue for finding neighbors from the retina. The projections regenerated under strobe illumination, like those regenerated without activity, had enlarged multiunit fields that averaged 33 ° in diameter. Arbors from these projections, stained by anterograde HRP transport, were again abnormal in much the same way as those from projections in which activity was blocked by TTX. The branches failed to form tight clusters and remained somewhat scattered.

A second anatomical study now supports the original electrophysiological findings. Cook and Rankin (1986) found that the distribution of cells in retina labeled by point injections of HRP in tectum normally shrinks to a tight cluster by 75 days, but in fish regenerating under strobe illumination (5 Hz) the cluster remains much larger.

The strobe and dark rearing experiments furnish important evidence that the TTX effects are not merely the result of abnormal metabolism. In particular they make unlikely the possibility that the TTX effect was caused by the deficit in axonal transport that was demonstrated by Edwards and Grafstein (1983). The high and low levels of activity in the strobe and dark reared fish are unlikely to be associated with similar decrements in axonal transport. The experiments also show that sharpening depends upon the pattern of activity, presumably the aforementioned correlation in firing between neighbors in the retina.

3.3 Sensitive Periods for Strobe and TTX

The periods of sensitivity for strobe and TTX are very similar and correspond to the period of synaptogenesis. The first synapses are made as the fibers enter the tectum at 14–16 days (Murray and Edwards 1982), and either strobe or TTX treatment applied before that time did not prevent the sharpening. Blocking from 14 to 34 days, however, was extremely effective, producing multiunit fields of almost 40 °. Synaptogenesis continues at a declining rate until at least 90 days, as determined by counts of synaptic density (Murray and Edwards 1982). Eisele and Schmidt (1988) have now assessed the effects of TTX and strobe applied for 2 week periods starting at later times. Either treatment begun at 35 days caused the multiunit fields which had already sharpened to become large again. The size of this effect became progressively smaller the later the strobe or TTX treatment was begun, and no effect was seen after 125 days or in the mature projection, where synaptogenesis should be minimal. The results suggest a parallel between the rate at which synapses are added and the degree of disorder produced by manipulating activity.

Neither TTX nor strobe appear to cause a decreased rate of synaptogenesis as judged by the field potentials elicited by optic nerve shock. A similar lack of effect of TTX was noted concerning neuromuscular synapse formation in vitro (Obata 1977). In fish tectum, the amplitude of the field potentials from eyes blocked by TTX or strobed during regeneration was not decreased (Schmidt et al. 1983; Eisele and Schmidt 1988). For the small field potentials in early regeneration, amplitude is likely to be a reasonably sensitive measure of the numbers of synapses formed, and the result therefore suggests an interference with the retinotopic deployment of the synapses rather than with the gross numbers made.

4 The Proposed Model

4.1 *Details of the Model*

A proposed model for sharpening is based upon the correlated firing of neighboring ganglion cells and the resultant summation of their EPSP's in the postsynaptic cells (Schmidt and Edwards 1983). The model (Fig. 3) also assumes that the arbors in the initial diffuse projection are larger than normal and have a high degree of overlap. For neighboring ganglion cells sharing common postsynaptic cells within the area of overlap of their arbors and firing with a high degree of correlation, there would often be a summation of their postsynaptic potentials. This

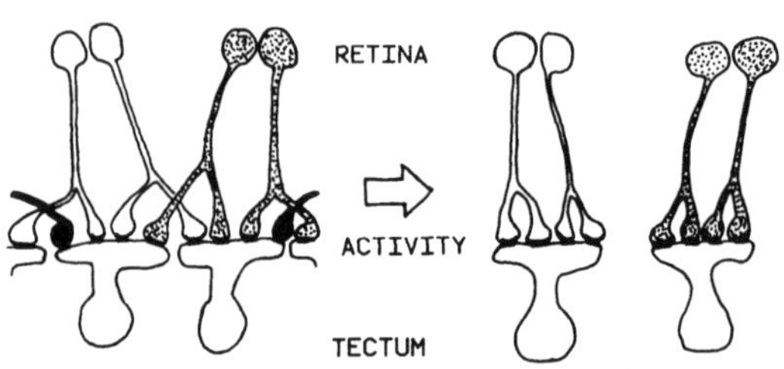

Fig. 3. A highly simplified model for stabilization of retinotopic synapses. On the *left*, two sets of neighboring ganglion cells from two retinal areas regenerate arbors contacting several tectal cells. Corelated activity will stabilize contacts of neighbors made onto the same postsynaptic cell, leading to the sharpened version at the *right*. (Schmidt and Eisele 1985)

summation would make these synapses more effective than those of fibers firing out of synchrony (from distant ganglion cells) and could lead to further increases in synaptic gain (potentiation). Finally, if the most effective synapses are differentially stabilized and retained (Hebb 1949; Changeux and Danchin 1976) the correlated activity would stabilize those that are retinotopically appropriate. Both blocking activity and the global synchronization of activity produced by strobe illumination would prevent the differential stabilization, in the first case because no synapses were stabilized and in the second case because all were equally stabilized due to the same pattern of activity. Specifically the model proposes two steps: first a potentiation of synaptic response (increased gain) of existing synapses, and second a slower anatomical change such that weaker synapses are removed and replaced by growth and elaboration of the stronger synapses. The morphological changes (above) produced by strobe synchrony support the second step, and the evidence below supports the first step by showing that synchronous activity, by producing summation of responses, can alter the gain of synapses.

4.2 Long-Term Potentiation and NMDA Receptors

During regeneration, there is a greatly increased capacity for potentiation of retinotectal synaptic transmission, assessed by recording tectal field potentials elicited by supramaximal stimulation of the optic nerve (Schmidt 1987). In the normal optic pathway, field potentials are very stable averaging about 4 mV in the retinal terminal layer. Moderate or high frequency trains of stimuli produce only very small increases in amplitude (20%–30% max). At 20 to 35 days postcrush, the regenerating projection gives much smaller (0.1–1.0 mV) responses that fatigue rapidly. However, administering as few as 6 to 20 stimuli at 0.1 Hz typically results in a 100% or more increase in amplitude. The cumulative increases after multiple potentiations are very large and stable for many hours. These long-term potentiations appear to be mediated by N-methyl D-aspartate (NMDA) receptors as in hippocampus (Harris et al. 1984), since application of 2-amino-5-phosphono valerate (APV) at 25 μM, which did not significantly decrement individual responses, prevented the increase in synaptic gain. Thus, this highly synchronous activity that produces a maximum summation of postsynaptic responses can produce large and long-lasting changes in synaptic gain. One can therefore speculate that smaller gain changes could be produced by the highly correlated activity of neighboring ganglion cells over the weeks that sharpening occurs under normal visual conditions.

The next step in linking this potentiation of coactive inputs to the sharpening process is to test whether chronic infusions of the NMDA receptor antagonist APV can block retinotopic sharpening. Preliminary results suggest that it does. Chronic APV application also prevents the segregation of retinal afferents into eye specific stripes in dually innervated tecta of three eyed frogs (Cline et al. 1987). This segregation process, previously shown to be activity-dependent (Meyer 1982; Boss and Schmidt 1984; Stryker and Harris 1986), is also thought to be driven by correlated activity between neighboring ganglion cells within each eye (but not between eyes).

NMDA receptors have special properties that enable them to generate a signal for the control of synaptic strength. The opening of the channels that they control requires not only the binding of the transmitter (presumably glutamate) but also a simultaneous depolarization of the cell by other inputs. When open, the channels allow an inward calcium current (MacDermott et al. 1986), that acts as a second messenger. The new findings suggest that a single synaptic mechanism has properties that allow it to control changes in synaptic gain as well as the structural refinements that occur during development, postlesion plasticity and learning processes.

5 Conclusion

Many questions have yet to be answered about the exact operation of these processes in retinotopic sharpening. Some involve the interaction with nonvisual inputs, either maps from other sensory modalities or modulatory activating systems. Others involve details of circuitry: the roles of lateral inhibition, of recurrent facilitatory loops (within tectum and through N. Isthmi), and of dendritic geometry in postsynaptic control of potentiation and structural refinements. Finally, there is the question of the nature of the changes underlying potentiation and synaptic stabilization, including the possible need for a signal from the postsynaptic cell that tells the presynaptic terminal that its input was correlated and is to be strengthened and retained. These are fundamental questions for neural development and for learning and memory as well as for neural plasticity following lesions.

References

Adamson JR, Burke J, Grobstein P (1984) Reestablishment of the ipsilateral oculotectal projection after optic nerve crush in the frog: evidence for synaptic remodelling during regeneration. J Neurosci 4:2635–2649

Arnett DW (1978) Statistical dependence between neighboring retinal ganglion cells in goldfish. Exp Brain Res 32:49–53

Bonhoeffer F, Huf J (1982) In vitro experiments on axon guidance demonstrating an anterior-posterior gradient on the tectum. EMBO J 1:427–430

Boss VC, Schmidt JT (1984) Activity and the formation of ocular dominance patches in dually innervated tectum of goldfish. J Neurosci 4:2891–2905

Changeux JP, Danchin A (1976) Selective stabilization of developing synapses as a mechanism for the specification of neuronal networks. Nature 264:705–712

Cline HT, Debski EA, Constantine-Paton M (1987) N-Methyl-D-aspartate receptor antagonist desegregates eye-specific stripes. P.N.A.S.U.S.A. 84:4342–4345

Cook JE, Rankin ECC (1986) Impaired refinement of the regenerated retinotectal projection of the goldfish in stroboscopic light: a quantitative HRP study. Exp Brain Res 63:421–430

Easter SS (1985) The continuous formation of the retinotectal map in goldfish with special attention to axonal pathway. In: Edelman GM, Gall WE, Cowan MW (eds) The molecular bases of neural development. Wiley, New York

Edwards DL, Grafstein B (1983) Intraocular tetrodotoxin in goldfish hinders optic nerve regeneration. Brain Res 269:1–14

Eisele LE, Schmidt JT (1988) Activity sharpens the regenerated retinotectal projection in goldfish: sensitive period for strobe illumination and lack of effect on synaptogenesis and on ganglion cell receptive field properties. J Neurobiol 19:395–411.

Fraser S (1985) Cell interactions involved in neuronal patterning: An experimental and theoretical approach. In: Edelman GM, Gall WE, Cowan MW (eds) The molecular bases of neural development. Wiley, New York
Fujisawa H (1987) Mode of growth of retinal axons within the tectum of *Xenopus* tadpoles and implications in the ordered neuronal connections between the retina and tectum. J Comp Neurol 260:127–139
Fujisawa H, Tani N, Watanabe K, Ibata Y (1982) Branching of regenerating retinal axons and preferential selection of appropriate branches for specific neuronal connection in the newt. Dev Biol 90:43–57
Ginsberg KS, Johnsen JA, Levine MW (1984) Common noise in the firing of neighboring ganglion cells in goldfish retina. J Physiol (Lond) 351:433–444
Harris EW, Ganong AH, Cotman CW (1984) Long term potentiation in the hippocampus involves activation of N-methyl-D-aspartate receptors. Brain Res 323:132–137
Hebb DO (1949) The organization of behavior. Wiley, New York
Humphrey MF, Beazley LD (1982) An electrophysiological study of early retinotectal projection patterns during optic nerve regeneration in *Hyla moorei*. Brain Res 239:595–602
MacDermott AB, Mayer ML, Westbrook GL, Smith SJ, Barker JL (1986) NMDA receptor activation increases cytoplasmic calcium concentration in cultured spinal cord neurons. Nature 321:519–522
Meyer RL (1980) Mapping the normal and regenerating retinotectal projection of goldfish with autoradiographic methods. J Comp Neurol 189:273–289
Meyer RL (1982) Tetrodotoxin blocks the formation of ocular dominance columns in goldfish. Science 218:589–591
Meyer RL (1983) Tetrodotoxin inhibits the formation of refined retinotopography in goldfish. Dev Brain Res 6:293–298
Murray M, Edwards M (1982) A qunatitative study of the reinnervation of goldfish optic tectum following optic nerve crush. J Comp Neurol 209:363–373
Murray M, Sharma S, Edwards M (1982) Target regulation of synaptic number in the compressed retinotectal projection of goldfish. J Comp Neurol 209:374–385
Obata K (1977) Development of neuromuscular transmission in culture with a variety of neurons and tetrodotoxin. Brain Res 119:141–150
Rankin ECC, Cook JE (1986) Topographic refinement of the regenerating retinotectal projection of the goldfish in standard laboratory conditions: a qunatitative WGA-HRP study. Exp Brain Res 63:432–446
Sachs GM, Jacobson M, Caviness VS (1986) Postnatal changes in arborization patterns of murine retinocollicular axons. J Comp Neurol 246:395–408
Sakaguchi DS, Murphey RK (1985) Map formation in the developing *Xenopus* retinotectal system: an examination of ganglion cell terminal arborizations. J Neurosci 5:3228–3245
Schmidt JT (1982) The formation of retinotectal projections. Trends Neurosci 5:111–116
Schmidt JT (1985) Factors involved in retinotopic map formation: complementary roles for membrane recognition and activity dependent synaptic stabilization. In: Edelman GM, Gall WE, Cowan WM (eds) The molecular bases of neural development. Wiley, New York
Schmidt JT (1987) Increased potentiation of postsynaptic response correlates with sensitive period during optic nerve regeneration in goldfish. Neurosci Abstr 13 (in press)
Schmidt JT, Edwards DL (1983) Activity sharpens the map during the regeneration of the retinotectal projection in goldfish. Brain Res 269:29–39
Schmidt JT, Eisele LE (1985) Stroboscopic illumination and dark rearing block the sharpening of the retinotectal map in goldfish. Neuroscience 14:535–546
Schmidt JT, Edwards DL, Stuermer CAO (1983) The reestablishment of synaptic transmission by regenerating optic axons in goldfish: time course and effects of blocking activity by intraocular injection of tetrodotoxin. Brain Res 269:15–27
Schmidt JT, Buzzard M, Turcotte J, Tieman DG (1988) Staining of regenerated optic arbors in goldfish tectum: Progressive changes in immature arbors and a comparison of mature regenerated arbors with normal arbors. J Comp Neurol 269:565–591.
Schneider GE, Rava L, Sachs GM, Jhaveri S (1981) Widespread branching of retinotectal axons: transient in normal development and anomalous in adults with neonatal lesions. Neurosci Abstr 7:732

Sperry RW (1963) Chemoaffinity in the orderly growth of nerve fiber patterns and connections. Proc Natl Acad Sci USA 50:703–709
Stryker MP, Harris WA (1986) Binocular impulse blockade prevents the formation of ocular dominance columns in cat visual cortex. J Neurosci 6:2117–2133
Stuermer CAO, Easter SS (1984) A comparison of the normal and regenerated retinotectal pathways of goldfish. J Comp Neurol 223:57–76
Willshaw DJ, Malsburg C von der (1976) How patterned neural connections can be set up by self-organization. Proc R Soc Lond B Biol Sci 194:421–445

VII. Lesions to Specific CNS Subsystems: Somatosensory Cortex, Motor Cortex, Basal Ganglia

Plasticity in Nonprimary Somatosensory Cortex of Adult Monkeys

T. P. Pons, P. E. Garraghty, and M. Mishkin
Laboratory of Neuropsychology, National Institute of Mental Health, Bethesda, Maryland 20892, USA

It has long been recognized that cortical maps, the topographically organized central representations of various sets of peripheral sensory receptors, can undergo extensive reorganization following central or peripheral injuries in neonatal or infant mammals (e.g., Sherman and Spear 1982; Kaas et al. 1983). Only during the past decade, however, has it become evident that cortical maps of adult mammals are likewise capable of a remarkable degree of reorganization following peripheral nerve injury (e.g., Kaas et al. 1983; Van der Loos and Welker 1985; Wall et al. 1985). These latter studies have demonstrated conclusively that nerve ligations, nerve crushes, and digit amputations result in the topographic reorganization of primary somatosensory cortex even in adult primates (e.g., Paul et al. 1972; Merzenich et al. 1983a, b, 1984; Wall et al. 1983).

Although these nerve manipulations in adult animals result in the reorganization of primary somatosensory cortex, their effects on nonprimary somatosensory areas have yet to be examined. Similarly, the effects on nonprimary somatosensory areas have yet to be investigated following damage to primary somatosensory cortex. Until recently, there was no reason to believe that elimination or removal of one cortical somatosensory area would have any effect on the somatotopic representation in another area. This was due, in part, to the belief that each cortical area with a well defined somatotopic map received its major source of activation directly from the thalamus (e.g., Jones and Powell 1970) and that the precisely organized thalamocortical projection was the major determinant of the cortical map. Within the past year, however, evidence has become available indicating that the cutaneous activation of neurons in the second somatosensory area (SII) is dependent upon a projection from postcentral cortex rather than on a direct projection from the thalamus (Friedman and Murray 1986; Pons et al. 1987a). This new evidence raises new questions, such as: What is the fate of the neurons in SII that are initially deactivated following removal of the postcentral representation of specific body parts? Do these neurons remain unresponsive to tactile stimulation, or do they have the capacity to reorganize and represent body parts whose representations are left intact? In this chapter we briefly review experiments demonstrating that in adult primates the SII body representation is capable of undergoing substantial topographic reorganization following damage to postcentral somatosensory cortex.

We used standard multiunit mapping procedures to study the topographic organization of the second somatosensory area (SII) of anesthetized macaques that

H. Flohr (Ed.)
Post-Lesion Neural Plasticity

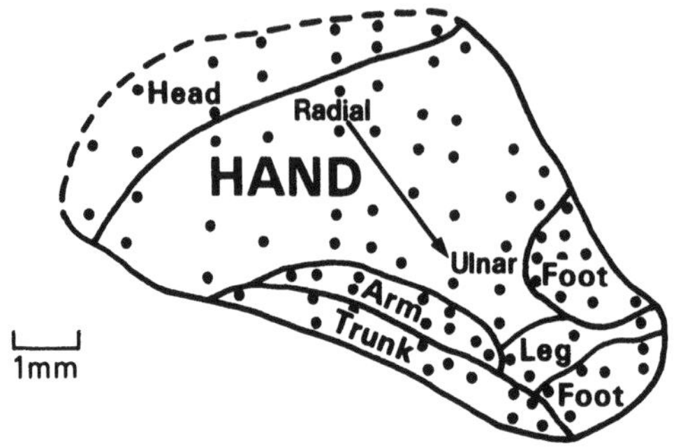

Fig. 1. A flattened map of the SII region showing somatotopic features as determined from electrophysiological recordings in five intact hemispheres. *Dots* indicate recording sites on which the maps are based. The *dashed line* indicates the border between the SII and postcentral head representations based solely on cytoarchitectural criteria. Note the large representation of the hand compared to that of any other body part

had received lesions of the entire representation of the hand in postcentral somatosensory cortex (area 3a, 3b, 1 and 2) either 12–24 h before the recording (acute survival) or 6–8 weeks before (chronic survival). Receptive fields for cells were determined by tactile stimulation with camel hair brushes, Von Frey hairs, and fine glass probes at 200–300 recording sites spaced across the entire SII representation. Small electrolytic marker lesions were placed at strategic points in the cortex to aid in later reconstruction of the data. The receptive field data were then used to construct maps indicating both the location and somatotopic relationship among the body part representations determined from individual recording sites (see Pons et al. 1987b for further details on map reconstruction). The pattern of topographic organization seen in SII of the acute and chronic survival animals was then compared to the pattern seen in SII of control monkeys with intact hemispheres.

Before presenting and interpreting the results obtained from the experimental animals, we must briefly describe the normal somatotopic organization found in SII of intact animals. Figure 1 illustrates several consistent features of the normal organization in SII, such as the location, distribution, and areal extent of the representations of various body parts. It can be seen that the representation of the body starts with the head rostrally in SII and then proceeds to the hand, arm, trunk, and leg before ending with the foot caudally. Much of SII is devoted to the representation of the hand. The representation of the foot is relatively small compared to that of the hand; in the cases studied, it never occupied more than 15% of the SII representation. Similarly, the representations of the trunk, arm, and leg always comprised a comparatively small portion of SII.

The SII head representation abuts the postcentral head representation, making it difficult to distinguish the precise border between these two cortical areas. The dashed line in Fig. 1 is an approximation of the border based solely on cytoarchitectural differences between the two cortical regions. Although it is possible to accurately identify the border between the postcentral and SII head representations by our electrophysiological mapping methods, such a goal was beyond the scope of the present study. It was sufficient for our purposes that the representation of the hand was always found immediately caudal to the head representation in the intact hemispheres, and thus the caudal border of the head representation served as a landmark to help identify the expected location of the hand representation in the experimental hemispheres.

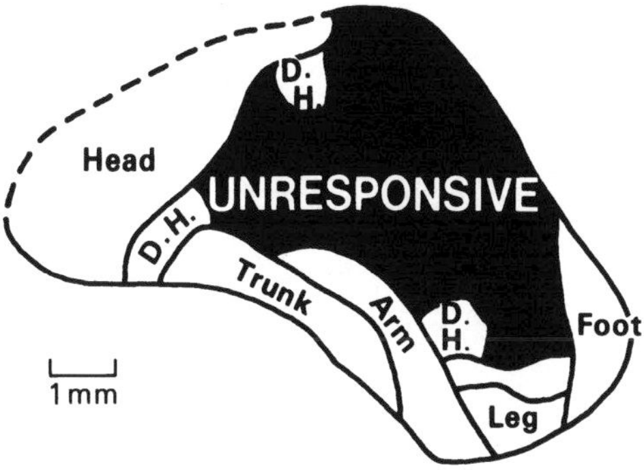

Fig. 2. A flattened map of the somatotopic organization of SII obtained 24 h after removal of the entire postcentral hand representation. Note the absence of somatic activity in the expected location of the hand representation in SII. Note also the normal location and extent of the representation of body parts other than the hand (cf. Fig. 1). Similar results were obtained in two additional cases. *D. H.* dorsal hand. Other conventions as in Fig. 1.

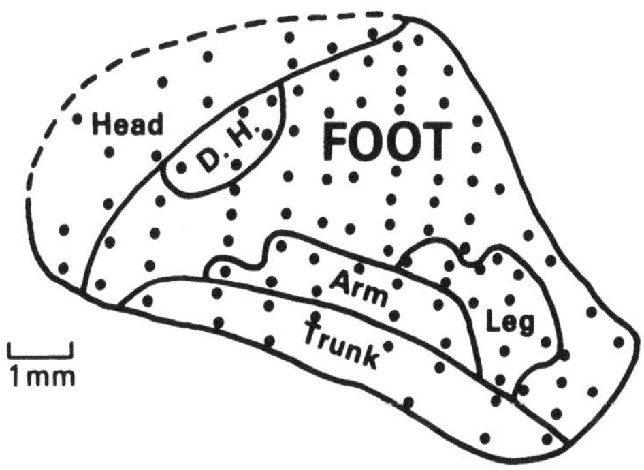

Fig. 3. A flattened map of the somatotopic organization seen in SII 7 weeks after removal of the entire postcentral hand representation. Note the expansion of the foot representation into the expected location of the hand representation (cf. Fig. 1). Similar results were obtained in three additional chronic cases. Other conventions as in Fig. 1

Figure 2 illustrates the somatotopy observed in SII 24 h after the removal of the postcentral hand representation in a representative case. Note that only the representation of the hand was "deactivated" in SII and not the representations of the head, trunk, arm, leg, or foot, all of which appeared to be completely normal as judged by both their relative location and their extent (the deactivation aspects of these results have been dealt with in detail in Pons et al. 1987a). We should also stress that the region of SII cortex normally occupied by the representation of the hand was completely silent and could not be activated by higher amplitude stimulation. After obtaining similar results in two additional acute survival animals, we wondered what the fate of the deactivated cortex would be over a longer survival period. That is, would the zone in SII formerly representing the hand continue to remain silent? Or would the deactivated zone eventually become reactivated, and, if so, would it again represent the hand or would it instead undergo some type of somatopic reorganization?

To answer these questions, we recorded from the SII region in four hemispheres from which we had removed the postcentral hand representations 6–8 weeks earlier. The results from a typical case are illustrated in Fig. 3, which shows that cortex in the expected location of the hand representation was now responsive once again, but not, as might have been expected, to stimulation of the hand; instead it was now responsive to stimulation of the foot! Equally surprising is that the representation of the foot expanded throughout almost the entire area that had formerly represented the hand; that is, the areal extent of the foot representation in the operated hemispheres was very nearly equal to the area of the combined representations of the foot and hand in the intact hemispheres. Further, there were no zones within the initially deactivated region of SII that remained unresponsive to somatic stimulation of some body part, at least none that we could detect using our experimental methods. This reorganization, in which an initially unresponsive region became completely filled in, took place over a linear cortical distance of more than 3 mm (cf. Fig. 2 and 3). Surprisingly again, threshold levels determined with Von Frey hairs were completely normal across the entire reorganized zone in SII.

Whether the representations of body parts other than the foot undergo expansion in SII following postcentral lesions of the hand representation is not yet clear. As illustrated in Fig. 3, there was no obvious increase in the representation of the head, even though it is normally located immediately adjacent to that of the hand. Similarly, there was no clear-cut enlargement of the representations of the arm, trunk, or leg. The ablation of the hand representation in postcentral cortex extended into the representations of both the head and arm, and perhaps this extension prevented the representations of these particular body parts from expanding in the SII region. But such an explanation cannot account for the failure of the leg and trunk representations to expand substantially, since our lesions did not encroach upon their representations in postcentral cortex.

Why the foot representation expands into the cortex vacated by the hand representation to a greater extent than the adjacent representations of other body parts is unclear. One possibility is that a large portion of the SII hand representation is specialized for receiving and processing inputs from glabrous skin. Consequently, when this cortical area is deprived of its normal inputs from the glabrous hand, it may be more receptive to new inputs (axonal sprouting) or more likely to express old ones (delayed unmasking of preexisting inputs) from the glabrous foot than those from the adjoining representations of hairy body parts. Another possibility, which does not exclude the first, is that differential sensory stimulation of the intact representations results in a selective expansion of the foot representation. Perhaps the sole or glabrous portion of a monkey's foot receives more tactile stimulation during the course of everyday cage behaviors than do hairy body parts such as the leg and thigh. It has been suggested that the differential sensory stimulation provided by selective use of a given body part, such as of a single digit of the hand, increases the size of the representation of that body part in primary somatosensory cortex (Jenkins et al. 1984); similar dynamic determinants may extend to nonprimary cortical areas as well.

Previous studies on topographic reorganization, in which primary somatosensory cortex of adult monkeys was examined following peripheral nerve damage (Merzenich et al. 1983 a, b; Wall et al. 1986), left open the possibility that the changes measured in cortex were simply a reflection of changes occurring at subcortical levels. The findings presented here, taken together with our previous finding that SII is critically dependent upon projections from postcentral cortex for its somatic activation (Pons et al. 1987 a), strongly suggest that the topographic reorganization in SII following restricted postcentral damage occurs within the cortex itself (cf. Van der Loos and Dorfl 1978).

Of course it could be that the observed reorganization in SII following damage to postcentral cortex is due to a delayed unmasking of preexisting thalamic inputs. Evidence consistent with such a possibility comes from previous studies demonstrating changes in the receptive field characteristics and gross responsiveness of thalamic neurons following spinal cord lesions. It is known that the Ventral Posterior Medial nucleus (VPM) of rodents receives a projection from both the trigeminal nucleus principalis (PrV; Feldman and Kruger 1980; Lund and Webster 1967) and the subnucleus interpolaris (SpVI; Peschanski 1984). In an elegant set of experiments, Rhoades and his colleagues (Rhoades et al. 1987) demonstrated that although kainic acid lesions of SpVi had no appreciable effect on the receptive field characteristics or gross excitability of neurons in VPM, kainic acid injections into PrV resulted in an immediate and vast reduction in the number of neurons in VPM that could be driven by stimulation of the mystacial vibrissae. In contrast, recordings obtained 6 to 90 days following such kainic acid injections into PrV revealed a massive restoration in the number of cells in VPM that were responsive to stimulation of the vibrissae. But the response properties now resembled thos of SpVi neurons in that they could be activated by many vibrassae rather than only by one or two as before (which resembles the properties of PrV neurons). If, at this point, kainic acid was injected into SpVi of the animals that had received PrV lesions 90 days earlier, there was again a massive decrease in the recording frequency of cells responsive to stimulation of the vibrissae, supporting the supposition that SpVi had been the new source supplying the driving input to the previously unresponsive neurons in VPM. Thus, whereas the major source of somatic activation of neurons in VPM representing the vibrissae appears to be PrV, there is a secondary input from SpVi that is normally dormant. Destruction of PrV, however, does not result in the immediate unmasking of the preexisting inputs from SpVi. Instead, following damage to PrV, there is a gradual functional reorganization during which neurons in VPM that previously exhibited PrV-like characteristics develop and exhibit SpVi-like properties.

Although it is possible that the postsurgical reorganization we found in SII could likewise be due to a gradual functional reorganization of subcortical inputs, such an explanation seems unlikely for several reasons. First, following ablation of the entire representation of a given body part in postcentral somatosensory cortex, the somatotopically matched portion of the Ventral Posterior Nucleus (VP) undergoes virtually complete degeneration, thus eliminating one of the major potential sources of somatic activation of SII (Pons et al. 1987a). No alteration of the Ventral Posterior Inferior nucleus (VPI), which projects directly to SII, is seen following such removals. Second, VPI, unlike VPL (at least in raccoons), is densely packed with GABA-positive neurons (Lin et al. 1986), implying that it does not exert an excitatory influence on its projection targets. Therefore, since VPI is the only thalamic nucleus besides VP known to have connections with SII, it seems an unlikely source to account for the observed reorganizational changes. Third, if, nonetheless, thalamic inputs from VPI to SII did account for the topographic properties of the map following a lesion of postcentral cortex, then the representation of the hand in SII would not be expected to disappear permanently. Rather, the SII hand representation would be expected to be restored, but to now reflect VPI-like properties rather than postcentral-like properties. Yet, as we have seen, neither VPI nor any other thalamic nucleus is capable of sustaining the hand representation in the absence of postcentral somatosensory cortex (Pons et al. 1987a). The only other known inputs conveying somatic information to SII come from the cortex (Friedman and Murray 1986; Friedman et al. 1986). So, unless one assumes a thalamic source for which there is presently no evidence, the somatotopic reorganization in SII must be taking place at the level of cortex. Fourth and finally, unlike the reorganization in the Rhoades et al. (1987) study, the reorganization in SII following damage to postcentral cortex was not somatotopically matched. In the Rhoades et al. (1987) study, neurons that previously represented a single vibrissa came to represent multiple vibrissae. But although the receptive fields of individual neurons in VPM clearly expanded in the above study, the authors indicated that this expansion did not extend outside the territory innervated by the trigeminal nerve. In our experiments, by contrast, receptive fields did expand outside the territory of the nerves innervating the hand.

A comparison between the extent of reorganizational changes seen in the present experiments with the extent of those seen in primary somatosensory cortex following peripheral nerve damage suggests the possibility that higher-order areas of somatosensory cortex have a greater capacity for reorganizational changes than do primary areas. Following deactivation of as few as two digits of the hand, the reorganization of the hand representation in primary somatosensory cortex extended over a distance of only 0.5–2.0 mm (Merzenich et al. 1983a, b). Furthermore, in those studies, small unresponsive zones were found embedded within the reorganized region. In contrast, the reorganization seen in SII following removal of the entire postcentral hand representation extended across more than 5.0 mm of cortex, and no unresponsive zones of cortex within SII were detected.

It is interesting to ask whether our findings in the adult somatosensory system can be generalized to other sensory systems (cf. Mishkin 1979). We know of no published reports documenting long-term changes in nonprimary visual or auditory areas after damage to the respective primary cortical receiving zone. Given the evidence of extensive topographic reorganization in SII following damage to primary areas of somatosensory cortex, one might speculate that damage to primary areas in other sensory modalities would likewise result in a previously unsuspected degree of reorganization in the higher-order cortical stations of those modalities.

References

Feldman SG, Kruger L (1980) An axonal transport study of the ascending projection of medial lemniscal neurons in the rat. J Comp Neurol 192:427–454

Friedman DP, Murray EA (1986) Thalamic connectivity of the second somatosensory area and neighboring somatosensory fields of the lateral sulcus of the macaque. J Comp Neurol 252:348–373

Friedman DP, Murray EA, O'Neill JB, Mishkin M (1986) Cortical connections of the somatosensory fields of the lateral sulcus of macaques: evidence for a corticolimbic pathway for touch. J Comp Neurol 252:323–347

Jenkins WM, Merzenich MM, Ochs MT (1984) Behaviorally controlled differential use of restricted hand surfaces induce changes in the cortical representation of the hand in area 3b of adult owl monkeys. Neurosci Abstr 10:665

Jones EG, Powell TPS (1970) Connexions of the somatic sensory cortex of the rhesus monkey. III. Thalamic connexions. Brain 93:37–56

Kaas JH, Merzenich MM, Killackey HP (1983) The reorganization of somatosensory cortex following peripheral nerve damage in adult and developing mammals. Annu Rev Neurosci 6:325–356

Lin C-S, Lu SM, Standage GP, Doetsch GS (1986) Glutamic acid decarboxylase, cholecystokinin and somatostatin immunoreactivities of somatosensory cortex and thalamus of the raccoon. Neurosci Abstr 12:1437

Loos H van der, Dorfl J (1978) Does the skin tell the somatosensory cortex how to construct a map of the periphery. Neurosci Lett 7:23–30

Loos H van der, Welker E (1985) Development and plasticity of somatosensory brain maps. In: Rowe M, Willis WD Jr (eds) Development, organization, and processing in somatosensory pathways. Liss, New York, pp 53–67

Lund RD, Webster KE (1967) Thalamic afferents from the spinal cord and trigeminal nuclei: an experimental anatomical study in the rat. J Comp Neurol 130:313–324

Merzenich MM, Kaas JH, Wall J, Nelson RJ, Sur M, Felleman D (1983a) Topographic reorganization of somatosensory cortical areas 3b and 1 in adult monkeys following restricted deafferentation. Neuroscience 8:33–55

Merzenich MM, Kaas JH, Wall JT, Sur M, Nelson RJ, Felleman DJ (1983b) Progression of change following median nerve section in the cortical representation of the hand in areas 3b and 1 in adult owl and squirrel monkeys. Neuroscience 10:639–665

Merzenich MM, Nelson RJ, Stryker MP, Cynader MS, Schoppmann A, Zook JM (1984) Somatosensory cortical map changes following digit amputation in adult monkeys. J Comp Neurol 224:591–605

Mishkin M (1979) Analogous neural models for tactual and visual learning. Neuropsychologia 17:139–151

Paul RL, Goodman H, Merzenich MM (1972) Alterations in mechanoreceptor input to Brodmann's areas 1 and 3 of the postcentral hand area of *Macaca mulatta* after nerve section and regeneration. Brain Res 36:1–19

Peschanski M (1984) Trigeminal afferents to the diencephalon in the rat. Neuroscience 12:465–487

Pons TP, Garraghty PE, Friedman DP, Mishkin M (1987a) Physiological evidence for serial processing in somatosensory cortex. Science 237:417–420

Pons TP, Wall JT, Garraghty PE, Cusick CG, Kaas JH (1987b) Consistent features of the representation of the hand in area 3b of macaque monkeys. Somatosens Res 4:309–331

Rhoades RW, Belford GR, Killackey HP (1987) Receptive-field properties of rat ventral posterior medial neurons before and after selective kainic acid lesions of the trigeminal brain stem complex. J Neurophysiol 57:1577–1600

Robinson CJ, Burton H (1980) Somatotopographic organization in the second somatosensory area of *M. fascicularis*. J Comp Neurol 192:43–67

Sherman SM, Spear PD (1982) Organization of visual pathways in normal and visually deprived cats. Physiol Rev 62:738–855

Wall JT, Felleman DJ, Kaas JH (1983) Recovery of normal topography in the somatosensory cortex of monkeys after nerve crush and regeneration. Science 221:771–773

Wall JT, Kaas JH, Felleman DJ (1985) Functional recovery in the somatosensory cortex of monkeys with regenerated nerves: replication versus reintegration. In: Rowe M, Willis WD Jr (eds) Development, organization, and processing in somatosensory pathways. Liss, New York, pp 277–287

Wall JT, Kaas JH, Sur M, Nelson RJ, Felleman DJ, Merzenich MM (1986) Functional reorganization in somatosensory cortical areas 3b and 1 of adult monkeys after median nerve repair: possible relationships to sensory recovery in humans. J Neurosci 6:218–233

Plasticity and Function of Associational Input to the Motor Cortex

H. Asanuma and M. Bornschlegl

The Rockefeller University, 1230 York Ave., New York, N.Y. 10021, USA

1 Introduction

It has been shown that there are cortical efferent zones within the depth of the motor cortex which when stimulated produce contraction of a particular muscle (Asanuma and Rosén 1972). Each efferent zone receives peripheral input related to contraction of the target muscle constituting a closed loop circuit between the cortex and the periphery (Rosén and Asanuma 1972). Part of this sensory input arrives at the motor cortex directly from the thalamus and part indirectly via the sensory cortex (Asanuma et al. 1979). The direct input ascends through the dorsal column (DC) to the thalamus and the indirect input ascends through the DC and the spinothalamic tract (Asanuma et al. 1980). Section of the dorsal column (Wall 1970) or ablation of the sensory cortex (Travis and Woolsey 1956) produces some motor deficit, but function recovers nearly completely within a week or two. However, when the sensory cortex was removed after dorsal column section, the animal was paralyzed (Asanuma and Arissian 1984). We were interested in this observation because the results suggested there was reorganization of the projection from the sensory cortex to the motor cortex following elimination of the direct input. In this series of experiments, however, we used thalamic lesion instead of DC section.

2 Methods

As the first step, we tried to reproduce the symptom of DC section by partial thalamic lesion. It is known (Asanuma et al. 1979) that the direct input to the motor cortex comes from the border area between n. ventralis posterolateralis pars oralis (VPLo) and pars caudalis (VPLc) (Olszewski 1952). Cynomolgus monkeys *(Macaca fascicularis)* were used. Kainic acid was injected with a Hamilton syringe into this border area guided by a stereotaxix atlas (Szabo and Cowan 1984). Figure 1 shows an example of the lesion revealed by later histological examination. The lesioned area covered n. ventralis anterior (VA), most of n. ventralis lateralis (VL), almost all of VPLo, the rostral VPLc and the subthalamic area. On the next day of the injection, this monkey showed involuntary irregular jerky movements of the contralateral arm with a frequency of about 3/s. When food

H. Flohr (Ed.)
Post-Lesion Neural Plasticity

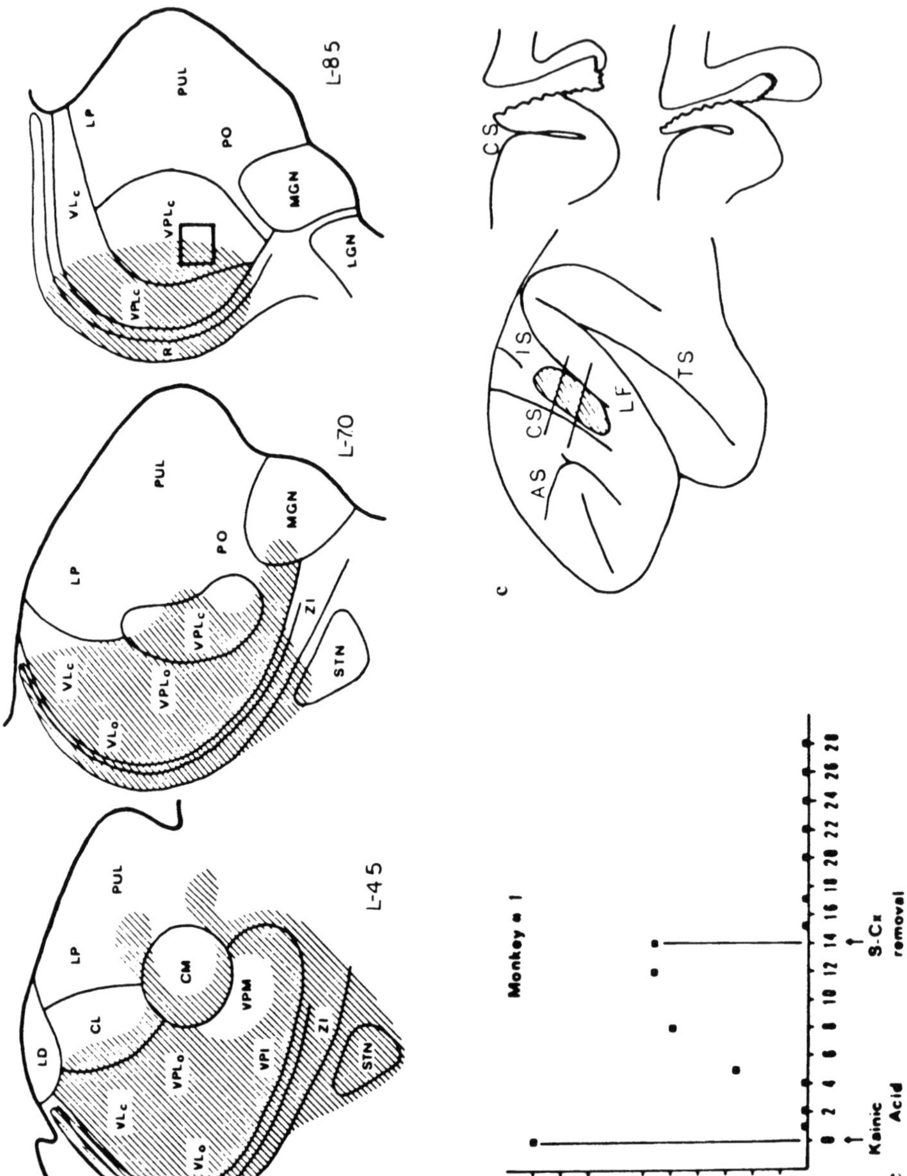
a
LD
LP
PUL
CL
CM
VLc
VPLo
VPM
VLo
VPI
ZI
STN
L-45
LP
PUL
PO
MGN
VLc
VPLo
VPLc
VLo
ZI
STN
L-70
LP
PUL
PO
VLc
VPLc
MGN
LGN
R
L-85
b
%
100
80
60
40
20
Monkey # 1
0 2 4 6 8 10 12 14 16 18 20 22 24 26 28
Kainic Acid
S-Cx removal
c
CS
IS
CS
AS
LF
TS

Fig. 1a–c

was given during the resting phase of this hyperkinesia, he could not reach the target. When food was placed into the hand, he could not bring it to the mouth. These symptoms disappeared rapidly and after 5 days, the monkey appeared normal in the cage. The motor skill of the hand was examined with our routine method which is described elsewhere (Asanuma and Arissian 1984). Briefly, the monkey was trained to sit in a monkey chair and to pick up a small piece of food from a hole in a food board which was turning at various speeds. The hand skill was measured by the maximum speed with which the monkey could pick up the food in 50% of the trial. Figure 1 B shows the recovery process of the hand skill. The monkey could not pick up the food at all during the initial 4 days, but the hand skill started recovering on the 5th day and rapidly returned to 60% of the normal skill within 2 weeks. This was the same time course as recovery after DC section. Then the monkey was reanesthetized and the postcentral gyrus was removed by section as shown in Fig. 1 C. The lesion included areas 1, 2 and most of 3 b but area 3 a and most of area 5 were intact. When the monkey recovered from the anesthesia, ataxia, dysmetria and hyperkinesia did not appear, but there was a severe loss of coordination in hand movements. The monkey could reach the food placed on a stationary foodboard, but could not pick it up. When the food was placed into the hand, he dropped it on the way to the mouth. This symptom persisted during the period of observation which lasted for 2 weeks. Similar observation was made in the other monkey in which the border area was lesioned. However, when both VPLo and VPLc were lesioned, the monkey was paralyzed and the motor function never recovered. The results suggested that the sensory input to the sensory cortex was necessary for the recovery of the function. Parallel with this experiment, Mackel performed the following experiment. It has been known that the motor deficit following cerebellar lesion recovers as time passes, but the recovered function is decompensated by the following dorsal column section (Carpenter and Correll 1961). Mackel (1978) removed the sensory cortex of the monkey in which the cerebellum was lesioned previously. He observed that the previously compensated function was decompensated by that removal. The results altogether supported our interpretation that lesion of the anterior thalamus (VL, VPLo) induced reorganization of the projection from the sensory cortex to the motor cortex to compensate the lost function, because removal of the sensory cortex by itself did not produce motor deficits (Tarvis and Woolsey 1956),

Fig. 1. a Extent of kainic acid lesion on saggital planes at L-4.5, 7.0 and 8.5. Lesions included STN, VL, VPLo and rostral VPLc. Abbreviations: *CL* n. ventralis lateralis; *CM* n. centralis medianum; *LGN* n. geniculatus lateralis; *LP* n. lateralis posterior; *MGN* n. geniculatus medialis; *PO* posterior complex; *PUL* n. pulvinaris; *STN* subthalamic nucleus; *VLc* n. ventralis lateralis pars caudalis; *VPI* n. ventralis posterior inferior; *VPLc* n. ventralis posterolateralis pars caudalis; *VPLo* n. ventralis posterolateralis pars oralis; *VPM* n. ventralis posterior medialis; *ZI* zona incerta. **b** Recovery of handskill following kainic acid injection to the thalamus and subsequent removal of the sensory cortex. The handskill was examined by the ability to pick up food from a rotating food board. *Ordinate* percent of control speed with which the monkey could pick up the food in 50% of the trials. *Abscissa* days after operation. **c** Location of cortical lesion and histological reconstruction of the lesion at the planes marked by lines on the *left*. The lesion included areas 1, 2 and most of 3 b. *AS* ansate sulcus; *CS* central sulcus; *IS* intraparietal sulcus; *LF* lateral fissure; *TS* temporal sulcus

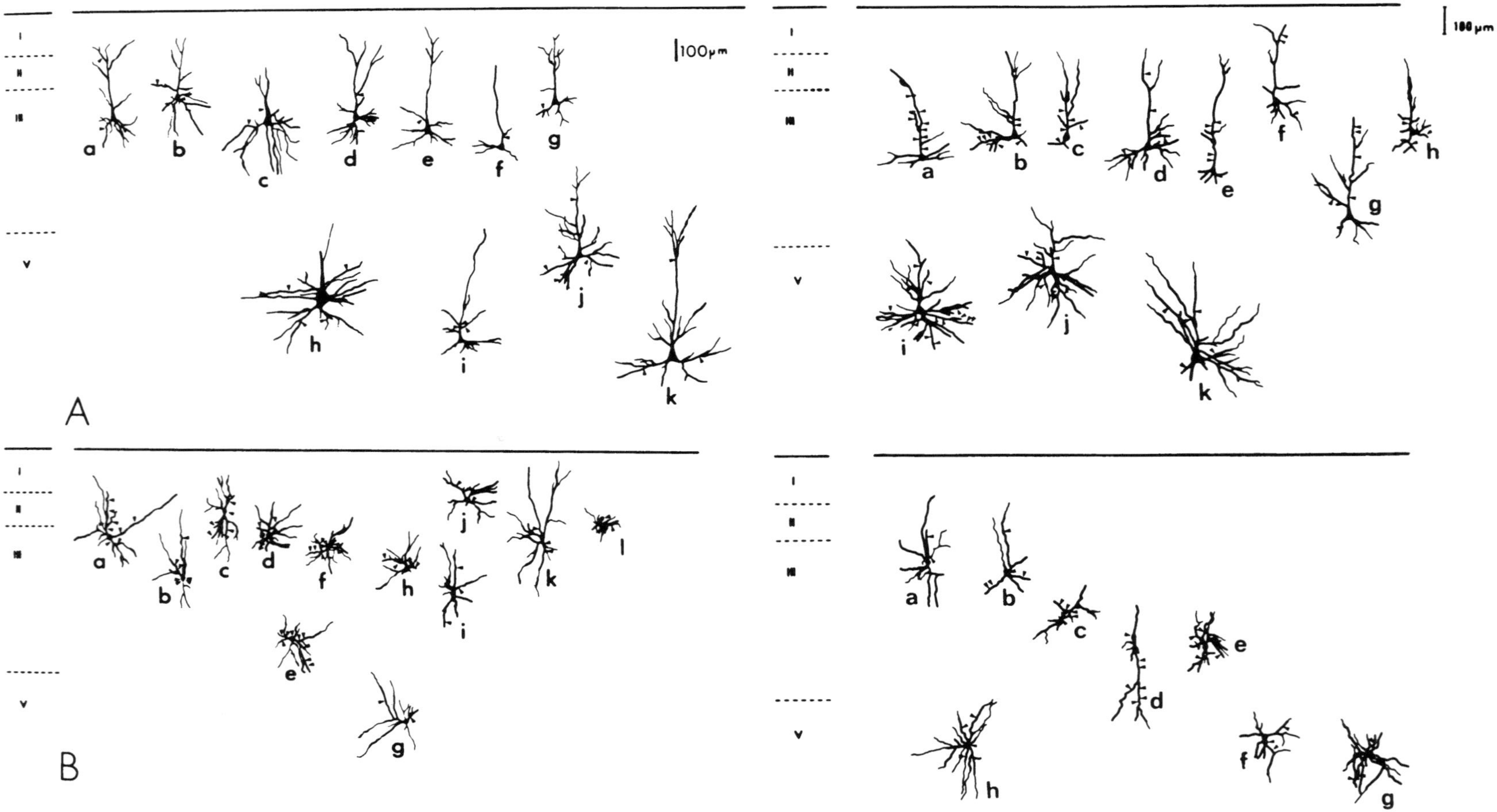

Fig. 2. Distribution of degenerating terminals of cortico-cortical fibers on identified postsynaptic neurons in the motor cortex in normal (*left*) and thalamic lesioned (*right*) cats. *Upper* (**A**) terminals contacting pyramidal cells. *Lower* (**B**) terminals contacting stellate cells. *Roman numerals* on the *left*

but when it was removed after compensation, it produced severe deficits. Similar results were obtained with cats trained to pick up food from a rotating beaker (unpublished observation).

A natural question was what kind of reorganization took place during the recovery period. We have reported physiological changes following the thalamic lesion (Asanuma et al. 1985) in the cat. These were: (1) increased efficiency of synaptic transmission at the terminal of the projection fibers from the sensory cortex to the motor cortex. (2) There was an expansion of the projection to the deep layer (V) of the motor cortex. In the normal cat (Kosar et al. 1985), sensory cortex stimulation produced monosynaptic EPSPs only in the cells located in the superficial layers (II and III). To obtain further insight into the reorganization, we have examined anatomical changes in the projection following thalamic lesion in the cat (Ichikawa et al. 1985; Ichikawa and Asanuma 1987). This was done in the following way. First, in the control experiments, area 2 of the sensory cortex was removed and 5 days later, distribution of the degenerating terminals synapsing on pyramidal and stellate cells in the motor cortex was examined. For this purpose, a small block of motor cortex was removed after 5 days and stained with Golgi method and pyramidal and stellate cells within the block were selected. Then serial ultrathin sections were cut and distribution of degenerating terminals on these cells were examined using an electron microscope. Then the same procedure was repeated with cats in which the border area between VL and VPL was lesioned 2 months prior to the terminal experiments. It should be noted that this border area corresponds to that of VPLo and VPLc in the monkey. Figure 2 shows the results. As shown on the left, the density of the terminals synapsing on the stellate cells was much higher than that on the pyramidal cells. On the stellate cells, there was an average of 8.4 degenerating synapses per unit length of dendrites, whereas on the pyramidal cells only 2.1 degenerating synapses on the corresponding length. Normally, 95% of the identified synapses on the stellate cells were either axo-dendritic or axo-somatic, whereas 50% of those on the pyramidal cells were on the spines. This configuration was changed drastically, as shown on the left. The density of the terminals on the stellate cells remained similar to the control, but that on the pyramidal cells was greatly increased. The average density per unit length of dendrites increased to 4.7 although half of these were still on the dendritic spines. The results demonstrated that there was a drastic change of the projection following the thalamic lesion.

3 Discussion and Conclusion

Among various inputs to the cerebral cortex, the associational input is the most numerous, yet least studied system. In fact, except for visual, auditory, somatic sensory and motor areas, the most dominant input comes through the association fibers. So far, this system has been studied primarily anatomically and physiological properties are not yet known. In the motor cortex, the major input has been said to come from the VL nucleus, but it also receives powerful input from the sensory and premotor areas (Jones et al. 1978); the exact ratio between the thalamic and associational inputs is not known. It is likely that the associational input is playing some unknown important role in the execution of movements. In fact, the present experiments revealed that near-total ablation of the VL and VPLo did not abolish the motor function, but the function was jeopardized when the input from the sensory cortex was removed. On the other hand, it is known that removal of this input by ablation of the sensory cortex produces least motor deficit (Tatton et al. 1975; Travis and Woolsey 1956) making speculation about the function of this system difficult. It has been shown in the cat that the projection from area 3a of the sensory cortex to the motor cortex is highly topographycally organized (Asanuma et al. 1981; Grant et al. 1975). Projection fibers from a small part of area 3a innervate one or a few groups of motor cortical neurons forming a columnar shape, suggesting that the activity of a sensory column is related to the activity of one or a few efferent zones in the motor cortex. We have recently found that the projection from area 2 to the motor cortex in the cat is similar to the projection from area 3a to the motor cortex (ms in preparation). How does this type of projection fit the present observation that this system is important for the recovery of the function? It is possible that specific input to a specific group of motor cortical neurons, i.e., a cortical efferent zone, is important for the execution of skilled movements. It is known that the projection fibers from the thalamus have special arborization and innervate a group of sensory and motor cortical neurons located close together (Asanuma et al. 1974; Lorente de No 1949). Therefore, it is likely that this specific input is provided by the thalamic input in normal animals. However, when this thalamic input is interrupted, the associational input is reinforced to substitute the lost function.

A question then arises as to what is the function of the association input in normal animals. One of the specific feature of the central nervous system is the redundancy of the same function. Therefore, it is possible that the same function is represented by multiple systems such as the thalamic and associational input. If this is the case, then the association fibers to the motor cortex functions only at the time of emergency. However, it is unlikely that this is the only function of this system. During the course of our previous study in which the monkey was trained without the sensory cortex (Asanuma and Arissian 1984), we received an impression that the monkey had difficulty in learning the motor skill. Encouraged by this observation, we examined whether there was a long-lasting potentiation (LTP) of synaptic transmission in this system which could be the basis of motor learning. It was found, to our surprise and delight, that tetanic stimulation of area 2 produced powerful LTP of the synaptic potentials in the motor cortex in the cat (Sakamoto et al. 1987). The results altogether indicated that this system constitutes the substrate of adaptive function which is known to exist in the mammalian central nervous system.

References

Asanuma H, Arissian K (1984) Experiments on functional role of peripheral input to motor cortex during voluntary movements in the monkey. J Neurophysiol 52:212–227

Asanuma H, Rosén I (1972) Topographical organization of cortical efferent zones projecting to distal forelimb muscles in the monkey. Exp Brain Res 14:243–256

Asanuma H, Fernandes J, Scheibel ME, Scheibel AB (1974) Characteristics of projections from the nucleus ventralis lateralis to the motor cortex in the cats: an anatomical and physiological study. Exp Brain Res 20:315–330
Asanuma H, Larsen KD, Yumiya H (1979) Direct sensory pathway to the motor cortex: a basis of cortical reflexes. In: Asanuma H, Wilson VJ (eds) Integration in the nervous system. Igaku-shoin, Tokyo New York, pp 223–238
Asanuma H, Larsen KD, Yumiya H (1980) Peripheral input pathways to the monkey motor cortex. Exp Brain Res 38:349–355
Asanuma H, Waters RS, Yumiya H (1981) Physiological properties of neurons projecting from area 3a to area 4 of feline cerebral cortex. J Neurophysiol 48:1048–1057
Asanuma H, Kosar E, Tsukahara N, Robinson H (1985) Modification of the projection from the sensory cortex to the motor cortex following the elimination of thalamic projections to the motor cortex in cats. Brain Res 345–79–86
Carpenter MB, Correll JW (1961) Spinal pathways mediating cerebellar dyskinesia in rhesus monkey. J Neurophysiol 24:535–551
Grant G, Landgren S, Silfvenius H (1975) Columnar distribution of U-fibers from the postcruciate cerebral projection area of the cat's group I muscle afferents. Exp Brain Res 24:57–74
Ichikawa M, Asanuma H (1987) Reorganization of the projection from the sensory cortex to the motor cortex following elimination of the thalamic projection to the motor cortex in cats; Golgi, electron microscope and degeneration study. Brain Res (in press)
Ichikawa M, Arissian K, Asanuma H (1985) Distribution of corticocortical and thalamocortical synapses on identified motor cortical neurons in the cat: Golgi, electron microscopic and degeneration study. Brain Res 345:87–101
Jones EG, Coulter JD, Hendry SH (1978) Intracortical connectivity of architechtonic fields in the somatic sensory, motor and parietal cortex of monkeys. J Comp Neurol 181:291–348
Kosar E, Waters RS, Tsukahara N, Asanuma H (1985) Anatomical and physiological properties of the projection from the sensory cortex to the motor cortex in normal cats: the difference between corticocortical and thalamocortical projections. Brain Res 345:68–78
Lorente de No R (1949) Cerebral cortex: architecture, intercortical connections, motor projection. In: Fulton JF (ed) Physiology of the nervous system, 3rd ed. Oxford Univ Press, New York, pp 274–298
Mackel R (1978) The role of the monkey sensory cortex in the recovery from cerebellar injury. Exp Brain Res 66:638–652
Olszewski J (1952) The thalamus of the *Macaca mulatta*. Kager, Basel
Rosén I, Asanuma H (1972) Peripheral afferent inputs to the forelimb area of the monkey motor cortex: input-output relations. Brain Res 14:257–273
Sakamoto T, Porter LL, Asanuma H (1987) Long-lasting potentiation of synaptic potentials in the motor cortex produced by stimulation of the sensory cortex in the cat: a basis of motor learning. Brain Res 413:360–364
Szabo J, Cowan WM (1984) A stereotaxic atlas of the brain of the cynomolgus monkey *(Macaca fascicularis)*. J Comp Neurol 222:265–300
Tatton EG, Forner SD, Gerstein GL, Chambers WW, Liu CM (1975) The effect of postcentral cortical lesions on motor responses to sudden upper limb displacements in monkeys. Brain Res 96:108–113
Travis AM, Woolsey CN (1956) Motor performance of monkeys after bilateral partial and total cerebral decortications. Am J Phys Med 35:273–289
Wall PD (1970) The sensory and motor role of impulses travelling in the dorsal columns towards cerebral cortex. Brain 93:505–524

Anatomical and Electrophysiological Studies of Aberrant Corticorubral Fibers Induced by Lesions of the Cerebral Cortex in Kitten

F. Murakami [1], S. Higashi [1], E. Kosar [2], and Y. Fujito [3]

[1] Department of Biophysical Engineering, Faculty of Engineering Science, Osaka University, Toyonaka 560, Osaka, Japan
[2] Monell Chemical Senses Center, 3500 Market Street, Philadelphia, Pennsylvania 19104, USA
[3] Department of Physiology, Sapporo Medical College, Nishi 17-chome, Minami Ichijo, Sapporo 060, Japan

1 Introduction

Brain lesions placed during early development have been thought to induce more axonal sprouting and synaptogenesis than do lesions made later in life (Cotman and Nadler 1978; Tsukahara 1981). Early brain damage is often associated with abnormal projections, which are not observed after adult brain damage.

Lesions of unilateral sensori-motor cortex of the kitten performed at less than 2 months after birth result in pronounced excitatory postsynaptic potentials (EPSP) in rubrospinal neurons, in response to stimulation of the contralateral sensori-motor cortex or the cerebral peduncle (Tsukahara and Fujito 1981; Tsukahara et al. 1983). This was interpreted to reflect collateral sprouting of ipsilaterally projecting corticofugal fibers which form new synapses with the neurons in the contralateral red nucleus (RN). Morphological studies demonstrated that there are aberrant crossed corticorubral projections in the neonatally hemispherectomized rat (Nah and Leong 1976a, b) and kitten (Villablanca et al. 1982). The present chapter seeks to address two issues. First: the possibility that crossed corticorubral projections are present during early development and that lesions placed in the cortex arrest the retraction of crossed projections. The second issue concerns the properties of synaptic transmission of the aberrant crossed projections. Although it has already been shown that the crossed corticorubral projection is functionally effective its properties of synaptic transmission have not yet been extensively studied.

2 Crossed Corticorubral Projections in Neonatal Cat

2.1 *Experiments Utilizing Anterograde Transport of Horseradish Peroxidase (HRP)*

If the aberrant crossed projections were the result of arrested retraction of crossed projections, it should be possible to find crossed projections in the neonatal kitten.

To test this possibility HRP was injected using a Hamilton syringe into the sensorimotor cortex of 7–30-day-old kittens and cats which had received unilateral ablations of the sensorimotor cortex (Kosar et al. 1985). The ablations were

H. Flohr (Ed.)
Post-Lesion Neural Plasticity

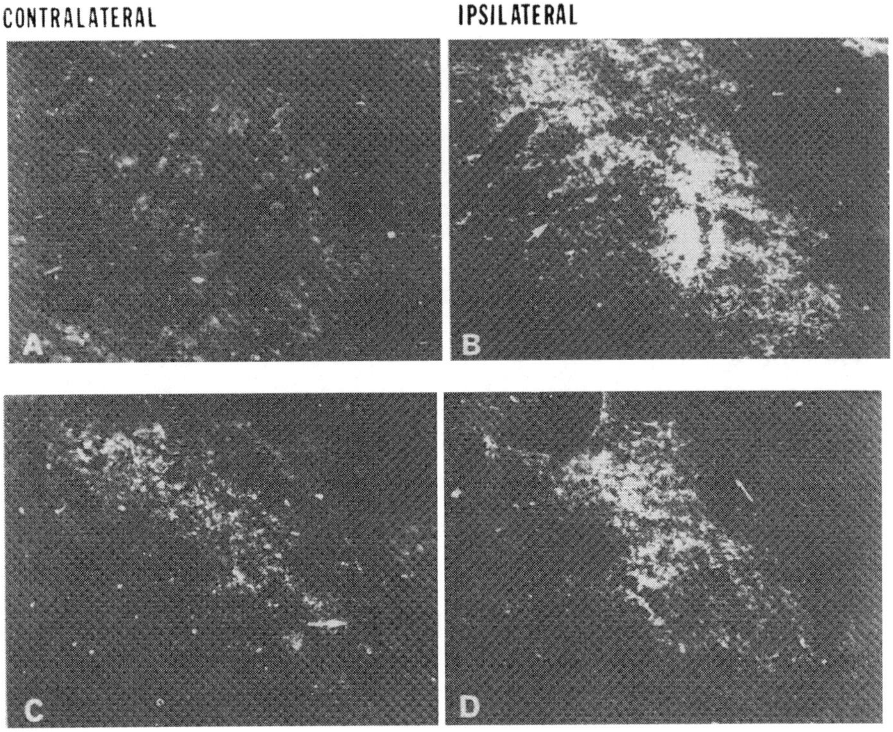

Fig. 1. A and **B** Sections through the contralateral and ipsilateral red nucleus of an intact kitten following unilateral injections of HRP into the sensori-motor cortex. The contralateral red nucleus is devoid of anterograde labeling (**A**) whereas the ipsilateral red nucleus (**B**) contains both labeled terminals and fibers. **C** and **D** anterograde corticorubral labeling resulting from unilateral injections of HRP into the intact motor cortex of kittens in which the opposite sensori-motor cortex had previously been ablated. Both the contralateral and ipsilateral red nuclei contain anterograde labeling. Dark-field illumination. *Arrows* point to labeled fibers

performed by suction 3.5–4 weeks after birth and the animals were allowed to survive for an additional 4 months. The animals were perfused intracardially 3–5 days later. Sections containing the RN were reacted with tetramethylbenzidine (TMB) (Mesulam 1982), coverslipped and observed under dark-field illumination. Figure 1 A and B shows photomicrographs of contralateral (A) and ipsilateral RN (B), respectively. Abundant labeling of both terminals and fibers can be seen in the RN ipsilateral to the injection. By contrast, the contralateral RN was devoid of labeling. In the animals in which the sensorimotor cortex had previously been ablated on one side, injections of HRP into the intact sensorimotor cortex resulted in terminal labeling within both ipsilateral (Fig. 1 D) and contralateral RN (Fig. 1 C).

2.2 *Experiment Utilizing Anterograde Transport of Phaseolus vulgaris Leucoagglutinin*

2.2.1 Labeled Fibers in the Red Nucleus

Plant lectin from the red kidney bean, *Phaseolus vulgaris* leucoagglutinin (PHA-L), is known to be an excellent anterograde tracer (Gerfen and Sawchenko 1984). PHA-L, when delivered iontophoretically into the CNS with fine tipped micropipets and then localized immunohistochemically, discretely labels cell bodies and their dendrites in a small area near the site of injection, along with their axons and terminal specializations. Hence, PHA-L has been successfully applied to various loci in the central nervous system as an anterograde tracer (e.g., Carlsen and Heimer 1986; Keller et al. 1985; Watts et al. 1987).

PHA-L was injected iontophoretically into the left pericruciate cortex of two groups of animals: seven intact control kittens 2–7 weeks of age and six kittens in which the sensorimotor cortex had previously been ablated unilaterally. The ablations were performed by suction 3–4.5 weeks after birth and the animals were allowed to survive for an additional 1–6 weeks. PHA-L, 2.5% in 0.05 M Tris-buffered saline (TBS), was injected iontophoretically through glass micropipets with a tip of 20–40 μm. After survival times of 7–14 days, the animals were deeply anesthetized and perfused transcardially. The brains were postfixed for 2 h and the brains stem containing the RN was dissected out and cut coronally. The sections were incubated for 12–14 h in affinity purified anti-PHA (L+E) serum and then processed for immunohistochemistry following the avidin-biotin complex (ABC) method.

In the RN ipsilateral to the injection, the labeled fibers were generally seen throughout the rostrocaudal extent of the RN in most of the animals. The fibers branch extensively and form varicosities which appear in both an en passant and terminal fashion along the trajectory of the axons (Fig. 2A). In the RN contralateral to the injection some labeled fibers were also observed, although the fibers were much more sparsely distributed than in the ipsilateral RN. They appeared to branch less than those observed ipsilaterally. Labeled fibers were seen in the contralateral RN in all of the intact kittens.

In the animals which underwent the cortical ablation, labeled fibers were also observed in the ipsilateral RN (Fig. 2B). However, in contrast to the RN of the normal kitten, many labeled fibers were seen in the RN contralateral to the cortical injection (Fig. 2C). The fibers branched as extensively as those observed in the ipsilateral RN and formed varicosities along the trajectory of the axons.

2.2.2 Increase of Crossed Corticorubral Fibers

Although labeled fibers were found in the RN contralateral to the injection in every kitten, those fibers were much more numerous in lesioned animals than in intact ones.

Figure 3A shows drawings of labeled fibers in the RN of two normal kittens. These drawings are representative of the corticorubral projections in these animals. The labeled fibers in the contralateral RN (A, right row) are much fewer

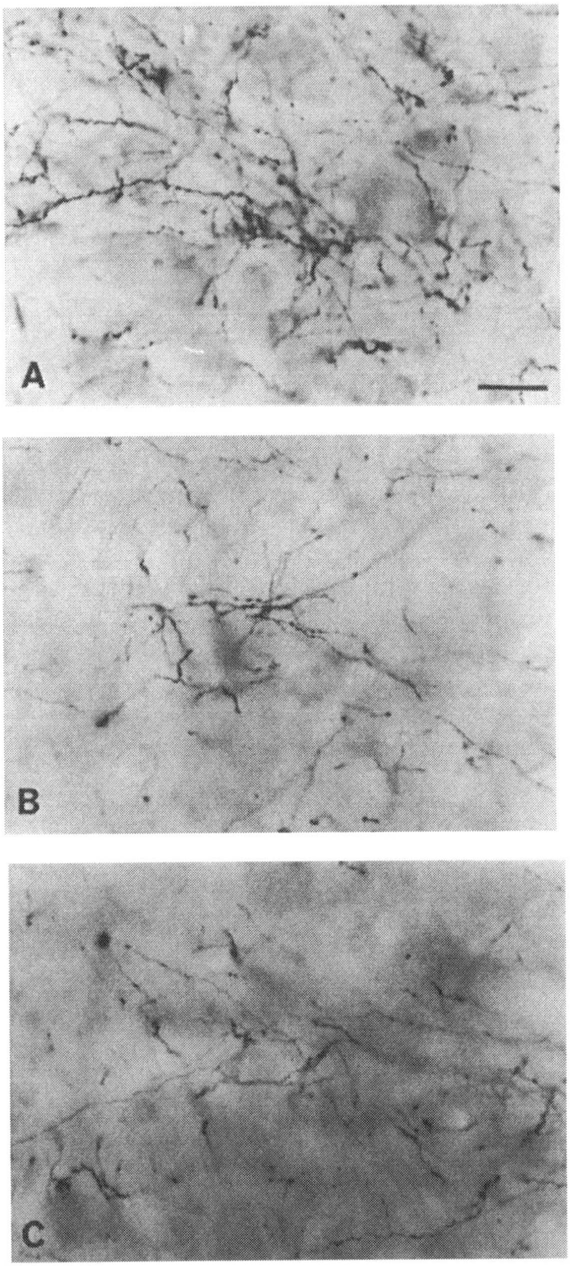

Fig. 2. Corticorubral fibers labeled with PHA-L. **A** Labeled fibers observed in the RN ipsilateral to the cortical injection in an intact animal. Many fibers with and without varicosities can be seen. Perfused 1 month after birth. **B** Labeled fibers in the red nucleus ipsilateral to the injection site. In this animal the sensorimotor cortex was ablated unilaterally 3 weeks after birth and animal was sacrificed 26 days later. PHA-L was injected into the intact sensorimotor cortex 14 days prior to perfusion. **C** Labeled fibers in the red nucleus contralateral to that shown in **B**. Note the abundance of labeled fibers. Although the fibers were more extensively labeled in the ipsilateral red nucleus in the animal shown in **A** than that in **C**, much fewer labeled fibers were found on the contralateral side of this intact animal. Bar: 25 μm

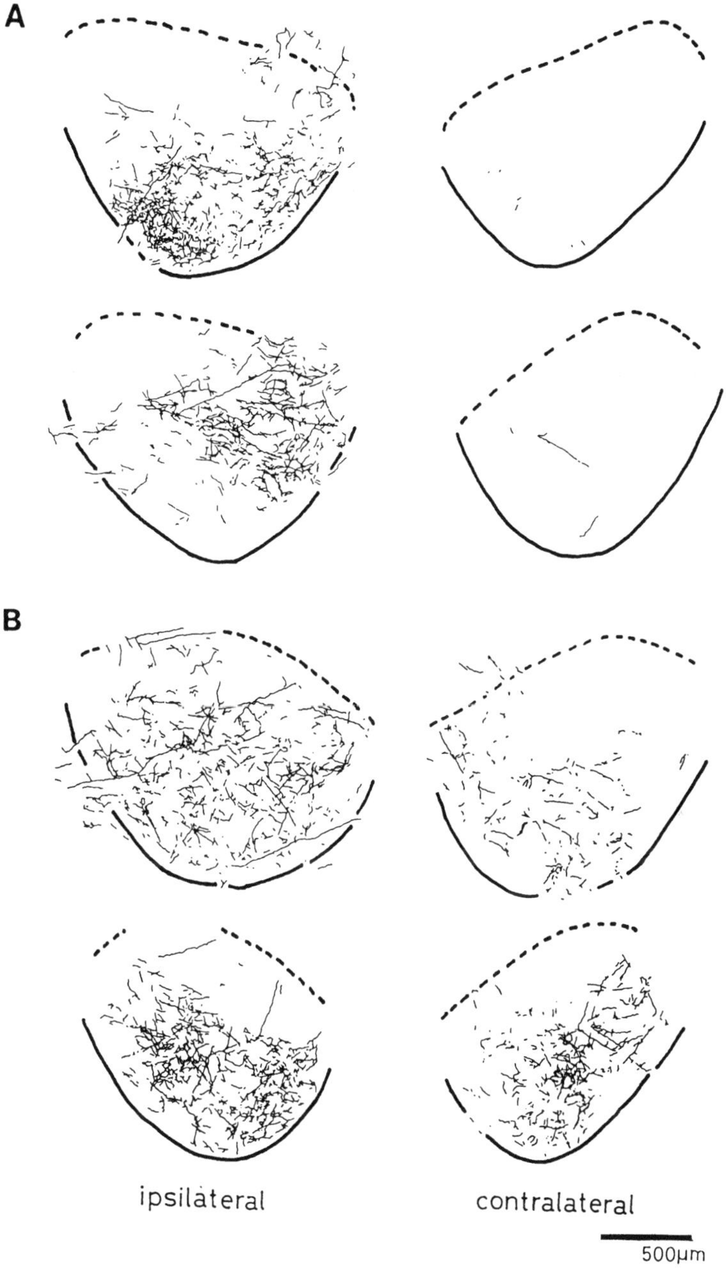

Fig. 3. Labeled corticofugal fibers in the red nucleus of intact kittens following injection of PHA-L into the sensorimotor cortex. *Left row* red nucleus ipsilateral to the injection. *Right row* red nucleus contralateral to the injection. **A** camera lucida drawings of labeled fibers in the red nucleus of two intact animals, 7 (*upper*) and 3 weeks (*lower*) of age. Frontal sections. **B** Labeled corticofugal fibers in the red nucleus of kittens which underwent unilateral cortical ablations. Camera lucida drawings of frontal sections. Sensorimotor cortex was ablated at 3 (*uper*) and 4 weeks (*lower*) after birth. Note that the contralateral labeling is almost comparable to the ipsilateral labeling

than those observed ipsilaterally (A, left row). By contrast, in the animals which underwent cortical lesions the number of labeled fibers observed in the RN contralateral to the injection sites (Fig. 3 B, right row) was almost comparable to those observed ipsilaterally (Fig. 3 B, left row).

It seemed that labeled fibers in the red nucleus were topographically arranged in both normal and lesioned animal; injections lateral to the cruciate sulcus resulted in extensive labelings at the medial portion of the RN and those into the medial part of the postcruciate sulcus resulted in extensive labelings at the ventral portion of the RN.

2.3 Electrophysiological Study of Crossed Corticorubral Projections

To examine whether the crossed corticorubral fibers are collaterals of the pyramidal tract, a collision test was performed. Stimulation of the contralateral CP (c-CP, A 8.0, L 5.5, H-4) was followed by stimulation of the contralateral pyramidal tract (c-PT, P 4.5, L 1, H-10) at various intervals while recording intracellularly from RN cells as shown in Fig. 4D (Kosar et al. 1985). The EPSP's elicited by stimulation of the c-PT began to show a depression at a stimulus interval of 1.8 ms and were reduced by 50% in amplitude at a stimulus interval of 1.6 ms. A complete collision was observed at an interval of 1.4 ms. The interval of 1.6 ms can be accounted for solely by the conduction time from the c-CP to the c-PT plus the absolute refractory period of corticorubral fibers. Thus, it is likely that the EPSP's elicited by stimulation of the c-CP were mediated, at least in part, by collaterals of the pyramidal tract fibers.

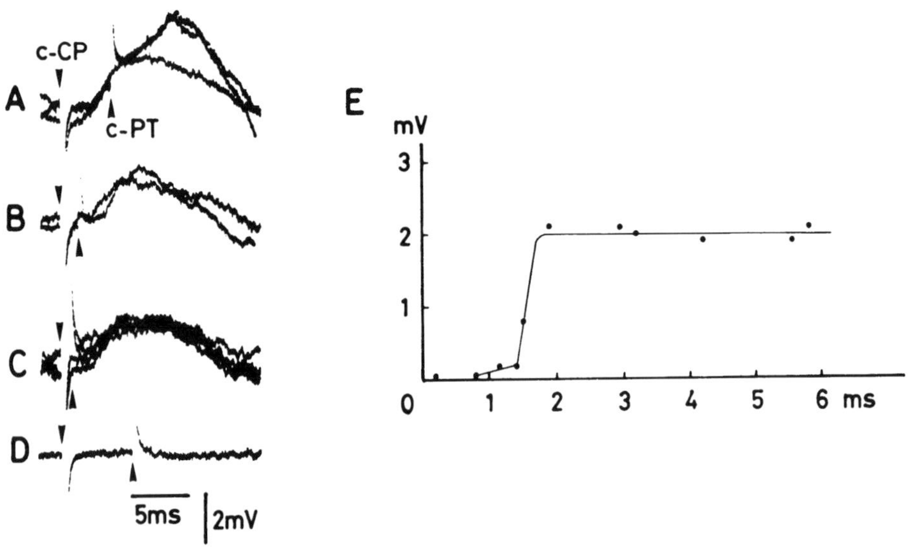

Fig. 4. Collision of the EPSP elicited by stimulation of contralateral cerebral peduncle (c-CP) and contralateral pyramidal tract (c-PT). From **A** to **C**, the stimulus interval was gradually reduced. **D** extracellular recording showing that there is no noticeable field potential. **E** the relation between the amplitude of the c-PT EPSP and the stimulus interval

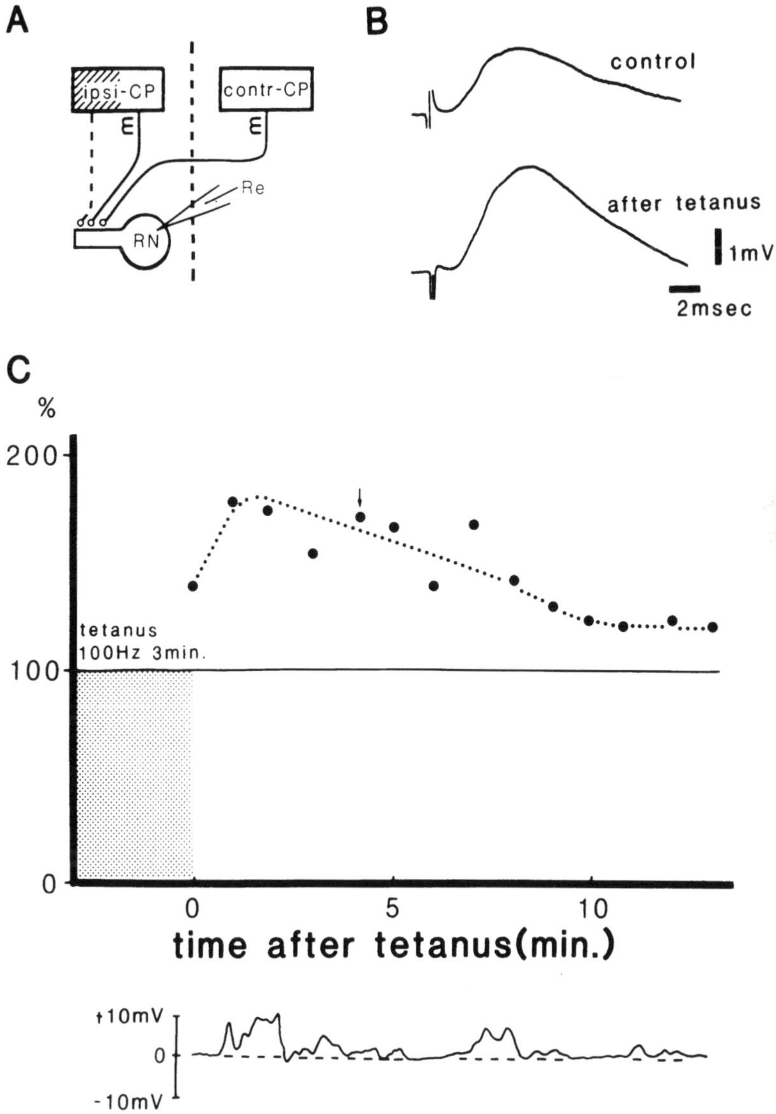

Fig. 5. Potentiation of ipsilaterally induced corticorubral EPSP following tetanic stimulation applied to ipsilateral cerebral peduncle. **A** Experimental arrangement. Intracellular recording was made from a red nucleus neuron of a cat whose sensorimotor cortex had been partially ablated unilaterally (indicated by *half-shaded column*). Stimulating electrode was placed at the ipsilateral cerebral peduncle. *ipsi-CP* ipsilateral cerebral peduncle; *contr-CP* contralateral cerebral peduncle; *RN* red nucleus neuron; *Re* recording electrode. *Dotted line* in the middle represents midline. **B** Averaged (20 times) records of ipsilaterally induced CP-EPSP. *Upper trace* shows control record before tetanic stimulation (100 Hz, 3 min) applied to CP. Lower trace shows the EPSP after tetanic stimulation. **C** Time course of the amplitude change of the ipsilaterally induced CP-EPSP following tetanic stimulation applied to ipsilateral CP. The amplitude (*ordinate*) is expressed as the percentage of the amplitude of the EPSP against that of the control value measured before tetanic stimulation. The point indicated by the *arrow* corresponds to the record shown in **B**. At the bottom is shown the intracellularly recorded membrane potential, after tetanus, as expressed by the value relative to that before tetanus. Note that the potentiation of the amplitude is not due to the hyperpolarization of the membrane potential

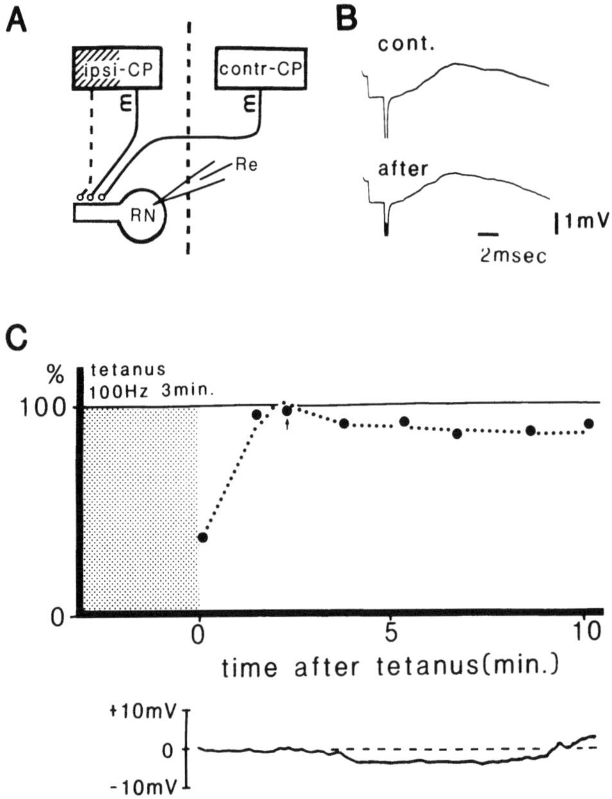

Fig. 6. Tetanic stimulation applied to contralateral cerebral peduncle. **A** Experimental arrangement. Intracellular recording was made from the red nucleus of a cat whose sensorimotor cortex had been partially ablated unilaterally. See legend to Fig. 4 for abbreviations in the figure. **B** Averaged (20 times) records of contralaterally induced CP-EPSP. **C** Time course of the amplitude of contralaterally induced CP-EPSP following tetanic stimulation. Note that the contralaterally induced EPSP is slightly reduced following tetanus. For detailed explanation of **B** and **C**, see legend to Fig. 5

In the experiment illustrated in Fig. 5, intracellular recordings were made in RN cells in a preparation which had undergone partial ablation of the sensorimotor cortex. It was possible to elicit EPSP's by stimulation of either ipsilateral or contralateral CP. Figure 5 shows the potentiation of the ipsilateral CP-EPSP's, following tetanic stimulation (100 Hz, 3 min). The amplitude of the EPSP's slowly increased and attained a peak after a few minutes and this potentiation lasted for more than 10 min. Figure 6 shows the time course of the amplitude of the corticorubral EPSP elicited in the same RN cells shown in Fig. 5, after similar tetanic stimulation of the c-CP. In contrast to the ipsilaterally evoked EPSP's, there was no potentiation of the amplitude of the EPSP following contralateral stimulation.

3 Discussion

Crossed corticorubral fibers are present in neonatal intact kittens, although the number is few. The presence of a crossed corticorubral projection has also been suggested in adult cats using silver impregnation methods (Rinvik 1965; Sadum 1975). The failure to find crossed projections by reacting anterogradely transported HRP with TMB could be due to the difficulty in discriminating a small number of labeled terminals from nonspecific precipitates (Mesulam 1982).

It is widely accepted that there are exuberant projections in the CNS and that they retract during normal development (Cowan et al. 1984; O'Leary et al. 1981; Purves and Lichtman 1985; Stanfied et al. 1982). Early brain damage could prevent the retraction. According to Leonard and Goldberger (1987), there are abundant crossed corticorubral fibers soon after birth. However, in our laboratory, extensive contralateral labeling was not observed in intact animals even when they were sacrificed at an earlier age than that at which the cortical lesions were performed in experimental animals. The number of labeled fibers was always much more numerous in the contralateral RN of lesioned animals than of intact ones. The aberrant crossed projections observed here are therefore at least partly ascribed to the increase of crossed fibers. The aberrant projection observed in the rat was similarly interpreted (Nah et al. 1980).

Collision experiments demonstrate that the crossed fibers are, at least in part, collaterals of pyramidal tract fibers. It is unknown whether a single cortical neuron innervates the RN of both sides, but in one case a single labeled fiber could be traced, by reconstruction from serial sections, from the contralateral RN to the ipsilateral RN. A further study is needed to confirm the possibility of bilateral innervation of the RN by a single corticorubral neuron.

A post-tetanic potentiation of corticorubral EPSP's similar to that observed here was described in adult cats. The reason for the difference in the post-tetanic potentiation between ipsilateral and contralateral CP-EPSP is unclear. One possible explanation could be insufficient quantities of available neurotransmitter within the newly formed synapses, as a result of their possible immaturity. More detailed physiological and anatomical studies are needed to assess this possibility.

In conclusion, unilateral ablation of the sensorimotor cortex results in an increase in crossed corticorubral fibers. Those fibers are at least in part collaterals of pyramidal tract fibers.

References

Carlsen J, Heimer L (1986) The projection from the parataenial thalamic nucleus, as demonstrated by the *Phaseolus vulgaris*-leucoagglutinin (PHA-L) method, identifies a subterritorial organization of the ventral striatum. Brain Res 374:375–379

Cotman CW, Nadler JV (1978) Reactive synaptogenesis in the hippocampus. In: Cotman CW (ed) Neuronal plasticity. Raven, New York, pp 227–271

Cowan WM, Fawcett JW, O'Leary DDM, Stanfield BB (1984) Regressive events in neurogenesis. Science 225:1258–1265

Gerfen CR, Sawchenko PE (1984) An anterograde neuroanatomical tracing method that shows the detailed morphology of neurons, their axons and terminals: Immunohistochemical localization of an axonally transported plant lectin, *phaseolus vulgaris* leucoagglutinin (PHA-L). Brain Res 290:219–238

Keller A, White EL, Cipolloni PB (1985) The identification of thalamocortical axon terminals in barrels of mouse SmI cortex using immunohistochemistry of anterogradely transported lectin (*Phaseolus vulgaris* leucoagglutinin). Brain Res 343:159–165

Kosar E, Fujito Y, Murakami F, Tsukahara N (1985) Morphological and electrophysiolgical study of sprouting of corticorubral fibers after lesions of the contralateral cerebrum in kitten. Brain Res 347:217–224

Leonard CT, Goldberger ME (1987) Consequences of damage to the sensorimotor cortex in neonatal and adult cats. II. Maintenance of exuberant projections. Dev Brain Res 32:15–30

Mesulam M-M (1982) Principles of horseradish peroxidase neurohistochemistry and their applications for tracing neuronal pathways – axonal transport, enzyme histochemistry and light microscopic analysis. In: Mesulam M-M (ed) Tracing neural connections. Willey, New York, pp 1–151

Murakami F, Tsukahara N, Fujito Y (1976) Physiological properties of the newly formed cortico-rubral synapses of red nucleus neurons due to collateral sprouting. Brain Res 103:147–151

Nah SH, Leong SK (1976a) Bilateral corticofugal projection to the red nucleus after neonatal lesions in the albino rat. Brain Res 107:433–436

Nah SH, Leong SK (1976b) An ultrastructural study of the anomalus corticorubral projection following neonatal lesions in the albino rat. Brain Res 111:162–166

Nah SH, Ong LS, Leong SK (1980) Is sprouting the result of a persistent neonatal connection? Neurosci Lett 19:39–44

O'Leary DDM, Stanfield BB, Cowan WM (1981) Evidence that the early postnatal restriction of the cells of origin of the callosal projection is due to the elimination of axon collaterals rather than to the death of neurons. Dev Brain Res 1:607–617

Purves D, Lichman JW (1985) Rearrangement of developing neuronal connections. In: Principles of neural development. Sinauer, Sunderland, pp 271–300

Rinvik E (1965) The corticorubral projection in the cat. Further observations. Exp Neurol 12:278–291

Sadun A (1975) Differential distribution of cortical terminations in the cat red nucleus. Brain Res 99:145–151

Stanfield BB, O'Leary DDM, Fricks C (1982) Selective collateral elimination in early postnatal development restricts cortical distribution of rat pyramidal tract neurons. Nature 298:371–373

Tsukahara N (1981) Synaptic plasticity in the mammalian central nervous system. Annu Rev Neurosci 4:351–379

Tsukahara N, Fujito Y (1981) Neuronal plasticity in the newborn and adult feline red nucleus. In: Flohr H, Precht W (eds) Lesion-induced neuronal plasticity in sensorimotor systems. Springer, Berlin Heidelberg New York, pp 64–74

Tsukahara N, Fujito Y, Kubota M (1983) Specificity of the newly-formed corticorubral synapses in the kitten red nucleus. Exp Brain Res 51:45–56

Villablanca JR, Olmstead CE, Sonnier BJ, McAllister JP, Gomez F (1982) Evidence for a crossed corticorubral projection in cats with one cerebral hemisphere removed neonatally. Neurosci Lett 33:241–246

Watts AG, Swanson LW, Sanchez-Watts G (1987) Efferent projections of the suprochiasmatic nucleus. I. Studies using anterograde transport of *Phaseolus vulgaris* leucoagglutinin in the rat. J Comp Neurol 258:204–229

Parallels in Behavioral and Neural Plasticity Induced by Unilateral Vibrissae Removal and Unilateral Lesion of the Substantia Nigra

J. P. Huston, H. Steiner, R. K. W. Schwarting, and S. Morgan
Institute of Physiological Psychology, University of Düsseldorf, Universitätsstraße 1, D-4000 Düsseldorf, FR Germany

1 Introduction

A prominent facial feature of mammals is the presence of long tactile hairs referred to as "vibrissae" or "whiskers". In fact, such tactile sense organs are found in almost all mammals except *Homo sapiens* (Ahl 1986). They are employed in a great variety of behaviors (see Gustafson and Felbain-Keramidas 1977 and Ahl 1986 for reviews of vibrissae function). For example, in a novel environment rats whisk their vibrissae over surfaces and edges in a repetitive manner (Vincent 1912; Welker 1964). This behavior is thought to serve for sampling tactile stimuli (Welker 1964; Zucker and Welker 1969; Shipley 1974). There is evidence that such tactile information from the vibrissae is analyzed in a complex form, which may even allow "shape" or "pattern" perception (Shipley 1974). The importance of the vibrissae system for the rat is also reflected by its large cortical representation. About 20% of the somatosensory cortex has been found to respond to stimulation of the vibrissae region of the face (Welker 1971).

We have employed the unilateral removal of the vibrissae as a tool to examine ensuing behavioral asymmetries in relation to concomitant changes in the central nervous system. In this paper we review a series of studies that show striking parallels between the effects of unilateral removal of vibrissae and unilateral injection of the neurotoxin 6-hydroxydopamine (6-OHDA) into the substantia nigra (SN) on the animal's behavior, and on the anatomy and biochemistry of the brain of the adult rat. These parallels are important from several perspectives. One, they indicate possible functional and anatomical links between the vibrissae system and the basal ganglia (BG). Two, they raise the possibility that unilateral removal of the vibrissae can serve as a model or substitute for the direct surgical interventions in the brain that are commonly employed to (a) study the effects of damage to the nigrostriatal pathway, for example, as a model of Parkinsonism, (b) investigate the neuropharmacology of catecholamine systems in the brain, and (c) investigate the dynamics of recovery from brain lesions.

1.1 The Vibrissae-Trigeminal System

The appearance, number and array of the vibrissae may greatly vary between species (Ahl 1986), and even between strains (Welker and Van der Loos 1986). In rats, tactile hairs are found upon the upper lip (mystacium), the lower lip, chin,

H. Flohr (Ed.)
Post-Lesion Neural Plasticity

brow, and cheek (Vincent 1913; Greene 1963). The mystacial vibrissae on each side of the snout (30–35 in the rat) are arranged in five (sometimes six) horizontal rows, each containing five to nine large vibrissae (Vincent 1913; Zucker and Welker 1969; Welker 1971).

Each vibrissa has its origin in a specialized follicle (Andres and von Düring 1973). The shaft of the vibrissa acts as a lever upon nerve endings of the follicle. Displacement of the vibrissa relative to its follicle and receptors results in a signal which is carried by the infraorbital branch of the trigeminal nerve to the trigeminal nuclear complex (Vincent 1913; Greene 1963; Arvidsson 1982). An electrophysiological study of single units in the trigeminal ganglion of the rat revealed that first-order trigeminal neurons can encode specific parameters of mechanical vibrissa displacement, including, among others, peripheral location, direction of deflection, onset, termination, amplitude, velocity, duration and temporal pattern (Zucker and Welker 1969). Processing of this information may allow analysis of the texture of the environment (Shipley 1974).

From the trigeminal nuclear complex, vibrissae information is transmitted to the contralateral ventrobasal complex of the thalamus (Smith 1973; Waite 1973a, b), and thence to the somatosensory cortex in the same hemisphere. In the rat (Welker 1971, 1976) and mouse (Woolsey and Van der Loos 1970; Van der Loos and Woolsey 1973) each mystacial vibrissa projects to an anatomically distinct group of cells called "barrel" in layer IV of the somatosensory cortex. These barrels match the arrangement of the vibrissae in the rat's face in terms of number and relative location. A somatotopic organization is also observed in the subcortical stations of the vibrissae-trigeminal system (Zucker and Welker 1969; Waite 1973a; Shipley 1974; Arvidsson 1982).

The importance of vibrissal sensory information for the anatomical organization of the brain has clearly been shown in the trigeminal system. The development of the distinct architecture of this system is critically dependent on sensory input from the vibrissae. Removing this input in neonatal rat and mouse by destruction of vibrissae follicles results in gross distortions in the anatomical organization of the trigeminal system (Van der Loos and Woolsey 1973; Killackey et al. 1976; Killackey and Belford 1979; Belford and Killackey 1979). This phenomenon seems to be age-dependent, since major anatomical alterations have only been found after follicle damage in the first postnatal week (Killackey and Belford 1979; Jeanmonod et al. 1981; Kaas et al. 1983). However, in the adult rat, removal of vibrissae has been shown to lower metabolic activity in the corresponding barrels (Durham and Woolsey 1978).

1.2 Functional Relationship Between the Vibrissae-Trigeminal System and the Basal Ganglia

We became interested in the vibrissae tactile system when we found evidence for a functional relationship between this sensory system and the BG. Manipulating the neural activity of the SN by lesion or by injection of neurotransmitter agonists resulted in a changed responsiveness to tactile stimulation in the vibrissae region of the face (Huston et al. 1980; Welzl and Huston 1981; Welzl et al. 1982). Unilateral intranigral injection of a GABA agonist, substance P and a metenkephalin analog led to an increased sensitivity of the side of the face contralateral to the injection (reviewed in Welzl et al. 1984). This effect was manifested in the perioral biting reflex (PBR): touching the face or the vibrissae around the mouth by a probe on the side contralateral to the injection resulted in vigorous biting of the probe. A systemic injection of the dopamine (DA) receptor agonist apomorphine (APO) elicited the PBR bilaterally. This response was also lateralized after a unilateral SN lesion (by 6-OHDA): animals which circled contraversively under APO also displayed a much stronger PBR upon stimulation of the side of the face contralateral to the lesion (Huston et al. 1980).

Recently, we obtained additional evidence for a relationship between the vibrissae system and BG structures. In experiments designed to examine neural plasticity in the BG in relation to experimentally induced behavioral changes, we found evidence for plasticity in SN efferents subsequent to various behavioral and surgical interventions (reviewed in Huston et al. 1987). Interestingly, there were striking similarities in neural and behavioral plasticity after injection of 6-OHDA into the SN and after removal of the vibrissae (reviewed below).

Having demonstrated functional interactions between the vibrissae-trigeminal system and the BG, it would be important to know how these systems interconnect anatomically and physiologically. Electrophysiological studies have shown that DA neurons of the SN respond to sensory input from various modalities (Chiodo et al. 1979; Romo and Schultz 1985; Strecker and Jacobs 1985; Rasmussen et al. 1986), including trigeminal stimulation (e.g., Harper et al. 1979; Chiodo et al. 1980). The route by which trigeminal sensory input reaches the DA pathway, however, remains unclear. Since the somatosensory cortex projects to the striatum (e.g., Wise and Jones 1977; Malach and Graybiel 1987), one bridge could be via cortico-striato-nigral projections. (A visual cortico-striato-nigral pathway has recently been demonstrated by Faull et al. 1986). Cortico-striatal projections may also directly influence DA transmission in the striatum by presynaptic regulation of the DA release from nigrostriatal nerve terminals (Romo et al. 1986). Several subcortical routes from the trigeminal nuclear complex to the BG have also been suggested, including: a direct trigeminostriatal projection (Yasui et al. 1987); a trigemino-pontine-nigral projection (Schneider 1986); a trigemino-tecto-thalamo-striatal axis (Krauthamer et al. 1987). Additionally, a raphé serotonergic influence on DA metabolism has been shown (e.g., Juorio and Greenshaw 1986; De Simoni et al. 1987); the raphé nuclei also receive input from the trigeminal nuclear complex (Lovick et al. 1977). However, the length of the latencies of trigeminally evoked SN responses may favor an oligosynaptic extrastriatal route to the SN (Harper et al. 1979).

Thus, there are numerous possibilities to account for a functional relationship between the vibrissae system and the BG. In the following sections we compare the effects of unilateral vibrissotomy and SN lesion by 6-OHDA at various levels.

2 6-OHDA- Versus Vibrissotomy-Induced Plasticity

2.1 *Anatomical Plasticity*

An anatomical projection found to change after a unilateral SN lesion and after hemivibrissotomy is the interhemispheric nigrostriatal projection. This connection was probably first described by Royce (1978) in the cat. Subsequent observations confirmed this tract in the rat (Veening et al. 1980; Fass and Butcher 1981; Collingridge 1982; Loughlin and Fallon 1982; Pritzel et al. 1983 a), cat (Fisher et al. 1984) and primate (Parent et al. 1983). Like its homolateral counterpart (Anden et al. 1964; Ungerstedt 1971 a), the crossed nigrostriatal projection is mainly catecholaminergic, most probably dopaminergic (Altar et al. 1983; Fallon et al. 1983; Pritzel et al. 1983 b; Consolazione et al. 1985). The proportion of interhemispheric projecting neurons was estimated to be less than 5% of the uncrossed projection (Altar et al. 1983; Fallon et al. 1983; Fisher et al. 1984; Morgan et al. 1985, 1986; Douglas et al. 1987). The interhemispheric fibers were found to decussate near their site of origin in the ventral mesencephalon (Steiner and Morgan, unpublished results; Douglas et al. 1987) and, like the uncrossed fibers, course rostrally via the medial forebrain bundle (Morgan and Huston 1988). Whereas the homolateral nigrostriatal projection has been shown to be topographically organized (Faull and Mehler 1978; Beckstead et al. 1979; Veening et al. 1980), little is known about such an organization in the crossed projection. However, we have recently found evidence for some dissociation between the two projections in the rostrocaudal distribution of the cells of origin (Morgan et al. 1985, 1986), as well as in the mediolateral distribution of the terminals in the striatum (Morgan and Huston 1988).

Unilateral injection of the neurotoxin 6-OHDA into the SN causes a degeneration of the ascending dopamine neurons (Ungerstedt 1971 b). Animals with such lesions exhibit sensorimotor asymmetries, including turning behavior toward the side of the injection and a loss of responsiveness to sensory stimulation on the body surface contralateral to the lesion (Ungerstedt 1971 b; Ljungberg and Ungerstedt 1976; Marshall 1979). Subsequent to the lesion, rats may recover from the induced sensorimotor asymmetries (Ungerstedt 1971 b; Marshall 1979; Steiner et al. 1988).

We examined the crossed nigrostriatal projection in rats at the point in time when the behavioral asymmetries were no longer apparent or had decreased. Animals showing recovery from asymmetry received an injection of the tract tracer horseradish peroxidase (HRP) or fluorescent tracers into the striatum on the side of the lesion. An analysis of their brains revealed evidence for an increased projection from the intact SN to the denervated striatum: these animals had more retrogradely labeled cells in the contralateral SN than normal rats (Pritzel et al. 1983 a).

Subsequently, we obtained evidence that such plasticity in crossed efferents of the SN could also be induced by peripheral interventions, which included vibrissae removal (Sabel et al. 1984). Moreover, we then found that removal of the vibrissae alone can alter the crossed nigrostriatal projection (Huston et al. 1986). In this experiment rats had their vibrissae shaved off on one side of the face daily for 10 consecutive days. At the end of this period, HRP was applied to the stria-

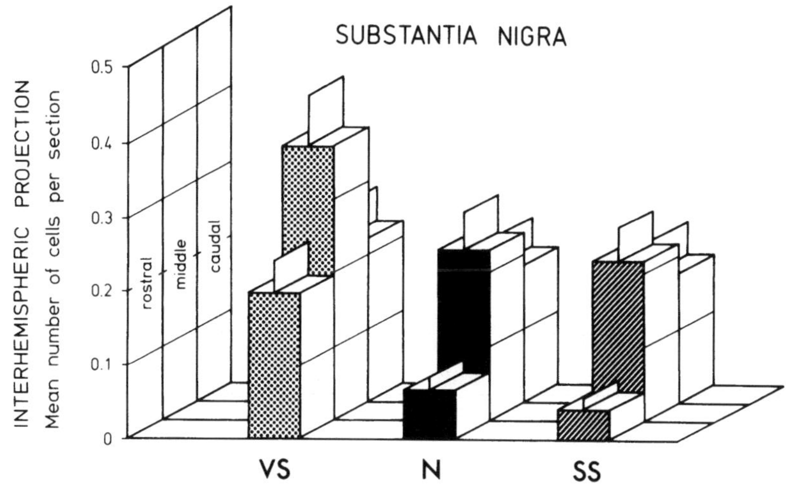

Fig. 1. The mean (+SEM) number of labeled cells per 50 µm section found in the rostral, middle and caudal parts of the substantia nigra contralateral to the striatum into which horseradish peroxidase (HRP) was deposited. *VS* vibrissae side: cell counts from unilaterally shaved animals with HRP applied to the striatum on the side opposite to the vibrissae removal; *N* normals: cell counts from controls with intact vibrissae; *SS* shaved side: cell counts from rats with HRP applied to the same side as vibrissae removal

tum either on the same side as, or the side opposite to the vibrissae removal. Normal control animals received the tracer either in the right or left striatum. The SN of both hemispheres were then examined for labeled cells. It was found that animals with HRP applied to the striatum on the side opposite to the vibrissae removal (i.e., the vibrissae side) had significantly more labeled neurons in the SN contralateral to the HRP deposition than rats with the tracer applied on the same side as vibrissae removal or normal controls (Fig. 1). No differences between the groups were found for the cell counts of the SN ipsilateral to the tracer application, i.e., in the homolateral nigrostriatal projection. Thus, after 10 days of unilateral vibrissae removal, evidence was found for an increase in the crossed nigrostriatal projection, which terminates in the striatum of the hemisphere opposite to the vibrissotomy, i.e., the hemisphere primarily affected by the sensory deprivation (see Sect. 1.1).

The change in uptake of HRP by the cells which give rise to the crossed nigrostriatal projection could be attributable to several different effects. It has been shown that uptake and retrograde transport of HRP are related to the activity of synapses and transmitter release (Kristensson 1978; Nishino et al. 1979; Eisenman and Azmitia 1982), as well as the number of synapses present in a HRP deposition site (Jones 1975). Thus, the change in cell-labeling found here could reflect alterations in synaptic activity and/or number of terminals.

It can be seen that after both interventions – unilateral 6-OHDA SN lesion and vibrissotomy – the crossed nigrostriatal projection shows similar changes. In the first case the striatum deafferented by the lesion received an apparently increased projection from the contralateral hemisphere. In the latter case, the striatum of the hemisphere partly deafferented by the sensory deprivation also received an apparently increased projection from the contralateral hemisphere.

2.2 Pharmacological Parallels

It has been shown before that rats develop a behavioral supersensitivity to DA agonists after degeneration of the nigrostriatal DA system (Ungerstedt 1971c). Ungerstedt (1971c) explained this phenomenon in terms of a denervation-induced increase in sensitivity of postsynaptic DA receptors in the striatum ipsilateral to the lesion. Numerous studies support this hypothesis (e.g., Feltz and de Champlain 1972; Creese et al. 1977; Mishra et al. 1980; Neve et al. 1982).

This supersensitivity is reflected by a reversing effect of DA receptor agonists on sensorimotor asymmetries of animals with unilateral DA depletion. As mentioned above, rats with unilateral SN lesions spontaneously circle toward the side of the lesion and respond less to stimulation of the contralateral body surface (Ungerstedt 1971b; Ljungberg and Ungerstedt 1976; Marshall 1979) (see Fig. 2). A systemic injection of the DA receptor agonist apomorphine, however, reverses the direction of sensorimotor asymmetry: the animals now circle toward the opposite (intact) side (Ungerstedt 1971c) and react preferentially upon stimulation of this side (Huston et al. 1980; Schallert et al. 1983).

Having found similar plasticity in the nigrostriatal projection after a 6-OHDA injection and after vibrissae removal, we wanted to know whether or not unilateral removal of the vibrissae would also lead to sensorimotor asymmetries, and whether such asymmetries could be influenced by APO (Steiner et al. 1986). We employed a new behavioral paradigm called "peritaxis" (described in detail in Steiner et al. 1988). This paradigm exploits the natural tendency of the rat to "ex-

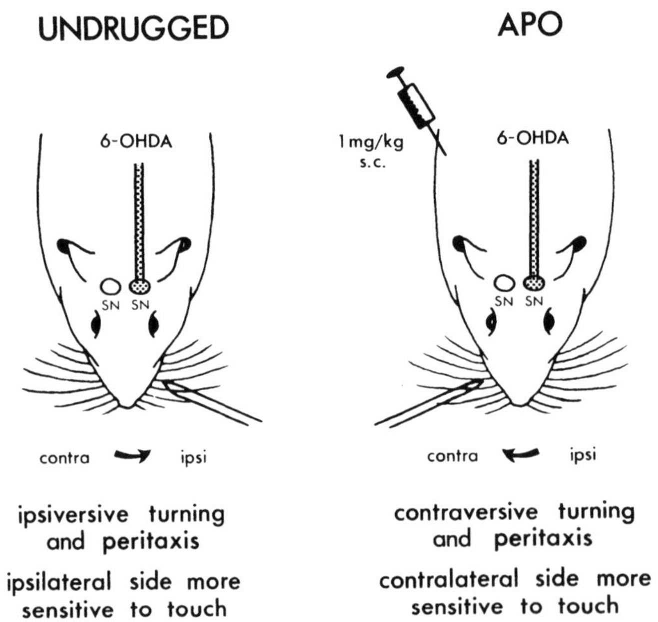

Fig. 2. Behavioral effects (9–11 days postlesion) of a unilateral injection of 6-OHDA into the substantia nigra (*SN*) of rats without (*UNDRUGGED*) and with additional challenge of dopamine receptors by subcutaneous injection of the dopamine receptor agonist apomorphine (*APO*)

plore" a new environment, a behavior which involves use of the vibrissae. When introduced in an open field, a rat will mainly locomote along the wall during the initial phase of testing (Schiorring 1979; Huston and Bures 1983). We measured the time rats moved along the wall of an open field with respect to the side of vibrissae removal. The behavior of two groups of rats was compared. One group had the vibrissae shaved on one side of the face for 10 days. The other group was sham-shaved for 10 days and had the vibrissae removed only 4 h before testing. Initially, the animal's spontaneous behavior was analyzed during a 5-min baseline period. One hour later the rat was injected with APO, and its drug-induced behavior was recorded.

Unilateral vibrissae removal resulted in an asymmetry in peritaxis: in the undrugged state both groups tended to scan the wall with the side of the face with intact vibrissae. Under the influence of APO the rats which were shaved for 4 h also spent significantly more time scanning the wall with the sensory intact side. In contrast, rats with vibrissae removed for 10 days now showed a reversed peritaxis asymmetry: they spent more time scanning the wall with the shaved side of the face (Steiner et al. 1986). Thus, unilateral removal of the vibrissae does lead to a sensorimotor asymmetry, and this asymmetry can be reversed by apomorphine after 10 days of vibrissotomy (Fig. 3A).

A comparison of the effects of apomorphine on rats after a 6-OHDA injection into the SN or vibrissotomy again suggests similar plasticity in DA transmission

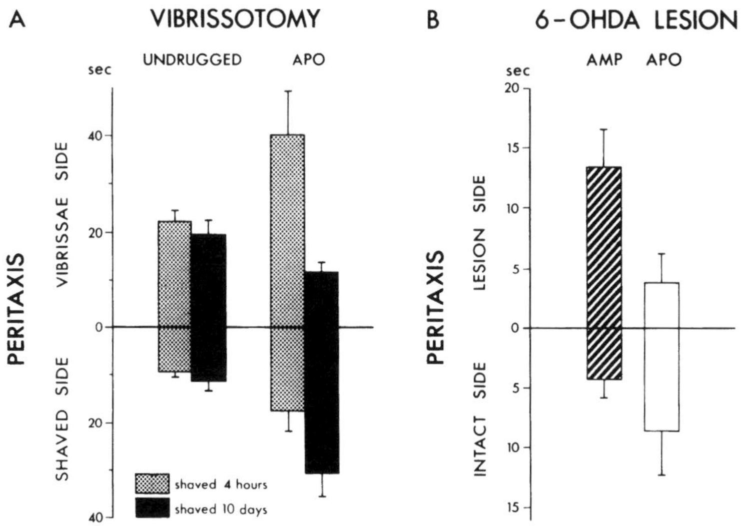

Fig. 3. Peritactic behavior of rats in 5-min test periods **A** after unilateral removal of the vibrissae (*VIBRISSOTOMY*) and **B** after a unilateral injection of 6-OHDA into the substantia nigra. **A** The mean (+SEM) duration of peritaxis with the shaved and the vibrissae side in groups of rats tested 4 h after shaving (*stippled bars*) or after 10 days of daily vibrissae removal (*black bars*). Rats were first tested undrugged. One hour later they were tested again under the influence of apomorphine (APO, 0.5 mg/kg, s.c.). **B** The mean (+SEM) duration of peritaxis with the side contralateral (*INTACT SIDE*) and ipsilateral (*LESION SIDE*) to the intranigral injection of 6-OHDA, tested after application of amphetamine (AMP, 1 mg/kg, i.p.; *hatched bars*) and apomorphine (APO, 1 mg/kg, s.c.; *open bars*) 9–11 days after the lesion

after both interventions. In the lesion model, the reversing effect of APO on sensorimotor asymmetries has been related to stimulation of supersensitive DA receptors in the striatum ipsilateral to the lesion (Ungerstedt 1971 c). It has yet to be determined whether or not a similar phenomenon (i.e., deafferentation-induced receptor supersensitivity) can account for the reversal in peritaxis seen after vibrissae removal. Interestingly, we recently also found peritaxis asymmetries after unilateral 6-OHDA injections, and these asymmetries could be reversed by apomorphine (Fig. 3 B; Steiner et al. 1988).

2.3 Plasticity in Dopamine Metabolism

Unilateral degeneration of DA neurons in the SN leads to a loss of DA in the striatum ipsilateral to the lesion. Extensive experimentation has indicated that this asymmetry in DA activity may partly be counteracted by metabolic adaptations in surviving DA terminals: neurons spared from the lesion may increase their rate of transmitter synthesis and release subsequent to the lesion (e.g., Hefti et al. 1980; Zigmond et al. 1984; Dravid et al. 1984).

To examine whether vibrissae removal would influence the DA metabolism in the striatum, we analyzed the levels of DA and its metabolites DOPAC and HVA in vibrissotomized rats (Steiner et al., in preparation). In Table 1 it can be seen that the animals with the vibrissae removed for 4 h had asymmetries in the levels of DA, DOPAC and HVA: they had a smaller transmitter and metabolite concentration in the striatum contralateral to the vibrissae removal, i.e., in the hemisphere primarily affected by the sensory deprivation. The rats shaved for 10 days did not exhibit such asymmetries. It remains to be determined whether similar plasticity as found in residual terminals after SN lesions or other central changes (for example, an increase in interhemispheric DA transmission, as de-

Table 1. Levels of the neurotransmitter DA and its metabolites DOPAC and HVA in the striatum ipsi- and contralateral to the side of vibrissae removal (μg/g brain tissue; mean + SEM)

	Striatum					
	DA		DOPAC		HVA	
4-h group						
Shaved side	11.213 + 0.164	p = 0.010	0.783 + 0.050	p = 0.053	0.431 + 0.028	p = 0.005
Vibrissae side	10.362 + 0.332		0.724 + 0.034		0.399 + 0.027	
10-day group						
Shaved side	9.769 + 0.185	p = 0.335	0.646 + 0.037	p = 0.348	0.358 + 0.033	p = 0.377
Vibrissae side	9.858 + 0.286		0.638 + 0.035		0.356 + 0.035	

The neurochemical analysis by means of high-performance liquid chromatography was performed in groups of rats tested after being unilaterally shaved only once (4-h group) and after 10 days of daily vibrissae removal (10-day group). P-values were obtained from t-tests for dependent samples.

scribed above) can account for the reduced asymmetries found after 10 days of vibrissae removal.

2.4 Recovery from Sensorimotor Asymmetries

In the postoperative period rats often recover spontaneously from SN lesion-induced sensorimotor asymmetries: ipsiversive turning behavior decreases, and orientation to sensory stimulation on the side contralateral to the lesion increases (Ungerstedt 1971 b; Ljungberg and Ungerstedt 1976; Marshall 1979; Neve et al. 1982; Steiner et al. 1985; Bonatz et al. 1987). Additionally, in parallel to turning behavior the lesion-induced peritaxis asymmetry decreases as well (Fig. 4; Steiner et al. 1988).

To determine whether rats would recover from the peritaxis asymmetry induced by vibrissae removal, we analyzed the intra-session pattern of peritaxis (Fig. 5). The peritaxis asymmetry was strongest during the first minutes of the 5-min testing period (when the animals showed most exploratory activity). In contrast to the rats shaved for 4 h, the animals shaved for 10 days did not exhibit significant differences in peritaxis between the vibrissae side and shaved side. Thus, our findings indicate recovery from peritaxis asymmetry in the rats that underwent vibrissae removal for 10 days.

Recovery of function after SN lesions has been related to DA activity remaining in the affected striatum. In contrast to animals which did not show recovery from lesion-induced sensorimotor asymmetries, rats which had recovered were shown to have an incomplete striatal DA depletion (Marshall 1979; Dravid et al. 1984; Carey 1986; Steiner et al. 1988). Pre- and postsynaptic plasticity (receptor supersensitivity, increase in transmitter turnover, enhancement in interhemispheric innervation) have been suggested to play a role in behavioral recovery in

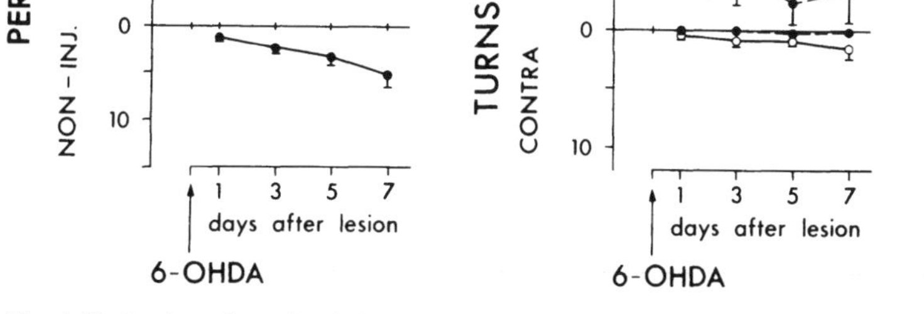

Fig. 4. Peritaxis and turning behavior during 7 days after intranigral injection of 6-OHDA (5-min test periods). **A** The mean (+SEM) duration of peritaxis with the side contralateral (*NON-INJ.*) and ipsilateral (*INJECTED SIDE*) to the 6-OHDA injection. **B** The mean (±SEM) number of ½-turns (*open circles*) and full turns (*full circles*) with a diameter of less than 20 cm contraversive (*CONTRA*) and ipsiversive (*IPSI*) to the side of lesion

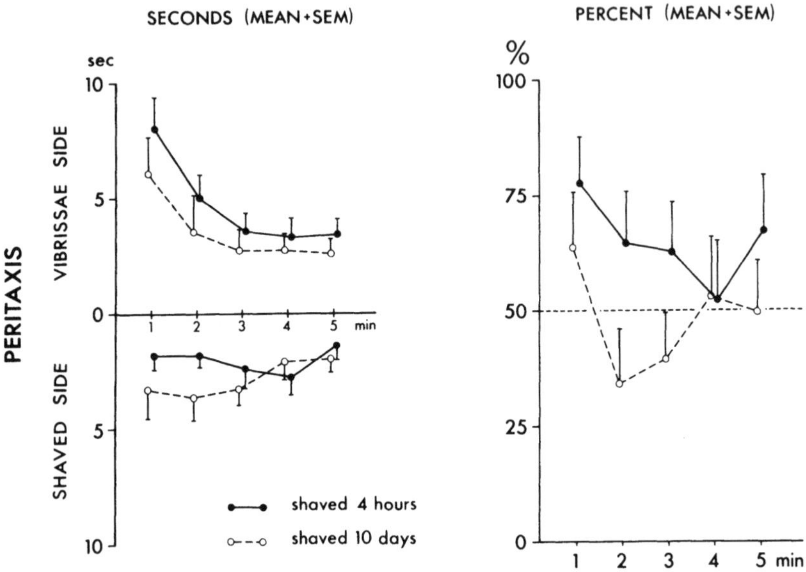

Fig. 5. Peritatic behavior after unilateral vibrissae removal within a 5-min test period as measured in a minute-by-minute analysis. *Left side* the mean (+SEM) duration of peritaxis with the shaved and vibrissae side in groups of rats tested 4 h after shaving (*full circles*) or after 10 days of daily vibrissae removal (*open circles*). *Right side* the corresponding percentage asymmetry score (mean+SEM): duration of peritaxis with the vibrissae side × 100 divided by the total duration of peritaxis

the lesion model (reviewed in Marshall 1984; Huston et al. 1987). Adaptations in receptor sensitivity and transmitter turnover are probably mediated by local feedback mechanisms (Hefti et al. 1980); and their link to behavioral adaptation remains unknown. In the case of the alterations in interhemispheric efferents of the SN, however, we have found evidence for a dependency on feedback from the animal's behavior (Morgan et al. 1983). Therefore, they could reflect changes linked to the animal's learning to compensate for induced asymmetries. It will be interesting to determine whether or not the neural alterations seen after vibrissotomy are related to the rat's learning to adapt to the condition of unilateral vibrissae.

3 Conclusions

Table 2 summarizes the main points we have emphasized to illustrate parallels in the behavioral and neural changes that occur after unilateral vibrissotomy and unilateral 6-OHDA injection into the SN.

First of all, the results clearly demonstrate a functional relationship between the vibrissae-trigeminal system and the basal ganglia. Secondly, they indicate the existence of a number of similarities in the effects of vibrissotomy and 6-OHDA SN lesion at the behavioral and neural levels. Here the question arises as to the degree of equivalence of the two interventions. Our results are certainly suggestive

Table 2. Summary of parallels in behavioral and neural plasticity found after unilateral lesion of the substantia nigra and unilateral vibrissae removal

Measure	Unilateral 6-OHDA lesion of the substantia nigra	Unilateral vibrissae removal
Crossed nigro-striatal projections	Increased cell-labeling in the projection which terminates in the denervated striatum (7 days after lesion)	Increased cell-labeling in the projection which terminates in the striatum of the deafferented hemisphere (after 10 days of vibrissae removal)
DA metabolism	Decrease in DA and metabolites in denervated striatum	Decrease in DA and metabolites in the striatum of the deafferented hemisphere after 4 hours but not after 10 days of vibrissae removal
Pharmacology: effects of apomorphine	Reversal in direction of asymmetries in turning, responsiveness and peritaxis	Reversal in direction of asymmetry in peritaxis
Spontaneous behavior	Asymmetry in peritactic behavior which recovers	Asymmetry in peritactic behavior which recovers

of an isomorphism in their influence on basal ganglia function, which raises the possibility that hemivibrissotomy can serve as a model for the 6-OHDA lesion.

The use of hemivibrissotomy as an intervention to study changes in brain and behavior promises obvious advantages over conventional experimental methods used to interfere with central nervous system function. (a) Hemivibrissotomy is easy to perform. (b) It does not necessitate the surgical destruction of brain tissue. (c) After hemivibrissotomy both brain hemispheres are available for analysis and comparison, thus, allowing use of interhemispheric control procedures, which have been so powerful in demonstrating the lateralized effects of various experimental manipulations. (d) The intervention is reversible. (e) Thorough knowledge exists about the central organization of the mystacial vibrissae system. (f) The importance of the vibrissae for behavior makes it likely that behavioral adaptation that occurs after manipulation of the vibrissae reflects learning processes. If so, hemivibrissotomy may also have utility for the study of behavioral and neural plasticity relevant to the understanding of mechanisms of learning.

Acknowledgements. This work was supported by grant Hu 306/3-3 from the Deutsche Forschungsgemeinschaft.

References

Ahl AS (1986) The role of vibrissae in behavior: a status review. Vet Res Commun 10:245–268

Altar A, Neve KA, Loughlin SE, Marshall JF, Fallon JH (1983) The crossed mesostriatal projection: neurochemistry and developmental response to lesion. Brain Res 279:1–8

Anden N-E, Carlsson A, Dahlström A, Fuxe K, Hillarp N-A, Larsson K (1964) Demonstration and mapping out of nigro-neostriatal dopamine neurons. Life Sci 3:523–530

Andres KH, Düring M von (1973) Morphology of cutaneous receptors. In: Iggo A (ed) Somatosensory System. Springer, Berlin Heidelberg New York, pp 3–28 (Handbook of Sensory Physiology, vol 2)
Arvidsson J (1982) Somatotopic organization of vibrissae afferents in the trigeminal sensory nuclei of the rat studied by transganglionic transport of HRP. J Comp Neurol 211:84–92
Beckstead RM, Domesick VB, Nauta WJH (1979) Efferent connections of the substantia nigra and ventral tegmental area in the rat. Brain Res 175:191–217
Belford GR, Killackey HP (1979) The development of vibrissae representation in subcortical trigeminal centers of the neonatal rat. J Comp Neurol 188:63–74
Bonatz AE, Steiner H, Huston JP (1987) Video image analysis of behavior by microcomputer: categorization of turning and locomotion after 6-OHDA injection into the substantia nigra. J Neurosci Meth 22:13–26
Carey RJ (1986) Relationship of changes in spontaneous motor activity to spontaneous circling in rats with unilateral 6-hydroxydopamine lesions of the substantia nigra. Exp Neurol 92:591–600
Chiodo LA, Caggiula AR, Antelman SM, Lineberry CG (1979) Reciprocal influences of activating and immobilizing stimuli on the activity of nigrostriatal dopamine neurons. Brain Res 176:385–390
Chiodo LA, Antelman SM, Caggiula AR, Lineberry CG (1980) Sensory stimuli alter the discharge rate of dopamine (DA) neurons: evidence for two functional types of DA cells in the substantia nigra. Brain Res 189:544–549
Collingridge GL (1982) Electrophysiological evidence for the existence of crossed nigrostriatal fibers. Experientia (Basel) 38:812–813
Consolazione A, Bentivoglio M, Goldstein M, Toffano G (1985) Evidence for crossed catecholaminergic nigrostriatal projections by combining wheat germ agglutinin-horseradish peroxidase retrograde transport and tyrosine hydroxylase immunocytochemistry. Brain Res 338:140–143
Creese I, Burt DR, Snyder SH (1977) Dopamine receptor binding enhancement accompanies lesion-induced behavioral supersensitivity. Science 197:596–598
De Simoni MG, Dal Toso G, Fodritto F, Sokola A, Algeri S (1987) Modulation of striatal dopamine metabolism by the activity of dorsal raphe serotonergic afferences. Brain Res 411:81–88
Douglas R, Kellaway L, Mintz M, Wageningen G van (1987) The crossed nigrostriatal projection decussates in the ventral tegmental decussation. Brain Res 418:111–121
Dravid A, Jaton AL, Enz A, Frei P (1984) Spontaneous recovery from motor asymmetry in adult rats with 6-hydroxydopamine-induced partial lesions of the substantia nigra. Brain Res 311:361–365
Durham D, Woolsey TA (1978) Acute whisker removal reduces neuronal activity in barrels of mouse Sml cortex. J Comp Neurol 178:629–644
Eisenman JS, Azmitia EC (1982) Physiological stimulation enhances HRP marking of salivary neurons in rats. Brain Res Bull 8:73–78
Fallon JH, Wang C, Kim Y, Canepa N, Loughlin S, Seroogy K (1983) Dopamine- and cholecystokinin-containing neurons of the crossed mesostriatal projection. Neurosci Lett 40:233–238
Fass B, Butcher LL (1981) Evidence for a crossed nigrostriatal pathway in rats. Neurosci Lett 22:109–113
Faull RLM, Mehler WR (1978) The cells of origin of nigrotectal, nigrothalamic and nigrostriatal projections in the rat. Neuroscience 3:989–1002
Faull RLM, Nauta WJH, Domesick VB (1986) The visual cortico-striato-nigral pathway in the rat. Neuroscience 19:1119–1132
Feltz P, Champlain J de (1972) Enhanced sensitivity of caudate neurones to microiontophoretic injections of dopamine in 6-hydroxydopamine treated cats. Brain Res 43:601–605
Fisher RS, Shiota C, Levine MS, Hull CD, Buchwald NA (1984) Subcortical crossed axonal projections to the caudate nucleus of the cat: a double-labelling study. Neurosci Lett 51:25–30
Greene EC (1963) Anatomy of the Rat. Hafner, New York
Gustafson JW, Felbain-Keramidas SL (1977) Behavioral and neural approaches to the function of the mystacial vibrissae. Psychol Bull 84:477–488

Harper JA, Labuszewski T, Lidsky TI (1979) Substantia nigra unit responses to trigeminal sensory stimulation. Exp Neurol 65:462–470
Hefti F, Melamed E, Wurtman RJ (1980) Partial lesions of the dopaminergic nigrostriatal system in rat brain: biochemical characterization. Brain Res 195:123–137
Huston JP, Bures J (1983) Innate and motivated behavior. In: Bures J, Buresova O, Huston JP (eds) Techniques and Basic Experiments for the Study of Brain and Behavior. Elsevier, Amsterdam, pp 77–133
Huston JP, Nef B, Papadopoulos G, Welzl H (1980) Activation and lateralization of sensorimotor field for perioral biting reflex by intranigral GABA agonist and by systemic apomorphine in the rat. Brain Res Bull 5:745–749
Huston JP, Morgan S, Lange KW, Steiner H (1986) Neuronal plasticity in the nigrostriatal system of the rat after unilateral removal of vibrissae. Exp Neurol 93:380–389
Huston JP, Morgan S, Steiner H (1987) Plasticity in crossed efferents from the substantia nigra as related to behavioral reorganization. In: Schneider JS, Lidsky TI (eds) Basal Ganglia and Behavior: Sensory Aspects of Motor Functioning. Huber, Bern, pp 89–102
Jeanmonod D, Rice FL, Van der Loos H (1981) Mouse somatosensory cortex: alterations in the barrelfield following receptor injury at different early postnatal ages. Neuroscience 6:1503–1535
Jones EG (1975) Possible determinants of the degree of retrograde neuronal labeling with horseradish peroxidase. Brain Res 85:249–253
Juorio AV, Greenshaw AJ (1986) The effect of raphé nuclei lesions on striatal tyramine concentration and dopamine turnover in the rat. Neurochem Res 11:687–693
Kaas JH, Merzenich MM, Killackey HP (1983) The reorganization of somatosensory cortex following peripheral nerve damage in adult and developing mammals. Annu Rev Neurosci 6:325–356
Killackey HP, Belford GR (1979) The formation of afferent patterns in the somatosensory cortex of the neonatal rat. J Comp Neurol 183:285–304
Killackey HP, Belford G, Ryugo R, Ryugo DK (1976) Anomalous organization of thalamocortical projections consequent to vibrissae removal in the newborn rat and mouse. Brain Res 104:309–315
Krauthamer GM, Yamasaki DS, Rhoades RW (1987) Does the neostriatum self-regulate its sensory input? The role of the superior colliculus. In: Schneider JS, Lidsky TI (eds) Basal Ganglia and Behavior: Sensory Aspects of Motor Functioning. Huber, Bern, pp 17–26
Kristensson K (1978) Retrograde transport of macromolecules in axons. Annu Rev Pharmacol Toxicol 18:97–110
Ljungberg T, Ungerstedt U (1976) Sensory inattention produced by 6-hydroxydopamine-induced degeneration of ascending dopamine neurons in the brain. Exp Neurol 53:585–600
Loughlin SE, Fallon JH (1982) Mesostriatal projections from ventral tegmentum and dorsal raphe: cells project ipsilaterally or contralaterally but not bilaterally. Neurosci Lett 32:11–16
Lovick TA, West DC, Wolstencroft JH (1977) Interactions between brain stem raphe nuclei and the trigeminal nuclei. In: Anderson DJ, Matthews B (eds) Pain in the Trigeminal Region. Elsevier, Amsterdam, pp 307–317
Malach R, Graybiel AM (1987) The somatic sensory corticostriatal projection: patchwork of somatic sensory zones in the extrastriosomal matrix. In: Schneider JS, Lidsky TI (eds) Basal Ganglia and Behavior: Sensory Aspects of Motor Functioning. Huber, Bern, pp 11–16
Marshall JF (1979) Somatosensory inattention after dopamine-depleting intracerebral 6-OHDA injections: spontaneous recovery and pharmacological control. Brain Res 177:311–324
Marshall JF (1984) Brain function: neural adaptations and recovery from injury. Annu Rev Psychol 35:277–308
Mishra RK, Marshall AM, Varmuza SL (1980) Supersensitivity in rat caudate nucleus: effects of 6-hydroxydopamine on the time course of dopamine receptor and cyclic AMP changes. Brain Res 200:47–57
Morgan S, Huston JP, Pritzel M (1983) Effects of reducing sensory-motor feedback on the appearance of crossed nigro-thalamic projections and recovery from turning induced by unilateral substantia nigra lesions. Brain Res Bull 11:721–727
Morgan S, Steiner H, Huston JP (1985) Relationship between interhemispheric nigrostriatal projections and the direction of rotational behavior induced by amphetamine. Exp Neurol 90:149–161

Morgan S, Steiner H, Rosenkranz C, Huston JP (1986) Dissociation of crossed and uncrossed nigrostriatal projections with respect to site of origin in the rat. Neuroscience 17:609–614
Morgan S, Huston JP (1988) Interhemispheric projections from the substantia nigra to the caudate-putamen and ventromedial thalamic nucleus as depicted by the anterograde transport of 3H-leucine. Brain Res Bull (in press)
Neve KA, Kozlowski MR, Marshall JF (1982) Plasticity of neostriatal dopamine receptors after nigrostriatal injury: relationship to recovery of sensorimotor functions and behavioral supersensitivity. Brain Res 244:33–44
Nishino H, Ono T, Sasaki K, Nishino A, Muramoto K (1979) Retrograde transport of horseradish peroxidase in sciatic nerve of rats and dystrophy mice. Neurosci Lett 14:1–6
Parent A, Mackey A, De Bellefeuille L (1983) The subcortical afferents to caudate nucleus and putamen in primate: a fluorescence retrograde double labeling study. Neuroscience 10:1137–1150
Pritzel M, Huston JP, Sarter M (1983 a) Behavioral and neuronal reorganization after unilateral substantia nigra lesions: evidence for increased interhemispheric nigrostriatal projections. Neuroscience 9:879–888
Pritzel M, Sarter M, Morgan S, Huston JP (1983 b) Interhemispheric nigrostriatal projections in the rat: bifurcating nigral projections and loci of crossing in the diencephalon. Brain Res Bull 10:385–390
Rasmussen K, Strecker RE, Jacobs BL (1986) Single unit response of noradrenergic, serotonergic and dopaminergic neurons in freely moving cats to simple sensory stimuli. Brain Res 369:336–340
Romo R, Schultz W (1985) Prolonged changes in dopaminergic terminal excitability and short changes in dopaminergic neuron discharge rate after short peripheral stimulation in monkey. Neurosci Lett 62:335–340
Romo R, Cheramy A, Godeheu G, Glowinski J (1986) In vivo presynaptic control of dopamine release in the cat caudate nucleus. III. Further evidence for the implication of corticostriatal glutamatergic neurons. Neuroscience 19:1091–1099
Royce GJ (1978) Cells of origin of subcortical afferents to the caudate nucleus: a horseradish peroxidase study in the cat. Brain Res 153:465–475
Sabel BA, Pritzel M, Morgan S, Huston JP (1984) Interhemispheric nigrothalamic projections and behavioral recovery following turning behavior induced by unilateral peripheral sensory and motor restriction. Exp Neurol 83:49–61
Schallert T, Upchurch M, Wilcox RE, Vaughn DM (1983) Posture-independent sensorimotor analysis of inter-hemispheric receptor asymmetries in neostriatum. Pharmacol Biochem Behav 18:753–759
Schiorring E (1979) An open field study of stereotyped locomotor activity in amphetamine-treated rats. Psychopharmacology 66:281–287
Schneider JS (1986) Interactions between the basal ganglia, the pontine parabrachial region, and the trigeminal system in cat. Neuroscience 19:411–425
Shipley MT (1974) Response characteristics of single units in the rat's trigeminal nuclei to vibrissa displacements. J Neurophysiol 37:73–90
Smith RL (1973) The ascending fiber projections from the principal sensory trigeminal nucleus in the rat. J Comp Neurol 148:423–446
Steiner H, Morgan S, Huston JP (1985) Effect of forebrain commissurotomy on recovery from unilateral 6-OHDA lesions of the substantia nigra and circling induced by apomorphine. Behav Brain Res 17:245–249
Steiner H, Huston JP, Morgan S (1986) Apomorphine reverses direction of asymmetry in facial scanning after 10 days of unilateral vibrissae removal in rat: vibrissotomy-induced denervation supersensitivity? Behav Brain Res 22:283–287
Steiner H, Bonatz AE, Huston JP, Schwarting R (1988) Lateralized wall-facing versus turning as measures of behavioral asymmetries and recovery of function after injection of 6-hydroxydopamine into the substantia nigra. Exp Neurol 99:556–566
Strecker RE, Jacobs BL (1985) Substantia nigra dopaminergic unit activity in behaving cats: effect of arousal on spontaneous discharge and sensory evoked activity. Brain Res 361:339–350

Ungerstedt U (1971 a) Stereotaxic mapping of the monoamine pathways in the rat brain. Acta Physiol Scand Suppl 367:1–48
Ungerstedt U (1971 b) Striatal dopamine release after amphetamine or nerve degeneration revealed by rotational behaviour. Acta Physiol Scand Suppl 367:49–68
Ungerstedt U (1971 c) Postsynaptic supersensitivity after 6-hydroxydopamine induced degeneration of the nigrostriatal dopamine system. Acta Physiol Scand Suppl 367:69–93
Van der Loos H, Woolsey TA (1973) Somatosensory cortex: structural alterations following early injury to sense organs. Science 179:395–398
Veening JG, Cornelissen FM, Lieven PAJM (1980) The topical organization of the afferents to the caudatoputamen of the rat. A horseradish peroxidase study. Neuroscience 5:1253–1268
Vincent SB (1912) The function of the vibrissae in the behavior of the white rat. Behav Monogr 1:4–81
Vincent SB (1913) The tactile hair of the white rat. J Comp Neurol 23:1–36
Waite PME (1973 a) Somatotopic organization of vibrissal responses in the ventro-basal complex of the rat thalamus. J Physiol 228:527–540
Waite PME (1973 b) The responses of cells in the rat thalamus to mechanical movements of the whiskers. J Physiol 228:541–561
Welker C (1971) Microelectrode delineation of fine grain somatotopic organization of SmI cerebral neocortex in albino rat. Brain Res 26:259–275
Welker C (1976) Receptive fields of barrels in the somato-sensory neocortex of the rat. J Comp Neurol 166:173–190
Welker E, Van der Loos H (1986) Quantitative correlation between barrel-field size and the sensory innervation of the whiskerpad: a comparative study in six strains of mice bred for different patterns of mystacial vibrissae. J Neurosci 6:3355–3373
Welker WI (1964) Analysis of sniffing of the albino rat. Behaviour 22:223–244
Welzl H, Huston JP (1981) Sensitization of the perioral biting reflex by intranigral GABA agonist after detelencephalization. Neurosci Lett 21:351–354
Welzl H, Flack HG, Huston JP (1982) Contraversive circling and facilitation of the perioral biting reflex by injection of substance P or D-Ala2-met-enkephalinamide into the substantia nigra. Behav Neural Biol 34:104–108
Welzl H, Schwarting R, Huston JP (1984) Substantia nigra efferents and afferents in the control of the perioral biting reflex. In: Bandler R (ed) Modulation of Sensorimotor Activity During Alterations in Behavioral States. Alan R Liss, New York, pp 415–429
Wise SP, Jones EG (1977) Cells of origin and terminal distribution of descending projections of the rat somatic sensory cortex. J Comp Neurol 175:129–158
Woolsey TA, Van der Loos H (1970) The structural organization of layer IV in the somatosensory region (S I) of mouse cerebral cortex: the description of a cortical field composed of discrete cytoarchitectonic units. Brain Res 17:205–242
Yasui Y, Itoh K, Mizuno N (1987) Direct projections from the caudal spinal trigeminal nucleus to the striatum in the cat. Brain Res 408:334–338
Zigmond MJ, Acheson AL, Stachowiak MK, Stricker EM (1984) Neurochemical compensation after nigrostriatal bundle injury in an animal model of preclinical parkinsonism. Arch Neurol 41:856–861
Zucker E, Welker WI (1969) Coding of somatic sensory input by vibrissae neurons in the rat's trigeminal ganglion. Brain Res 12:138–156

VIII Lesions to Specific CNS Subsystems: Auditory System, Olfactory System

Axonal Regeneration After Transection of the Commissure of the Inferior Colliculus

I. Taniguchi and H. Tanaka

Medical Research Institute, Tokyo Medical and Dental University, 1-5-45 Yushima, Bunkyo-ku, Tokyo 113, Japan

1 Introduction

Regenerative responses to damage in the central nervous system (CNS) can be elicited by two types of mechanisms. One is regenerative sprouting or axonal regeneration from damaged neurons and the other is collateral sprouting from intact neurons. Collateral sprouting occurs widely in the CNS, as well as in the peripheral nervous system (Raisman 1969; Moore et al. 1971; Lynch et al. 1973). On the other hand, in the mammalian CNS, axonal regeneration is usually abortive. Since Cajal's description in 1928, the capacity for axonal regeneration of transected spinal cord has been extensively studied. Such studies have shown that the newly formed axonal sprouts could not continue growth across the transected site because of obstruction caused by the formation of glial scars. Conditions in the glial environment of injured fibers have a decisive role in successful axonal elongation in the CNS (Kawaguchi et al. 1981; Benfey and Aguayo 1982).

In previous studies (Taniguchi and Saito 1982), we observed axonal regeneration in the commissure of the interior colliculus (CIC) of mice following its transection. In the lesioned structure of the CIC, no glial scars were observed.

While evidence concerning morphological re-organization of the CNS is accumulating, little is known about the consequent recovery of function at the cellular level. The purpose of this study was to examine whether the regenerated CIC axons could re-establish synapses with neurons in the inferior colliculus (IC) or not. If the synapses were re-established, our concern was their functional features. This approach should provide useful data related to the problem of functional recovery.

2 General Procedures

2.1 Histological Experiment

The subjects in all experiments described here were 4-week-old albino mice (ICR strain, Institute for Cancer Research) weighing about 20 g. The CIC was totally transected under pentobarbital anesthesia (3 mg/100 g body weight).

At various postoperative survival intervals, approximately 0.5 μl of a 30% solution of horseradish peroxidase (HRP) in distilled water was injected into the left

H. Flohr (Ed.)
Post-Lesion Neural Plasticity

IC. All animals survived for 24 h after the injection. Frozen sections of the brain were made at 50 μm thickness and incubated with 3,3′-diaminobenzidine and hydrogen peroxidase according to the procedure described by LaVail and LaVail (1972).

2.2 Physiological experiment

The CIC transection of the mice was made at the age of 4 weeks and they were used for the physiological experiment at least 4 weeks later. The control experiment was performed using 2-month-old intact animals.

Extracellular recordings were obtained from a single neuron or a few neurons of the IC with a glass-coated elsiloy microelectrode (Suzuki and Azuma 1976). Action potentials from a single neuron was isolated with a level detector for computer processing. To stimulate the IC neurons, pure tone bursts (5 ms rise and fall times, 40 ms duration) were used. These tone bursts were delivered from a loudspeaker in front of the animal's head. The best frequency at the minimum threshold (BF) was determined for each IC neuron, and the impulse discharges from the neuron to acoustic and/or electric stimuli were represented in the form of peri-stimulus-time (PST) histograms.

Another elsiloy electrode was inserted into the opposite IC to stimulate the CIC fibers with an electrical stimulus (0.1 ms duration, 20–100 μA). If the CIC fibers have synaptic connections with the IC neurons, the latter will respond to the electrical stimulus with the postsynaptic potentials. Their auditory responses may also be affected by the electrical stimulus.

Since the CIC fibers reciprocally innervate the IC, they can be electrically stimulated in two directions: orthodromic and antidromic. To examine the direction of the impulse conduction, the impulse collision technique was used. That is, electrical stimulation was triggered by spontaneous discharges from the opposite IC. If the CIC fiber was electrically stimulated in the antidromic direction, its impulse will collide in the CIC fiber with the spontaneous discharge impulse.

It is possible that the IC neurons will respond to the electrical stimulation by current spread or by different pathways other than the CIC. To examine this, we made a transection between the two electrodes for recording and stimulating after acoustical identification of the IC neurons and examined the effect of the electrical stimulation on them.

At the end point of each successful penetration, a lesion was made by passing an electric current of a few μA for a few seconds through the microelectrodes. Brain sections of 50 μm were made and stained using the Klüver-Barrera method.

3 The Time Course of Axonal Regeneration in the CIC After its Transection

When the HRP injection was made into the dorsal part of the IC in normal animals, a dense band of fibers was observed in the CIC (Fig. 1 A). They run in parallel toward the opposite IC and their terminals reach a symmetrical region corresponding to the injection site. The short fibers of the CIC terminate at the dorsal

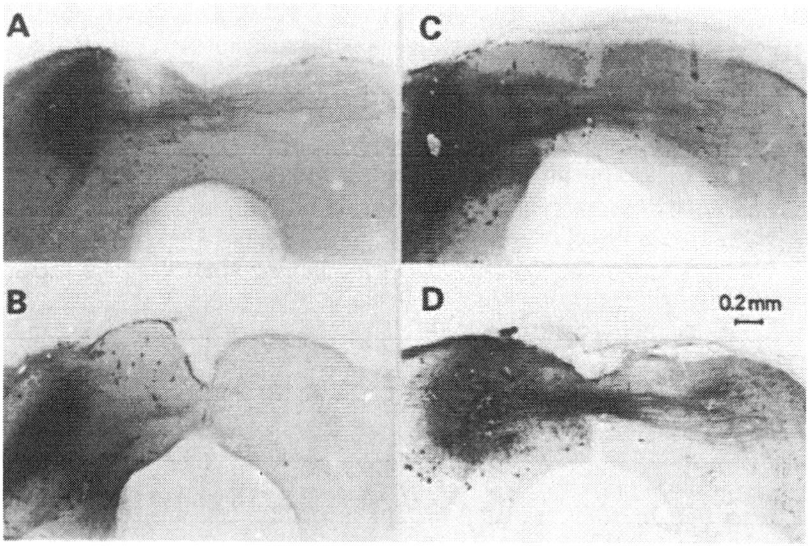

Fig. 1. The time course of axonal regeneration in the mouse CIC after its transection. HRP was injected into the left IC. **A** Coronal section through the IC of a control animal. The labeled fibers cross the CIC towards the opposite IC. **B** Within 2 weeks after transection, the labeled fibers do not cross the lesion. **C** The regenerating fibers have reached the opposite IC 3 weeks after transection. **D** The regenerating axons with the laminar alignment end at the appropriate destination 4 weeks after transection. (Taniguchi and Saito 1982)

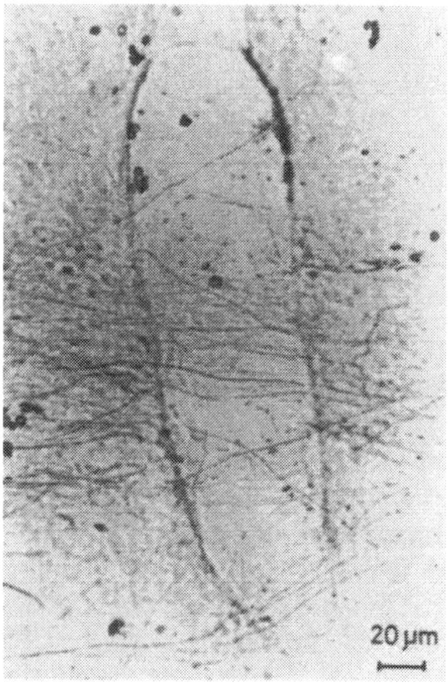

Fig. 2. Regenerating axons around the lesion in the CIC 15 weeks after transection. Note the large axons in disarray passing through the partial contact defect of the CIC. (Taniguchi and Saito 1982)

half of the contralateral IC (Woollard and Harpman 1940). A few HRP-positive cell somata are also found at the region where the fibers terminate. These findings suggest that HRP was transported in two ways, with either anterograde or retrograde axonal flow, and that the CIC fibers innervate the IC reciprocally.

By 2 weeks after transection of the CIC, the neurons have sent some fibers tipped with growth cones to the border of the lesion as shown in Fig. 1 B. They could not grow beyond the lesion although the incision had healed.

Three weeks after transection, axonal outgrowth was found beyond the lesion (Fig. 1 C). No glial scars were observed. Their terminals were confined to a definite symmetrical region corresponding to the injection site. A few HRP-positive cells were found also in the contralateral IC. They were labeled by the retrograde transport of HRP.

Figure 1 D shows a dense band of regenerating axons 4 weeks after the transection. Parallel alignment of the fibers is obvious and the fibers terminate at their appropriate destinations. The diameters of the regenerating axons are several times as large as the normal ones.

If the spatial gap due to the incision still remained, the sprouting axons usually could not cross the gap of the lesion. However, in the animal with a partial contact defect at the center of the CIC, the fibers with a large diameter were found passing through the defect in an irregular way (Fig. 2).

4 Function of the CIC Projection to the IC

Little is known about the function of the CIC fibers. It was believed that their influence on IC neurons was insignificant and that these neurons should receive the dominant input through the lateral lemniscus (LL). Unexpectedly, the CIC fibers were found to have a strong influence on the IC neurons.

Figure 3 shows the inhibitory influence of the CIC fiber on the IC neuron in the intact animal. This IC neuron exhibited on- and off-responses to the BF sound stimulus (6.9 kHz) (A). When an electrical stimulus was added to the sound, auditory responses were clearly inhibited (B). To the electrical stimulus, the neuron responded with a shorter latency than to the sound stimulus (C). Such an inhibitory response pattern was found in about 65% of the neurons which were responsive to acoustic stimuli (N = 51). A few neurons (10%) showed summational or facilitative responses to the combination of acoustic and electrical stimuli. The remaining 25% were undefined for their response properties to the electrical stimulation.

The experiment to examine the electrical effects by different paths other than the CIC on the IC neurons showed that such a possibility could be ruled out. In other words, the responses to acoustic stimuli could still be recorded even after acute transection of the CIC, but the responses to electrical stimuli could not.

The impulse collision test demonstrated that in almost all cases the impulses responding to electrical stimulus were conducted orthodromically along the CIC fibers. This means that the impulses recorded from the IC were postsynaptically evoked.

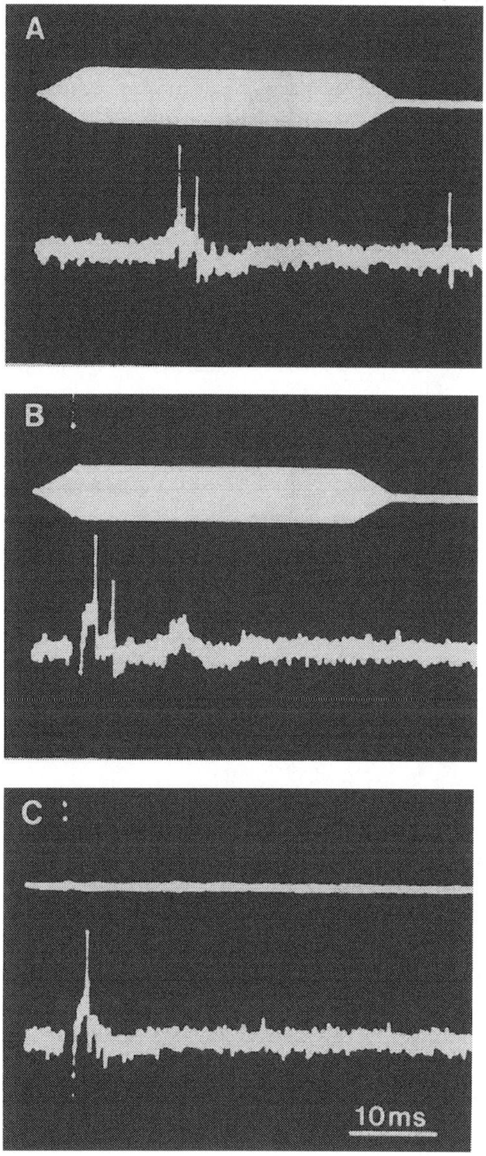

Fig. 3. The impulse discharges from a single IC neuron of the control animal. **A** The sound stimulus of BF, 6.9 kHz at 56 dB SPL (*upper trace*) evoked the impulses (*lower trace*). **B** The auditory response was inhibited by the addition of an electrical stimulus. **C** The impulse discharges to the electrical stimulus alone

5 Function of Re-Established Synapses

The same technique as described for the intact animals was applied to the animals operated upon in order to examine whether the synapses were re-established or not. If the auditory responses from the IC neurons in the animals operated upon are modulated by electrical stimulation, as seen in intact animals, it suggests that the synapses are re-established in the IC.

Results similar to those seen in the intact animals as described above were obtained. Many IC neurons responsive to acoustic stimuli also initiated impulse dis-

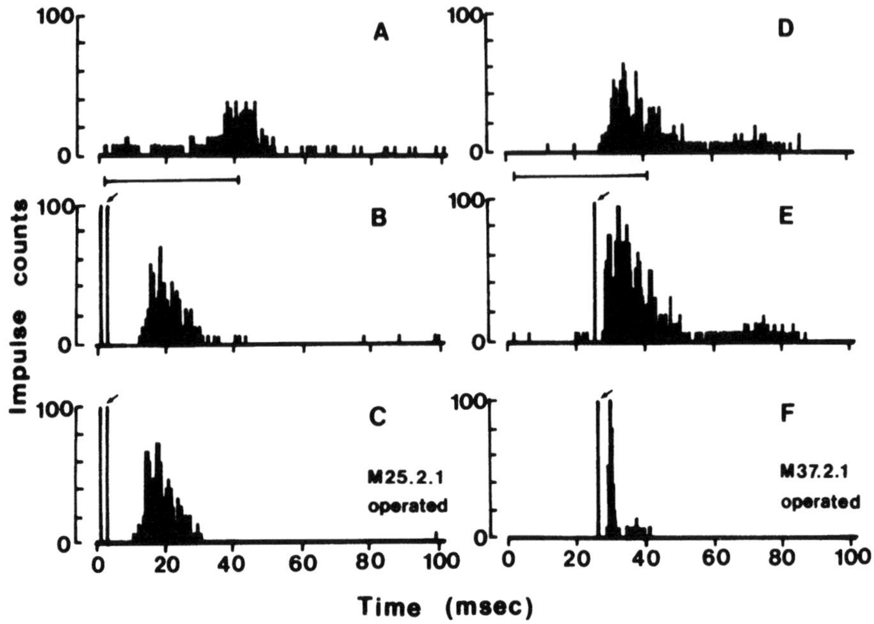

Fig. 4. PST histograms representing the different response patterns obtained from the two IC neurons in the animals operated upon. **A, D** The response to the sound stimulus alone. **B, E** The response to the acoustic and electrical stimuli combination. The auditory response was inhibited in B when the electrical stimulus was added, while in E the facilitative response was evoked. **C, F** The response to the electrical stimulus alone. The horizontal bars indicate sound stimulation. The *arrows* in **B, C, E** and **F** indicate artifacts due to electrical stimulation

charges to electrical stimuli. When the electrical stimulus was added to the acoustic stimulus, the auditory responses were modulated.

Figure 4 shows two representative types of response patterns in the form of PST histograms. One was obtained from the neuron whose auditory responses were inhibited by the electrical stimulus (Fig. 4 A, B, C). The other is from the neuron which showed summational responses to the combination of acoustic and electrical stimuli (Fig. 4 D, E, F).

Approximately 53% of the neurons responsive to the acoustic stimulus (N= 102) were suppressed by the electrical stimulus, while about 9% of the neurons showed summational or facilitative responses to the combination stimulus. The remaining 38% were undefined with respect to their response properties to the electrical stimulation.

6 Comparison of Synaptic Features in Operated and Unoperated Animals

6.1 Proportion of Excitatory and Inhibitory Synapses

Two types of synapses were found in both operated mice and unoperated controls. In the neuronal response patterns, we could not find any differences between the two different subject groups. In addition to these two grops, there was

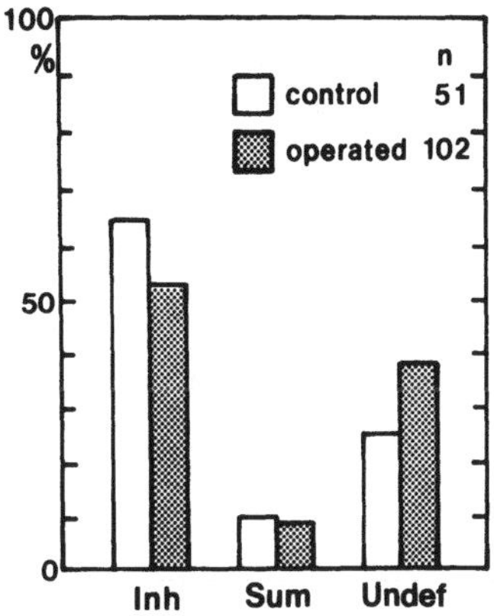

Fig. 5. The proportions of the IC neurons having different synaptic features in the operated mice and the unoperated controls. *Inh* inhibition; *Sum* summation or facilitation; *Undef* undefined

another type of neuron in which the synaptic features could not be identified. This group belonging to the IC neurons were clearly reponsive to the acoustic stimuli, but it was difficult to determine their responsiveness to the electrical stimuli.

Although we could not find any characteristic difference in the response patterns of the IC neurons between the control animals and the animals operated upon, there was a statistical difference in the proportion of the neurons with undefined synaptic features between the two subject groups (Fig. 5). In other words, the proportion of such a type of neuron in the animals operated upon is larger than that in the control animals.

For further analytical comparison, the chi-square test was applied to the two neuronal populations of operated and unoperated animals. The calculated χ^2 was $\chi^2 = 201.7 > \chi_0^2$ (0.01), thus we can conclude that the proportion of the two populations is different.

The large proportion of the undefined group in the animals undergoing the operation suggests difficulty in re-establishing the synapses.

6.2 Distribution of BF's

We identified the IC neurons in operated and unoperated mice by acoustic stimulation, and determined their BF's. The BF's of the neurons, except the neuron group with undefined synaptic features, are distributed in the frequency range from 3 to 18 kHz.

The mean BF values were 8.2 kHz for the operated animals and 8.8 kHz for the unoperated controls. The chi-square test applied to the distribution of the BF's also demonstrated a difference between the two subject groups.

The mean BF value for the operated animals is slightly lower than that for the unoperated controls. Many previous studies on the tonotopical organization of the IC in different species have shown that neurons having lower BF's are located in the more dorsal layers of the IC (Rose et al. 1963; Merzenich and Reid 1974). Such an organization was also found in our study on mice IC. These results suggest that the neurons identified in the operated animals are distributed more superficially than in the unoperated controls. The regenerating CIC axons will most likely cease to grow if they encounter the appropriate targets at a closer location.

The mean value of the BF's for the functionally undefined neuron group is 9.9 kHz for the operated animals and 10.8 kHz for the unoperated controls. These BF values are higher than those for the functionally defined neuron group. Tonotopically, the higher BF indicates that the neurons are located in the deeper layer. Those neurons may intrinsically not have the input from the opposite IC through the CIC. This is in agreement with the evidence that the short CIC projection ends at the dorsal half of the IC (Woollard and Harpman 1940). Therefore, this is probably why we had difficulty in determining the synaptic features of those neurons.

7 Re-Established Neural Networks

Some statistical differences were found between the two neuronal populations of the operated mice and the unoperated controls. However, with respect to the response patterns, many neurons in the animals operated upon were similar to those in the control animals.

Interestingly, the majority of the neurons studied had inhibitory synapses. This indicates that at least an inhibitory interneuron exists between the CIC terminals and the IC neurons. Such an interneuron is probably activated by a collateral from the CIC fiber. Thus, the excitatory input from the LL and the excitatory and inhibitory inputs from the opposite IC through the CIC may converge on an IC neuron.

8 Conclusions

Although there were some statistical differences in the synaptic features of the IC neurons between operated and unoperated animals, the re-established synapses seem to be properly re-organized into newly established neural circuits which may be similar to those in intact animals. This means the restitution of synaptic integrity occurs even after mechanical transection of the axons within the mammalian CNS.

Since our experiments were conducted in young mice, certain embryological characteristics may be involved. Therefore there is most likely a limit to generalizations that can be deduced from these results. However, the evidence presented here may provide new data about functional recovery at the neuronal level.

References

Benfey M, Aguayo AJ (1982) Extensive elongation of axons from rat brain into peripheral nerve grafts. Nature 296:150–152

Cajal SR (1928) Degeneration and regeneration of the nervous system. Oxford Univ Press, London

Kawaguchi S, Miyata H, Kawamura M, Harada Y (1981) Morphological and electrophysiological evidence for axonal regeneration of axotomized cerebellothalamic neurons in kitten. Neurosci Lett 25:13–18

LaVail JH, LaVail MM (1972) Retrograde axonal transport in the central nervous system. Science 176:1416–1417

Lynch G, Deadwyler S, Cotman C (1973) Postlesion axonal growth produces permanent functional corrections. Science 180:1364–1366

Merzenich MM, Reid MD (1974) Representation of the cochlea within the inferior colliculus of the cat. Brain Res 77:397–415

Moore RY, Björklund A, Stenevi U (1971) Plastic changes in the adrenergic innervation of the rat septal area in response to denervation. Brain Res 33:13–35

Raisman G (1969) Neuronal plasticity in the septal nuclei of the adult rat. Brain Res 14:25–48

Rose JE, Greenwood DD, Goldberg JM, Hind JE (1963) Some discharge characteristics of single neurons in the inferior colliculus of the cat. I. Tonotopical organization, relation of spike-counts to tone intensity, and firing patterns of single elements. J Neurophysiol 26:295–320

Suzuki H, Azuma M (1976) A glass-insulated "elsiloy" microelectrode for recording unit activity in chronic monkey experiments. Electroencephalogr Clin Neurophysiol 41:93–95

Taniguchi I, Saito N (1982) Plastic reorganization after transection of the commissure of the inferior colliculus. Proc Jpn Acad 58:181–184

Woollard HH, Harpman JA (1940) The connexions of the inferior colliculus and of the dorsal nucleus of the lateral lemniscus. J Anat 74:441–458

Post-Lesion Plasticity in the Auditory System of the Cricket

K. Schildberger and F. Huber
Max-Planck-Institut für Verhaltensphysiologie, D-8131 Seewiesen, FR Germany

1 Introduction

For a long time it was believed that insects, unlike vertebrates, do not form synaptic connections after sensory deprivation different to those normally established during embryonic and postembryonic development. Here, we present results from lesion experiments in the auditory system of crickets that show changes in morphology and physiology of some identified auditory neurons which could account for compensation phenomena observed in phonotactic behavior after sensory deprivation.

The ears of the cricket are located in the tibiae of the forelegs. The axons of the auditory sensory cells originate in the auditory organ, extend through the leg, enter the prothoracic ganglion and terminate there ipsilaterally in an area termed auditory neuropil. Because both ears exhibit mirror-imaged directional characteristics it is believed that a comparison of the excitation of both ears within the CNS initiates a turning command toward the side most strongly stimulated. According to this binaural hypothesis crickets with only one ear functioning should be unable to find the sound source but should circle toward the side of the functioning ear.

2 Phonotaxis with One Ear

By measuring the course and velocity of a phonotactically responding animal on a treadmill (see Weber et al. 1981), we can determine the angle of a freely walking animal in relation to sound direction. A normal two-eared cricket walks toward the direction of an attractive sound source by meandering around sound direction with maximal deviations of about $\pm 40^{\circ}$ (Figs. 1, 2). Some animals with one foreleg amputated or the auditory nerve transected after the imaginal molt can maintain, at least for some time, a stable walking course depending on sound direction (Fig. 1, see also Huber 1987; Huber et al. 1984; Schmitz 1986). An animal tested for phonotactic performance five days after transection of one auditory nerve (Fig. 2A) still showed a loudspeaker-dependent mean walking course, but course stability was weak, as expressed in the broad range of walking angles. The mean walking angle deviated to the side of the remaining ear. When tested 12 days after nerve transection, course stability was improved and the deviation from

H. Flohr (Ed.)
Post-Lesion Neural Plasticity

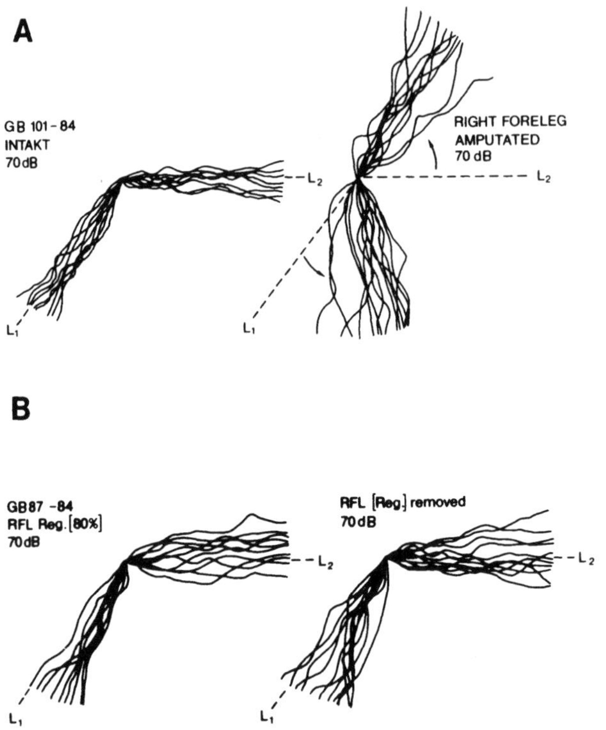

Fig. 1. A Phonotactic performance of a two-eared adult cricket on a walking compensator during simulated calling song (70 dB, 5 kHz) before (*left*) and after amputation of the right foreleg; L_1, L_2 mark loudspeaker positions, *arrows* indicate the shift of tracking angle; **B** walking traces of an adult female with the right foreleg (*RFL*) amputated in the 6th larval instar and with subsequent regeneration (*left*) and after amputation of the regenerated leg (*right*). (After Huber 1987)

sound direction decreased. The auditory nerve was not regenerated 12 days after the transection. Thus it seems that the auditory system can be readjusted after sensory deprivation to enable sound localization with one ear.

If a cricket is able to adjust its auditory system, it might be possible to disturb this process. We blocked one or both ears by waxing the tympana on the day of the imaginal molt and let the animals live for 6 weeks under different sound conditions (Klose and Schildberger 1988).

With one ear blocked and a calling song presented permanently for 6 weeks at 80 dB SPL, the animal exhibited a weak course stability, often circled and deviated to the side of the ear which was not blocked (Fig. 2B). After removal of the wax, the test on the same day revealed a weak course stability, circling and deviation from sound direction, but now to the side of the previously waxed ear. Course stability was improved 3 days after wax removal, but an error angle still remains. Thus, not only the loss of an ear can be, at least partly, compensated, but also long-term activation and deactivation of the auditory system can influence the localization process which formerly was believed as innate, rather stereotyped and based on a hard wired network.

With a foreleg amputated during postembryonic development some of the adult animals showed a course stability nearly as good as intact crickets (Fig. 1B). In contrast to intact crickets, however, these animals always deviated to the intact side, with an error angle depending on sound intensity (Fig. 2C). Moreover, the intensity range within which stable phonotactic walking courses occur was narrowed compared to two-eared animals (Schmitz et al. 1988). What mechanisms

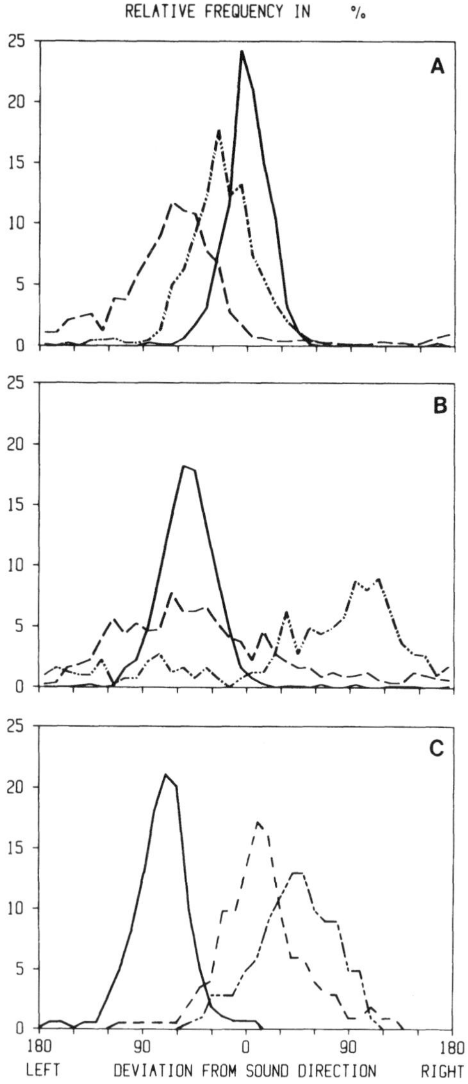

Fig. 2. Phonotactic performance of female crickets to simulated calling song shown as percentual distribution of walking angles relative to the speaker direction. **A** Adult female before (——), 5 days (– – –) and 12 days (–· · –) after transsection of the right auditory nerve cut; **B** with tympana of the left foreleg waxed and with permanent presentation of calling song for 6 weeks before the test; performance after sound treatment with the left leg waxed (–· · –· · –), after removal of the wax on the same day (– – –) and 3 days later (——). **C** Performance of two adult females in which the right (——) respectively the left foreleg (– – – 60 dB, –· · – 70 dB) has been amputated in the penultimate larval instar

could be responsible for the behavioral compensation after the loss of one ear during postembryonic development?

3 Auditory Organ

Can the ear be regenerated and reconnected to central neurons? Insects are able to regenerate body appendages during postembryonic development. Leg regeneration takes at least two larval instars to start. Depending on the time of the amputation, the leg can regenerate up to normal length (Schmitz et al. 1988). Its motor coordination is not impaired; walking speed and courses do not differ from normal animals. However, in the regenerated foreleg, some internal structures are

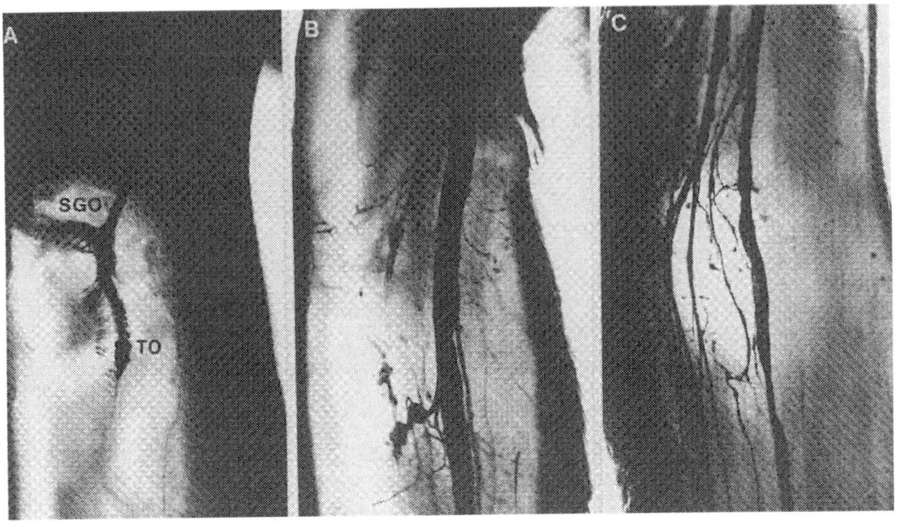

Fig. 3. Retrograde cobalt fills of the auditory nerve reveal a tympanal (*TO*) and a subgenual organ (*SGO*) in a normal adult leg (**A**); but no tympanal organ in adult legs of animals with one foreleg amputated in the 2nd larval instar and subsequently regenerated (**B, C**)

missing or disorganized. Normally there are two tracheal branches within the tibia of the foreleg, an outer one contacting the posterior tympanum and an inner one to which the auditory organ is attached (Michel 1974; Eibl 1978). In the regenerated tibia this geometry is distorted; more than two tracheal branches are developed, smaller in diameter and topographically dislocated. Moreover, in several dozens of regenerated legs we never observed an auditory organ (Fig. 3), even if the leg could regenerate to its full length after amputation in the first larval instar. The nerve – homologous in position to the auditory nerve of intact animals – either runs with only few branches through the whole tibia or spreads out and innervates scattered sensory cells. Such six-legged animals possess only one auditory organ and some of them can localize a sound source nearly as well as normal animals (cf. Fig. 1 B).

4 Central Auditory Neurons

In the prothoracic ganglion interneurons were identified which receive input from the auditory receptor cells (Wohlers and Huber 1982). Figure 4 shows typical features of two of these cell types. A local interneuron, the omeganeuron ON1, connects the auditory neuropils of both hemispheres and another auditory neuron, AN2, which ascends up to the brain. Both neurons are excited by the ear ipsilaterally to their main dendritic fields and inhibited by the contralateral ear. Major excitatory input arrives from the ear on the cell body side for the ON1 and on the side contralateral to the soma for AN2. Each of these neurons has a mirror-imaged partner cell originating in the other ganglion half with corresponded input.

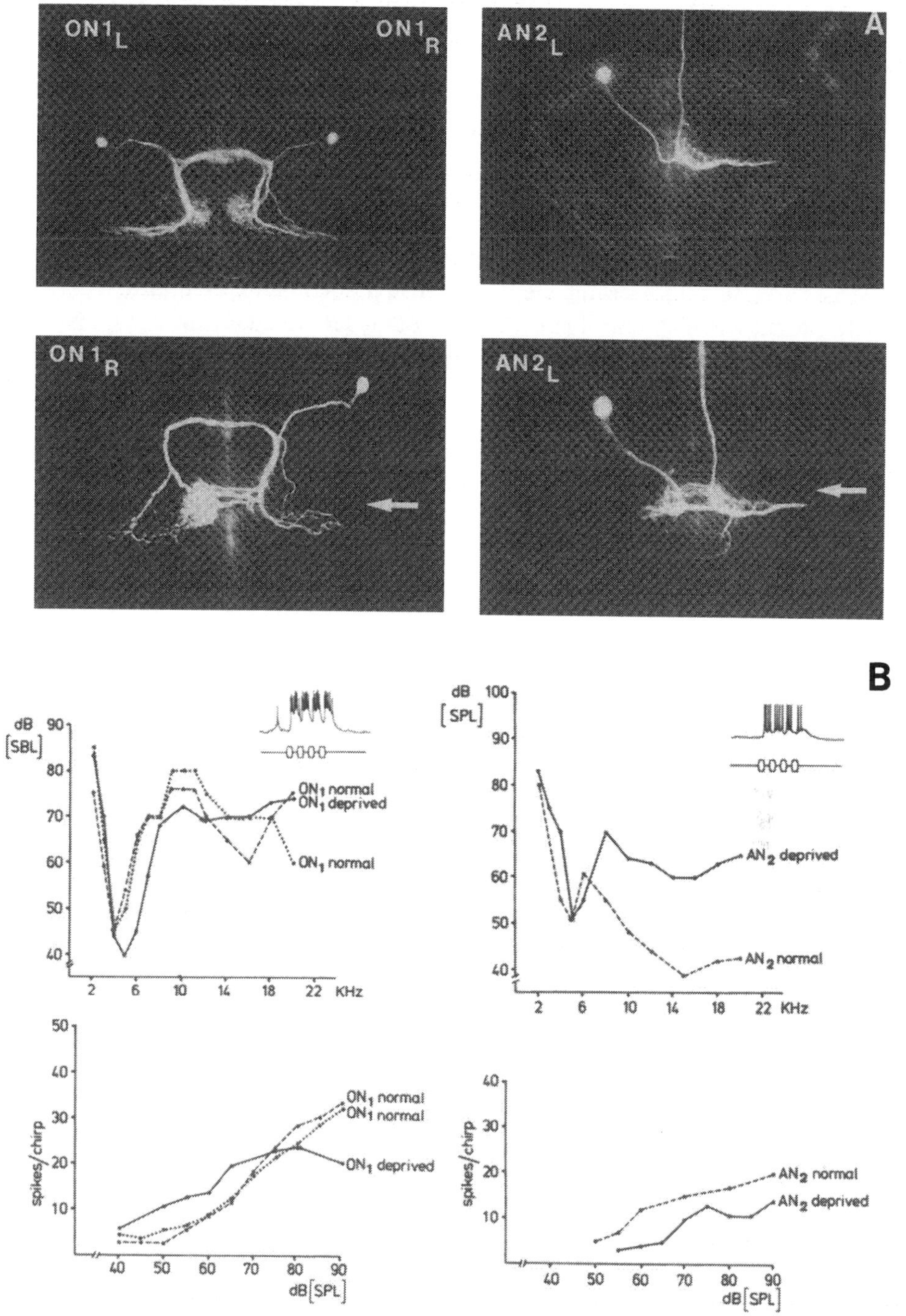

Fig. 4. Auditory neurons of the prothoracic ganglion of normal and unilaterally deafferented animals. **A** Morphology of the mirror-imaged omeganeurons ON_1 and one neuron AN_2 of normal animals (*top*) and of animals with the right foreleg amputated (*arrow*) in a larval instar (*bottom*). **B** Auditory threshold curves of ON_1 and AN_2 of normal and unilaterally deprived animals (*top*), intensity response functions at 5 kHz of the same neurons (*bottom*). (After Schildberger et al. 1986)

In animals with a leg amputated in a larval instar, the neurons with the main dendritic field on the deafferented side (arrow in Fig. 4) sprout dendrites across the ganglionic midline into the auditory neuropil of the intact side (Schildberger et al. 1986). These dendrites form new connections which are functional. ON1 is now excited by the remaining contralateral ear which normally inhibits the neuron. Moreover, the functional characteristics of the ON1 with sprouted dendrites seem to be restored because absolute sensitivity and the shape of the auditory threshold curves of the neurons closely resemble that of ON1 in normal animals. The same seems to be true for the AN2 except for a loss of sensitivity for high frequency sound. These results confront us with an apparent paradox. There is only one ear left, the central pathways of both sides are excited by the same ear, and a presumed comparison of the excitation of both sides should provide no useful information for sound localization, but some of these animals walk oriented.

5 A Possible Model

We should like to propose a model that could solve this contradiction (Fig. 5 a). In the case of an interneuron being connected to only one ear, the directional sensitivity of that ear can be read out from the intensity response function of the interneuron. This is so because the effective sound pressure level is ambivalent with regard to intensity and direction. Therefore, in one-eared animals sound intensity – completely encoded in the intensity response function of the corresponding neuron – is a measure for sound direction. If we transfer this situation to animals where central neurons have grown dendrites to reach auditory input from the remaining ear, some predictions can be made. If excitation of a neuron on the intact side is compared within the CNS with the one of the partner neuron on the deprived side and transferred into a turning command, different walking courses will occur. With the neuron on the intact side always stronger excited than its partner neuron on the deprived side, the animal should circle toward the intact side, or to the opposite side if the deprived and newly connected neuron is more strongly excited. There would be no stable walking courses in both conditions. However, if the intensity response functions of the two neurons intersect then there exists an intensity range, i.e., a certain sound direction, where both neurons are equally excited. At this point a stable walking course is predicted. Localization with this mechanism is only possible in a restricted range of sound intensities. At high sound pressure levels both neurons will be more strongly excited than at the intensity of the intersection point, independent of sound direction.

The existing behavioral data do not contradict the model. The stability of walking courses of one-eared animals decreases at high sound intensities and the animals circle toward the intact side. Compared to normal animals one-eared crikkets exhibit oriented walking only at a narrower range of intensities, mostly at lower and moderate intensities (Schmitz et al. 1988). At the neuronal level (Fig. 5 B) there is evidence that intensity response functions of mirror-imaged cells intersect in animals exhibiting phonotactic tracking with one ear (Schildberger and Kleindienst 1988).

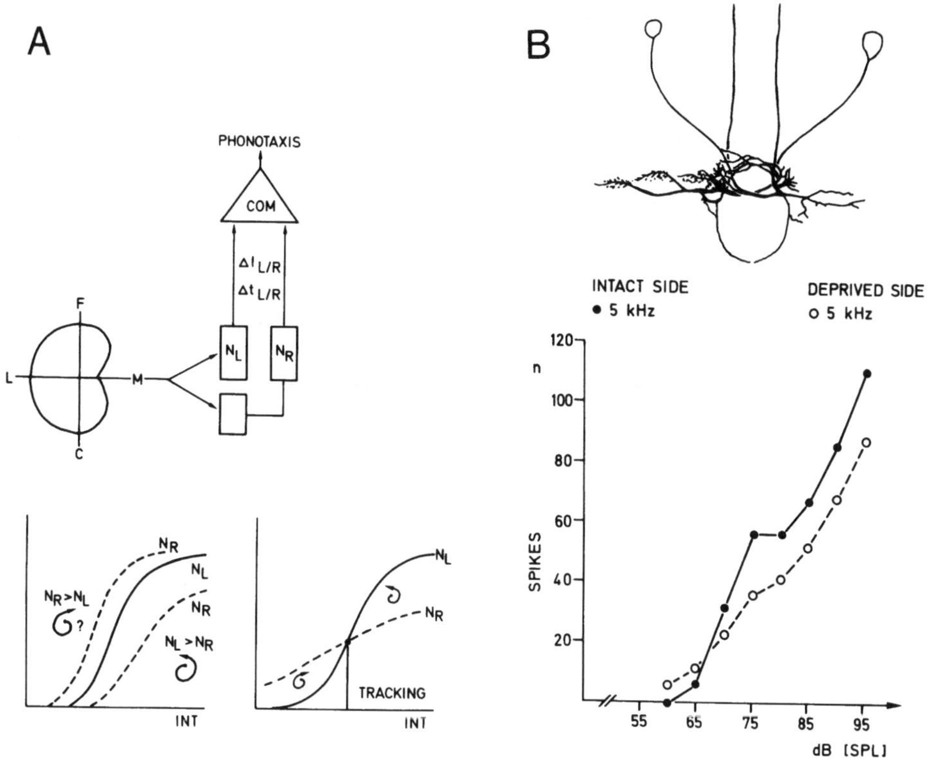

Fig. 5. One possible mechanism of phonotaxis with one ear **A**; *top* the model includes the directional sensitivity of the remaining ear (cardoid *left*), the connections of auditory neurons of the intact (N_L) and deprived (N_R) side with the remaining ear, a central comparison (*COM*) of the excitation of both sides (I or t *L*/*R*) and a resulting turning command; *bottom* consequences of different intensity response functions of both sides for behavior. **B** *Top* a pair of AN_2 neurons of the intact and deprived side of the same individual; *bottom* corresponding intensity response functions of the two neurons at 5 kHz. (Wohlers unpublished)

6 Regenerated Ears

So far we considered animals with a regenerated leg missing a functional ear. However, if the leg is amputated not at the coxa trochanter but at the femur tibia joint, the tibia regenerates with an ear (Fig. 6). Position, numbers of sensory cells and their arrangement in the auditory organ, as well as tracheae and tympana, are similar to those of normal animals. The central projections of the regenerated sensory axons resemble in shape and location those of the other, intact, side. Moreover, auditory interneurons of the side of the regenerated ear show no obvious morphological and physiological differences compared to normal cells, especially no dendritic sprouting across the midline. Finally, the regenerated ear must have been functional because the animals show stable walking courses dependent on sound direction, and circle after amputation of the regenerated ear.

These results raise a developmental problem. Dendritic sprouting in neurons occurs 7 days after leg amputation in adults (Schmitz in preparation). Regener-

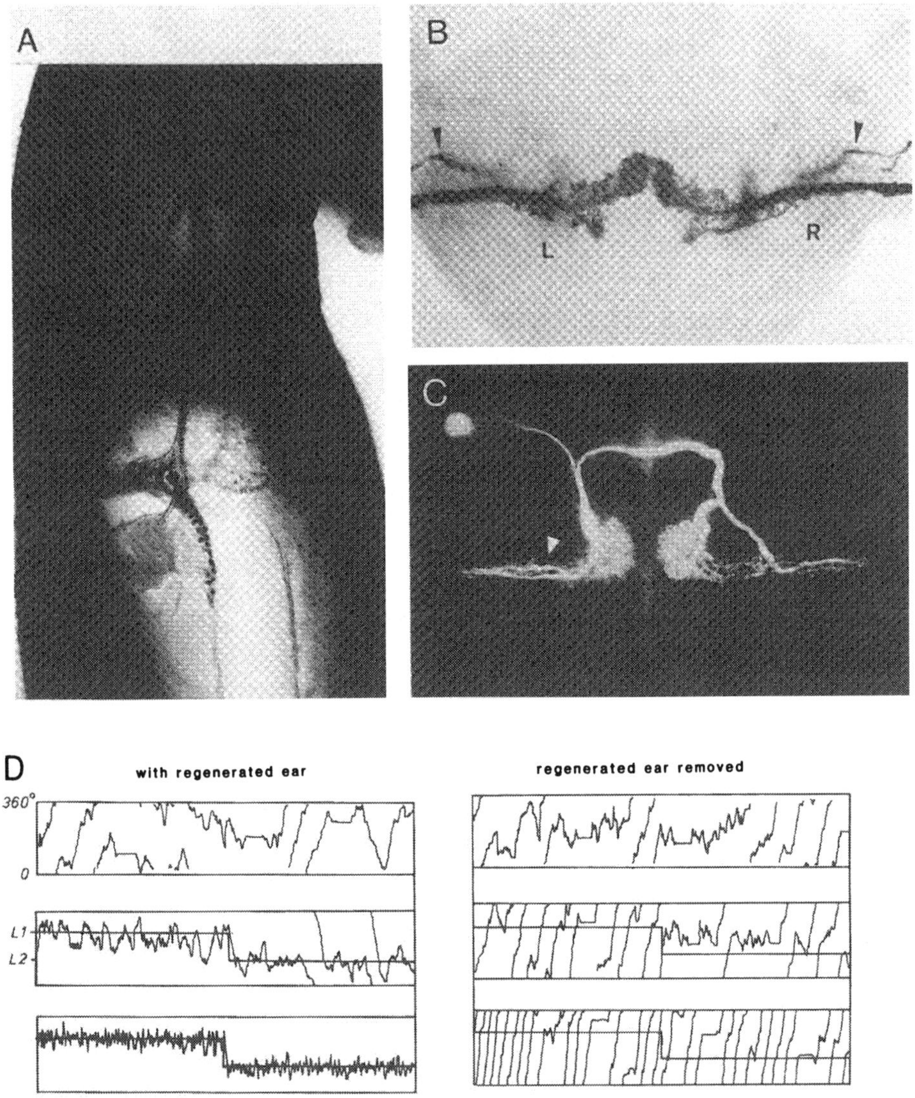

Fig. 6. Regeneration of a functional ear in animals with unilateral amputation at the femur-tibia joint in the 6th larval instar. **A** Retrograde cobalt fill of the auditory nerve shows a completely regenerated ear. **B** Anterograde cobalt fill of the auditory nerves of both ears in the same individual, the central projections of the regenerated auditory sensory fibers (left ganglion half, *L*) resemble in shape and location those of the intact partner leg (*R*), *arrow heads* mark stains of fibers from probably other mechanosensory systems. **C** Omeganeuron ON1 with the cell body and dendritic field on the side of regenerated ear (*left*, marked by an *arrowhead*) is similar in morphology to normal ON1s. **D** Phonotacic walking before (*left*) and after amputation (*right*) of the regenerated leg indicate that the regenerated ear was functional; *straight horizontal lines* mark loudspeaker positions (L1, L2), walking angles are plotted versus time

ation of the ear takes at least 3 molts (about 6 weeks) after amputation. Although studies of the morphology of sensory cells and interneurons remain to be done in developing and regenerating animals, it seems likely that central neurons exhibit dendritic sprouting already before the regenerating sensory axons again reach their final targets. But there are no transmidline sproutings of interneurons of adult animals with regenerated ear.

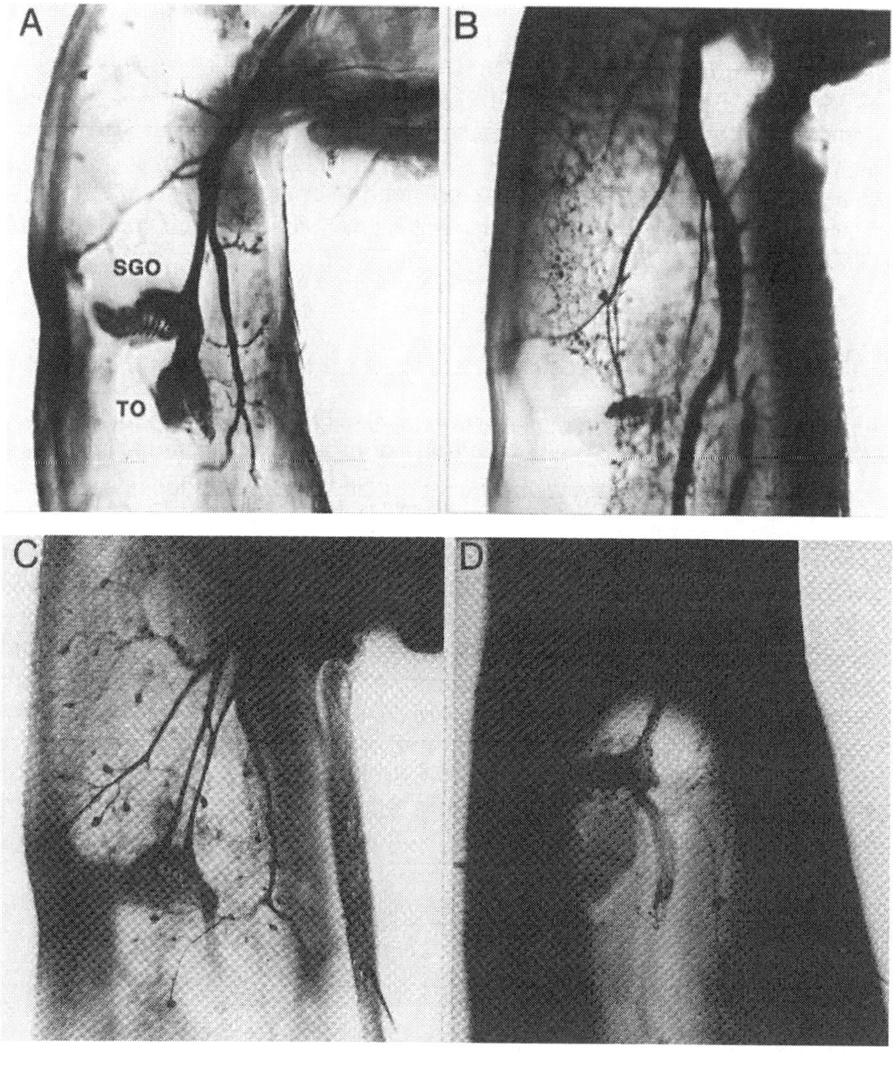

Fig. 7. Transplantations of fore- and middle leg tibiae; **A** the auditory and the subgenual organ of a prothoracic leg in the 6th larval instar in which the transplantations have been done. **B** Nervous and sensory organization within the tibia of an adult middle leg where the tibia was transplanted from the foreleg in the 6th larval instar, note that the auditory organ is missing but a subgenual organ present. **C** Normal adult middle leg tibia for comparison. **D** Fully differentiated auditory organ in an adult foreleg with the middle leg tibia transplanted to the foreleg femur (cf. Fig. 3 A); parts of the organ only lightly stained

One could presume that these sproutings occurred during postembryonic development, contacted sensory afferents of the other – normal – side, were disconnected and replaced by connections with the delayed ingrowing afferents of the regenerating ear. Thus, insect auditory neurons are not only able to form behaviorally useful new connections after unilateral deprivation, but may also disconnect and reorganize already formed synapses.

The limiting factor for ear regeneration cannot be time alone. Even if the leg was amputated near the coxa-trochanter region in the 1st larval instar (12 weeks before imaginal molt) it regenerated to full length without an ear. With amputation at the femur tibia joint in the 7th larval instar (about 4 weeks before imaginal molt) the leg regenerated to 80% of its normal length with a complete and functional ear. This suggests cells or structures within the prothoracic segment, particularly within the femural regions, which may induce transformation of tibial epidermal cells into auditory sensory cells and their proper location. Some evidence for the importance of the prothoracic segment for ear differentiation is provided by transplantation experiments.

7 Tibia Transplantations

In the 6th larval instar most of the sensory cells of the auditory organ are differentiated and their axons have reached the final projection area in the prothoracic ganglion, though the ear is not sensitive to sound because a tympanum is not yet formed (Fig. 7 A; see also Ball and Young 1974). If the tibia of the foreleg of such a larva is transplanted onto the femur of the ipsilateral midleg, no auditory organ could be found weeks later in the adult middle leg, but a subgenual organ was present (Fig. 7 B) as in the normally developed middle leg (Fig. 7 C). The reciprocal transplantation of a midleg tibia with no auditory organ onto an ipsilateral femur of the foreleg revealed an ear in the adult (Fig. 7 D). Obviously, there are factors in the prothoracic segment or the proximal parts of the foreleg which favor ear differentiation, factors that are missing in the mesothoracic segment or leg. Absence of a previously present ear in a transplanted foreleg tibia may even point to factors within the mesothoracic segment that suppress ear development.

Acknowledgements. We thank M. Klose, M. L. Obermayer, S. Schmaderer, B. Schmitz and D. W. Wohlers for providing unpublished results.

References

Ball EE, Young D (1974) Structure and development of the auditory system in the prothoracic leg of the cricket, *Teleogryllus commodus* II. Postembryonic development. Z Zellforsch 147:313:324

Eibl E (1978) Morphology of sense organs in the proximal parts of the tibiae of *Gryllus campestris* L. and *Gryllus bimaculatus* de Geer. Zoomorpholy 89:185–205

Huber F (1987) Plasticity in the auditory system of crickets: phonotaxis with one ear and neuronal reorganization within the auditory pathway. J Comp Physiol A 161:583–604

Huber F, Kleindienst HU, Weber T, Thorson J (1984) Auditory behavior of the cricket III. Tracking of male calling song by surgically and developmentally one-eared females, and the curious role of the anterior tympanum. J Comp Physiol A 155:725–738

Klose M, Schildberger K (1988) Sound localization after longterm acoustic stimulation. In: Elsner N, Barth FG (eds) Sense Organs. Interfaces between environment and behaviour. Beiträge zur 16. Göttinger Neurobiologentagung, Abstract 142, G Thieme, Stuttgart New York

Michel K (1974) Das Tympanalorgan von *Gryllus bimaculatus* DeGeer (Saltatotria Gryllidae) Z Morphol Tiere 77:285–315

Schildberger K, Wohlers DW, Schmitz B, Kleindienst HU, Huber F (1986) Morphological and physiological changes in central auditory neurons following unilateral foreleg amputation in larval crickets. J Comp Physiol A 158:291–300

Schildberger K, Kleindienst HU (1988) Sound localization in one-eared crickets. In: Elsner N, Barth FG (eds) Sense Organs. Interfaces between environment and behaviour. Beiträge zur 16. Göttinger Neurobiologentagung, Abstract 18, G Thieme, Stuttgart New York

Schmitz B (1986) Compensation of unilateral hearing deficiencies in the acoustic orientation of crickets. In: Elsner N, Rathmayer W (eds) Sensomotorik identifizierte Neurone. Beiträge zur 14. Göttinger Neurobiologentagung, Abstract 66, p 124

Schmitz B, Kleindienst HU, Schildberger K, Huber F (1988) Acoustic orientation in adult, female crickets (Gryllus bimaculatus de Geer) after unilateral foreleg amputation in the larva. J Comp Physiol A 162:715–728

Weber T, Thorson J, Huber F (1981) Auditory behavior of the cricket. I. Dynamics of compensated walking and discrimination paradigms on the Kramer treadmill. J Comp Physiol 141:215–232

Wohlers DW, Huber F (1982) Processing of sound signals by six types of neurons in the prothoracic ganglion of the cricket *Gryllus campestris* L. J Comp Physiol 146:161–173

Peripheral and Central Post-Lesion Plasticity in the Olfactory System of the Goldfish: Behavior and Morphology

H. P. Zippel[1], D. L. Meyer[2] and M. Knaust[1]

[1] Physiologisches Institut der Universität, Humboldtallee 23, D-3400 Göttingen, FR Germany
[2] Anatomisches Institut der Universität, Kreuzbergring 36, D-3400 Göttingen, FR Germany

1 Introduction

The enormous regenerative capacity of primary olfactory receptor neurons of vertebrates has been demonstrated by the findings of a number of investigations. The majority of publications are based on neuroanatomical (see, e.g., the numerous publications by Graziadei et al.), and the minority on physiological findings (electroolfactogram, single cell recordings; see, e.g., Simmons and Getchell 1981 a, b). In the publications cited and in those of other investigators, there is general agreement that olfactory mucosa receptors degenerate following axotomy, and are later replaced by new receptor cells which differentiate from basal cells. Axons of the new olfactory receptors grow out and terminate in the glomerular layer at the surface of the olfactory bulb, where they make synaptic contact with relay neurons. Whilst studies of morphological and physiological aspects have demonstrated a highly developed regenerative capacity, related functional and behavioral investigations have been rare in comparison with the number of similar studies on other sensory and motor systems. Thus, until now, one can only assume that the regenerated sensory neurons, together with their central connections, might be functionally and topographically reorganized in such a way as to enable the animal to appropriately behave and orientate in its olfactory environment.

Hence the present behavioral and neuroanatomical investigations in goldfish have been designed (1) to examine the relationship between morphological changes and changes in the perception of spontaneously preferred olfactory (food) stimuli before and after inflicting lesions in the olfactory system, (2) to test the animals' ability to respond to a stimulus learned preoperatively following regeneration, (3) to investigate the animals' ability to discriminate between two odors, one of which had been rewarded preoperatively ("specific" regeneration), and (4) to test for functional and morphological regeneration after lesions in peripheral and in central olfactory pathways, and after partial ablation of olfactory bulbs.

Accordingly, at various time intervals before and after lesions to the olfactory system, behavioral and morphological comparisons were made between operated and intact, sham-operated specimens.

H. Flohr (Ed.)
Post-Lesion Neural Plasticity

Material and Methods

2.1 Surgery

The fish were anesthetized with MABA (formerly Tricain or MS-222, Sandoz; 300 mg in 1000 ml tap water) and the cranium overlying the olfactory bulbs and tracts was removed. Lesions made for the behavioral experiments were always bilateral, for histological investigations unilateral, and were performed in the following series (Fig. 1 A): (1) Intracranial receptor axotomy, i.e., loss of receptor input to the olfactory bulb; (2) Rostral bulbectomy, i.e., extirpation of the rostral half of the olfactory bulbs: loss of receptor input to the olfactory bulb and of approximately 50% of the olfactory bulb neurons; (3) Caudal bulbectomy, i.e., extirpation of the caudal half of the olfactory bulbs: complete loss of bulbar output to telencephalic nuclei, of efferent, centrifugal pathways to olfactory bulbs, and of approximately 50% of bulbar neurons.

2.2 Training and Testing Apparatus

Details of the standardized, semi-automatic apparatus for a shock-free training have been described elsewhere (Zippel et al. 1981), and Fig. 1 B outlines the essential points. During the entire pre- and post-operative test and training period the animals were maintained in groups of two in separate aquaria under a 12 h light/12 h dark cycle. For feeding, testing, or training purposes, opaque plastic funnels were suspended at each end of the tank, the tips of which were perforated with holes through which the food-reward (Tubifex worms) was delivered during training.

The olfactory stimuli were (1) Tubifex extract (concentrations based on 1 g wet weight), which is spontaneously preferred by the fish. The animals' reactions to these stimuli were always recorded before training, and Tubifex extract served as the test stimulus for determining the time course of functional regeneration. Details of olfactory and gustatory thresholds for the various Tubifex extracts will be presented with the results. (2) Amino acids and synthetic odors which do not induce responses, i.e., which are neither spontaneously preferred nor actively avoided by the animals. These stimuli served as training stimuli, and again, details will be presented with the results.

All experiments were preceded by a habituation period (4–5 weeks) during which the animals learned to associate the funnels with food. By the end of this habituation period, a complex behavioral repertoire had developed and the following behavioral patterns were recorded: (1) Funnel Orientation (FO) – the time (s) spent by the animals in the immediate vicinity of the funnels,

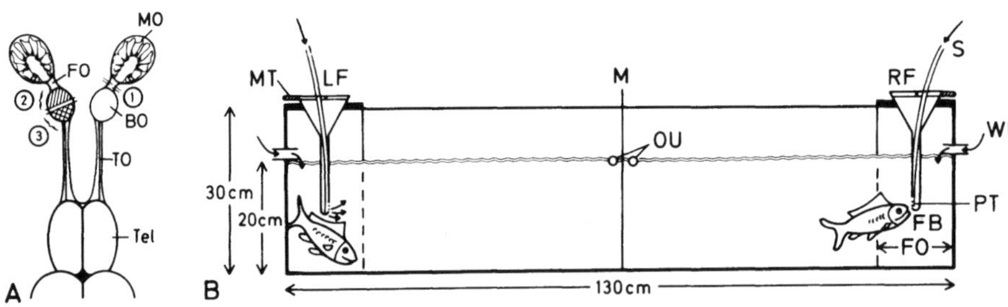

Fig. 1. Schematic representation of the olfactory system lesions (**A**) and of the testing and training apparatus (**B**). Operations for the behavioral series were always made bilaterally (for details see text); **A** *1* receptor axotomy; *2* rostral bulbectomy; *3* caudal bulbectomy; *MO* Mucosa olfactoria; *FO* Fila olfactoria; *BO* Bulbus olfactorius; *TO* Tractus olfactorius; *Tel* Telencephalon; **B** *FO* funnel orientation; *FB* funnel biting (automatically recorded by a photocell system, the light beam of which was broken during biting of the funnel by moving a flat metal tongue = *MT* affixed to the rim of the funnel); *LF, RF* left and right funnels; *S* stimulus application (200 µl/s, 12 ml/min) through plastic tubes to the perforated tip (*PT*) of the funnel; stimulus concentrations mentioned in the text are maximum concentrations at the tip of the funnel; *WI* tap water inlet; *OU* tap water outlet in the middle (*M*) of the tanks' back-screen

and (2) Funnel Biting (FB) – the number of bites at each of the funnels, which may be interpreted as a specifically orientated food expectation. FB was registered automatically by a photocell system, while FO was recorded manually by operating a time switch. Final data and statistical analyses were performed using a PDP-11/34 computer.

2.3 Training and Testing Procedure

At the onset of each session, a *control registration* (3 min duration) was made of the animals' reaction to the funnels in the absence of stimuli. In the immediately succeeding *test* or *training period* (also of 3 min duration), the animals' behavior during stimulus application was recorded.

At no time during or immediately following a *test session* was the animals' behavior reinforced, in contrast to the *training sessions* which were immediately followed by positive reinforcement with the animals being fed Tubifex worms through the appropriate funnel (interval reinforcement). The training or test stimulus was always applied through the funnel to which less activity was orientated during the control registration (i.e., on the "negative"/"disliked" side of the tank). The training experiments fall into two categories: (1) training to a single stimulus, and (2) discrimination training, during which the animals learned to differentiate odors qualitatively. During discrimination training a "positive" (i.e., reinforced, through the spontaneously disliked funnel) stimulus was simultaneously offered with a "negative" (i.e., nonreinforced, competing) stimulus, applied through the funnel spontaneously preferred by the specimens. Statistical evaluation of the pre- and postoperative behavior was made using the Wilcxon rank test.

2.4 Neuroanatomy

The behavioral tests were accompanied by histological investigations. These consisted of HRP studies performed on long-term regenerated fish, or on animals shortly after the behavioral recovery of function. HRP (0.1–0.3 µl, Boehringer, grade III) was injected (1) into one nostril (which was then closed with Histoacryle) to demonstrate receptor terminals in the olfactory bulb, and (2) into the olfactory bulb or into the telencephalon to investigate the presence of regenerated central connections. Survival time was 3 days at 22 °C water temperature. Thereafter the animals were anesthetized and transcardially perfused with saline, followed by a 2% solution of glutaraldehyde in 0.05 M phosphate buffer (pH 7.4). The brains were postfixed in the same fixative. To avoid freezing artifacts in later stages, the brains were then soaked in 30% sucrose for 8–12 h. Thereafter sections were cut in a cryostat at 20 µm. The peroxidase reaction was performed with bencidine dihydrochloride (for details, see v. Bartheld et al. 1984), and the sections were counterstained with Safranine and neutral red.

3 Results

3.1 General Remarks

Behavioral studies on goldfish, in this case pre-training and training experiments, using olfactory stimuli are very time-consuming (Zippel 1970; Zippel and Voigt 1982), and several months of daily training and testing were necessary before the operations could be performed. Test and training sessions were conducted once per day only in order to maintain a suitable level of motivation. The data presented here concentrate on the postoperative test period. Spontaneous reactions to various stimuli and detailed training results will be published elsewhere.

3.2 Functional Regeneration After Training to a Single Olfactory Stimulus

From previous studies it is evident that peripheral and central pathways in the goldfish olfactory system regenerate functionally and morphologically (receptor axotomy: Zippel et al. 1970; olfactory tract and anterior commissure: Zippel and Westerman 1970). More recent studies have shown that after complete bilateral bulbectomy, receptor cells in the olfactory mucosa are present following long term regeneration (Zippel et al. 1981). The outgrowing receptor axons form more or less bulb-like "neuromas" in place of the former olfactory bulbs. Neither neurons nor synapses are present within such "neuromas" (Graziadei personal communication). The permanent behavioral deficit (i.e., no return to the preoperative olfactory thresholds at any time, Zippel et al. 1981) can therefore be sufficiently explained by he fact that no connections are found between the olfactory mucosa and the telencephalon (Graziadei personal communication). The present investigation concentrated on partial olfactory bulbectomy in order to examine the relationship between changes in the perception of olfactory stimuli and morphological changes after receptor axotomy, as well as after extirpation of the rostral half of the olfactory bulb (Rostral Bulbectomy, RBE); or after dissection of the olfactory tracts, and extirpation of the caudal halves of the olfactory bulbs (Caudal Bulbectomy, CBE).

In the first series, 21 groups of goldfish were trained using amyl acetate (10^{-5} M) as the reinforced stimulus; zhe animals' spontaneous reactions to different concentrations of Tubifex-food-extracts (10^{-5}–10^{-9} g/l) were also tested. In accordance with earlier findings (Zippel and Voigt 1982), a nearly linear decline in funnel orientation and funnel biting was recorded with decreasing concentrations of the stimulus, the olfactory threshold concentration (i.e., no behavioral change in comparison with the preceding stimulus-free control registration) lying below 10^{-8} g/l. Furthermore, generalization tests to taste stimuli were made with acetic acid (10^{-6} and 10^{-5} M), a stimulus to which goldfish can easily be trained (Zippel et al. 1981).

After 45 days of training, the 21 groups were subdivided into three collectives: seven groups served as sham-operated controls, in seven the rostral half (RBE) and in the remaining seven the caudal half (CBE) of the olfactory bulbs was extirpated bilaterally. After the sham operation the control animals' behavior to the training stimulus and low concentrations of Tubifex extract did not change; some tests are depicted in Fig. 2 C. Immediately after partial bulbectomy, the operated animals did not respond to the training stimulus (amyl acetate, 10^{-5} M, RBE: Fig. 2 A, day 3; CBE: Fig. 2 B, day 7). However, when the stimulus concentration increased to concentrations above the taste threshold (Zippel and Voigt 1982), a generalized positive behavior was noticed (RBE, days 7 and 8; CBE, days 10, 11 and 13).

As we expected, a comparatively slow time course of regeneration (or no functionally complete recovery at all) after partial bulbectomy, generalizing experiments to taste stimuli were made with acetic acid from postoperative day 9 onwards. However, when acetic acid was applied in a 10^{-5} molar concentration (i.e., below the taste threshold; Zippel et al. 1981) on day 9 postoperative, the rostrally bulbectomized animals responded positively, and positive reactions were also re-

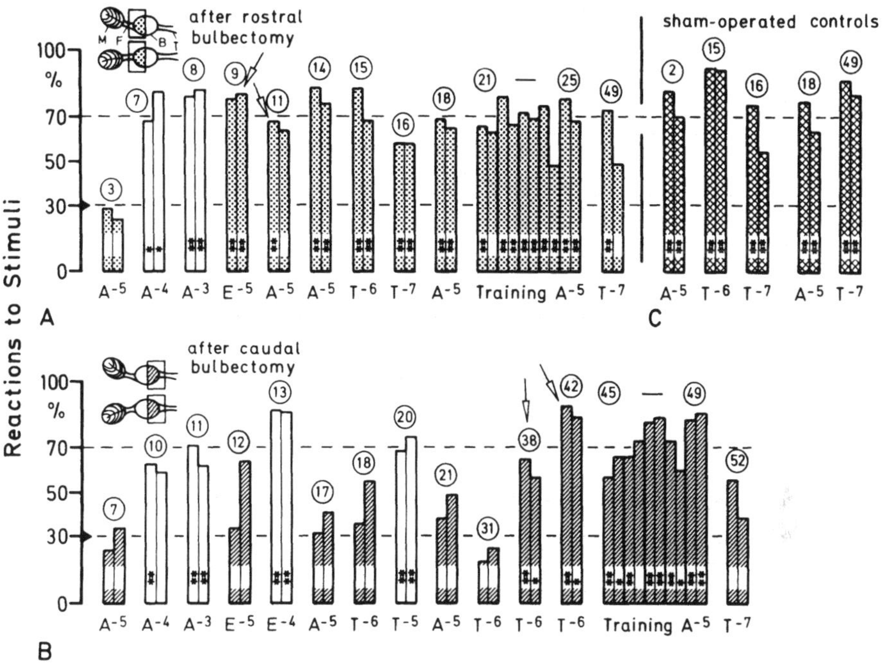

Fig. 2. Behavior of 21 groups of goldfish after bilateral rostral (**A** RBE; 7 groups) or caudal (**B** CBE; 7 groups) bulbectomy in comparison with sham operated (**C** 7 groups) controls. Pre-operative training stimulus: amyl acetate 10^{-5} M (A^{-5}); pre-operative test stimuli: acetic acid 10^{-5}, 10^{-4} M (E^{-5}, E^{-4}) and various concentrations of Tubifex extract (*T* indices for concentrations in g wet weight per liter). Post-operative responses are expressed as mean values (in percent) to the respective stimulus; funnel orientation in the left and funnel biting in the right half of the respective column. Reactions close to or above the 70% level represent a significant positive behavior. The *broken line* at 30% indicates the mean values for the reactions to the spontaneously disliked funnel during the stimulus-free control registration (for details see methods). *Open columns* stimulus concentrations above "taste" threshold; *circled numbers* time in days after the lesion; *arrows* time of functional recovery. Statistical significance was tested (Wilcoxon rank test) by comparing the FO and the FB values to the test or training stimulus, with the FO and FB values at the same funnel during the stimulus-free control registration. *Two asterisks* (inserted into the respective column) significant positive reactions at the 1% level; *one asterisk* significant at the 5% level; *no asterisk* no significant difference between test and control behavior

corded during presentation of the training stimulus on day 11. The fact that a functional regeneration had taken place in the RBE collective after this extremely short postoperative period became obvious in the subsequent tests with no significant differences in behavior between the sham-operated controls and the RBE collective being found following application of the training stimulus or low-concentrated Tubifex extracts (Fig. 2A, 2C; days 15, 16, 18). During a postoperative training period (Fig. 2A, days 2–25) the behavior of the rostrally bulbectomized specimens was similar to that noted immediately before the operation.

In contrast to the RBE collective, the CBE animals' behavior remained unchanged (in comparison with the stimulus-free control registration) over a period of 6 weeks when the stimuli were offered in concentrations below the taste threshold (Fig. 2 B, days 12, 17, 18, 21, 31), whereas concentrations above the taste threshold elicited positive reactions (Fig. 2 B, days 10, 11, 13, 20). A return to the preoperative behavior was first recorded on postoperative day 38, and a highly positive response to the low concentration Tubifex-extract was found by day 42 (i.e., 6 weeks postoperative). Thereafter these animals' behavior was also comparable to their pre-operative training behavior when the training stimulus was presented in five subsequent training sessions (Fig. 2 B, days 45–49).

Partial bulbectomies were repeated in a second series of experiments on ten groups of goldfish trained to Na-glutamate (10^{-5} M). In this series the RBE-collective was investigated postoperatively with low concentrations of Tubifex food extracts (10^{-7} g/l) only [since the stimuli applied at days 7 and 8 (Fig. 2 A) were above the taste threshold in the first series] and, in this case, a return to pre-operative olfactory thresholds was observed from day 8 onwards. In contrast, the CBE collective again showed little responsiveness for approximately 6 weeks before gradually regaining the pre-operative sensitivity.

In conclusion, the above data indicate that a functional recovery occurs after partial bulbectomy, although at different time intervals for the different collectives. The assumption that the functional regeneration was due to an at least partial reconstruction of the pre-operative peripheral and central connections, was investigated in the neuroanatomical part of this study.

3.3 Morphological Regeneration After Partial Bulbectomy

The initial series of HRP-studies was performed on fish from the first behavioral experiments (Fig. 2) described above. The brains were investigated 20 months after partial bulbectomy, and compared with preparations from intact animals. HRP was always injected unilaterally, with the contralateral side serving as a control for possible unspecific reactions. Macroscopically it was evident that the spherical shape of the olfactory bulb was reestablished in the RBE- and CBE collective, although the diameter of the regenerated bulbs was about half the size of those found in intact animals. In RBE animals connections from the olfactory mucosa were similar to those of intact animals. After HRP injection into the nostril, an orderly projection was found similar to that in the intact animals (Fig. 3). Following ablation of the caudal halves of the olfactory bulbs, however, the appearance was different. In these cases the olfactory bulbs contained an inner sack surrounded by granulated tissue, with the incoming fibers terminating inside this tissue. In no case were fibers found traveling through the olfactory bulb and tract directly to the telencephalon.

In order to demonstrate the central connections [1], HRP was injected into the olfactory bulb and into the telencephalon (Fig. 4) in animals previously subjected to caudal bulbectomy. Following olfactory bulb injection, fibers were found in the ipsilateral telencephalon (Fig. 4 B), and after injection into the telencephalon retrogradely labeled mitral cells were present in the ipsilateral olfactory bulb (Fig. 4 C).

---

[1] These experiments were made by Ch. v. Bartheld, whose cooperation is gratefully acknowledged.

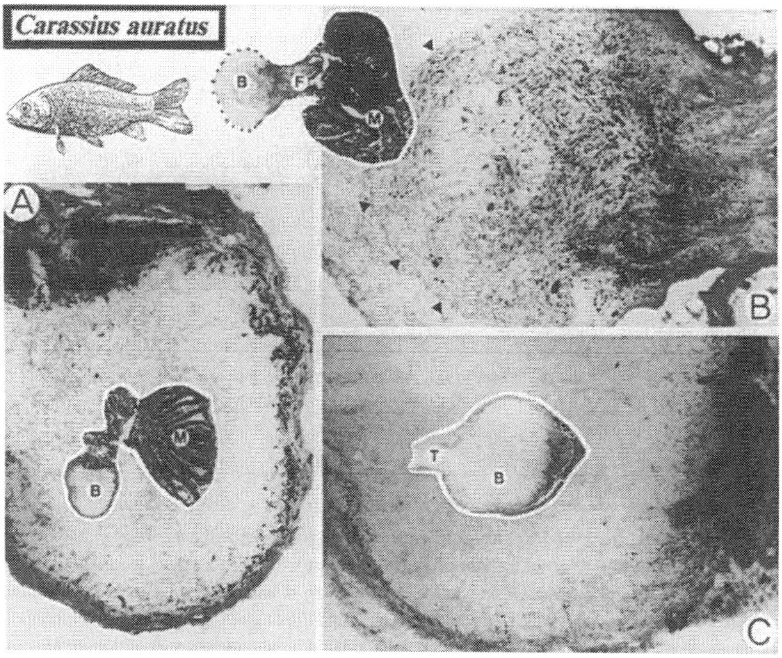

Fig. 3. HRP studies 20 months after partial bulbectomy. HRP injection into nostril. **A** After rostral bulbectomy **B** After caudal bulbectomy; *arrowheads* indicate HRP-labeled terminals at the surface of the olfactory bulb (*B*); **C** intact control; *M* Mucosa olfactoria; *T* Tractus olfactorius

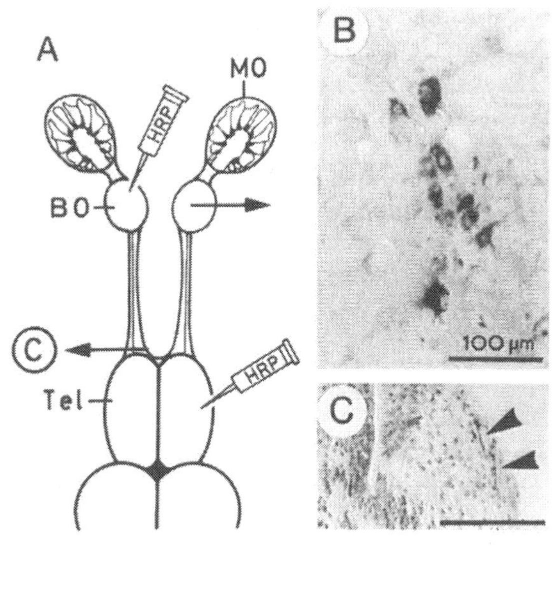

Fig. 4. HRP studies after caudal bulbectomy. **A** Sites of injection; *MO* Mucosa olfactoria; *BO* Bulbus olfactorius; *T* Tractus olfactorius; *Tel* Telencephalon **B** Labeled mitral cells in the olfactory bulb after HRP injection into the telencephalon **C** Labeled fibers in the telencephalon after injection into the olfactory bulb

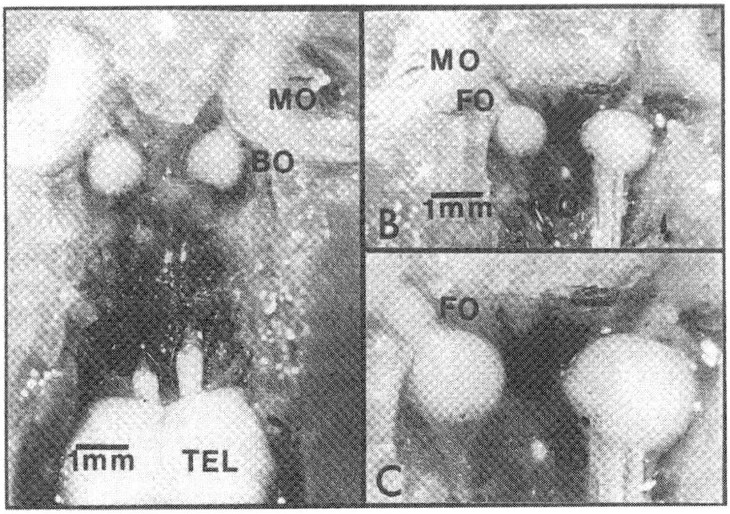

Fig. 5. Preparation sites shortly after functional recovery from partial bulbectomy. Nine weeks after caudal (**A**) and 20 days after rostral (**B, C**) bulbectomy. In both cases the spherical shape of the olfactory bulbs has regenerated. In the CBE animals, in which the olfactory tracts were cut directly behind the bulb, the tracts have shrunken to roughly 25% of their initial lengths. In this situs the thin and transparent regenerated pathways can only be seen at the rostral end of the stump. In the RBE animals the regenerated fila olfactoria (*FO*) are comparable with those in intact animals. For abbreviations see Fig. 4

In a second series, neuroanatomical investigations were performed as soon as behavioral tests demonstrated a return to pre-operative level (RBE animals 20 days, and CBE animals 9 weeks postoperative). From the photographs taken during preparation (Fig. 5) it is evident that shortly after the functional recovery, a surprisingly high level of regeneration had been achieved, with the spherical shape of the olfactory bulb being reestablished nearly completely in both collectives. In RBE animals (Fig. 5 B, C) the connections between the olfactory mucosa and the olfactory bulbs were not significantly different to those of intact animals, whereas in the CBE animals central connections between the olfactory bulbs and the telencephalon were extremely thin and much more "transparent" (Fig. 5 A). The HRP experiments demonstrated fiber connections between the olfactory mucosa and the olfactory bulb to be present in RBE and CBE specimens, with a pattern of innervation qualitatively similar to that found in the long-term regenerated material. A quantitative estimation is not possible at present.

3.4 Behaviorally "Specific" Regeneration

From the above series of experiments it is evident that after partial ablations of olfactory bulbs, and a subsequent regeneration of respective pathways, goldfish react positively to low threshold olfactory stimuli and to individual odors. In this series of experiments three questions were to be answered: (1) do animals, following lesions in the olfactory system, return to the original olfactory threshold after

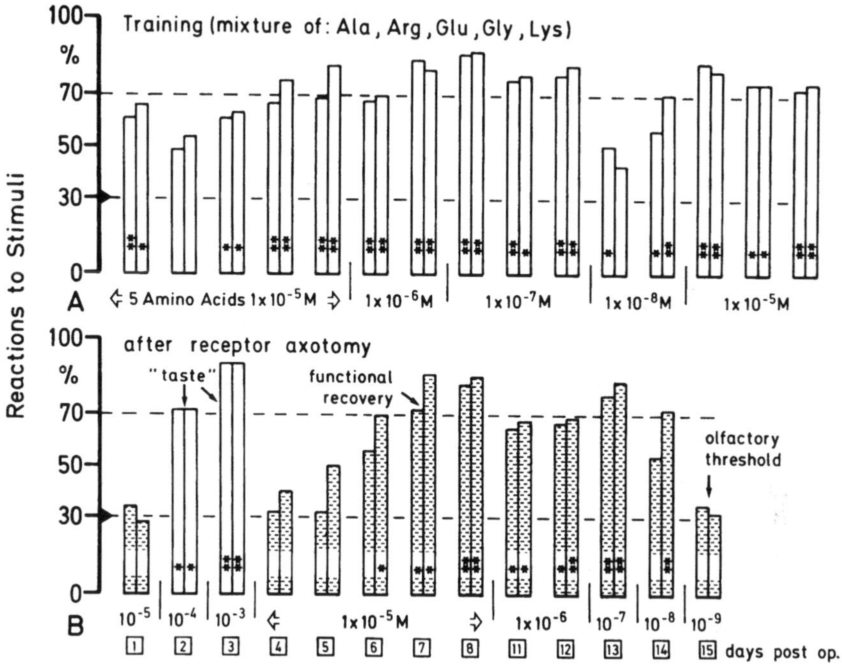

Fig. 6. Threshold investigations using seven groups of animals after training to a mixture of five amino acids (**A**), and after receptor axotomy (**B**). For details see legend to Fig. 2

regeneration? (2) Are they able to discriminate between similar odors? (3) Can they discriminate the initially learned odors from others, unknown to them? "Specific" regeneration of receptor axons, i.e., the animals' ability to discriminate between different olfactory stimuli, which have been learned during training, was first described in 1970 (Zippel et al.). In these experiments, the stimuli used were rather dissimilar synthetic odors (amyl acetate, coumarine, eugenol, morpholine), and threshold investigations were not performed.

We therefore reinvestigated the specific regeneration using (1) threshold investigations and, in a second series, (2) using experiments employing very similar olfactory stimuli (amino acids) following discrimination training. In the first series five amino acids behaviorally relevant for fish were used as the training stimuli (Ala, Arg, Glut, Gly, Lys 10^{-5} M). Figure 6 A demonstrates the final training period (after 45 training sessions) to a 10^{-5} M training stimulus, and the subsequent threshold investigations. At the 10^{-8} M concentration, the trained animals' behavior was still slightly positive, but the first of two tests was more or less indifferent, i.e., the stimulus concentration was close to the threshold. After three rewards with the training stimulus in its initial concentration the animals were lesioned. Immediately following receptor axotomy the fish responded positively only when the stimulus was applied in concentrations above the taste threshold (Fig. 6 B, day 2, 3). But by day 7 postoperative the functional recovery was quite evident. From day 8 onwards, when the stimulus concentration was lowered in log concentrations, the animals' behavior was comparable to the pre-operative reactions. Whereas they responded positively down to the 10^{-8} M concentration, they failed to respond to the 10^{-9} M stimulus intensity, findings which indicate full recovery of sensitivity following axonal regeneration.

A subsequent differentiation training was possible, however, only when the competing stimulus was very different from the training stimulus; i.e., while it was possible to train the animals to discriminate between "five amino acids", vs. amyl acetate (both 10^{-6} M), discrimination training with "five amino acids" (Ala, Arg, Lys, Glut, Gly) vs. Alanin 10^{-6} M was not effective. The question whether this failure was due to a difficulty in discriminating between the stimuli, to "overtraining" in the long-lasting series (124 test and training sessions), or to a form of regeneration not permitting "specific discrimination", was therefore investigated in the subsequent series.

In this final series ten groups of animals were discrimination trained to a bouquet of four amino acids (Ala, Arg, Lys, Glut, 10^{-6} M) vs. glycine (10^{-6} M) as the competing stimulus. The fact that 60 training sessions were necessary before the animals were capable of discriminating between the stimuli, and that up to the final training session there was a considerable fluctuation in individual groups' behavior, clearly demonstrates the animals' difficulty in discriminating between the stimuli applied. After 60 training sessions, our ten groups of animals were subdivided in two collectives of five each: five groups were sham-operated, and in five groups the olfactory nerves were bilaterally dissected close to the olfactory bulb. Thereafter the animals were neither behaviorally tested, nor fed through the funnels for 9 days, in order to avoid any extinction or reward for the sham-operated or the operated collective. On day 10 postoperative the animals were again confronted with the pre-operative training stimuli (Fig. 7) and both collectives responded positively to the training stimulus, i.e., a specific regeneration could be demonstrated in the operated animals even following training to an extremely difficult discrimination task.

During postoperative days 11–13 the behavioral response to the pre-operative training stimuli remained unchanged. The introduction of a new, less familiar stimulus (amyl acetate 10^{-6} M) neither resulted in a preference for the novel stimulus, nor did it cause indifferent behavior, as both the operated and the sham-operated animals responded positively to the training stimulus.

Both collectives were now trained to the four amino acids vs. amyl acetate (Fig. 7, days 14 to 17 post op.), and thereafter the formerly sham-operated animals were subjected to rostral bulbectomy. After 9 days without any behavioral test or funnel feeding the fishes' behavior was again investigated. During the first 2 days (10, 11 post RBE) glycine (10^{-6} M) was the competing stimulus and in both tests the reactions of the control fish (formerly axotomized) and of the rostral bulbectomized animals were positive, i.e., even after rostral bulbectomy and loss of half of the olfactory bulb specific regeneration occurred. During days 12–15 post RBE, amyl acetate (10^{-6} M) served as the competing stimulus. In the first test the control and the rostral bulbectomized animals' behavior was different from that recorded after the first operation (day 12 post RBE) with the introduction of the "unfamiliar" stimulus leading to some "uncertainty" in both groups of animals, and the recorded mean values dropping to roughly 60% for funnel orientation and funnel biting. Thereafter, i.e., as a result of the reinforcement of

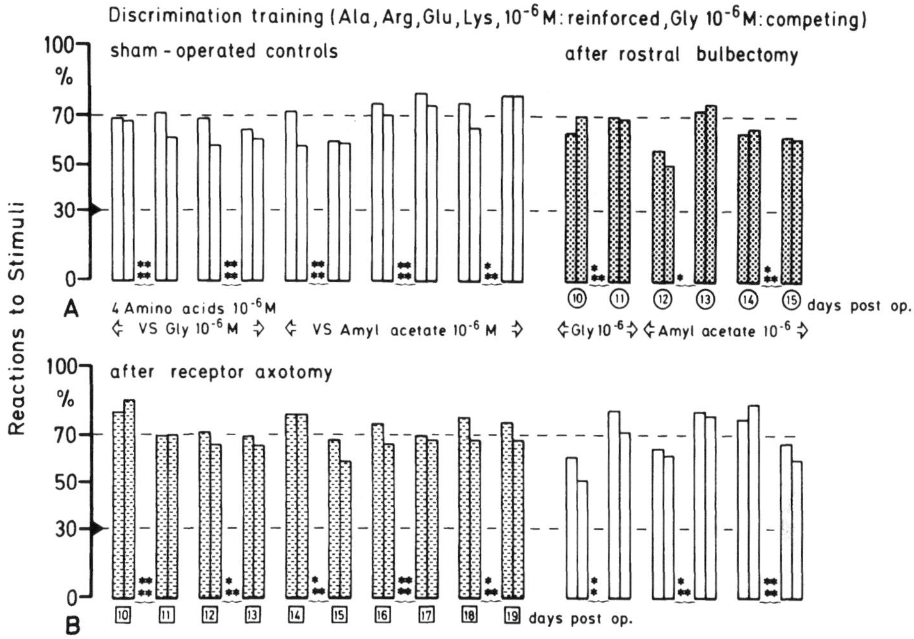

Fig. 7. Specific regeneration after discrimination training. Ten groups of animals were discrimination-trained in 60 daily sessions, five groups were then operated (receptor axotomy: **B**) and after a regeneration time of 10 days their behavior was compared to that of the other five sham-operated (**A**) groups (stimuli inserted in the figure). After discrimination-training with a new, unfamiliar, competing stimulus, the sham-operated animals were subjected to rostral bulbectomy, and following another regenerative period of 10 days, the RBE collectives' behavior was compared with the formerly axotomized animals' reactions. For statistical analysis the data of 2 successive days with identical stimuli were pooled, and the significance levels are given between the respective columns (*FO* left; *FB* right). For details see legend to Fig. 2

the four amino acids after test 12, the positive reactions recorded on days 13–15 demonstrate that both the collectives were capable of discriminating between the training stimulus and amyl acetate as significant positive reactions were found in both collectives. Thus a specific regeneration was also achieved after rostral bulbectomy. The slight behavioral deficit (roughly 10–15% in FO and FB) recorded in the RBE collective may have been caused by the loss of olfactory bulb neurons. A quantitative analysis of this behavioral deficit is not possible at present.

4 Discussion

Following partial bilateral ablation of the olfactory bulbs a return to the pre-operative behavior was found at different (but for the respective lesions characteristic) time intervals after the operation. Since, with the exception of some of our own previous investigations, no other extensive behavioral studies are available on regeneration in the olfactory system, the data presented here can only be compared with effects of partial lesions, morphological, and electrophysiological observations in other species.

From the post-lesion behavioral tests it is evident that immediately after a bilateral interruption of the primary olfactory input the animals are no longer able to respond to low-threshold stimuli. The question whether or not the positive, generalizing behavior to stimuli applied at higher concentration can be interpreted as a generalized reaction mediated by the gustatory system, is caused by the high-threshold trigeminal system or by both systems, must remain open at present. However, that the positive reactions recorded after regeneration are mediated via reconstructed olfactory connections, rather than by a sensitivity increase in the systems mentioned above, is evident from the following facts: a functional regeneration, i.e., the return to the pre-operative behavior occurs (1) with a different, and reproducable time course, and (2) after total bulbectomy (Zippel et al. 1981) the animals' behavior remains at a high-threshold level even after 12 months postoperative. In the latter case no connections were found between the neuroma formed in place of the former olfactory bulb, and the telencephalon.

The time course of regeneration is depicted schematically in Fig. 8. After receptor axotomy and following rostral bulbectomy in which the animals lost the peripheral input as well as approximately 50% of their bulbar relay- and interneurons, surprisingly short functional regeneration (7–10 days) was observed. At this time degenerative processes are at their maximum in all other previously studied vertebrates (e.g., trout: Evans and Hara 1985; frog: Burns et al. 1981; salamander: Simmons and Getchell 1981 a, b; Simmons et al. 1981; hamster: Samanen and Forbes 1984; Constanzo 1985; rat: Graziadei and Monti Graziadei 1980; monkey: Monti Graziadei et al. 1980). From previous (Zippel et al. 1970) observations, and from the HRP cases reported here, it is evident that innervation of the goldfish olfactory bulb is more or less comparable to that of intact animals at the time of functional recovery. The phenomenon of rapid regeneration in goldfish must remain unexplained at present as in haematoxylin-eosin stained sections of the olfactory epithelium of specimens previously subjected to receptor axotomy or total bulbectomy no obvious differences were present. Survival times studied were 6 h, 12 h, 24 h, 2, 4, 6, and 9 days. The question whether (1) in the goldfish receptors survive after surgery and regenerate their axons, or (2) undifferentiated receptors migrate to the surface of the olfactory mucosa and rapidly grow axons to the olfactory bulb, is currently being investigated using olfactory marker protein. The latter possibility however receives support from Ichikawas' and Uedas' (1977) ultrastructural observations. After receptor axotomy in goldfish the authors report degeneration of only two types of receptors. Another type (considered to be nonolfactory) was present in the olfactory mucosa and had occupied the postions where the two degenerated cell types were previously localized.

After receptor axotomy and after rostral bulbectomy a specific regeneration was found, i.e., not only a return of pre-operative responses to natural and to trained odor stimuli, but also a return to threshold concentrations. Behavior acquired in a difficult, long-lasting discrimination training is present after regeneration (i.e., a highly specific regeneration). This supports the hypothesis that the regenerating receptor axons either project to specific glomeruli (Doucette et al. 1983; Graziadei et al. 1986) and relay neurons, or are functionally specifically "programmed" at the time they mature and make synaptic contact with the relay

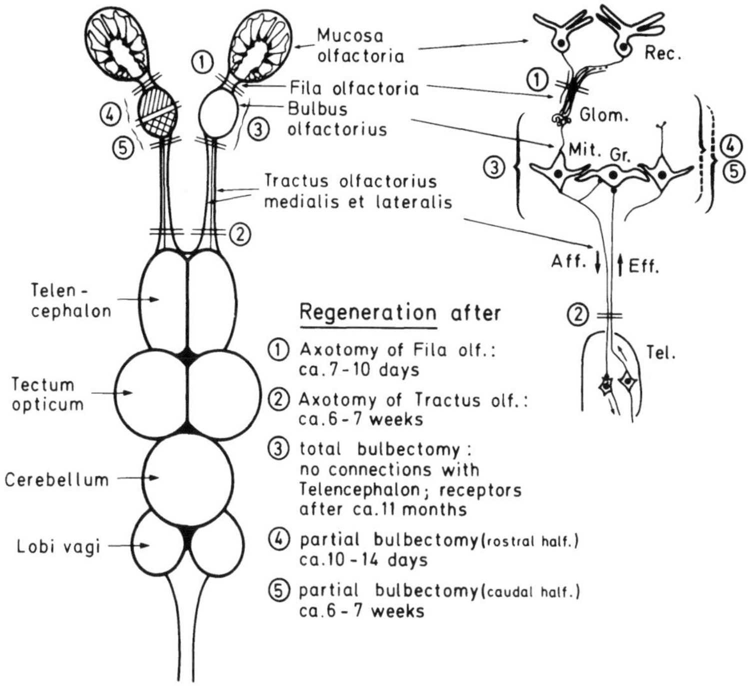

Fig. 8. A schematic presentation of peripheral and central lesions in the goldfish olfactory system. At the *left* a schema of the goldfish's brain showing the operation sites (always bilateral for the behavioral tests); the sketch on the *right* demonstrates the main pathways and main neuronal structures destroyed by the respective operations, and the *insert* presents a list of the time courses of regeneration

neurons in the olfactory bulb (which is apparently the case during development, Gesteland et al. 1982). Whether or not the synaptic input to the reduced number of mitral cells has increased, as recently demonstrated by Greer and Halasz (1987) in a mouse mutant (PCD), must remain open with regard to the rostrally and caudally bulbectomized animals. However, that a surgical reduction of the olfactory bulbs' tissue does not significantly reduce the animals acquired behavior, has recently been demonstrated by Hudson and Distel (1987) in rabbit pups.

So far only regeneration of the peripheral connections and recovery of behavior have been discussed. Virtually nothing is known about the central projections after surgical interruption of connections with the telencephalon. After complete bilateral bulbectomy (Fig. 8, 3) neither a functional (Zippel et al. 1981), nor a morphological (Grazaidei, personal communication) regeneration was found. This is in accordance with the findings of Butler et al. (1984), who reported that neonatally bulbectomized rats with new olfactory-neocortical connections were anosmic. After sectioning the olfactory tracts (Fig. 8, 2), goldfish could be trained to olfactory stimuli after a 6–7-week regeneration period, and a regeneration of the central connections was histologically demonstrated (Zippel and Westerman 1970). From the data presented here it is evident that even after bilateral caudal bulbectomy (Fig. 8, 5), a functional and a morphological regeneration occurs. An olfactory bulb of half the normal size and extremely thin connections are evidently sufficient for regaining pretraining behavior. Whether or not all the afferent and efferent pathways previously described for the goldfish (v. Bartheld et al. 1984) are regenerated still remains open. Furthermore, we have so far been unable to investigate "specific" regeneration after lesioning the olfactory tracts or the caudal olfactory bulbs as the "forgetting period" following differentiation training is shorter than the regeneration time of 6–7 weeks.

In conclusion, after lesioning the olfactory pathways a functional recovery occurs with the time course depending on the type of lesions. Where the distance between the regenerating fibers and the target neurons is short, (approximately 1 mm: receptor axotomy, rostral bulbectomy) an extremely short functional and morphological regeneration was found, whereas long-distance (4–5 mm) regenerations (tractotomy and caudal bulbectomy) needed 6–7 weeks for a behavioral and morphological reconstruction. This is, in part, in accordance with Evans and Hara's (1985) findings in trout (which have an extremely long olfactory nerve), in which they found a very slow regeneration, which was still electrophysiologically (EOG) incomplete after survival times of 230 days. The fact that in the goldfish experiments presented here the regeneration time courses after partial ablations of the olfactory bulbs were practically identical to the time courses of recovery after lesions of peripheral and central pathways, was unexpected.

Acknowledgement. We thank Dr. R. Hudson for critically reading the manuscript.

References

Bartheld CS von, Meyer DL, Fiebig E, Ebbesson SOE (1984) Central connections of the olfactory bulb in the goldfish, *Carassius auratus*. Cell Tissue Res 238:475–487

Burns SM, Mitchell JA, Getchell ML, Getchell TV (1981) Functional correlates of degeneration and renewal of cilia and knobs of olfactory receptor neurons in the frog. Chem Sens 6:307–315

Butler AB, Graziadei PPC, Monti Graziadei GA, Slotnick BM (1984) Neonatally bulbectomized rats with new olfactory-neocortical connections are anosmic. Neurosci Lett 48:247–254

Constanzo RM (1985) Neural regeneration and functional reconnection following olfactory nerve transection in hamster. Brain Res 361:258–266

Doucette JR, Kiernan JA, Flumerfelt BA (1983) The re-innervation of olfactory glomeruli following transection of primary olfactory axons in the central or peripheral nervous system. J Anat 137:1–19

Evans RE, Hara TJ (1985) The characteristics of the electro-olfactogram (EOG): its loss and recovery following olfactory nerve section in rainbow trout (*Salmo gairdneri*). Brain Res 330:65–75

Gesteland RC, Yancey RA, Farbman AI (1982) Development of olfactory receptor neuron selectivity in the rat fetus. Neuroscience 7:3127–3136

Graziadei PPC, Monti Graziadei GA (1980) Neurogenesis and neuron regeneration in the olfactory system of mammals. III. Deafferentation and reinnervation of the olfactory bulb following section of the fila olfactoria in rat. J Neurocytol 9:145–162

Graziadei PPC, Monti Graziadei GA (1986) Principles of organization of the vertebrate olfactory glomerulus: an hypothesis. Neuroscience 19:1025–1035

Greer CA, Halasz N (1987) Plasticity of dendrodendritic microcircuits following mitral cell loss in the olfactory bulb of the murine mutant Purkinje cell degeneration. J Comp Neurol 256:284–298

Hudson R, Distel H (1987) Regional autonomy in the peripheral processing of nipple-search odors. Brain Res 421:85–94

Ichikawa M, Ueda K (1977) Fine structure of the olfactory epithelium in the goldfish, *Carassius auratus*. Cell Tissue Res 183:445–455

Monti Graziadei GA, Karlan MS, Bernstein JJ, Graziadei PPC (1980) Reinnervation of the olfactory nerve in monkey (*Saimiri scuireus*). Brain Res 189:343–354

Samanen DW, Forbes WB (1984) Replication and differentiation of olfactory receptor neurons following axotomy in the adult hamster: a morphometric analysis of postnatal neurogenesis. J Comp Neurol 225:202–211

Simmons PA, Getchell TV (1981 a) Neurogenesis in olfactory epithelium: loss and recovery of transepithelial voltage transients following olfactory nerve section. J Neurophysiol 45:516–528

Simmons PA, Getchell TV (1981 b) Physiological activity of newly differentiated olfactory receptor neurons correlated with morphological recovery from olfactory nerve section in the salamander. J Neurophysiol 45:529–549

Simmons PA, Rafols JA, Getchell TV (1981) Ultrastructural changes in olfactory receptor neurons following olfactory nerve section. J Comp Neurol 197:237–257

Zippel HP (1970) Verhaltenskomponenten und Differenzierungsvermögen in der straffreien Geruchsdressur beim Goldfisch (*Carassius auratus*). Z Vgl Physiol 69:54–78

Zippel HP, Voigt R (1982) Neuronal correlates of olfactory behavior in goldfish. In: Hara TJ (ed) Chemoreception in fishes. Elsevier Scientific, Amsterdam, pp 181–199

Zippel HP, Westermann RA (1970) Geruchsdifferenzierungsvermögen der Karausche (*Carassius carassius*) nach funktioneller und histologischer Regeneration des Tractus olfactorius und der Commissura anterior. Z Vgl. Physiol 69:38–53

Zippel HP, Baumgarten R von, Westermann RA (1970) Histologische, funktionelle und spezifische Regeneration nach Durchtrennung der Fila olfactoria beim Goldfisch (*Carassius auratus*). Z Vgl Physiol 69:79–98

Zippel HP, Breipohl W, Schoon H (1981) Functional and morphological changes in fish chemoreception systems following ablation of the olfactory bulbs. In: Flohr H, Precht W (eds) Lesion-induced neuronal plasticity in sensorimotor systems. Springer, Berlin Heidelberg New York, pp 377–394

IX Pharmacology of CNS Injury

Melanocortins: A Potential Role in the Therapy of Neurological Disorders

P. De Koning [1], P. M. Edwards [2], and W. H. Gispen [1]

[1] Division of Molecular Neurobiology, Rudolf Magnus Institute for Pharmacology and Institute of Molecular Biology and Medical Biotechnology, University of Utrecht, Padualaan 8, NL-3584 CH Utrecht, The Netherlands

[2] Bogazici University, School of Arts and Sciences, Department of Biology, Bebek, Istanboel, Turkey

1 Introduction

Melanocortins are peptides related to the pituitary hormones ACTH and α-MSH. As pituitary hormones, both α-MSH and ACTH mediate adaptive responses to changes in the environment, either by changes in skin color or alterations in energy metabolism. These same peptides are also now known to be neuropeptides, that is they are both synthesized in, and have effects on, neurons. The proposal that nerve tissue can be considered as a target for melanocortins was suggested by De Wied from studies on the effects of these agents on behavioral parameters (for a review see De Wied and Jolles 1982). Acute (short-term) effects on the electrical functions of motor neurons have been shown in the cat spinal cord by Krivoy et al. (1985), frog neuromuscular preparations (Johnston et al. 1983) and in the motor system of immature and hypophysectomized rats (Gonzales and Strand 1981; Strand and Smith 1980; Smith and Strand 1981).

Many peptide hormones have long-term trophic influences on the target cells in which they also exert short-term functional effects. In nerve tissue, a trophic influence of melanocortins is evidenced by increased macromolecular synthesis under certain conditions (Dunn and Schotman 1981; Bijlsma et al. 1983a) and enhanced repair of damaged peripheral nerves (Bijlsma et al. 1984). A vast array of effects of melanocortins on neural function have been described and have been collectively described as an enhancement of neural plasticity (Gispen et al. 1986), thus permitting behavioral adaptations to environmental change. At the simplest and most direct level, neural plasticity can be seen as the ability to overcome direct damage to the nervous system, and it is this aspect of melanocortin action which is emphasized here. The ability of melanocortins to facilitate repair following mechanical damage (cut or crush lesion) in experimental models has been demonstrated by measuring the recovery in functional, electrophysiological and histological terms (Saint-Come and Strand 1985; Bijlsma et al. 1984; Verhaagen et al. 1987b; De Koning and Gispen 1987; Dekker et al. 1987a). The potential applicability of the effects of melanocortins in experimental models to clinical problems requires careful consideration. The need for therapeutic agents is desperate. Microsurgical methods for the repair of damaged nerves have probably now reached their ultimate level of sophistication and further improvements in the final outcome of the repair will only be achieved by chemotherapy. Still more urgent is the need for treatments of neurological damage caused by disease rather

H. Flohr (Ed.)
Post-Lesion Neural Plasticity

than mechanical trauma, for which no adequate therapy is currently available. Most of the evidence for a neurotrophic role for melanocortins comes from studies on rodents suffering from a well-defined mechanical lesion to the sciatic nerve. The clinical problems faced by neurologists are vastly different. Careful consideration of the nature of the active peptides, the characteristics of their effects in experimental animal models and analysis of the clinical problems are required for a successful transfer of melanocortins from the laboratory to the clinic, and this is the objective of this chapter.

1.1 Melanocortins

In the pituitary, melanocortins are formed by post-translational processing of a large polypeptide precursor, proopiomelanocortin (POMC). Nucleotide sequencing of the cloned cDNA for the precursor polypeptide allowed the complete amino acid sequence to be determined (Nakanishi et al. 1979). The precursor itself is biologically inactive, but contains the sequences of a large number of potentially active peptides, released by proteolytic cleavage of the precursor. The biologically active peptides that can be derived from POMC are shown in Fig. 1. Those peptides that are shaded are all neurotrophic in an animal model (Van der Zee et al. 1988) and contain a sequence of amino acids homologous to residues 4–10 of ACTH (Fig. 2). The synthetic peptide containing only these residues is

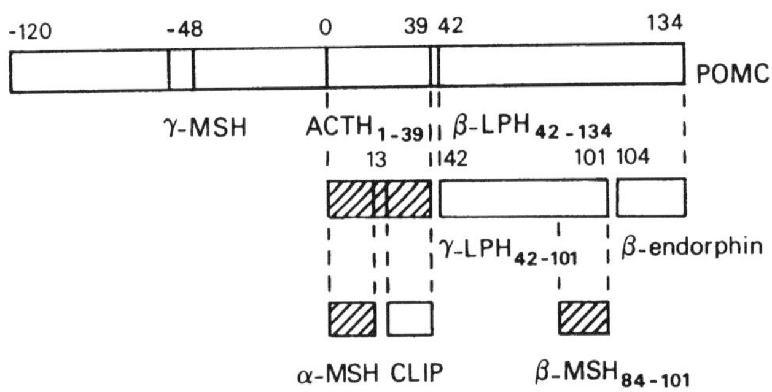

Fig. 1. Proopiomelanocortin and its naturally occurring cleavage products. The *shaded peptides* are all neurotrophic in an animal model

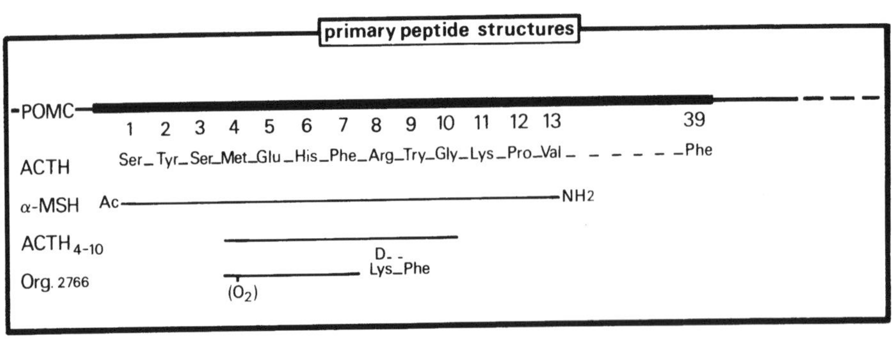

Fig. 2. Primary peptide structures of the ACTH derivatives and analogs most commonly used in the treatment of nervous tissue damage

also active. Of many modified melanocortins that have been synthesized, only Org. 2766 (a modified form of $ACTH_{4-9}$, see Fig. 2 for structure) has been shown to enhance nerve repair. This peptide is relatively resistant to metabolic breakdown. Both $ACTH_{4-9}$ and Org. 2766 are virtually devoid of the classical adrenocortical secretional activity and, for reasons other than those discussed in this chapter, have already been widely used in humans with no toxic side-effects (De Wied and Jolles 1982).

In common with a large number of biologically active peptides originally described in peripheral tissues, melanocortins have now been demonstrated in neurons in many brain regions (Palkovits 1984). These peptides are synthesized from the identical POMC precursor as is found in the anterior pituitary (Liotta et al. 1984). These neuronal melanocortins may act as neuromodulators, modifying electrical activity as has been shown in the neuromuscular system (see above). The presence of POMC-derived peptides in motorneurons of immature, but not adult mice suggests a possible role in normal neuronal growth (Haynes and Smith 1984, 1985). Other studies have also suggested that α-MSH is important in neuronal development (Swaab and Martin 1981). The normal maturational disappearance of α-MSH from motorneurons does not occur in dystrophic mice, possibly indicative of a damage-related expression of neuronal POMC (Haynes and Smith 1985). However, α-MSH is not produced in detectable amounts in response to damage in adult rat sciatic nerves (Verhaagen et al. 1988).

In the past, natural $ACTH_{1-39}$ and synthetic $ACTH_{1-24}$ were widely used in a variety of neurological diseases such as multiple sclerosis, myasthenia gravis, lesions and compressions of peripheral nerves and neuropathies. The contradictory results of these studies may be attributed to the corticotrophic activity of ACTH, since steroids and neurotropic peptides are known to have antagonistic effects on behavior and muscle. The use of shorter fragments of ACTH (notably $ACTH_{4-10}$ and Org. 2766) that have full neurotrophic activity (Bijlsma et al. 1981) but little or no corticotrophic activity will avoid this problem. The structure of melanocortins that have been tested to date has been based on the structure of ACTH. It has been suggested that peptides with similar activity might be produced by injury-induced proteolysis of the intermediate size neurofilament protein (Edwards et al. 1984). Since the full sequence of α-MSH is not present in the neurofilament protein (Shaw et al. 1985; Verhaagen et al. 1986), nonlinear structures may be considered. At present the putative neurofilament-derived, MSH-like peptides have not been isolated.

2 Neurological Disorders and Compensation

2.1 Mechanical Injury

The most obvious use for melanocortins is in the treatment of mechanical injury to nerves. In experimental animals, melanocortins influence the early response to injury by increasing the number of outgrowing sprouts and possibly also the time point at which regenerative sprouts form (Verhaagen et al. 1987a). The rate of outgrowth of the regenerating nerves is not affected. In keeping with this early action is the finding that melanocortin treatment exclusively during the 8 days after injury is sufficient for maximal peptide effects measured at later times (Edwards et al. 1984). Delayed peptide treatment is ineffective. Functional recovery in rats is measured by parameters that are crude in comparison to those assessed in patients and suggests a recovery period that is shortened by 1 or 2 days. If this were the only effect of melanocortins, the clinical benefit would be questionable. However, the results of electrophysiological experiments provide evidence that melanocortins also improve the quality of the nerve repair. In untreated rats, the motor nerve conduction velocity in the regenerated sciatic nerve remains depressed up to 200 days following a crush lesion. Treatment with melanocortins during the first 8 days after injury results in the return to normal values by 90 days (De Koning et al. 1987a). Similar long-term effects were seen on the recovery of sensory nerve conduction velocity. Melanocortin treatment also preferentially accelerates the reformation and stabilization of small motor units (Saint-Come and Strand 1985). Small motor units are less susceptible to fatigue than detailed larger motor units. The larger number of small motor units in peptide-treated animals also allows control of motor functions. Thus, there is evidence that melanocortins improve the quality of repair, rendering the nervous system in a more favorable position to withstand renewed damage and age-related changes. Remodeling of the neuromuscular junction occurs continuously throughout life and involves sprouting at the intraterminal, terminal, and collateral levels (Barker and Ip 1966). It seems reasonably well-established that there is an age-related loss of motorneurons (Tomlinson and Irving 1977). Muscle fibers which become denervated in this way are probably reinnervated by collateral sprouts, which results in fiber-type grouping and enlargement of remaining motor units (Jennekens et al. 1971). Since melanocortin therapy following a crush lesion results in the formation of more motor units as compared to untreated animals, the age-related compensatory processes would be expected to be sustained successfully over a longer period of time. We have preliminary evidence that treatment with Org. 2766 during the 8 days following an initial injury enhances recovery from a second, untreated injury (De Koning unpublished data). Evidence that melanocortin treatment of neurotoxin-induced damage improves the capacity to cope with a subsequent insult (see below) supports the proposal that the benefits of melanocortin treatment are much more significant than simply a slightly faster recovery time.

The best dosing regime and route of administration of melanocortins can ultimately be established only from clinical trials but some useful indicators can already be deduced from animal studies. Oral dosing has not been shown to be effective in the treatment of nerve damage (Bijlsma et al. 1983; Van der Zee et al. 1988). Repeated subcutaneous dosing and long-term treatment by means of mini-pumps have been used successfully (Dekker et al. 1987b). Several studies have shown an inverted bell-shaped relationship between dose of melanocortins and response (Gonzales and Strand 1981; Verhaagen et al. 1987a; De Koning et al. 1986). Thus too high a dose will be ineffective. The timing of treatment is also critical (Verhaagen et al. 1987a). A clinically interesting possible treatment of nerve injury where the site of injury is accessible is suggested by the discovery that local administration of melanocortins is effective in treating both nerve crush and transection (Edwards et al. 1986). Recent work in our laboratory (Van der Zee et al. 1988) has shown that coating the site of a nerve injury with a biodegradable matrix impregnated with melanocortins enhances nerve repair. These studies offer a hopeful application in the neurosurgical treatment of traumatic nerve transections and compression injuries.

2.2 Collateral Sprouting

In many disorders of the nervous system (e.g., motorneuron disorders and polyneuropathies) there exists the capacity to compensate to some degree by collateral sprouting of surviving neurons. Collateral sprouting is regulated by mechanisms which to some degree are similar to those active in peripheral sprouting but there are also important differences. Experimental evidence suggests that melanocortins, besides stimulating peripheral (regenerative) sprouting, also increase motorneuron collateral sprouting capacity (De Koning et al. 1988). The rat soleus muscle receives most of its innervation through the L5 root, while the remainder is conveyed in the L4 and L6 roots. Transection of the L5 root results in collateral sprouting from the remaining L4 and L6 motor units. In untreated rats, these remaining motor units can increase their muscle fibre number about fivefold, as estimated from the maximal twitch tension per motor unit. Treatment with Org. 2766 following L5 section increased the maximal extent to which one motor unit could sprout to about 7.5-fold. Histological examination of the same muscles in which the twitch tension had been measured revealed that the increase could be accounted for entirely by an increase in the number of innervated fibers per motor unit. The muscle fibers themselves were not affected by the melanocortin treatment, an observation which agrees with findings of Strand and co-workers (Strand and Kung 1980; Smith and Strand 1981). Thus, melanocortin treatment might be of benefit to patients with motorneuron disorders such as amyotrophic lateral sclerosis, in which collateral sprouting has to compensate for the loss of functional motor units. A clinical trial in which this issue will be investigated is already in progress.

2.3 Neuropathies

Enhancement by melanocortin treatment of compensatory collateral sprouting represents a possible means of ameliorating the consequences of the loss of functional neurons in disease. However, treatment of neuropathies in a manner that either diminishes the extent of the neuronal damage or encourages direct repair is a still more attractive prospect, especially in the long term. In order to investigate the possibility that melanocortins might be of benefit, animal models of toxic neuropathies are currently being investigated. Acrylamide is a cumulative toxin which causes a mainly peripheral neuropathy in rats and man. The intoxication causes degeneration of distal regions of nerve axons and recovery is brought about by regeneration from surviving proximal portions of the nerve (Suzuki and Pfaff 1979). The repair process is itself compromised, as evidenced by diminished regeneration following a mechanical injury (Griffin et al. 1976) and inhibited sprouting (Kemplay and Cavanagh 1984). Many of the biochemical changes that occur during acrylamide intoxication resemble those observed following axotomy (Bisby and Redshaw 1987). Thus the acrylamide model will be very interesting to study in relation to melanocortin in order to establish the potential for these agents in the treatment of neurological disorders.

Human cases of neuropathy caused by acrylamide are rare but neuropathies caused by cancer chemotherapeutic agents present a serious clinical problem. Cisplatin [cis diamine-dichloroplatinum (II)] is widely used as a cancer chemotherapeutic agent but its neurotoxicity limits its therapeutic application. The neurotoxic side-effects include peripheral, primarily sensory, neuropathy, retrobulbar neuritis, and neurosensory hearing loss. Neurophysiological studies show decreased sensory, but normal motor nerve conduction velocities with occasional signs of denervation. It has been difficult to set up an animal model in which cisplatin neuropathy could be properly evaluated, mainly because of animals dying from renal toxicity and other systemic effects. However, protection of the kidney by means of a diuretic have enabled us to produce a definitive peripheral neuropathy in the rat. The severity of the neuropathy was assessed by means of longitudinal sensory and motor nerve conduction velocity measurements (De Koning

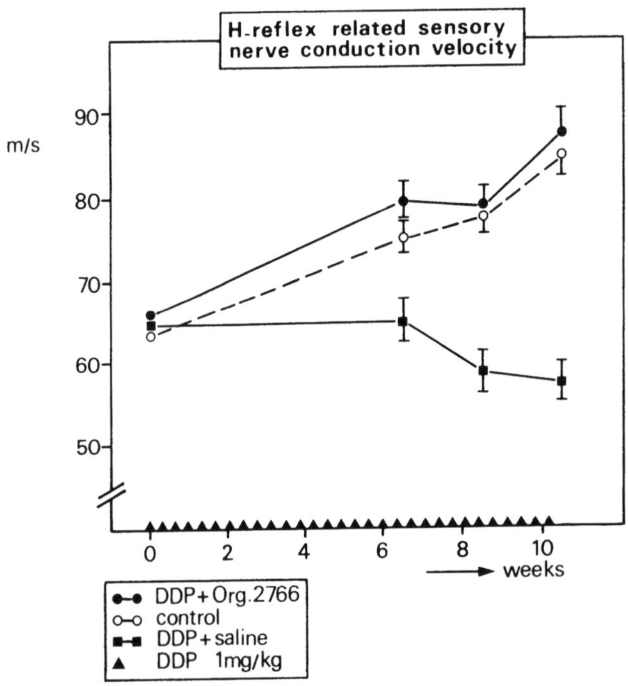

Fig. 3. Time course of the H-reflex related sensory nerve conduction velocity (HSNCV) in untreated rats (○), rats concurrently treated with cisplatin and Org. 2766 (●) and rats which concurrently received cisplatin and saline (■). Values ±SEM. The rats were about 200 g (approx. 12–13 weeks) at the onset of the experiment which accounts for the increase in HSNCV in the untreated animals as the HSNCV normally increases with age during the first 6 months of life. Cisplatin was administered 2 times a week on Tuesday and Friday (1 mg/kg body weight) while Org. 2766 was administered four times a week on Tuesday, Thursday, Friday and Sunday (10 μg/rat). The difference between cisplatin-saline and either cisplatin-Org. 2766 or control animals was highly significant (ANOVA: $p < 0.001$)

et al. 1987a). After a mean cumulative dose of cisplatin of 13 mg/kg body weight, the sensory conduction started to deteriorate until, at a mean cumulative dose of 20 mg/kg body weight, the velocity was only 68% of that in control animals (Fig. 3). The slowing occurred primarily distally. No slowing of the motor nerve conduction velocity was observed. When cisplatin-treated rats received concomitant injections of Org. 2766 throughout the intoxication period, no slowing of the sensory nerve conduction velocity occurred (Fig. 3). Thus Org. 2766 has a protective action against cisplatin neurotoxicity. This $ACTH_{4-9}$ analog does not alter the anti-tumor activity of cisplatin (De Koning et al. 1987b). Based on these encouraging results, a clinical trial on the protective effect of Org. 2766 against cisplatin neurotoxicity will be carried out among ovarian cancer patients.

Treatment with Org. 2766 of rats already neuropathic as a result of cisplatin treatment ameliorates the neuropathy even when cisplatin treatment is continued (Gerritsen van der Hoop et al. 1988). Furthermore, treatment with Org. 2766 during an initial cisplatin intoxication renders rats less susceptible to a subsequent exposure. This is reminiscent of the better quality of recovery from mechanical damage already discussed.

One of the most frequent polyneuropathies encountered in the clinic is that associated with diabetes mellitus. Preliminary studies on rats rendered diabetic by a single injection of streptozotocin indicate that in these rats slowing of peripheral nerve conduction velocities occurs, indicative of a diabetic neuropathy (Van der Zee et al. 1987). Studies are in progress to evaluate the possible beneficial effects of melanocortins in this respect.

3 Summary

The evidence that melanocortins improve the speed and quality of repair of mechanically damaged nerves in experimental animals is sufficient to make one optimistic about the value of these peptides in analogous clinical problems. Nontoxic, noncorticotropic peptides, which can be applied topically at the site of nerve injury, can be expected to provide a chemotherapeutic aid to the successful repair of damaged nerves in the clinic in the near future.

Successful treatment of multifocal damage, as in neuropathies, where such topical treatment is impossible, will depend on careful clinical trials to establish suitable dosage, treatment regimes, and mode of administration. As we have described, both timing and dosage are crucial factors in the effectiveness of ACTH-related peptide action. These phenomena are consistent with competition by ACTH-related peptides with more active endogenous substances. If the putative neurofilament-derived peptides can be isolated and characterized, a new family of peptides without these problems could be envisaged, but for the present this remains highly speculative.

To our knowledge, only one study of the treatment of human neurological disease by noncorticotrophic melanocortins has been published. Strand et al. (1976) reported that administration of a single dose of $ACTH_{4-10}$ to a patient with a hereditary form of an anterior horn cell disorder (M. Kugelberg-Welander) prevented the decline in muscle action potential on repetitive nerve stimulation. This effect may be related to acute effects of these peptides on neuromuscular function in experimental animals. It is not known whether there is any connection between the short-term improvements in nerve function and the long-term growth effects of melanocortins and both may be important in the treatment of neurological disorders.

References

Barker D, Ip MC (1966) Sprouting and degeneration of mammalian motor axons in normal and de-afferented skeletal muscle. Proc R Soc Biol 163:538–554

Bijlsma WA, Jennekens FGI, Schotman P, Gispen WH (1981) Effects of corticotrophin (ACTH) on recovery of sensorimotor function in the rat: structure-activity study. Eur J Pharmacol 76:73–79

Bijlsma WA, Schotman P, Jennekens FGI, Gispen WH (1983a) Regeneration and protein synthesis in the spinal cord the influence of ACTH-like peptides on cell-free protein synthesis. Neurosci Lett 38:297–302

Bijlsma WA, Schotman P, Jennekens FGI, Gispen WH, De Wied D (1983b) The enhanced recovery of sensorimotor function in rats is related to the melanotropic moiety of ACTH/MSH neuropeptides. Eur J Pharmacol 92:231–236

Bijlsma WA, Jennekens FGI, Schotman P, Gispen WH (1984) Neurotrophic factors and regeneration in the peripheral nervous system. Psychoneuroendocrinology 9:199–215

Bisby MA, Redshaw JD (1987) Acrylamide neuropathy: changes in the composition of proteins of fast axonal transport resemble those observed in regenerating axons. J Neurochem 48:924–928

Dekker AJAM, Gispen WH, Wied D de (1987a) Axonal regeneration, growth factors and neuropeptides. Life Sci 41:1667–1687

Dekker AJAM, Princen MM, Nijs H de, Leede LGJ de, Broekkamp CLE (1987b) Acceleration of recovery from sciatic nerve damage by the ACTH(4–9) ANALOG Org. 2766; different routes of administration. Peptides 8:1057–1059

Dunn AJ, Schotman P (1981) Effects of ACTH and related peptides on cerebral RNA and protein synthesis. Pharmacol Ther 12:353–372

Edwards PM, Zee CEEM van der, Verhaagen J, Schotman P, Jennekens FGI, Gispen WH (1984) Evidence that the neurotrophic actions of α-MSH may derive from its ability to mimic the actions of a peptide formed in degenerating nerve stumps. J Neurol Sci 64:333–340

Edwards PM, Kuiters RRF, Boer GJ, Gispen WH (1986) Recovery from peripheral nerve transection is accelerated by local application of α-MSH by means of Accurel polypropylene tubes. J Neurol Sci 74:171–176

Gerritsen van der Hoop R, Koning P de, Neijt JP, Jennekens FGI, Gispen WH (1988) Efficacy of the neuropeptide Org. 2766 in the prevention and treatment of cisplatin-induced neurotoxicity in rats. Eur J Cancer Clin Oncol 24:637–642

Gispen WH, Isaacson RL, Spruijt BM, Wied D de (1986) Melanocortins, neural plasticity and aging. Prog Neuro-Psychopharmacol Biol Psychiatry 10:415–426

Gonzales ER, Strand FL (1981) Neurotropic action of MSH/$ACTH_{4-10}$ on neuromuscular function in hypophysectomized rats. Peptides 2:107–113

Griffin JW, Price DR, Drachman DB (1976) Impaired axonal regeneration in acrylamide intoxication. J Neurobiol 8:355–370

Haynes LW, Smith ME (1984) The action of proopriomelanocortin peptides at the developing neuromuscular junction. TIPS 5:165–168

Haynes LW, Smith ME (1985) Presence of immunoreactive α-melanotropin and β-endorphin in spinal motorneurons of the dystrophic mouse. Neurosci Lett 58:13–18

Jennekens FGI, Tomlinson BE, Walton JN (1971) Histochemical aspects of five limb muscles in old age. J Neurol Sci 14:250–276

Johnston MF, Kravitz EA, Meiri H, Rahaminoff R (1983) Adrenocorticotropic hormone causes long-lasting potentiation of transmitter release from motor nerve terminals. Science 220:1071–1072

Kemplay S, Cavanagh JB (1984) Effects of acrylamide and other sulfhydryl compounds in vivo and in vitro on staining of motor nerve terminals by the zinc iodide-osmium technique. Muscle Nerve 7:94–100

Koning P de, Gispen WH (1987) Org. 2766 improves functional and electrophysiological aspects of regenerating sciatic nerve in the rat. Peptides 8:415–422

Koning P de, Brakkee JH, Gispen WH (1986) Methods for producing a reproducible crush in the sciatic and tibial nerve of the rat and rapid and precise testing of return of sensory function: beneficial effects of melanocortins. J Neurol Sci 74:237–246

Koning P de, Neijt JP, Jennekens FGI, Gispen WH (1987a) Evaluation of cis-diaminechloroplatinum (II) (cisplatin) neurotoxicity in rats. Toxicol Appl Pharmacol 89:81–87

Koning P de, Neijt JP, Jennekens FGI, Gispen WH (1987b) Org. 2766 protects from cisplatin-induced neurotoxicity in rats. Exp Neurol 97:746–750

Koning P de, Verhaagen J, Sloot W, Jennekens FGI, Gispen WH (1988) Org. 2766 stimulates collateral sprouting in rat soleus muscle following partial denervation. Muscle & Nerve, in press

Krivoy WA, Couch JR, Stewart JA (1985) Modulation of spinal synaptic transmission by β-melanocyte stimulating hormone (β-MSH) Psychoneuroendocrinology 10:1–3–108

Liotta AS, Advis JP, Krause JK, McKelvy JF, Krieger DT (1984) Demonstration of in vivo synthesis of adult rat brain. J Neuroscience 4:956–965

Nakanishi S, Inoue A, Kita T, Nakamura M, Chang ACY, Cohen SN, Numa S (1979) Nucleotide sequence of cloned cDNA for bovine corticotropin-B-lipotropin precursor. Nature 278:423–427

Palkovits M (1984) Distribution of neuropeptides in the central nervous system: a review of biochemical mapping studies. Prog Neurobiol 23:151–189

Saint-Come C, Strand FL (1985) ACTH/MSH_{4-10} improves motor unit reorganization during peripheral nerve regeneration in the rat. Peptides 6:77–83

Shaw G, Fisher S, Weber K (1985) α-MSH and neurofilament explanation for neurofibrillary staining with α-MSH antibodies. FEBS Lett 181:343–346

Smith CM, Strand FL (1981) Neuromuscular response of the immature rat to ACTH/MSH_{4-10}. Peptides 2:197–206

Strand FL, Kung TT (1980) ACTH accelerates recovery of neuromuscular function following crushing of peripheral nerve. Peptides 1:135–138

Strand FL, Smith CM (1980) LPH, ACTH, MSH and motor systems. Pharmacol Ther 11:509–533

Strand FL, Cayer A, Gonzalez E, Stoboy (1976) Peptide enhancement of neuromuscular function: animal and clinical studies. Pharmcol Biochem Behav 5:179–187

Suzuki K, Pfaff LD (1979) Acrylamide neuropathy in rats. Acta Neuropathol 24:197–213

Swaab DF, Martin JT (1981) Functions of α-melanotropin and other opiomelanocortin peptides in lavior, intrauterine growth and brain development. Peptides of the pars intermedia. Pitman Medical, London (Ciba Foundation Symposium 81) pp 196–217

Tomlinson BE, Irving D (1977) The number of limb motor neurons in the human lumbosacral cord throughout life. J Neurol Sci 34:213–219

Verhaagen J, Edwards PM, Schotman P, Jennekens FGI, Gispen WH (1986) Characterization of epitopes shared by α-melanocyte stimulating hormone (α-MSH) and the 150 kDa neurofilament protein (NF150); relationship to neurotrophic sequences. J Neurosci Res 16:589–600

Verhaagen J, Edwards PM, Jennekens FGI, Schotman P, Gispen WH (1987a) Early effect of an ACTH(4–9) analog (Org. 2766) on regenerative sprouting demonstrated by the use of neurofilament-binding antibodies isolated from a serum raised by α-MSH immunization. Brain Res 404:142–150

Verhaagen J, Edwards PM, Jennekens FGI, Gispen WH (1987b) Pharmacological aspects of the influence of melanocortins on the formation of regenerative peripheral nerve sprouts. Peptides 8:581–584

Verhaagen J, Edwards PM, Gispen WH (1988) Damaged rat peripheral nerves do not contain detectable amounts of α-MSH. J. Neurosci Res 19:14–18

Wied D de, Jolles J (1982) Neuropeptides derived from Proopiocortin: behavioral, physiological and neurochemical effects. Physiol Rev 62:976–1059

Zee CEEM van der, Edwards PM, Koning P de, Gispen WH (1987) Beneficial effects of melanocortins on axonal regeneration: possibility for treatment of diabetic-induced peripheral neuropathy. Neurosci Lett Suppl 22, p 254

Zee CEEM van der, Brakkee JH, Gispen WH (1988) α-MSH and Org. 2766 in peripheral nerve regeneration: different routes of delivery. Eur J Pharmacol 147:351–357

Sprouting Pattern and B-50 Phosphorylation in Regenerating Sciatic Nerve Respond to ACTH Peptides

F. L. Strand, L. A. Zuccarelli, B. Kirschenbaum, and R. E. Frischer
Center for Neuroscience, New York University, Washington Square, New York, USA

1 Introduction

Nerve sprouting in regeneration occurs as a restorative process following nerve crush or section. During regeneration, the administration of ACTH or ACTH peptide fragments such as αMSH (ACTH 1–13), ACTH 4–10, or an analogue of ACTH 4–9 (Org. 2766), stimulates the formation of nerve sprouts. The accelerated return of motor and sensory function following treatment with ACTH peptides (Strand and Kung 1980; Bijlsma et al. 1983) is due to an earlier initiation of sprouting following crush denervation (Verhaagen et al. 1986a). Reinnervated motor units formed under the influence of early peptide treatment are considerably smaller, thus permitting finer motor control, than in saline-treated animals (Saint-Come and Strand 1985, 1987). This suggests an organizational role for these peptides on the pattern by which regenerating axons reinnervate skeletal muscle.

While regeneration is often seen as a recapitulation of ontogeny, there are important differences. One of these is that sprouting during development occurs in virgin territory: nerve terminals innervate selective muscle fibers to form the initial endplates. Regenerating nerve fibers return to these preformed endplates, their pathway being influenced by the previously innervated muscle fibers. In both regenerating and developing peripheral motor nerves, axonal sprouting may be nodal, terminal or both (Brown 1984). In our studies on the developing neuromuscular system, sprouting is seen as an increased arborization of preterminal branches entering the muscle (Acker et al. 1984) and increased terminal branching within the endplate itself (Fig. 1), processes accelerated by ACTH/MSH peptides (Rose et al. 1988). This led to the question as to whether these peptides would have similar effects on regenerating nerves entering denervated endplates. We report here on quantitative morphological changes in endplate parameters from reinnervated skeletal muscle 15 days after crush denervation as related to the recovery of motor and sensory function.

An interesting biochemical correlation has been found between the increased synthesis of the membrane protein B-50, also known as GAP-43 and protein F1 (Zwiers et al. 1980; Nelson and Routtenberg 1985) and nerve growth in both regenerating and developing neurons (Verhaagen et al. 1986a; Jacobson et al. 1985; Skene et al. 1986). Protein B-50 is greatly increased in growth cone membranes and its phosphorylation is sensitive to ACTH peptides (Zwiers et al. 1976; Gispen

H. Flohr (Ed.)
Post-Lesion Neural Plasticity

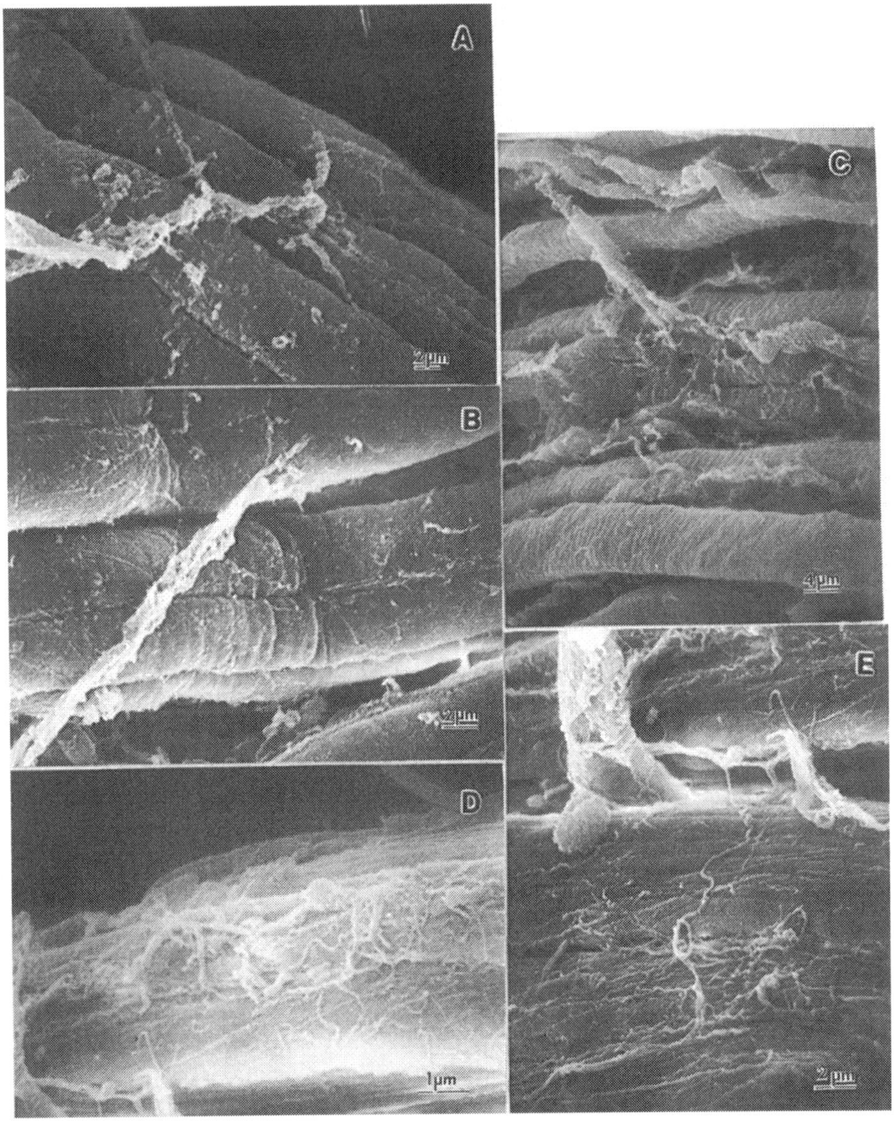

Fig. 1. Scanning electron micrographs of endplate regions from the EDL muscles of 15-day-old rats. **A** Saline-treated neonate. The nerve can be seen branching into narrower nerve terminals along the muscle fiber. ×3420 **B** ORG 2766-treated (10 μg/kg) neonate. The nerve terminal shows a marked reduction in branching. ×3610 **C** ACTH 4–10-treated (10 μg/kg) neonate showing extensive nerve terminal branching. ×2000 **D** ORG 2766-treated (0.01 μg/kg) neonate with arborization of nerve terminals. ×5000 **E** Higher magnification of ORG 2766-treated (0.01 μg/kg) neonate showing considerable nerve terminal branching at the endplate. ×10500

et al. 1979). These findings prompted us to study the fate of B-50 protein during regenerative axonal outgrowth as influenced by the administration of ACTH peptides.

2 Methods

2.1 Animals

Mature male Sprague-Dawley rats (180–200 g) were housed individually, maintained on a 12 h dark : 12 h light cycle and fed rat chow and water ad lib. Elizabethan rat collars were placed on the animals from day 4 to day 12 to prevent self-mutilation. The collars were removed only for testing. All surgeries were performed between 10 am and 12 pm and all injections were administered during these times. Surgery was performed under anesthesia with 8% chloral hydrate (0.6 ml/0.1 kg i.p.). The right EDL muscle was denervated by crushing the right sciatic nerve with a #5 Dumont forceps with a uniformly filed tip, producing a crush 1 mm wide approximately 6 mm below the sciatic notch and the wound closed with surgical staples. Immediately after the operation, treatment commenced with the i.p. administration of one of the following;

1. 0.9% saline
2. ACTH 4–10 (10 μg)
3. αMSH (10 μg)

This treatment was repeated every 48 h until the time of death (6, 15 and 21 days after denervation).

2.2 Morphological Techniques

The EDL muscle was fixed in situ and prepared for light microscopy (LM) using a combined silver-cholinesterase stain which facilitated the quantitative measurements of morphological parameters. These included total length of nerve terminal branching and cholinesterase-positive endplate area and perimeter. Experimental preparations were observed with an Olympus BH 100X oil immersion objective and measurements performed with a Bioquant image analysis system. Student's t-test was used for statistical evaluation. Data are expressed as mean and standard error of the mean (SEM).

2.3 Functional Studies

Return of Motor Function. The walking pattern of rats was observed on days 6, 8, 10, 12, 14, 16, 18, 19, 20, and 21 following nerve crush using a method based on that of De Medinaceli et al. (1982). The hind feet were lightly rubbed along the surface of a pad of India ink, the animals placed at one end of a narrow, 76-cm-long walkway which had a sheet of white paper lying on the bottom and a cover producing a dark area at the far end. As the rats walked toward the dark area, their feet produced sharp prints on the white paper. From these prints the following parameters were measured: (1) distance between toes of one foot to toes of opposite foot, (2) distance between second and fourth toes, (3) distance between first and fifth toes, (4) distance between tip of third toe to the proximal tip of metatarsals. The print length represents the degree of reinnervation of the calf muscles, whereas the toespread measurements represents the degree to which the small muscles of the feet have been reinnervated.

Return of Sensory Function. A graded analysis of sensory function as described by De Koning et al. (1986) was utilized. A series of small graded electric currents (0.05, 0.1, 0.2, 0.3, 0.4, 0.5, and 0.6 mA) was applied to the footsole of the rat and the strength of current which induced retraction of the foot recorded.

2.4 B-50 Phosphorylation

Sample Preparation. On day 6 following surgery, the sciatic nerves were exposed and frozen in situ with liquid nitrogen. A 5-mm section distal to the site of the crush was removed, as was a 5-mm section from a comparable site on the contralateral nerve. These sections were stored at −80 °C prior to phosphorylation. In preparation for phosphorylation, each section was homogenized in 10 × volume 0.32 M sucrose, 5.0 mM Hepes buffer pH 7.3. The crude homogenate was centrifuged at 5000 g to give a pellet with nuclear and cell debris which was discarded. A small quantity of the remaining homogenate was used to determine protein concentration by the method of Ramagli and Rodriquez (1985), modified by using the lysis buffer described below. All samples were then normalized to 0.5 μg/μL.

Phosphorylation. 100 μl containing 50 μg protein per sample were phosphorylated using the method of Hoch et al. (1984). $ATP^{\gamma 32}$ (100 μCi) at a concentration of 40 μM/g protein plus 2.0 mM magnesium chloride was used to initiate the reaction. The reaction was stopped after 30 s by the addition of 10 μl of stop solution containing at final concentration 1.0 mM EDTA, 1.0 mM sodium tetrathionate (Paulson 1980) and 5.0 mM ATP, and by immediate immersion in liquid nitrogen. Samples were lyophilized and resuspended in a lysis buffer containing 2% NP40, 4% ampholyte and 100 mM dithiotreitol and 9.5 M urea (Rimpilainen and Righetti 1985).

Electrophoresis. Two-dimensional polyacrylamide gel electrophoresis was performed using a modification of O'Farrell's procedure (1975). Two significant modifications of the first dimension isoelectric focusing tube (IEF) gel preparation were the use of a 3.5% T and 25% C_{bis} and the attachment of the IEF gel to the second dimension gel using a stacking gel buffer containing 1.0% SDS and 1.0% melted agarose. SDS-PAGE slab gels for the second dimension separations were prepared using buffers and acrylamide solutions containing 0.1% SDS (O'Farrell 1975). Exponential gradient gels 8%–18% acrylamide were cast and electrophoresed (Marotta et al. 1978). For the molecular weight calibration of 2D gels, IEF lysis buffer was used to dilute 1:100 Pharmacia low and high molecular weight calibration proteins. This was polymerized in the same agarose buffer which sealed the IEF tube gels to the second dimension. The distances migrated by the standard proteins in calibration tubes were the same as those subjected to isoelectric focusing. Bio Rad Silver Stain Kit was used for silver staining of the gels. Garrison's technique (1983) was followed for autoradiography.

3 Results

3.1 Morphology

Peptide treatment markedly ($p < 0.001$) increases nerve terminal branching within the endplate 15 days after nerve crush, with α-MSH treatment being more impressive than ACTH 4–10 (Table 1 and Fig. 2). This table also shows that whereas endplate perimeter and endplate area are effectively increased by α-MSH, treatment with ACTH 4–10 does not affect these parameters.

3.2 Return of Function

Motor function as measured by the partial print length appears to return more rapidly in peptide-treated animals than in the saline-treated controls. This difference is seen most clearly at days 14 and 16: by day 18 there is no discernable difference between any of the animal groups. Toe-spreading is greater in peptide-treated rats at day 14 than in the saline controls; again, this difference is eliminated by day 19.

Sensory Function. The most lateral aspect of the foot sole, innervated by the sciatic nerve, does not respond to electrical stimulation before day 14. ACTH- and α-MSH-treated animals are the first to respond with a contraction of the toes

Table 1. Effect of administration of 10 μg peptide/48 h on endplate parameters from EDL muscles 15 days after crush lesion

Parameters	Saline	ACTH 4-10	α-MSH
Nerve terminal branching (μm)	256.6 ± 4.3	336.8 ± 12.5***	369.3 ± 17.5***
Perimeter (μm)	232.0 ± 6.0	234.8 ± 5.9	255.7 ± 4.7**
Area (μm^2)	3501.2 ± 200.3	3572.7 ± 188.5	4091.9 ± 152.4*

n = 36 means ± SEM; * $p<0.05$; ** $p<0.01$; *** $p<0.001$.

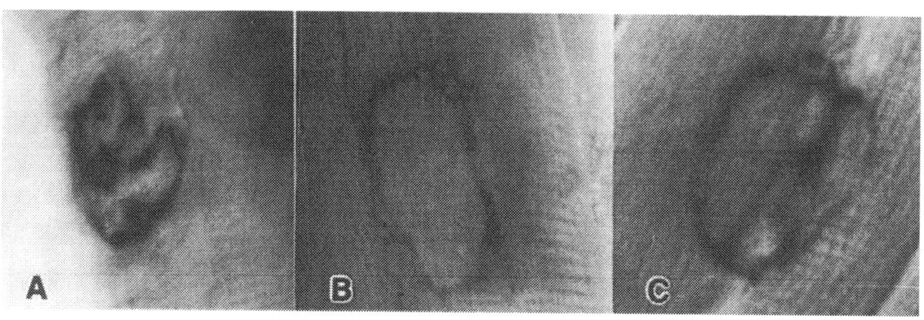

Fig. 2. Light micrographs of silver-cholinesterase stained endplates at the denervated EDL muscle. **A** α-MSH-treated; **B** saline-treated; **C** ACTH 4–10-treated. Increased density and complex branching are clearly evident in the peptide-treated endplates, **A** and **C.** The saline-treated sample, **B,** reveals an open, simple endplate interior

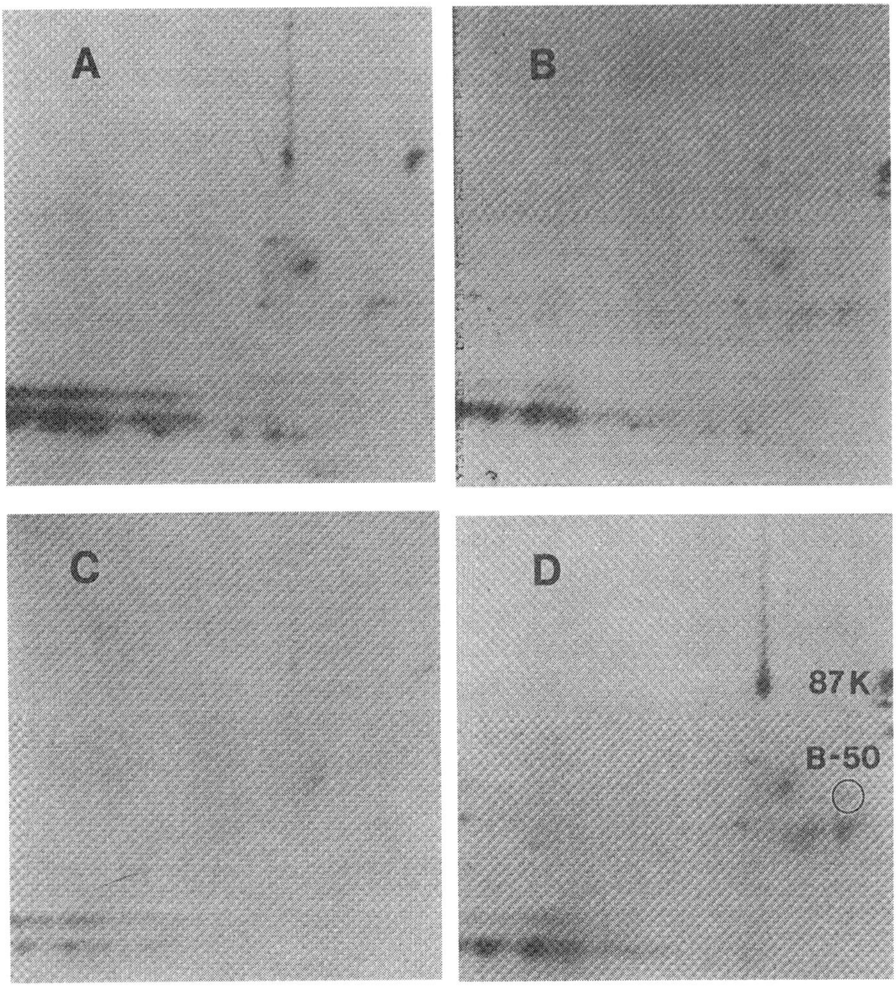

Fig. 3. Autoradiographs of 2-D gels of phosphorylated proteins from α-MSH- and saline-treated crushed and contralateral intact nerves. **A** saline-intact nerve; **B** saline-crushed nerve; **C** α-MSH intact nerve; **D** α-MSH-crushed nerve. The presence of phosphorylated B-50 protein (*circled*) can be seen only in the autoradiograph made from α-MSH-treated crushed nerve

closest to the point of stimulation. In this preliminary study, limb retraction does not appear to be significantly affected by either peptide.

It is possible that a more precise localization of the point of stimulation might elicit clearer distinctions.

3.3 Phosphorylation of B-50

Figure 3 is a compilation of four two-dimensional gels. Phosphorylated B-50 can only be seen in the gel from the crushed regenerated nerve from animals treated with αMSH (Fig. 3 D); it is not apparent in the contralateral peptide treated nerve or in the crushed saline nerve or its uncrushed counterpart.

4 Discussion

Our quantitative morphological results clearly show that both ACTH 4–10 and αMSH increase the length of nerve terminal branching within the endplate of denervated EDL muscles 15 days after nerve crush. However, only αMSH increases the endplate perimeter and area. This is in accord with our results on the effects of ACTH 4–10 on the developing neuromuscular junction: this peptide significantly increases nerve terminal branching but has no effect on the other two parameters (Rose et al. 1988; Frischer and Strand 1988).

In studies of regenerating peripheral motor nerves, axonal sprouting appears to be mainly collateral in nature. In the developing neuromuscular system, sprouting is seen as an increased arborization of preterminal branches entering the muscle and increased length and complexity of the terminals within the endplate (Frischer and Strand 1988). As both these sprouting processes are augmented by ACTH/MSH peptides, the question arises as to whether these peptides wield an organizational effect on the pattern by which the regenerating axons reinnervate skeletal muscle to form new motor units. We now have convincing evidence that ACTH 4–10 acts as a modulator of neuromuscular maturation, with the ultimate effect (either acceleration or deceleration) depending upon the time during gestation or neonatal life during which the developing system is exposed to this peptide (Rose and Strand 1987, 1988).

The importance of the pattern of nerve sprouting is indicated by electrophysiological studies that show that there are important qualitative differences between motor units reinnervated under peptide influence and saline-treated controls (Saint-Come and Strand 1985; Strand and Smith 1986). Peptide-treated reinnervated muscles have a greater number of motor units, with a size distribution more nearly that of intact muscles: saline-treated reinnervated muscles consist of few, large motor units. Thus ultimate control of movement in reinnervated muscle is better in the peptide-treated animals.

Other fundamental questions arise from these observations. Recent studies indicate a complex interaction, age-dependent, between nerve and muscle fibers in the formation of motor unit types (fast vs. slow) in development (Sanes 1987). Are the same cues responsible for the size of the motor units? Are these interactions replayed or modified during regeneration? Muscle fibers produce neuronotrophic factors (O'Brien and Fischbach 1986) which may be of vital importance in the ultimate matching of nerve fiber type and degree of branching with muscle fiber type and number innervated. We are now undertaking further studies to follow the pattern of regenerating nerve terminals as they construct new motor units.

Our results permit us to see phosphorylated B-50 in crushed regenerating nerve treated with αMSH, not apparent in crushed saline-treated controls. Thus the role of αMSH in regeneration previously described (Edwards and Gispen, 1985; De Graan et al. 1986) is significantly expanded. Crushing the nerve increases both MSH-like substances and B-50 protein within the regenerating sprouts (Edwards et al. 1984), administration of αMSH stimulates the outgrowth of nerve fibers following nerve crush (Verhaagen et al. 1986b). It has been suggested (Gispen et al. 1985; Nelson and Routtenberg 1985), that the degree of phosphorylation of B-50 contributes to the persistence of synaptic plasticity and improved transmitter release in CNS preparations. The presence of phosphorylated B-50 in vitro in α-MSH-treated crushed regenerating sciatic nerves found in this study points to the possibility that MSH peptides in the peripheral nervous system can act in the phosphatidyl-inositol mechanism responsible for the phosphorylation of B-50. If the peptide is acting at the regulatory level of intracellular calcium which is involved with the subsequent formation of cytoskeletal proteins, then the correlation between MSH peptides and the accelerated, enhanced growth during regeneration and development may be established.

While B-50 is abundant in the sprouts of regenerating neurons, it is synthesized, accumulated, and phosphorylated at much higher levels in developing neurons. Its synthesis peaks during the first 2 weeks of neonatal life, a period of active nerve outgrowth and synaptogenesis (Jacobson et al. 1986; Skene et al. 1986). It is during this period of neonatal life that the infant neuromuscular system is responsive to ACTH peptides which accelerate its maturation (Frischer et al. 1985; Saint-Come et al. 1982, 1985).

While ACTH 4–10 vigorously improves motor performance and increases resistance to fatigue, its effect on endplate area and perimeter is not as significant as the effect of α-MSH on these parameters. ACTH 4–10 has neurotrophic actions which appears to be restricted to the aspect of nerve terminal branching only. Since α-MSH increases endplate perimeter and area, as well as terminal branching of regenerating nerves, a stronger structural basis for improved neurotransmitter release and fine motor control is in place sooner than in the untreated denervated controls. The concomitant increase in B-50 induced by α-MSH supplies a biochemical basis for the enhanced neuromuscular performance demonstrated by our earlier work (Strand et al. 1976).

It is not merely the rapidity with which a muscle is reinnervated that is important in clinical studies. Of equal or greater importance is the quality of the reinnervation which will determine the ultimate level of motor control. In light of this and the results of this experiment, α-MSH is a strong candidate for regulating the pattern by which motor units are restructured after denervation.

Acknowledgements. We are grateful to Dr. Monroe Yoder for advice and assistance with the light and scanning microscopy. This study was supported by The Council for Tobacco Research and by Grant BRSG RR07062 awarded by the Biomedical Research Support Grant Program of Research Resources, National Institutes of Health. We also thank Organon for ACTH 4–10 (Org. 0163) and for αMSH.

References

Acker GR, Frischer R, Strand FL (1984) ACTH peptide neuromodulation in the developing neuromuscular system as seen through three different perspectives. Ann NY Acad Sci 435:370–375

Bijlsma WA, Schotman P, Jenekens FGI, Gispen WH, Wied D de (1983) The enhanced recovery of sensorimotor function in rats is related to the melanotropic moiety of ACTH/MSH neuropeptides. Eur J Pharmacol 9:231–236

Brown MC (1984) Sprouting of motor nerves in adult muscle: a recapitulation of ontogeny. TINS 7(1):10–14

Edwards PM, Zee CEEM van der, Verhaagen J, Schotman P, Jennekens FGI, Gispen WH (1984) Evidence that the neurotrophic actions of α-MSH may derive from its ability to mimick the actions of a peptide formed in degenerating nerve stumps. J Neurol Sci 64:333–340

Frischer RE, El-Kawa N, Strand FL (1985) ACTH peptides as organizers of neuronal patterns in development: maturation of the rat neuromuscular junction as seen by scanning electron microscopy. Peptides 6(2):13–19

Frischer RE, Strand FL (1988) ACTH peptides stimulate motor nerve sprouting in development. Exp Neurol 100:531–541

Garrison JC (1983) Measurement of hormone-stimulated protein phosphorylation in intact cells. Methods in enzymology 99:22–36

Gispen WH, Zwiers H, Wiegant VM, Schotman P, Wilson JE (1979) The behaviorally active neuropeptide ACTH as neurohormone and neuromodulator: the role of cyclic nucleotides and membrane phosphoproteins. Adv Exp Med Biol 116:119–224

Gispen WH, Zwiers H, Wiegant VM, Schotmann P, Wilson JE (1985) The role of phosphoprotein B-50 in phosphoinositide metabolism in brain synaptic plasma membranes. In: Bleasdale JE, Hasuer G, Eichberg J (eds) Inositol and phosphoinositides. Human, New Jersey, pp 399–413

Graan PNE de, Oestreicher AB, Schrama LH, Gispen WH (1986) Phosphoprotein B-50: localization and function. In: Gispen WH, Routtenberg A (eds) Progress in brain research, vol 69. Elsevier Science, Amsterdam, pp 37–50

Hoch DB, Wilson JE, Dingledine RJ (1984) Factors affecting the phosphorylation of a 41,000 Dalton protein in hippocampal subcellular fractions. Brain Res 302:117–124

Jacobson RD, Virag I, Skene JHP (1986) A protein associated with axon growth, GAP-43, is widely distributed and developmentally regulated in rat CNS. J Neurosci 5:1843–1855

Koning P de, Gispen WH (1987) Org 2766 improves functional and electrophysiological aspects of regenerating sciatic nerve in the rat. Peptides 8:415–422

Koning P de, Brakee JH, Gispen WH (1986) Methods for producing a reproducible crush in the sciatic nerve of the rat and the rapid and precise testing of return of sensory function: beneficial effects of melanocortins. J Neurol Sci 74:237–246

Marotta CA, Harris JL, Gilbert JM (1978) Characterization of multiple forms of brain tubulin subunits. J Neurochem 30:1431–1440

Medinaceli LW de, Freed J, Wyatt RJ (1982) An index of the functional condition of rat sciatic nerve based on measurements made from walking tracks. Exp Neurol 77:634-643

Nelson RB, Routtenberg A (1985) Characterization of protein F1 (47 kDa, 4.5 pI): a kinase C substrate directly related to neural plasticity. Exp Neurol 89:213–224

O'Brien RJ, Fischbach GD (1986) Isolation of embryonic chick motoneurons and their survival in vitro. J Neurosci 6:3265–3274

O'Farrell PH (1975) High resolution two-dimensional electrophoresis of proteins. J Biol Chem 250:4007–4021

Paulson JR (1980) Sulfhydryl reagents prevent dephosphorylation and proteolysis of histones in isolated HeLa metaphase chromosomes. Eur J Biochem 111:189–197

Ramagli LS, Rodriquez LV (1985) Quantitation of microgram amounts of protein in two-dimensional polyacrylamide gel electrophoresis sample buffer. Electrophoresis 6:559–563

Rimpilainen MA, Righetti PG (1985) Membrane protein analysis by isoelectric focusing in immobilized pH gradients. Electrophoresis 6:419–422

Rose KJ, Strand FL (1987) Response of the developing neuromuscular system of the rat to nicotine and the neurotropic peptide fragment ACTH/MSH 4–10. Ann NY Acad Sci 494:319–322

Rose KJ, Strand FL (1988) Mammalian neuromuscular development accelerated with early but slowed with late gestational administration of ACTH peptide. Synapse 2:200–204

Rose KJ, Frischer RE, King JA, Strand FL (1988) Neonatal neuromuscular parameters vary in susceptibility to postnatal ACTH 4–10 administration. Peptides 9:151–156

Saint-Come C, Strand FL (1985) ACTH 4–10 improves motor unit performance during peripheral nerve regeneration. Peptides 6(1):77–83

Saint-Come C, Strand FL (1987) ACTH 4–9 analog (Org. 2766) improves qualitative and quantitative aspects of nerve regeneration. Peptides 9 SI(1):215–221

Saint-Come C, Acker GR, Strand FL (1982) Peptide influences on the development and regeneration of motor performance. Peptides 3:439–499

Saint-Come C, Acker GR, Strand FL (1985) Development and regeneration of motor systems under the influence of ACTH peptides. Psychoneuroendo 10:445–459

Sanes JR (1987) Cell lineage and the origin of muscle fiber types. TINS 10(6):219–221

Skene JHP, Jacobson RD, Snipes GJ, McGuire CB, Norden JJ, Freeman JA (1986) A protein induced during nerve growth (GAP-43) is a major component of growth-cone membranes. Science 233:783–785

Strand FL, Kung T (1980) ACTH accelerates recovery of neuromuscular function following crushing of peripheral nerve. Peptides 1:135–138

Strand FL, Smith CM (1986) LPH, ACTH, MSH and motor systems. Neuropeptides and behavior, vol 1. In: Wied D de, Gispen WH, Wimersma Greidanus TjB van (eds) CNS effects of ACTH, MSH and opioid peptides. Pergamon, New York, pp 245–271

Strand FL, Cayer AM, Gonzalez E, Stoboy H (1976) Peptide enhancement of neuromuscular function: animal and clinical studies. Pharmacol Biochem Behav 5:179–187

Verhaagen J, Edwards PM, Jenekens FGI, Schotman P, Gispen WH (1986a) α-Melanocyte-stimulating hormone stimulates the outgrowth of myelinated nerve fibers after peripheral nerve crush. Exp Neurol 92:451–454

Verhaagen J, Hooff COM van, Edwards PE et al. (1986b) The kinase C substrate protein B-50 and axonal regeneration. Brain Res Bull 17:734–741

Zwiers H, Veldhuis D, Schotman P, Gispen WH (1976) ACTH, cyclic nucleotides and brain protein phosphorylation in vitro. J Neurochem Res 1:669–677

Zwiers H, Schotman P, Gispen WH (1980) Purification and some characteristics of an ACTH-sensitive protein kinase and its substrate protein in rat brain membranes. J Neurochem 34:1689–1699

ACTH (4–10) and ACTH (1–24) Influence Neurite Outgrowth and Neural Development of Fetal Rat Cerebral Cultures

C. Richter-Landsberg, I. Bruns, and H. Flohr
Department of Neurobiology, University of Bremen, D-2800 Bremen, FR Germany

1 Introduction

The central and peripheral nervous systems are targets for a group of ACTH/MSH-like peptides, collectively termed melanocortins. These peptides and some of their (noncorticotrophic) fragments and analogs, such as ACTH (4–10) and ACTH (4–9), α-MSH and Org 2766, affect numerous neurophysiological functions, in particular, those underlying memory and learning (for reviews see Gispen and Zwiers 1985, De Wied and Jolles 1982). Recent evidence indicates that these peptides also influence lesion-induced plastic processes in both the central and peripheral nervous system.

Beneficial effects of melanocortins on functional recovery from CNS damage have been observed in various lesion paradigms. They enhance recovery from vestibular (Flohr and Lüneburg 1982; Flohr et al. 1985; Lüneburg and Flohr 1987; Igarashi et al. 1985), septal (Isaacson and Poplawski 1983, 1985), hippocampal (Hannigan and Isaacson 1985) and parafascicular (Nyakas et al. 1985) lesions.

Studies of the peripheral nervous system indicate that melanocortins enhance recovery of sensory and motor functions after nerve crush or section. They influence axon regeneration and the restoration of synapses at the peripheral target (Gonzales and Strand 1981; Bijlsma et al. 1981, 1983, 1984; Edwards et al. 1984; Verhaagen et al. 1986).

A variety of neurotrophic factors positively influence regenerative processes (Bijlsma et al. 1984). In this respect, however, the mechanism of action of ACTH-like neuropeptides has not yet been resolved. On the basis of studies on effects of ACTH fragments on neuronal RNA and protein metabolism, it has been suggested that these peptides exert neurotrophic functions (Dunn and Schotman 1986; Bijlsma et al. 1984).

To further elucidate the potential neurotrophic role of melanocortins, we have used a tissue culture approach. The effect of ACTH (4–10) and ACTH (1–24) on neuronal growth and development of cultured embryonic rat cerebral cells was investigated.

Dissociated fetal rat cerebral cells in culture undergo morphological and biochemical differentiation and express distinct features of the developing central nervous system, such as myelinogenesis, synaptogenesis, and neurite outgrowth (Yavin and Yavin 1977). This culture system consists of a mixed population of neuronal and glia cells (Bock et al. 1980); and astroglia, as identified by immuno-

H. Flohr (Ed.)
Post-Lesion Neural Plasticity

labeling with antisera against glial acidic fibrillary protein, are observed to develop and mature after the first week in culture (Richter-Landsberg 1988). Serum deprivation and the use of poly L-lysine-coated culture dishes support the survival of neurons and the elaboration and maintenance of an extensive neuritic network (Yavin and Yavin 1980). Thus, a system is provided to study the effects of extracellular factors on the developmental regulation and cellular interactions of neuronal cells.

2 Methods

2.1 Cell Culture

Cerebral hemispheres from 10–12 16-day-old rat embryos were pooled and dissociated mechanically as described previously (Yavin and Yavin 1974). Cells were seeded on poly L-lysine-coated culture dishes in Eagle's basal medium (BME) supplemented with 20% fetal calf serum. After 2 h the culture medium was replaced by BME containing 0.5% fetal calf serum with or without ACTH 4–10 or ACTH 1–24 as indicated. Fresh ACTH was supplemented every second day. Cell morphology was observed by phase contrast microscopy.

2.2 Indiret Immunofluorescent Staining

Cultures were grown on poly L-lysine-coated glass coverslips. After washing with Ca^{2+}, Mg^{2+}-free phosphate buffered saline (PBS), cells were fixed with 3% paraformaldehyde for 30 min. For cell surface staining with anti D2 protein antiserum (AbD2), cultures were incubated with antisera (AbD2 1 : 100) for 45 min at room temperature. After washing with PBS, cells were incubated with the appropriate second antiserum (45 min, RT). For cytoskeletal staining, cells were pretreated for 30 min with 0.1% Triton X-100 in PBS and then incubated with anti-neurofilament anti-serum (AbNF) (at a dilution of 1 : 100 of the polyclonal and 1 : 3 of the monoclonal antiserum) and antigliafilament antiserum (AbGFAP) (at a dilution of 1 : 200 of the polyclonal and 1 : 3 of the monoclonal antiserum). The second antibodies were fluorescein-isothiocyanate (FITC)-conjugated goat anti-rabbit or mouse for the D2-, neurofilament- and gliafilamentstaining. Fluorescent labeling was monitored using a Zeiss epifluorescence microscope equipped with an automatic camera.

2.3 Acetylcholinesterase Activity

Acetylcholinesterase activity (AChE) was determined according to Ellman et al. (1961). One unit of enzyme activity is defined as the transformation of one nmole substrate per min at 20 °C. Specific activity is expressed as units of enzyme activity per mg of total cell protein. Numbers represent the mean value of four experiments.

2.4 Incorporation of Amino Acids

For radioactive labeling cells were incubated in the presence or absence of ACTH (4–10, and 1–24, at 10^{-8}M) and for the last 24 h in the presence of [^{35}S]-methionine (10 μCi/ml). At the end of the incubation period, cells were washed with phosphate-buffered saline (PBS) removed and homogenized in PBS in the cold. Total bound radioactivity was determined in aliquots after precipitation with 10% trichloroacetic acid. Protein concentrations were determined according to Bradford (1976), using bovine serum albumin as a standard.

3 Results

3.1 Phase Contrast Microscopy

Cerebral hemispheres of 16-day-old fetal rats were prepared as described before (Yavin and Yavin 1974, 1980) and single cell suspensions were seeded on poly L-lysine coated culture dishes and incubated in low serum-containing (0.5%) medium for up to 3 weeks. Under these conditions cells were firmly attached within 30–40 min and during the next 24 h had formed small clusters of cells, extending thin neuritic processes (Yavin et al. 1987). During the following 2 weeks in culture, neuronal cells formed large aggregates interconnected by bundles of fibers, and the extensive neuritic network was covering the whole culture dish (Fig. 1). When ACTH (1–24, and 4–10) was added to the culture medium 2 h after cellular plating and resupplied every second day, after 1 week in vitro an increase in the density of the neuritic network and the number and size of the cellular aggregates were observed, as compared to the untreated control cultures. At the same time, the fasciculation of neurites to large neuritic bundles of extensive length was greatly enhanced (Fig. 1). These morphological effects were observed with ACTH-fragments at a concentration range of 10^{-6}–10^{-8}M. Maximal effects were observed at a concentration of 10^{-7}M. ACTH-peptides exerted their morphological action during the first and second week in culture, after which time cultures grown in the absence of ACTH had reached the same state of differentiation.

3.2 Immunocytochemical Staining

Indirect immunofluorescent staining with antiserum against purified tubulin protein (AbTub) and antiserum against purified neurofilament proteins (AbNF) revealed that ACTH neuropeptides induced the formation of a neurotubule- and neurofilament-containing neuritic network (Figs. 2, 3 a, b). Figure 2 shows that in the presence of ACTH-fragments microtubule-containing neurites were preferentially associated to fascicles (Fig. 2 b, arrows), and the neuronal aggregates have been formed to a greater extent. Cells in the aggregates, the neuritic processes, and the large bundles are of neuronal origin, as identified by indirect immunofluorescent staining with AbNF (Fig. 3 a, b). To further identify neurons and their filamentous network, cultures were labeled with antiserum against purified rat D2-protein (AbD2), a neuron-specific cell surface glycoprotein, which is immu-

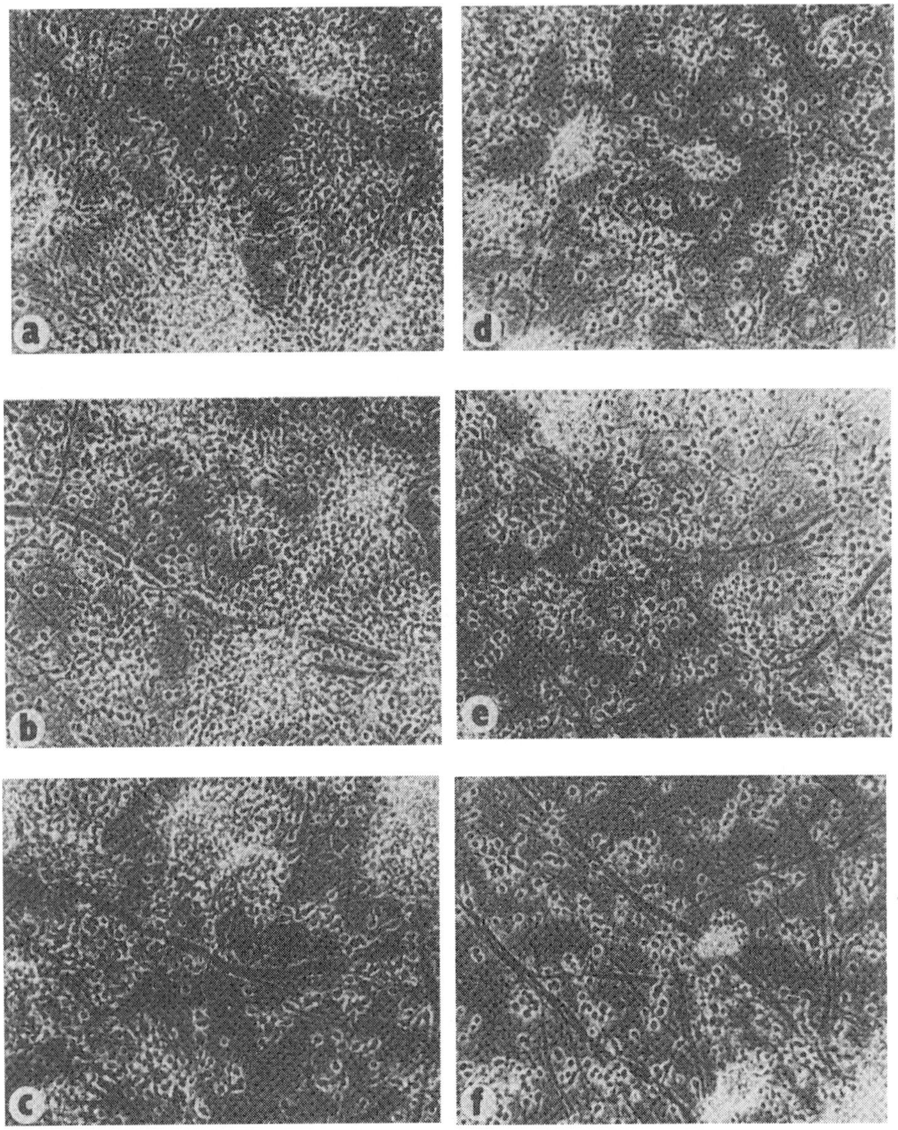

Fig. 1. Effect of ACTH on the development of rat cerebral cells in culture. Phase microscopical images of 7-day- (**a–c**) and 10-day (**d–f**) -old cultures are seen. Control cultures (**a, d**) were incubated in low serum-containing BME (see Sect. 2), and for ACTH treatment culture medium was supplemented with $ACTH_{1-24}$ (**b, e**) or $ACTH_{4-10}$ (**c, f**), at a concentration of 10^{-8}M. ACTH was added 2 h after plating and supplemented fresh every other day. Bar = 50 μm

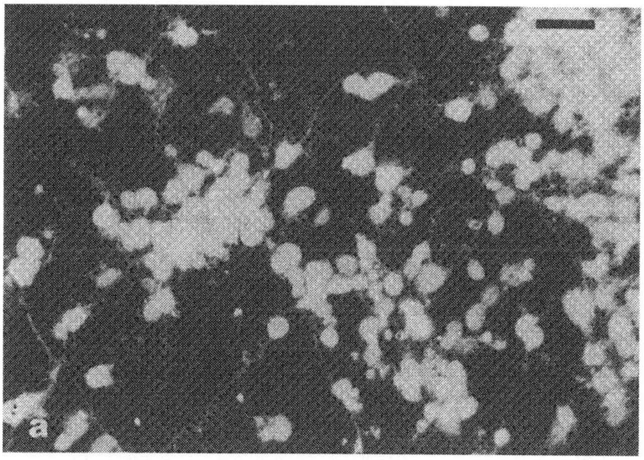

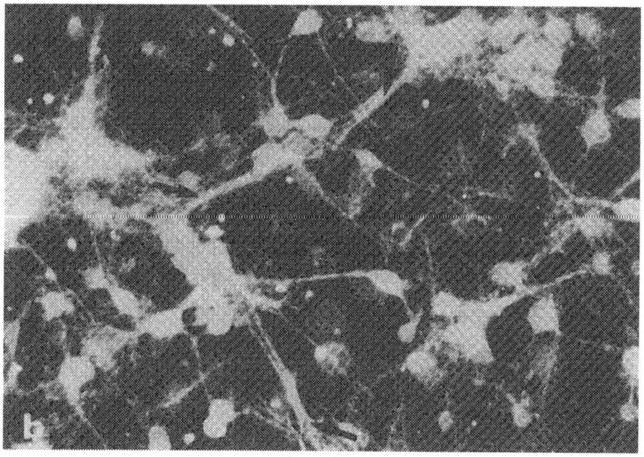

Fig. 2. Immunofluorescent staining of rat cerebral cells with AbTub after 11 days in vitro. The fluorescent images of a control culture (**a**) and after treatment with $ACTH_{4-10}$ (10^{-8}M) (**b**) are seen. ACTH-treatment induces the formation of microtubule containing neurites and the neuritic fasciculation (**b,** *arrows*). Bar = 50 μm

nologically related to the cell adhesion molecule N-CAM (Jørgensen et al. 1980). The large cellular aggregates, the neuritic network covering the culture dish, as well as the large neuritic bundles were intensely stained with AbD2 (Fig. 3c, d).

Astroglia-like cells appeared approximately after 1 week in culture and were identified by indirect immunofluorescent procedure using antisera raised against purified glia filament acidic protein (AbGFAP). Both ACTH-fragments seemed to accelerate the maturation of astrocytes in this culture system. Already during the second week in vitro, ACTH-treated cultures contained glia cells with astrocytic processes with thickened fibrillary ends, and a relatively high number of cells had formed an extensive glia filament-containing network, filling out the cell somata (Fig. 4b). Control cultures at this time mostly expressed branched processes and the typical perikarya were rarely formed (Fig. 4a).

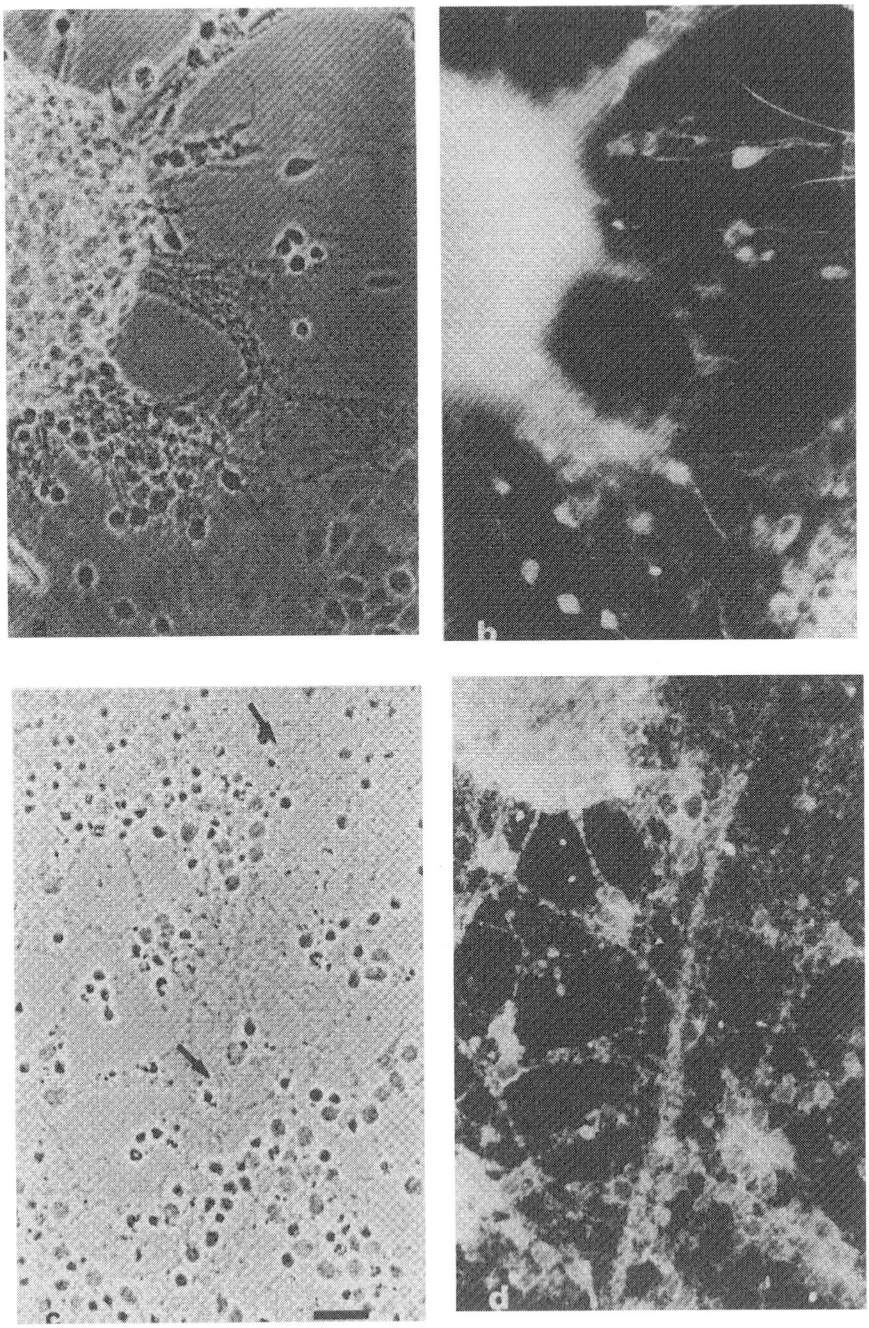

Fig. 3. Immunofluorescent staining of $ACTH_{4-10}$ (10^{-8}M)-treated rat cerebral cells after 12 days in culture. Immunofluorescent images after staining with AbNF (**b**) or AbD2 (**d**) and the corresponding phase images (**a, c**) are shown. *Arrows* (**c, d**) point to ACTH-induced AbD2-positive neuritic bundles. Bar = 50 μm

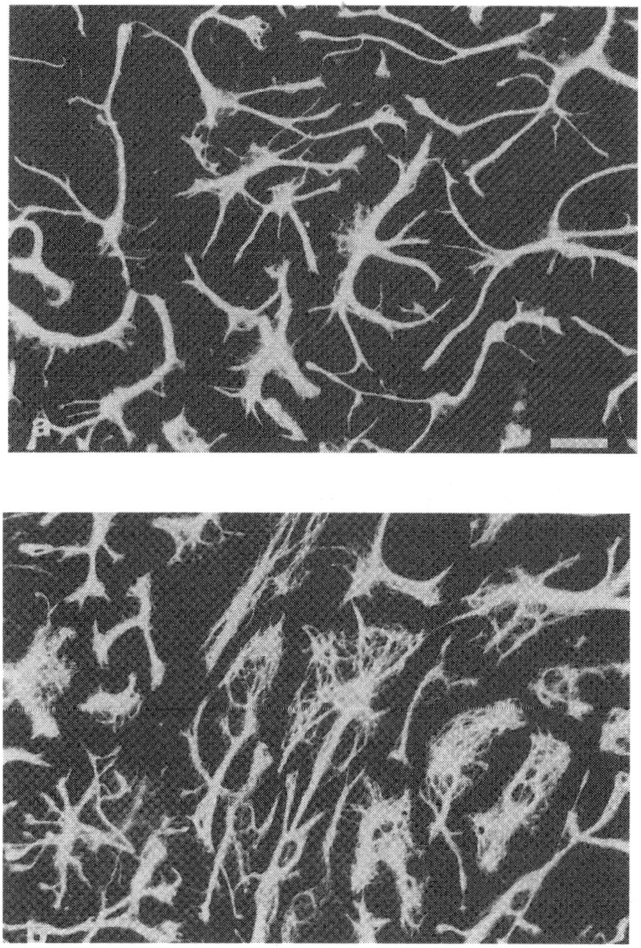

Fig. 4. Effect of ACTH on astroglia morphology. Immunofluorescent images after staining with AbGFAP of 12-day-old cultures incubated in the absence (panel **a**) or presence of $ACTH_{4-10}$ (10^{-8}M) (panel **b**) are seen. Bar = 50 µm

3.3 Protein Synthesis and Acetylcholinesterase Activity

Incorporation of [^{35}S]-methionine into acid-insoluble proteins of rat cerebral cultures during differentiation was analyzed. Figure 5 indicates that long-term treatment with ACTH did not affect the incorporation of [^{35}S]-methionine. Thus, ACTH did not increase the amount of newly synthesized proteins. In addition, no changes in the protein patterns of control and ACTH-treated cultures were seen, as analyzed by one-dimensional SDS-polyacrylamid-gelelectrophoresis (data not shown).

However, ACTH-fragments transiently affected acetylcholin-esterase activity (AChE) in this culture system. The maturation of cerebral cells in vitro was ac-

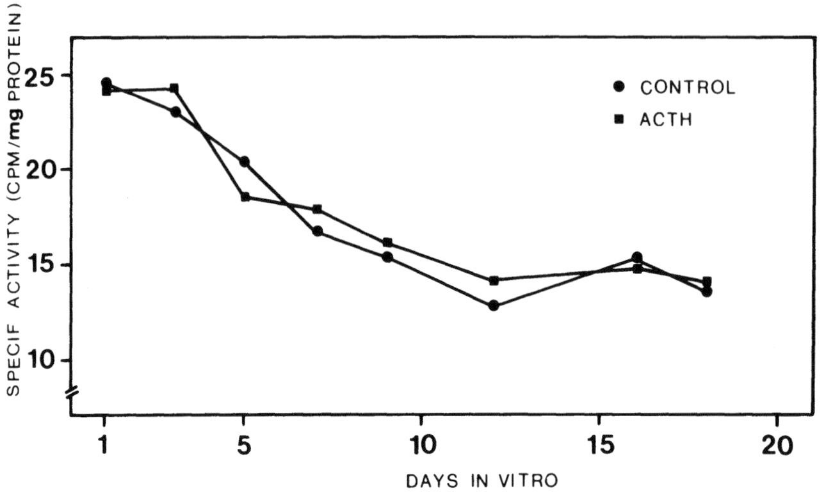

Fig. 5. Effect of ACTH on the incorporation of [^{35}S]-methionine into cellular proteins of rat cerebral cells during development in vitro. (●——●) untreated control cultures, (■——■) cultures were treated with $ACTH_{4-10}$ (10^{-8}M). Each *point* represents the mean value of a triplicate culture from two separate experiments

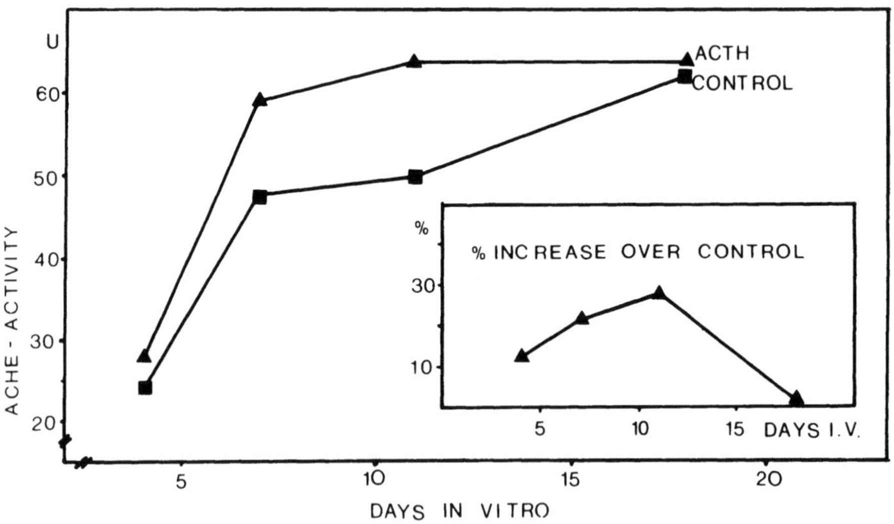

Fig. 6. Effect of ACTH on acetylcholinesterease activity during neuronal development in vitro. AChE was determined in control cultures (■——■) and after chronic tresatment with $ACTH_{4-10}$ (10^{-8}M) (▲——▲). Values represent the mean of a duplicate culture from two independent experiments

companied by a steady increase in AChE activity (Fig. 6). During the second week in culture after the continuous treatment with ACTH-fragments (10^{-8}M) AChE activity was increased by 30% as compared to the untreated controls (Fig. 6). After 2–3 weeks, however, no differences in the enzyme level of control and ACTH-treated cultures were observed (Fig. 6).

4 Discussion

Long-term treatment of dissociated rat cerebral cells in vitro with ACTH (4–10) or ACTH (1–24) at low concentrations (10^{-8}M) modulated neuronal development and differentiation. During the second week in culture, an increase in the density of the neuritic network and a promotion of the fasciculation of neurites to bundles of extensive length and an increase in the size and number of neuronal aggregates was observed (Figs. 1–3). Immunofluorescent staining demonstrated that the ACTH-promoted neuritic bundles were of neuronal origin, since they were positively stained with AbD2 and were highly enriched in microtubules and neurofilaments (Figs. 2 and 3).

Microtubules as part of the cytoskeleton are involved in the organization and maintenance of cell shape and are principal structural components of neurites (for review see Littauer and Ginzburg 1985). Microtubule-associated proteins (MAP's) have been shown to promote microtubule assembly and the state of phosphorylation of MAP's affect the initiation and elongation steps of microtubule assembly and stability (Pallas and Solomon 1982). In addition, microtubule-microtubule interactions and the binding of microtubules to other constituents of the cytoskeleton, such as neurofilaments, have been shown to be regulated through phosphorylation of cytoskeletal proteins (Leterrier et al. 1982).

Since ACTH does not seem to induce the rate of synthesis of specific proteins, in particular, no increase in tubulin and actin synthesis was observed (Edwards et al. 1985), it might exert its effects through modulating the phosphorylation of cytoskeletal proteins. Alteration of the phosphorylation of cytoskeletal proteins might lead to an increased intraorganellar binding, which possibly results in the formation of neuritic bundles. The initiation of neurite outgrowth does not necessarily require the new synthesis of tubulin but can be carried out by using the intracellular pool of soluble tubulin (Littauer and Ginzburg 1985). ACTH influences the activity of protein kinases in the brain and thus modulates the phosphorylation of a variety of substrate proteins (Wiegant et al. 1985; Gispen and Zwiers 1985; Flohr et al. 1985). Its effects on the phosphorylation of cytoskeletal proteins, however, remain to be investigated.

Another possibility to explain the enhancement in neuritic fasciculation could be the induction of neural cell adhesion molecules (Edelman 1984). NgCAM, which is immunologically related to NILE glycoprotein and the L1-antigen (Bock et al. 1985), is involved in neurite fasciculation (Stallcup and Beasley 1985; Fischer et al. 1986). However, indirect immunoprecipitation experiments of ACTH-treated and untreated [^{3}H]-fucose-labeled cell extracts of embryonic rat cerebral cells during development (analyzed from day 4 to 18) did not show any differences in NILE glycoprotein concentration (Richter-Landsberg unpublished).

The development of rat cerebral cells in vitro is accompanied by a steady increase in AChE-activity, possibly reflecting the development of cholinergic neurons. ACTH fragments accelerate the increase in AChE activity and the maximal level is observed one week earlier than in the untreated control cultures. AChE metabolizes acetylcholine and is specifically associated with cholinergic neurons (Levey et al. 1983). It has been observed that prenatal administration of ACTH neuropeptides results in an accelerated maturation of the neuromuscular junction (Strand and Smith 1980). It might be possible that immature neurons or only neurons at a specific stage of development are susceptible to the action of neuropeptides. Evidence has accumulated that ACTH neuropeptides influence neurotransmitter metabolism (for review see Gispen and Zwiers 1985). In particular, ACTH increased the synthesis of acetylcholine in brain slices (Torda 1952), and accelerated choline release (Strand and Smith 1980). A recent study (Azmitia and Kloet 1987), using a tissue culture approach, indicated that ACTH also enhanced the maturation of serotonergic neurons.

Daval et al. (1983) have described the influence of ACTH (1–24) on neurite outgrowth and neuronal survival of isolated neurons from chick embryo cerebral hemispheres, indicating its function as a neurotrophic agent and a survival factor. Our data showed that also the shorter ACTH-fragment (4–10) was active. In the culture system used in our studies, however, the morphological effects of ACTH were exerted only during a specific period of culture development, and after 3 weeks in vitro ACTH-treated cultures resembled the untreated controls in morphological appearance and AChE activity. Thus, ACTH trophically influenced and modulated growth of rat CNS neurons, but was not a necessary requirement for neuronal culture development and neuronal survival.

Our experiments indicated that chronic treatment of neuronal cultures with ACTH fragments also influenced growth and development of astroglia cells (Fig. 4). Cultures incubated in the presence of ACTH fragments contain astroglia cells with thick fibrillary GFAP-containing processes and fibrillary somata at earlier times in vitro than the untreated controls. In addition, glia cell processes in the presence of ACTH appeared shorter and with thickened fibrillary, "growth cone"-like ends (Fig. 4). After 3 weeks, however, no difference in astroglia cell appearance in both culture types was observed. Whether this was the effect of ACTH neuropeptides on glia cells directly, or was caused by soluble factors released by ACTH-stimulated neurons, is subject of further investigations in our laboratory.

Acknowledgements. This work was supported by grants of the Deutsche Forschungsgemeinschaft (Ri 384/2 to C.R-L., and F1 139/2-1 to H.F.).

References

Azmitia EC, Kloet ER de (1987) ACTH neuropeptide stimulation of serotonergic neuronal maturation in tissue culture: modulation by hippocampal cells. Prog Brain Res 72:311–318

Bijlsma WH, Jennekens FGI, Schotman P, Gispen WH (1981) Effects of corticotropin (ACTH) on recovery of sensorymotor function in the rat: structure activity study. Eur J Pharmacol 76:73–79

Bijlsma WH, Jennekens FGI, Schotman P, Gispen WH (1983) Stimulation by $ACTH_{4-10}$ of nerve fiber regeneration following sciatic nerve crush. Muscle Nerv 6:104–112

Bijlsma WH, Jennekens FGI, Schotman P, Gispen WH (1984) Neurotrophic factors and regeneration in the peripheral nervous system. Psychoneuroendocrinology 9:199–215

Bock E, Yavin Z, Jørgensen OS, Yavin E (1980) Nervous system specific proteins in developing rat cerebral cells in culture. J Neurochem 35:1297–1302

Bock E, Richter-Landsberg C, Faissner A, Schachner M (1985) Demonstration of immunochemical identity between the nerve growth factor-inducible large external (NILE) glycoprotein and the cell adhesion molecule L1. EMBO J 4:2765–2768

Bradford MM (1976) A rapid and sensitive method for the quantification of microgram quantities of protein utilizing the principle of protein-dye binding. Anal Biochem 72:248–254
Daval JL, Louis JC, Gerard MJ, Vincendon G (1983) Influence of adrenocorticotropic hormone on the growth of isolated neurons in culture. Neurosci Lett 36:299–204
Dunn AJ, Schotman P (1986) Effects of ACTH and related peptides on cerebral RNA and protein synthesis. In: Wied D de, Gispen WH, Wimersa Greidanus TJB van (eds) Neuropeptides and behavior, vol I. CNS effects of ACTH, MSH and opioid peptides. Pergamon, New York, pp 165–187
Edelman GM (1984) Expression of cell adhesion molecules during embryogenesis and regeneration. Exp Cell Res 161:1–16
Edwards PM, Zee CEEM van der, Verhaagen J, Schotman P, Jennekens FGI, Gispen WH (1984) Evidence that the neurotrophic actions of α-MSH may derive from its ability to mimic the actions of a peptide formed in degenerating nerve stumps. J Neurol Sci 64:333–340
Edwards PM, Verhaagen J, Spierings T, Schotman P, Jennekens FGI, Gispen WH (1985) The effect of $ACTH_{4-10}$ on protein synthesis, actin and tubulin during regeneration. Brain Res Bull 15:267–272
Ellman GL, Courtney KD, Andres V Jr, Featherstone RM (1961) A new and rapid colorimetric determination of acetylcholinesterase activity. Biochem Pharmacol 7:88–95
Fischer DR, Künemund V, Schachner M (1986) Neurite outgrowth patterns in cerebellar microexplant cultures are affected by antibodies to the cell surface glycoprotein L1. J Neurosci 6:605–612
Flohr H, Lüneburg U (1982) Effects of $ACTH_{4-10}$ on vestibular compensation. Brain Res 248:169–173
Flohr H, Lüneburg U, Richter-Landsberg C (1985) ACTH/MSH-like neuropeptides and lesion-induced plastic processes. In: Keller EL, Zee DS (eds) Adaptive processes in visual and oculomotor systems. Pergamon, New York, pp 409–416
Gispen WH, Zwiers H (1985) Behavioral and neurochemical effects of ACTH. In: Lajtha A (ed) Handbook of neurochemistry, vol 8, 2nd edn. Plenum, New York, pp 375–412
Gonzales ER, Strand FL (1981) Neurotrophic action of MSH/ACTH 4–10 in neuromuscular function in hypophysectomized rats. Peptides 2:107–113
Hannigan JH, Isaacson RL (1985) The effects of Org. 2766 on the performance of sham, neocortical, and hippocampal-lesioned rats in a food search task. Pharmacol Biochem Behav 23:1019–1027
Igarashi M, Ishikawa K, Ishii M, Schmidt KH (1985) Effect of ACTH (4–10) on equilibrium compensation after unilateral labyrinthectomy in the squirrel monkey. Eur J Pharmacol 119:239–242
Isaacson RL, Poplawsky (1983) An ACTH 4–9 analog (Org. 2766) speeds recovery from septal hyperemotionality in the rat. Behav Neural Biol 39:52–59
Isaacson RL, Poplawski A (1985) ACTH 4–10 produces a transient decrease in septal hyperemotionality in the rat. Behav Neurol Biol 43:109–113
Jørgensen OS, Delouvée A, Thiery JP, Edelman GM (1980) The nervous system specific protein D2 is involved in adhesion among neurites from cultured rat ganglia. FEBS Lett 111:39–42
Leterrier JF, Liem RKH, Shelanski ML (1982) Interactions between neurofilaments and microtubule-associated proteins: a possible mechanism for intraorganellar bridging. J Cell Biol 95:982–986
Levey AI, Wainer BH, Mufson EJ, Mesulam EJ (1983) Co-localization of acetylcholinesterase and choline acetyltransferase in the rat cerebrum. Neuroscience 9:9–22
Littauer U, Ginzburg I (1985) Expression of microtubule proteins in the brain. In: Zomzely-Neurath C, Walker WA (eds) Gene expression in brain. Wiley, New York, pp 125–156
Lüneburg U, Flohr H (1988) Effects of melanocortins on vestibular compensation. In: Vestibulospinal control of posture and movement. In: Pompeiano O, Allum JHJ (eds) Prog Brain Res, vol 76. Elsevier, Amsterdam, pp 421–429
Nyakas C, Veldhuis HD, Wied D de (1985) Beneficial effect of chronic treatment with Org. 2766 and αMSH on impaired reversal learning of rats with bilateral lesions of the parafiscular area. Brain Res Bull 15:257–265
Pallas D, Solomon F (1982) Cytoplasmic microtubule associated proteins: phosphorylation at novel sites is correlated with their incorporation into assembled microtubules. Cell 30:407–414

Richter-Landsberg C (1988) NILE glycoprotein in developing rat cerebral cells in culture. Cell Tissue Res 225:181–190
Stallcup WB, Beasley LL (1985) Involvement of the nerve growth factor-inducible large external (NILE) glycoprotein in neurite fasciculation in primary cultures of rat brain. Proc Natl Acad Sci USA 82:1276–1280
Strand FL, Smith CM (1980) LPH, ACTH, MSH and motor systems. Pharmacol Ther 11:509–533
Torda C (1952) Effects of a single injection of corticotropin (ACTH) on ammonium ion and acetylcholine content of the brain. Am J Physiol Ther 11:509–533
Verhaagen J, Edwards PM, Jennekens FGI, Schotman P, Gispen WH (1986) α-Melanocyte-stimulating hormone stimulates the outgrowth of myelinated nerve fibers after peripheral nerve crush. Exp Neurol 92:451–454
Wied D de (1969) Effects of peptide hormones on behavior. In: Ganong WF, Martini L (eds) Frontiers in neuroendocrinology. Oxford University Press, New York, pp 97–140
Wied D de, Jolles J (1982) Neuropeptides derived from proopiocortin: behavioral, physiological, and neurochemical effects. Physiol Rev 62:976–1059
Wiegant VM, Oestreicher AB, Zwiers H, Gispen WH (1985) ACTH and brain cAMP and phosphoproteins. In: Wied D de, Gispen WH, Wimersa Greidanus TJB van (eds) Neuropeptides and behavior, vol I. CNS effects of ACTH, MSH and opioid peptides. Pergamon, New York, pp 189–210
Yavin E, Yavin Z (1974) Attachment and culture of dissociated cerebral cells from rat embryo cerebral hemispheres on polylysine-coated surfaces. J Cell Biol 62:540–546
Yavin Z, Yavin E (1977) Synaptogenesis and myelinogenesis in dissociated cerebral cells from rat embryo on polylysine coated surfaces. Exp Brain Res 29:137–147
Yavin Z, Yavin E (1980) Survival and maturation of cerebral neurons on poly (L-lysine)-surfaces in the absence of serum. Dev Biol 75:454–459
Yavin E, Gil S, Guroff G, Hama T, Richter-Landsberg C (1987) Glycoconjugate metabolism, accretion and release during neurite outgrowth in normal and transformed cells of neuronal origin. In: Althaus HH, Seifert W (eds) Glial-neuronal communication in development and regeneration. Springer, Berlin Heidelberg New York, pp 303–321

Comparative Effect of Some Neurotropic Agents on Balance Compensation After Unilateral and Bilateral (Two-Staged) Labyrinthectomy in Squirrel Monkeys

M. Igarashi, M. Ishii, K. Ishikawa, and T. Himi

Dept. of Otorhinolaryngology and Communicative Sciences, Baylor College of Medicine
Houston, Texas 77030–3498, USA

1 Introduction

The dynamic repair process of body equilibrium after vestibular injury is a typical representation of nervous system plasticity. The body equilibrium system needs to re-establish a coordinated sensorimotor balance. For this purpose, the input from the remaining vestibular organs is believed to carry the primacy for balance compensation with its receptor characteristics. In addition, afferents from other sensory modalities are mobilized, modulated, and efficiently integrated at many levels of the central nervous system, particularly after the total vestibular loss; during the Bechterew's compensation. Thus, the global and behavioral measurements in animal models is advantageous, for it can assess the integrated function of the entire neural system.

It is clinically important to use any noninvasive procedure which can help the initiation, acquisition, and maintenance of balance compensation after vestibular injury. For example, the application of running-type exercise demonstrated clear acceleration of balance regainment in squirrel monkeys (Igarashi et al. 1981). Also, it has been reported that various neuropharmacological agents can enhance the balance repair, and investigations characterize the symptom modification in many animal models along the course of balance compensation (Bienhold et al. 1981; Schaefer and Meyer 1981; Flohr and Lüneburg 1982). Identifying the effective drugs, particularly in the experimental primate model, is very important for its possible generalization to man.

2 Methods

Experimental subjects used in this study series were healthy squirrel monkeys *(Saimiri sciureus)* of both sexes, approximately 2–3 years old, weighing about 700–850 g.

In order to measure behavioral locomotor balance function, the squirrel monkey platform runway test (Igarashi 1974) was used. This test measures the degree and direction of gait deviation in a natural over-ground run. The test apparatus consisted of two cubical end-boxes, connected by a 4-m-long enclosed straight runway, with longitudinally installed pressure-sensors (in a parallel fashion) under the floor. The monkeys were trained to shuttle the end-boxes; thereafter,

H. Flohr (Ed.)
Post-Lesion Neural Plasticity

they were shaped for running only on the center line until they reached the pre-procedural criterion with acceptable repeatability. The animals ran ten shuttles per trial day.

For the recording of horizontal spontaneous nystagmus (SN), the squirrel monkeys were restrained, and a standard DC electro-nystagmogram was obtained for 3 min in the dark. The maximum SPEV of the day was calculated based on analysis of five nystagmic beats with most high speeds. The eye speed calibration was obtained pre-operatively using 30 °/s constant speed optokinetic stimulus.

Sterile labyrinthectomies were performed under general anesthesia induced by intra-peritoneal injection of pentobarbital sodium, 25 mg/kg. A combined trans-meatal and trans-mastoidal approach to the vestibular end organs was used. Details were described elsewhere (Igarashi et al. 1981). Unilateral labyrinthectomy (UL) was placed on the left ear. The post-operative study for 90–100 days confirmed a complete recovery of equilibrium function. Then, the contralateral labyrinthectomy was placed on the right side.

Three monkeys each were assigned to the ACTH-(4–10) 250 µg/kg group, ACTH-(4–10) 500 µg/kg group, and control physiological saline group. Daily injections were given subcutaneously, immediately following the scheduled daily tests, for a 28-day period starting on the day of the UL and bilateral labyrinthectomy (BL). The locomotor ataxiametry was done daily and the SN was recorded every other day. For the analysis, ataxia data were pooled for every 3 days, and nystagmus data were pooled for every 6 days, except for the first 3 post-operative days. Two-way analysis of variance was used.

For the three animals in the TRH group, an osmotic mini-pump was implanted subcutaneously on the back immediately after UL and BL. The pump was filled with TRH in physiological saline solution that produced a calculated infusion rate of 250 µg/kg/day for 14 days. On the 14th postoperative day after the scheduled test, the mini-pump was removed.

3 Results

3.1 Stages of Balance Compensation

When the balance compensation of over-ground straight run was studied in squirrel monkeys after UL, about day 12 was the transition point from acute compensation stage to chronic compensation maintenance stage. This was found also in static oculomotor balance function. The speed of balance regainment prior to this transition point was faster than that in the subsequent stage. This transition of stages seemed to exist after two-staged BL, but it was less clear.

3.2 Compensation of Locomotor Balance Function

The results of locomotor ataxiametry are summarized in Fig. 1. After UL, both drugs contributed significantly to the reduction of deviation (Fig. 1 A and C). After BL, ACTH-(4–10) showed a significant effect (Fig. 1 B). The TRH group depicted some positive but nonsignificant effect (Fig. 1 D).

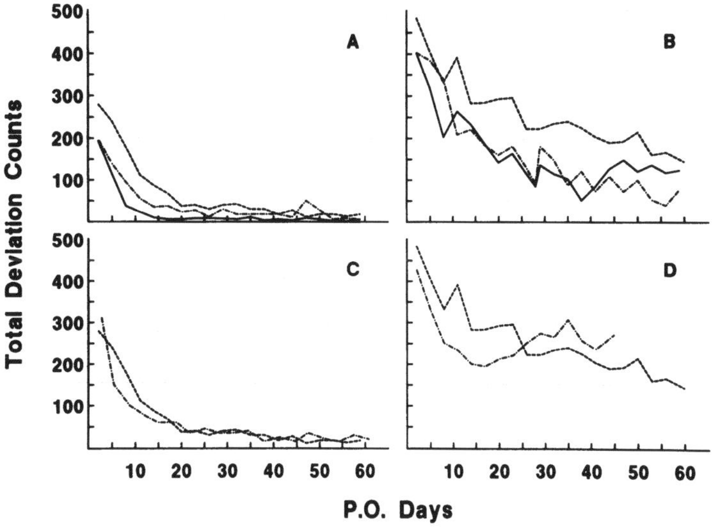

Fig. 1. Total locomotor deviation counts measured by squirrel monkey platform runway test. **A** and **C** After unilateral labyrinthectomy. **B** and **D** After second-stage (bilateral) labyrinthectomy. *Solid lines* in **A** and **B** ACTH-(4–10) 500 μg/kg(day. *Dash* and *dot lines* in **A** and **B** ACTH-(4–10) 250 μg/kg/day. *Dash* and *dot lines* in **C** and **D** TRH 250 μg/kg/day. *Dashed lines* in **A–D** no drug controls

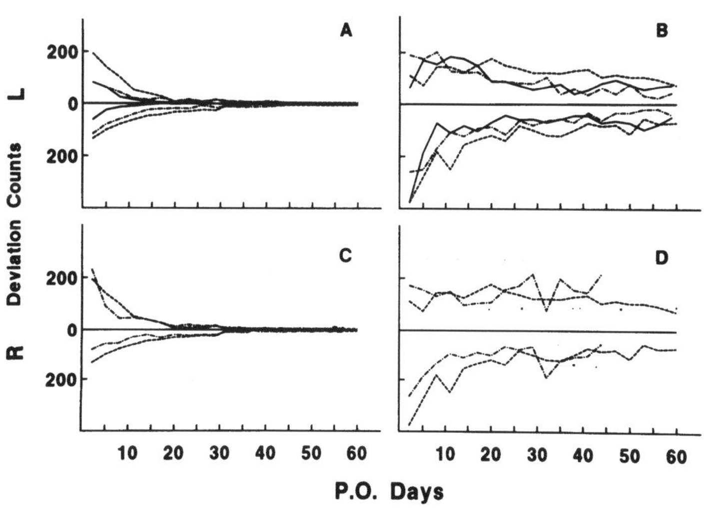

Fig. 2. Deviation counts plotted separately to left (*L*) and right (*R*). **A–D** and all lines represent the same experimental conditions as in Fig. 1

Unlike the situation after UL, all monkeys after BL showed severe difficulty and variability in locomotor performance. The time required to regain the preoperative performance criterion level was extremely long, and the decay of locomotor deviation was very slow.

When the direction of ataxia was analyzed (Fig. 2), there was a trend showing that the drug's effectiveness appeared slightly more on the deafferentated side; left side after UL, and right side after contralateral labyrinthectomy.

3.3 Compensation of Oculomotor Balance Function

Figure 3 displays the results of SN studies. Both drugs after UL, and ACTH-(4–10) after BL contributed significantly to the reduction of SPEV. Unlike the severe locomotor imbalance, SPEV of SN after BL was low and reduced to the lower level quickly. This could be reflectory to the lowered balance level of vestibular nuclei activities after UL.

Regarding the compensation pattern of SN after UL, TRH seemed to show more effect during the drug administration than that produced by ACTH-(4–10) (Fig. 3 A and C), even though the direct comparison may not be possible due to the difference in dose-effectiveness relationship. After BL, the difference between the two drugs was unremarkable (Fig. 3 B and D).

With the injections of ACTH-(4–10), the SN, both after UL and BL, showed a retained SPEV reduction after drug withdrawal during the chronic compensation maintenance stage. With TRH, there was a clear reduction during the acute compensation stage, particularly after UL. After withdrawal of the drug, however, rebound increases were found both after UL and BL, and thereafter there was no difference between the experimental and control groups. Therefore, the working mechanism of TRH should be mostly a nonorganic nature. In contrast, with ACTH-(4–10), SN did not show any rebounding phenomenon after withdrawal.

3.4 Unilateral vs. Bilateral Lesion

After UL, both drugs showed effectiveness both in vestibulo-spinal and vestibulo-oculomotor balance function during the acute compensation stage. This was clear in the vestibulo-oculomotor balance function with TRH, whereas ACTH-(4–10) showed extended effectiveness during the chronic compensation maintenance

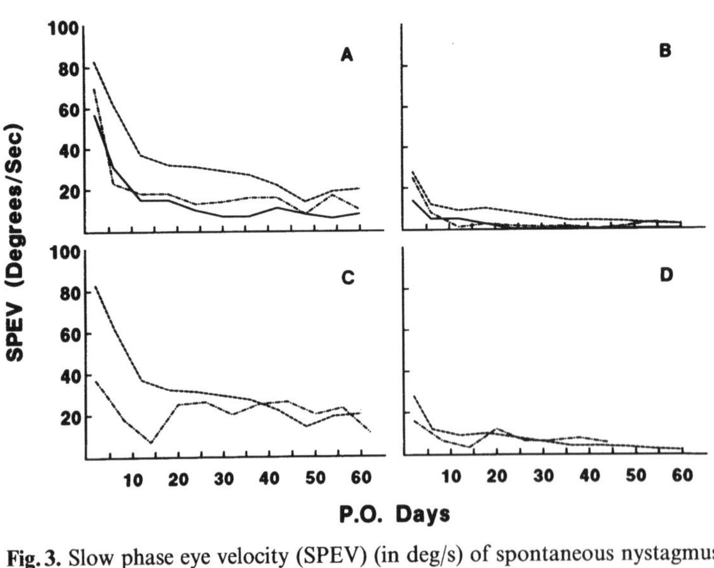

Fig. 3. Slow phase eye velocity (SPEV) (in deg/s) of spontaneous nystagmus. **A–D** and all lines represent the same experimental conditions as in Fig. 1

stage. After BL, ACTH-(4–10) depicted retained effectiveness during the chronic stage, whereas TRH showed its effect mostly in the acute compensation stage.

Generally, the drug's effect was clearer in static oculomotor balance after UL, and in reduction of locomotor deviation after BL.

4 Discussion

The squirrel monkey platform runway test is to evaluate the balance function in coordinated over-ground locomotion basically propelled by the hind legs, which has a dominant linkage to the lateral vestibulo-spinal tract. Therefore, the post-UL analysis of neuro-transmitter dynamics in this neural channel should correlate well with balance function. Indeed, our squirrel monkey study on the dynamics of GABA-like immunoreactivity in lateral vestibular nucleus (Thompson et al. 1986) showed a good correlation with locomotor balance regainment during the acute compensation stage. Thus, the positive drug effect found in locomotor balance function in this present study should be related, at least in part, to that neural network.

Flohr et al. (1981) showed that Bechterew's phenomenon in the frog was prevented following commissurectomy, thus confirming Bechterew's compensation is related to the translateral neural modulation. Therefore, any positive drug effect found in the present study after BL should be related to that neuroanatomical circuitry.

ACTH-(4–10) given into the frog's dorsal lymph sac (Flohr and Lüneburg 1982) enhanced the balance compensation process (in head deviation) after UL. When we used this drug via the systemic route to squirrel monkeys, the balance compensation was enhanced both in vestibulo-spinal and vestibulo-oculomotor function, both after UL and BL (Igarashi et al. 1985; Ishii and Igarashi 1987). Thus, ACTH-(4–10) could penetrate through the barrier, at least to a certain extent, in this animal model, and worked on the common neural network of these two subsystems, probably vestibular nuclei.

With ACTH-(4–10), the monkeys showed less variability in the locomotor balance data, indicating the existence of the attentive effect of this agent. According to de Wied (1965, 1969), ACTH and its analogs influence learning, and acquisition and maintenance of adaptive behavior. Also, these neuropeptides regulate the spinal excitability (Strand et al. 1976; Krivoy and Zimmerman 1977; Gonzales and Strand 1981).

It has been known that any sort of alerting maneuver such as noise application increases the SPEV of SN in animals after UL. Therefore, the SN "reduction" by ACTH-(4–10) should indicate that the drug's effectiveness is more on neural repair mechanism, overcoming arousal increase. Also, no clear rebound increase was found in SN after the termination of ACTH-(4–10) injections, thus indicating that the organic repair of involved neural network is solidified by the time of drug withdrawal.

After BL, the effectiveness of ACTH-(4–10) to the plastic remodeling underlying the balance compensation process can work only with other modality inputs. ACTH may influence the production of brain cyclic AMP (Wiegant et al. 1979), synthesis of brain proteins and RNA, and activation of adenylate cyclase (Schotman et al. 1976, 1980; Dunn and Schotman 1981). Also, ACTH-(4–10) enhances the growth of regenerating axons of peripheral nerves and brings a more rapid recovery of functions (Strand and Kung 1980; Bijlsma et al. 1981).

TRH is broadly distributed in the central nervous system, and exerts behavioral effects (Yarbrough 1979; Jackson 1982). Our study of squirrel monkey locomotometry indicated that the dopanergic system was involved in mediating the TRH's effect (Ishii and Igarashi 1986). Indeed, the dopanergic-releasing action of TRH and its analogs have been reported (Heal and Green 1979; Miyamato et al. 1979; Bennet et al. 1981). Thus, TRH could increase arousal (Stanton et al. 1980), and then motor activity which allows squirrel monkeys to perceive more error signals, and thus, to promote neurochemical modification. The effectiveness of voluntary movement in the vestibular compensation was already demonstrated through the physical exercise and other sensorimotor activation studies (Lacour et al. 1976; Igarashi et al. 1981; Petrosini 1983).

It was reported that TRH is metabolized with a very short half-life (Bassiri and Utiger 1973; Brewster and Waltham 1981). In the present study, after removing the mini-pump from the monkeys, the SPEV of SN showed a clear rebound. Thus, TRH was working more in short-term plasticity of motor output, and less in organic repair.

The concentration of TRH is reported to be high in the cerebellum (Winters et al. 1974). Furthermore, Pacheco et al. (1981) have shown uneven distribution of TRH in the cerebellum; highly concentrated in the paraflocculi, flocculi, nodulus and deep cerebellar nuclei, but not in the cerebellar hemispheres and vermis. It has been known that the flocculus and nodulus are closely linked to vestibulo-oculomotor system (Ito et al. 1974; Haddad et al. 1977; Courjon et al. 1982; Igarashi and Ishikawa 1985). Even though recovery patterns of SN and locomotor ataxia are not directly comparable to each other, TRH showed more effective control on SN, possibly through the neural network involving the flocculus and nodulus.

In summarizing, the administration of ACTH-(4–10) or TRH modified the characteristics and time courses of vestibulo-spinal and vestibulo-oculomotor balance compensation both after UL and BL. However, there seemed to be some difference about their working mechanisms. In general, ACTH-(4–10) appeared to contribute more actively to the organic repair or remodeling of the neural network related to the motor output system, whereas TRH could contribute more to short-term functional plasticity in motor outputs.

Acknowledgements. This study was supported in part by NINCDS grant NS-10940 and NASA grant NAG-2-289. Technical assistance by K. Schmidt, C. Castillo, P. Martin, M. Del Rosario, R. Buentello, S. Charlton and J. Brown is deeply appreciated.

References

Bassiri RM, Utiger RD (1973) Metabolism and excretion of exogeneous thyrotropin-releasing hormone in humans. J Clin Invest 52:1616–1619

Bennet GW, Marsden CA, Metcalf G et al. (1981) Analogues of thyrotropin-releasing hormone (TRH) stimulate in vitro release of endogenous dopamine from rat brain regions. Br J Pharmacol 74:227–228

Bienhold H, Abeln W, Flohr H (1981) Drug effects on vestibular compensation. In: Flohr H, Precht W (eds) Lesion-induced neuronal plasticity in sensorimotor systems. Springer, Berlin Heidelberg New York, p 265

Bijlsma WA, Jennekins FGI, Schotman P, Gispen WH (1981) Corticotrophin (ACTH) like peptides stimulate peripheral nerve regeneration. In: Mohn van Hof (ed) Functional recovery from brain damage. Elsevier, Amsterdam, p 441

Brewster D, Waltham K (1981) TRH degradation rates vary widely between different animal species. Biochem Pharmacol 30:619–622

Courjon JH, Flandrin JM, Jeannerod M et al. (1982) The role of the flocculus in vestibular compensation after hemilabyrinthectomy. Brain Res 239:251–257

Dunn AJ, Schotman P (1981) Effects of ACTH and related peptides on cerebral RNA and protein synthesis. Pharmacol Ther 12:353–372

Flohr H, Lüneburg U (1982) Effects of ACTH-(4–10) on vestibular compensation. Brain Res 248:169–173

Flohr H, Bienhold H, Abeln W, Macskovics I (1981) Concepts of vestibular compensation. In: Flohr H, Precht W (eds) Lesion-induced neuronal plasticity in sensorimotor systems. Springer, Berlin Heidelberg New York, p 153

Gonzales ER, Strand FL (1981) Neurotropic action of MSH/ACTH-(4–10) on neuromuscular function in hypophysectomized rats. Peptides 2(1):107

Haddad GM, Friendlich AR, Robinson DA (1977) Compensation of nystagmus after VIIIth nerve lesions in vestibulo-cerebellectomized cats. Brain Res 135:192–196

Heal DJ, Green AR (1979) Administration of thyrotropin releasing hormone (TRH) to rats releases dopamine in N. accumbens but not N. caudatus. Neuropharmacology 18:23–31

Igarashi M (1974) Squirrel monkey platform runway test. Acta Oto-Laryngol 77:284–288

Igarashi M, Ishikawa K (1985) Post-labyrinthectomy balance compensation with preplacement of cerebellar vermis lesion. Acta Oto-Laryngol 99:452–458

Igarashi M, Levy JK, O-Uchi T et al. (1981) Further study of physical exercise and locomotor balance compensation after unilateral labyrinthectomy in squirrel monkeys. Acta Oto-Laryngol 92:101–105

Igarashi M, Ishikawa K, Ishii M et al. (1985) Effect of ACTH-(4–10) on equilibrium compensation after unilateral labyrinthectomy in the squirrel monkey. Eur J Pharmacol 119:239–242

Ishii M, Igarashi M (1986) Effect of thyrotropin-releasing hormone on vestibular compensation in primates. Am J Oto-Laryngol 7:177–180

Ishii M, Igarashi M (1987) Effect of ACTH-(4–10) on Bechterew's compensation in squirrel monkeys. ORL 49:87–92

Ito M, Shiida T, Yagi N et al. (1974) Visual influence on rabbit's horizontal vestibulo-ocular reflex presumably effected via the cerebellar flocculus. Brain Res 65:170–174

Jackson I (1982) Thyrotropin-releasing hormone. N Engl J Med 306:145–155

Krivoy WA, Zimmermann E (1977) An effect of α-melanocyte stimulating hormone (α-MSG) on motoneurons of cat spinal cord. Eur J Pharmacol 46:315–322

Lacour M, Roll JP, Appaix M (1976) Modifications and development of spinal reflexes in the alert baboon (*Papio papio*) following an unilateral vestibular neurotomy. Brain Res 133:255–269

Miyamato M, Narumi S, Nagai Y et al. (1979) Thyrotropin-releasing hormone: hyperactivity and mesolimbic dopamine system in rats. Jpn J Pharmacol 29:335–347

Pacheco MF, McKelvy JF, Woodward DJ et al. (1981) TRH in the rat cerebellum: I. Distribution and concentration. Peptides 2:277–282

Petrosini L (1983) Compensation of vestibular symptoms in hemilabyrinthectomized guinea pigs: role of the sensorimotor activation. Behav Brain Res 8:335–342

Schaefer KP, Meyer DL (1981) Aspects of vestibular compensation in guinea pigs. In: Flohr H, Precht W (eds) Lesion-induced neuronal plasticity in sensory motor systems. Springer, Berlin Heidelberg New York, p 197

Schotman P, Reigh MEA, Wimersma Greidanus TB van, Gispen W, Wied D de (1976) Hypothalamic and pituitary peptide hormones and the central nervous system. With special references to the neurochemical effects of ACTH. In: Gispen (ed) Molecular and functional neurobiology. Elsevier, Amsterdam, p 309

Schotman P, Heuven-Nolson D van, Gispen WH (1980) Protein synthesis in a cell-free system from rat brain sensitive to ACTH-like peptides. J Neurochem 34:1661–1670
Stanton TL, Winokur A, Beckman AL (1980) Reversal of natural CNS depression by TRH action in the hippocampus. Brain Res 181:470–475
Strand FL, Kung TT (1980) ACTH accelerates recovery of neuromuscular function following crushing of peripheral nerve. Peptides 1:135–138
Strand FL, Cayer A, Gonzales E, Stoboy H (1976) Peptide enhancement of neuromuscular function: Animal and clinical studies. Pharmacol Biochem Behav 5(1):179–187
Thompson GC, Igarashi M, Cortez AM (1986) GABA imbalance in squirrel monkey after unilateral vestibular end-organ ablation. Brain Res 370:182–185
Wied D de (1965) The influence of the posterior and intermediate lobe of the pituitary and pituitary peptides on the maintenance of a conditioned avoidance response in rats. Int J Neuropharmacol 4:157–167
Wied D de (1969) Effects of peptide hormones on behavior. In: Ganong, Martini (eds) Frontiers in neuroendocrinology, vol 1. Oxford University Press, New York, p 97
Wiegant VM, Dunn AK, Schotman P, Gispen WH (1979) ACTH-like neurotrophic peptides: possible regulators of rat brain cyclic AMP. Brain Res 168:565–584
Yarbrough GG (1979) On the neuropharmacology of thyrotropin releasing hormone (TRH) Prog Neurobiol 12:291–312

Glucocorticoids and Lesion-Induced Plasticity

E. R. de Kloet, J. A. M. van Eekelen, and M. A. Seger
Rudolf Magnus Institute for Pharmacology, Medical Faculty, University of Utrecht, Padvalaan 8, NL-3584 CH Utrecht, The Netherlands

1 Introduction

Glucocorticoids (GC) are secreted by the adrenal cortex in response to stressful stimuli. These include disturbances in the environment, psychological factors, metabolic insults and tissue damage. The GC elevate blood glucose level, promote fluid excretion, alter vascular tone, have an immunosuppressive and anti-inflammatory action, and influence cognition, mood and affect (Selye 1976). GC actions are catabolic and inhibit growth, cell division, protein synthesis, and the uptake of glucose and amino acids. GC control synthesis, release and action of intercellular mediators such as prostaglandins, leukotrenes, lymphokines and a variety of other peptides, neurotransmitters, and hormones that are released as immediate mediators of the stress response. It has been postulated that these GC actions all suppress the immediate defense mechanisms to stress, which would themselves be damaging if left uncontrolled (Munck et al. 1984). In line with this view is the notion of Tausk (1951) that "Cortisol treatment is appropriate, where the defensive reactions of the organism cause more damage than the agent against which they defend". Metaphorically, the action of GC could be considered "as limiting water damage caused by the fire brigade" (Tausk 1951). However, when an elevated level of GC persists, then the catabolic effects of the steroid prevail with detrimental consequences for the organism, e.g., steroid diabetes, hypertension, myopathy, and cognitive disturbances.

GC are secreted as the last step in the chain of the humoral events that are organized in the hypothalamus-pituitary-adrenal (HPA) system. Ascending neural circuits transduce the stress stimulus to hypothalamic paraventricular neurons (PVN), that release the corticotrophin-releasing factor (CRF) into the portal vasculature. CRF activates the pro-opiomelanocortin (POMC) cells in the anterior pituitary and subsequently ACTH stimulates GC secretion. GC feeds back on brain and pituitary to suppress the stress-activated HPA system. It is, therefore, not surprising that GC receptors or Type 2 receptors are present in abundance in the CRF neurons and in the anterior pituitary POMC cells. Monoaminergic and peptidergic neurons of the neural stress circuitry also contain high amounts of Type 2 GC receptors. The hippocampus, one of the limbic structures, contains Type 2 receptors in abundance in the granular neurons of the dentate gyrus and the pyramidal neurons. In addition, glial cells contain Type 2 receptors (Reul and De Kloet 1985; Fuxe et al. 1985; Van Eekelen et al. 1987).

H. Flohr (Ed.)
Post-Lesion Neural Plasticity

The brain and pituitary also contain a mineralocorticoid receptor or Type 1 receptor. In the brain this receptor has an almost exclusive localization in neurons of the limbic system (hippocampus, dorsal septum). A peculiar property of these neuronal Type 1 receptors is that they respond with stringent specificity to very low doses of the naturally occurring GC. Accordingly, the Type 1 receptors are thought to be involved in tonic control (synchronization and coordination) of circadian-driven brain mechanisms, rather than in feedback actions mediated by Type 2 on stress-induced responses (De Kloet et al. 1986; De Kloet and Reul 1987; McEwen et al. 1986; Funder and Sheppard 1987). Recently, the primary structure of Type 1 and Type 2 corticoid receptors has been resolved (Hollenberg et al. 1985; Gustaffson et al. 1987; Arrizza et al. 1987).

In this chapter, we will focus on the role of GC in morphological, chemical, and functional responses to denervation of the hippocampus and the intermediate lobe of the pituitary (IL). One of the consequences is the induction of the Type 2 receptors which develops slowly after the lesion. This is in particular a remarkable phenomenon for the IL, which after denervation contains Type 2 receptors that become active in the regulation of POMC gene expression and biosynthesis.

2 Glucocorticoids and Recovery from Lesions in the Hippocampus

It has been postulated that high blood levels of GC have a progressive catabolic effect on neurons, which results in increased vulnerability to neurotoxic and metabolic insults. This hypothesis is based on experiments involving exposure of hippocampal neurons to various excitotoxins. These neurons appeared more damaged in animals treated with high doses of GC than in adrenalectomized (ADX) animals (Sapolski and Pulsinelli 1985). In addition, elevated GC level in blood also increased ischemic damage to hippocampal neurons, while the reverse was observed after ADX (Sapolski and Pulsinelli 1985).

GC have been shown to influence reactive growth in the hippocampus after lesions of the entorhinal cortex. When analyzed 2 weeks after the lesion, high doses of GC administered daily to these rats had suppressed outgrowth and sprouting of commissural-associational fibers in the dentate gyrus and had enhanced the hypertrophy of glial cells (Scheff et al. 1980; Scheff and Cotman 1982). Subsequent studies have revealed that in young rats circulating levels of GC comparable to those found in the aged rat already resulted in marked suppression of the adrenergic sprouting response following lesion of the fimbria-fornix bundle (De Kosky et al. 1984). They have suggested that high GC levels are an important factor that contribute to the slowing down of nerve regeneration with advancing age. The role of glial cells in the GC action resulting in reduced sprouting is conjectural. However, neuronal degeneration and astrocyte hypertrophy is characteristic of the aging hippocampus (Landfield et al. 1978). Peripheral nerve regeneration also slowed down after GC administration (Strand and Smith 1986). How GC might act is not known. Either they may affect axonal transport of factors regulating axonal growth, or have a catabolic influence on cell metabolism.

While high doses of GC may impair recovery of brain function following damage, low levels of GC seem to be permissive for regeneration. ADX, for instance, suppresses homotypical sprouting of serotonergic fibers in the hippocampus following lesions of the cingulum bundle and normalization occurs following replacement with physiological doses of corticosterone (Zhou and Azmitia 1985). These effects of GC on sprouting of damaged neurons are reminiscent of the developmental effects of GC. High amounts of GC given post-natally depress neurogenesis, while constant low levels are an essential requirement for "normal" post-natal brain development as expressed in growth and differentiation of monoaminergic systems (Bohn 1984; Bohn et al. 1986).

3 Lesion-Induced Glucocorticoid Receptor Plasticity in the Hippocampus

In a recent series of experiments we have used unilateral removal of the dorsal hippocampus (HX) as a model system to study GC receptor plasticity in the remaining contralateral hippocampus lobe (see Fig. 1). In the case of unilateral ablation of the dorsal hippocampus, the innervation of commissural-associational fibers to the contralateral lobe is removed. Morphological observations on the dentate gyrus have shown that as a consequence astrocytes hypertrophy and nerve terminals degenerate. The appearance of the scavenging glial cells and the disappearance of nerve terminals during the first 5 days is followed by a period of very active synaptogenesis from 6 to 15 days post-lesion. Regeneration then

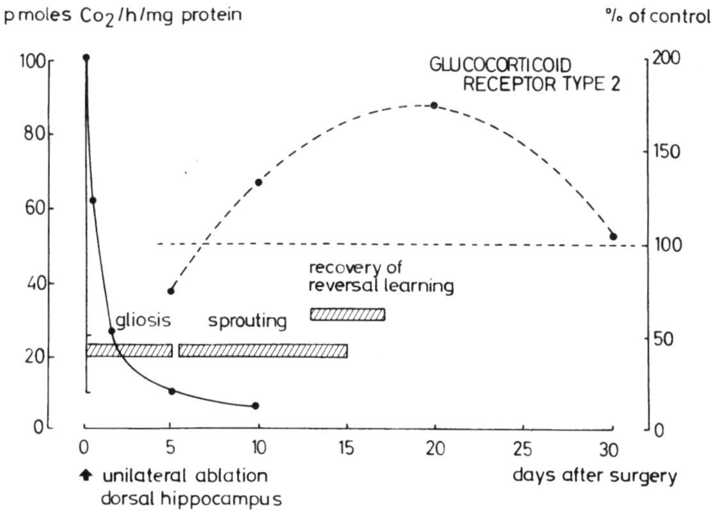

Fig. 1. Time course of effect of unilateral removal of dorsal hippocampus on morphological, biochemical and behavioural responses associated with contralateral hippocampus lobe. Pronounced gliosis at day 1–5 and most vigorous sprouting of commissural associational fibers at day 6–15 post-lesion. (McWilliams and Lynch 1980). Recovery of reversal learning in a T-maze occurs between day 13–17 depending on the presence of adrenocortical secretion (de Ronde et al. 1986). A tenfold increase in ornithine decarboxylase (ODC) activity is observed at 6 h post-lesion (de Kloet et al. 1983a). Glucocorticoid Type 2 receptors are decreased 25% in capacity at 5 days and increased 75% at 20 days post-lesion (data are given as % of control-B_{max} of Scatchard analyses)

slowly continued and was almost complete after 2 months (McWilliams and Lynch 1979). In the next paragraphs we will discuss two biochemical responses that take place following the HX lesion.

The first response regards ornithine decarboxylase (ODC) which is the rate-limiting enzyme in the biosynthesis of polyamines. The enzyme can be considered as a marker for protein synthesis, cell division and growth and it has been proposed, therefore, as an index of enhanced cell proliferation in the brain (Kohsaka et al. 1982; Goertz 1979). ODC responds to unilateral HX with a rapid rise in activity in the contralateral lobe reaching a tenfold higher level at 6 h after the lesion, and then slowly returning to baseline levels during the next 2 days. In the absence of the adrenals the effect of HX on ODC was more potent and more prolonged, lasting up to 5 days post-HX. Chronic administration of low doses of dexamethasone to ADX rats attenuates the HX-induced ODC response in these rats. On the other hand, high doses of dexamethasone induce ODC activity. From these experiments it can be concluded that after HX, ODC is a marker of the biochemical responses taking place in the contralateral lobe and that GC have a dual effect on the enzyme, i.e., attenuation of lesion-induced ODC rise, and independently, induction of ODC (De Kloet et al. 1983 a). The actual cellular localization of the ODC response is not known, but the time course of the morphological events and the ODC response suggests requirement of activation of a glial cell Type 2 GC receptor.

The second response involving GC regards plasticity in the Type 2 GC receptor system itself. Type 2 receptor capacity showed a decrease the first week after HX, followed by a compensatory increase in the subsequent 3 weeks. Maximal increase was 72% at 3 weeks post-lesion, reaching pre-surgery level again at the fourth week. The lesion-induced rise in Type 2 receptor number was transient in animals with an intact pituitary-adrenal system, but appeared to remain for up to 3 months in ADX rats (Nyakas et al. 1981; De Ronde et al. 1986). After HX, reversal learning behavior in a T-shaped maze was disturbed during the first 2 weeks, but restored the following week. ADX seemed to facilitate recovery of the deficit. Thus, unilateral HX induces a period of considerable plasticity in Type 2 receptors in particular during the period of active synaptogenesis. This receptor plasticity coincides with the recovery of a lesion-induced deficit in hippocampus-associated behavior.

Similar relationships between Type 2 receptor capacity and behavioral performances were observed after selective lesioning of distinct cell fields of the hippocampus with local kainic acid administration (Nyakas et al. 1983; Handelmann and Olton 1981). The lesion-induced Type 2 receptor plasticity has also been observed following selective lesioning of neurotransmitter pathways innervating the hippocampus. Two weeks after lesioning of the raphe-hippocampus serotonin pathway with the neurotoxin 5,7 dihydroxytryptamine, the number of corticosteroid receptors in the hippocampus showed a twofold increase (de Kloet et al. 1983 b). In other cells containing steroid receptors, similar effects of neurotransmitters have been described. For instance, changes in noradrenergic input alter the number of progesterone receptors in the hypothalamus (Nock et al. 1981; Blaustein and Lether 1987) and of estrogen receptors in the pineal gland (Cardinali et al. 1975). Thus, neurotransmitters may influence the receptor content of steroid-sensitive cells, and this may constitute an important mechanism by which neurotransmitters modulate steroid-dependent cellular events.

4 Glucocorticoid Receptor Plasticity in the Pituitary: Functional Consequences

A striking example of Type 2 receptor plasticity was recently discovered in the intermediate lobe of the rat pituitary. As pointed out before, the IL contains cells which synthesize POMC, as is the case in the adjacent corticotrophes of the anterior lobe (AL). Both POMC-producing cell types differ, however, in the regulation of POMC gene expression, biosynthesis, and release, as well as in the profile and biological activity of the end-products released (Eipper and Mains 1980). POMC synthesis in the AL is stimulated by CRF, inhibited by GC and the AL contains an abundance of Type 2 GC receptors. In contrast, the IL has no detectable GC receptors, does not respond to circulating GC, and is instead under control of dopaminergic, GABA-ergic and serotonergic fibers arising from the hypothalamus. Dopamine and GABA tonically inhibit the release of IL products, while serotonin has a stimulatory effect (Jones et al. in press).

It has been shown that the Type 2 receptor could be induced following denervation, i.e., transection of the pituitary stalk (Antakly et al. 1985). Induction of the receptor was also observed in pituitary grafts to the kidney capsule and in IL monolayer cultures. It was subsequently suggested that the Type 2 receptor induction may be related to removal of the hypothalamic dopaminergic input to the IL cells, since dopamine inhibited the expression of the receptor in primary culture (Antakly et al. 1987).

These findings prompted us to investigate whether the induced Type 2 receptor is involved in the regulation of POMC gene expression. In order to reduce the chance of possible infarcts or tissue necrosis, deafferentation of the anterior hypothalamus was chosen, which is known to sever the neural input to the IL. This lesion led to a gradual induction of the GC receptor. The GC receptor was visualized with a monoclonal antibody, which detects the steroid-activated receptor (Van Eekelen et al. 1987). The time course of receptor induction was similar to that observed previously in the brain. Thus, maximal immunostaining was observed at 3 weeks after the lesion. Two series of experiments were performed. First, 1 week after the deafferentation the animals were killed and the neurointermediate lobes (NILs) were incubated in vitro for 24 h with tritiated amino acids. A two- to fourfold increase of incorporation of tritium into HPLC-separated POMC peptides was observed with no concomittant change in the POMC peptide content of the NIL (Seger et al. 1987a). Adrenalectomy attenuated the lesion-induced increase in POMC peptide labeling, which was increased again upon chronic corticosterone replacement (see Table 1).

Secondly, the POMC mRNA was measured in the NIL after hypothalamic deafferentation followed by ADX, with and without GC replacement. As shown in Table 1, lesion and ADX induced a decrease in POMC mRNA, which slowly developed over a period of 3 weeks post-surgery and which can be restored to normal by chronic corticosterone replacement. As can be expected, the lesion itself

Table 1. Effect of lesion and chronic GC treatment on POMC gene expression and biosynthesis

Treatment	POMC mRNA in		POMC mRNA in		POMC biosynthesis in neurointermediate lobe
	Anterior 1 week	Pituitary 3 weeks	Intermediate 1 week	Pituitary 3 weeks	
A) Sham-lesion − GC	100 ± 16	100 ± 18	100 ± 10	100 ± 19	100 ± 21
B) Sham-lesion + GC	84 ± 10	3 ± 1**	135 ± 20	62 ± 8	95 ± 16
C) Lesion − GC	88 ± 11	27 ± 8**	120 ± 18	28 ± 5**	34 ± 2*
D) Lesion + GC	51 ± 5**	21 ± 5	124 ± 21	70 ± 10**	82 ± 14**

All animals were adrenalectomized 3 days after lesion and received 100 mg corticosterone (+GC) or cholesterol (−GC) pellets subcutaneously. Significance levels were determined by Student's t-test as follows: B or C vs. A, D vs. C (* $p<0.05$, ** $p<0.01$). The POMC mRNA levels represent the ratio of POMC mRNA to actin mRNA, which served as an internal control for the dot-blot assay. The POMC biosynthesis index was calculated from the incorporation of tritiated amino acid into HPLC-separated POMC peptides, as described in Seger et al. 1987a. Both sets of data are expressed as the percent ± SE relative to group A which was set at 100%.

also reduced POMC mRNA levels in the AL since neurosecretion of CRF and several other ACTH secretogogues is interrupted. However, the effect of ADX and corticosterone replacement, i.e., a decrease in POMC mRNA levels, persisted in the AL following the hypothalamic lesion. Therefore, while corticosterone replacement for 3 weeks slowly decreases POMC mRNA level in the AL, the corticosterone replacement induced much to our surprise a pronounced increase in the POMC mRNA level in the NIL after lesion (see Table 1; Seger et al. 1987 b).

These findings are reminiscent of recent observations regarding the regulation of enkephalin biosynthesis by glucocorticoids. Denervation of the adrenal chromaffin cells increases pro-enkephalin mRNA and enkephalin content of the medulla. This effect is attenuated following hypophysectomy or when the medullae were grown in tissue culture in the absence of corticosteroids. Indeed, subsequent experiments showed that corticosteroids displayed a tonic permissive effect on enkephalin biosynthesis, that becomes apparent upon alterations in nerve impulse flow (LaGamma and Adler 1987; Yoburn et al. 1987). Interestingly, the co-localized adrenalin appears also to be under the control of GC via a stimulating effect on tyrosine hydroxylase (TH) and phenyl ethanolamine transferase (PNMT) (Wurtman and Axelrod 1966). Accordingly, the differential control of adrenalin (effector) and enkephalin (modulator) by GC and depolarizing stimuli suggests a finely tuned regulatory mechanism.

Taken together, these novel findings have dramatically altered our current understanding of GC action. Denervation induces Type 2 receptors and responsivity to GC in tissues normally not responsive to the steroid hormone. Moreover, GC may positively regulate the POMC gene in the IL, which is on the contrary subjected to negative control in the AL. A number of exciting questions can now be addressed. For instance, is corticosterone affecting another structural element of the POMC gene in the IL or is regulation exerted rather by tissue specific trans-acting factors? Is the induction of GC receptors and the enhanced synthesis of trophic signals (see Gispen et al. this Vol.) released after processing of POMC a characteristic of regenerating and developing tissue? Recent evidence suggests that during development the Type 2 GC receptor may indeed be transiently induced in some neurons and then disappears when neural input to the receptor containing cells is established (Van Eekelen et al. 1987; Rosenfeld et al. 1988 a, b).

5 Concluding Remarks

In this chapter we have discussed three important aspects of GC action on neural tissue after lesion. First, the response of cells to the lesion depends on the concentration of GC. High levels of circulating GC have a catabolic influence on neurons, in particular those in the hippocampus. Moreover, high levels of GC results in a down-regulated Type 2 receptor number, which provides a smaller feedback signal of the GC to restore homeostasis. Chronically elevated GC levels are, therefore, detrimental for neuronal survival during excitotoxic and metabolic challenges.

Second, high levels of GC impair sprouting and regeneration. Accordingly, when GC treatment is applied for management of brain edema or inflammatory processes, the potential negative effects of chronically high GC on recovery from brain damage should be an important consideration in clinical practice. High GC levels are encountered in patients suffering from Cushing's disease or after exposure to chronic GC therapy. In addition, chronic inability to cope with environmental challenges (stress) or aging result in elevated GC level. Such conditions may be expected to impair neuronal survival, outgrowth, and sprouting of damaged neurons. On the other hand, growth and differentiation of, for instance, monoaminergic systems, are stimulated by low physiological amounts of GC. Apparently, some of these developmentally dependent trophic (permissive) influences of corticosterone, i.e., on the sprouting of serotonin neuron return during adulthood.

The third aspect is concerned with the induction of the Type 2 GC receptor in the denervated cells. The phenomenon has been observed so far in hippocampus and NIL, but may well be a general molecular response to damage. The response is quite selective for the Type 2 receptor, which is the receptor that mediates the feedback action of GC. Although in hippocampus cells the Type 1 receptor appears to be co-localized with Type 2 (Van Eekelen et al. in press), the density of Type 1 – which is the GC receptor involved in tonic control of brain function (see de Kloet and Reul 1987) – shows only small lesion-induced plasticity. The induction of the Type 2 receptor has a slow onset. Receptor increase is highest during the period of most active synaptogenesis. In intact animals the receptor increase is transient. ADX results in persistently elevated receptor level. It should be emphasized that the receptor plasticity is observed post-synaptically (see Fig. 2). Our most recent data suggest a function of newly induced receptor

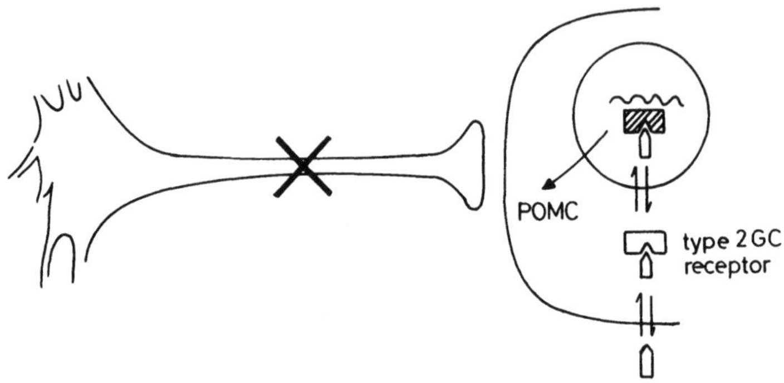

Fig. 2. Long-term (1–4 weeks) post-lesion plasticity of Type 2 GC receptor and POMC gene expression in denervated cell. In hippocampus Type 2 GC receptors increase in number. In the intermediate lobe of the pituitary Type 2 GC receptors are induced, while tissue was previously devoid of receptor. Receptors respond to GC with stimulation of POMC gene expression (cf. blockade of POMC gene expression in the anterior lobe). Positive control of GC on POMC gene expression is thought to make trophic peptides available in the denervated cells. *GC* glucocorticoid; *POMC* pro-opiomelanocortin; □ activated receptor binds to DNA and stimulates gene transcription

in recovery from tissue damage. Increased number of receptors will permit a larger GC signal to the denervated cells at the time reinervation progresses. Such a molecular response seems at odds with the (supraphysiological) above-mentioned detrimental effects of GC on axonal sprouting. However, the local postsynaptic change in receptor concentration may provide a more selective and physiological GC signal for tissue recovery. Such a signal may be needed in response to higher energy demand of the denervated cell in the process of re-innervation.

The most dramatic example of Type 2 receptor induction is in the IL, a tissue which is before denervation devoid of these receptors and not directly under control of GC. Following denervation the Type 2 GC receptors are induced and become functional in the regulation of POMC gene expression. In contrast, however, to the AL where GC suppresses POMC synthesis the steroid enhances POMC gene expression in the IL. Future research will reveal whether the receptor induction and the stimulation of the synthesis of trophic POMC peptides (see Gispen this Vol.) may be implicated in the repair of neuronal innervation and be required for survival of the denervated cell.

References

Antakly T, Sasaki A, Liotta AS, Palkovits M, Krieger DT (1985) Induced expression of the glucocorticoid receptor in the rat intermediate pituitary lobe. Science 229:277–279

Antakly T, Mercille S, Cote JP (1987) Tissue specific dopaminergic regulation of the glucocorticoid receptor in the rat pituitary. Endocrinology 120:1558–1562

Blaustein JD, Lether B (1987) Noradrenergic regulation of cytosol estrogen receptors in female rat hypothalamus: possible role of α_2-noradrenergic receptors. Brain Res 404:51–57

Bohn MC (1984) Glucocorticoid induced teratologies of the nervous system. In: Yanai J (ed) Neurobehavioural teratology. Elsevier, Amsterdam, pp 365–387
Bohn MC, Goldstein M, Black IB (1986) Expression and development of phenylethanol amine N-methyl transferase (PNMT) in rat brain stem: studies with glucocorticoids. Dev Biol 114:180–193
Cardinali DP, Nagle AC, Rosner JM (1975) Control of estrogen and androgen receptors in the rat pineal gland by catecholamine neurotransmitters. Life Sci 16:93–106
Eekelen JAM van, Kiss JZ, Westphal HM, Kloet ER de (1987) Immunocytochemical study on the intracellular localization of the Type 2 glucocorticoid receptor in the rat brain. Brain Res 436:120–128
Eekelen JAM van, Rosenfeld P, Levine S, Westphal HM, Kloet ER de (1987) Postnatal disappearance of glucocorticoid receptor immunoreactivity in the suprachiasmatic nucleus of the rat. Neurosci Res 1:129–133
Eipper BA, Mains RE (1980) Structure and biosynthesis of proadrenocorticotropin/endorphin and related peptides. Endocr Rev 1:1–27
Funder JW, Sheppard K (1987) Adrenocortical steroids and the brain. Ann Rev Physiol 49:397–411
Fuxe K, Wikström AC, Ökret S et al. (1985) Mapping of glucocorticoid receptor immunoreactive neurons in the rat tel- and diencephalon using a monoclonal antibody against rat liver glucocorticoid receptor. Endocrinology 117:1803–1812
Goertz B (1979) Effect of polyamines on cell-free protein synthetizing systems from rat cerebral cortex, cerebellum, and liver. Brain Res 173:125–135
Gustaffson JA, Carlstedt-Duke J, Poellinger L et al. (1987) Biochemistry, molecular biology and physioslogy of the glucocorticoid receptor. Endocr Rev 8:185–234
Handelmann GE, Olton DS (1981) Spatial memory following damage to hippocampal CA_3 pyramidal cells with kainic acid: impairment and recovery with preoperative training. Brain Res 217:41–58
Hollenberg SM, Weinberger C, Ong ES et al. (1985) Primary structure and expression of a functional human glucocorticoid receptor cDNA. Nature 318:635–640
Jones M, Gillham B Factors involved in the regulation of ACTH/βLPH. Physiol Rev (in press)
Kloet ER de, Reul JMHM (1987) Feedback action and tonic influence of corticosteroids on brain function: a concept arising from the heterogeneity of brain receptor systems. Psychoneuroendocrinology 12:85–105
Kloet ER de, Angelucci L, Kovacs GL, Versteeg DHG, Bohus B (1983 a) The glucocorticoid receptor system in rat hippocampus: a site of integration between body and mind. In: Endroczi E, Wied D de, Angelucci L, Scapagnini U (eds) Integrative neurohumoral mechanisms. Elsevier, Amsterdam, pp 147–155
Kloet ER de, Cousin MA, Veldhuis HD, Voorhuis THD, Lando D (1983 b) Glucocorticoids modulate the response of ornithine decarboxylase to unilateral removal of the dorsal hippocampus. Brain Res 275:91–98
Kloet ER de, Reul JMHM, Ronde FSW de, Bloemers M, Ratka A (1986) Function and plasticity of brain corticosteroid receptor systems: action of neuropeptides. J Steroid Biochem 25:723–731
Kohsaka S, Heacock AM, Klunger PD, Porta R, Agranoff B (1982) Dissociation of enhanced ornithine decarboxylase activity and optic nerve regeneration in goldfish. Dev Brain Res 4:149–156
Kosky ST de, Scheff SW, Cotman CW (1984) Elevated corticosterone levels: a possible cause of reduced axon sprouting in aged animals. Neuroendocrinology 38:33–38
LaGamma EF, Adler JE (1987) Glucocorticoids regulate adrenal opiate peptides. Mol Brain Res 388:125–131
Landfield PW, Waymire PC, Lynch G (1978) Hippocampal aging and adrenocorticoids: quantitative correlation. Science 202:1098–1102
McEwen BS, Kloet ER de, Rostene W (1986) Adrenal steroid receptors and action in the nervous system. Physiol Rev 66:1121–1188
McWilliams R, Lynch G (1979) Terminal proliferation in the partially deafferentiated dentate gyrus: time courses for the appearance and removal of degeneration and the replacement of lost terminals. J Comp Neurol 187:191–198

Munck A, Guyre PM, Holbrook NJ (1984) Physiological function of glucocorticoids in stress and their relation to pharmacological action. Endocr Rev 5:25–44
Nock BJ, Blaustein JD, Feder HH (1981) Changes in noradrenergic transmission alter the concentration of cytoplasmic progestin receptors in hypothalamus. Brain Res 207:371–396
Nyakas CS, Kloet ER de, Veldhuis HD, Bohus B (1981) Corticosterone binding capacity increases in contralateral hippocampus after partial unilateral hippocampectomy. Neurosci Lett 21:339–343
Nyakas CS, Kloet ER de, Veldhuis HD, Bohus B (1983) Hippocampal corticosterone receptors and novelty-induced behavioural activity: effect of kainic acid lesion in the hippocampus. Brain Res 288:219–228
Reul JMHM, Kloet ER de (1985) Two receptor systems for corticosterone in rat brain: microdistribution and differential occupation. Endocrinology 117:2505–2511
Ronde FSW de, Kloet ER de, Nyakas C (1986) Cortisteroid receptor plasticity and recovery of a deficient hippocampus-associated behaviour after unilateral (dorsal) hippocampectomy. Brain Res. 374:219–226
Rosenfeld P, Eekelen JAM van, Levine S, Kloet ER de (1988 a) Ontogeny of the glucocorticoid receptor: an immunocytochemical study. Dev Brain Res 42:113–118
Rosenfeld P, Sutanto W, Levine S, Kloet ER de (1988 b) Ontogeny of Type 1 and Type 2 corticosteroid receptors in the rat hippocampus. Dev Brain Res 42:119–127
Sapolski R (1985) A mechanism for glucocorticoid toxicity in the hippocampus: increased neuronal vulnerability to metabolic insults. J Neurosci 5:1228–1232
Sapolski R, Pulsinelli W (1985) Glucocorticoids potentiate ischemic injury to neurons: therapeutic implication. Science 229:1397–1399
Sapolski R, Krey LC, McEwen BS (1986) The neuroendocrinology of stress and aging: The glucocorticoid cascade hypothesis. Endocr Rev 7:284–301
Scheff SW, Cotman CW (1982) Chronic glucocorticoid therapy alters axon sprouting in the hippocampal dentate gyrus. Exp Neurol 76:644–654
Scheff SW, Benardo LS, Cotman CW (1980) Hydrocortisone administration retards axon sprouting in the rat dentate gyrus. Exp Neurol 68:195–201
Seger MA, Kiss JZ, Tol HHM van, Burbach JPH (1987 a) Biosynthesis of pro-opiomelanocortin is increased in the rat intermediate pituitary following denervation. Neurosci Lett 83:149–155
Seger MA, Eekelen JAM van, Kiss JZ, Burbach JPH, Kloet ER de (1987 b) Stimulation of pro-opiomelanocortin gene expression by glucocorticoids in the denervated rat intermediate pituitary gland. Neuroendocrinology 47:350–357
Selye H (1976) Stress in health and disease. Butterworth, London
Strand FL, Smith CM (1986) LPH, ACTH, MSH, and motor systems. In: Wied D de, Gispen WH, Wimersma Greidanus TJB van (eds) Neuropeptides and behaviour, vol 2. Pergamon, Oxford, pp 245–272
Tausk M (1951) Hat die Nebennierenrinde tatsächlich eine Verteidigungsfunktion? In: Das Hormon. Organon, Oss, The Netherlands 3:1–4
Wurtman R, Axelrod J (1966) Control by enzymatic synthesis of adrenaline in the adrenal medulla by adrenal cortisol steroids. J Biol Chem 241:2301–2305
Yoburn BC, Franklin SO, Calvano SE, Inturris CH (1987) Regulation of rat adrenal medullary enkephalins by glucocorticoids. Life Sci 40:2495–2503
Zhou FC, Azmitia EC (1985) The effect of adrenalectomy and corticosterone on homotypical collateral sprouting of serotonergic fibers in the hippocampus. Neurosci Lett 54:111–116

Ganglioside Treatment and Post-Lesion Neural Plasticity. A Unifying Theory on Ganglioside Mechanism of Action

A. Gorio [1] and A. M. Di Giulio [2]

[1] Institute of Pharmocological Sciences, Faculty of Pharmacy, Via Balzaretti 9, and
[2] Department of Medical Pharmacology, Faculty of Medicine, Via Vanvitelli 32, University of Milan, Milan, Italy

1 Introduction

The study of the mechanisms underlying nervous system repair following lesions is one of the key problems of modern neuroscience. The anatomical repair with synapse reformation and other forms of restoration of functions not necessarily accompanied by neuronal regeneration and/or sprouting are named under the general term of post-lesion neuronal plasticity. In this report we give a rapid critical review of some of the most relevant experimental results obtained in the past years using gangliosides as potential agents capable of enhancing neuronal repair processes. A variety of approaches, ranging from in vitro to in vivo models, were used to assess the effects of these agents, aiming to achieve a clear experimental result and an understanding of their mechanism of action.

Gangliosides were discovered some 40 years ago by Klenk and co-workers as constituents of neurons; however, their tissue distribution is more diffused and at least 60 different ganglioside molecules are now well known (Gorio 1986). Subcellular studies have shown that gangliosides are major cell membrane contituents mainly present in the outer layer with the hydrophobic moiety embedded into the lipid bilayer and the sugar portion protruding into the extracellular fluids (Gorio 1986). Their strategic location has suggested a role as cell membrane receptors, being capable of interacting with a variety of molecules (Gorio 1986). In the case of cholera toxin, the monosialoganglioside Gm1 acts as the unique receptor for the toxin; such specificity is high and displayed even by exogenous gangliosides incorporated into the cell membrane from the incubating medium (Gorio 1986). These latter results suggested that exogenously incorporated gangliosides may be capable of acting like endogenous.

Specific expression of different gangliosides occurs during developmental stages of various brain areas in normal and pathological conditions, suggesting a possible role of these substances. On the other hand, stronger evidence is required to support these conclusions.

2 Ganglioside Treatment and Neuronal Regeneration

In vitro experiments using a variety of cell lines or primary neuronal cultures showed that the addition of gangliosides to the culture media enhanced neuronal

H. Flohr (Ed.)
Post-Lesion Neural Plasticity

differentiation, neuritogenesis, and synaptogenesis (Gorio 1986). Such morphological effects were suggestive of a direct interaction of exogenous gangliosides with the processes underlying neuronal development. PC12 Pheochromocytoma cells were employed for examining in details the above phenomena (Ferrari et al. 1983). PC12 cells differentiate forming a neurite network after addition of nerve growth factor (NGF) to the culture medium. Gangliosides failed to affect PC12 growth by themselves; however, neuritogenesis was increased in a dose-dependent manner when the glycosphingolipids were added with NGF. Neurite regeneration was also increased by the treatment. Binding experiments with radioactively labeled NGF showed that ganglioside treatment did not in any way increase or change NGF incorporation by PC12 cells. Therefore the effects of gangliosides cannot be ascribed to an enhancement of the trophic factor receptor number per PC12 cell, but rather to an enhancement of the capacity of these cells to respond to the specific growth factor.

The in vivo results obtained by treating experimental animals with gangliosides during muscle reinnervation are in agreement with the above conclusions. Morphological, biochemical, and electrophysiological techniques were used for assessing the process of muscle reinnervation following complete or partial denervation. The extensor digitorum longus muscle was denervated by crushing the sciatic nerve, standard intracellular electrophysiological recordings revealed that the rate and extent of sprouting and polyneuronal innervation were enhanced by ganglioside treatment (Gorio et al. 1980). The process of synaptic repression was unaltered and the rate of axonal regeneration as well. The extent of peripheral nerve sprouting was also enhanced in the case of soleus muscle reinnervation following partial denervation. The index of sprouting was significantly increased (Gorio et al. 1983). The iris reinnervation by noradrenergic axons was also accelerated after chemical sympathectomy (Jonsson et al. 1984). The apparent results are twofold, indicating that ganglioside treatment enhanced the dynamics and the extent of nerve sprouting and that the effect was not restricted to a single neuronal kind, as a variety of other observations have suggested (Gorio 1986).

Specific lesions of neuronal pathways in the central nervous system have given further insights on ganglioside action upon damaged neurons. Noradrenergic and serotoninergic axons in the CNS of newborn and adult animals were lesioned with 6-hydroxidopamine and 5–7 dihydroxitryptamine, respectively. Newborn rats were lesioned with an intraperitoneal injection of the designated neurotoxin then treated with GM1 ganglioside for the subsequent 4 days. In both cases, the primary degeneration occurred normally in the most distal 5-HT projections (i.e., cerebral cortex and spinal cord), but 2 weeks later regenerating fibers were already evident in the frontal cortex and proximal portions of the spinal cord of GMI-treated animals. The regenerative wave proceeded gradually until normal 5-HT levels and fiber distribution were observed. Saline-treated animals displayed a similar pattern of regeneration with a much retarded start (Jonsson et al. 1984; Fusco et al. 1985; Gorio 1986). Since regeneration was observed 10 days after ganglioside treatment had ceased, these results suggested that ganglioside acted by preventing retrograde regeneration of the lesioned neurons, allowing a faster regeneration rather than stimulating regeneration directly. Similar conclusions were drawn from experiments in adult rats concerning intracerebral lesions of noradrenergic fibers with 6-OHDA and GM1 ganglioside treatment (Kojima et al. 1984). On the other hand, there are also many direct experimental data that support this hypothesis about ganglioside mode of action. The serotoninergic pathways experiments, mentioned above, show that in addition to promoting regeneration of the lesioned fibers, ganglioside treatment partially prevented the hyperinnervation (pruning effect) of the areas closer to the cell bodies located in the raphe (Jonsson et al. 1984). Therefore ganglioside treatment did not enhance the aberrant neurite outgrowth characteristic of the pruning effect. We have tackled specifically this issue by studying the effect of GM1 treatment on the hyperinnervation of the frontal cortex by dopaminergic and serotoninergic axons after noradrenaline depletion with 6-OHDA in newborn animals (Jonsson et al. 1984). The experimental animals were treated with GM1 ganglioside from day 1 to 4 post-natally, then several weeks later the serotoninergic and dopaminergic innervation of the frontal cortex was monitored. It was found that ganglioside treatment fully prevented the hyperinnervation process by both monoaminergic pathways. These results are reinforcing the suggestion that in the central nervous system gangliosides may act by affecting the lesioned axons and/or neurons. If the ganglioside stimulatory effect was exerted upon any axonal system in growth, the treatment should have stimulated hyperinnervation of the cortex. On the other hand, peripheral nerve sprouting is enhanced by ganglioside treatment (Gorio 1986).

3 Ganglioside Treatment and Prevention of Neuronal Functional Decay

In vivo and in vitro ganglioside treatment can affect the capacity of neurons to withstand insults otherwise lethal. A variety of experimental strategies were developed to evaluate such a protective mechanism. Neuromuscular preparations or brain slices were incubated in vitro with gangliosides or dissected out after the animals were pre-treated for 3 days with a daily injection of gangliosides. The phrenic-diaphragm preparations were bathed in a saline solution lacking potassium ions for several hours; this condition causes a blockade of the Na-pump and a depolarization, which is accompanied by an influx of sodium and a loss of potassium ions. When the bathing medium is adjusted to the correct potassium concentration, the Na-pump is suddenly activated with repolarization of the resting potential and gives rise to a transient hyperpolarization which is directly correlated with the Na-pump activity. Preincubation of the neuromuscular preparation with gangliosides for 2 h prior to potassium re-admission enhanced significantly the Na-pump activity. However, if gangliosides were added several hours (6–8 h at least) prior to potassium readmission or the animals were pretreated, the depolarization was absent and no activation of the pump was ever observed upon normalization of potassium ion levels (Vyskocil et al. 1985). These data suggested a protective effect of long-term ganglioside treatments, while short-term treatments may cause enhancement of membrane enzyme activity. Similar protective effects were obtained in the case of in vitro hypoxia (Vyskocil et al. 1985). Hippocampal slices were prepared and incubated for several hours with gangliosides or saline, then subjected to depolarization induced by hypoxia. The ganglioside treatment significantly reduced the extent of depolarization if the incubation period was at least 6 h long (Janigro et al. 1984). Therefore these results suggested a decay-preventive effect of ganglioside administration upon membrane functions. In vivo treatment with gangliosides is also capable of preventing membrane enzyme deactivation occurring during the re-oxygenation following the ischemic period of slice preparation. The activity of Na-K ATPase and Ca-Mg ATPase decreases significantly and the mitochondria swell greatly due to the intracellular calcium loading. In the case of ganglioside treated animals the membrane enzyme deactivation is fully prevented and the number of swollen mitochondria is reduced by 80%. These data indicate a cell membrane protective effect of gangliosides, which subsequently is reflected in an intracellular preservation as indicated by the reduced calcium load and mitochondrial swelling.

4 Ganglioside Treatment and Trans-Synaptic Degeneration

The effects of ganglioside treatment upon trans-synaptic degenerative events were studied by monitoring the intraspinal degenerative processes occurring in the dorsal horn of the spinal cord several days after peripheral nerve lesions. We have clearly shown that section of the sciatic nerve causes met-enkephalinergic inter-

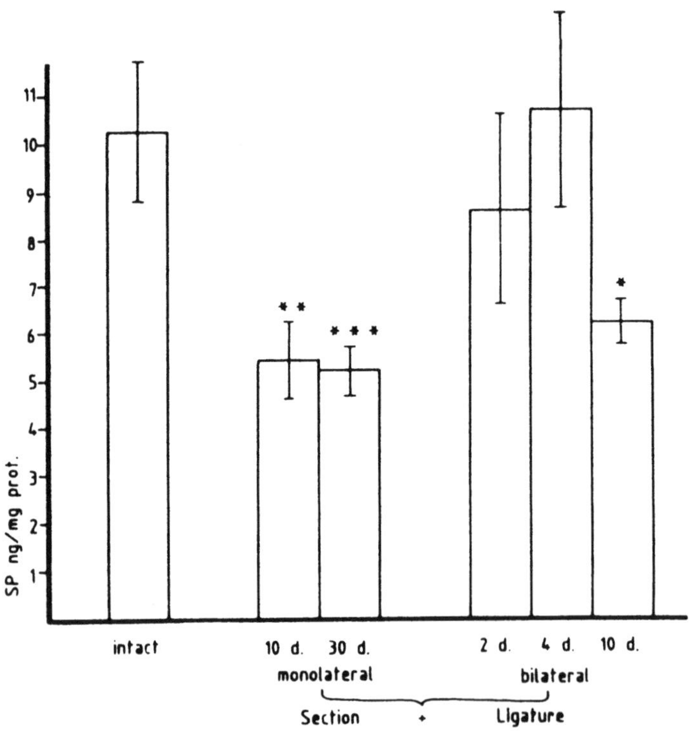

Fig. 1. The histogram shows the levels of substance P in the lumbar spinal cord in unlesioned rats and in experimental groups at various time after mono- or bilateral resection of the sciatic nerve. There is a significant drop of substance P levels at 10 and 30 days post-lesioning. The peptide loss is similar for both mono- and bilateral lesion

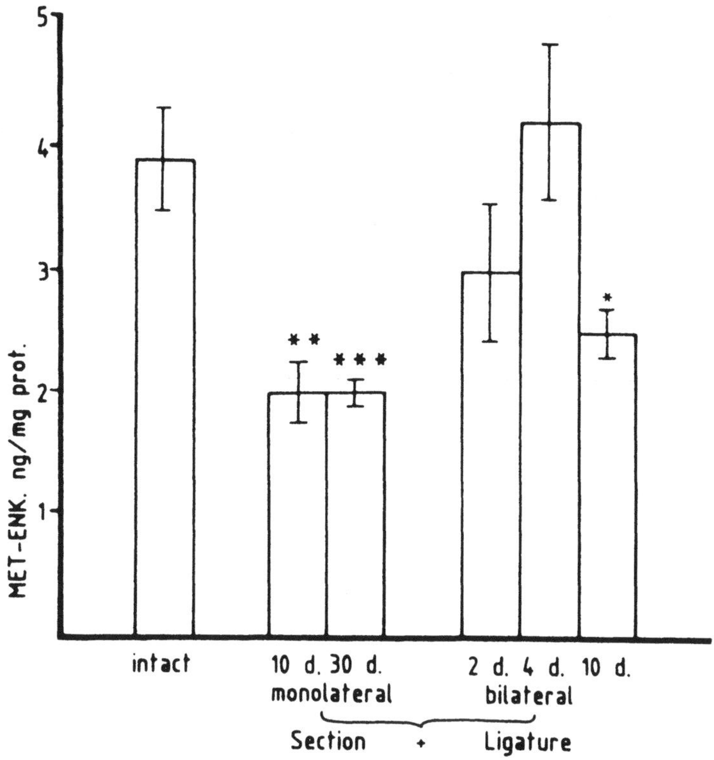

Fig. 2. The histogram shows the levels of met-enkephalin in the lumbar spinal cord of rats from control unlesioned groups and from animal groups with mono- or bilateral resection of the sciatic nerve. The met-enkephalin loss is highly significant at 10 days and remains similar at 30 days

neurons atrophy in addition to a significant degeneration of the sensory input (Di Giulio et al. 1985). Substance P is present in high amounts in the axon tips of the sensory input to the substantia gelatinosa of the dorsal horn, which stained as a band by substance P antibodies. Ten–fifteen days after lesioning the sciatic nerve, the substance P stain decreases drastically (Di Giulio et al. 1985) and the peptide levels measured with the radioimmunoassay technique are reduced significantly (Fig. 1). Met-enkephalin interneurons present mainly in the lamina II of the substantia gelatinosa also undergo atrophy, as shown by radioimmunoassay, which indicates a significant reduction of peptide levels following the lesion (Fig. 2). The peptidergic alterations are restricted to the dorsal horn, as demonstrated by radioimmunoassay (Fig. 3) and immunocytochemistry with specific antibodies (Di Giulio et al. 1985). These intraspinal degenerative atrophies are preceded by a transient increase of serotonin metabolism in the lumbar segment of the spinal cord (Di Giulio et al. 1987). Such an increase is observed within 24 h from lesioning the nerve and maintained for several days, then prior the peptidergic loss it returns to normal levels (Fig. 4). To assess whether serotonin release was involved in the degenerative process, a group of animals was subjected to parachlorophenilalanine treatment for depleting serotonin (Fig. 5) and then underwent surgery

GROUP	DORSAL HORN		VENTRAL HORN	
	SP	ME	SP	ME
	ng/mg prot. ± S.E.			
CONTROL	30.2±2.4	4.0±0.21	4.8±0.53	1.6±0.17
LESION	21.1±1.8**	2.7±0.17***	4.3±0.50	1.6±0.16

Fig. 3. Schematic illustration of how the peptidergic loss occurring 10 days after sciatic nerve resection is restricted to the dorsal horn

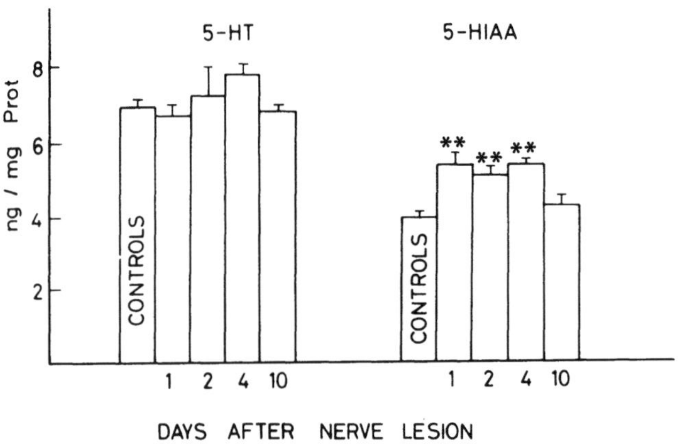

Fig. 4. It is shown that within 24 h after sciatic nerve resection the serotonin turnover, measured as 5-HIAA levels, is significantly increased in the lumbar spinal cord. The lesion effect on serotonin is transient and subsides at 10 days' post-lesioning. (Di Giulio et al. 1987)

for lesioning the peripheral nerve. It was found that in the animals depleted of serotonin, the sciatic nerve lesions induced substance P loss but failed to cause the trans-synaptic degenerative atrophy of the met-enkephalinergic interneurons (Fig. 6). In conclusion peripheral nerve lesions trigger intraspinal rearrangements which may be regulated by the descending serotoninergic pathways.

The effect of ganglioside treatment was tested in such an experimental paradigm. Animals were lesioned by sectioning the sciatic nerve and then treated with gangliosides; it was observed that the lesion caused the substance P loss but the

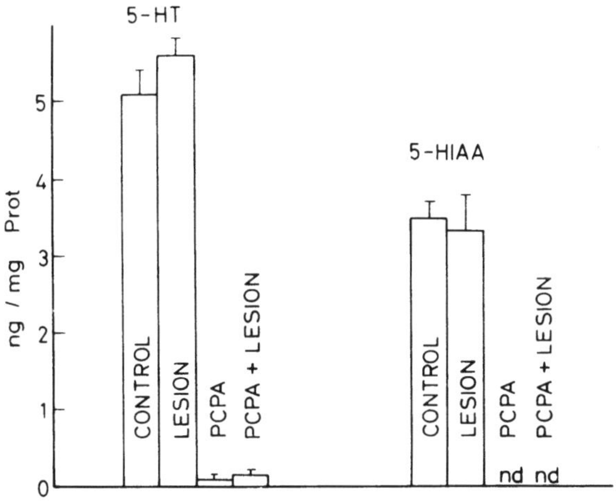

Fig. 5. Treatment with para-chlorophenilalanine depletes serotonin and 5-HIAA in the lumbar spinal cord of control and lesioned animals. Treatment began 3 days before sciatic nerve resection and repeated every 3 days subsequently. The dosage of p-CPA was 300 mg/kg for the first treatment and subsequently was reduced to 100 mg/kg. Animals were killed 20 days after sciatic nerve lesion. (Di Giulio et al. 1987)

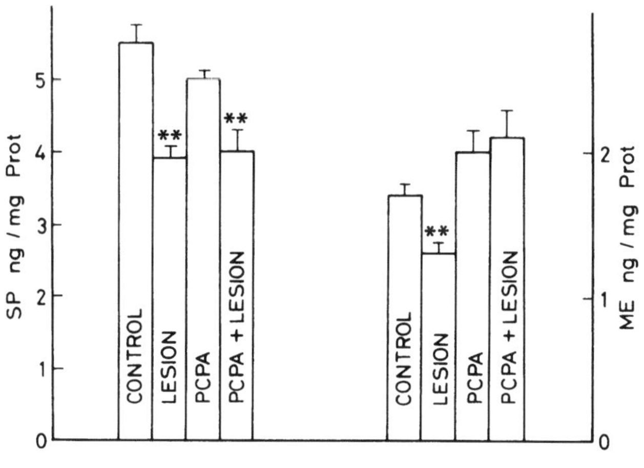

Fig. 6. The significant substance P loss is present in the lumbar spinal cord of lesioned animals also in the case of p-CPA treatment. Conversely, the met-enkephalin loss is fully prevented by the serotonin depletion induced by p-CPA in the sciatic nerve lesioned animals. (Di Giulio et al. 1987)

trans-synaptic atrophy of the met-enkephalinergic interneurons was fully prevented by the treatment. In other words, gangliosides administration did not affect the primary sensory degeneration directly lesioned during the nerve resection but successfully prevented the central consequences of such a process. This effect could be relevant for the basic understanding of ganglioside activity, but also for a putative antinociceptive property of these agents in the case of the phantom limb syndrome being the enkephalinergic interneurons of the dorsal horn an important relay of the nociceptive circuit.

5 Conclusions

There are two basic observations on the effects of ganglioside treatment deriving from the results above summarized: one is that the treatment with these agents of regenerating and/or sprouting peripheral neurons causes an enhancement of the sprouting process, while the effects observed in the CNS seem to relate to a protective effect upon membrane alterations underlying retrograde degeneration and neuronal loss. It is conceivable that the same basic molecular mechanism is responsible for the various responses observed with the different experimental paradigms. We postulate that a possible candidate mechanism has an effect upon an excessive calcium loading by neurons. It has been demonstrated that neuronal injury causes intracellular calcium accumulation; similarly sprouting axons, in vivo and in vitro as well, show calcium spikes at their growing tips. We postulate that a modulation of the excessive calcium entering the sprouting axons may stabilize the newly formed sprout and so prevent retraction of the bud which often occurs in normal conditions. The result is a net enhancement of sprouting and neuritogenesis. On the other hand, protection from excessive calcium loading may be responsible for the prevention of retrograde degeneration, trans-synaptic degeneration, and enhancement of neuronal survival.

We have sufficient results to support these conclusions, which are suggested by both phrenic-diaphragm and hippocampal slice experiments, indicating protection from ionic loading, ultrastructural preservation and prevention of membrane function deactivation. Recent results obtained by us are showing that in the case of chronic irreversible ischemia, a particular schedule of treatment with gangliosides limits the calcium loading to the first 24 h, during the following days no further calcium accumulation is observed.

References

Ferrari G, Fabris M, Gorio A (1983) Gangliosides enhance neurite outgrowth in PC12 cells. Dev Brain Res 8:215–221

Fusco M, Dona' M, Hallmann H, Jonsson G, Gorio A (1985) Enhancement of 5-HT fiber regrowth in the spinal cord by GM1 treatment. J Neurosci Res 15:467–480

Giulio AM di, Borella F, Mantegazza P, Hong JS, Panozzo C, Zanoni R, Gorio A (1985a) Structural and biochemical alterations in the dorsal horn of the spinal cord caused by peripheral nerve lesions. Peptides 6:249–256

Giulio AM di, Borella F, Parenti M et al. (1985b) Transganglionic intraspinal alterations of opioid peptides caused by lesions to peripheral nerves. In: Biggio G, Spano PF, Toffano G, Gessa GL (eds) Modulation of central and peripheral transmitters function. Liviana, Padova, Italy, pp 537–544

Giulio AM di, Mantegazza P, Dona' P, Gorio A (1985c) Peripheral nerve lesions cause simultaneous alterations of substance P and enkephalin levels in the spinal cord. Brain Res 342:405–408

Giulio AM di, Tenconi B, Mannavola A, Mantegazza P, Schiavinato A, Gorio A (1987) Spinal cord interneuron degenerative atrophy caused by perypheral nerve lesions is prevented by serotonin depletion. J Neurosci Res (in press)

Gorio A (1986) Ganglioside enhancement of neuronal differentiation, plasticity and repair. Clin Neurobiol 2:241–296

Gorio A, Carmignoto G, Facci L, Finesso M (1980) Motor nerve sprouting induced by ganglioside treatment. Possible implications for gangliosides on neuronal growth. Brain Res 191:236–241

Gorio A, Marini P, Zanoni R (1983) Muscle reinnervation. III. Motorneuron sprouting capacity, enhancement by ganglioside treatment. Neuroscience 8:417–429

Janigro D, Gregorio F di, Vyskocil F, Gorio A (1984) Gangliosides dual mode of action: a working hypothesis. J Neurosci Res 12:499–509

Jonsson G, Gorio A, Hallman H, Janigro D, Kojima H, Luthman J, Zanoni R (1984) Effect of Gm1 ganglioside on developing and mature serotonin and noradrenaline neurones lesioned by selective neurotoxins. J Neurosci Res 12:459–475

Kojima H, Gorio A, Janigro D, Jonsson G (19849 GM1 ganglioside enhances regrowth of noradrenaline nerve terminals in rats cerebral cortex lesioned by the toxin 6-Hydroxidopamine. Neuroscience 13:1011–1022

Vyskocil F, Gregorio F di, Gorio A (1985) The facilitating effect of gangliosides on the electrogenic (Na-K) pump and the resistance of the membrane potential to hypoxia in neuromuscular preparations. Pfluegers Arch 403:1–6

Role of Gangliosides in Functional Recovery of Damaged Nervous System

A. Consolazione, G. Vantini, S. Seren, G. Toffano, and A. Leon
Fidia Research Laboratories, Via Ponte della Fabbrica, 3/A, 35031 Abano Terme (PD), Italy

1 Introduction

Lesion and transplantation studies conducted during the last decade have shown that the adult mammalian brain possesses a remarkable capacity for positively adapting to injury. Neuronotrophic interactions and/or activation of novel or usually quiescent processes have been implicated and agents or conditions capable of affecting these processes are currently been examined in an attempt to decrease damage (i.e., sparing) and/or improve restoration of function (i.e., functional recovery) of damaged neuronal tissues.

Among factors that may be able to modulate lesion-induced adaptative morphofunctional neuronal responses, gangliosides are today receiving increasing attention. In particular, studies from our and other laboratories have shown that the systemic administration of monosialoganglioside GM1, or its inner ester derivative, the siagoside, is effective in enhancing neuronal repair in experimental in vivo paradigms mimicking pathological situations of the central nervous system (CNS). Approaches for investigating an effect of chronic ganglioside administration essentially derive from evidence indicating that such molecules may play a prominent role in the processes of neuritogenesis and neurite repair, both in vitro and in vivo. Although their primary mechanism of action is still obscure, numerous in vitro studies indicate that the observed effects are presumably related to facilitation of membrane-mediated transfer of "trophic" information and/or reduction of membrane-mediated neurotoxic events.

The present chapter will briefly summarize the information today available with regard to the potential therapeutical efficacy of monosialoganglioside in facilitating recovery of the adult CNS following traumatic, toxic, or ischemic experimentally induced brain damage. It also briefly discusses the possible mechanisms underlying the GM1 effects in vivo.

2 Gangliosides and in Vitro Studies

Gangliosides, sialic acid containing glycosphingolipids, have long been recognized as characteristic cell surface constituents of mammalian cells, in particular neuronal cells. Although these molecules have been identified as receptors for bacterial toxins and viruses, relatively little is known about their normal function

H. Flohr (Ed.)
Post-Lesion Neural Plasticity

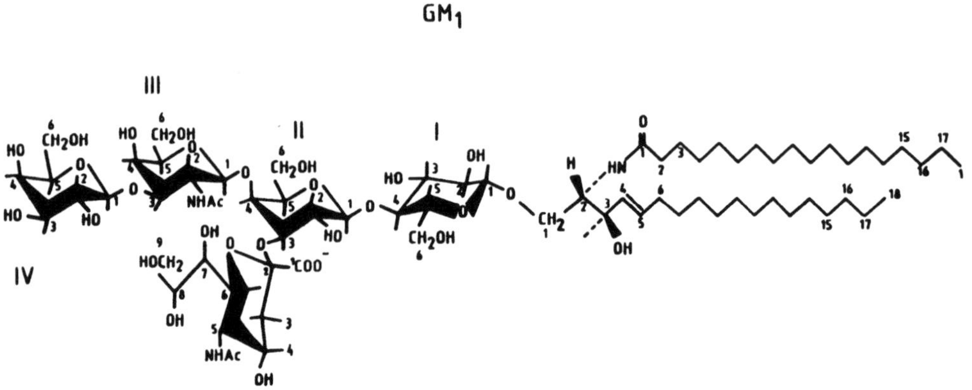

Fig. 1. Chemical sturcture of GM1 monosialoganglioside

in neuronal membranes (for review see Ando 1983). In this regard, several lines of evidence are now available suggesting a link between endogenously occurring gangliosides, including GM1 (Fig. 1), and plasticity of developing and mature CNS neurons (for a recent review see Tettamanti et al. 1986).

In addition, in vitro studies have shown that exogenously supplied gangliosides facilitate survival, as well as morphological and biochemical differentiation of cultured neuronal cells. For example, the addition of mixed gangliosides, or GM1 alone, to the culture medium has been shown to increase synapse formation in spinal cord-muscle preparations (Obata et al. 1977) and in neuroblastoma cells (Spoerri 1983). In this latter system, neurite outgrowth has also been shown to be enhanced by the gangliosides (Roisen et al. 1981; Leon et al. 1983; Byrne et al. 1983). Furthermore, utilizing PC12 pheochromocytoma cells, a cell line that produces neurites in response to nerve growth factor (NGF), it was observed that the gangliosides, including GM1, enhance neurite outgrowth only in the presence of NGF (Ferrari et al. 1983; Katoh-Semba et al. 1984). A similar synergistic effect between neuronotrophic factors (NTF) and gangliosides has been confirmed also in primary neuronal systems, which are strictly dependent on NGF, or other NTF(s) for survival and/or neurite elongation (Leon et al. 1984; Doherty et al. 1985; Skaper et al. 1985; Dal Toso et al. 1988; Leon et al. 1988).

Interestingly, recent research has shown that exogenously supplied gangliosides are also able to decrease effects of potential neurotoxic agents (e.g., glutamate) in cultured neuronal cells. In particular, addition of gangliosides to cerebellar granular cells in culture inhibits glutamate-induced translocation of protein kinase C (Vaccarino et al. 1987). Thus, it appears that exogenously supplied gangliosides, at least in vitro, show dual effects: they are able both to facilitate neuronal cell responsiveness to NTF's and to decrease neuronal cell responsiveness to neurotoxic agents. Although the underlying molecular mechanisms are still obscure, it is known that both these effects involve stable association of the ganglioside molecule with the neuronal cell surface (Facci et al. 1984).

3 Gangliosides and CNS Repair Processes

Prompted by the above evidence, the possible effect of the monosialoganglioside on CNS repair following injury was evaluated. The systemic administration of GM1 has been reported to be effective in ameliorating outcome, both temporally and qualitatively, following several types of experimentally induced brain damage in adult animals.

For example, following partial unilateral hemitransection of the nigrostriatal pathway in adult rats, repeated administration of GM1 was found to facilitate functional recovery of dopaminergic function in the lesioned striatum and substantia nigra (Toffano et al. 1983; 1984). This effect was associated with enhanced survival of nigral dopaminergic cell bodies (Agnati et al. 1983; Consolazione et al. 1987) as well as with reduction in the imbalance of striatal energy metabolism and blood flow occurring between the striatum of the lesioned and unlesioned side (Agnati et al. 1985). Analogously, GM1 was shown to facilitate maintenance of cell viability and recovery of specific biochemical markers following mechanical or electrolytical injury to the cholinergic neurons of the basal forebrain (Oderfeld-Nowak et al. 1984; Casamenti et al. 1985; Cuello et al. 1986; Sofroniew et al. 1986). Facilitatory effects of GM1 have also been observed following neurotoxic lesions. GM1 administration restores striatal dopamine content following MPTP treatment in mice (Hadjiconstantinou et al. 1986), and it is also capable of partly counteracting the degenerative effects induced by 5,7-dihydroxytryptamine and 6-hydroxydopamine on serotonergic and noradrenergic systems, respectively (Jonsson et al. 1984; Fusco et al. 1986).

Interestingly, in the above-mentioned lesion paradigms, the effects of chronic GM1 treatment were assessed in the late phases following the injury and, when tested, were associated with behavioral recovery. This indicates that the chronic GM1 treatment is efficacious in facilitating long-term recovery of function following an acute CNS insult. However, recent evidence suggests that the monosialoganglioside is also effective in the early phases following CNS damage. For example, Li et al. (1986) reported that GM1 treatment reduces behavioral asymmetry and loss of striatal Na^+/K^+ ATPase activity at very early time points following unilateral hemitransection of the nigrostriatal pathway. These latter effects were dependent on the time interval elapsing between the injury and the first ganglioside administration. Furthermore, short-term effects of GM1 have been studied in different experimental models of cerebral ischemia. In these cases, the ganglioside was shown to limit cerebral electrolytic imbalance and edema and to improve blood flow at the early phases of experimental stroke (Tanaka et al. 1986; Karpiak et al. 1987). In addition, the ganglioside-treated animals showed less severe morphological damage occurring in defined brain areas, as well as less severe neurological deficits (Aldinio et al. 1987; Greenberg et al. 1987). In some cases, the siagoside, utilized because of its greater penetrability across the blood brain barrier (Ghidoni et al. 1986), was more efficacious than GM1 (Cahn et al. 1986).

Therefore, the above results indicate that treatment with monosialoganglioside has also beneficial effects at acute time points following brain damage, including cerebral ischemia. The capability of GM1 to facilitate functional recovery following CNS insults thus seems, at least in part, to involve also a reduction of secondary degenerative processes occurring at early post-lesion times. Conceivably, this decreased secondary degeneration results in sparing of neuronal tissue not affected by the primary insult, an effect dependent on early initiation of GM1 treatment.

4 GM1 in Vivo Effects: Mechanistic Consideration

Traumatic or ischemic CNS injury is often associated with primary and secondary neuronal damage and consequent cell death. Evidence is today available indicating that the neuronal cell death, secondary to the primary insult, may be related to an abnormal ratio between endogenously occurring neurotoxic and neuronotrophic influences. In addition, as previously mentioned, there is evidence that gangliosides can affect in vitro neuronal cell responsiveness to both neurotoxic and neuronotrophic influences.

As concerns neurotoxic influences, there is evidence today suggesting that the excitatory amino acids, massively released following an ischemic insult, are most probably the cause of secondary death of neurons receiving an excitatory input (Rothman and Olney 1986). For example, recent investigations have shown that glutamate antagonists markedly decrease the selective ischemic neuronal cell death occurring in the CA1 area of the hippocampus (Rothman and Olney 1986). A similar effect has recently been observed also following ganglioside treatment (Aldinio et al. 1987). As such, and in accord with the previously mentioned in vitro studies, it may be hypothesized that the GM1-induced prevention of secondary CNS neuronal damage may be related with reduction of the effects of neurotoxic agents.

As concerns NTF influences, it is now known that they are present in the adult brain (Varon et al. 1984) and that they are essential components in the regulation of neuronal cell survival and repair after an insult. It has been hypothesized that a neuronotrophic deficit (perhaps due to an increased trophic need) may occur following brain damage. In addition, although increases of neuronotrophic activity have been reported to occur after brain injury (Nieto-Sampedro et al. 1982), these increases appear to be either insufficient or to occur at a time subsequent to most neuronal cell death. Consequently, and again in accord with the in vitro studies, a current working hypothesis is that the GM1-induced facilitation of CNS functional recovery in vivo may be related with facilitation of neuronal response to neuronotrophic signals, per se inadequate at early post-lesion times. This hypothesis is further supported by recent in vivo studies showing that GM1 treatment is able to enhance NGF effects both in the peripheral and central nervous system (Cuello et al. 1987; Vantini et al. 1988).

In conclusion, it is feasible that the ganglioside can have dual effects also in vivo. The GM1-induced antagonism of neurotoxic effects may be involved, at least in part, in the prevention of secondary neuronal cell damage, whereas the GM1-induced facilitation of neuronotrophic effects may well underlie the GM1 capability to facilitate functional recovery.

References

Agnati L, Fuxe K, Calzà L, Benfenati F, Cavicchioli L, Toffano G, Goldstein M (1983) Ganglioside increase the survival of lesioned nigral dopamine neurons and favour the recovery of dopaminergic synaptic function in striatum of rats by collateral sprouting. Acta Physiol Scand 119:347–363

Agnati LF, Fuxe K, Benfenati F et al. (1985) Effects of ganglioside GM1 treatment on striatal glucose metabolism blood flow and protein phosphorylation of the rat. Acta Physiol Scand 125:43–53

Aldinio C, Seren S, Lazzaro A et al. (1987) Ischemia-induced neuronal cell injury: effects of the inner ester derivative of GM1 ganglioside. Soc Neurosci, 17th Annual Meeting, Abs 418.9

Ando S (1983) Gangliosides in the nervous system. Neurochem Int 5:507–537

Byrne MC, Ledeen RW, Roisen FJ, Yorke G, Scalafani JR (1983) Ganglioside-induced neuritogenesis: verification that gangliosides are the active agents and comparison of molecular species. J Neurochem 41:1214–1222

Cahn J, Borzeix MG, Toffano G (1986) Effect of GM1 ganglioside and of its inner ester derivative in a model of transient cerebral ischemia in the rat. In: Tettamanti G, Ledeen RW, Sandhoff K, Nagai Y, Toffano G (eds) Gangliosides and neuronal plasticity. Liviana, Padova, pp 435–443

Casamenti F, Bracco L, Bartolini L, Pepeu G (1985) Effects of ganglioside treatment in rats with a lesion of the cholinergic forebrain nuclei. Brain Res 338:45–52

Consolazione A, Aldinio C, Cavicchioli L, Leon A, Toffano G (1987) Gangliosides as a tool for studying the repair of adult mammalian brain after injury. In: Vernadakis A (ed) Model systems of development and aging of the nervous system. Martinus Nijhoff, The Hague, Netherland, pp 443–452

Cuello AC, Stephen PH, Tagari PC (1986) Retrograde changes in the nucleus basalis of the rat, caused by cortical damage, are prevented by exogenous ganglioside GM1. Brain Res 376:373–377

Cuello AC, Maysinger D, Garofalo L, Tagari PC, Stephen PH, Pioro E, Piotte M (1987) Influence of gangliosides and nerve growth factor on the plasticity of forebrain cholinergic neurons. In: Fuxe K, Agnati LF (eds) Receptor-receptor interactions. Mac Millan, London, England, pp 62–75

Dal Toso R, Giorgi O, Soranzo C et al. (1988) Development and survival of neurons in dissociated fetal mesencephalic serum-free cultures: I effects of cell density and of an adult mammalian striatal-derived neuronotrophic factor (SDNF). J Neurosci 8:733–745

Doherty O, Dickson JG, Flanigan TP, Walsh FS (1985) Ganglioside GM1 does not initiate but enhances neurite regeneration of nerve growth factor-dependent sensory neurones. J Neurochem 44:1259–1265

Facci L, Leon A, Toffano G, Sonnino S, Ghidoni R, Tettamanti G (1984) Promotion of neuritogenesis in mouse neuroblastoma cells by exogenous ganglioside. Relationship between the effect and the cell association of ganglioside GM1. J Neurochem 42:299–305

Ferrari G, Fabris M, Gorio A (1983) Gangliosides enhance neurite outgrowth in PC12 cells. Dev Brain Res 8:215–221

Fusco M, Donà M, Tessari F, Hallman H, Jonsson G, Gorio A (1986) GM1 ganglioside counteracts selective neurotoxin-induced lesion of developing serotonin neurons in rat spinal cord. J Neurosci Res 15:467–479

Ghidoni R, Trinchera M, Venerando B, Fiorilli A, Tettamanti G (1986) Metabolism of exogenous GM1 and related glycolipids in the rat. In: Tettamanti G, Ledeen RW, Sandhoff K, Nagai Y, Toffano G (eds) Gangliosides and neuronal plasticity. Liviana, Padova, pp 183–200

Greenberg JH, Komatsumoto S, Dara E, Tanaka K, Hiwkey W, Toffano G, Reivich M (1987) The effects of GM1 in focal ischemia in the cat. J Cereb Blood Flow Metab 7:S180

Hadjiconstantinou M, Rossetti ZL, Paxton RC, Neff NH (1986) Administration of GM1 ganglioside restores the dopamine content in striatum after chronic treatment with MPTP. Neuropharmacology 25:1075–1077

Jonsson G, Gorio A, Hallman H, Janigro D, Kojima H, Luthman J, Zanoni R (1984) Effects of GM1 ganglioside on developing and mature serotonin and noradrenaline neurons lesioned by selective neurotoxins. J Neurosci Res 12:459–475

Karpiak SE, Li YS, Mahadik SP (1987) Gangliosides (GM1 and AGF2) reduce mortality due to ischemia: protection of membrane function. Stroke 18:184–187

Katoh-Semba R, Skaper SD, Varon S (1984) Interaction of GM1 ganglioside with PC12 pheochromocytoma cells: serum and NGF-dependent effects on neuritic growth (and proliferation). J Neurosci Res 12:299–310

Leon A, Facci L, Benvegnù D, Toffano G (1982) Morphological and biochemical effects of gangliosides in neuroblastoma cells. Dev Neurosci 5:108–114

Leon A, Benvegnù D, Dal Toso R, Presti D, Facci L, Giorgi O, Toffano G (1984) Dorsal root ganglia and nerve growth factor: a model for understanding the mechanism of GM1 effects on neuronal repair. J Neurosci Res 12:277–287

Leon A, Dal Toso R, Presti D et al. (1988) Development and survival of neurons in dissociated fetal mesencephalic serum-free cell cultures: II modulatory effects of gangliosides. J Neurosci 8:746–753

Li YS, Mahadik SP, Rapport MM, Karpiak SE (1986) Acute effects of GM_1 ganglioside: reduction in both behavioral asymmetry and loss of Na^+-K^+ ATPase after nigrostriatal transection. Brain Res 377:292–297

Nieto-Sampedro M, Lewis ER, Cotman CW et al. (1982) Brain injury causes a time dependent increase in neuronotrophic activity at the lesion site. Science 217:860–861

Obata K, Oide M, Handa S (1977) Effects of glycolipids on in vitro development of neuromuscular junction. Nature 266:369–371

Oderfeld-Nowak B, Skup M, Ulas J, Jezierska M, Gradkowska M, Zaremba M (1984) Effect of GM1 ganglioside treatment on postlesion responses of cholinergic enzymes in rat hippocampus after various partial deafferentations. J Neurosci Res 12:409–420

Roisen FJ, Bartfeld H, Nagele R, Yorke G (1981) Ganglioside stimulation of axonal sprouting in vitro. Science 214:577–578

Rothman SM, Olney JW (1986) Glutamate and the pathophysiology of hypoxic-ischemic brain damage. Ann Neurol 19:105–111

Skaper SD, Katoh-Semba R, Varon S (1985) GM1 ganglioside accellerates neurite outgrowth from primary peripheral and central neurons under selective culture conditions. Dev Brain Res 23:19–26

Sofroniew MV, Pearson RCA, Cuello AC, Tagari PC, Stephens PH (1986) Parenterally administered GM1 ganglioside prevents retrograde degeneration of cholinergic cells of the rat basal forebrain. Brain Res 398:393–396

Spoerri PE (1983) Effects of gangliosides on the in vitro development of neuroblastoma cells: an ultrastructural study. Int J Dev Neurosci 6:383–391

Tanaka K, Dora E, Urbanics R, Greenberg JH, Toffano G, Reivich M (1986) Effect of the ganglioside GM1 on cerebral metabolism, microcirculation, recovery kinetics of ECoG and histology, during the recovery period following focal ischemia in cats. Stroke 17:1170–1178

Tettamanti G, Ledeen RW, Sandhoff K, Nagai Y, Toffano G (eds) (1986) Gangliosides and neuronal plasticity. Liviana, Padova

Toffano G, Savoini G, Moroni F, Lombardi G, Calzà L, Agnati LF (1983) GM1 ganglioside stimulates the regeneration of dopaminergic neurons in the central nervous system. Brain Res 261:163–166

Toffano G, Agnati LF, Fuxe K, Aldinio C, Consolazione A, Valenti G, Savoini G (1984) Effect of GM1 ganglioside treatment on the recovery of dopaminergic nigro-striatal neurons after different types of lesion. Acta Physiol Scand 122:313–321

Vaccarino F, Guidotti A, Costa E (1987) Sphingoglycolipid inhibition of glutamate-mediated protein kinase C translocation and activation. PNAS 84:8707–8711

Vantini G, Fusco M, Bigon E, Leon A (1988) GM1 ganglioside potentiates the effect of nerve growth factors in preventing vinblastine-induced sympathectomy in newborn rats. Brain Res 448:252–258

Varon S, Manthorpe M, Williams LR (1984) Neuronotrophic and neurite-promoting factors and their clinical potentials. Dev Neurosci 6:73–100

Central and Cerebellar Norepinephrine Depletion and Vestibulo-ocular Reflex (VOR) Adaptation

J.G. McElligott[1] and W. Freedman[2]

[1] Department of Pharmacology, Temple University School of Medicine, Philadelphia, PA 19140, USA

[2] Department of Electrical and Computer Engineering, Drexel University, Philadelphia, PA 19104, USA

1 Introduction

Adaptation of the vestibulo-ocular reflex (VOR) presents an attractive model for investigating neural plasticity in the central nervous system because a great deal is known about the neural organization that is responsible for its operation. Modifications in the gain of the VOR can be brought about by a mismatch of visual and vestibular inputs. This can be accomplished by modifying vision with reversing prisms (Gonshor and Melvill Jones 1976), magnifying or reducing lenses (Miles and Fuller 1974; Demer and Robinson 1982), or by altering the normal visual environment in other ways to produce a conflict between the two senses (Ito et al. 1974; Maioli and Precht 1984). The result is an adjustment of the gain of the VOR that allows this reflex to continue to carry out its primary function, i.e., to maintain an image of visual interest on the fovea of the eye in the presence of head movements. This reflex has a well-established time course during which adaptation takes place. Modifications to the gain of the VOR occur within a relatively short period (1–2 h) and can be maintained over an extended period of time (days to weeks).

Experimental evidence has been put forth to either affirm (Ito 1984) or deny (Miles and Lisberger 1981; Demer and Robinson 1982) that the cerebellum is the neuronal site where these modifications take place. Lesion studies (Ito et al. 1979; Ito and Miyashita 1975; Ito et al. 1974; Robinson 1976; Demer and Robinson 1982) have been one of the more popular techniques used in an effort to identify the site of the neural changes that underlie VOR adaptation. Yet the question remains largely unanswered. However, there appears to be common agreement that damage or removal of the flocculus of the cerebellum will prevent gain adaptation of the VOR.

1.1 The Noradrenergic System

Recent work has implicated norepinephrine (NE) as an important transmitter and modulator that facilitates learning, adaptation, as well as developmental and behavioral changes. The small pontine nucleus locus coerulus (LC) and its many projections are a major source of central NE. Crow (1968) suggested that NE released in the cerebral cortex from the terminals of the dorsal noradrenergic bundle (DNB) may serve to change the synaptic structure of cortical circuits

H. Flohr (Ed.)
Post-Lesion Neural Plasticity

which could encode the learning experience. Kety (1970) hypothesized that brain NE may facilitate the consolidation of learning by stimulating protein synthesis or by some other trophic processes occurring at recently activated synapses.

Moreover, an extensive number of studies have attempted to show that NE is essential during early development for the formation of functional connections in the visual cortex. Both electrophysiological (Kasamatsu and Pettigrew 1976) as well as behavioral (Gordon et al. 1986) evidence indicate that depletion of NE with 6-OHDA will negate the effect of visual experience in monocularly occluded kittens during the critical development period. However, this work has not been without controversy, since a number of investigators (Daw et al. 1985a, b; Trombley et al. 1986) have not been able to repeat all the findings of the inital studies by Kasamatsu and his co-workers. However, Kasamatsu and Shirokawa (1985) provided additional evidence that the 6-OHDA induced deficit is due to NE depletion since they can produce the same results using a noradrenergic beta-blocker. Furthermore, recent work (Bear and Singer 1986) has implicated acetylcholine (Ach) along with norepinephrine as being responsible for the deficits brought about by 6-OHDA induced depletions.

1.2 Cerebellar Noradrenergic System

The cerebellum, which has been implicated as an important site involved in VOR adaptation, also has a rich noradrenergic innervation which is derived entirely from the locus coeruleus (LC) (Bloom et al. 1971; Hoffer et al. 1973; Kobayashi et al. 1974; Olson and Fuxe 1971). The proximal dendrites of Purkinje cells, within the molecular layer of the cortex, receive nearly all of the innervation. Gilbert (1975) has postulated that this innervation is necessary for motor responses to be learned. Experimental evidence has shown that adaptive locomotor changes (Watson and McElligott 1983, 1984) are associated with cerebellar NE reductions. More specifically, one of these studies (Watson and McElligott 1984) indicated that the ability to learn specific locomotor tasks correlates directly with cerebellar NE concentration.

One of the important issues that has arisen in the investigation of VOR adaptation has been to identify the site or sites in the neural circuitry where the neural changes occur that are responsible for this phenomenon. Discovery of these sites will facilitate our ability to understand the underlying processes for VOR adaptation and possibly for neural plastic changes in general.

We wished to investigate the influence of norepinephrine on VOR adaptation and to localize where in the brain this influence is being manifested. In particular, we have attempted to determine if a generalized depletion of central stores of NE has a detrimental effect on VOR adaptation and then to investigate if a more selective depletion of NE, specifically in the cerebellum, also adversely affected VOR adaptive gain changes.

2 Methods

A preformed ocular search coil (Judge et al. 1980) was implanted under the conjunctiva of female cats (n = 10) in order to measure their eye movements. These

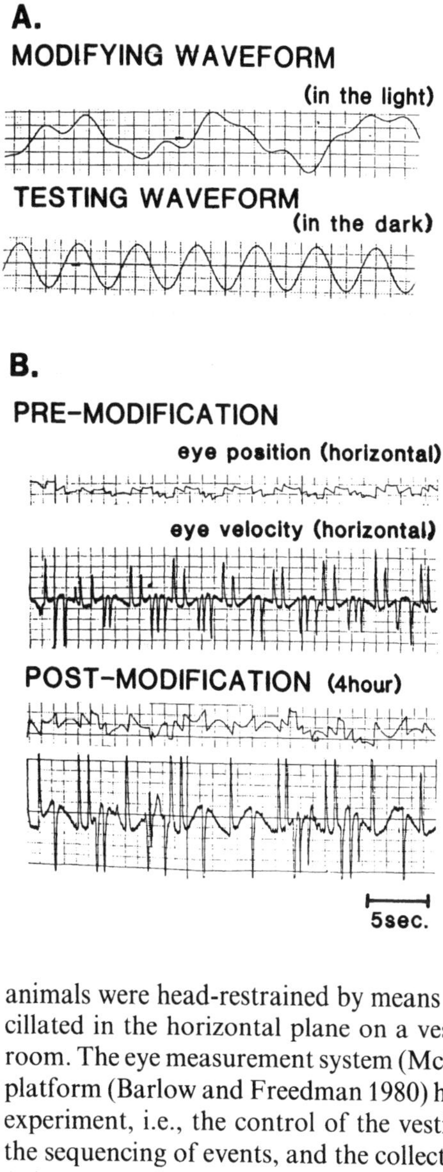

Fig. 1. a A complex waveform (*Modifying Waveform*), produced by the computer which is constructed by a linear combination of several sinusoidal frequencies (0.07 to 0.4 Hz). This signal drives the platform and drum (180 ° out of phase with each other) during VOR modification. **b** Measurement of the VOR gain is carried out in the dark using a 0.25 Hz (±10 °) signal (*Testing Waveform*). Horizontal eye position and velocity signals for a test animal at the beginning (*Pre-Modification*) and after 4 h (*Post-Modification*) of exposure to the adaptive VOR gain increase paradigm

animals were head-restrained by means of an acrylic head block implant and oscillated in the horizontal plane on a vestibular testing platform in a light-proof room. The eye measurement system (McElligott et al. 1979) and the vestibular test platform (Barlow and Freedman 1980) have been described previously. The entire experiment, i.e., the control of the vestibular platform and optokinetic display, the sequencing of events, and the collection of data, was under computer control (Digital Equipment Corp. PDP 11/10).

Each animal, while visually fixating on an object of interest, was oscillated at the test frequency (0.25 Hz ± 10 °). Visual fixation served as a reference against which all VOR values were measured and as a control for the integrity of the visual system after depletion of norepinephrine.

In order to increase the gain of the VOR, the platform oscillated each animal in a pseudo-random fashion (sum of four sinusoids between 0.07–0.4 Hz; Fig. 1 a) as the illuminated optokinetic drum was rotated 180 ° out of phase with the table. Each animal was constantly monitored by the experimenter to ensure that it was in an alert state. The number of saccades made by an animal during the modification periods was counted and used as an estimate of its level of alertness.

At the end of each 15 min period, the lights were extinguished and the test animal was rotated at a single frequency (0.25 Hz ± 10 °; Fig. 1 a) in the dark for 32 s to measure the gain and phase angle of the VOR. This procedure was repeated 16 times over 4 h. The signals of table, optokinetic drum position, as well as horizontal eye position and velocity were recorded on a pen recorder (Brush Model# 680). Table and eye position signals were also entered into a computer and stored on a magnetic hard disk for later analysis.

The animal was alerted by loud noises or sounds made during alternate test periods (i.e., at the 0.5 h, 1.0 h, 1.5 h, etc.). During the other test periods (i.e., at the 0.25 h, 0.75 h, 1.25 h, etc.) no arousing stimuli were presented. VOR data were analyzed only for the eight activated test periods.

The horizontal eye position signals were analyzed on a PDP 11/73 computer by a set of programs which removed the fast phases (i.e., saccades) and reconstructed the slow phase eye position signal. The gain and phase angle of the remaining slow phase of the horizontal eye position signal with respect to the head and platform position output signal were computed for the 32 s VOR test. The program picked out the "best" 40% of data in the 32 s test period where "best" was defined as the section of the reconstructed data in which the peak-to-peak value of the horizontal eye signal was the largest. Fast Fourier transforms (FFT) of the reconstructed eye signal and the head position signal were calculated. The ratio (FFT-eye/FFT-head) was computed to yield the VOR gain and phase angle at the test frequency (0.25 Hz).

After modification of the VOR, each animal was returned to the colony for a period of 1 to 2 weeks in order for the VOR gain to return to its normal value. Central stores of norepinephrine were then depleted by intra-cisternal injection of 6-OHDA. Solutions of 6-OHDA were prepared within a cold nitrogen atmosphere to prevent auto-oxidation. The 6-OHDA was dissolved in 0.9% sterile saline with 0.1 mg/ml ascorbic acid added as an antioxidant. In order to provide maximum preferential depletion of central stores of NE and to minimize interference with dopaminergic systems, 6-OHDA was administered to the first ten cats in three doses (100 μl; 0.4–0.8 mg) spaced over a 7 day period, as has been described previously (Watson and McElligott 1983).

For the second experiment, we wished to obtain a selective lesion of cerebellar NE without any depletion to other areas of the central nervous system. Thus, in three animals bilateral infusions of the toxin, 6-OHDA, were introduced directly into the coeruleo-cerebellar noradrenergic pathway. The toxin was infused through microdialysis probes (free base concentration = 120 ng/μl; flow rate = 1 μl/min; duration = 1.5 h) that were stereotaxically placed in the coeruleo-cerebellar pathway of a ketamine anesthetized animal. These probes (Bioanalytic Systems Inc. Part #MF5140; dialysis tube, length = 3 mm, o.d. = 200 μm) allow a more widespread diffusion of the toxin than is possible using conventional intracerebral injection probes.

Before the resumption of behavioral testing, a post-infusion recovery period of 2 weeks was allowed for 6-OHDA uptake and for subsequent degeneration of the catecholamine system to occur. The VOR modification test procedure was then repeated.

At the conclusion of testing, neurochemical analysis was carried out in order to verify the extent of catecholamine depletion. The cats were anesthetized (ketamine 50 mg/kg) and placed in a sterotaxic instrument. The cerebrum and cerebellum were exposed by performing a craniotomy. Tissue samples from the cerebellum, visual cortex, and brainstem were collected.

The samples were rapidly excised and placed on a glass plate over ice to minimize the oxidation of catecholamines during dissection. The excised tissue was wrapped in aluminum foil and immediately quick-frozen by immersion in liquid nitrogen. It was kept briefly over solid carbon dioxide until subsequent storage at $-90\,^{\circ}$ F. Quantitative determinations of norepinephrine and dopamine were made using high performance liquid chromatography with electrochemical detection (HPLC-ED) (Watson and McElligott 1983).

For the first experiment which involved the more generalized NE depletion by intra-cisternal injection of 6-OHDA, there was a reduction of cerebellar NE to 21% of the control values. A significant depletion of NE also occurred in the visual cortex (31% of control). The brain stem showed some decrement in NE which was not statistically significant (n.s., Student's t-test). Dopamine levels were not significantly affected. For the second part of this study, when infusion was made directly into the cerebellum by a bilateral infusion into the coeruleo-cerebellar pathway, cerebellar NE was reduced to 25% of controls without any depletion of NE or DA in other regions of the brain.

3 Results

3.1. VOR Modification Before and After Central Norepinephrine Depletion

3.1.1 VOR Modification Before 6-OHDA Injection

Ten head-restrained cats were oscillated at a test frequency of $0.25\ \mathrm{Hz} \pm 10\,^{\circ}$ in the horizontal plane via a hydraulically driven vestibular platform in the light to test their ability to fixate on an object of visual interest and then in the dark to measure the initial gain and phase of the VOR. The average gain of the VOR relative to that obtained during the visual fixation in the light for all the animals was 0.86 ± 0.20 with a range from 0.48 to 1.11 for the ten animals tested.

Each animal was oscillated in a pseudo-random fashion (modifying waveform; Fig. 1 a) as an illuminated optokinetic drum was rotated 180 degrees out of phase with the table in order to increase the gain of the VOR. VOR testing was carried out every 15 min in the dark at the test frequency ($0.25\ \mathrm{Hz} \pm 10\,^{\circ}$). However, the VOR data were analyzed only for those periods where each animal was alerted by loud noises or sounds which occurred at the end of each 30 min segment during the 4 h of modification.

There was an increase in the average VOR gain for the ten cats tested (Figs. 1 b and 2) over the first 1.5 h, which was followed by a period of elevated gain (1.27 ± 0.05) during the remainder of the modification period. There was no change in the phase angle (av. $= -176.7 \pm 6.5$) at the test frequency (0.25 Hz) over this period.

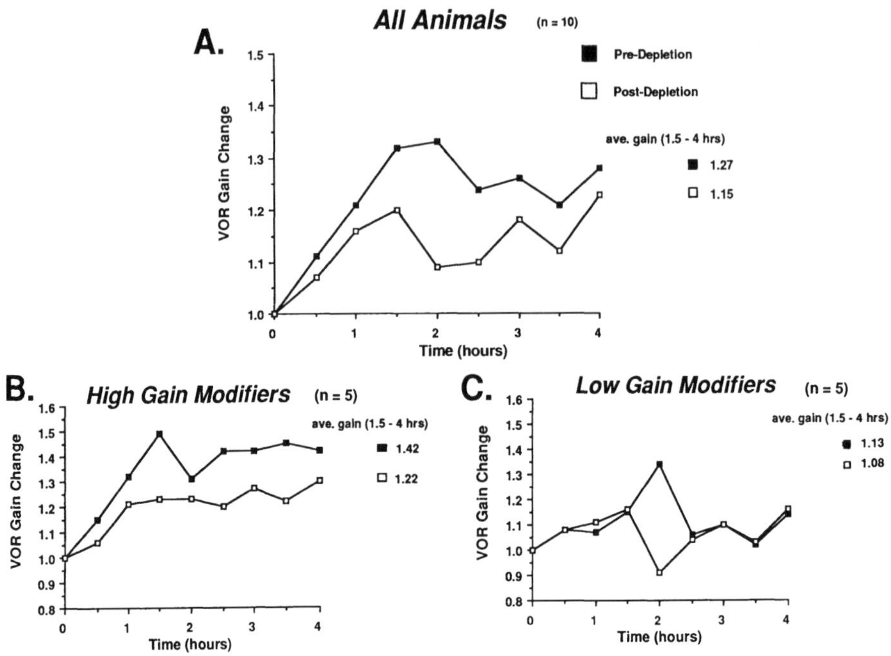

Fig. 2. This figure depicts the adaptive VOR gain increases over a 4 h modification period (**a**) averaged for all the animals (n = 10) before (*dark boxes*) and after (*light boxes*) administration of intra-cisternal injection of 6-OHDA which depletes central nervous system NE. The normalized average gain achieved was significantly reduced from 1.27 ± 0.05 to 1.15 ± 0.06 after 6-OHDA treatment (two-way ANOVA $p \leqq 0.05$). In **b** and **c**, the data are presented for these same animals when they are divided into high and low gain modifiers respectively. It is evident that the gain is significantly reduced in the high gain modifiers after 6-OHDA treatment and that little effect is observed in those animals classified as low gain modifiers

3.1.2 VOR Modification After 6-OHDA Injection

One to two weeks after initial modification and testing, the same ten animals were injected intra-cisternally with 6-OHDA to reduce NE levels in the central nervous system. Following a post-injection period of 2 weeks, during which time central stores of NE were depleted, the animals were retested for their ability to increase the gain of their VOR. The post 6-OHDA VOR gain increase (open squares, Fig. 2a) was reduced significantly over the entire time course of the modification (two-way ANOVA $p \leqq 0.05$). The animals achieved an average gain of only 1.15 ± 0.06. The phase angles both before and after 6-OHDA treatment were similar, and did not change as a result of this treatment nor as a consequence of the VOR modification.

There was a wide variety of gain changes manifested by the ten cats prior to depletion (range = 0.98 to 1.62; Fig. 3; dark squares). If these animals were divided into low and high modifier groups, then the effect of NE depletion on changes in the VOR gain became much more pronounced. A high gain modifier was defined to be any animal which achieved an average gain $\geqq 1.25$. For these high gain cats (n = 5; Fig. 2b), the average pre 6-OHDA gain change achieved was

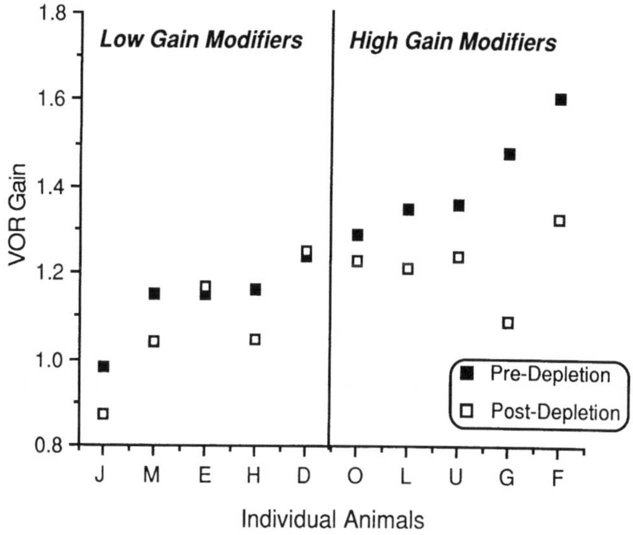

Fig. 3. There were a wide variety of gain changes manifested by each of the ten cats in the normal state prior to depletion (dark boxes; range = 0.98 to 1.62). This figure depicts in rank order the VOR gain changes achieved by individual animals after 4 h of exposure to the adaptive VOR gain increase paradigm. These animals can be divided into low and high gain modifier groups. A high gain modifier was defined to be a cat that achieved an average gain ≧1.25. High gain modifiers show a greater decrement in VOR gain increase after depletion

1.42. After 6-OHDA treatment, the average gain achieved was 1.22, which was a statistically significant reduction (two-way ANOVA $p \leqq 0.005$). The low gain cats (gain < 1.25; n = 5; Fig. 2c) achieved an average gain of 1.13 prior to the 6-OHDA treatment and 1.08 after treatment. No statistically significant increase in gain (n.s.; two-way ANOVA) was observed either before or after 6-OHDA treatment.

The average absolute gain of the VOR relative to that obtained during visual fixation was approximately the same before (0.86 ± 0.20) and after (0.90 ± 0.14) 6-OHDA treatment. In addition, no significant difference in this regard was noted between the high and low gain modifiers either before or after the 6-OHDA injection.

3.1.3 Control for Attentiveness and Arousal Level

Decrements in the ability of cats to increase VOR gain which were observed after 6-OHDA treatment could be attributed to a lack of attentiveness during the modification period or a decrease in the arousal level during the VOR testing periods. This could also explain the lack of significant increases in VOR gain in the low gain modifier animals. As a control, we attempted to keep the animals alert during the modification and test periods. Alert cats make many saccades to explore their environment whereas less attentive and drowsy animals make few of these rapid eye movements. Thus, as an index of attentiveness, counts of saccadic eye movements were made throughout the 4 h period.

The average number of saccades made during the initial modification period (i.e., during the first 15 min period when no arousing stimuli were presented) was equal both before and after 6-OHDA injection. This indicates that NE depletion did not severely alter the animals' overall arousal or attention level. Since no attempt was made by the experimenter to activate the animal during the initial period, the number of saccades made spontaneously during the initial modification period served as the index to gauge the initial level of arousal when VOR modification commenced.

An attempt was made to keep the level of saccade production during each of the subsequent 15 min modification periods approximately equal to that obtained spontaneously during the initial modification period. The data for the average number of saccades in the high gain modifiers emitted both before and after 6-OHDA administration were not significantly different and were relatively constant (approximately 30/min) during the period (n.s.; two-way ANOVA). Similar results were also obtained for the number of saccades emitted before and after 6-OHDA treatment in the low gain modifiers (n.s.; two-way ANOVA).

As an additional control, similar VOR gain modifications were carried out after each animal was injected with amphetamine (5 mg, S.C.). This produced a highly alerted animal during the entire modification period. Under amphetamine, a decrement in the ability of the high gain modifiers to produce an increase in the gain of the VOR after central NE depletion was also noted (McElligott and Freedman 1987).

3.2 VOR Modification Before and After Selective Cerebellar Norepinephrine Depletion

Studies have been initiated to localize more specifically the region in the central nervous system where the depletion of NE via 6-OHDA may be having an effect on the modification of the VOR gain. We have carried out experiments similar to those described in the first part of the study, but infused 6-OHDA directly into the noradrenergic fibers as they course through the cerebellum. In our initial experiments, we have found that bilateral infusion of 6-OHDA into the coeruleo-cerebellar pathway via a microdialysis probe will produce a selective depletion of NE in the cerebellum without producing a decrease in other parts of the central nervous system.

This study was carried out in three cats in a manner similar to that of the previous experiment but VOR gain modification was conducted over a 2.5 h period since it had been shown previously that the animals achieved the final short term modification after a 1.5 h period. In this experiment, animals were alerted at each VOR testing period (i.e., every 15 min). Data for two adjacent periods were averaged together and plotted at 30 min intervals (Fig. 4).

As can be seen from Fig. 4, for these three animals there was an average increase in VOR gain to 1.30 for initial adaptation prior to cerebellar depletion by intra-cerebellar NE depletion. When these animals underwent 2.5 h of VOR modification after NE depletion, the average VOR gain did not change (average VOR gain change = 0.98). However, when the gain changes for the individual animals are examined (Fig. 5), it is clear that as in the previous study, there were high (Cats

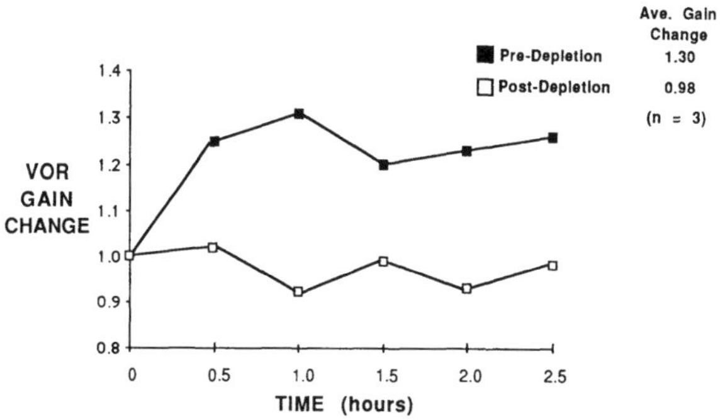

Fig. 4. This figure depicts the adaptive VOR gain increases over a 2.5 h modification period averaged for the three animals before (*dark boxes*) and after (*light boxes*) administration of bilateral intra-cerebellar injection of 6-OHDA which exclusively depletes cerebellar NE. The normalized average gain achieved prior to depletion was 1.30. No gain change (0.98) was observed for the group following the selective cerebellar NE depletion

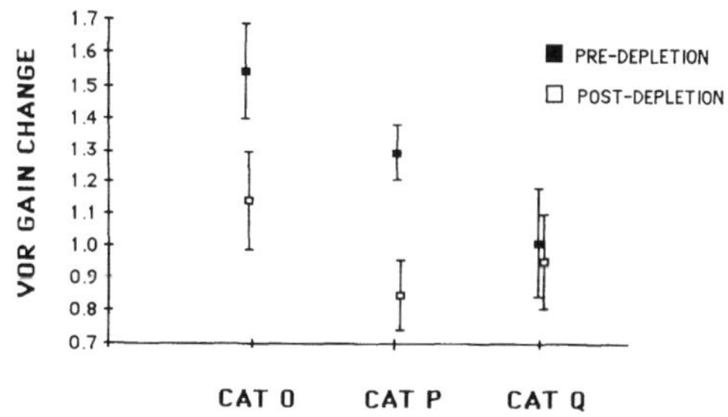

Fig. 5. This figure depicts the VOR gain changes achieved by the three animals after 2.5 h of exposure to the adaptive VOR gain increase paradigm before (*dark boxes*) and after (*light boxes*) cerebellar NE depletion. Prior to depletion these three cats manifested different VOR gain changes like that observed in the first experiment (Fig. 3). As in the previous experiment, high gain modifiers (Cats *O* and *P*) show significant decrements in VOR gain increase after depletion

O and P) and low gain modifiers (Cat Q). The most significant decrement in the ability to modify VOR gain occurred in the high gain modifiers.

4 Discussion

4.1 Central Norepinephrine Depletion and VOR Adaptation

Over a 4 h period, cats that were kept alert increased the gain of their horizontal VOR when rotated in the horizontal plane in a manner that was equal in magnitude but opposite in direction (180 ° out of phase) to that of a surrounding optokinetic drum. This increase in gain occurred during the first 1.5 h of modification and remained elevated over the following period of continued modification. When these same animals were intra-cisternally injected with 6-OHDA which depleted central stores of NE, there was an overall reduction in the VOR gain increase over the same time period. This phenomenon also occurred in these same animals when kept aroused and in a highly alerted state following a subcutaneous injection of amphetamine.

The magnitude of the VOR gain increase prior to depletion was different for each of the ten animals. Individual animals could manifest a significant increase in VOR gain (Cat F; 1.62) or could produce no increase (Cat J; 0.98) at all (Fig. 3). The most severe deficit in VOR adaptation after NE depletion occurred in those animals (high gain modifiers) that produced the greatest gain increase prior to 6-OHDA injection. The animals that manifested little or no VOR adaptation before the depletion (low gain modifiers), showed no significant decrement in adaptation (average VOR gain change before NE depletion = 1.13; after NE depletion = 1.08). For these animals it would be difficult to measure a decrement in VOR adaptation, since they manifested poor, if any, adaptation prior to depletion.

It is well known that VOR gain can be quite variable and that low or non optimal VOR gains occur if an animal is not fully alerted or activated. There were several indications that the animals maintained adequate levels of attention and that the decreases in VOR gain change after 6-OHDA were not due to decreases in activation or arousal levels during either the 15 min modifying or the 32 s testing periods.

Firstly, we have taken as a level of attentiveness the number of spontaneous saccades made by the animal during the initial 15 min modification period. Using this index, we find that equal levels of attention were maintained throughout all the modification periods both before and after NE depletion. Secondly, injection of amphetamine produced a fully alerted animal during the modification and test periods. Thus, the decrease in VOR gain change after 6-OHDA could not be attributed to a lack of attentiveness. Furthermore, the low gain modifiers had saccade counts that were not statistically different from those of the high gain modifiers, which indicates comparable alertness levels for these two groups of animals.

A study by Keller and Smith (1983) also showed that depletion of NE by intracisternal 6-OHDA injection hinders adaptive VOR gain decreases in cats. Our study confirmed their previously reported result and showed that NE depletion impeded adaptive VOR gain increases as well. Furthermore, our study also demonstrated that this reduction in adaptation was not just due to a decrease in attention or arousal levels.

It is also important to note that the doses of 6-OHDA administered to the animals did not cause any overt behavioral or other physical signs that can accompany large doses of 6-OHDA, such as motor disturbances, weight loss, sham rage, and lack of grooming behavior (Jonsson et al. 1972). In fact, it was not possible to detect any overt change in the normal behavior of the animals after the injection. We attribute this to the fact that the 6-OHDA injections were given in smaller multiple doses which tend to preserve the DA systems while still depleting the NE systems (Watson and McElligott 1983).

These experiments indicate that NE plays an important role in the mechanism of VOR adaptation. NE is an ubiquitous neurotransmitter which has a wide spread distribution in all cortical areas of the brain and, in addition, is found in many subcortical areas. For example, there is a substantial NE projection to the inferior olive with little or no evidence for its presence in the vestibular nuclei or abducens nuclei (Kobayashi et al. 1974; Moore and Bloom 1979). There are also many other oculomotor structures in the brain stem that receive NE innervation (Moore and Bloom 1979). Given the route of the injection of 6-OHDA and the widespread depletion of NE, it is not possible in this experiment to localize precisely the site(s) where NE may be acting.

4.2 Cerebellar Norepinephrine Depletion and VOR Adaptation

In the first experiment when NE was depleted by intra-cisternal administration of 6-OHDA, the most severe depletion occurred in the cerebellum. Thus, a next logical experiment was to deplete cerebellar NE without affecting transmitter concentration in other areas of the nervous system. When this experiment was carried out in three animals (Fig. 4), the results showed that selective cerebellar NE depletion impeded VOR gain modification in a manner similar to that found when there was a more generalized CNS noradrenergic depletion (Fig. 2). Two out of the three animals showed a significant decrement in VOR adaptation after selective cerebellar NE depletion. The third animal in this experiment was similar to the low gain modifiers in the first experiment and showed no adaptive VOR gain changes before or after depletion.

Norepinephrine has been shown to act as an important neuromodulatory agent in the cerebellum. NE whether iontophoretically applied (Siggins et al. 1971) or released consequent to LC stimulation (Hoffer et al. 1973), produces a depression of spontaneous activity in Purkinje cells. Furthermore, Purkinje cell responsiveness to excitation via climbing and mossy fiber afferents, and to inhibition by basket and stellate cells is augmented (Freedman et al. 1977). Thus, NE selectively improves the signal-to-noise ratio of evoked versus spontaneous activity, enhancing the sensitivity of cerebellar circuitry to both excitatory and inhibitory afferent synaptic inputs (Ebner and Bloedel 1981; Moises and Woodward 1980; Moises et al. 1979). Low doses of NE having no direct depressant action also produce an increased sensitivity to evoked activity. A recent study (McElligott et al. 1986) has shown that climbing fiber enhancement which was originally studied by Ebner and Bloedel (Ebner and Bloedel 1981; Ebner et al. 1983) is significantly diminished when the NE level in the cerebellum is reduced. Climbing fiber enhancement is the phenomenon whereby somatosensory mossy fiber signals to Purkinje cells are enhanced in the presence of a concomitant climbing fiber response.

Woodward et al. (1979) have postulated that cerebellar NE is not directly involved with specific or detailed information transfer, but rather that NE serves in the cerebellum and throughout the central nervous system as a neuromodulator, altering postsynaptic responsiveness to conventional neurotransmitter systems. Gilbert (1975) has proposed that increased discharge of locus coeruleus cells releases large quantities of NE at Purkinje cell synapses to signal a favorable outcome worthy of consolidation. Thus, enhanced short-term changes of mossy fiber input to the Purkinje cells effected by the climbing fibers could be turned into long-term synaptic changes.

A number of functional behavioral studies have implicated cerebellar NE in neural plasticity. For example, deficits in the ability to learn novel locomotor tasks were directly correlated with selective NE lesions in the cerebellum (Watson and McElligott 1984). In addition, recovery of function after cerebral cortical lesions has been facilitated by infusion of NE into the cerebellum (Boyeson et al. 1986). These studies lead us to believe that the cerebellum is a good candidate area where NE may be manifesting its influence on VOR adaptation.

Recently, Miyashita and Watanabe (1984) have shown that depletion of serotonin (5-HT) and norepinephrine by intraventricular injection of 5,7 dihydroxytryptamine (5,7-DHT) results in the loss of VOR adaptation in pigmented rabbits. This study also reported that there is no loss in VOR adaptation after injection of 6-OHDA which depletes cerebellar NE but not 5-HT. Thus, Miyashita and Watanabe conclude that 5-HT and not NE plays an important role in maintaining adaptative modifiability of the VOR.

This is not in agreement with the results presented in our work or in another study by Keller and Smith (1983) which implicate NE as being responsible for a reduction of VOR adaptability in the cat. This disagreement could be attributed to the species difference between the rabbit and the cat or may stem from the fact NE was depleted to a greater extent in the studies carried out in the cat.

In conclusion, our experiments have shown that the ability to produce adaptive VOR gain increases is impeded in cats after 6-OHDA induced NE depletion. Two separate experiments, the first after a general nervous system NE depletion and the second after a more localized cerebellar NE depletion, show that the ability to produce adaptive VOR gain increases is hindered. Thus, the neuronal mechanism which controls adaptive VOR gain changes could be mediated by cerebellar norepinephrine. Further experimentation to test this hypothesis is now underway.

Acknowledgements. It is with pleasure that we acknowledge the technical assistance and enthusiasm of Troy Carter during the course of these experiments. This work was supported by grants from the National Science Foundation (BNS-8410231) and from NIH (SO7-RR05417).

References

Barlow D, Freedman W (1980) Human cervico-ocular reflex in the adult. Acta Oto-Laryngol 89:487–496

Bear MF, Singer W (1986) Modulation of visual cortical plasticity by acetylcholine and noradrenaline. Nature 320:172–176

Bloom F, Hoffer BJ, Siggins GR (1971) Studies on norepinephrine containing afferents to Purkinje cells of a rat cerebellum. I. Localization of fibers and their synapses. Brain Res 25:501–521

Boyeson MG, Krobert KA, Hughes JM (1986) Norepinephrine infusions into the cerebellum facilitate recovery from sensori-motor cortex injury in the rat. Soc Neurosci Abst 12:#304.13, 1120

Crow TJ (1968) Cortical synapses and reinforcement: a hypothesis. Nature 219:736–737

Daw NW, Videen TO, Parkinson D, Rader K (1985a) DSP-4 depletes noradrenaline in kitten visual cortex without altering the effects of monocular deprivation J Neurosci 5:1925–1933

Daw NW, Videen TO, Rader RK, Robertson T, Coscia CJ (1985b) Substantial reduction of noradrenalin in kitten visual cortex by intraventricular injections of 6-hydroxydopamine does not always prevent ocular dominance shifts after monocular deprivation. Exp Brain Res 59:30–35

Demer JL, Robinson DA (1982) Effects of reversible lesions and stimulation of olivocerebellar system on vestibulo-ocular reflex plasticity. J Neurophysiol 47:1084–1107
Ebner JT, Bloedel JR (1981) Role of climbing fiber afferent input in determining responsiveness of Purkinje cells to mossy fiber inputs. J Neurophysiol 45:962–971
Ebner T, Qi-Xiang Y, Bloedel JR (1983) Increase in Purkinje cell gain associated with naturally activated climbing fiber input. J Neurophysiol 50:205–219
Freedman R, Hoffer BJ, Woodward DJ, Puro DA (1977) A functional role for the adrenergic input to the cerebellar cortex: interaction of norepinephrine with activity evoked by mossy and climbing fibers. Exp Neurol 55:269–288
Gilbert PFC (1975) How the cerebellum could memorize movements. Nature 254:688–689
Gonshor A, Melvill Jones G (1976) Extreme vestibulo-ocular adaptation induced by prolonged optical reversal of vision. J Physiol (Lond) 256:381–384
Gordon B, Moran J, Trombley P, Soyke J (1986) Visual behavior of monocularly deprived kittens treated with 6-hydroxydopamine. Brain Res 389:21–29
Hoffer BJ, Siggins GR, Oliver AP, Bloom FE (1973) Activation of the pathway from locus coeruleus to rat cerebellar Purkinje neurons: pharmacological evidence for central adrenergic inhibition. J Pharmacol Exp Ther 184:553–569
Ito M (1984) The cerebellum and neural control. Raven, New York
Ito M, Jastreboff PJ, Miyashita Y (1979) Adaptive modification of the rabbit's horizontal vestibulo-ocular reflex during sustained vestibular and optokinetic stimulation. Exp Brain Res 37:17–30
Ito M, Miyashita Y (1975) The effects of chronic destruction of the inferior olive upon visual modification of the horizontal vestibulo-ocular reflex of rabbits. Proc Jpn Acad 51:716–720
Ito M, Shiiga T, Yagi N, Yamamoto M (1974) The cerebellar modification of rabbits' horizontal vestibulo-ocular reflex induced by sustained head rotation combined with visual stimulation. Proc Jpn Acad 50:85–89
Jonsson G, Malmfors T, Sachs C (1972) Effects of drugs on the 6-hydroxydopamine induced degeneration of adrenergic nerves. Res Commun Chem Pathol Pharmacol 3:543–556
Judge SJ, Richmond BJ, Chu FC (1980) Implantation of magnetic search coils for measurement of eye position: an improved method. Vision Res 20:535–553
Kasamatsu T, Pettigrew JD (1976) Depletion of brain catecholamines: failure of ocular dominance shift after monocular occlusion in kittens. Science 194:206–209
Kasamatsu T, Shirokawa T (1985) Involvement of β-adrenoreceptors in the shift of ocular dominance after monocular deprivation. Exp Brain Res 59:507–514
Keller EL, Smith MJ (1983) Suppressed visual adaptation of the vestibulo-ocular reflex in catecholamine depleted cats. Brain Res 258:323–327
Kety SS (1970) The biogenic amines in the central nervous system: their possible roles in arousal, emotion, and learning. In: Schmidt FO (ed) The Neurosciences: Second Study Program. Rockefeller University Press, New York, pp 324–336
Kobayashi RM, Palkovits M, Kopin IJ, Jacobowitz DM (1974) Biochemical mapping of the noradrenergic nerves arising from the rat locus coeruleus. Brain Res 77:269–279
Maioli C, Precht W (1984) The horizontal optokinetic nystagmus in the cat. Exp Brain Res 55(3):494–506
McElligott JG, Ebner TJ, Bloedel JR (1986) Reduction of cerebellar norepinephrine alters climbing fiber enhancement of mossy fiber input to the Purkinje cell. Brain Res 397:245–252
McElligott JG, Freedman W (1988) Vestibulo-ocular reflex adaptation in cats before and after depletion of norepinephrine. Exp Brain Res 69:509–521
McElligott JG, Loughnane M, Mays L (1979) The use of synchronous demodulation for the measurement of eye movements by means of an ocular magnetic search coil. IEEE Trans Biomed Eng 26:370–374
Miyashita Y, Watanabe E (1984) Loss of vision-guided adaptation of the vestibulo-ocular reflex after depletion of brain serotonin in the rabbit. Neurosci Lett 51:177–182
Miles FA, Fuller JH (1974) Adaptive plasticity of the vestibulo-ocular response of the rhesus monkey. Brain Res 80:512–516
Miles FA, Lisberger SA (1981) Plasticity in the vestibulo-ocular reflex: a new hypothesis. Annu Rev Neurosci 4:273–299

Moises HC, Woodward DJ (1980) Potentiation of GABA inhibitory action in cerebellum by locus coeruleus stimulation. Brain Res 182:327–344

Moises HC, Woodward DJ, Hoffer BJ, Freedman R (1979) Interactions of norepinephrine with Purkinje cell responses to putative amino acid neurotransmitters applied by microiontophoresis. Exp Neurol 64:493–515

Moore RY, Bloom FE (1979) Central catecholamine neuron systems: anatomy and physiology of norepinephrine and epinephrine systems. Annu Res Neurosci 2:113–168

Olson L, Fuxe K (1971) On the projections from the locus coeruleus noradrenaline neurons: the cerebellar innervation. Brain Res 28:165–171

Robinson DA (1976) Adaptive gain control of the vestibulocular reflex to the cerebellum. J Neurophysiol 39:954–969

Siggins GR, Hoffer BJ, Bloom FE (1971) Studies on norepinephrine containing afferents to Purkinje cells of rat cerebellum: III. Evidence for mediation of norepinephrine effect by cyclic 3′,5′-adenosine monophosphate. Brain Res 25:535–553

Trombley P, Allen EE, Soyke J, Blaha CD, Lane RF, Gordon B (1986) Doses of 6-hydroxydopamine sufficient to deplete norepinephrine are not sufficient to decrease plasticity in the visual cortex. J Neurosci 6:266–273

Watanabe E (1984) Neuronal events correlated with long term adaptation of the horizontal vestibulo-ocular reflex in the primate flocculus. Brain Res 297:169–174

Watson M, McElligott JG (1983) Six-hydroxydopamine and the acquisition and performance of specific locomotor tasks in rats. Pharmacol Biochem Behav 18:927–934

Watson M, McElligott JG (1984) Cerebellar norepinephrine depletion and impaired acquisition of specific locomotor tasks in rats. Brain Res 296:129–138

Woodward DJ, Moises HC, Waterhouse BD, Hoffer BJ, Freedman R (1979) Modulatory actions of norepinephrine in the central nervous system. Fed Proc 38:2109–2116

Sensorimotor Activity and Metabolic Factors in Vestibular Compensation

M. Lacour, L. Ez-Zaher, and C. Xerri

Université Aix-Marseille I, Département de Psychophysiologie, U.A.-C.N.R.S. N° 372, Centre de St. Jérôme, Avenue E. Normandie Niemen, 13397 Marseille Cedex 13, France

1 Introduction

Dramatic impairments of postural, locomotor and oculomotor functions are shown following unilateral vestibular neurectomy or hemilabyrinthectomy in many species. Functional recovery develops with time, leading to a total or near complete compensation of the sensorimotor deficits. It is generally assumed that these disturbances are due to the strong imbalance between the spontaneous firing rate and the dynamic response properties of the vestibular nuclei neurons on both sides (Mc Cabe and Ryu 1969; Shimazu and Precht 1966; Precht 1974; Xerri et al. 1983; Lacour et al. 1985). During the time course of vestibular compensation, symmetrical resting rates are observed and the dynamic neuronal characteristics of thc deafferented vestibular cells are partially restored (Precht et al. 1966; Pompeiano et al. 1984; Ried et al. 1984). These neurophysiological data are closely correlated with the biochemical changes observed in many central nervous structures, including the deafferented vestibular nuclei. Modifications in the energy metabolism of vestibular nuclei located on the lesioned side were shown in the rat (Llinas and Walton 1979) and in the frog (Flohr et al. 1981). By using the deoxyglucose method and by comparing the distribution of radioactivity in uncompensated and compensated animals, these authors pointed to an increase of the glucose uptake within the deafferented vestibular nuclei during the chronic stage of compensation. Assuming that metabolic activity measured as glucose uptake closely parallels functionally significant activity (Sokoloff 1977), this method provides also a good index of activity in the central nervous system.

It was proposed that vestibular compensation could be considered as a sensorimotor relearning process requiring the subject's activity (Xerri and Lacour 1980; Lacour 1981). The effect of physical exercise upon nystagmus (Igarashi et al. 1975) and locomotor balance (Igarashi et al. 1981) was demonstrated in the hemilabyrinthectomized squirrel monkey. The role of sensorimotor activity in postural and equilibrium function recovery was also well established in unilateral vestibular neurectomized monkey (Lacour et al. 1976) and cat (Xerri and Lacour 1980; Lacour 1984). These latter studies have clearly shown that sensorimotor restriction applied in the early stages of compensation prevented and strongly delayed the recovery process. It was therefore concluded that active sensorimotor exploration was needed to achieve normally developing and well-adapted compensations. Sensorimotor learning processes may also induce modifications in the

H. Flohr (Ed.)
Post-Lesion Neural Plasticity

energy metabolism in nervous structures implicated in functional recovery. Considering now that brain-lesioned patients have to lie in their bed sometimes for a rather long period of time, therefore behaving like sensorimotor-restrained human beings, it is of obvious clinical interest to find pharmacologic principles capable of counterbalancing the negative effects associated with the lack of sensorimotor activity, by potentiating the plasticity mechanisms.

The aim of the present study was to analyze the effect of an extract of *Ginkgo biloba* (EGb 761) on the postural and locomotor balance recovery after unilateral vestibular neurectomy in the cat. This extract, composed of flavonoids and *Ginkgo biloba*-specific flavone glycoside esters and terpenes, has effects on capillary permeability, ion transport, and cerebral energy metabolism. It produces changes in brain energy metabolism in the rat, characterized by an improvement of glucose and deoxyglucose uptake and an increase in the cerebral level of ATP (Rapin and Le Poncin-Lafitte 1979; Karcher et al. 1984), leading to protective effects against cerebral edema and hypoxia. The effects of EGb on the recovery process after unilateral vestibular neurectomy was assessed here by comparing the time course of recovery in sensorimotor restrained and unrestrained cats, submitted or not to a postoperative EGb-treatment.

2 Material and Methods

These experiments were performed on 14 adult cats, weighing between 3 and 4 kg, submitted to a unilateral vestibular neurectomy performed on the left side. The animals were divided into four groups including two experimental and two control groups.

The two experimental groups (N = 8 cats) were daily treated with 50 mg/kg/day of an extract of *Ginkgo biloba* (EGb 761) until 30 days postoperative. This standardized extract was the same as that used in the commercial preparation, Tanakan (IPSEN Laboratory, France). One group (group 1; n = 5 cats) was composed of animals which underwent a 7-day sensorimotor restriction (SMR) applied just after the vestibular lesion. Such a SMR is known to produce the strongest effects on vestibular compensation, by preventing and delaying postural and equilibrium recovery (cf. Lacour et al. 1976; Xerri and Lacour 1980). In the other group (group 2; n = 3 cats) animals were kept in their usual laboratory environment after surgery, and allowed to move freely.

In the two control groups (N = 6 cats), animals were not treated at all with EGb 761; but some of them (group 3; n = 3) were submitted to the same SMR as group 1 cats, while the remaining cats (group 4; n = 3) could actively behave in their environment, as group 2 animals.

The effect of the extract of *Ginkgo biloba* (EGb) on postural and locomotor balance recovery was assessed by comparing the time-course of compensation in these four groups of cats: treated and free-moving (T-F) or sensorimotor restrained (T-SMR) cats, untreated and free-moving (NT-F) or sensorimotor restrained (NT-SMR) cats. In the SMR groups, clinical observations and behavioral tests have begun just immediately after the end of the SMR application (9th postoperative day), while they were performed as soon as the first post-lesion day in the free-moving groups.

Recovery of postural and locomotor balance functions was described by means of behavioral observations and quantified by using the rotating beam test.

2.1 Vestibular Neurectomy

Unilateral vestibular neurectomy was performed on the left side, under sterile conditions and visual control through a dissecting microscope. The vestibular nerves were sectioned at a post-ganglion level after mastoidectomy, partial destruction of the bony labyrinth and surgical exposure of the internal auditory canal. The classical, locomotor, and oculomotor (nystagmus) disorders were used as criteria indicating the effectiveness of the lesion.

2.2 Sensorimotor Restriction (SMR)

A 7 days, SMR was applied during the first postoperative week (from 2nd to 9th postoperative days). The cat was placed in a small cage provided with wire-netting (40 × 23 × 24 cm) and containing a thick rubber. Under these experimental conditions, the animal had normal vision but it could neither stand erect nor move inside the cage.

2.3 Clinical Observations

Measurements of postural and locomotor disorders were daily performed and subsequent compensation was quantified using observation platforms. Four clinical states of recovery were determined on the basis of the number of falls to the operated side per unit of time, and of the degree of postural-locomotor asymmetry. Postural-locomotor asymmetries were quantified by measuring the deviation of the cat trajectory during walking (in cm/m) and the surface of the support (in cm^2) occupied by the four legs of the animal when it stood erect. Degree of head-tilt and tonus of fore- and hindlimbs were subjectively appreciated using four-point scales. State I of recovery corresponded to the stage of maximal disorders observed just after surgery. By convention, states II and III were defined as intermediate states corresponding to 30% and 60% recoveries, respectively. State IV represented an almost complete compensation of all the disorders. These rough observations have constituted rather complementary investigations, able to provide useful informations to be correlated with the data from the following well quantified behavioral test.

2.4 Locomotor Equilibrium Recovery

Locomotor equilibrium recovery was studied using a behavioral test described in previous papers (rotating beam: cf. Xerri and Lacour 1980). It consists of two compartments connected by a cylindric beam (3 m long, 12 cm diameter, 1.2 m above ground) which could be rotated along its longitudinal axis with variable speeds (0 to 35 m/min). The cats were trained before surgery to walk on the rotating beam, and we measured their maximum performance. This maximum performance (Max P) corresponded to the highest speed of rotation of the beam which did not lead to a fall for four consecutive trials. Animals reached Max P within 8 to 12 training sessions (1 h duration per day). Maximum performance was very comparable from one cat to another (range: 21 to 35 m/min). Data obtained after surgery were expressed for each cat in percent of its maximal preoperative value.

2.5 *Spontaneous Vestibular Nystagmus*

Unilateral vestibular neurectomy induces a strong spontaneous vestibular nystagmus characterized by eye movements with fast phases beating toward the intact side, and slow phases directed to the lesioned side. The eye movements were recorded using DC-electrodes (Ag-AgCl) chronically implanted in the horizontal axis of the eyeballs. Frequency of the nystagmus was expressed by the total number of beats per unit of time (10 s) as a function of the postoperative time. EOG recordings were only made in the free-moving groups of cats, in both EGb-treated and untreated groups, until total compensation. Spontaneous vestibular nystagmus recovery was studied both in the light and in total darkness. These experiments were performed every day. Total duration of recordings was about 15 min in order to prevent variations due to change in the level of alertness.

3 Results

3.1 *Clinical Observations*

Figure 1 shows the time-course of postural disorders compensation in the four groups of cats. A strong effect of EGb is observed in both free and SMR cats, characterized by faster recovery curves as compared to the untreated cats.

In the treated and free-moving cats (T-F), postural recovery rapidly develops within the first postoperative week. An almost complete compensation of the deficits is seen 9 to 10 days after surgery. On the contrary, a slower time-course is found in the untreated and free-moving group (NT-F), as shown by the slope of the recovery curve. In this latter group, the initial disorders corresponding to state I of recovery are observed during a longer period of time (4 days), while near full compensation (state IV) is achieved only after around 3 weeks postoperative. Finally, recovery leading to normal postural behavior in the free-moving cats is established about 10 days earlier in the EGb-treated cats.

Similar conclusions can be drawn when comparing the two SMR groups. Recovery is faster in the treated animals (T-SMR: 16th postoperative day, on the average) than in the untreated ones (NT-SMR: 24th postoperative day). A rather similar benefit of 8 days is observed in this treated group. Interestingly enough is the observation that postural deficits are more attenuated just at the end of the restriction period in the T-SMR cats as compared to the NT-SMR animals. In this latter group, the SMR prevents postural recovery, as already described in our previous papers (Lacour et al. 1976: monkey; Xerri and Lacour 1980: cat), and animals exhibit the same syndrome as that seen just after surgery. The lack of sen-

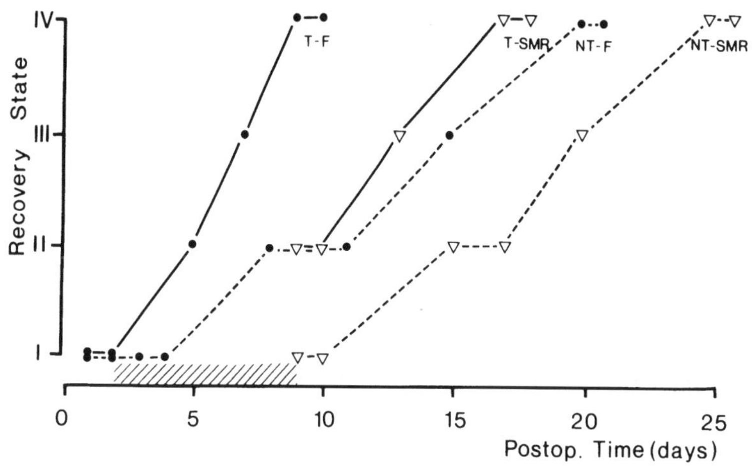

Fig. 1. Effects of the extract of *Ginkgo biloba* on postural deficits compensation in free-moving and sensorimotor restrained cats. Mean recovery curves of posturo-locomotor disorders in the four groups of cats. Group 1 (n = 5) was treated and sensorimotor restrained (T-SMR); group 2 (n = 3) was composed of treated and free-moving cats (T-F); group 3 (n = 3) was untreated and sensorimotor restrained (NT-SMR) while group 4 (n = 3) was made of untreated and free-moving animals (NT-F). On the ordinate: clinical state of recovery. State I corresponds to maximal disorders; states II and III to intermediate states with 30% and 60% recovery, respectively, and state IV to a near complete recovery. On the abscissae: postoperative time in days. The 7-day period of sensorimotor restriction (groups 1 and 3) is represented as the hatched area (from the 2nd to the 9th postoperative day). Comparison between treated (*solid lines*) and untreated (*dashed lines*) groups, free-moving (*black points*) and sensorimotor restrained (*triangles*) cats show that the more rapid recovery is found when both activity and treatment are present (*T-F*); neither activity nor treatment leads to the strongest delayed compensation (*NT-SMR*)

sorimotor activity has less dramatic effects under pharmacological treatment with the extract of *Ginkgo biloba*.

The whole of these data indicate that the best postural recovery curve is found when both activity of the animal and EGb-treatment are given, and the worst one when none of these two factors is present. Intermediate and very identical patterns of compensation are observed in the groups of cats submitted to either postoperative activity or pharmacologic treatment.

3.2 Locomotor Equilibrium Recovery

In all the groups of cats, locomotor balance recovery develops as postural deficits are nearly fully compensated. As expected, the cats remain unable to walk on the rotating beam, even in the lowest range of rotation speed, until postural asymmetry was strongly attenuated. More precisely, postural symmetry was not regained before 5 and 9 days in the treated and untreated free-moving groups, respectively, before 12 and 15 postoperative days in the treated and untreated sensorimotor restrained groups, respectively.

The more rapid recovery is again found in the cats submitted to a postoperative EGb-treatment and allowed to behave actively (T-F) in their environment just after surgery (Fig. 2). Their maximal performance reaches 50% of the control

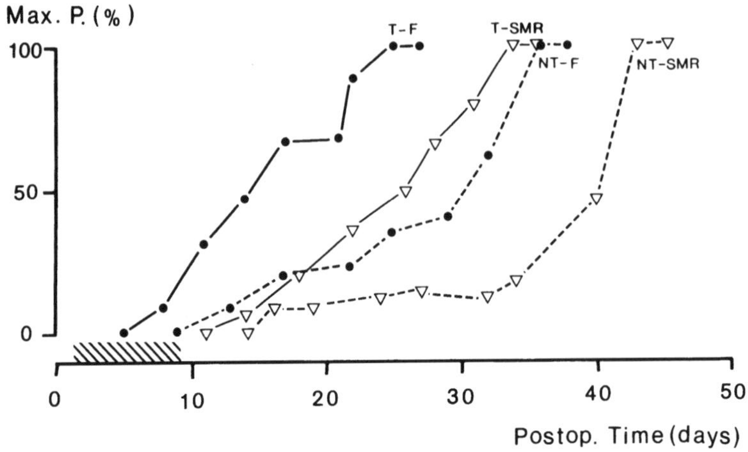

Fig. 2. Effects of the extract of *Ginkgo biloba* on locomotor equilibrium recovery in free-moving and sensorimotor restrained cats. Mean recovery curves of the maximal performance of the cats on the rotating beam as a function of the postoperative time in days (*abscissae*). The *Max. P.* (*ordinate*) is expressed in percent of the preoperative values. Same conventions as in Fig. 1: *T-F* treated and free-moving cats; *T-SMR* treated and sensorimotor restrained cats; *NT-F* untreated and free-moving cats; *NT-SMR* untreated and sensorimotor restrained cats. Period of sensorimotor restriction is indicated as the hatched area. Note again that the better recovery is found in the T-F group and the worst in the NT-SMR one

preoperative level as soon as 14 days postoperative, and 100% at the end of the third postoperative week. On the contrary, a significant and important delay is observed in the untreated free-moving group of cats (NT-F), with 50% and 100% of the control level of maximal performance reached around 30 and 35 postoperative days, respectively.

Delay in the locomotor balance recovery is also more prominent in the sensorimotor restrained and untreated group (NT-SMR), since equilibrium recovery remains at a very low level of performance (10%) until more than 1 month after surgery. Complete recovery is only observed around 45 postoperative days. On the contrary, locomotor balance recovery is more rapidly achieved in the T-SMR cats. Moreover, recovery develops more regularly in this latter group with averaged half and full compensation occurring 25 and 32 days postoperative, respectively.

Finally, the same recovery patterns are observed when considering either postural or locomotor equilibrium recovery. The association of EGb-treatment with sensorimotor activity leads to the fastest recovery, within 25 days, while a 17 days' delay is induced when none of these two parameters is present. Once again, only one of the parameters (pharmacological treatment or activity) leads to similar patterns of recovery, with intermediate time-courses with respect to the T-F and NT-SMR groups of cats.

3.3 Oculomotor Recovery

Post-lesion spontaneous horizontal eye movements were recorded in total darkness and in the light in the two groups of free-moving cats. Figure 3 shows the

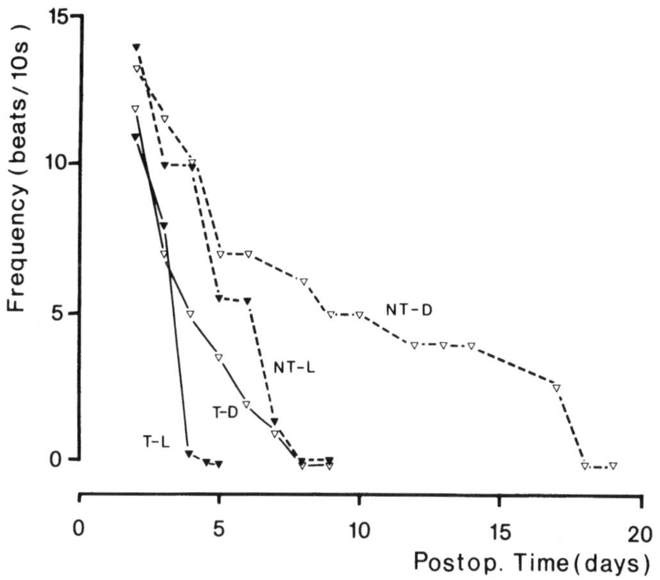

Fig. 3. Effects of the extract of *Ginkgo biloba* on vestibular nystagmus compensation in the free-moving cats. Mean recovery curves of oculomotor function in the treated (n = 3; *solid lines*) and untreated (n = 3; *dashed lines*) groups of cats. Frequency of vestibular nystagmus was measured in the light (*filled triangles*) and in total darkness (*open triangles*). It was expressed by the number of beats per 10 s (*ordinate*) as a function of the postoperative time in days (*abscissae*). Disappearance of nystagmus is more rapid in the treated group of cats in both conditions (light and darkness)

nystagmus frequency (in beats per 10 s) in these two experimental conditions as a function of the postoperative time. Compensation of spontaneous nystagmus develops more rapidly in both light and darkness in the EGb-treated animals.

In the treated cats, the frequency of vestibular nystagmus observed after surgery is around 11–12 beats/10 s in light as well as in darkness. It rapidly declines to zero within 4 and 8 days, when eye movements were recorded in the light and in darkness, respectively. In the untreated group, compensation develops much more slowly. It is achieved in around 8 days when recordings are made in the light, but it requires a very long period of time when tested in darkness (18 days). These delays are classically observed in unilateral vestibular neurectomized cats, but what is interesting is that they are considerably shortened in the EGb-treated cats. This indicates that the extract of *Ginkgo biloba* accelerates the compensation of all the deficits induced by the vestibular lesion.

4 Discussion

The present study brings strong experimental support in favor of the role of an extract of *Ginkgo biloba* in compensating the vestibular deficits after unilateral vestibular neurectomy in the cat. Postoperative EGb-treatment acts on the whole vestibular syndrome by accelerating postural, locomotor balance, and oculomotor recovery. This very positive effect of EGb on vestibular compensation is observed both in the sensorimotor restrained and free-moving groups of cats; but the better pattern of recovery is always found when postoperative EGb-treatment is associated with postoperative sensorimotor activity of the animal (T-F group). In these cats, a 20–25 days' benefit is noticed in the recovery time course, as compared to the restrained and untreated group (NT-SMR). Intermediate patterns of recovery are found when only treatment or activity are present (T-SMR and NT-F groups). These data can be summarized as follows:

T-F > T-SMR = NT-F > NT-SMR.

Moreover, a time benefit also occurs in the treated group of cats when considering compensation of vestibular nystagmus in both light (T-L) and darkness (T-D). Compensation is more rapidly achieved under EGb, even in the absence of visual information, as stated by the following conclusion:

T-L > T-D = NT-L > NT-D.

The extract of *Ginkgo biloba* (EGb 761) is a standardized substance containing flavone glycoside esters together with specific terpene lactones (bilobalid and ginkgolides: cf. Major 1967; Chatterjee and Trunzler 1981). It was demonstrated that EGb application improved the cerebral blood flow in experimental ischemic conditions and enhanced the tolerance of cerebral energy metabolism against hypoxia (Le Poncin-Lafitte et al. 1980; Karcher et al. 1984). It was also found that EGb normalized glucose and ATP levels (Karcher et al. 1985). This latter pharmacological effect is of interest, since biochemical changes, like increased energy metabolism as measured by glucose uptake, were recorded in central nervous structures involved in vestibular compensation in the rat (Llinas and Walton 1979) and in the frog (Flohr et al. 1981). The beneficial effect of EGb can be interpreted in terms of such a pharmacological property, since increased metabolic activity was frequently shown in functionally reorganizing neural networks.

Increase in energy metabolism is also commonly observed in central nervous structures involved in learning processes. A metabolic mapping of sensorimotor functions has been recently scheduled in the human cerebral cortex during various tasks like somatosensory discrimination (Roland and Larsen 1976), visual discrimination (Roland and Skinhoj 1981), motor sequence tests (Roland et al. 1980), mental tasks (Roland and Friberg 1985), etc. Metabolic data from these functional mapping studies show a close correlation between regional cerebral oxygen consumption, regional cerebral blood flow and regional cerebral glucose consumption (Roland 1985). These variables provide a good quantitative estimate of local neuronal and glial metabolism, the increase of which mainly reflects the synaptic activity in both intrinsic neurons and terminal of afferent neurons.

According to these observations, the EGb could rebalance the spontaneous firing rate of vestibular neurons on both sides, by accelerating the restoration of neural activity in the deafferented vestibular nuclei. Disappearance of the spontaneous vestibular nystagmus, occurring earlier in the treated cats, can be considered as a good index of central reequilibration and of the influence of the extract of *Ginkgo biloba* on this neural process. Return to symmetrical activities in vestibular nuclei (Mc Cabe and Ryu 1969; Precht et al. 1966; Pompeiano et al. 1984;

Ried et al. 1984) is classically evoked as a key process in vestibular compensation. It underlies postural compensation and allows recalibration of the dynamic vestibular reflexes (Precht et al. 1980; Lacour 1981; Xerri and Lacour 1983). We are now testing this hypothesis by recording the extracellular unit activity of vestibular neurons on the deafferented side in both untreated and EGb-treated unilateral vestibular neurectomized cats.

A better functional recovery is observed in the free-moving cats as compared to the sensorimotor restrained group, a result already demonstrated in our previous studies in the monkey (Lacour et al. 1976) and in the cat (Xerri and Lacour 1980), and confirmed by the present investigation. Changes in energy metabolism, due to activity, could also explain this faster recovery. Behavioral activity feeds the central nervous system with various reafferent signals which would induce metabolic changes and, therefore, modifications in the synaptic activity of the implicated structures, leading to functional reorganizations of the deafferented or impaired systems. It seems reasonable to suggest that such metabolic mechanisms, which are basically involved in all learning processes in the normal brain and in functional recovery in the injured brain, could be potentiated by the EGb-treatment. If similar cellular processes are activated by either sensorimotor activity or treatment with the extract of *Ginkgo biloba,* one could expect a better recovery when associating the two experimental variables. Interestingly, the present studies support this hypothesis. The recovery patterns are delayed when only one variable is available (treatment or sensorimotor activity) and much slower with none of them.

If true, this conception does not rule out other cellular mechanisms. It was shown in rat posterior pituitary that activity-dependent energy metabolism reflected sodium pump activity (Mata et al. 1980). Such cellular correlates are also proposed in the neuronal models of associative learning (Byrne 1987). Interestingly enough, it has been reported that EGb modifies Na^+ flux across the cell membrane (Etienne et al. 1982) and, in this way, protects the erythrocytes against osmotic shock. A curative effect of EGb was also found in thriethyltin-induced cerebral edema, probably due to an improvement of the astroglial reaction (Otani et al. 1986; Sancesario and Kreutzberg 1986).

Whatever the mode of action of EGb (energy metabolism, enzymatic activity, ion transport, etc.), the present investigations clearly demonstrate that postoperative treatment with an extract of *Ginkgo biloba* strongly accelerates the vestibular compensation process. This substance has beneficial effects on both restrained and unrestrained groups of cats. It could act as a substitute of behavioral activity and potentiate the cellular mechanisms of neural plasticity. It might therefore not only supply the lack of sensorimotor activity in patients who have to rest in bed after surgery or disease, but also help those who can behave actively. Moreover, assuming that plasticity mechanisms are basically identical in all types of CNS injury, we believe that EGb treatment might be used as valuable therapy in the general frame of brain function recovery.

Acknowledgements. This work was supported by grants from U.A. CNRS 372, and from IPSEN Laboratory (France) which provided us the standardized extract of Ginkgo Biloba (EGb 761) used in the commercial preparation, Tanakan. The authors wish to thank M.R. Ibrahim for typing the manuscript and L. Agate for preparing the photographs.

References

Byrne JH (1987) Cellular analysis of associative learning. Physiol Rev 67:329–439

Chatterjee SS, Trunzler G (1981) Neue Ergebnisse aus der *Ginkgo* Forschung. Arztez Naturheilverf 22:593–604

Etienne A, Hecquet F, Clostre F, Defeudis FV (1982) Comparaison des effets d'un extrait de *Ginkgo biloba* et de la chlorpromazine sur la fragilité osmotique, in vitro, d'érithrocytes de rat. J Pharmacol 13:291–298

Flohr H, Bienhold H, Abeln W, Macskovics I (1981) Concepts of vestibular compensation. In: Flohr H, Precht W (eds) Lesion-induced neuronal plasticity in sensori-motor systems. Springer, Berlin Heidelberg New York, pp 153–172

Igarashi M, Alford BR, Kato Y, Levy JK (1975) Effect of physical exercise upon nystagmus and locomotor dysequilibrium after labyrinthectomy in experimental primates. Acta Oto-Laryngol 79:214–220

Igarashi M, Levy JK, O-Uchi T, Reschke MF (1981) Further study of physical exercise and locomotor balance compensation after unilateral labyrinthectomy in squirrel monkeys. Acta Oto-Laryngol 92:101–105

Karcher L, Zagerman P, Krieglstein J (1984) Effect of an extract of *Ginkgo biloba* on rat brain energy metabolism in hypoxia. Naunyn-Schmiedebergs Arch Pharmacol 327:31–35

Karcher L, Chatterjee SS, Gabard B, Krieglstein J (1985) Extract of *Ginkgo biloba* and triethyltin toxicity: biochemical studies in brain. Iupar 9th International Congress of Pharmacology

Lacour M (1981) Contribution à l'étude de la restauration des fonctions posturo-cinétiques après labyrinthectomie chez le singe et le chat. Thèse Doctorat es-Sciences, Marseille, pp 152

Lacour M (1984) Restauration fonctionnelle et réapprentissage: exemple de la compensation vestibulaire. In: Toupet M (ed) XVIIIth Symp Eng, Ipsen, pp 93–106

Lacour M, Roll JP, Appaix M (1976) Modifications and development of spinal reflexes in the alert baboon (*Papio papio*) following an unilateral vestibular neurotomy. Brain Res 113:255–269

Lacour M, Manzoni D, Pompeiano O, Xerri C (1985) Central compensation of vestibular deficits. III. Response characteristics of lateral vestibular neurons to roll tilt after contralateral labyrinth deafferentation. J Neurophysiol 54:988–1005

Llinas R, Walton K (1979) Vestibular compensation: a distributed property of the central nervous system. In: Asanuma H, Wilson VJ (eds) Integration in the nervous system. Igashu-Shoin, Tokyo New York, pp 145–166

Major RT (1967) The *Ginkgo,* the most ancient living tree. Science 157:1270–1273

Mata M, Fink DJ, Gainer H et al. (1980) Activity-dependent energy metabolism in rat posterior pituitary reflects sodium pump activity. J Neurochem 34:213–215

Mc Cabe BF, Ryu JH (1969) Experiments on vestibular compensation. Laryngoscope 79:1728–1736

Otani M, Chatterjee SS, Gabard B, Kreutzberg GW (1986) Effect of an extract of *Ginkgo biloba* on triethyltin-induced cerebral edema. Acta Neuropathol 69:54–65

Pompeiano O, Xerri C, Gianni S, Manzoni D (1984) Central compensation of vestibular deficits. II. Influence of roll tilt on different-size lateral vestibular neurons after ipsilateral labyrinth deafferentation. J Neurophysiol 52:18–38

Poncin-Lafitte MC le, Rapin J, Rapin JR (1980) Effects of Ginkgo-biloba on changes induced by quantitative cerebral microembolization in rats. Arch Int Pharmacodyn Ther 243:236–244

Precht W (1974) Characteristic of vestibular neurons after acute and chronic labyrinthine destruction. In: Kornhuber HH (ed) Vestibular system, Handbook of sensory physiology, vol VI. Springer, Berlin Heidelberg New York, pp 451–462

Precht W, Shimazu H, Markham CH (1966) A mechanism of central compensation of vestibular function following hemilabyrinthectomy. J Neurophysiol 29:996–1010

Precht W, Maioli C, Dieringer N, Cochran S (1981) Mechanisms of compensation of the vestibulo-ocular reflex after vestibular neurotomy. In: Flohr H, Precht W (eds) Lesion-induced neuronal plasticity in sensorimotor systems. Springer, Berlin Heidelberg New York, pp 222–230

Rapin JR, Poncin-Lafitte M le (1979) Modèle expérimental d'ischémie cérébrale. Action préventive de l'extrait de *Ginkgo*. Sémin Hôp (Paris) 55:2047–2050

Ried S, Maioli C, Precht W (1984) Vestibular nuclear neuron activity in chronically hemilabyrinthectomized cats. Acta Oto-Laryngol 98:1–13

Roland PE (1985) Cortical organization of voluntary behavior in man. Hum Neurobiol 4:155–167

Roland PE, Larsen B (1976) Focal increase of cerebral blood flow during stereognostic testing in man. Arch Neurol 33:551–558

Roland PE, Friberg L (1985) Localization of cortical areas activated by thinking. J Neurophysiol 53:1219–1243

Roland PE, Skinhoj E (1981) Extrastriate cortical areas activated during visual discrimination in man. Brain Res 222:166–171

Roland PE, Larsen B, Lassen NA, Skinhoj E (1980) Supplementary motor area and other cortical areas in organization of voluntary movements in man. J Neurophysiol 43:118–136

Sancessario G, Kreutzberg GW (1986) Stimulation of astrocytes affects cytotoxic brain edema. Acta Neuropathol 72:3–14

Schaefer KP, Meyer DL (1973) Compensatory mechanisms following labyrinthine lesions in the guinea-pig. A simple model of learning. In: Zippel HP (ed) Memory and transfer of information. Plenum, New York, pp 203–232

Shimazu H, Precht W (1966) Inhibition of central vestibular neurons from the contralateral labyrinth and its mediating pathway. J Neurophysiol 29:467–492

Sokoloff L (1977) Relation between physiological function and energy metabolism in the central nervous system. J Neurochem 29:13–26

Xerri C, Lacour M (1980) Compensation des déficits posturaux et cinétiques après neurectomie vestibulaire unilatérale chez le chat. Rôle de l'activité sensorimotrice. Acta Oto-Laryngol 90:414–424

Xerri C, Lacour M, Manzoni D, Pompeiano O (1983) Behavioral aspects and central neuronal events in vetibular compensation. In: Horn H (ed) Multimodal convergences in sensory system. Fischer, Stuttgart, pp 291–301

Role of the Ca^{2+} Entry Blocker Flunarizine in Vestibular Compensation

E. Tolu, O. Mameli, F. Melis, and M. A. Caria
Institute of Human Physiology, Sassari University Medical School,
v. le S. Pietro 43/B, 07100 Sassari, Italy

1 Introduction

The Ca^{2+} entry blocker flunarizine is widely used for treating patients with vertigo of vestibular origin (Holmes et al. 1984, for review), but there are few experimental studies about the effects of the drug on vestibular apparatus, although the vestibular depressant effect of flunarizine has been indirectly observed by some workers. Observations performed in humans (Oosterveld et al. 1974a; Mangabeira-Albernaz et al. 1978) and experimental animals (Oosterveld 1974b; Shiozaki et al. 1982) have shown that this drug provokes significant reduction in frequency and peak velocity of the slow component of eye nystagmus, suggesting that flunarizine exerts a depressant effect on vestibular reflexes by reducing vestibular receptor excitability. In fact, by blocking Ca^{2+} fluxes through the cellular membranes, the drug reduces the magnitude of the mechano-electrical transduction in the vestibular receptors, which are Ca^{2+}-dependent (Corey and Hudspeth 1979; Jørgensen 1983). Depression of receptor excitability in the intact labyrinth could justify the observation that repeated flunarizine administration in hemilabyrinthectomized guinea pigs accelerates vestibular compensation (Tolu and Mameli 1984). In hemilabyrinthectomized animals, the asymmetrical activity of the vestibular nuclear complexes of the two sides is considered responsible for the appearance of motor and postural vestibular deficits. Recovery of symmetrical activity in the vestibular nuclei, characterized by increased activity in the vestibular nuclei of the deafferentated side and/or reduced activity of the intact one, represents the crucial condition for vestibular compensation (McCabe et al. 1972; Precht 1974; Azzena et al. 1976, 1977). The reduction of the receptor excitability of the intact side induced by flunarizine could facilitate the re-establishment of equilibrium between the vestibular nuclear activities of both sides, thus accelerating the compensatory processes. However, compensation of the vestibular syndrome is a complex phenomenon that involves different nervous structures, for example the vestibular nuclei (Igarashi et al. 1978), the intervestibular fibers (Flohr et al. 1981; Galiana et al. 1984), the cerebral cortex (Di Giorgio 1939; Giretti 1971; Schaefer and Meyer 1981), the cerebellum (Azzena et al. 1979; Schaefer and Meyer 1981), the precerebellar nuclei such as the inferior olive (Azzena et al. 1979; Mameli and Tolu 1986) and the lateral reticular nuclei (Azzena et al. 1981) and the spinal cord (Azzena 1969; Azzena et al. 1976, 1977, 1981; Tolu et al. 1980). These structures participate in vestibular compensation through a

H. Flohr (Ed.)
Post-Lesion Neural Plasticity

substitution process, by exciting the vestibular nuclei of the deafferentated side and by inhibiting the vestibular nuclei of the intact one. The accelerated compensation observed in the animals treated with flunarizine could, therefore, depend not only upon reduction of the receptor excitability but also upon an involvement of other compensatory structures. Since flunarizine significantly influences the activities of the vestibular nuclei and cerebellum (Tolu and Mameli 1987), the experiments were set up to analyze the effect of flunarizine on vestibular compensation.

2 Method

The experiments were performed on 80 fasted guinea pigs. Twenty animals, under ether anesthesia, were hemilabyrinthectomized by injecting a 50% solution of mineral oil and chloroform into the middle ear (Simonelli 1923; Azzena 1969). The time course of vestibular compensation was studied by measuring the frequency and duration of the eye nystagmus and head deviation degrees, every hour throughout 24 h after labyrinthine lesion. A model 7 Grass polygraph connected to silver electrodes, applied at the outer canthi was used to record the eye nystagmus. The method for measuring head deviation, described in a previous paper (Tolu and Mameli 1984), was as follows:

A small mirror was fixed on the center of the forehead of the animal which was held on a table within a plastic box, with the emerging head free to move. Around the table was placed a half graduated arc scale (0 °–90 °–0 °) with a runner on its upper edge that carried a mobile light. The animal's body was positioned so that the spinal cord lay along the 0 °–0 ° line traced on the table. The mobile light was directed towards the mirror which reflected back onto the graduated scale giving the degree of head deviation (Fig. 1 A–B). The time course of the compensatory process was analyzed in the controls (10 animals) and in animals treated every 6 h for 24 h with flunarizine (10 animals). The general experimental procedure for the remaining sixty guinea pigs was as follows:

Under ketamine anesthesia (50 mg/kg i.p.) the animals were tracheotomized, immobilized with Fazadon, artificially ventilated and fixed in a stereotaxic frame. Tidal volume was adjusted to maintain normal the expired CO_2. Body temperature, rectally measured, was maintained at 37.5 ± 0.5 °C. ECG and femoral artery pressure were monitored throughout the experiment. A catheter was introduced, through the esophagus, into the stomach for drug administration. A craniotomy exposed the cerebellar surface. Extracellular spontaneous activity and evoked potentials were recorded by means of a tungsten microelectrode insulated except at the tip (900–1000 kOhm resistance), advanced by an electronic microdrive into the cerebellar cortex (lamella V) or the vestibular nuclei. The spontaneous activity of the Purkinje and vestibular cells was recorded and stored on magnetic tape for later analysis. Single spikes, converted into standard pulses through a window discriminator, were fed to a computer (Nicolet 1072) for constructing the frequency distribution histograms (FDHs) and the cumulative frequency distributions (CFDs) during 102.4 s of analysis. The CFDs were constructed by adding to every address of memory the counts of all previous addresses. For each unit,

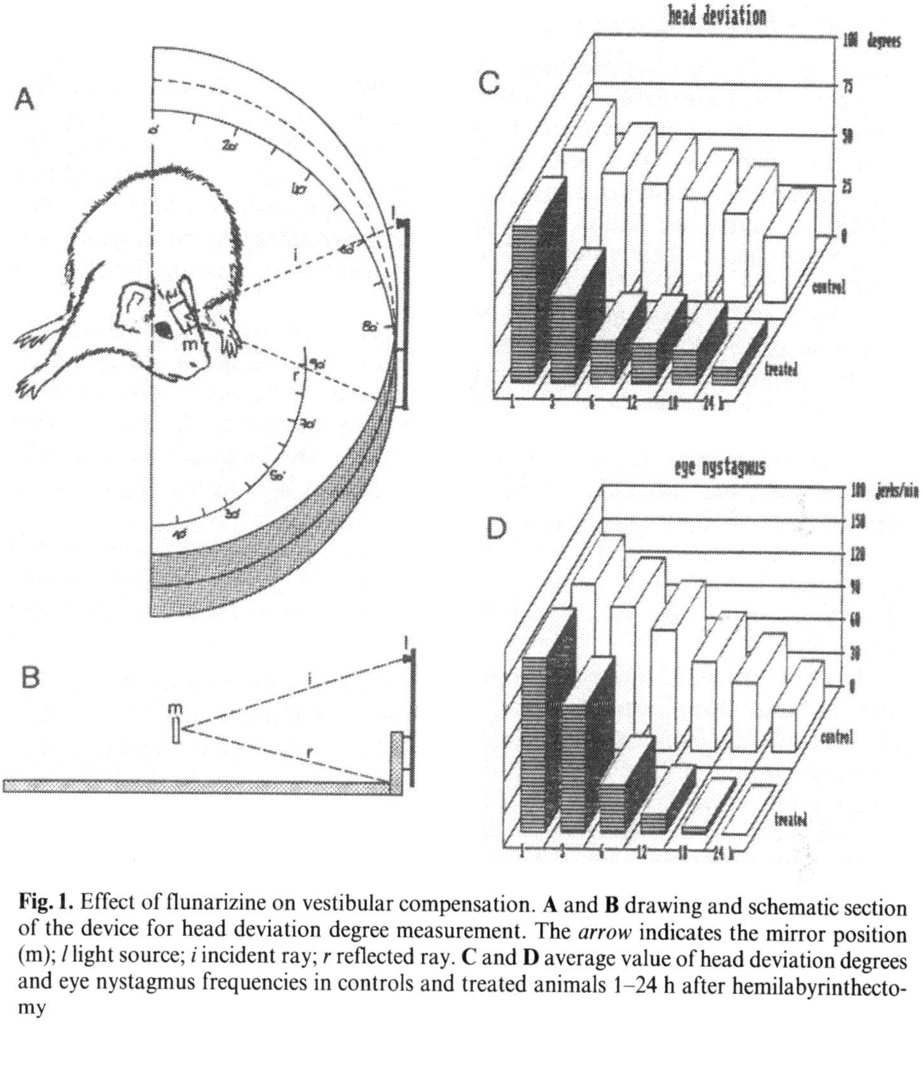

Fig. 1. Effect of flunarizine on vestibular compensation. **A** and **B** drawing and schematic section of the device for head deviation degree measurement. The *arrow* indicates the mirror position (m); *l* light source; *i* incident ray; *r* reflected ray. **C** and **D** average value of head deviation degrees and eye nystagmus frequencies in controls and treated animals 1–24 h after hemilabyrinthectomy

the CFDs so obtained were used to compare the spontaneous firing rate recorded before and after drug administration. The electrophysiological studies were performed in two distinct sets of experiments.

2.1 First Set (30 Experiments)

The spontaneous activity of single vestibular cells was analyzed in three different groups of animals: (1) normal guinea pigs (10 experiments); (2) bilaterally labyrinthectomized (10 experiments) and (3) labyrinthectomized-cerebellectomized animals (10 experiments). In the 2nd and 3rd groups, bilateral labyrinthine lesions were performed 1 month before the experiments. In the 3rd group, the additional cerebellectomy was performed about 1 h before the recordings. In all experiments, the recordings were performed before and after drug administration.

2.2 *Second Set (30 Experiments)*

In all animals, the epitympanic recessus was opened and a pair of silver wires (50 μm diam., insulated except for the tips) were respectively inserted into the ampulla of the horizontal or vertical semicircular canal and into the surrounding bone. The wires were then fixed in place with dental cement and connected to a model S8 Grass stimulator. The evoked response of the vestibular nuclei was elicited by cathodal stimulation of the ipsilateral labyrinthine receptors (duration: 0.1 ms; frequency: 3 Hz; intensity: twice the threshold for evoking the N_1 wave potential).

The trend of spontaneous and evoked activities of single Purkinje cells (P-cells), localized in the pars intermedia of the anterior lobe, with parallel variations of the spontaneous and evoked vestibular nuclear cells, were analyzed in 15 guinea pigs. Responses of P-cells to electrical stimulation of the ipsilateral radial nerve (double shocks, 0.1 ms duration, 3 Hz) and vestibular cells to electrical stimulation of the ipsilateral labyrinth were analyzed by constructing the post-stimulus time histograms (PSTHs) and the relative CFDs.

The evoked vestibular potentials were evaluated in eight normal animals and in seven cerebellectomized animals. Cerebellectomy was performed 1 h before the recordings by suction of the cerebellum, but leaving intact the cerebellar peduncles.

In all experiments, the spontaneous and evoked activities of vestibular nuclei and cerebellar cortex were analyzed before (basal conditions) and 60–180 min after flunarizine administration (5 mg/kg/os suspended in gum arabic). It had been previously established that this dosage of the drug was the most suitable for reducing the vestibular deficits of a hemilabyrinthectomized animal (Tolu and Mameli 1984). At the end of the experiments, an electrolytic lesion marked the recorded site for later histological control, and the animals were killed with a barbiturate overdose. The statistical significance of the differences between results obtained in basal conditions and after drug administration was elaborated with Student's t-test for mean comparisons.

3 Results

3.1 *Effects of Flunarizine on Time-Course of Vestibular Compensation*

After hemilabyrinthectomy a marked head deviation of $75 \pm 4\,^{\circ}$ ($\bar{x} \pm SD$) with spontaneous eye nystagmus of 150.62 ± 17.5 jerks/min was observed. In the control animals, these vestibular deficits gradually diminished, but 24 h after operation a distinct head deviation (32.6 ± 3.4) and spontaneous nystagmus (38.2 ± 7.1) were still observed. On the contrary, in the animals treated with flunarizine recovery occurred more rapidly than in the controls. Twenty four hours after hemilabyrinthectomy, not only had the spontaneous nystagmus disappeared but head deviation was significantly reduced to $8.4 \pm 2.8\,^{\circ}$ (Fig. 1 C–D).

3.2 *Effects of Flunarizine on Spontaneous and Evoked Activity of the Vestibular and Purkinje Cells*

The analysis of the spontaneous vestibular cell activity was performed in normal, bilaterally labyrinthectomized and labyrinthectomized-cerebellectomized animals. In each animal group there was an appreciable change in the spontaneous firing rate at 60 min after flunarizine administration. At 90–120 min, a maximal depression was reached and at 180 min a partial recovery was observed (Fig. 2A). The percentage variation of the mean spontaneous firing rate of all vestibular cells tested in each animal group, showed significant differences ($P < 0.001$) between basal conditions and following drug administration. The Student's t-test for paired observations also showed significant differences between groups ($P < 0.001$), particularly at 90 min. In normal animals, the depressant effect of the drug was more marked and recovery was less complete than in bilaterally labyrinthectomized and in labyrinthectomized-cerebellectomized animals, as reported in Fig. 2B.

The spontaneous firing rate was analyzed in two animals of each group without drug administration; no significant variations were observed.

Concomitant recordings of spontaneous and evoked activities in the cerebellar cortex and vestibular nuclei following flunarizine administration are reported in Fig. 3. The spontaneous firing rate of the P-cell as well as the evoked response markedly increased after drug treatment. A parallel depression of the sponta-

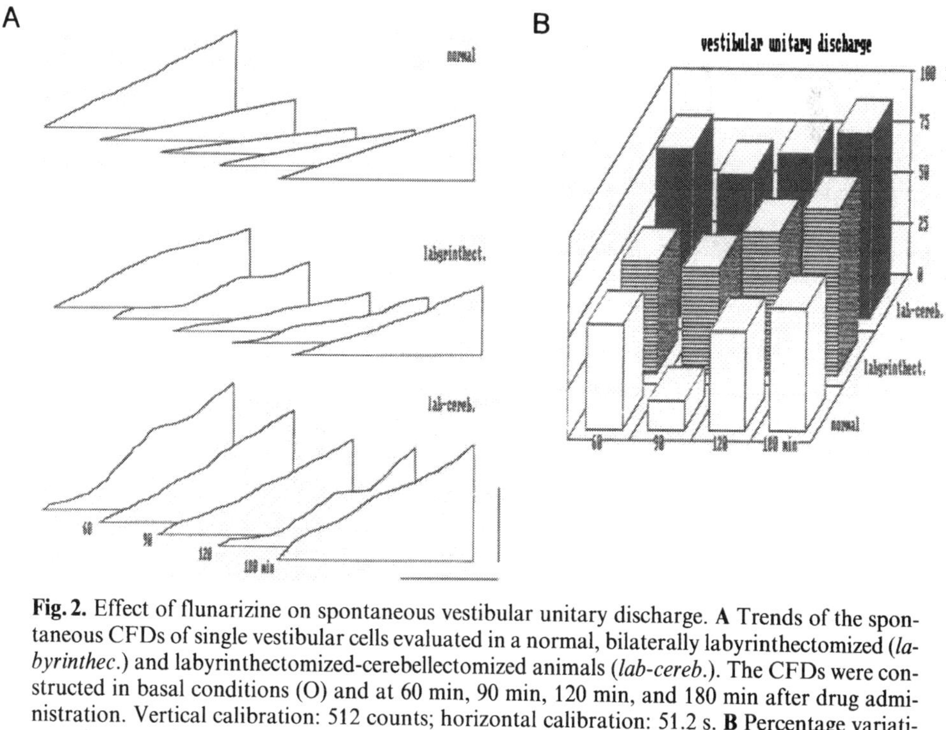

Fig. 2. Effect of flunarizine on spontaneous vestibular unitary discharge. **A** Trends of the spontaneous CFDs of single vestibular cells evaluated in a normal, bilaterally labyrinthectomized (*labyrinthec.*) and labyrinthectomized-cerebellectomized animals (*lab-cereb.*). The CFDs were constructed in basal conditions (O) and at 60 min, 90 min, 120 min, and 180 min after drug administration. Vertical calibration: 512 counts; horizontal calibration: 51.2 s. **B** Percentage variations of averaged spontaneous vestibular unitary discharge of all cells tested following flunarizine treatment. Comparison is made with basal conditions defined as 100%

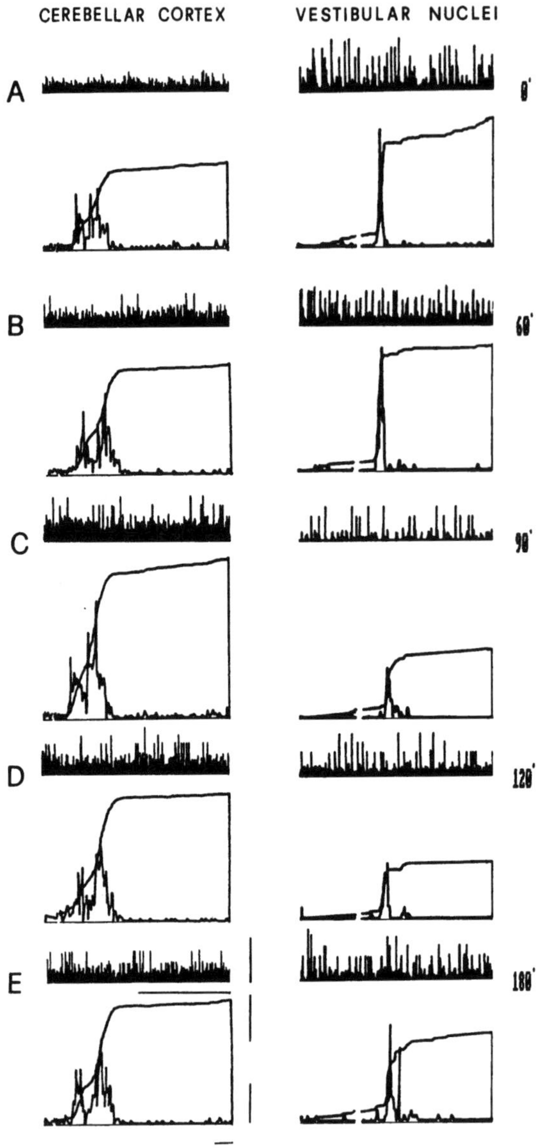
CEREBELLAR CORTEX
VESTIBULAR NUCLEI
A
B
C
D
E
0'
60'
90'
120'
180'

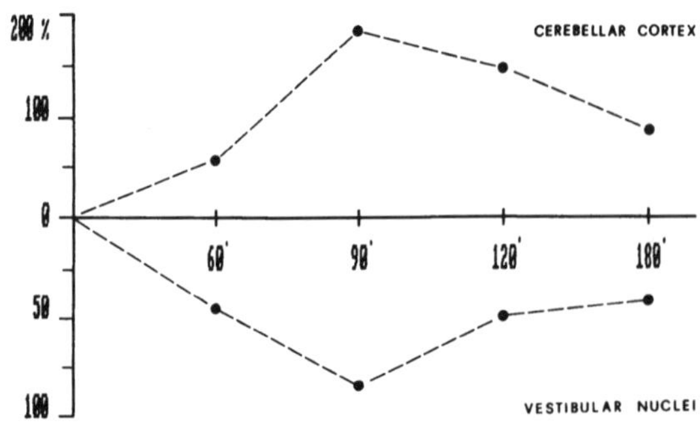
200 %
100
0
50
100
60'
90'
120'
180'
CEREBELLAR CORTEX
VESTIBULAR NUCLEI

neous and evoked vestibular nuclear activities was also observed. Diagrammatic representation of the trend percentage variation of the mean spontaneous frequencies of all tested neurons, 60–180 min after drug administration, is reported in the lower part of Fig. 3. Comparison is made with basal conditions (100%) defined as O. The spontaneous firing rate of P-cells increased by 54% at 60 min, 180% at 90 min, 145% at 120 min, and 85.2% at 180 min after drug administration, whereas the spontaneous firing rate of vestibular cells decreased by 45.6%, 86.1%, 50.6%, and 41.8%.

During drug treatment, differences between mean values of spontaneous firing rates and evoked response amplitudes were always highly significant ($P<0.001$) in the cerebellar cortex and vestibular nuclei. The excitation of the cerebellar cortex was significantly correlated to the vestibular nuclear depression with a statistical significance of 1‰.

3.3 Effects of Flunarizine on Evoked Field Potentials in Vestibular Nuclei

Experiments were performed in normal and cerebellectomized animals. In basal conditions (Fig. 4A), the vestibular potentials generated by electrical stimulation of the ipsilateral labyrinthine receptors were characterized by an early positive or positive-negative wave (P), followed by a large negative wave (N_1) and a delayed negative wave (N_2). The mean latency expressed as mean ± SD was 0.68 ± 0.20 ms for P, 1.08 ± 0.28 ms for N_1 and 2.43 ± 0.29 ms for N_2. In the intact animal, both N_1 and N_2 components were depressed in amplitude after drug administration (Fig. 4B-E). In particular, 90–120 min after drug administration the amplitude of the N_1 wave decreased by about 40%. A partial recovery was observed at 180 min. The N_2 potential was almost completely abolished and the amplitude remained at low level (about 20% of the basal value) during the experiment. Lengthening in latency (mostly of the N_2 wave) and greater duration (mostly of the N_1 wave) were also observed. Latency of the N_1 wave was 1.1 ms in basal conditions, 1.4 ms at 60–90th min, 1.8 ms at the 120th min and 1.5 ms at the 180th min. Correspondingly, the N_1 duration was 1.1 ms, 2.8 ms, 2.5 ms, and 2.1 ms. The N_2 latency increased from 2.3 ms in basal conditions to 3.9 ms at the 60th min, 5.3 ms at 90th min, 5.6 ms at 120th min and 5.3 ms at the 180th min.

In the cerebellectomized animal, the vestibular potentials were similar to those elicited in the intact animal. Drug administration depressed the N_1 and potenti-

Fig. 3. Effect of flunarizine on spontaneous and evoked P-cell and vestibular cell activities. Contemporaneous recordings from the pars intermedia of lamella V and medial vestibular nucleus were performed in basal conditions (**A**) and 60 min (**B**), 90 min (**C**), 120 min (**D**) and 180 min (**E**) after drug administration. In each time recording, the upper traces report the spontaneous FDHs; the lower traces report the PSTHs and CFDs of the P-cell and vestibular cell responses. The PSTHs and CFDs were constructed by 100 consecutive electrical stimuli applied at the ipsilateral radial nerve and labyrinth, respectively. The *gaps* show the moment of stimulation. Vertical calibrations: 10 counts for FDHs; 10 counts for PSTHs and 1 count added/stimulus for CFDs. These values are reduced to half for vestibular recordings. Horizontal calibrations: 51.2 s for FDHs; 7 ms and 1.4 ms for PSTHs and CFDs of P-cell and vestibular cell responses, respectively. Trends of the mean percentage increase in spontaneous firing rate of P-cells and the mean percentage decrease in that of vestibular cells are reported in the lower diagram. Comparison is made with basal conditions (100%) defined as O

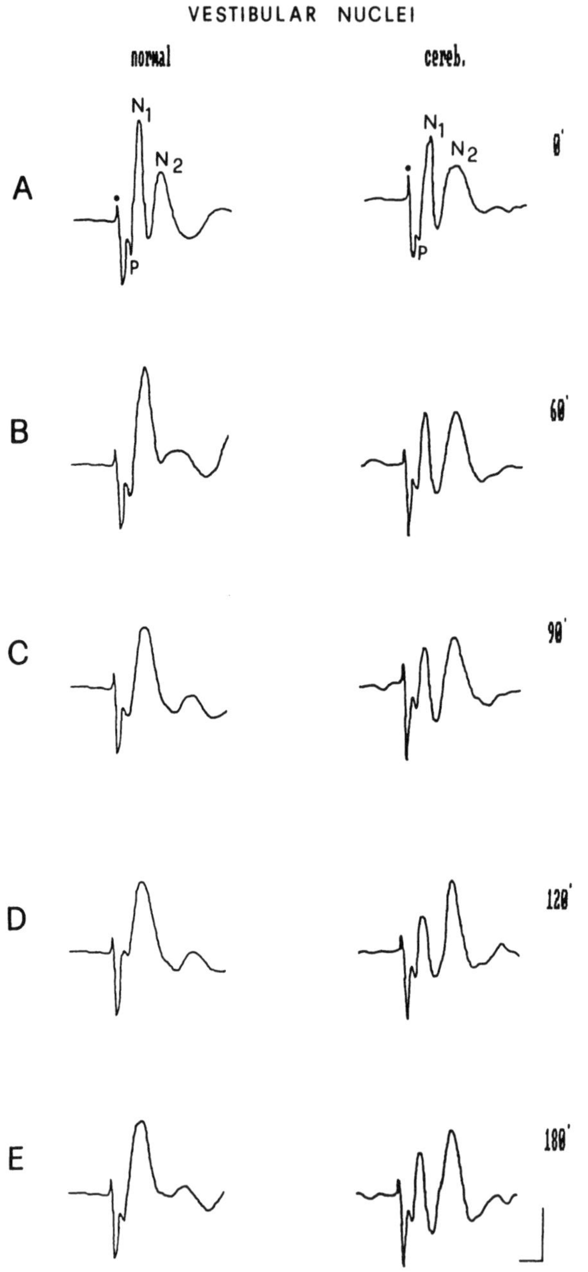

Fig. 4. Effect of flunarizine on evoked vestibular potentials. **A–E** Field potentials evoked in the medial vestibular nuclei of a normal and a cerebellectomized animal by electrical stimulation of the ipsilateral labyrinth. **A** Basal conditions; **B** 60 min; **C** 90 min; **D** 120 min and **E** 180 min after flunarizine administration. The *dots* indicate shock artifacts. Vertical calibration: 0.5 mV; horizontal calibration: 2 ms

ated the N_2 wave. Maximal depression of the N_1 response (about 55% of the basal amplitude) was reached 90–120 min after drug administration and partial recovery occurred at the 180th min. The N_2 potential amplitude, on the other hand, linearly increased, reaching about 160% of basal value at the 180th min. Significant variations in the latency or duration of vestibular potentials were not observed.

In all experiments, the differences between field potentials measured in basal conditions and following flunarizine treatment were highly significant ($P<0.001$).

4 Discussion

The present results agree with our previous findings that flunarizine administration significantly accelerates vestibular compensation in hemilabyrinthectomized guinea pigs (Tolu and Mameli 1984).

Significant differences in the spontaneous firing rate reduction, observed between normal and bilaterally labyrinthectomized animals, suggest that the drug prevalently depresses the vestibular receptor excitability. In the labyrinthectomized-cerebellectomized animals, the inhibitory effect of flunarizine on spontaneous vestibular cell activity was always present but less pronounced than in the bilaterally labyrinthectomized animals, suggesting that the drug directly depresses vestibular cell activity.

The contemporaneous recording of the cerebellar and vestibular activities showed that the most important effect of flunarizine in depressing vestibular activity occurred through a strong excitation of the cerebellar cortex. Confirmation could be seen in the significant reduction of the vestibular potential that occurred in normal animals with respect to the cerebellectomized ones.

Flunarizine selectively blocks the Ca^{2+} slow channels (voltage-dependent ion channels) increasing its effect during depolarization of the cell membrane as well as during exalted activity of the tissue (use-dependence) (Ehara and Daufmann 1978; Bolton 1979, Holmes et al. 1984, for review). The specific mechanism of the drug in depressing vestibular activity is not yet clear. In neuronal tissue, modulation of Ca^{2+} channels could be an important factor in the regulation of neurotransmitter release from presynaptic nerve terminals. Depression of spontaneous and evoked vestibular cell activity, amplitude reduction, and increase of latency of N_1 and N_2 field potentials suggest that the drug blocks the excitatory impulses at synaptic level in the region of the vestibular nuclei. This action could also arise from the antihistaminic effect exerted by flunarizine (Awouters et al. 1975).

Much less is known about the mechanism with which flunarizine increases excitability of the P-cells in the cerebellar cortex. The P-cells are known to receive two classes of excitatory inputs: climbing and parallel fibers. The pattern of P-cell response to climbing fibers activation was better studied in guinea pig cerebellar slices (Llinàs and Sugimori 1980 a, b). P-cell excitation was characterized by a fast somatic Na-dependent spike, followed by a dendritic Ca-dependent depolarization. The latter was followed by a Ca- and voltage-dependent K conductance which generated a fall in the dendritic excitability. On blocking the Ca current or using a Ca-free bath, the P-cell increases the spontaneous firing rate up to complete inactivation by continuous depolarization. From these experiments it appears that the slow Ca-K system represents an important modulation of the P-cell excitability. That is, flunarizine could block selectively but not completely the slow Ca channels, with consequent increase in P-cell excitability. However, this hypothesis does not exclude other possible drug action, such as interference with intracerebellar circuitry to release the P-cell from the inhibitory controls.

The question which arises is: through what mechanisms does the drug influence the time course of the vestibular compensation?

After hemilabyrinthectomy, the asymmetrical output from the vestibular nuclei of both sides becomes dependent upon the uncontested activity of the intact labyrinth. Depression of the receptor excitability by pharmacological aids could then facilitate the recovery of the balance between the vestibular nuclear complex of the two sides.

Flohr et al. (1981) showed that metabolic activity of the vestibular nuclei of the intact side greatly increased, and it may well be that the drug inhibitory effect correspondingly increases due to use-dependence. Such a direct inhibitory effect of flunarizine, particularly upon the vestibular nuclei of the intact side, could contribute in balancing the output of the vestibular nuclear complexes.

Involvement of the cerebellum in vestibular compensation has been well known for some time (Azzena et al. 1979; Schaefer and Meyer 1981). In hemilabyrinthectomized and compensated animals, the lesion of olivo-cerebellar and reticolo-cerebellar afferents or functional inactivation of the cerebellar cortex induced the release of the vestibular symptoms (Azzena et al. 1979, 1981; Mameli and Tolu 1986). Functional inactivation of the cerebellar cortex by cooling induced an asymmetrical behavior of the vestibular nuclei characterized by exalted activity in the deafferentated side and exalted-depressed activity in the intact side (Azzena et al. 1979). These different influences of the cerebellum on the vestibular nuclei of the two sides could depend upon the asymmetrical volleys of the spino-cerebellar afferents. In fact, after hemilabyrinthectomy, the distorted posture assumed by the animal asymmetrically excites the extero- and proprioceptors to supply spatial error information along the spinal cord. These impulses depress the vestibular cells of the intact side and exalt the vestibular cells of the deafferentated side as well as asymmetrically exciting the cerebellar cortex (Azzena et al. 1976, 1977, 1979, 1981; Mameli and Tolu 1986). The strong spinal excitation of the vestibular nuclei in the deafferentated side could, through crossed commissural connections, depress the vestibular nuclei of the intact side to equilibrate the activity between the vestibular complexes. The cerebellum, asymmetrically influenced by impulses from the spinal afferents, effects much more control than the intact vestibular nuclei over the excitation level of the deafferentated nuclei, as observed following functional inactivation of the cerebellar cortex (Azzena et al. 1979). Therefore, the increase in cerebellar excitability induced by flunarizine could potentiate the modulation of the excitatory fluxes directed at restoring the "normal" activity of the deafferentated vestibular nuclei.

In conclusion, flunarizine treatment in hemilabyrinthectomized animals accelerates vestibular compensation probably through three different mechanisms: (1) by depressing the receptor excitability of the intact labyrinth; (2) by directly inhibiting the activity of the vestibular nuclei, particularly of the intact side; (3) by increasing the excitability of the cerebellar cortex which in turn modulates the vestibular nuclear activity.

Acknowledgements. This work was supported by a grant of the Italian Ministero della Pubblica Istruzione.

References

Awouters F, Niemegeers JE, Janssen PAJ (1975) Cinnarizine and flunarizine potent inhibitors of anaphylactic shock in guinea pigs. Arch Int Pharmacodyn Ther 217:38–43

Azzena GB (1969) Role of the spinal cord in compensating the effects of hemilabyrinthectomy. Arch Ital Biol 65:43–53

Azzena GB, Mameli O, Tolu E (1976) Vestibular nuclei of hemilabyrinthectomized guinea pigs during decompensation. Arch Ital Biol 114:389–398

Azzena GB, Mameli O, Tolu E (1977) Vestibular units during decompensation. Experienta (Basel) 33:234–235

Azzena GB, Mameli O, Tolu E (1979) Cerebellar contribution in compensating the vestibular function. In: Granit R, Pompeiano O (eds) Reflex control of posture and movement. Elsevier Biomed, Amsterdam, North Holland, pp 599–606 (Progress in Brain Research, vol 50)

Azzena GB, Tolu E, Mameli O (1981) The lateral reticular nucleus. Role in vestibular compensation. In: Flohr H, Precht W (eds) Lesion-induced neuronal plasticity in sensorimotor systems. Springer, Berlin Heidelberg New York, pp 254–264

Bolton TB (1979) Mechanisms of action of transmitters and other substances on smooth muscle. Physiol Rev 59:606–718

Corey DP, Hudspeth AJ (1979) Ionic basis of the receptor potential in a vertebrate hair cell. Nature 281:675–677

Di Giorgio AM (1939) Effetti di lesioni unilaterali della corteccia cerebrale sui fenomeni di compenso da emislabirintazione. Atti Accad Fisiol Fac Med Siena 2:382–384

Ehara T, Daufmann R (1978) The voltage- and time-dependent effects of (−)-verapamil on the slow inward current in isolated cat ventricular myocardium. J Pharmacol Exp Ther 207:49–55

Flohr H, Bienhold H, Abeln W, Macskovics I (1981) Concepts of vestibular compensation. In: Flohr H, Precht W (eds) Lesion-induced neuronal plasticity in sensorimotor systems. Springer, Berlin Heidelberg New York, pp 153–172

Galiana HL, Flohr H, Melvill Jones GA (1984) A reevaluation of intervestibular nuclear coupling: its role in vestibular compensation. J Neurophysiol 51:242–259

Giretti ML (1971) Spinal compensation of the cerebral release phenomena. Exp Neurol 30:459–466

Holmes B, Brogden RN, Heel RC, Speight TM, Avery GS (1984) Flunarizine. A review of its pharmacodynamic and pharmacokinetic properties and therapeutic use. Drugs 27:6–44

Igarashi M, Levy JK, Reschke MF, Kubo T, Watson T (1978) Locomotor dysfunction after surgical lesions in the unilateral vestibular nuclei region in squirrel monkeys. Arch Oto-Rhino-Laryngol 221:89–95

Jørgensen F (1983) Influence of Ca^{2+} on the mechanosensitivity of the hair cells in the lateral line organs of *Necturus maculosus*. Acta Physiol Scand 118:423–431

Llinàs R, Sugimori M (1980a) Electrophysiological properties of in vitro Purkinje cell somata in mammalian cerebellar slices. J Physiol (Lond) 305:171–195

Llinàs R, Sugimori M (1980b) Electrophysiological properties of in vitro Purkinje cell dendrites in mammalian cerebellar slices. J Physiol (Lond) 305:197–213

Mameli O, Tolu E (1986) Involvement of the inferior olivary neurons in vestibular compensation. Brain Res Bull 16:739–743

Mangabeira-Albernaz PL, Ganança MM, Novo NF, Rodrigues de Paiva E (1978) Flunarizine and cinnarizine as vestibular depressants. A statistical study. ORL 40:92–100

McCabe BF, Ryu JH, Sekitani T (1972) Further experiments on vestibular compensation. Laryngoscope 82:381–396

Oosterveld WJ (1974a) Pilot evaluation of flunarizine (R 14950) in vertigo. A double blind trial. Can J Otolaryngol 3:284–292

Oosterveld WJ (1974b) Vestibular pharmacology of flunarizine compared to that of cinnarizine. ORL 36:157–164

Precht W (1974) Characteristics of vestibular neurons after acute and chronic labyrinthine destruction. In: Kornhuber HH (ed) Vestibular system, Part 2. Springer, Berlin Heidelberg New York, pp 451–462 (Handbook of Sensory Physiology, vol VI/2)

Schaefer KP, Meyer DL (1981) Aspects of vestibular compensation in the guinea pigs. In: Flohr H, Precht W (eds) Lesion-induced neuronal plasticity in sensorimotor systems. Springer, Berlin Heidelberg New York, pp 197–207

Shiozaki S, Kubo K, Shuto K, Nakamizo N (1982) Effects of flunarizine on nystagmus and cochlear blood flow. Folia Pharmacol Jpn 80:429–439

Simonelli G (1923) Un metodo per la distruzione chimica del labirinto. Arch Fisiol 21:231–233

Tolu E, Mameli O (1984) Pharmacological improvement of vestibular plasticity. Effects of a Ca^{2+}-antagonist agent. Pharmacol Res Commun 16:1161–1173

Tolu E, Mameli O (1987) Direct and indirect action of flunarizine on vestibular activity. Pharmacol Res Commun 19:241–254

Tolu E, Mameli O, Azzena MT, Azzena GB (1980) Dynamic responses of vestibular cells during spinal decompensation. Physiol Behav 25:637–640

Effect of Substance P on Vestibular Compensation

U. Lüneburg and H. Flohr
Department of Neurobiology, University of Bremen, D-2800 Bremen 33, FR Germany

1 Introduction

Since the discovery of Substance P (SP) by von Euler and Gaddum (1931) and its identification as an undecapeptide by Chang and Leeman (1970), SP has been one of the most thoroughly studied neuropeptides in the nervous system. The peptide has a wide and heterogeneous distribution in both the central and peripheral nervous system of mammalian and submammalian species (including amphibia) (Lembeck and Zetler 1962) and belongs, by definition, to a distinct family of peptides, the tachykinins, which all share a common C-terminal sequence of five amino acids (Erspamer 1981). As a neurotransmitter and/or neuromodulator SP subserves multiple physiological roles (Krivoy et al. 1979; Nicoll et al. 1980). Besides its functional involvement in nociception and diverse behavioral patterns such as grooming, ingestion, and locomotion (for review see Huston and Stäubli 1981; Nemeroff et al. 1984), SP may also participate in plastic processes as revealed in various learning paradigms. Several studies have demonstrated that intracerebroventricular or peripheral administration of SP induces dose-dependent effects on operant conditioning, generally promoting the acquisition of the learned behavior (Melo and Graeff 1975; Goldstein and Malick 1977; Graeff and Arisawa 1978; Stein and Belluzzi 1979). In avoidance-conditioning peripheral administration of SP enhanced the acquisition and retention of the task (Wetzel and Matthies 1982; Schlesinger et al. 1983a; Pelleymounter et al. 1986; Tomaz and Huston 1986), effects which could not be confirmed by Hecht et al. (1979) and Luttinger et al. (1982), however. Effects of local administration of SP into the CNS are dependent on the site of injection: administered to the lateral hypothalamus or the magnocellular basal nucleus the peptide facilitates the responding in different avoidance paradigms (Huston and Stäubli 1981; Kafetzopoulos et al. 1986), whereas avoidance performance is reduced by injections into the substantia nigra or the amygdala (Huston and Stäubli 1981; Sagen and Routtenberg 1981; Stäubli and Huston 1985). Similar effects could be observed after intracerebroventricular (Pelleymounter et al. 1986) and central (Gaffori et al. 1984) application of SP fragments. In addition, SP attenuates the degree of retrograde amnesia induced by electroconvulsive shock or cycloheximide on conditioned avoidance responding (Schlesinger et al. 1983b).

Despite this obvious involvement of SP in mechanisms underlying adaption to environmental changes, there does not exist, to our knowledge, any evidence

H. Flohr (Ed.)
Post-Lesion Neural Plasticity

for an influence of SP on *lesion-induced* learning processes, i.e., on those plastic reactions responsible for functional recovery following damage of the CNS or its sensory or motor interface. The purpose of the present study was therefore to investigate the action of SP on a lesion-induced plastic process. As a paradigm vestibular compensation was employed, i.e., the functional recovery following unilateral vestibular lesions (Schaefer and Meyer 1974; Flohr et al. 1981; Precht and Dieringer 1985). The experimental subjects were grass frogs *(Rana temporaria)*. In this species, unilateral destruction of the labyrinth or vestibular nerve causes two groups of functional deficits: Firstly, asymmetries in the tonic influence on posture, resulting in a curvature of the longitudinal body axis, head deviation around the longitudinal axis, and alterations in the extensor/flexor tonus; secondly, an impairment of the dynamic vestibular reflexes dependent on bilateral inputs such as VOR and VCR. The initial postoperative stage of severe symptoms is followed by a gradual, but remarkably complete recovery of function. In the present investigation, the time course of compensation of the main postural symptom, head deviation, was chosen as a measure of pharmacological effects. In *Rana temporaria* unilateral labyrinthectomy leads to an initial head deviation of 33 ± 4 S.D. degrees. The time course of compensation follows an exponential function ($\tau \approx 20$ days).

2 Methods

The experiments were carried out on common grass frogs *(Rana temporaria)* weighing 16–28 g. The animals were maintained in groups of seven at room temperature in a moist environment and fed with crickets. Unilateral labyrinthectomy was performed under MS 222 anesthesia using the extracranial approach as described by Ewald (1892). To quantify the time-course of vestibular compensation, head deviation was measured daily using specially adopted video equipment. SP was dissolved in frog Ringer in appropriate concentrations and the various doses were injected in a volume of 17 ml/kg into the dorsal lymph sac. Control animals were sham-treated with frog Ringer. SP was administered in the following doses: 50, 100, 250, 400, 1000, and 4000 µg/kg. The treatment consisted of 24 daily injections beginning on the first postoperative day.

3 Results

Chronic postoperative treatment of hemilabyrinthectomized grass frogs with SP significantly accelerates vestibular compensation in a dose-dependent manner (Figs. 1 and 2). With the lowest tested dose (50µg/kg/day) only a slight and delayed effect is observable. With doses between 100 and 4000 µg/kg/day a pronounced accelerating effect is obtained, the onset of which is remarkably fast (see Figs. 1 and 2). The peptide effect is most marked in the earliest stage of compensation and declines thereafter, suggesting that SP exerts its action by modifying

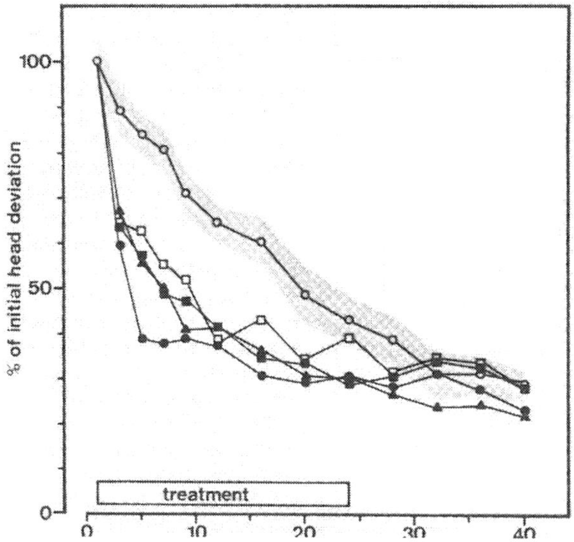

Fig. 1. Effect of Substance P on vestibular compensation following unilateral labyrinthectomy in the grass frog. The peptide was daily administered in four different doses from the 1st to the 24th postoperative day. ○ Ringer-treated control group (n = 42), mean values ± S.E. (*shaded area*); ▲250, □400, ●1000, and ■4000 μg/kg/day of Substance P (n = 14 in all groups). *Ordinate* mean values of head deviation in percent of initial head deviation (measured on the first postoperative day). *Abscissa* time in days

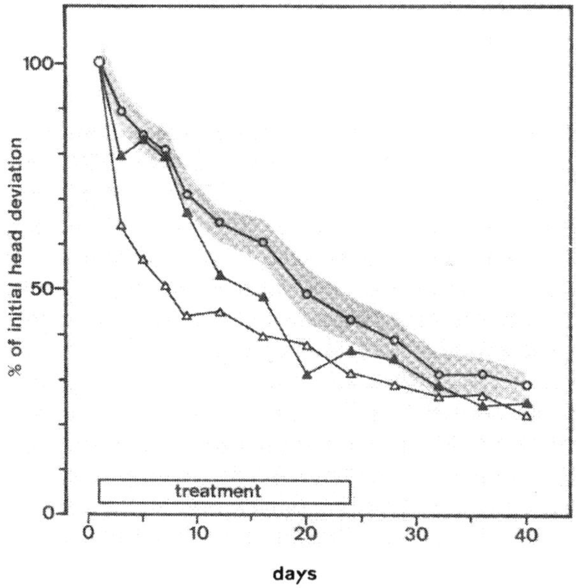

Fig. 2. Effect of Substance P on the course of vestibular compensation. Substance P was administered daily in two different doses from the 1st to the 24th postoperative day. ○Ringer-treated control group (n = 42), mean values ± S.E. (*shaded area*). ▲50 and △100 μg/kg/day of Substance P (n = 14 in both groups). *Ordinate* mean values of head deviation in percent of initial head deviation (measured on the first postoperative day). *Abscissa* time in days

the initial response of the vestibular system to unilateral deafferentation. In Fig. 3 the various treated groups are compared on the basis of half-compensation time, i.e., the time over which the initial head deviation (measured at the first postoperative day) is reduced by 50%. Sham-treated control animals reach this criterion in about 20 days; for SP-treated animals this period is shortened dose-dependently by 50%–75%.

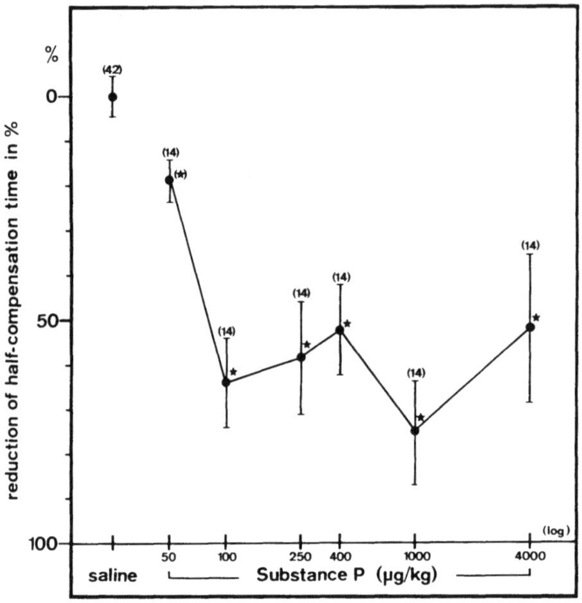

Fig. 3. Survey of efficacies of different doses of Substance P in facilitating vestibular compensation following unilateral labyrinthectomy in the grass frog. The relative effects of the various tested doses are compared on the basis of half-compensation time, i.e., the time over which the initial head deviation is reduced by 50% (see text). The reduction of half-compensation time is expressed in percentage (±S.E.) of half-compensation time of the control, which is defined as 0% (±S.E.). Number of animals in *brackets*. Original data were tested for significance by WELCH's t-test, two-tailed analysis. $(*) = 2P < 0.01$, $* = 2P < 0.001$, peptide versus control

4 Discussion

The restitution of function following unilateral labyrinthectomy does not involve regeneration of the peripheral vestibular end organ or nerve (Magnus 1924), but consists of a reorganization of the remaining central vestibular circuits. A prerequisite for the recovery from asymmetrical posture is generally assumed to be the restitution of the symmetrical activity in the vestibular nuclei of both sides (Dieringer and Precht 1979a, b; Flohr et al. 1981; Precht and Dieringer 1985). One hypothesis proposes that a reactivation of neurons in the deafferented nucleus is brought about by collateral sprouting of remaining inputs and subsequent formation of new synapses at vacated sites. Reactive synaptogenesis has been observed as a response to partial denervation in different CNS areas and, indeed, it has been suggested to be the decisive mechanism in the compensatory rewiring of neuronal circuitry following CNS damage (Cotman 1978). SP may promote such lesion-induced synapse formation since there is evidence that SP has neurotrophic properties: in cultured neuroblastoma cells and peripheral ganglia the peptide induces neurite outgrowth and differentiation (Narumi and Maki 1978; Lindner et al. 1980; Mitta et al. 1984; Lindner 1984; Lindner et al. 1986). These findings are supported by observations by Jonsson and Hallman (1982a, b) and Nakai and Kasamatsu (1984), who have presented evidence that SP counteracts the neurodegenerative effects of 6-hydroxydopamine (6-OHDA) supposedly by stimulating reinnervation of 6-OHDA-treated areas. It remains questionable, however, whether the remarkably fast onset of the action of SP observed in the present experiment can be explained by assuming trophic effects of SP on collateral sprouting.

For the above-mentioned effects of SP in classical learning paradigms, Huston and Stäubli (1981) have forwarded the hypothesis that SP activates specific reinforcement systems which control adaptive neural changes. This concept cannot easily be extended to the results observed here in a lesion-induced plastic process. However, with regard to the striking similarities between the action of SP in the present paradigm and in learning processes which occur in response to environmental changes, it is tempting to assume that both processes depend on similar or identical mechanisms. In this view, then, vestibular compensation is not or not only due to local regenerative events caused by the lesion (such as collateral sprouting), but rather equivalent to those forms of experience-dependent plasticity in the vestibular system that occur in the absence of lesions, e.g., those that follow a manipulation of visual experience.

Acknowledgements. This work was supported by grant no. FL 139/1-2 of the Deutsche Forschungsgemeinschaft.

References

Chang MM, Leeman SE (1970) Isolation of a sialogogic peptide from bovine hypothalamic tissue and its characterization as Substance P. J Biol Chem 245:4784–4790

Cotman CW (ed) (1978) Neuronal plasticity. Raven, New York

Dieringer N, Precht W (1979a) Mechanisms of compensation for vestibular deficits in the frog. I. Modification of the excitatory commissural system. Exp Brain Res 36:311–328

Dieringer N, Precht W (1979b) Mechanisms of compensation for vestibular deficits in the frog. II. Modification of inhibitory pathways. Exp Brain Res 36:329–341

Erspamer V (1981) The tachykinin peptide family. TINS 4:267–271

Euler US von, Gaddum JH (1931) An unidentified depressor substance in certain tissue extracts. J Physiol 72:74–87

Ewald JR (1892) Physiologische Untersuchungen über das Endorgan des N. octavus. Bergmann, Wiesbaden

Flohr H, Bienhold H, Abeln W, Macskovics I (1981) Concepts of vestibular compensation. In: Flohr H, Precht W (eds) Lesion-induced neuronal plasticity in sensorimotor systems. Springer, Berlin Heidelberg New York, pp 153–172

Gaffori O, Stewart JM, DeWied D (1984) Influence of Substance P and fragments on passive avoidance behavior. Experientia (Basel) 40:89–91

Goldstein JM, Malick JB (1977) Effect of Substance P on medial forebrain bundle self-stimulation in rats following intracerebral administration. Pharmacol Biochem Behav 7:475–478

Graeff FG, Arisawa EAL (1978) Effect of intracerebroventricular bradykinin, angiotensin II, and substance P on multiple fixed-interval fixed-ratio responding in rabbits. Psychopharmacology 57:89–95

Hecht K, Oehme P, Poppei M, Hecht T (1979) Conditioned-reflex learning of normal juvenile and adult rats exposed to the action of Substance P and of a Substance P analogue. Pharmazie 34:419–423

Huston JP, Stäubli U (1981) Substance P and its effects on learning and memory. In: Martinez JL, Jensen RA, Mesing RB, Rigter H, McGaugh JL (eds) Endogenous peptides and learning and memory processes. Academic Press, London, pp 521–540

Jonsson G, Hallman H (1982a) Substance P counteracts neurotoxin damage on norepinephrine neurons in rat brain during ontogeny. Science 215:75–77

Jonsson G, Hallman H (1982b) Substance P modifies the 6-hydroxydopamine-induced alteration of postnatal development of central noradrenaline neurons. Neuroscience 7:2909–2918

Kafetzopoulos E, Holzhäuer M-S, Huston JP (1986) Substance P injected into the region of the nucleus basalis magnocellularis facilitates performance of an inhibitory avoidance task. Psychopharmacology 90:281–283

Krivoy WA, Couch JR, Henry JL, Stewart JM (1979) Synaptic modulation by Substance P. Fed Proc 38:2344–2347
Lembeck F, Zetler G (1962) Substance P: a polypeptide of possible physiological significance, especially within the nervous system. Int Rev Neurobiol 4:159–215
Lindner G (1984) Über die Wirkung von N- und C-terminalen Sequenzen der Substanz P auf nichtneuronale Zellen in vitro (Nervengewebekultur). Z Mikrosk Anat Forsch 98:107–118
Lindner G, Große G, Oehme P, Jentzsch KD (1980) Beitrag zur Wirkung von Substanz P auf die Nervenfaserregeneration in der Gewebekultur. Z Mikrosk Anat Forsch 94:661–668
Lindner G, Große G, Jentsch KD, Oehme P, Wenzel M (1986) Wirkung von Derivaten des Substanz P – Dipeptides (3–4) auf das Nervenfaserwachstum in der Gewebekultur. Z Hirnforsch 27:639–649
Luttinger D, Nemeroff CB, Prange AJ (1982) The effects of neuropeptides on discrete-trial conditioned avoidance responding. Brain Res 237:183–192
Magnus R (1924) Körperstellung. Springer, Berlin
Melo JC, Graeff FG (1975) Effect of intracerebroventricular bradykinin and related peptides on rabbit operant behavior. J Pharmacol Exp Ther 193:1–10
Mitta M, Nakamura M, Kohne J et al. (1984) Growth of central Substance P containing neurons into superior cervical ganglia transplanted in the spinal cord of adult rats. Brain Res 324:134–137
Nakai K, Kasamatsu T (1984) Accelerated regeneration of central catecholamine fibers in cat occipital cortex: effects of substance P. Brain Res 323:374–379
Narumi S, Maki Y (1978) Stimulatory effects of Substance P on neurite extension and cyclic AMP levels in cultured neuroblastoma cells. J Neurochem 30:1321–1326
Nemeroff GB, Kalivas PW, Golden RN, Prange AJ (1984) Behavioral effects of hypothalamic hypophysiotropic hormones, neurotensin, substance P, and other neuropeptides. Pharmacol Ther 24:1–56
Nicoll RA, Schenker C, Leeman SE (1980) Substance P as a neurotransmitter candidate. Annu Rev Neurosci 3:227–268
Pelleymounter MA, Fischer Q, Schlesinger K, Hall M, Armey P de, Stewart J (1986) The effect of Substance P and its fragments on passive avoidance retention and brain monoamine activity. Behav Brain Res 21:119–127
Precht W, Dieringer N (1985) Neuronal events paralleling functional recovery (compensation) following peripheral vestibular lesions. In: Berthoz A, Melvill Jones G (eds) Adaptive mechanisms in gaze control. Elsevier, Amsterdam, pp 251–268
Sagen J, Routtenberg A (1981) Specific anatomical and synaptic sites of neuropeptide action in memory formation. In: Martinez JL, Jensen RA, Messing RB, Rigter H, McGaugh JL (eds) Endogenous peptides and learning and memory processes. Academic Press, London, pp 541–561
Schaefer K-P, Meyer DL (1974) Compensation of vestibular lesions. In: Kornhuber KH (ed) Handbook of sensory physiology, vol IV/2. Springer, Berlin Heidelberg New York, pp 463–490
Schlesinger K, Lipsitz DU, Peck PL, Pelleymounter MA, Stewart JM, Chase TN (1983a) Substance P enhancement of passive and active avoidance conditioning in mice. Pharmacol Biochem Behav 19:655–661
Schlesinger K, Lipsitz DU, Peck PL, Pelleymounter MA, Stewart JM, Chase TN (1983b) Substance P reversal of electroconvulsive shock and cycloheximide-induced retrograde amnesia. Behav Neural Biol 39:30–39
Stäubli U, Huston JP (1985) Central action of Substance P: possible role in reward. Behav Neural Biol 43:100–108
Stein L, Belluzzi JD (1979) Brain endorphins: possible role in reward and memory formation. Fed Proc 38:2468–2472
Tomaz C, Huston JP (1986) Facilitation of conditioned inhibitory avoidance by post-trial peripheral injection of Substance P. Pharmacol Biochem Behav 25:469–472
Wetzel W, Matthies H (1982) Effect of Substance P on the retention of a brightness discrimination task in rats. Acta Biol Med Ger 41:647–652

Subject Index

abnormal innervation 145
abnormal pathways 495
acetylcholine receptor 126
- muscarinic 62
acetylcholinesterase 62, 621
acoustic neurinoma 294
ACTH 4–10 605, 611, 612, 628–633
acute vestibular syndrome 397, 403
adrenocorticotropic hormone (ACTH) 595
- fragments 615
age dependent recovery 157
anatomical reorganization 217
apolipoproteins 50
areae 17 and 18 187
association cortex 514, 516
association input 523
astrocytes 57, 58, 62
astroglia 619
auditory system 567
autoradiography 176
axon trees 139
axonal arborization 139
- conservation of 139, 144
- quantification of 141
axonal regeneration 129, 555–562
- time course of 426, 556–558
axonal transport 113

balance compensation 627, 628, 631
behavioral compensation 565

Ca^{2+} entry blocker
- and vestibular activity 687
calmodulin 113
catecholamines 661, 662
cDNA, protein F1 97–99
cell migration 73
cerebellar defect 431
cerebellar hemispherectomy 203
- and development 203ff
cerebellar lesion 331, 443, 447, 448, 450
cerebello-cerebral response 426
cerebellofugal projection
- regeneration of 423
- uncrossed ipsilateral 425
cerebellothalamic connections 204
cerebellothalamic projection 422
cerebellum 203, 213–220, 314, 315, 661, 662, 692
cerebral ischemia 657
cerebralcortical lesions 443, 445, 446
cerebrorubral connections 204, 208
cervical proprioception 329
cervicoocular reflex 281, 321ff
cholesterol 85, 89–91
chromatolysis 57
collateral sprouting 35, 37, 223, 353, 702
commissural fiber system 345–346
competition 126, 127
- of neurones for trophic factors 30, 31
computer models
- of synaptic stabilization 13
conditioning lesion 113
corpus callosum 187, 189
cortex 511
- somatosensory 511
cortical maps 511
corticorubral projection 527
corticospinal tract 166, 167, 224, 227
cosmos 936, 782, 1129 332
cricket phonotactic behavior 565
crossed corticorubral fibers 529
cutaneous nerves 37

deafferentation 482, 486
- hindlimb 249
decompensation 316ff
dopamine 539–547
dorsal root ganglia (DRG) 86–89
dorsolateral pontine nucleus (DLPN) 414, 417

early lesions 157
endplate 605, 609
enucleation 173
equilibrium recovery 677
evoked vestibular potentials 694
excitatory amino acids 481, 482, 656–658
eye movements 293

facial nerve 57
flocculus 315
flunarizine 687
fluorescent dyes 130
fluorescent latex beads 90

GABA immunocytochemistry 437
gangliosides 645 ff, 655
Ginkgo biloba 676 ff
glial cells 49
– modulation factors 51
glial fibrillary acidic protein (GFAP) 57
glucocorticoids 635–642
– and brain 636–638
– and hippocampus 636–638
– and neuropeptides 641, 642
– and pituitary 639–642
– and receptors 635–642
glycosphingolipids 646
GM1 ganglioside 645, 646
growth associated protein 43 (GAP43) 103
growth cone phosphoprotein pp46 105
growth cones 245–247, 351–353
growth factors 36, 37

hemicerebellectomy 213–220
hemidecortication 157
hemilabyrinthectomy 321, 381, 393 ff
hemisection 223
hemispherectomy 165
hippocampus 636–638
hopping reaction 159
6-hydroxydopamine (6-OHDA) 537–547, 647, 662, 664

ibotenic acid 413, 447
image analysis 174
increased synaptic strength 30, 31
infant lesion effect 213, 219
inferior colliculus
– commissures of the 555–562
– function of 558–562
inferior olive 206

kainic acid 519
kinase C 103

labyrinthectomy 305, 313 ff, 345–355, 627–631, 633
labyrinthine input on
– locus coeruleus neurons 263
– Renshaw cells 266
lateral geniculate nucleus 481
leg regeneration 567
locomotion 214, 249, 252, 253
locomotor balance function 628
locus coeruleus 259
long-term potentiation (LTP) 94–97, 505

macrophage 85, 86, 88–91
magnetic field 65
melanocortins 595, 596
melanocyte stimulating hormone (MSH) 595, 605, 611, 612
microglia 58, 63
microgravity 331, 332, 334
microtubules 619
monocular deprivation 455
motor behavior 249
motor cortex 519
motor deficit 521
motor recovery 214–217, 223

neonatal brain injury 165
nerve development 605, 606, 611
nerve growth factor (NGF) 36, 37, 116, 656–658
nerve regeneration 35, 37, 605, 611, 612
– and motor function 609
– and sensory function 609
– in the CNS 49
neural injury treatment 645
neurite outgrowth 617
neuromuscular junction 26
neuronal environmental modifications 49, 50
neuronal injury 49
neuronotoxicity 656–658
neuronotrophic factors 656–658
neuropeptides 641, 642
neurotrophic factors 615
N-methyl-DL-aspartic acid (NMDA) 417
– receptors 505
noradrenergic coeruleospinal system 259
norepinephrine 661, 662
nucleus praepositus hypoglossi 387

occipital lobectomy 305
ocular dominance 188
oculomotor balance function 630
oculomotor system 321, 323–327, 443

olfactory bulbectomy 580 ff
- caudal 580 ff
- rostral 580 ff
olfactory system 577 ff
one-ear tracking model 571
open loop 414
optic chiasm 189
optic nerve regeneration 499
optokinetic nystagmus 455, 458 ff
optokinetic pathways 386
optokinetic reflexes (OKR) 381 ff, 455, 456
optokinetic stimulation 455
ornithine decarboxylase 637

packing density 174
paraflocculus 314
partial denervation 11
path- and homefinding 489
pathfinding 234–237, 240–247
Pcd mice 431
phagocytosis 88–91
phaseolus vulgaris leucoagglutinin 529
phosphoinositides 104
phospholipase C 104
polyneuronal innervation 4, 11
pontine nuclei 206
post-tetanic potentiation 534
postural reflexes 252
posture 214
pressure block 473
pronase injection into single axons 25
proopiomelanocortin 596, 639–641
protein B-50 103, 605, 610–612
protein F1 93, 105
protein kinase C 93
pruning effect 146
pseudo-dendrites 62
pursuit eye movements 443, 444
pursuit recovery 416

rat cerebral cells 615
reactive synaptogenesis 702
recalibration 406–407
recovery 305
- from unilateral labyrinthine loss 287–288
- of motor behavior 249
- optic tract 473
recurrent inhibition 260
red nucleus 205, 527, 529, 532
re-established synapses 559–562
regeneration 3, 57, 63, 233–247
- central olfactory connections 582
- functional, olfactory 580 ff
- in the CNS 473
- of cerebellofugal projection 423
- olfactory bulb 580 ff
- olfactory nerve 584
reinforcement systems 703
reinnervation 11
removal of one excitatory axon 29
removal of one inhibitory axon 28
Renshaw cells 260
retinal axons 489
- regeneration of 489
retinal slip 462
retinotectal projection 499
retinotectal system 494
rewiring of central neurons 568
rhodamine-labeled lipoprotein 88–91

scar 240–245
Schwann cell 85–91, 116
semicircular canal plugging 322
sensorimotor asymmetries 540–545
sensorimotor cortex 528, 529
sensorimotor restriction 677
sensory reinnervation 37
sensory substitution 357, 359
sheath cell-axon interaction 65
slow component b (Scb) 114
smooth pursuit 413
somatotopic reorganization 513
spared root preparation 252
spinal afferents 378
spinal cord 223, 233–247, 249
spinal decompensation 378
split brain 187
spontaneous nystagmus 305
sprouting 3, 93, 249 ff
striatum 539–547
Substance P 699
substantia nigra 537–547
supersensitivity 481, 487
synapse elimination 11, 123–126
synapse formation 3
synaptic basal lamina 126
synaptic specificity 237–239
synaptic stabilization 504
synaptic stripping 58, 63
synaptogenesis 646

tachykinins 699
target recognition 494
tectal lesions 139
territorial specificity of trophic effects 32
thyreotropin releasing hormone (TRH) 628–634
tibia transplantation 574
tissue shrinkage
- due to deafferentation 348, 353

transneuronal atrophy 482, 487
transplantation 130
transsynaptic degeneration 648
trophic factors
– muscle derived 13
tubulin 113

vagal nerve 62
vermis 315
vestibular compensation 313 ff, 342, 345–355, 357, 371, 381 ff, 393 ff, 675 ff, 695, 699
vestibular cortex 371
vestibular deafferentation 378
vestibular lesions 293, 321, 331
vestibular nerve 373
vestibular neurons 376
vestibular nuclei 431, 675 ff, 691
vestibular nystagmus 678
vestibular receptors 687
vestibular zone 345 ff
vestibuloocular reflex (VOR) 281–288, 293 ff, 305, 321 ff, 381 ff, 661
– adaptation of 281–283, 287–288, 305
– voluntary modulation of 286–288
vestibulospinal reflexes
– gain regulation 268
vibrissae 537–547
visual cortex 173, 187, 188
visual discrimination 158
visual smooth pursuit 281, 288
visual substitution 406
visual-vestibular interaction 387, 394, 406–407
vital dyes 126

weaver mice 431
wheat germ agglutinin-horseradish peroxidase (WGA–HRP) 223, 224, 227

SPRINGER NATURE

GPSR Compliance

The European Union's (EU) General Product Safety Regulation (GPSR) is a set of rules that requires consumer products to be safe and our obligations to ensure this.

If you have any concerns about our products, you can contact us on ProductSafety@springernature.com

In case Publisher is established outside the EU, the EU authorized representative is:

Springer Nature Customer Service Center GmbH
Europaplatz 3
69115 Heidelberg, Germany

FSC
www.fsc.org
MIX
Papier | Fördert gute Waldnutzung
FSC® C083411

Kolibri 360 GmbH
Ferdinand-Jühlke-Straße 7
99095 Erfurt, Deutschland
produktsicherheit@kolibri360.de